Counciling

Monday 10:00 - 11:00
Tuesday 2:00 - 3:00
Wednesday 1:10 - 2:00
Thursday 9:10 - 10:00
Friday 10:10 - 11:00

Monday   10:00 —

Friday   10:00

SIR ISAAC NEWTON

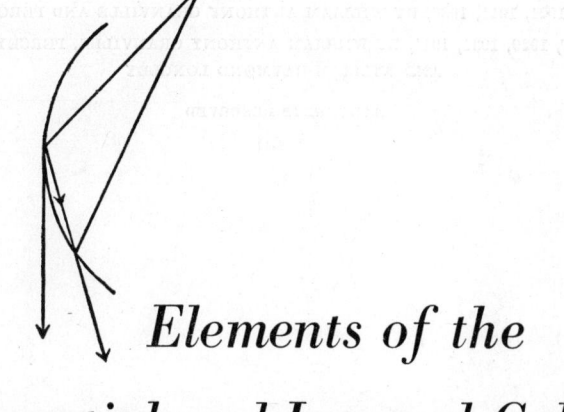

Elements of the Differential and Integral Calculus

NEW REVISED EDITION

WILLIAM ANTHONY GRANVILLE
PERCEY F. SMITH
WILLIAM RAYMOND LONGLEY
YALE UNIVERSITY

GINN AND COMPANY
BOSTON · NEW YORK · CHICAGO · ATLANTA · DALLAS
PALO ALTO · TORONTO · LONDON

© COPYRIGHT, 1957, BY WILLIAM RAYMOND LONGLEY

PHILIPPINES COPYRIGHT, 1958, BY WILLIAM RAYMOND LONGLEY

COPYRIGHT, 1904, 1911, 1939, BY WILLIAM ANTHONY GRANVILLE AND PERCEY F. SMITH

COPYRIGHT, 1929, 1934, 1941, BY WILLIAM ANTHONY GRANVILLE, PERCEY F. SMITH AND WILLIAM RAYMOND LONGLEY

ALL RIGHTS RESERVED

158.11

D

PREFACE

In the preface to the revised edition it was pointed out that a chapter on Hyperbolic Functions had been added and that cylindrical coordinates had been employed to broaden the applications of double integration. These features have been retained in the New Revised Edition, where the only significant change in the text occurs in the treatment of the total differential in Article 227.

Following the Table of Integrals there has been added a list of the linear differential equations, with their solutions, which occur most frequently in applications to mechanics.

The most important changes have been made in the problems, several of which have been replaced by others to increase the interest of the student, and new ones have been added in most of the chapters. The additional problems, collected from prize examinations, have been retained for the use of superior students.

Answers to many of the problems are given in the text. Some of the answers are purposely omitted in order to give the student greater self-reliance in checking his work. Teachers who desire answers to the other problems should communicate with the publishers.

The undersigned wishes to acknowledge the debt which he and thousands of teachers and students owe to the late Professor William A. Granville, the author of the original book, and to the late Professor Percey F. Smith who, as editor and co-author of later editions, has kept Granville's Calculus up to date for so many years.

WILLIAM R. LONGLEY

CONTENTS

DIFFERENTIAL CALCULUS

CHAPTER I. COLLECTION OF FORMULAS 1

Formulas from elementary algebra and geometry, 1 · Formulas from plane trigonometry, 2 · Formulas from plane analytic geometry, 3 · Formulas from solid analytic geometry, 4 · Greek alphabet, 6

CHAPTER II. VARIABLES, FUNCTIONS, AND LIMITS 7

Variables and constants, 7 · Interval of a variable, 7 · Continuous variation, 7 · Functions, 8 · Independent and dependent variables, 8 · Notation of functions, 8 · Division by zero excluded, 9 · Graph of a function; continuity, 10 · Limit of a variable, 10 · Limiting value of a function, 11 · Theorems on limits, 11 · Continuous and discontinuous functions, 12 · Infinity (∞), 13 · Infinitesimals, 17 · Theorems concerning infinitesimals and limits, 17

CHAPTER III. DIFFERENTIATION 19

Introduction, 19 · Increments, 19 · Comparison of increments, 20 · Derivative of a function of one variable, 21 · Symbols for derivatives, 22 · Differentiable functions, 23 · General rule for differentiation, 23 · Interpretation of the derivative by geometry, 25

CHAPTER IV. RULES FOR DIFFERENTIATING ALGEBRAIC FORMS 28

Importance of the General Rule, 28 · Differentiation of a constant, 29 · Differentiation of a variable with respect to itself, 29 · Differentiation of a sum, 30 · Differentiation of the product of a constant and a function, 30 · Differentiation of the product of two functions, 30 · Differentiation of the product of n functions, n being a fixed number, 31 · Differentiation of a function with a constant exponent. The Power Rule, 32 · Differentiation of a quotient, 32 · Differentiation of a function of a function, 37 · Differentiation of inverse functions, 38 · Implicit functions, 39 · Differentiation of implicit functions, 40

CHAPTER V. VARIOUS APPLICATIONS OF THE DERIVATIVE . 42

Direction of a curve, 42 · Equations of tangent and normal; lengths of subtangent and subnormal, 43 · Maximum and minimum values of a function; introduction, 47 · Increasing and decreasing functions. Tests, 50 · Maximum and minimum values of a function; definitions, 52 · First method for examining a function for maximum and minimum values. Working rule, 53 · Maximum or minimum values when $f'(x)$ becomes infinite and $f(x)$ is continuous, 55 · Maximum and minimum values. Applied problems, 57 · Derivative as the rate of change, 64 · Velocity in rectilinear motion, 65 · Related rates, 67

CONTENTS

CHAPTER VI. SUCCESSIVE DIFFERENTIATION AND APPLICATIONS 73

Definition of successive derivatives, 73 · Successive differentiation of implicit functions, 73 · Direction of bending of a curve, 75 · Second method for testing for maximum and minimum values, 76 · Points of inflection, 79 · Curve-tracing, 81 · Acceleration in rectilinear motion, 83

CHAPTER VII. DIFFERENTIATION OF TRANSCENDENTAL FUNCTIONS. APPLICATIONS 86

Formulas for derivatives; second list, 86 · The number e. Natural logarithms, 87 · Exponential and logarithmic functions, 89 · Differentiation of a logarithm, 90 · Differentiation of the exponential function, 91 · Differentiation of the general exponential function. Proof of the Power Rule, 92 · Logarithmic differentiation, 93 · The function $\sin x$, 97 · Theorem, 98 · Differentiation of $\sin v$, 99 · The other trigonometric functions, 99 · Differentiation of $\cos v$, 100 · Proofs of formulas XV–XIX, 100 · Comments, 101 · Inverse trigonometric functions, 105 · Differentiation of arc $\sin v$, 106 · Differentiation of arc $\cos v$, 106 · Differentiation of arc $\tan v$, 107 · Differentiation of arc ctn v, 108 · Differentiation of arc sec v and arc csc v, 108 · Differentiation of arc vers v, 109

CHAPTER VIII. APPLICATIONS TO PARAMETRIC EQUATIONS, POLAR EQUATIONS, AND ROOTS 115

Parametric equations of a curve. Slope, 115 · Parametric equations. Second derivative, 119 · Curvilinear motion. Velocity, 120 · Curvilinear motion. Component accelerations, 121 · Polar coördinates. Angle between the radius vector and the tangent line, 123 · Lengths of polar subtangent and polar subnormal, 126 · Real roots of equations. Graphical methods, 128 · Second method for locating real roots, 130 · Newton's method, 131

CHAPTER IX. DIFFERENTIALS 136

Introduction, 136 · Definitions, 136 · Approximation of increments by means of differentials, 137 · Small errors, 138 · Formulas for finding the differentials of functions, 140 · Differential of the arc in rectangular coördinates, 142 · Differential of the arc in polar coördinates, 144 · Velocity as the time-rate of change of arc, 146 · Differentials as infinitesimals, 146 · Order of infinitesimals. Differentials of higher order, 148

CHAPTER X. CURVATURE. RADIUS AND CIRCLE OF CURVATURE 149

Curvature, 149 · Curvature of a circle, 149 · Formulas for curvature; rectangular coördinates, 150 · Special formula for parametric equations, 151 · Formula for curvature; polar coördinates, 151 · Radius of curvature, 152 · Railroad or transition curves, 152 · Circle of curvature, 153 · Center of curvature, 157 · Evolutes, 158 · Properties of the evolute, 162 · Involutes and their mechanical construction, 163 · Transformation of derivatives, 166

CONTENTS

CHAPTER XI. THEOREM OF MEAN VALUE AND ITS APPLICATIONS 169

Rolle's Theorem, 169 · Osculating circle, 170 · Limiting point of intersection of consecutive normals, 171 · Theorems of Mean Value (Laws of the Mean), 172 · Indeterminate forms, 174 · Evaluation of a function taking on an indeterminate form, 174 · Evaluation of the indeterminate form $\frac{0}{0}$, 175 · Evaluation of the indeterminate form $\frac{\infty}{\infty}$, 178 . Evaluation of the indeterminate form $0 \cdot \infty$, 178 · Evaluation of the indeterminate form $\infty - \infty$, 179 · Evaluation of the indeterminate forms 0^0, 1^∞, ∞^0, 180 · The Extended Theorem of Mean Value, 182 · Maxima and minima treated analytically, 182

INTEGRAL CALCULUS

CHAPTER XII. INTEGRATION; RULES FOR INTEGRATING STANDARD ELEMENTARY FORMS 187

Integration, 187 · Constant of integration. Indefinite integral, 189 · Rules for integrating standard elementary forms, 190 · Formulas (3), (4), (5), 193 · Proofs of (6) and (7), 198 · Proofs of (8)–(17), 200 · Proofs of (18)–(21), 203 · Proofs of (22) and (23), 211 · Trigonometric differentials, 213 · Integration of expressions containing $\sqrt{a^2 - u^2}$ or $\sqrt{u^2 \pm a^2}$ by trigonometric substitution, 221 · Integration by parts, 223 · Comments, 227

CHAPTER XIII. CONSTANT OF INTEGRATION 229

Determination of the constant of integration by means of initial conditions, 229 · Geometrical signification of the constant of integration, 229 · Physical signification of the constant of integration, 233

CHAPTER XIV. THE DEFINITE INTEGRAL 237

Differential of the area under a curve, 237 · The definite integral, 237 · Calculation of a definite integral, 239 · Change in limits corresponding to change in variable, 240 · Calculation of areas, 241 · Area when the equations of the curve are given in parametric form, 242 · Geometrical representation of an integral, 244 · Approximate integration; trapezoidal rule, 245 · Simpson's rule (parabolic rule), 247 · Interchange of limits, 249 · Decomposition of the interval of integration of the definite integral, 250 · The definite integral a function of its limits, 250 · Improper integrals. Infinite limits, 250 · Improper integrals. When $y = \phi(x)$ is discontinuous, 251

CHAPTER XV. INTEGRATION A PROCESS OF SUMMATION . . 254

Introduction, 254 · The Fundamental Theorem of Integral Calculus, 254 · Analytical proof of the Fundamental Theorem, 257 · Areas of plane curves; rectangular coördinates, 258 · Areas of plane curves; polar coördinates, 262 · Volumes of solids of revolution, 265 · Length of a curve, 271 · Lengths of plane curves; rectangular coördinates, 272 · Lengths of plane curves; polar coördinates, 274 · Areas of surfaces of revolution, 277 · Solids with known parallel cross sections, 283

CONTENTS

CHAPTER XVI. FORMAL INTEGRATION BY VARIOUS DEVICES ... 289

Introduction, 289 · Integration of rational fractions, 289 · Integration by substitution of a new variable; rationalization, 296 · Binomial different'als, 299 · Conditions of rationalization of the binomial differential, 302 · Transformation of trigonometric differentials, 303 · Miscellaneous substitutions, 305

CHAPTER XVII. REDUCTION FORMULAS. USE OF TABLE OF INTEGRALS ... 307

Introduction, 307 · Reduction formulas for binomial differentials, 307 · Reduction formulas for trigonometric differentials, 312 · Use of a table of integrals, 315

CHAPTER XVIII. CENTROIDS, FLUID PRESSURE, AND OTHER APPLICATIONS ... 320

Moment of area; centroids, 320 · Centroid of a solid of revolution, 323 · Fluid pressure, 325 · Work, 328 · Mean value of a function, 333

DIFFERENTIAL AND INTEGRAL CALCULUS

CHAPTER XIX. SERIES ... 338

Definitions, 338 · The geometric series, 339 · Convergent and divergent series, 340 · General theorems, 341 · Comparison tests, 342 · Cauchy's test-ratio test, 345 · Alternating series, 347 · Absolute convergence, 348 · Summary, 348 · Power series, 350 · The binomial series, 353 · Another type of power series, 354

CHAPTER XX. EXPANSION OF FUNCTIONS ... 357

Maclaurin's series, 357 · Operations with infinite series, 362 · Differentiation and integration of power series, 365 · Approximate formulas derived from Maclaurin's series, 367 · Taylor's series, 369 · Another form of Taylor's series, 371 · Approximate formulas derived from Taylor's series, 372

CHAPTER XXI. ORDINARY DIFFERENTIAL EQUATIONS ... 375

Differential equations — order and degree, 375 · Solutions of differential equations. Constants of integration, 376 · Verification of the solutions of differential equations, 377 · Differential equations of the first order and of the first degree, 378 · Two special types of differential equations of higher order, 387 · Linear differential equations of the second order with constant coefficients, 390 · Applications. Compound-interest law, 399 · Applications to problems in mechanics, 402 · Linear differential equations of the nth order with constant coefficients, 407

CHAPTER XXII. HYPERBOLIC FUNCTIONS ... 414

Hyperbolic sine and cosine, 414 · Other hyperbolic functions, 415 · Table of values of the hyperbolic sine, cosine, and tangent. Graphs, 415 · Hyperbolic functions of $v + w$, 417 · Derivatives, 420 · Relations to the equilateral hyperbola, 420 · Inverse hyperbolic functions, 423 · Derivatives (continued), 425 · Telegraph line, 428 · Integrals, 430 · Integrals (continued), 432 · The gudermannian, 435 · Mercator's Chart, 438 · Relations between trigonometric and hyperbolic functions, 440

CONTENTS

CHAPTER XXIII. PARTIAL DIFFERENTIATION 444

Functions of several variables. Continuity, 444 · Partial derivatives, 445 · Partial derivatives interpreted geometrically, 446 · The total differential, 449 · Approximation of the total increment. Small errors, 451 · Total derivatives. Rates, 455 · Change of variables, 457 · Differentiation of implicit functions, 458 · Derivatives of higher order, 462

CHAPTER XXIV. APPLICATIONS OF PARTIAL DERIVATIVES 466

Envelope of a family of curves, 466 · The evolute of a given curve considered as the envelope of its normals, 469 · Tangent line and normal plane to a skew curve, 471 · Length of arc of a skew curve, 473 · Normal line and tangent plane to a surface, 475 · Geometric interpretation of the total differential, 477 · Another form of the equations of the tangent line and normal plane to a skew curve, 480 · Law of the Mean, 482 · Maxima and minima of functions of several variables, 483 · Taylor's theorem for functions of two or more variables, 488

CHAPTER XXV. MULTIPLE INTEGRALS 491

Partial and successive integration, 491 · Definite double integral. Geometric interpretation, 492 · Value of a definite double integral taken over a region S, 497 · Plane area as a definite double integral. Rectangular coördinates, 498 · Volume under a surface, 501 · Directions for setting up a double integral, 503 · Moment of area and centroids, 503 · Theorem of Pappus, 504 · Center of fluid pressure, 506 · Moment of inertia of an area, 508 · Polar moment of inertia, 510 · Polar coördinates. Plane area, 512 · Problems using polar coördinates, 514 · General method for finding the areas of curved surfaces, 517 · Volumes found by triple integration, 521 · Volumes using cylindrical coördinates, 524

CHAPTER XXVI. CURVES FOR REFERENCE 531

CHAPTER XXVII. TABLE OF INTEGRALS 538

INDEX 553

GOTTFRIED WILHELM LEIBNITZ

DIFFERENTIAL CALCULUS

CHAPTER I

COLLECTION OF FORMULAS

1. Formulas from elementary algebra and geometry. For the convenience of the student we give in Arts. 1–4 the following lists of formulas. We begin with algebra.

(1) **Quadratic** $\quad Ax^2 + Bx + C = 0.$

Solution. 1. By factoring: Factor $Ax^2 + Bx + C$, set each factor equal to zero, and solve for x.

2. By completing the square: Transpose C, divide by the coefficient of x^2, add to both members the square of half the coefficient of x, and extract the square root.

3. By the formula $\quad x = \dfrac{-B \pm \sqrt{B^2 - 4AC}}{2A}.$

Nature of the roots. The expression $B^2 - 4AC$ beneath the radical in the formula is called the *discriminant*. The two roots are real and unequal, real and equal, or imaginary, according as the discriminant is positive, zero, or negative.

(2) **Logarithms**

$\log ab = \log a + \log b. \qquad \log a^n = n \log a. \qquad \log 1 = 0.$

$\log \dfrac{a}{b} = \log a - \log b. \qquad \log \sqrt[n]{a} = \dfrac{1}{n} \log a. \qquad \log_a a = 1.$

(3) **Binomial theorem** (n being a positive integer)

$(a+b)^n = a^n + na^{n-1}b + \dfrac{n(n-1)}{\lfloor 2} a^{n-2}b^2 + \dfrac{n(n-1)(n-2)}{\lfloor 3} a^{n-3}b^3 + \cdots$

$\qquad\qquad + \dfrac{n(n-1)(n-2)\cdots(n-r+2)}{\lfloor r-1} a^{n-r+1}b^{r-1} + \cdots.$

(4) **Factorial numbers.** $n! = \lfloor n = 1 \cdot 2 \cdot 3 \cdot 4 \cdots (n-1)n.$

In the following formulas from elementary geometry, r or R denotes radius, a altitude, B area of base, and s slant height.

(5) **Circle.** Circumference $= 2\pi r$. Area $= \pi r^2$.

(6) **Circular sector.** Area $= \tfrac{1}{2} r^2 \alpha$, where $\alpha =$ central angle of the sector measured in radians.

1

(7) **Prism.** Volume $= Ba$.
(8) **Pyramid.** Volume $= \frac{1}{3} Ba$.
(9) **Right circular cylinder.** Volume $= \pi r^2 a$. Lateral surface $= 2\pi r a$. Total surface $= 2\pi r(r + a)$.
(10) **Right circular cone.** Volume $= \frac{1}{3} \pi r^2 a$. Lateral surface $= \pi r s$. Total surface $= \pi r(r + s)$.
(11) **Sphere.** Volume $= \frac{4}{3} \pi r^3$. Surface $= 4\pi r^2$.
(12) **Frustum of a right circular cone.** Volume $= \frac{1}{3} \pi a(R^2 + r^2 + Rr)$. Lateral surface $= \pi s(R + r)$.

2. Formulas from plane trigonometry. Many of the following formulas will be found useful.

(1) **Measurement of angles.** There are two common methods of measuring angular magnitude; that is, there are two unit angles.

Degree measure. The unit angle is $\frac{1}{360}$ of a complete revolution and is called a *degree*.

Circular measure. The unit angle is an angle whose subtending arc is equal to the radius of that arc, and is called a *radian*.

The relation between the unit angles is given by the equation

$$180 \text{ degrees} = \pi \text{ radians } (\pi = 3.14159 \cdots),$$

the solution of which gives

$$1 \text{ degree} = \frac{\pi}{180} = 0.0174 \cdots \text{ radian}; \quad 1 \text{ radian} = \frac{180}{\pi} = 57.29 \cdots \text{ degrees.}$$

From the above definition we have

$$\text{Number of radians in an angle} = \frac{\text{subtending arc}}{\text{radius}}.$$

These equations enable us to change from one measurement to another.

(2) **Relations**

$$\text{ctn } x = \frac{1}{\tan x}; \quad \sec x = \frac{1}{\cos x}; \quad \csc x = \frac{1}{\sin x}.$$

$$\tan x = \frac{\sin x}{\cos x}; \quad \text{ctn } x = \frac{\cos x}{\sin x}.$$

$$\sin^2 x + \cos^2 x = 1; \quad 1 + \tan^2 x = \sec^2 x; \quad 1 + \text{ctn}^2 x = \csc^2 x.$$

(3) **Formulas for reducing angles**

Angle	Sine	Cosine	Tangent	Cotangent	Secant	Cosecant
$-x$	$-\sin x$	$\cos x$	$-\tan x$	$-\text{ctn } x$	$\sec x$	$-\csc x$
$90° - x$	$\cos x$	$\sin x$	$\text{ctn } x$	$\tan x$	$\csc x$	$\sec x$
$90° + x$	$\cos x$	$-\sin x$	$-\text{ctn } x$	$-\tan x$	$-\csc x$	$\sec x$
$180° - x$	$\sin x$	$-\cos x$	$-\tan x$	$-\text{ctn } x$	$-\sec x$	$\csc x$
$180° + x$	$-\sin x$	$-\cos x$	$\tan x$	$\text{ctn } x$	$-\sec x$	$-\csc x$
$270° - x$	$-\cos x$	$-\sin x$	$\text{ctn } x$	$\tan x$	$-\csc x$	$-\sec x$
$270° + x$	$-\cos x$	$\sin x$	$-\text{ctn } x$	$-\tan x$	$\csc x$	$-\sec x$
$360° - x$	$-\sin x$	$\cos x$	$-\tan x$	$-\text{ctn } x$	$\sec x$	$-\csc x$

COLLECTION OF FORMULAS

(4) Functions of $(x+y)$ and $(x-y)$

$$\sin(x+y) = \sin x \cos y + \cos x \sin y.$$
$$\sin(x-y) = \sin x \cos y - \cos x \sin y.$$
$$\cos(x+y) = \cos x \cos y - \sin x \sin y.$$
$$\cos(x-y) = \cos x \cos y + \sin x \sin y.$$
$$\tan(x+y) = \frac{\tan x + \tan y}{1 - \tan x \tan y}. \quad \tan(x-y) = \frac{\tan x - \tan y}{1 + \tan x \tan y}.$$

(5) Functions of $2x$ and $\tfrac{1}{2}x$

$$\sin 2x = 2 \sin x \cos x; \quad \cos 2x = \cos^2 x - \sin^2 x; \quad \tan 2x = \frac{2 \tan x}{1 - \tan^2 x}.$$

$$\sin \frac{x}{2} = \pm \sqrt{\frac{1 - \cos x}{2}}; \quad \cos \frac{x}{2} = \pm \sqrt{\frac{1 + \cos x}{2}}; \quad \tan \frac{x}{2} = \pm \sqrt{\frac{1 - \cos x}{1 + \cos x}}.$$

$$\sin^2 x = \tfrac{1}{2} - \tfrac{1}{2} \cos 2x; \quad \cos^2 x = \tfrac{1}{2} + \tfrac{1}{2} \cos 2x.$$

(6) Addition theorems

$$\sin x + \sin y = 2 \sin \tfrac{1}{2}(x+y) \cos \tfrac{1}{2}(x-y).$$
$$\sin x - \sin y = 2 \cos \tfrac{1}{2}(x+y) \sin \tfrac{1}{2}(x-y).$$
$$\cos x + \cos y = 2 \cos \tfrac{1}{2}(x+y) \cos \tfrac{1}{2}(x-y).$$
$$\cos x - \cos y = -2 \sin \tfrac{1}{2}(x+y) \sin \tfrac{1}{2}(x-y).$$

(7) Relations for any triangle

Law of sines. $\qquad \dfrac{a}{\sin A} = \dfrac{b}{\sin B} = \dfrac{c}{\sin C}.$

Law of cosines. $\qquad a^2 = b^2 + c^2 - 2bc \cos A.$

Formulas for area. $\qquad K = \tfrac{1}{2} bc \sin A.$

$$K = \frac{\tfrac{1}{2} a^2 \sin B \sin C}{\sin(B+C)}.$$

$$K = \sqrt{s(s-a)(s-b)(s-c)}, \quad \text{where} \quad s = \tfrac{1}{2}(a+b+c).$$

3. Formulas from plane analytic geometry. The more important formulas are given in the following list.

(1) Distance between two points $P_1(x_1, y_1)$ and $P_2(x_2, y_2)$

$$d = \sqrt{(x_1 - x_2)^2 + (y_1 - y_2)^2}.$$

Slope of $P_1 P_2$. $\qquad m = \dfrac{y_1 - y_2}{x_1 - x_2}.$

Midpoint. $\qquad x = \tfrac{1}{2}(x_1 + x_2), \quad y = \tfrac{1}{2}(y_1 + y_2).$

(2) Angle between two lines

$$\tan \theta = \frac{m_1 - m_2}{1 + m_1 m_2}.$$

(For parallel lines, $m_1 = m_2$; for perpendicular lines, $m_1 m_2 = -1$.)

(3) Equations of straight lines

Point-slope form. $\qquad y - y_1 = m(x - x_1).$

Slope-intercept form. $\qquad y = mx + b.$

Two-point form. $\qquad \dfrac{y - y_1}{x - x_1} = \dfrac{y_2 - y_1}{x_2 - x_1}.$

Intercept form. $\qquad \dfrac{x}{a} + \dfrac{y}{b} = 1.$

(4) Perpendicular distance from the line $Ax + By + C = 0$ to $P_1(x_1, y_1)$

$$d = \frac{Ax_1 + By_1 + C}{\pm \sqrt{A^2 + B^2}}.$$

(5) Relations between rectangular and polar coördinates

$$x = \rho \cos \theta, \quad y = \rho \sin \theta, \quad \rho = \sqrt{x^2 + y^2}, \quad \theta = \arctan \frac{y}{x}.$$

(6) Equation of the circle

Center $(h, k).$ $\qquad (x - h)^2 + (y - k)^2 = r^2.$

(7) Equations of the parabola

Vertex the origin. $\qquad y^2 = 2\,px,$ focus $(\tfrac{1}{2}\,p, 0).$

$\qquad\qquad\qquad\qquad x^2 = 2\,py,$ focus $(0, \tfrac{1}{2}\,p).$

Vertex $(h, k).$ $\qquad (y - k)^2 = 2\,p(x - h),$ axis $y = k.$

$\qquad\qquad\qquad\qquad (x - h)^2 = 2\,p(y - k),$ axis $x = h.$

Axis the y-axis. $\qquad y = Ax^2 + C.$

(8) Equations of other curves

Ellipse with center at the origin and with foci on the x-axis. $(a > b)$

$$\frac{x^2}{a^2} + \frac{y^2}{b^2} = 1.$$

Hyperbola with center at the origin and with foci on the x-axis.

$$\frac{x^2}{a^2} - \frac{y^2}{b^2} = 1.$$

Equilateral hyperbola with center at the origin and with the coördinate axes for asymptotes.

$$xy = C.$$

See also Chapter XXVI.

4. Formulas from solid analytic geometry. Some of the more important formulas are given.

(1) Distance between $P_1(x_1, y_1, z_1)$ and $P_2(x_2, y_2, z_2)$

$$d = \sqrt{(x_1 - x_2)^2 + (y_1 - y_2)^2 + (z_1 - z_2)^2}.$$

(2) Straight line

Direction cosines: $\cos \alpha$, $\cos \beta$, $\cos \gamma$.
Direction numbers: a, b, c.

Then
$$\frac{\cos \alpha}{a} = \frac{\cos \beta}{b} = \frac{\cos \gamma}{c}.$$

$$\cos^2 \alpha + \cos^2 \beta + \cos^2 \gamma = 1.$$

$$\cos \alpha = \frac{a}{\pm \sqrt{a^2 + b^2 + c^2}},$$

$$\cos \beta = \frac{b}{\pm \sqrt{a^2 + b^2 + c^2}},$$

$$\cos \gamma = \frac{c}{\pm \sqrt{a^2 + b^2 + c^2}}.$$

For the line joining (x_1, y_1, z_1) and (x_2, y_2, z_2)

$$\frac{\cos \alpha}{x_2 - x_1} = \frac{\cos \beta}{y_2 - y_1} = \frac{\cos \gamma}{z_2 - z_1}.$$

(3) Two lines

Direction cosines: $\cos \alpha$, $\cos \beta$, $\cos \gamma$; $\cos \alpha'$, $\cos \beta'$, $\cos \gamma'$.
Direction numbers: a, b, c; a', b', c'.
If θ = angle between the lines,

$$\cos \theta = \cos \alpha \cos \alpha' + \cos \beta \cos \beta' + \cos \gamma \cos \gamma',$$

$$\cos \theta = \frac{aa' + bb' + cc'}{\sqrt{a^2 + b^2 + c^2} \sqrt{a'^2 + b'^2 + c'^2}}.$$

Parallel lines. $\quad \dfrac{a}{a'} = \dfrac{b}{b'} = \dfrac{c}{c'}.$

Perpendicular lines. $\quad aa' + bb' + cc' = 0.$

(4) **Equations of the straight line with direction numbers** a, b, c **passing through** (x_1, y_1, z_1)

$$\frac{x - x_1}{a} = \frac{y - y_1}{b} = \frac{z - z_1}{c}.$$

(5) **Plane.** For the plane $Ax + By + Cz + D = 0$ the coefficients A, B, C are the direction numbers of a line perpendicular to the plane.

Equation of a plane passing through (x_1, y_1, z_1) and perpendicular to the line with direction numbers A, B, C.

$$A(x - x_1) + B(y - y_1) + C(z - z_1) = 0.$$

(6) Two planes

Equations:
$$Ax + By + Cz + D = 0,$$
$$A'x + B'y + C'z + D' = 0.$$

Direction numbers of the line of intersection:

$$BC' - CB', \quad CA' - AC', \quad AB' - BA'.$$

DIFFERENTIAL CALCULUS

If θ = angle between the planes, then

$$\cos\theta = \frac{AA' + BB' + CC'}{\sqrt{A^2 + B^2 + C^2}\sqrt{A'^2 + B'^2 + C'^2}}.$$

(7) Cylindrical coördinates.* The distance z of a point $P(x, y, z)$ from the XY-plane and the polar coördinates (ρ, θ) of its projection $(x, y, 0)$ on the XY-plane are called the *cylindrical coördinates* of P. The cylindrical coördinates of P are written (ρ, θ, z).

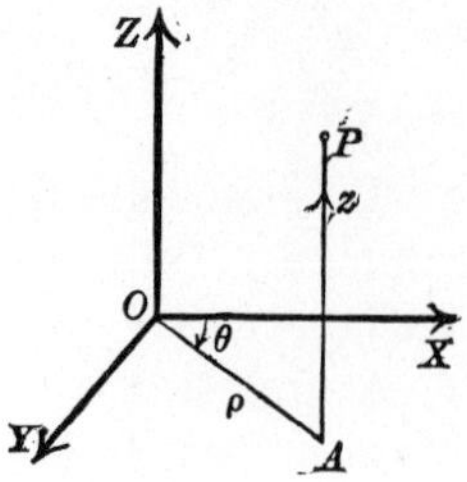

If the rectangular coördinates of P are x, y, z, then, from the definitions and the figure, we have

$$x = \rho\cos\theta, \quad y = \rho\sin\theta, \quad z = z;$$

$$\rho^2 = x^2 + y^2, \quad \theta = \arctan\frac{y}{x}.$$

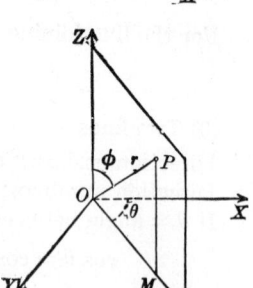

(8) Spherical coördinates.* The radius vector r of a point P, the angle ϕ between OP and the z-axis, and the angle θ between the projection of OP on the XY-plane and the x-axis are called the *spherical coördinates* of P. ϕ is called the *colatitude* and θ the *longitude*. The spherical coördinates of P are written (r, ϕ, θ).

If the rectangular coördinates of P are x, y, z, then, from the definitions and the figure, we have

$$x = r\sin\phi\cos\theta, \quad y = r\sin\phi\sin\theta, \quad z = r\cos\phi;$$

$$r^2 = x^2 + y^2 + z^2, \quad \theta = \arctan\frac{y}{x}, \quad \phi = \arctan\frac{\sqrt{x^2 + y^2}}{z}.$$

5. Greek alphabet

Letters	Names	Letters	Names	Letters	Names
A α	Alpha	I ι	Iota	P ρ	Rho
B β	Beta	K κ	Kappa	Σ σ s	Sigma
Γ γ	Gamma	Λ λ	Lambda	T τ	Tau
Δ δ	Delta	M μ	Mu	Υ υ	Upsilon
E ϵ	Epsilon	N ν	Nu	Φ ϕ	Phi
Z ζ	Zeta	Ξ ξ	Xi	X χ	Chi
H η	Eta	O o	Omicron	Ψ ψ	Psi
Θ θ	Theta	Π π	Pi	Ω ω	Omega

*For a discussion of cylindrical and spherical coördinates see Smith, Gale, and Neelley's "New Analytic Geometry, Revised Edition" (Ginn and Company), pp. 320-322

CHAPTER II

VARIABLES, FUNCTIONS, AND LIMITS

6. Variables and constants. A *variable* is a quantity to which an unlimited number of values can be assigned in an investigation. Variables are denoted usually by the later letters of the alphabet.

A quantity whose value is fixed in any investigation is called a *constant*.

Numerical or *absolute constants* retain the same values in all problems; as 2, 5, $\sqrt{7}$, π, etc.

Arbitrary constants are constants to which numerical values may be assigned, and they retain these assigned values throughout the investigation. They are usually denoted by the earlier letters of the alphabet.

Thus, in the equation of a straight line,

$$\frac{x}{a} + \frac{y}{b} = 1,$$

x and y are the variable coördinates of a point moving along the line, while the arbitrary constants a and b are the intercepts, for which definite values are assumed.

The *numerical* (or *absolute*) *value* of a constant a, as distinguished from its algebraic value, is represented by $|a|$. Thus, $|-2| = 2 = |2|$. The symbol $|a|$ is read "the numerical value of a."

7. Interval of a variable. Very often we confine ourselves to a portion only of the number system. For example, we may restrict our variable so that it shall take on only values lying between a and b. Also, a and b may be included, or either or both excluded. We shall employ the symbol $[a, b]$, a being less than b, to represent the numbers a, b, and all the numbers between them, unless otherwise stated. This symbol $[a, b]$ is read "the interval from a to b."

8. Continuous variation. A variable x is said to vary continuously through an interval $[a, b]$ when x *increases* from the value a to the value b in such a manner as to assume all values between a and b in the order of their magnitudes, or when x *decreases* from $x = b$ to $x = a$, assuming in succession all intermediate values. This may be illustrated geometrically by the diagram on page 8.

The origin being at O, lay off on the straight line the points A and B corresponding to the numbers a and b. Also, let the point P correspond to a particular value of the variable x. Evidently the interval $[a, b]$ is represented by the segment AB. As x varies continuously through the interval $[a, b]$, the point P generates the segment AB when x increases, or the segment BA when x decreases.

9. Functions. When two variables are so related that the value of the first variable is determined when the value of the second variable is given, then the first variable is said to be a function of the second.

Nearly all scientific problems deal with quantities and relations of this sort, and in the experiences of everyday life we are continually meeting conditions illustrating the dependence of one quantity on another. For instance, the *weight* a man is able to lift depends on his *strength*, other things being equal. Similarly, the *distance* a boy can run may be considered as depending on the *time*. Or we may say that the *area* of a square is a function of the *length of a side*, and the *volume* of a sphere is a function of its *diameter*.

10. Independent and dependent variables. The second variable, to which values may be assigned at pleasure within limits depending on the particular problem, is called the *independent variable*, or *argument*; and the first variable, whose value is determined when the value of the independent variable is given, is called the *dependent variable*, or *function*.

Frequently, when we are considering two related variables, it is in our power to fix upon either as the independent variable; but having once made the choice, no change of independent variable is allowed without certain precautions and transformations. For example, the area of a square is a function of the length of its side. Conversely, the length of a side is a function of the area.

11. Notation of functions. The symbol $f(x)$ is used to denote a function of x, and is read f *of* x. In order to distinguish between different functions, the prefixed letter is changed, as $F(x)$, $\phi(x)$, $f'(x)$, etc.

During any investigation a functional symbol indicates the same law of dependence of the function upon the variable. In the simpler cases this law takes the form of a series of analytical operations upon the variable. Hence, in such a case, the functional symbol will indicate the same operations or series of operations applied to different values of the variable. Thus, if
$$f(x) = x^2 - 9x + 14,$$
then
$$f(y) = y^2 - 9y + 14.$$

Also $f(b+1) = (b+1)^2 - 9(b+1) + 14 = b^2 - 7b + 6,$
$f(0) = 0^2 - 9 \cdot 0 + 14 = 14,$
$f(-1) = (-1)^2 - 9(-1) + 14 = 24,$
$f(3) = 3^2 - 9 \cdot 3 + 14 = -4.$

12. Division by zero excluded. The quotient of two numbers a and b is a number x such that $a = bx$. Obviously, division by zero is ruled out by this definition. For if $b = 0$, and we recall that any number times zero equals zero, we see that x does not exist unless $a = 0$. But, in this case, x may be any number whatever. The forms

$$\frac{a}{0}, \quad \frac{0}{0},$$

are, therefore, meaningless.

Care should be taken not to divide by zero inadvertently. The following fallacy is an illustration.

Assume that $\qquad a = b.$
Then, evidently, $\qquad ab = a^2.$
Subtracting b^2, $\qquad ab - b^2 = a^2 - b^2.$
Factoring, $\qquad b(a-b) = (a+b)(a-b).$
Dividing by $a-b$, $\qquad b = a+b.$
But $\qquad a = b;$
therefore $\qquad b = 2b,$
or $\qquad 1 = 2.$

The absurd result is due to the fact that we divided by $a - b = 0$.

PROBLEMS

1. Given $f(x) = x^3 - 5x^2 - 4x + 20$, show that
$$f(1) = 12, \quad f(5) = 0, \quad f(0) = -2f(3), \quad f(7) = 5f(-1).$$

2. If $f(x) = 4 - 2x^2 + x^4$, find $f(0), f(1), f(-1), f(2), f(-2).$

3. If $F(\theta) = \sin 2\theta + \cos \theta$, find $F(0), F(\tfrac{1}{2}\pi), F(\pi).$

4. Given $f(x) = x^3 - 5x^2 - 4x + 20$, show that
$$f(t+1) = t^3 - 2t^2 - 11t + 12.$$

5. Given $f(y) = y^2 - 2y + 6$, show that
$$f(y+h) = y^2 - 2y + 6 + 2(y-1)h + h^2.$$

6. Given $f(x) = x^3 + 3x$, show that
$$f(x+h) - f(x) = 3(x^2+1)h + 3xh^2 + h^3.$$

7. Given $f(x) = \dfrac{1}{x}$, show that $f(x+h) - f(x) = -\dfrac{h}{x^2 + xh}.$

8. Given $\phi(z) = 4^z$, show that $\phi(z+1) - \phi(z) = 3\phi(z).$

9. If $\phi(x) = a^x$, show that $\phi(y) \cdot \phi(z) = \phi(y+z)$.

10. Given $\phi(x) = \log \dfrac{1-x}{1+x}$, show that
$$\phi(y) + \phi(z) = \phi\left(\dfrac{y+z}{1+yz}\right).$$

11. Given $f(x) = \sin x$, show that
$$f(x+2h) - f(x) = 2 \cos(x+h) \sin h.$$

HINT. Use (6), p. 3.

13. Graph of a function; continuity. Consider the function x^2, and let

(1) $\qquad y = x^2.$

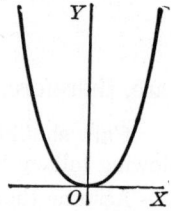

This relation gives a value of y for any value of x; that is, y *is defined* by (1) for all values of the independent variable. The locus of (1), a parabola (see figure), is called the *graph* of the function x^2. If x varies continuously (Art. 8) from $x = a$ to $x = b$, then y will vary continuously from $y = a^2$ to $y = b^2$, and the point $P(x, y)$ will move continuously along the graph from the point (a, a^2) to (b, b^2). Also, a and b may have any values. We then say, "the function x^2 is continuous for all values of x."

Consider the function $\dfrac{1}{x}$, and let

(2) $\qquad y = \dfrac{1}{x}.$

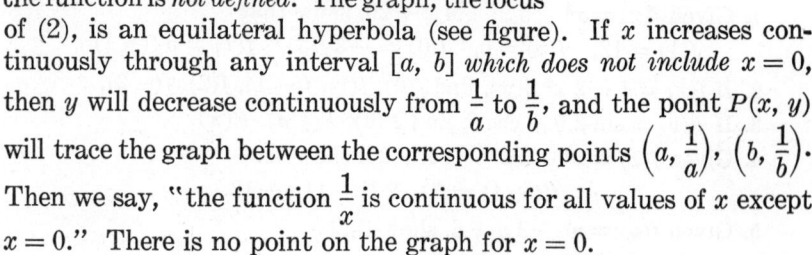

This equation gives a value of y for any value of x except $x = 0$ (Art. 12). For $x = 0$ the function is *not defined*. The graph, the locus of (2), is an equilateral hyperbola (see figure). If x increases continuously through any interval $[a, b]$ *which does not include* $x = 0$, then y will decrease continuously from $\dfrac{1}{a}$ to $\dfrac{1}{b}$, and the point $P(x, y)$ will trace the graph between the corresponding points $\left(a, \dfrac{1}{a}\right)$, $\left(b, \dfrac{1}{b}\right)$. Then we say, "the function $\dfrac{1}{x}$ is continuous for all values of x except $x = 0$." There is no point on the graph for $x = 0$.

These examples illustrate the concept of continuity of a function. A definition is given in Art. 17.

14. Limit of a variable. The idea of a variable approaching a limit occurs in elementary geometry in establishing a formula for the area of a circle. The area of a regular inscribed polygon with any number of sides n is considered, and n is then assumed to increase indefinitely.

The variable area then approaches a limit, and this limit is defined as the area of the circle. In this case the variable v (the area) increases constantly, and the difference $a - v$, where a is the area of the circle, diminishes and ultimately becomes less than any preassigned number, however small.

The relation illustrated is made precise by the

DEFINITION. *The variable v is said to approach the constant l as a limit when the successive values of v are such that the numerical value of the difference $v - l$ ultimately becomes and remains less than any preassigned positive number, however small.*

The relation defined is written $\lim v = l$. For convenience, we shall use the notation $v \to l$, read, "v approaches l as a limit," or, more briefly, "v approaches l." (Some authors use the notation $v \doteq l$.)

ILLUSTRATIVE EXAMPLE. Let the values of v be

$$2 + 1,\ 2 + \frac{1}{2},\ 2 + \frac{1}{4},\ \cdots,\ 2 + \frac{1}{2^n},\ \cdots,$$

without end. Then, obviously, $\lim v = 2$, or $v \to 2$.

If we mark on a straight line, as in Art. 8, the point L corresponding to the limit l, and from L lay off on each side a length ϵ, *however small*, then the points determined by v will ultimately all lie within the segment corresponding to the interval $[l - \epsilon,\ l + \epsilon]$.

15. Limiting value of a function. In applications, the situation that usually arises is this. We have a variable v, and a given function z of v. The independent variable v assumes values such that $v \to l$. We then have to examine the values of the dependent variable z, and, in particular, determine if z approaches a limit. If there is a constant a such that $\lim z = a$, then the relation described is written

$$\lim_{v \to l} z = a,$$

read, "the limit of z, as v approaches l, is a."

16. Theorems on limits. In calculating the limiting value of a function, the following theorems may be applied. Proofs are given in Art. 20.

Suppose u, v, and w are functions of a variable x, and suppose that

$$\lim_{x \to a} u = A,\quad \lim_{x \to a} v = B,\quad \lim_{x \to a} w = C.$$

Then the following relations hold.

(1) $\qquad \lim\limits_{x \to a} (u + v - w) = A + B - C.$

(2) $\qquad \lim\limits_{x \to a} (uvw) = ABC.$

(3) $\qquad \lim\limits_{x \to a} \dfrac{u}{v} = \dfrac{A}{B},$ if B is not zero.

Briefly, in words, *the limit of an algebraic sum, of a product, or of a quotient is equal, respectively, to the same algebraic sum, product, or quotient of the respective limits, provided, in the last named, that the limit of the denominator is not zero.*

If c is a constant (independent of x) and B is not zero, then, from the above,

(4) $\quad \lim_{x \to a} (u + c) = A + c, \quad \lim_{x \to a} cu = cA, \quad \lim_{x \to a} \frac{c}{v} = \frac{c}{B}.$

Consider some examples.

1. Prove $\lim_{x \to 2} (x^2 + 4x) = 12.$

Solution. The given function is the sum of x^2 and $4x$. We first find the limiting values of these two functions. By (2),

$$\lim_{x \to 2} x^2 = 4, \text{ since } x^2 = x \cdot x.$$

By (4), $\qquad \lim_{x \to 2} 4x = 4 \lim_{x \to 2} x = 8.$

Hence, by (1), the answer is $4 + 8 = 12$.

2. Prove $\lim_{z \to 2} \dfrac{z^2 - 9}{z + 2} = -\dfrac{5}{4}.$

Solution. Considering the numerator, $\lim_{z \to 2} (z^2 - 9) = -5$, by (2) and (4). For the denominator, $\lim_{z \to 2} (z + 2) = 4$. Hence, by (3), we have the required result.

17. Continuous and discontinuous functions. In Ex. 1 of the preceding article, where it was shown that

$$\lim_{x \to 2} (x^2 + 4x) = 12,$$

we observe that the answer is the *value* of the function for $x = 2$. That is, the limiting value of the function when x approaches 2 as a limit is equal to the value of the function for $x = 2$. The function is said to be *continuous* for $x = 2$. The general definition is as follows.

DEFINITION. A function $f(x)$ is said to be *continuous* for $x = a$ if the limiting value of the function when x approaches a as a limit is the value assigned to the function for $x = a$. In symbols, if

$$\lim_{x \to a} f(x) = f(a),$$

then $f(x)$ is continuous for $x = a$.

The function is said to be *discontinuous* for $x = a$ if this condition is not satisfied.

Attention is called to the following two cases of common occurrence.

VARIABLES, FUNCTIONS, AND LIMITS

CASE I. As an example illustrating a simple case of a function continuous for a particular value of the variable, consider the function
$$f(x) = \frac{x^2 - 4}{x - 2}.$$

For $x = 1$, $f(x) = f(1) = 3$. Moreover, if x approaches 1 as a limit, the function $f(x)$ approaches 3 as a limit (Art. 16). Hence the function is continuous for $x = 1$.

CASE II. The definition of a continuous function assumes that the function is already defined for $x = a$. If this is not the case, however, it is sometimes possible to assign such a value to the function for $x = a$ that the condition of continuity shall be satisfied. The following theorem covers these cases.

Theorem. *If $f(x)$ is not defined for $x = a$, and if*
$$\lim_{x \to a} f(x) = B,$$
then $f(x)$ will be continuous for $x = a$ if B is assumed as the value of $f(x)$ for $x = a$.

Thus, the function
$$\frac{x^2 - 4}{x - 2}$$
is not defined for $x = 2$ (since then there would be division by zero). But for every other value of x,
$$\frac{x^2 - 4}{x - 2} = x + 2;$$
and
$$\lim_{x \to 2}(x + 2) = 4;$$
therefore
$$\lim_{x \to 2} \frac{x^2 - 4}{x - 2} = 4.$$

Although the function is not defined for $x = 2$, if we arbitrarily assign to it the value 4 for $x = 2$, it becomes continuous for this value.

A function $f(x)$ is said to be continuous in an interval when it is continuous for all values of x in this interval.*

In the calculus we have to calculate frequently the limiting value of a function of a variable v when v approaches as a limit a value a lying in an interval in which the function is continuous. This limiting value is the value of the function for $v = a$.

18. Infinity (∞). If the numerical value of a variable v ultimately becomes and remains greater than any preassigned positive number,

---

* In this book we shall deal only with functions which are in general continuous, that is, continuous for all values of x, with the possible exception of certain isolated values, our results in general being understood as valid only for those values of x for which the function in question is actually continuous.

however large, we say *v becomes infinite*. If v takes on only positive values, it becomes positively infinite; if negative values only, it becomes negatively infinite. The notation used for the three cases is

$$\lim v = \infty, \quad \lim v = +\infty, \quad \lim v = -\infty.$$

In these cases v does not approach a limit as defined in Art. 14. The notation $\lim v = \infty$, or $v \to \infty$, must be read "v becomes infinite," and not "v approaches infinity."*

We may now write, for example,

$$\lim_{x \to 0} \frac{1}{x} = \infty,$$

meaning that $\frac{1}{x}$ becomes infinite when x approaches zero.

Referring to Art. 17, it appears that if

$$\lim_{x \to a} f(x) = \infty,$$

that is, if $f(x)$ becomes infinite as x approaches a as a limit, then $f(x)$ is discontinuous for $x = a$.

A function may have a limiting value when the independent variable becomes infinite. For example,

$$\lim_{x \to \infty} \frac{1}{x} = 0.$$

And, in general, if $f(x)$ approaches the constant value A as a limit when $x \to \infty$, we use the notation of Art. 17 and write

$$\lim_{x \to \infty} f(x) = A.$$

Certain special limits occur frequently. These are given below. The constant c is not zero.

	Written in the form of limits	*Abbreviated form often used*
(1)	$\lim\limits_{v \to 0} \dfrac{c}{v} = \infty.$	$\dfrac{c}{0} = \infty.$
(2)	$\lim\limits_{v \to \infty} cv = \infty.$	$c \cdot \infty = \infty.$
(3)	$\lim\limits_{v \to \infty} \dfrac{v}{c} = \infty.$	$\dfrac{\infty}{c} = \infty.$
(4)	$\lim\limits_{v \to \infty} \dfrac{c}{v} = 0.$	$\dfrac{c}{\infty} = 0.$

---

* On account of the notation used and for the sake of uniformity, the expression $v \to +\infty$ is sometimes read "v approaches the limit plus infinity." Similarly, $v \to -\infty$ is read "v approaches the limit minus infinity," and $v \to \infty$ is read "v, in numerical value, approaches the limit infinity."

This phraseology is convenient, but the student must not forget that infinity is **not a** limit, for infinity is not a number at all.

VARIABLES, FUNCTIONS, AND LIMITS

These special limits are useful in finding the limiting value of the quotient of two polynomials when the variable becomes infinite. The following example will illustrate the method.

ILLUSTRATIVE EXAMPLE. Prove $\lim\limits_{x \to \infty} \dfrac{2x^3 - 3x^2 + 4}{5x - x^2 - 7x^3} = -\dfrac{2}{7}$.

Solution. Divide numerator and denominator by x^3, the highest power of x present in either. Then we have

$$\lim_{x \to \infty} \frac{2x^3 - 3x^2 + 4}{5x - x^2 - 7x^3} = \lim_{x \to \infty} \frac{2 - \dfrac{3}{x} + \dfrac{4}{x^3}}{\dfrac{5}{x^2} - \dfrac{1}{x} - 7}.$$

The limit of each term in numerator or denominator containing x is zero, by (4). Hence, by (1) and (3) of Art. 16 we obtain the answer. In any similar case the first step is therefore as follows.

Divide both numerator and denominator by the highest power of the variable occurring in either.

If u and v are functions of x, and if

$$\lim_{x \to a} u = A, \quad \lim_{x \to a} v = 0,$$

and if A is not zero, then

$$\lim_{x \to a} \frac{u}{v} = \infty.$$

This notation provides for the exceptional case of (3), Art. 16, when $B = 0$ and A is not zero. See also Art. 20.

PROBLEMS

Prove each of the following statements.

1. $\lim\limits_{x \to \infty} \dfrac{5 - 2x^2}{3x + 5x^2} = -\dfrac{2}{5}$.

Proof. $\lim\limits_{x \to \infty} \dfrac{5 - 2x^2}{3x + 5x^2} = \lim\limits_{x \to \infty} \dfrac{\dfrac{5}{x^2} - 2}{\dfrac{3}{x} + 5}$.

[Dividing both numerator and denominator by x^2.]

The limit of each term in numerator and denominator containing x is zero, by (4). Hence, by (1) and (3), Art. 16, we obtain the answer.

2. $\lim\limits_{x \to \infty} \dfrac{4x + 5}{2x + 3} = 2$.

3. $\lim\limits_{t \to 0} \dfrac{4t^2 + 3t + 2}{t^3 + 2t - 6} = -\dfrac{1}{3}$.

4. $\lim\limits_{h \to 0} \dfrac{x^2 h + 3xh^2 + h^3}{2xh + 5h^2} = \dfrac{x}{2}$.

DIFFERENTIAL CALCULUS

5. $\lim\limits_{h \to \infty} \dfrac{3h + 2xh^2 + x^2h^3}{4 - 3xh - 2x^3h^3} = -\dfrac{1}{2x}$.

6. $\lim\limits_{k \to 0} \dfrac{(2z + 3k)^3 - 4k^2z}{2z(2z - k)^2} = 1$.

7. $\lim\limits_{y \to \infty} \dfrac{4y^2 - 3}{2y^3 + 3y^2} = 0$.

8. $\lim\limits_{x \to \infty} \dfrac{6x^3 - 5x^2 + 3}{2x^3 + 4x - 7} = 3$.

9. $\lim\limits_{x \to \infty} \dfrac{a_0 x^n + a_1 x^{n-1} + \cdots + a_n}{b_0 x^n + b_1 x^{n-1} + \cdots + b_n} = \dfrac{a_0}{b_0}$.

10. $\lim\limits_{x \to 0} \dfrac{a_0 x^n + a_1 x^{n-1} + \cdots + a_n}{b_0 x^n + b_1 x^{n-1} + \cdots + b_n} = \dfrac{a_n}{b_n}$.

11. $\lim\limits_{x \to \infty} \dfrac{ax^4 + bx^2 + c}{dx^5 + ex^3 + fx} = 0$.

12. $\lim\limits_{x \to \infty} \dfrac{ax^4 + bx^2 + c}{dx^3 + ex^2 + fx + g} = \infty$.

13. $\lim\limits_{s \to a} \dfrac{s^4 - a^4}{s^2 - a^2} = 2a^2$.

14. $\lim\limits_{h \to 0} \dfrac{(x+h)^n - x^n}{h} = nx^{n-1}$. \hspace{2em} ($n$ = positive integer)

15. $\lim\limits_{x \to 2} \dfrac{x^2 + x - 6}{x^2 - 4} = \dfrac{5}{4}$.

16. $\lim\limits_{h \to 0} \dfrac{\sqrt{x+h} - \sqrt{x}}{h} = \dfrac{1}{2\sqrt{x}}$.

Proof. The limiting value cannot be found by substituting $h = 0$, for we then obtain (Art. 12) the indeterminate form $\dfrac{0}{0}$. We then transform the expression in a suitable manner as indicated below, namely, rationalize the numerator.

$$\dfrac{\sqrt{x+h} - \sqrt{x}}{h} \times \dfrac{\sqrt{x+h} + \sqrt{x}}{\sqrt{x+h} + \sqrt{x}} = \dfrac{x+h-x}{h(\sqrt{x+h} + \sqrt{x})} = \dfrac{1}{\sqrt{x+h} + \sqrt{x}}.$$

Hence $\quad \lim\limits_{h \to 0} \dfrac{\sqrt{x+h} - \sqrt{x}}{h} = \lim\limits_{h \to 0} \dfrac{1}{\sqrt{x+h} + \sqrt{x}} = \dfrac{1}{2\sqrt{x}}$.

17. Given $f(x) = x^2$, show that
$$\lim\limits_{h \to 0} \dfrac{f(x+h) - f(x)}{h} = 2x.$$

18. Given $f(x) = ax^2 + bx + c$, show that
$$\lim\limits_{h \to 0} \dfrac{f(x+h) - f(x)}{h} = 2ax + b.$$

19. Given $f(x) = \dfrac{1}{x}$, show that
$$\lim\limits_{h \to 0} \dfrac{f(x+h) - f(x)}{h} = -\dfrac{1}{x^2}.$$

20. If $f(x) = x^3$, find $\lim\limits_{h \to 0} \dfrac{f(x+h) - f(x)}{h}$.

VARIABLES, FUNCTIONS, AND LIMITS

19. Infinitesimals. A variable v which approaches zero as a limit is called an *infinitesimal*. This is written (Art. 14)

$$\lim v = 0 \quad \text{or} \quad v \to 0,$$

and means that the numerical value of v ultimately becomes and remains less than any preassigned positive number, however small.

If $\lim v = l$, then $\lim (v - l) = 0$; that is, *the difference between a variable and its limit is an infinitesimal.*

Conversely, *if the difference between a variable and a constant is an infinitesimal, then the variable approaches the constant as a limit.*

20. Theorems concerning infinitesimals and limits. In the following considerations all variables are assumed to be functions of the same independent variable and to approach their respective limits as this variable approaches a fixed value a. The constant ϵ is a preassigned positive number, as small as we please, but not zero.

We first prove four theorems on infinitesimals.

I. *An algebraic sum of n infinitesimals is an infinitesimal, n being a fixed number.*

For the numerical value of the sum will become and remain less than ϵ when the numerical value of each infinitesimal becomes and remains less than $\dfrac{\epsilon}{n}$.

II. *The product of a constant c by an infinitesimal is an infinitesimal.*

For the numerical value of the product will be less than ϵ when the numerical value of the infinitesimal is less than $\dfrac{\epsilon}{|c|}$.

III. *The product of n infinitesimals is an infinitesimal, n being a fixed number.*

For the numerical value of the product will become and remain less than ϵ when the numerical value of each infinitesimal becomes and remains less than the nth root of ϵ.

IV. *If $\lim v = l$, and l is not zero, then the quotient of an infinitesimal i by v is also an infinitesimal.*

For we can choose a positive number c, numerically less than l, such that the numerical value of v ultimately becomes and remains greater than c, and also such that the numerical value of i becomes and remains less than $c\epsilon$. Then the numerical value of the quotient will become and remain less than ϵ.

Proofs of the theorems of Art. 16. Let

(1) $\qquad u - A = i, \ v - B = j, \ w - C = k.$

Then i, j, k are functions of x, and each approaches zero as $x \to a$; that is, they are infinitesimals (Art. 19). From equations (1), we obtain

(2) $$u + v - w - (A + B - C) = i + j - k.$$

The right-hand member is an infinitesimal by theorem I above. Hence, by Art. 19,

(3) $$\lim_{x \to a} (u + v - w) = A + B - C.$$

From (1), we have $u = A + i$, $v = B + j$. By multiplication and transposing AB we get

(4) $$uv - AB = Aj + Bi + ij.$$

By the above theorems I–III the right-hand member is an infinitesimal, and hence

(5) $$\lim_{x \to a} uv = AB.$$

The proof is readily extended to the product uvw.

Finally, we may write

(6) $$\frac{u}{v} - \frac{A}{B} = \frac{A + i}{B + j} - \frac{A}{B} = \frac{Bi - Aj}{B(B + j)}.$$

The numerator in (6) is an infinitesimal by theorems I and II. By (3) and (4), $\lim B(B + j) = B^2$. Hence, by theorem IV, the right-hand member in (6) is an infinitesimal, and

(7) $$\lim_{x \to a} \frac{u}{v} = \frac{A}{B}.$$

Hence the statements in Art. 16 are proved.

PROBLEMS

Using equations (1) and the methods indicated above, prove the following statements.

1. $\lim_{x \to a} u^3 = A^3.$

2. $\lim_{x \to a} uvw = ABC.$

3. $\lim_{x \to a} \sqrt{u} = \sqrt{A}.$

CHAPTER III

DIFFERENTIATION

21. Introduction. We shall now proceed to investigate the manner in which a function changes in value as the independent variable changes. The fundamental problem of the differential calculus is to establish a measure of this change in the function with mathematical precision. It was while investigating problems of this sort, dealing with continuously varying quantities, that Newton* was led to the discovery of the fundamental principles of the calculus, the most scientific and powerful tool of the modern mathematician.

22. Increments. The *increment* of a variable in changing from one numerical value to another is the *difference* found by subtracting the first value from the second. An increment of x is denoted by the symbol Δx, read "delta x." The student is warned against reading this symbol "delta times x."

Evidently this increment may be either positive or negative † according as the variable in changing increases or decreases. Similarly,

Δy denotes an increment of y,
$\Delta \phi$ denotes an increment of ϕ,
$\Delta f(x)$ denotes an increment of $f(x)$, etc.

If in $y = f(x)$ the independent variable x takes on an increment Δx, then Δy will denote the corresponding increment of the function $f(x)$ (or dependent variable y).

The increment Δy is always to be reckoned from the definite initial value of y corresponding to the arbitrarily fixed initial value of x from which the increment Δx is reckoned. For instance, consider the function

$$y = x^2.$$

---

* Sir Isaac Newton (1642–1727), an Englishman, was a man of the most extraordinary genius. He developed the science of the calculus under the name of Fluxions. Although Newton had discovered and made use of the new science as early as 1670, his first published work in which it occurs is dated 1687, having the title "Philosophiae Naturalis Principia Mathematica." This was Newton's principal work. Laplace said of it, "It will always remain preëminent above all other productions of the human mind." See frontispiece.

† Some writers call a *negative increment* a *decrement*.

DIFFERENTIAL CALCULUS

Assuming $x = 10$ for the initial value of x fixes $y = 100$ as the initial value of y.

Suppose x increases to $x = 12$, that is, $\Delta x = 2$;
then $\quad y$ increases to $y = 144$, and $\Delta y = 44$.
Suppose x decreases to $x = 9$, that is, $\Delta x = -1$;
then $\quad y$ decreases to $y = 81$, and $\Delta y = -19$.

In the above example, y increases when x increases and y decreases when x decreases. The corresponding values of Δx and Δy have like signs. It may happen that y decreases as x increases, or the reverse; in either case Δx and Δy will then have opposite signs.

23. Comparison of increments. Consider the function

(1) $$y = x^2.$$

Assuming a fixed initial value for x, let x take on an increment Δx. Then y will take on a corresponding increment Δy, and we have

$$y + \Delta y = (x + \Delta x)^2,$$

or $\quad y + \Delta y = x^2 + 2x \cdot \Delta x + (\Delta x)^2.$
Subtracting (1), $\quad y \quad\quad = x^2$
(2) $\quad\quad\quad \Delta y = \quad\quad 2x \cdot \Delta x + (\Delta x)^2$

we get the increment Δy in terms of x and Δx.

To find the ratio of the increments, divide both members of (2) by Δx, giving

$$\frac{\Delta y}{\Delta x} = 2x + \Delta x.$$

If the initial value of x is 4, it is evident (Art. 16) that

$$\lim_{\Delta x \to 0} \frac{\Delta y}{\Delta x} = 8.$$

Let us carefully note the behavior of the ratio of the increments of x and y as the increment of x diminishes.

Initial Value of x	New Value of x	Increment Δx	Initial Value of y	New Value of y	Increment Δy	$\frac{\Delta y}{\Delta x}$
4	5.0	1.0	16	25.	9.	9.
4	4.8	0.8	16	23.04	7.04	8.8
4	4.6	0.6	16	21.16	5.16	8.6
4	4.4	0.4	16	19.36	3.36	8.4
4	4.2	0.2	16	17.64	1.64	8.2
4	4.1	0.1	16	16.81	0.81	8.1
4	4.01	0.01	16	16.0801	0.0801	8.01

DIFFERENTIATION

It is apparent that as Δx decreases, Δy also diminishes, but their ratio takes on the successive values 9, 8.8, 8.6, 8.4, 8.2, 8.1, 8.01, illustrating the fact that the value of $\frac{\Delta y}{\Delta x}$ can be brought as near to 8 as we please by making Δx sufficiently small. Therefore

$$\lim_{\Delta x \to 0} \frac{\Delta y}{\Delta x} = 8.$$

24. Derivative of a function of one variable. The fundamental definition of the differential calculus is as follows.

The derivative* of a function is the limit of the ratio of the increment of the function to the increment of the independent variable, when the latter increment varies and approaches zero as a limit.

When the limit of this ratio exists, the function is said to be *differentiable*, or to *possess a derivative*.

The above definition may be given in a more compact form *symbolically* as follows. Given the function

(1) $$y = f(x),$$

and consider x to have a fixed value.

Let x take on an increment Δx; then the function y takes on an increment Δy, the new value of the function being

(2) $$y + \Delta y = f(x + \Delta x).$$

To find the increment of the function, subtract (1) from (2), giving

(3) $$\Delta y = f(x + \Delta x) - f(x).$$

Dividing both members by the increment of the variable, Δx,

(4) $$\frac{\Delta y}{\Delta x} = \frac{f(x + \Delta x) - f(x)}{\Delta x}.$$

The limit of the right-hand member when $\Delta x \to 0$ is, from the definition, the *derivative* of $f(x)$, or by (1), of y, and is *denoted by the symbol* $\frac{dy}{dx}$. Therefore

(A) $$\frac{dy}{dx} = \lim_{\Delta x \to 0} \frac{f(x + \Delta x) - f(x)}{\Delta x}$$

defines the *derivative of y [or $f(x)$] with respect to x.*

From (4) we get also

$$\frac{dy}{dx} = \lim_{\Delta x \to 0} \frac{\Delta y}{\Delta x}.$$

---

* Also called the *differential coefficient* or the *derived function*.

Similarly, if u is a function of t, then

$$\frac{du}{dt} = \lim_{\Delta t \to 0} \frac{\Delta u}{\Delta t} = \text{derivative of } u \text{ with respect to } t.$$

The process of finding the derivative of a function is called *differentiation*.

25. Symbols for derivatives. Since Δy and Δx are always finite and have definite values, the expression

$$\frac{\Delta y}{\Delta x}$$

is really a fraction. The symbol

$$\frac{dy}{dx},$$

however, is to be regarded *not as a fraction but as the limiting value of a fraction*. In many cases it will be seen that this symbol does possess fractional properties, and later on we shall show how meanings may be attached to dy and dx, but for the present the symbol $\frac{dy}{dx}$ is to be considered as a whole.

Since the derivative of a function of x is in general also a function of x, the symbol $f'(x)$ is also used to denote the derivative of $f(x)$. Hence, if
$$y = f(x),$$
we may write
$$\frac{dy}{dx} = f'(x),$$
which is read "the derivative of y with respect to x equals f prime of x." The symbol

$$\frac{d}{dx}$$

when considered by itself is called the *differentiating operator*, and indicates that any function written after it is to be differentiated with respect to x. Thus,

$\frac{dy}{dx}$ or $\frac{d}{dx} y$ indicates the derivative of y with respect to x;

$\frac{d}{dx} f(x)$ indicates the derivative of $f(x)$ with respect to x;

$\frac{d}{dx}(2x^2 + 5)$ indicates the derivative of $2x^2 + 5$ with respect to x.

y' is an abbreviated form of $\frac{dy}{dx}$.

The symbol D_x is used by some writers instead of $\frac{d}{dx}$. If, then,
$$y = f(x),$$

we may write the identities

$$y' = \frac{dy}{dx} = \frac{d}{dx} y = \frac{d}{dx} f(x) = D_x f(x) = f'(x).$$

It must be emphasized that the variable, in the essential step of letting $\Delta x \to 0$, is Δx, and not x. The value of the latter is assumed fixed from the start. To emphasize that $x = x_0$ throughout, we may write

$$f'(x_0) = \lim_{\Delta x \to 0} \frac{f(x_0 + \Delta x) - f(x_0)}{\Delta x}.$$

26. Differentiable functions. From the Theory of Limits it is clear that if the derivative of a function exists for a certain value of the independent variable, the function itself must be continuous for that value of the variable.

The converse, however, is not always true, functions having been discovered that are continuous and yet possess no derivative. But such functions do not occur often in applied mathematics, and *in this book only differentiable functions are considered*, that is, functions that possess a derivative for all values of the independent variable save at most for isolated values.

27. General Rule for Differentiation. From the definition of a derivative it is seen that the process of differentiating a function $y = f(x)$ consists in taking the following distinct steps.

General Rule for Differentiation*

First Step. *In the function replace x by $x + \Delta x$, and calculate the new value of the function, $y + \Delta y$.*

Second Step. *Subtract the given value of the function from the new value and thus find Δy (the increment of the function).*

Third Step. *Divide the remainder Δy (the increment of the function) by Δx (the increment of the independent variable).*

Fourth Step. *Find the limit of this quotient when Δx (the increment of the independent variable) varies and approaches zero as a limit. This is the derivative required.*

The student should become thoroughly familiar with this rule by applying the process to a large number of examples. Three such examples will now be worked out in detail. Note that the theorems of Art. 16 are used in the Fourth Step, x being held constant.

* Also called the *Four-Step Rule*.

DIFFERENTIAL CALCULUS

ILLUSTRATIVE EXAMPLE 1. Differentiate $3x^2 + 5$.

Solution. Applying the successive steps in the General Rule, we get, after placing
$$y = 3x^2 + 5,$$

First Step.
$$\begin{aligned} y + \Delta y &= 3(x + \Delta x)^2 + 5 \\ &= 3x^2 + 6x \cdot \Delta x + 3(\Delta x)^2 + 5. \end{aligned}$$

Second Step.
$$\begin{aligned} y + \Delta y &= 3x^2 + 6x \cdot \Delta x + 3(\Delta x)^2 + 5 \\ y &= 3x^2 \hspace{4cm} + 5 \\ \hline \Delta y &= \hspace{1cm} 6x \cdot \Delta x + 3(\Delta x)^2 \end{aligned}$$

Third Step. $\quad \dfrac{\Delta y}{\Delta x} = 6x + 3 \cdot \Delta x.$

Fourth Step. In the right-hand member let $\Delta x \to 0$. Then by **(A)**
$$\frac{dy}{dx} = 6x. \text{ Ans.}$$

Or
$$y' = \frac{d}{dx}(3x^2 + 5) = 6x.$$

ILLUSTRATIVE EXAMPLE 2. Differentiate $x^3 - 2x + 7$.

Solution. Place $\quad y = x^3 - 2x + 7.$

First Step.
$$\begin{aligned} y + \Delta y &= (x + \Delta x)^3 - 2(x + \Delta x) + 7 \\ &= x^3 + 3x^2 \cdot \Delta x + 3x \cdot (\Delta x)^2 + (\Delta x)^3 - 2x - 2 \cdot \Delta x + 7. \end{aligned}$$

Second Step.
$$\begin{aligned} y + \Delta y &= x^3 + 3x^2 \cdot \Delta x + 3x \cdot (\Delta x)^2 + (\Delta x)^3 - 2x - 2 \cdot \Delta x + 7 \\ y &= x^3 \hspace{6cm} - 2x \hspace{2cm} + 7 \\ \hline \Delta y &= \hspace{1cm} 3x^2 \cdot \Delta x + 3x \cdot (\Delta x)^2 + (\Delta x)^3 \hspace{1cm} - 2 \cdot \Delta x \end{aligned}$$

Third Step. $\quad \dfrac{\Delta y}{\Delta x} = 3x^2 + 3x \cdot \Delta x + (\Delta x)^2 - 2.$

Fourth Step. In the right-hand member let $\Delta x \to 0$. Then by **(A)**
$$\frac{dy}{dx} = 3x^2 - 2. \text{ Ans.}$$

Or
$$y' = \frac{d}{dx}(x^3 - 2x + 7) = 3x^2 - 2.$$

ILLUSTRATIVE EXAMPLE 3. Differentiate $\dfrac{c}{x^2}$.

Solution. Place $\quad y = \dfrac{c}{x^2}.$

First Step. $\quad y + \Delta y = \dfrac{c}{(x + \Delta x)^2}.$

Second Step. $\quad y + \Delta y = \dfrac{c}{(x + \Delta x)^2}.$
$$y = \dfrac{c}{x^2}$$
$$\Delta y = \dfrac{c}{(x + \Delta x)^2} - \dfrac{c}{x^2} = \dfrac{-c \cdot \Delta x(2x + \Delta x)}{x^2(x + \Delta x)^2}.$$

Third Step. $\quad \dfrac{\Delta y}{\Delta x} = -c \cdot \dfrac{2x + \Delta x}{x^2(x + \Delta x)^2}.$

Fourth Step. In the right-hand member let $\Delta x \to 0$. Then by **(A)**
$$\frac{dy}{dx} = -c \cdot \frac{2x}{x^2(x)^2} = -\frac{2c}{x^3}. \text{ Ans.} \quad \left(y' = \frac{d}{dx}\left(\frac{c}{x^2}\right) = -\frac{2c}{x^3}.\right)$$

PROBLEMS

Use the General Rule in differentiating each of the following functions.

1. $y = 2 - 3x$. Ans. $y' = -3$.
2. $y = mx + b$. $y' = m$.
3. $y = ax^2$. $y' = 2ax$.
4. $s = 2t - t^2$. $s' = 2 - 2t$.
5. $y = cx^3$. $y' = 3cx^2$.
6. $y = 3x - x^3$. $y' = 3 - 3x^2$.
7. $u = 4v^2 + 2v^3$. $u' = 8v + 6v^2$.
8. $y = x^4$. $y' = 4x^3$.
9. $\rho = \dfrac{2}{\theta+1}$. $\dfrac{d\rho}{d\theta} = -\dfrac{2}{(\theta+1)^2}$.
10. $y = \dfrac{3}{x^2+2}$. $\dfrac{dy}{dx} = -\dfrac{6x}{(x^2+2)^2}$.
11. $s = \dfrac{t+4}{t}$. $\dfrac{ds}{dt} = -\dfrac{4}{t^2}$.
12. $y = \dfrac{1}{1-2x}$. Ans. $\dfrac{dy}{dx} = \dfrac{2}{(1-2x)^2}$.
13. $\rho = \dfrac{\theta}{\theta+2}$. $\dfrac{d\rho}{d\theta} = \dfrac{2}{(\theta+2)^2}$.
14. $s = \dfrac{At+B}{Ct+D}$. $\dfrac{ds}{dt} = \dfrac{AD-BC}{(Ct+D)^2}$.
15. $y = \dfrac{x^3+1}{x}$. $\dfrac{dy}{dx} = 2x - \dfrac{1}{x^2}$.
16. $y = \dfrac{1}{x^2+a^2}$. $\dfrac{dy}{dx} = -\dfrac{2x}{(x^2+a^2)^2}$.
17. $y = \dfrac{x}{x^2+1}$. $\dfrac{dy}{dx} = \dfrac{1-x^2}{(x^2+1)^2}$.
18. $y = \dfrac{x^2}{4-x^2}$. $\dfrac{dy}{dx} = \dfrac{8x}{(4-x^2)^2}$.

19. $y = 3x^2 - 4x - 5$.
20. $s = at^2 + bt + c$.
21. $u = 2v^3 - 3v^2$.
22. $y = ax^3 + bx^2 + cx + d$.
23. $\rho = (a - b\theta)^2$.
24. $y = (2-x)(1-2x)$.
25. $y = (Ax + B)(Cx + D)$.
26. $s = (a + bt)^3$.
27. $y = \dfrac{x}{a+bx^2}$.
28. $y = \dfrac{a+bx^2}{x^2}$.
29. $y = \dfrac{x^2}{a+bx^2}$.

28. Interpretation of the derivative by geometry. We shall now consider a theorem which is fundamental in all applications of the differential calculus to geometry. It is necessary to recall the definition of the *tangent line* to a curve at a point P on the curve. Let a secant be drawn through P and a neighboring point Q on the curve (see figure). Let Q move along the curve and approach P indefinitely. Then the secant will revolve about P, and its limiting position is the tangent line at P. Let

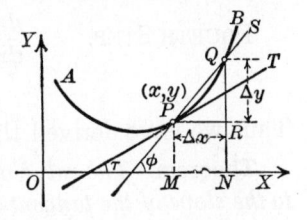

(1) $\qquad\qquad\qquad y = f(x)$

be the equation of a curve AB. This curve is the graph of $f(x)$ (see figure).

Now differentiate (1) by the General Rule and interpret each step geometrically from the figure (p. 25). We choose a point $P(x, y)$ on the curve, and a second point $Q(x + \Delta x, y + \Delta y)$ near P, also on the curve.

First Step. $\qquad y + \Delta y = f(x + \Delta x) \qquad = NQ$

Second Step. $\qquad y + \Delta y = f(x + \Delta x) \qquad = NQ$

$$\underline{y \qquad\quad = f(x) \qquad\qquad\ = MP = NR}$$
$$\Delta y = f(x + \Delta x) - f(x) \ = RQ$$

Third Step. $\qquad \dfrac{\Delta y}{\Delta x} = \dfrac{f(x + \Delta x) - f(x)}{\Delta x} = \dfrac{RQ}{MN} = \dfrac{RQ}{PR}$

$\qquad\qquad\qquad\quad = \tan \angle RPQ = \tan \phi$
$\qquad\qquad\qquad\quad =$ slope of secant line PQ.

At this point, therefore, we see that the ratio of the increments Δy and Δx equals the slope of the secant line drawn through the points $P(x, y)$ and $Q(x + \Delta x, y + \Delta y)$ on the graph of $f(x)$.

Let us examine the geometric meaning of the Fourth Step. The value of x is now regarded as fixed. Hence P is a *fixed point* on the graph. Also, Δx is to vary and approach zero as a limit. Obviously, therefore, the point Q is to move *along the curve* and approach P as a limiting position. The secant line drawn through P and Q will then turn about P and approach the tangent line at P as its limiting position. In the figure,

$$\phi = \text{inclination of the secant line } PQ,$$
$$\tau = \text{inclination of the tangent line } PT.$$

Then $\lim\limits_{\Delta x \to 0} \phi = \tau$. Assuming that $\tan \phi$ is a continuous function (see Art. 70), we have, therefore,

Fourth Step. $\qquad \dfrac{dy}{dx} = f'(x) = \lim\limits_{\Delta x \to 0} \tan \phi = \tan \tau,$

$\qquad\qquad\qquad\quad\ = $ *slope of the tangent line at P.*

Thus we have derived the important

Theorem. *The value of the derivative at any point of a curve is equal to the slope of the tangent line to the curve at that point.*

It was this tangent problem that led Leibnitz[*] to the discovery of the differential calculus.

---

[*] Gottfried Wilhelm Leibnitz (1646–1716) was a native of Leipzig. His remarkable abilities were shown by original investigations in several branches of learning. He was first to publish his discoveries in calculus in a short essay appearing in the periodical *Acta Eruditorum* at Leipzig in 1684. It is known, however, that manuscripts on Fluxions

DIFFERENTIATION

ILLUSTRATIVE EXAMPLE. Find the slopes of the tangents to the parabola $y = x^2$ at the vertex and at the point where $x = \frac{1}{2}$.

Solution. Differentiating by the General Rule (Art. 27), we get

(2) $\dfrac{dy}{dx} = 2x =$ slope of tangent line at *any* point (x, y) on curve.

To find slope of tangent at the vertex, substitute $x = 0$ in (2), giving
$$\dfrac{dy}{dx} = 0.$$

Therefore the tangent at the vertex has the slope zero; that is, it is parallel to the x-axis and in this case coincides with it.

To find the slope of the tangent at the point P, where $x = \frac{1}{2}$, substitute in (2), giving
$$\dfrac{dy}{dx} = 1;$$

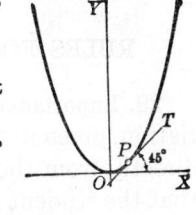

that is, the tangent at the point P makes an angle of $45°$ with the x-axis.

PROBLEMS

Find by differentiation the slope and inclination of the tangent line to each of the following curves at the point indicated. Verify the result by drawing the curve and the tangent line.

1. $y = x^2 - 2$, where $x = 1$.   Ans. 2; $63° 26'$.

2. $y = 2x - \frac{1}{2}x^2$, where $x = 3$.   4. $y = 3 + 3x - x^3$, where $x = -1$.

3. $y = \dfrac{4}{x-1}$, where $x = 2$.   5. $y = \dfrac{x}{x+1}$, where $x = 1$.

6. Find the point on the curve $y = 7x - 3x^2$ where the inclination of the tangent line is $45°$.   Ans. $(1, 4)$.

7. Find the points on the curve $3y = x^3 - 3x$ where the tangent line is parallel to the line $y = 3x$.   Ans. $(2, \frac{2}{3})$, $(-2, -\frac{2}{3})$.

In each of the three following problems find (a) the points of intersection of the given pair of curves; (b) the slope and inclination of the tangent line to each curve, and the angle between the tangent lines, at each point of intersection (see (2), p. 3).

8. $y = 1 - x^2$,   Ans. Angle of intersection $=$ arc tan $\frac{4}{3} = 53° 8'$.
$y = x^2 - 1$

9. $y = x^2$,   10. $y = x^3$,
$x + y - 2 = 0$. $y = 1 + x - x^2$.

11. Find the angle of intersection between the curves $9y = x^3$ and $y = 6 + 8x - x^3$ at the point $(3, 3)$.   Ans. $21° 27'$.

written by Newton were already in existence, and from these some claim Leibnitz got the new ideas. The decision of modern times seems to be that both Newton and Leibnitz invented the calculus independently of each other. The notation used today was introduced by Leibnitz.

CHAPTER IV

RULES FOR DIFFERENTIATING ALGEBRAIC FORMS

29. Importance of the General Rule. The General Rule for differentiation, given in the last chapter (Art. 27) is fundamental, being found directly from the definition of a derivative, and it is very important that the student should be thoroughly familiar with it. However, the process of applying the rule to examples in general has been found too tedious or difficult: consequently special rules have been derived from the General Rule for differentiating certain standard forms of frequent occurrence in order to facilitate the work.

It has been found convenient to express these special rules by means of formulas, a list of which follows. The student should not only memorize each formula when deduced, but should be able to state the corresponding rule in words.

In these formulas u, v, and w denote functions of x which are differentiable.

Formulas for Differentiation

I $\qquad \dfrac{dc}{dx} = 0.$

II $\qquad \dfrac{dx}{dx} = 1.$

III $\qquad \dfrac{d}{dx}(u + v - w) = \dfrac{du}{dx} + \dfrac{dv}{dx} - \dfrac{dw}{dx}.$

IV $\qquad \dfrac{d}{dx}(cv) = c\dfrac{dv}{dx}.$

V $\qquad \dfrac{d}{dx}(uv) = u\dfrac{dv}{dx} + v\dfrac{du}{dx}.$

VI $\qquad \dfrac{d}{dx}(v^n) = nv^{n-1}\dfrac{dv}{dx}.$

VI a $\qquad \dfrac{d}{dx}(x^n) = nx^{n-1}.$

VII $\qquad \dfrac{d}{dx}\left(\dfrac{u}{v}\right) = \dfrac{v\dfrac{du}{dx} - u\dfrac{dv}{dx}}{v^2}.$

RULES FOR DIFFERENTIATING ALGEBRAIC FORMS

VII a $$\frac{d}{dx}\left(\frac{u}{c}\right) = \frac{\frac{du}{dx}}{c}.$$

VIII $$\frac{dy}{dx} = \frac{dy}{dv} \cdot \frac{dv}{dx}, \text{ } y \text{ being a function of } v.$$

IX $$\frac{dy}{dx} = \frac{1}{\frac{dx}{dy}}, \text{ } y \text{ being a function of } x.$$

30. Differentiation of a constant. A function that is known to have the same value for every value of the independent variable is constant, and we may denote it by
$$y = c.$$

As x takes on an increment Δx, the function does not change in value, that is, $\Delta y = 0$, and
$$\frac{\Delta y}{\Delta x} = 0.$$

But $$\lim_{\Delta x \to 0} \frac{\Delta y}{\Delta x} = \frac{dy}{dx} = 0.$$

I $$\therefore \frac{dc}{dx} = 0.$$

The derivative of a constant is zero.

This result is readily foreseen. For the locus of $y = c$ is a straight line parallel to OX, and its slope is therefore zero. But the slope is the value of the derivative (Art. 28).

31. Differentiation of a variable with respect to itself

Let $$y = x.$$
Following the General Rule (Art. 27), we have

FIRST STEP. $\quad y + \Delta y = x + \Delta x.$

SECOND STEP. $\quad \Delta y = \Delta x.$

THIRD STEP. $\quad \dfrac{\Delta y}{\Delta x} = 1.$

FOURTH STEP. $\quad \dfrac{dy}{dx} = 1.$

II $\quad \therefore \dfrac{dx}{dx} = 1.$

The derivative of a variable with respect to itself is unity.

This result is readily foreseen. For the slope of the line $y = x$ is unity.

32. Differentiation of a sum

Let $$y = u + v - w.$$

By the General Rule,

FIRST STEP. $\quad y + \Delta y = u + \Delta u + v + \Delta v - w - \Delta w.$

SECOND STEP. $\quad \Delta y = \Delta u + \Delta v - \Delta w.$

THIRD STEP. $\quad \dfrac{\Delta y}{\Delta x} = \dfrac{\Delta u}{\Delta x} + \dfrac{\Delta v}{\Delta x} - \dfrac{\Delta w}{\Delta x}.$

Now (Art. 24),

$$\lim_{\Delta x \to 0} \frac{\Delta u}{\Delta x} = \frac{du}{dx}, \quad \lim_{\Delta x \to 0} \frac{\Delta v}{\Delta x} = \frac{dv}{dx}, \quad \lim_{\Delta x \to 0} \frac{\Delta w}{\Delta x} = \frac{dw}{dx}.$$

Hence, by (1), Art. 16,

FOURTH STEP. $\quad \dfrac{dy}{dx} = \dfrac{du}{dx} + \dfrac{dv}{dx} - \dfrac{dw}{dx}.$

III $\quad \therefore \dfrac{d}{dx}(u + v - w) = \dfrac{du}{dx} + \dfrac{dv}{dx} - \dfrac{dw}{dx}.$

A similar proof holds for the algebraic sum of any number of functions.

The derivative of the algebraic sum of n functions is equal to the same algebraic sum of their derivatives, n being a fixed number.

33. Differentiation of the product of a constant and a function

Let $$y = cv.$$

By the General Rule,

FIRST STEP. $\quad y + \Delta y = c(v + \Delta v) = cv + c\Delta v.$

SECOND STEP. $\quad \Delta y = c\Delta v.$

THIRD STEP. $\quad \dfrac{\Delta y}{\Delta x} = c\dfrac{\Delta v}{\Delta x}.$

Whence, by (4), Art. 16,

FOURTH STEP. $\quad \dfrac{dy}{dx} = c\dfrac{dv}{dx}.$

IV $\quad \therefore \dfrac{d}{dx}(cv) = c\dfrac{dv}{dx}.$

The derivative of the product of a constant and a function is equal to the product of the constant and the derivative of the function.

34. Differentiation of the product of two functions

Let $$y = uv.$$

By the General Rule,

FIRST STEP. $\quad y + \Delta y = (u + \Delta u)(v + \Delta v).$

RULES FOR DIFFERENTIATING ALGEBRAIC FORMS

Multiplied out, this becomes
$$y + \Delta y = uv + u\Delta v + v\Delta u + \Delta u \Delta v.$$

SECOND STEP. $\quad \Delta y = u\Delta v + v\Delta u + \Delta u \Delta v.$

THIRD STEP. $\quad \dfrac{\Delta y}{\Delta x} = u\dfrac{\Delta v}{\Delta x} + v\dfrac{\Delta u}{\Delta x} + \Delta u \dfrac{\Delta v}{\Delta x}.$

Applying (2) and (4), Art. 16, noting that $\lim\limits_{\Delta x \to 0} \Delta u = 0$, and hence that the limit of the product $\Delta u \dfrac{\Delta v}{\Delta x}$ is zero, we have

FOURTH STEP. $\quad \dfrac{dy}{dx} = u\dfrac{dv}{dx} + v\dfrac{du}{dx}.$

V $\quad \therefore \dfrac{d}{dx}(uv) = u\dfrac{dv}{dx} + v\dfrac{du}{dx}.$

The derivative of the product of two functions is equal to the first function times the derivative of the second, plus the second function times the derivative of the first.

35. Differentiation of the product of n functions, n being a fixed number. When both sides of **V** are divided by uv, this formula assumes the form

$$\frac{\dfrac{d}{dx}(uv)}{uv} = \frac{\dfrac{du}{dx}}{u} + \frac{\dfrac{dv}{dx}}{v}.$$

If, then, we have the product of n functions
$$y = v_1 v_2 \cdots v_n,$$
we may write

$$\frac{\dfrac{d}{dx}(v_1 v_2 \cdots v_n)}{v_1 v_2 \cdots v_n} = \frac{\dfrac{dv_1}{dx}}{v_1} + \frac{\dfrac{d}{dx}(v_2 v_3 \cdots v_n)}{v_2 v_3 \cdots v_n}$$

$$= \frac{\dfrac{dv_1}{dx}}{v_1} + \frac{\dfrac{dv_2}{dx}}{v_2} + \frac{\dfrac{d}{dx}(v_3 v_4 \cdots v_n)}{v_3 v_4 \cdots v_n}$$

$$= \frac{\dfrac{dv_1}{dx}}{v_1} + \frac{\dfrac{dv_2}{dx}}{v_2} + \frac{\dfrac{dv_3}{dx}}{v_3} + \cdots + \frac{\dfrac{dv_n}{dx}}{v_n}.$$

Multiplying both sides by $v_1 v_2 \cdots v_n$, we get

$$\frac{d}{dx}(v_1 v_2 \cdots v_n) = (v_2 v_3 \cdots v_n)\frac{dv_1}{dx} + (v_1 v_3 \cdots v_n)\frac{dv_2}{dx} + \cdots$$
$$+ (v_1 v_2 \cdots v_{n-1})\frac{dv_n}{dx}.$$

The derivative of the product of n functions, n being a fixed number, is equal to the sum of the n products that can be formed by multiplying the derivative of each function by all the other functions.

36. Differentiation of a function with a constant exponent. The Power Rule. If the n factors in the above result are each equal to v, we get

$$\frac{\dfrac{d}{dx}(v^n)}{v^n} = n\frac{\dfrac{dv}{dx}}{v}.$$

VI $$\therefore \frac{d}{dx}(v^n) = nv^{n-1}\frac{dv}{dx}.$$

When $v = x$, this becomes

VI a $$\frac{d}{dx}(x^n) = nx^{n-1}.$$

We have so far proved **VI** only for the case when n is a positive integer. In Art. 65, however, it will be shown that this formula holds true for any value of n, and we shall make use of this general result now.

The derivative of a function with a constant exponent is equal to the product of the exponent, the function with the exponent diminished by unity, and the derivative of the function.

This rule is called the *Power Rule*.

37. Differentiation of a quotient

Let $$y = \frac{u}{v}. \qquad (v \neq 0)$$

By the General Rule,

FIRST STEP. $$y + \Delta y = \frac{u + \Delta u}{v + \Delta v}.$$

SECOND STEP. $$\Delta y = \frac{u + \Delta u}{v + \Delta v} - \frac{u}{v} = \frac{v \cdot \Delta u - u \cdot \Delta v}{v(v + \Delta v)}.$$

THIRD STEP. $$\frac{\Delta y}{\Delta x} = \frac{v\dfrac{\Delta u}{\Delta x} - u\dfrac{\Delta v}{\Delta x}}{v(v + \Delta v)}.$$

Applying (1)–(4), Art. 16,

FOURTH STEP. $$\frac{dy}{dx} = \frac{v\dfrac{du}{dx} - u\dfrac{dv}{dx}}{v^2}.$$

VII $$\therefore \frac{d}{dx}\left(\frac{u}{v}\right) = \frac{v\dfrac{du}{dx} - u\dfrac{dv}{dx}}{v^2}.$$

RULES FOR DIFFERENTIATING ALGEBRAIC FORMS

The derivative of a fraction is equal to the denominator times the derivative of the numerator, minus the numerator times the derivative of the denominator, all divided by the square of the denominator.

When the denominator is constant, set $v = c$ in **VII**, giving

VII a
$$\frac{d}{dx}\left(\frac{u}{c}\right) = \frac{\frac{du}{dx}}{c}.$$

$$\left[\text{Since } \frac{dv}{dx} = \frac{dc}{dx} = 0.\right]$$

We may also get **VII a** from **IV** as follows:

$$\frac{d}{dx}\left(\frac{u}{c}\right) = \frac{1}{c}\frac{du}{dx} = \frac{\frac{du}{dx}}{c}.$$

The derivative of the quotient of a function by a constant is equal to the derivative of the function divided by the constant.

PROBLEMS*

Differentiate the following functions.

1. $y = x^3$.

Solution. $\frac{dy}{dx} = \frac{d}{dx}(x^3) = 3x^2$. Ans.   By VI a

$[n = 3.]$

2. $y = ax^4 - bx^2$.

Solution. $\frac{dy}{dx} = \frac{d}{dx}(ax^4 - bx^2) = \frac{d}{dx}(ax^4) - \frac{d}{dx}(bx^2)$ by III

$= a\frac{d}{dx}(x^4) - b\frac{d}{dx}(x^2)$ by IV

$= 4ax^3 - 2bx$. Ans.   By VI a

3. $y = x^{\frac{4}{3}} + 5$.

Solution. $\frac{dy}{dx} = \frac{d}{dx}(x^{\frac{4}{3}}) + \frac{d}{dx}(5)$ by III

$= \frac{4}{3}x^{\frac{1}{3}}$. Ans.   By VI a and I

4. $y = \frac{3x^3}{\sqrt[5]{x^2}} - \frac{7x}{\sqrt[8]{x^4}} + 8\sqrt[7]{x^3}$.

Solution. $\frac{dy}{dx} = \frac{d}{dx}(3x^{\frac{13}{5}}) - \frac{d}{dx}(7x^{\frac{1}{2}}) + \frac{d}{dx}(8x^{\frac{3}{7}})$ by III

$= \frac{39}{5}x^{\frac{8}{5}} + \frac{7}{3}x^{-\frac{4}{3}} + \frac{24}{7}x^{-\frac{4}{7}}$. Ans.   By IV and VI a

*When learning to differentiate, the student should have oral drill in differentiating simple functions.

5. $y = (x^2 - 3)^5$.

Solution. $\dfrac{dy}{dx} = 5(x^2 - 3)^4 \dfrac{d}{dx}(x^2 - 3)$ by VI

$[v = x^2 - 3, \text{ and } n = 5.]$

$= 5(x^2 - 3)^4 \cdot 2x = 10\,x(x^2 - 3)^4$.   Ans.

We might have expanded this function by the Binomial Theorem ((3), p. 1), and then applied III, etc., but the above process is to be preferred.

6. $y = \sqrt{a^2 - x^2}$.

Solution. $\dfrac{dy}{dx} = \dfrac{d}{dx}(a^2 - x^2)^{\frac{1}{2}} = \dfrac{1}{2}(a^2 - x^2)^{-\frac{1}{2}} \dfrac{d}{dx}(a^2 - x^2)$ by VI

$[v = a^2 - x^2, \text{ and } n = \tfrac{1}{2}.]$

$= \dfrac{1}{2}(a^2 - x^2)^{-\frac{1}{2}}(-2x) = -\dfrac{x}{\sqrt{a^2 - x^2}}$.   Ans.

7. $y = (3x^2 + 2)\sqrt{1 + 5x^2}$.

Solution. $\dfrac{dy}{dx} = (3x^2 + 2)\dfrac{d}{dx}(1 + 5x^2)^{\frac{1}{2}} + (1 + 5x^2)^{\frac{1}{2}}\dfrac{d}{dx}(3x^2 + 2)$ by V

$[u = 3x^2 + 2, \text{ and } v = (1 + 5x^2)^{\frac{1}{2}}.]$

$= (3x^2 + 2)\dfrac{1}{2}(1 + 5x^2)^{-\frac{1}{2}}\dfrac{d}{dx}(1 + 5x^2) + (1 + 5x^2)^{\frac{1}{2}} 6x$ by VI etc.

$= (3x^2 + 2)(1 + 5x^2)^{-\frac{1}{2}} 5x + 6x(1 + 5x^2)^{\frac{1}{2}}$

$= \dfrac{5x(3x^2 + 2)}{\sqrt{1 + 5x^2}} + 6x\sqrt{1 + 5x^2} = \dfrac{45x^3 + 16x}{\sqrt{1 + 5x^2}}$.   Ans.

8. $y = \dfrac{a^2 + x^2}{\sqrt{a^2 - x^2}}$.

Solution. $\dfrac{dy}{dx} = \dfrac{(a^2 - x^2)^{\frac{1}{2}}\dfrac{d}{dx}(a^2 + x^2) - (a^2 + x^2)\dfrac{d}{dx}(a^2 - x^2)^{\frac{1}{2}}}{a^2 - x^2}$ by VII

$= \dfrac{2x(a^2 - x^2) + x(a^2 + x^2)}{(a^2 - x^2)^{\frac{3}{2}}}$

[Multiplying both numerator and denominator by $(a^2 - x^2)^{\frac{1}{2}}$.]

$= \dfrac{3a^2x - x^3}{(a^2 - x^2)^{\frac{3}{2}}}$.   Ans.

Prove each of the following differentiations.

9. $\dfrac{d}{dx}(3x^4 - 2x^2 + 8) = 12x^3 - 4x$.

10. $\dfrac{d}{dx}(4 + 3x - 2x^3) = 3 - 6x^2$.

11. $\dfrac{d}{dt}(at^5 - 5bt^3) = 5at^4 - 15bt^2$.

12. $\dfrac{d}{dz}\left(\dfrac{z^2}{2} - \dfrac{z^7}{7}\right) = z - z^6$.

RULES FOR DIFFERENTIATING ALGEBRAIC FORMS 35

13. $\dfrac{d}{dx}\sqrt{v} = \dfrac{1}{2\sqrt{v}}\dfrac{dv}{dx}$.

14. $\dfrac{d}{dx}\left(\dfrac{2}{x} - \dfrac{3}{x^2}\right) = -\dfrac{2}{x^2} + \dfrac{6}{x^3}$.

15. $\dfrac{d}{dt}\left(2\,t^{\frac{4}{3}} - 3\,t^{\frac{2}{3}}\right) = \tfrac{8}{3}\,t^{\frac{1}{3}} - 2\,t^{-\frac{1}{3}}$.

16. $\dfrac{d}{dx}\left(2\,x^{\frac{3}{4}} + 4\,x^{-\frac{1}{4}}\right) = \tfrac{3}{2}\,x^{-\frac{1}{4}} - x^{-\frac{5}{4}}$.

17. $\dfrac{d}{dx}\left(x^{\frac{2}{3}} - a^{\frac{2}{3}}\right) = \tfrac{2}{3}\,x^{-\frac{1}{3}}$.

18. $\dfrac{d}{dx}\left(\dfrac{a + bx + cx^2}{x}\right) = c - \dfrac{a}{x^2}$.

19. $y = \dfrac{\sqrt{x}}{2} - \dfrac{2}{\sqrt{x}}$. $\qquad \dfrac{dy}{dx} = \dfrac{1}{4\sqrt{x}} + \dfrac{1}{x\sqrt{x}}$.

20. $s = \dfrac{a + bt + ct^2}{\sqrt{t}}$. $\qquad \dfrac{ds}{dt} = -\dfrac{a}{2\,t\sqrt{t}} + \dfrac{b}{2\sqrt{t}} + \dfrac{3\,c\sqrt{t}}{2}$.

21. $y = \sqrt{ax} + \dfrac{a}{\sqrt{ax}}$. $\qquad \dfrac{dy}{dx} = \dfrac{a}{2\sqrt{ax}} - \dfrac{a}{2\,x\sqrt{ax}}$.

22. $r = \sqrt{1 - 2\,\theta}$. $\qquad \dfrac{dr}{d\theta} = -\dfrac{1}{\sqrt{1 - 2\,\theta}}$.

23. $f(t) = (2 - 3\,t^2)^3$. $\qquad f'(t) = -18\,t(2 - 3\,t^2)^2$.

24. $F(x) = \sqrt[3]{4 - 9\,x}$. $\qquad F'(x) = -\dfrac{3}{(4 - 9\,x)^{\frac{2}{3}}}$.

25. $y = \dfrac{1}{\sqrt{a^2 - x^2}}$. $\qquad \dfrac{dy}{dx} = \dfrac{x}{(a^2 - x^2)^{\frac{3}{2}}}$.

26. $f(\theta) = (2 - 5\,\theta)^{\frac{3}{5}}$. $\qquad f'(\theta) = -\dfrac{3}{(2 - 5\,\theta)^{\frac{2}{5}}}$.

27. $y = \left(a - \dfrac{b}{x}\right)^2$. $\qquad \dfrac{dy}{dx} = \dfrac{2\,b}{x^2}\left(a - \dfrac{b}{x}\right)$.

28. $y = \left(a + \dfrac{b}{x^2}\right)^3$. $\qquad \dfrac{dy}{dx} = -\dfrac{6\,b}{x^3}\left(a + \dfrac{b}{x^2}\right)^2$.

29. $y = x\sqrt{a + bx}$. $\qquad \dfrac{dy}{dx} = \dfrac{2\,a + 3\,bx}{2\sqrt{a + bx}}$.

30. $s = t\sqrt{a^2 + t^2}$. $\qquad \dfrac{ds}{dt} = \dfrac{a^2 + 2\,t^2}{\sqrt{a^2 + t^2}}$.

31. $y = \dfrac{a - x}{n + x}$. $\qquad \dfrac{dy}{dx} = -\dfrac{2\,a}{(a + x)^2}$.

32. $y = \dfrac{a^2 + x^2}{a^2 - x^2}.$ $\dfrac{dy}{dx} = \dfrac{4 a^2 x}{(a^2 - x^2)^2}.$

33. $y = \dfrac{\sqrt{a^2 + x^2}}{x}.$ $\dfrac{dy}{dx} = -\dfrac{a^2}{x^2 \sqrt{a^2 + x^2}}.$

34. $y = \dfrac{x}{\sqrt{a^2 - x^2}}.$ $\dfrac{dy}{dx} = \dfrac{a^2}{(a^2 - x^2)^{\frac{3}{2}}}.$

35. $r = \theta^2 \sqrt{3 - 4\theta}.$ $\dfrac{dr}{d\theta} = \dfrac{6\theta - 10\theta^2}{\sqrt{3 - 4\theta}}.$

36. $y = \sqrt{\dfrac{1 - cx}{1 + cx}}.$ $\dfrac{dy}{dx} = -\dfrac{c}{(1 + cx)\sqrt{1 - c^2 x^2}}.$

37. $y = \sqrt{\dfrac{a^2 + x^2}{a^2 - x^2}}.$ $\dfrac{dy}{dx} = \dfrac{2 a^2 x}{(a^2 - x^2)\sqrt{a^4 - x^4}}.$

38. $s = \sqrt[3]{\dfrac{2 + 3t}{2 - 3t}}.$ $\dfrac{ds}{dt} = \dfrac{4}{(2 + 3t)^{\frac{2}{3}}(2 - 3t)^{\frac{4}{3}}}.$

39. $y = \sqrt{2 px}.$ $\dfrac{dy}{dx} = \dfrac{p}{y}.$

40. $y = \dfrac{b}{a}\sqrt{a^2 - x^2}.$ $\dfrac{dy}{dx} = -\dfrac{b^2 x}{a^2 y}.$

41. $y = \left(a^{\frac{2}{3}} - x^{\frac{2}{3}}\right)^{\frac{3}{2}}.$ $\dfrac{dy}{dx} = -\sqrt[3]{\dfrac{y}{x}}.$

Differentiate each of the following functions.

42. $f(x) = \sqrt{2x} + \sqrt[3]{3x}.$

43. $y = \dfrac{2 - x}{1 + 2x^2}.$

44. $y = \dfrac{x}{\sqrt{a - bx}}.$

45. $s = \dfrac{\sqrt{a + bt}}{t}.$

46. $r = \dfrac{\sqrt[3]{a + b\theta}}{\theta}.$

47. $y = x^2\sqrt{5 - 2x}.$

48. $y = x\sqrt[3]{2 + 3x}.$

49. $s = \sqrt{2t - \dfrac{1}{t^2}}.$

50. $y = (x + 2)^2 \sqrt{x^2 + 2}.$

51. $y = \dfrac{\sqrt{1 + 2x}}{\sqrt[3]{1 + 3x}}.$

In each of the following problems find the value of $\dfrac{dy}{dx}$ for the given value of x.

52. $y = (x^2 - x)^3$; $x = 3$. *Ans.* 540.

53. $y = \sqrt[3]{x} + \sqrt{x}$: $x = 64$. $\dfrac{1}{12}.$

RULES FOR DIFFERENTIATING ALGEBRAIC FORMS

54. $y = (2x)^{\frac{1}{3}} + (2x)^{\frac{2}{3}}$; $x = 4$. Ans. $\frac{5}{8}$.

55. $y = \sqrt{9 + 4x^2}$; $x = 2$. $\frac{8}{5}$.

56. $y = \dfrac{1}{\sqrt{25 - x^2}}$; $x = 3$. $\frac{3}{64}$.

57. $y = \dfrac{\sqrt{16 + 3x}}{x}$; $x = 3$. $-\frac{41}{90}$.

58. $y = x\sqrt{8 - x^2}$; $x = 2$. 0.

59. $y = x^2\sqrt{1 + x^3}$; $x = 2$. 20.

60. $y = (4 - x^2)^3$; $x = 3$.

63. $y = x\sqrt{3 + 2x}$; $x = 3$.

61. $y = \dfrac{x^2 + 2}{2 - x^2}$; $x = 2$.

64. $y = \sqrt{\dfrac{4x + 1}{5x - 1}}$; $x = 2$.

62. $y = \dfrac{\sqrt{5 - 2x}}{2x + 1}$; $x = \frac{1}{2}$.

65. $y = \sqrt{\dfrac{x^2 - 5}{10 - x^2}}$; $x = 3$.

38. Differentiation of a function of a function. It sometimes happens that y, instead of being defined directly as a function of x, is given as a function of another variable v, which is defined as a function of x. In that case y is a function of x through v and is called a *function of a function.*

For example, if
$$y = \frac{2v}{1 - v^2},$$
and
$$v = 1 - x^2,$$
then y is a function of a function. By eliminating v we may express y directly as a function of x, but in general this is not the best plan when we wish to find $\dfrac{dy}{dx}$.

If $y = f(v)$ and $v = \phi(x)$, then y is a function of x through v. Hence, when we let x take on an increment Δx, v will take on an increment Δv and y will also take on a corresponding increment Δy. Keeping this in mind, let us apply the General Rule simultaneously to the two functions
$$y = f(v) \quad \text{and} \quad v = \phi(x).$$

FIRST STEP. $y + \Delta y = f(v + \Delta v)$ $\quad v + \Delta v = \phi(x + \Delta x)$

SECOND STEP. $y + \Delta y = f(v + \Delta v)$ $\quad v + \Delta v = \phi(x + \Delta x)$
$\qquad\qquad\quad\; y \;\;\;\;\;\; = f(v)$ $\qquad\qquad v \;\;\;\;\;\;\; = \phi(x)$
$\qquad\qquad\overline{\;\Delta y = f(v + \Delta v) - f(v)\;}$ $\quad\overline{\;\Delta v = \phi(x + \Delta x) - \phi(x)\;}$

THIRD STEP. $\dfrac{\Delta y}{\Delta v} = \dfrac{f(v + \Delta v) - f(v)}{\Delta v},$ $\quad\dfrac{\Delta v}{\Delta x} = \dfrac{\phi(x + \Delta x) - \phi(x)}{\Delta x}.$

The left-hand members show one form of the ratio of the increment of each function to the increment of the corresponding variable, and the right-hand members exhibit the same ratios in another form. Before passing to the limit let us form a product of these two ratios, choosing the left-hand forms for this purpose.

This gives $\dfrac{\Delta y}{\Delta v} \cdot \dfrac{\Delta v}{\Delta x}$, which equals $\dfrac{\Delta y}{\Delta x}$.

Write this $\dfrac{\Delta y}{\Delta x} = \dfrac{\Delta y}{\Delta v} \cdot \dfrac{\Delta v}{\Delta x}.$

FOURTH STEP. When $\Delta x \to 0$, then also $\Delta v \to 0$. Passing to the limit,

(A) $\qquad\qquad\dfrac{dy}{dx} = \dfrac{dy}{dv} \cdot \dfrac{dv}{dx}.$ \hfill By (2), Art. 16

This may also be written

(B) $\qquad\qquad\dfrac{dy}{dx} = f'(v) \cdot \phi'(x).$

If $y = f(v)$ and $v = \phi(x)$, the derivative of y with respect to x equals the product of the derivative of y with respect to v by the derivative of v with respect to x.

39. Differentiation of inverse functions. Let y be given as a function of x by means of the relation

$$y = f(x).$$

It is often possible in the case of functions considered in this book to solve this equation for x, giving

$$x = \phi(y);$$

that is, we may also consider y as the independent and x as the dependent variable. In that case

$$f(x) \quad \text{and} \quad \phi(y)$$

are said to be *inverse functions*. When we wish to distinguish between the two it is customary to call the first one given the *direct function* and the second one the *inverse function*. Thus, in the examples which follow, if the second members in the first column are taken as the direct functions, then the corresponding members in the second column will be respectively the *inverse functions*.

$$y = x^2 + 1, \qquad\qquad x = \pm\sqrt{y-1}.$$
$$y = a^x, \qquad\qquad x = \log_a y.$$
$$y = \sin x. \qquad\qquad x = \arcsin y.$$

RULES FOR DIFFERENTIATING ALGEBRAIC FORMS

Let us now differentiate the inverse functions
$$y = f(x) \quad \text{and} \quad x = \phi(y)$$
simultaneously by the General Rule.

FIRST STEP. $\quad y+\Delta y = f(x+\Delta x) \qquad x+\Delta x = \phi(y+\Delta y).$

SECOND STEP. $\quad \dfrac{\begin{array}{l} y+\Delta y = f(x+\Delta x) \\ y = f(x) \end{array}}{\Delta y = f(x+\Delta x) - f(x)} \qquad \dfrac{\begin{array}{l} x+\Delta x = \phi(y+\Delta y) \\ x = \phi(y) \end{array}}{\Delta x = \phi(y+\Delta y) - \phi(y)}$

THIRD STEP. $\quad \dfrac{\Delta y}{\Delta x} = \dfrac{f(x+\Delta x) - f(x)}{\Delta x}, \quad \dfrac{\Delta x}{\Delta y} = \dfrac{\phi(y+\Delta y) - \phi(y)}{\Delta y}.$

Taking the product of the left-hand forms of these ratios, we get
$$\frac{\Delta y}{\Delta x} \cdot \frac{\Delta x}{\Delta y} = 1,$$
or
$$\frac{\Delta y}{\Delta x} = \frac{1}{\dfrac{\Delta x}{\Delta y}}.$$

FOURTH STEP. When $\Delta x \to 0$, then also, in general, $\Delta y \to 0$. Passing to the limit,

(C) $\qquad\qquad\qquad \dfrac{dy}{dx} = \dfrac{1}{\dfrac{dx}{dy}}, \qquad\qquad \text{by (3), Art. 16}$

or

(D) $\qquad\qquad\qquad f'(x) = \dfrac{1}{\phi'(y)}.$

The derivative of the inverse function is equal to the reciprocal of the derivative of the direct function.

40. Implicit functions. When a relation between x and y is given by means of an equation not solved for y, then y is called an *implicit function* of x. For example, the equation

(1) $\qquad\qquad\qquad x^2 - 4y = 0$

defines y as an implicit function of x. Evidently x is also defined by means of this equation as an implicit function of y.

It is sometimes possible to solve the equation defining an implicit function for one of the variables and thus obtain an **explicit function**. For instance, equation (1) may be solved for y, giving

$$y = \tfrac{1}{4} x^2,$$

showing y as an explicit function of x. In a given case, however, such a solution may be either impossible or too complicated for convenient use.

DIFFERENTIAL CALCULUS

41. Differentiation of implicit functions. When y is defined as an implicit function of x, it was explained in the last article that it might be inconvenient to solve for y in terms of x or x in terms of y (that is, to find y as an explicit function of x, or x as an explicit function of y).

We then follow the rule:

Differentiate the terms of the equation as given, regarding y as a function of x, and solve for $\dfrac{dy}{dx}$.

This process will be justified in Art. 231. Only corresponding values of x and y which satisfy the given equation may be substituted in the derivative.

Let us apply this rule in finding $\dfrac{dy}{dx}$ from

$$ax^6 + 2x^3y - y^7x = 10.$$

Then $\quad \dfrac{d}{dx}(ax^6) + \dfrac{d}{dx}(2x^3y) - \dfrac{d}{dx}(y^7x) = \dfrac{d}{dx}(10);$

$$6ax^5 + 2x^3\frac{dy}{dx} + 6x^2y - y^7 - 7xy^6\frac{dy}{dx} = 0;$$

$$(2x^3 - 7xy^6)\frac{dy}{dx} = y^7 - 6ax^5 - 6x^2y;$$

$$\frac{dy}{dx} = \frac{y^7 - 6ax^5 - 6x^2y}{2x^3 - 7xy^6}. \quad Ans.$$

The student should observe that in general the result will contain both x and y.

PROBLEMS

Find $\dfrac{dy}{dx}$ for each of the following functions.

1. $y = u^6, \; u = 1 + 2\sqrt{x}.$ $\qquad$ *Ans.* $\dfrac{dy}{dx} = \dfrac{6u^5}{\sqrt{x}}.$

2. $y = \sqrt{2u} - u^2, \; u = x^3 - x.$ $\qquad \dfrac{dy}{dx} = \left(\dfrac{1}{\sqrt{2u}} - 2u\right)(3x^2 - 1).$

3. $y = \dfrac{a-u}{a+u}, \; u = \dfrac{b-x}{b+x}.$ $\qquad \dfrac{dy}{dx} = \dfrac{4ab}{(a+u)^2(b+x)^2}.$

4. $y = u\sqrt{a^2 - u^2}, \; u = \sqrt{1-x^2}.$ $\qquad \dfrac{dy}{dx} = \dfrac{x(2u^2 - a^2)}{\sqrt{(a^2-u^2)(1-x^2)}}.$

5. $15x = 15y + 5y^3 + 3y^5.$ $\qquad \dfrac{dy}{dx} = \dfrac{1}{1+y^2+y^4}.$

6. $x = \sqrt{y} + \sqrt[3]{y}.$ $\qquad \dfrac{dy}{dx} = \dfrac{6y^{\frac{2}{3}}}{3y^{\frac{1}{6}}+2}.$

RULES FOR DIFFERENTIATING ALGEBRAIC FORMS 41

7. $y^2 = 2\,px$.
8. $x^2 + y^2 = r^2$.
9. $b^2x^2 + a^2y^2 = a^2b^2$.
10. $\sqrt{x} + \sqrt{y} = \sqrt{a}$.
11. $x^{\frac{2}{3}} + y^{\frac{2}{3}} = a^{\frac{2}{3}}$.
12. $x^3 - 3\,axy + y^3 = 0$.
13. $x^3 - 3\,xy^2 + y^3 = 10$.
14. $x + 2\sqrt{xy} + y = a$.
15. $x^2 + a\sqrt{xy} + y^2 = b^2$.
16. $x^4 + 4\,x^2y^2 + y^4 = 20$.
17. $ax^3 - 3\,b^2xy + cy^3 = 1$.
18. $\sqrt{\dfrac{y}{x}} + \sqrt{\dfrac{x}{y}} = 6$.

Find the slope of each of the following curves at the given point.

19. $x^2 + 2\,xy - 3\,y^2 + 11 = 0$; (2, 3). *Ans.* $\tfrac{5}{7}$.
20. $x^3 + 3\,x^2y + y^3 = 3$; (−1, 1). $\tfrac{1}{2}$.
21. $\sqrt{2}\,x + \sqrt{3}\,y = 5$; (2, 3).
22. $x^2 + 4\sqrt{xy} + y^2 = 25$; (1, 4).
23. $x^3 - axy + 3\,ay^2 = 3\,a^3$; (a, a).
24. $2\,x^2 + y\sqrt{xy} - y^2 = 132$; (8, 2).

25. Show that the parabolas $y^2 = 2\,px + p^2$ and $y^2 = p^2 - 2\,px$ intersect at right angles.

26. Show that the circle $x^2 + y^2 - 12\,x - 6\,y + 25 = 0$ is tangent to the circle $x^2 + y^2 + 2\,x + y = 10$ at the point (2, 1).

27. At what angle does the line $y = 2\,x$ cut the curve $x^2 - xy + 2\,y^2 = 28$?

28. If $f(x)$ and $\phi(y)$ are inverse functions, show that the graph of $\phi(x)$ may be found as follows: construct the graph of $-f(x)$ and rotate it around the origin 90° counterclockwise.

ADDITIONAL PROBLEMS

1. The vertex of the parabola $y^2 = 2\,px$ is the center of an ellipse. The focus of the parabola is an end of one of the principal axes of the ellipse, and the parabola and ellipse intersect at right angles. Find the equation of the ellipse. *Ans.* $4\,x^2 + 2\,y^2 = p^2$.

2. A circle is drawn with its center at $(2\,a, 0)$ and with a radius such that the circle cuts the ellipse $b^2x^2 + a^2y^2 = a^2b^2$ at right angles. Find the radius. *Ans.* $r^2 = \tfrac{3}{4}(3\,a^2 + b^2)$.

3. From any point P on an ellipse lines are drawn to the foci. Prove that these lines make equal acute angles with the normal at P.

4. Prove that the line $Bx + Ay = AB$ is tangent to the ellipse $b^2x^2 + a^2y^2 = a^2b^2$ if, and only if, $B^2a^2 + A^2b^2 = A^2B^2$.

5. Find the equation of the tangent to the curve $x^m y^n = a^{m+n}$ at any point. Prove that the portion of it intercepted between the axes is divided in the ratio $\dfrac{m}{n}$ at the point of contact. *Ans.* $my_1(x - x_1) + nx_1(y - y_1) = 0$.

6. If k is the slope of a tangent to the hyperbola $b^2x^2 - a^2y^2 = a^2b^2$, prove that its equation is $y = kx \pm \sqrt{a^2k^2 - b^2}$, and show that the locus of the points of intersection of the perpendicular tangents is $x^2 + y^2 = a^2 - b^2$.

CHAPTER V

VARIOUS APPLICATIONS OF THE DERIVATIVE

42. Direction of a curve. It was shown in Art. 28 that if

$$y = f(x)$$

is the equation of a curve (see figure), then

$$\frac{dy}{dx} = \text{slope of the line tangent to the curve at } P(x, y).$$

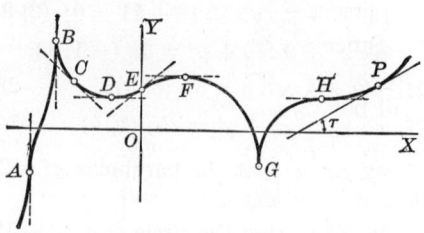

The *direction of a curve* at any point is defined as the direction of the tangent line to the curve at that point. Let $\tau =$ inclination of the tangent line. Then the slope $= \tan \tau$, and

$$\frac{dy}{dx} = \tan \tau = \text{slope of the curve at any point } P(x, y).$$

At points such as D, F, H, where the direction of the curve is parallel to the x-axis *and the tangent line is horizontal*,

$$\tau = 0; \text{ therefore } \frac{dy}{dx} = 0.$$

At points such as A, B, G, where the direction of the curve is perpendicular to the x-axis *and the tangent line is vertical*,

$$\tau = 90°; \text{ therefore } \frac{dy}{dx} \text{ becomes infinite}.$$

ILLUSTRATIVE EXAMPLE 1. Given the curve $y = \frac{x^3}{3} - x^2 + 2$ (see figure).

(a) Find the inclination τ when $x = 1$.
(b) Find τ when $x = 3$.
(c) Find the points where the direction of the curve is parallel to OX.
(d) Find the points where $\tau = 45°$.
(e) Find the points where the direction of the curve is parallel to the line $2x - 3y = 6$ (line AB).

Solution. Differentiating, $\frac{dy}{dx} = x^2 - 2x = \tan \tau$.

VARIOUS APPLICATIONS OF THE DERIVATIVE 43

(a) When $x=1$, $\tan \tau = 1 - 2 = -1$; therefore $\tau = 135°$. *Ans.*
(b) When $x=3$, $\tan \tau = 9 - 6 = 3$; therefore $\tau = 71° \, 34'$. *Ans.*
(c) When $\tau = 0$, $\tan \tau = 0$; therefore $x^2 - 2x = 0$. Solving this equation, we get $x = 0$ or 2. Substituting in the equation of the curve, we find $y = 2$ when $x = 0$, $y = \frac{2}{3}$ when $x = 2$. Hence the tangent lines at $C(0, 2)$ and $D(2, \frac{2}{3})$ are horizontal. *Ans.*
(d) When $\tau = 45°$, $\tan \tau = 1$; therefore $x^2 - 2x = 1$. Solving this equation, we get $x = 1 \pm \sqrt{2} = 2.41$ and -0.41, giving two points where the slope of the curve (or tangent) is unity.
(e) Slope of the given line $= \frac{2}{3}$; therefore $x^2 - 2x = \frac{2}{3}$. Solving, we get $x = 1 \pm \sqrt{\frac{5}{3}} = 2.29$ and -0.29, giving the abscissas of the points F and E where the direction of the given curve (or tangent) is parallel to the line AB.

Since a curve at any point has the same direction as its tangent line at that point, the angle between two curves at a common point will be the angle between their tangent lines at that point.

ILLUSTRATIVE EXAMPLE 2. Find the angle of intersection of the circles
(A) $\qquad x^2 + y^2 - 4x = 1$,
(B) $\qquad x^2 + y^2 - 2y = 9$.

Solution. Solving simultaneously, we find the points of intersection to be $(3, 2)$ and $(1, -2)$.

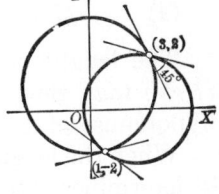

Let $m_1 =$ slope of the tangent to the circle A at (x, y),
and $\qquad m_2 =$ slope of the tangent to the circle B at (x, y).

Then from (A), $\qquad m_1 = \dfrac{dy}{dx} = \dfrac{2-x}{y}$, $\qquad$ by Art. 41

and from (B), $\qquad m_2 = \dfrac{dy}{dx} = \dfrac{x}{1-y}$. $\qquad$ By Art. 41

Substituting $x = 3$, $y = 2$, we have
$\qquad m_1 = -\frac{1}{2} =$ slope of tangent to (A) at $(3, 2)$.
$\qquad m_2 = -3 =$ slope of tangent to (B) at $(3, 2)$.

The formula for finding the angle θ between two lines whose slopes are m_1 and m_2 is
$$\tan \theta = \frac{m_1 - m_2}{1 + m_1 m_2}. \qquad (2), \text{Art. 3}$$

Substituting, $\qquad \tan \theta = \dfrac{-\frac{1}{2} + 3}{1 + \frac{3}{2}} = 1$; $\therefore \theta = 45°$. *Ans.*

This is also the angle of intersection at the point $(1, -2)$.

43. Equations of tangent and normal; lengths of subtangent and subnormal. The equation of a straight line passing through the point (x_1, y_1) and having the slope m is $\qquad y - y_1 = m(x - x_1)$. $\qquad$ (3), Art. 3

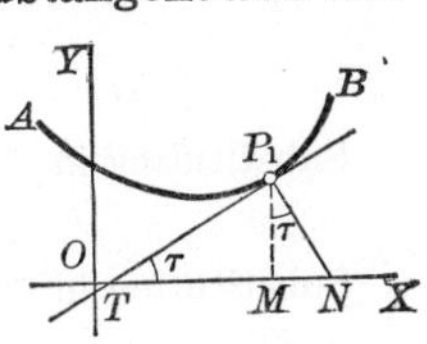

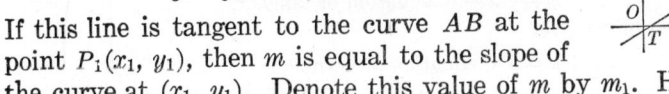

If this line is tangent to the curve AB at the point $P_1(x_1, y_1)$, then m is equal to the slope of the curve at (x_1, y_1). Denote this value of m by m_1. Hence at the point of contact $P_1(x_1, y_1)$ the *equation of the tangent line* TP_1 is

(1) $\qquad y - y_1 = m_1(x - x_1)$.

DIFFERENTIAL CALCULUS

The normal being perpendicular to the tangent, its slope is the negative reciprocal of m_1 ((2), Art. 3). And since it also passes through the point of contact $P_1(x_1, y_1)$, we have for the *equation of the normal* P_1N,

(2) $$y - y_1 = -\frac{1}{m_1}(x - x_1).$$

That portion of the tangent which is intercepted between the point of contact and OX is called the *length of the tangent* ($= TP_1$), and its projection on the x-axis is called the *length of the subtangent* ($= TM$). Similarly, we have the *length of the normal* ($= P_1N$) and the length of the subnormal ($= MN$).

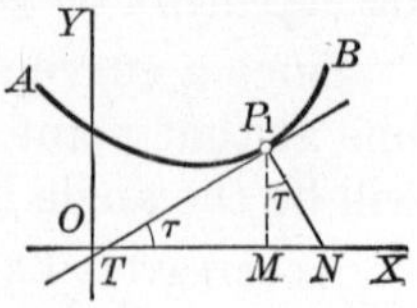

In the triangle TP_1M, $\tan \tau = m_1 = \dfrac{MP_1}{TM}$; therefore

(3) $$TM^* = \frac{MP_1}{m_1} = \frac{y_1}{m_1} = \text{length of subtangent}.$$

In the triangle MP_1N, $\tan \tau = m_1 = \dfrac{MN}{MP_1}$; therefore

(4) $$MN^* = m_1 MP_1 = m_1 y_1 = \text{length of subnormal}.$$

The length of the tangent (TP_1) and the length of the normal (P_1N) may then be found directly from the figure, each being the hypotenuse of a right triangle having two legs known.

When the length of subtangent or subnormal at a point on a curve is determined, the tangent and normal may easily be constructed.

PROBLEMS

1. Find the equations of tangent and normal and the lengths of subtangent, subnormal, tangent, and normal, at the point (a, a) on the cissoid $y^2 = \dfrac{x^3}{2a - x}$.

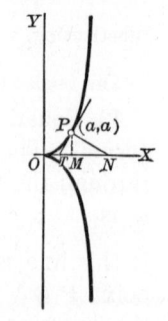

Solution. $\dfrac{dy}{dx} = \dfrac{3ax^2 - x^3}{y(2a - x)^2}.$

Substituting $x = a$, $y = a$, we have

$$m_1 = \frac{3a^3 - a^3}{a(2a - a)^2} = 2 = \text{slope of tangent}.$$

Substituting in (1) gives

$$y = 2x - a, \text{ equation of tangent}.$$

Substituting in (2) gives

$$2y + x = 3a, \text{ equation of normal}.$$

---

* If the subtangent extends to the right of T, we consider it positive; if to the left, negative. If the subnormal extends to the right of M, we consider it positive; if to the left, negative.

VARIOUS APPLICATIONS OF THE DERIVATIVE

Substituting in (3) gives $\quad TM = \dfrac{a}{2} =$ length of subtangent.

Substituting in (4) gives $\quad MN = 2\,a =$ length of subnormal.

Also, $PT = \sqrt{(TM)^2 + (MP)^2} = \sqrt{\dfrac{a^2}{4} + a^2} = \dfrac{a}{2}\sqrt{5} =$ length of tangent,

and $\quad PN = \sqrt{(MN)^2 + (MP)^2} = \sqrt{4\,a^2 + a^2} = a\sqrt{5} =$ length of normal.

Find the equations of the tangent and normal at the given point.

2. $y = x^3 - 3\,x$; $(2, 2)$. $\quad$ Ans. $9\,x - y - 16 = 0$, $x + 9\,y - 20 = 0$.

3. $y = \dfrac{2\,x + 1}{3 - x}$; $(2, 5)$. $\quad\quad\quad 7\,x - y - 9 = 0$, $x + 7\,y - 37 = 0$.

4. $2\,x^2 - xy + y^2 = 16$; $(3, 2)$.

5. $y^2 + 2\,y - 4\,x + 4 = 0$; $(1, -2)$.

6. Find the equations of the tangent and normal at (x_1, y_1) to the ellipse $b^2 x^2 + a^2 y^2 = a^2 b^2$.
$\quad$ Ans. $b^2 x_1 x + a^2 y_1 y = a^2 b^2$, $a^2 y_1 x - b^2 x_1 y = x_1 y_1 (a^2 - b^2)$.

7. Find the equations of the tangent and normal, and the lengths of the subtangent and subnormal, at the point (x_1, y_1) on the circle $x^2 + y^2 = r^2$.
$\quad$ Ans. $x_1 x + y_1 y = r^2$, $x_1 y - y_1 x = 0$, $-\dfrac{y_1^2}{x_1}$, $-x_1$.

8. Show that the subtangent to the parabola $y^2 = 2\,px$ is bisected at the vertex, and that the subnormal is constant and equal to p.

Find the equations of the tangent and normal, and the lengths of the subtangent and subnormal, to each of the following curves at the points indicated.

9. $ay = x^2$; (a, a). $\quad\quad$ Ans. $2\,x - y = a$, $x + 2\,y = 3\,a$, $\dfrac{a}{2}$, $2\,a$.

10. $x^2 - 4\,y^2 = 9$; $(5, 2)$. $\quad\quad 5\,x - 8\,y = 9$, $8\,x + 5\,y = 50$, $\tfrac{16}{5}$, $\tfrac{5}{4}$.

11. $9\,x^2 + 4\,y^2 = 72$; $(2, 3)$.

12. $xy + y^2 + 2 = 0$; $(3, -2)$.

13. Find the area of the triangle formed by the x-axis and the tangent and the normal to the curve $y = 6\,x - x^2$ at the point $(5, 5)$. $\quad$ Ans. $\tfrac{425}{8}$.

14. Find the area of the triangle formed by the y-axis and the tangent and the normal to the curve $y^2 = 9 - x$ at the point $(5, 2)$.

Find the angles of intersection of each of the following pairs of curves.

15. $y^2 = x + 1$, $x^2 + y^2 = 13$. $\quad\quad$ Ans. $109°\,39'$.

16. $y = 6 - x^2$, $7\,x^2 + y^2 = 32$.
$\quad$ Ans. At $(\pm 2, 2)$, $5°\,54'$; at $(\pm 1, 5)$, $8°\,58'$.

17. $y = x^2$, $y^2 - 3\,y = 2\,x$.

18. $x^2 + 4\,y^2 = 61$, $2\,x^2 - y^2 = 41$.

Find the points of contact of the horizontal and vertical tangents to each of the following curves.

19. $y = 5x - 2x^2$. Ans. Horizontal, $(\frac{5}{4}, \frac{25}{8})$.

20. $3y^2 - 6y - x = 0$. Vertical, $(-3, 1)$.

21. $x^2 + 6xy + 25y^2 = 16$. Horizontal, $(3, -1)$, $(-3, 1)$.
Vertical, $(5, -\frac{3}{5})$, $(-5, \frac{3}{5})$.

22. $x^2 - 8xy + 25y^2 = 81$.

23. $x^2 - 24xy + 169y^2 = 25$.

24. $169x^2 + 10xy + y^2 = 144$.

25. Show that the hyperbola $x^2 - y^2 = 5$ and the ellipse $4x^2 + 9y^2 = 72$ intersect at right angles.

26. Show that the circle $x^2 + y^2 = 8ax$ and the cissoid $(2a - x)y^2 = x^3$

(a) are perpendicular at the origin;

(b) intersect at an angle of 45° at two other points. (See figure in Chapter XXVI.)

27. Show that the tangents to the folium of Descartes $x^3 + y^3 = 3axy$ at the points where it meets the parabola $y^2 = ax$ are parallel to the y-axis. (See figure in Chapter XXVI.)

28. Find the equation of the normal to the parabola $y = 5x + x^2$ which makes an angle of 45° with the x-axis.

29. Find the equations of the tangents to the circle $x^2 + y^2 = 58$ which are parallel to the line $3x - 7y = 19$.

30. Find the equations of the normals to the hyperbola $4x^2 - y^2 = 36$ which are parallel to the line $2x + 5y = 4$.

31. Find the equations of the two tangents to the ellipse $4x^2 + y^2 = 72$ which pass through the point $(4, 4)$. Ans. $2x + y = 12$, $14x + y = 60$.

32. Show that the sum of the intercepts on the coördinate axes of the tangent line at any point to the parabola $x^{\frac{1}{2}} + y^{\frac{1}{2}} = a^{\frac{1}{2}}$ is constant and equal to a. (See figure in Chapter XXVI.)

33. Show that for the hypocycloid $x^{\frac{2}{3}} + y^{\frac{2}{3}} = a^{\frac{2}{3}}$ the portion of the tangent line at any point included between the coördinate axes is constant and equal to a. (See figure in Chapter XXVI.)

34. The equation of the path of a ball is $y = x - \dfrac{x^2}{100}$; the unit of distance is 1 yd., the x-axis being horizontal, and the origin being the point from which the ball is thrown. (a) At what angle is the ball thrown? (b) At what angle will the ball strike a vertical wall 75 yd. from the starting point? (c) If the ball falls on a horizontal roof 16 yd. high, at

VARIOUS APPLICATIONS OF THE DERIVATIVE 47

what angle will it strike the roof? (d) If thrown from the top of a building 24 yd. high, at what angle will the ball strike the ground? (e) If thrown from the top of a hill which slopes downward at an angle of 45°, at what angle will the ball strike the ground?

35. The cable of a suspension bridge hangs in the form of a parabola and is attached to supporting pillars 200 ft. apart. The lowest point of the cable is 40 ft. below the points of suspension. Find the angle between the cable and the supporting pillars.

44. Maximum and minimum values of a function; introduction. In a great many practical problems we have to deal with functions which have a greatest (maximum) value or a least (minimum) value,* and it is important to know what particular value of the variable gives such a value of the function.

For instance, suppose that it is required to find the dimensions of the rectangle of greatest area that can be inscribed in a circle of radius 5 inches. Consider the circle in the following figure. Inscribe any rectangle, as BD.

Let $CD = x$; then $DE = \sqrt{100 - x^2}$, and the area of the rectangle is evidently

(1) $$A = x\sqrt{100 - x^2}.$$

That a rectangle of maximum area must exist may be seen as follows. Let the base $CD\ (=x)$ increase to 10 inches (the diameter); then the altitude $DE = \sqrt{100 - x^2}$ will decrease to zero and the area will become zero. Now let the base decrease to zero; then the altitude will increase to 10 inches and the area will again become zero. It is therefore evident by intuition that there exists a greatest rectangle. By a careful study of the figure we might suspect that when the rectangle becomes a square its area would be greatest, but this would be guesswork. A better way would evidently be to plot the graph of the function (1) and note its behavior. To aid us in drawing the graph of (1), we observe that

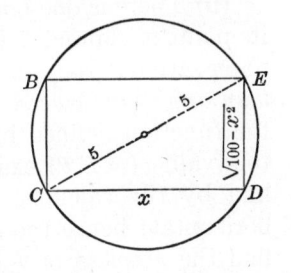

(a) from the nature of the problem it is evident that x and A must both be positive; and

(b) the values of x range from zero to 10 inclusive.

Now construct a table of values and draw the graph, as in the figure on page 48.

*There may be more than one of each, as illustrated on page 55.

DIFFERENTIAL CALCULUS

What do we learn from the graph?

x	A
0	0
1	9.9
2	19.6
3	28.6
4	36.6
5	43.0
6	48.0
7	49.7
8	48.0
9	39.6
10	0.0

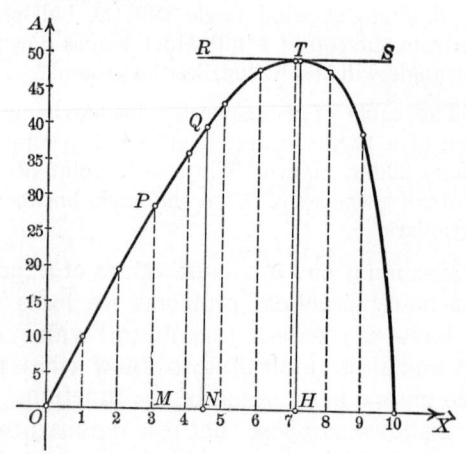

(a) If carefully drawn, we may find quite accurately the area of the rectangle corresponding to any value of x by measuring the length of the corresponding ordinate. Thus,

when $\quad\quad x = OM = 3$ in.,
then $\quad\quad A = MP = 28.6$ sq. in.;
and when $\quad x = ON = 4\frac{1}{2}$ in.,
then $\quad\quad A = NQ =$ about 39.8 sq. in. (found by measurement).

(b) There is one horizontal tangent (RS). The ordinate TH from its point of contact T is greater than any other ordinate. Hence this observation: *One of the inscribed rectangles has evidently a greater area than any of the others.* In other words, we may infer from this that the function defined by (1) has a *maximum value*. We cannot find this value ($= HT$) exactly by measurement, but it is very easy to find by the calculus. We observed that at T the tangent was horizontal; hence the slope will be zero at that point (Art. 42). To find the abscissa of T we then find the derivative of A with respect to x from (1), place it equal to zero, and solve for x. Thus we have

(1) $$A = x\sqrt{100 - x^2},$$

$$\frac{dA}{dx} = \frac{100 - 2x^2}{\sqrt{100 - x^2}}, \quad \frac{100 - 2x^2}{\sqrt{100 - x^2}} = 0.$$

Solving, $\quad\quad x = 5\sqrt{2}.$

Substituting, we get $\quad DE = \sqrt{100 - x^2} = 5\sqrt{2}.$

VARIOUS APPLICATIONS OF THE DERIVATIVE

Hence the rectangle of maximum area inscribed in the circle is a square of area
$$A = CD \times DE = 5\sqrt{2} \times 5\sqrt{2} = 50 \text{ sq. in.}$$

The length of HT is therefore 50.

Take another example. A wooden box is to be built to contain 108 cu. ft. It is to have an open top and a square base. What must be its dimensions in order that the amount of material required shall be a minimum; that is, what dimensions will make the cost the least?

Let x = length of side of square base in feet, and y = height of box.

Since the volume of the box is given, however, y may be found in terms of x. Thus,

$$\text{Volume} = x^2 y = 108; \quad \therefore y = \frac{108}{x^2}.$$

We may now express the number ($= M$) of square feet of lumber required as a function of x as follows.

Area of base $= x^2$ sq. ft.,

and Area of four sides $= 4xy = \dfrac{432}{x}$ sq. ft. Hence

(2) $$M = x^2 + \frac{432}{x}$$

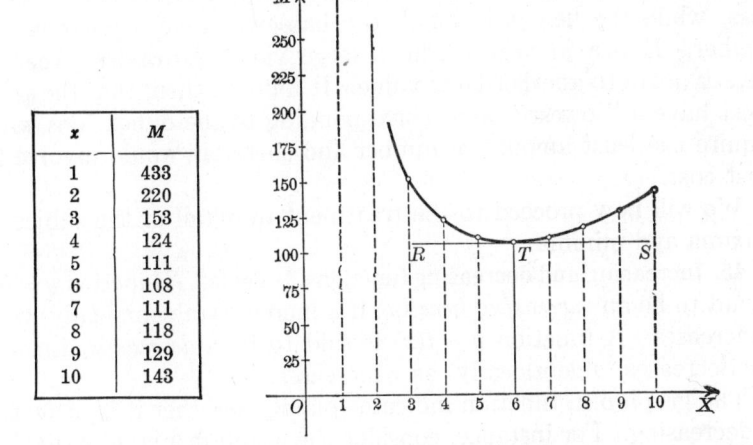

x	M
1	433
2	220
3	153
4	124
5	111
6	108
7	111
8	118
9	129
10	143

is a formula giving the number of square feet required in any such box having a capacity of 108 cu. ft. Draw a graph of (2), as in the figure.

What do we learn from the graph?

(a) If carefully drawn, we may measure the ordinate corresponding to any length ($= x$) of the side of the square base and so determine the number of square feet of lumber required.

(b) There is one horizontal tangent (RS). The ordinate of its point of contact T is less than any other ordinate. Hence this observation: *One of the boxes evidently takes less lumber than any of the others.* In other words, we may infer that the function defined by (2) has a *minimum value*. Let us find this point on the graph exactly, using the calculus. Differentiating (2) to get the slope at any point, we have

$$\frac{dM}{dx} = 2x - \frac{432}{x^2}.$$

At the lowest point T the slope will be zero. Hence

$$2x - \frac{432}{x^2} = 0;$$

that is, when $x = 6$ the least amount of lumber will be needed.

Substituting in (2), we see that this is

$$M = 108 \text{ sq. ft.}$$

The fact that a least value of M exists is also shown by the following reasoning. Let the base increase from a very small square to a very large one. In the former case the height must be very great and therefore the amount of lumber required will be large. In the latter case, while the height is small, the base will take a great deal of lumber. Hence M varies from a large value, grows less, then increases again to another large value. It follows, then, that the graph must have a "lowest" point corresponding to the dimensions which require the least amount of lumber and therefore would involve the least cost.

We will now proceed to the treatment in detail of the subject of maxima and minima.

45. Increasing and decreasing functions.* Tests. A function $y = f(x)$ is said to be an *increasing function* if y increases (algebraically) when x increases. A function $y = f(x)$ is said to be a *decreasing* function if y decreases (algebraically) as x increases.

The graph of a function indicates plainly whether it is increasing or decreasing. For instance, consider the graph in Fig. *a*, p. 51.

---

* The proofs given here depend chiefly on geometric intuition. The subject of maxima and minima will be treated analytically in Art. 125.

VARIOUS APPLICATIONS OF THE DERIVATIVE

As we move along the curve from left to right the curve is *rising*; that is, as x increases the function ($= y$) increases. Obviously, Δy and Δx agree in sign.

On the other hand, in the graph of Fig. b, as we move along the curve from left to right the curve is *falling*; that is, as x increases, the function ($= y$) always decreases. Clearly, Δy and Δx have opposite signs.

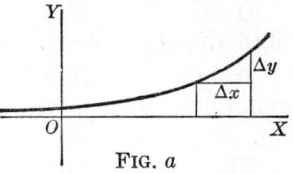

FIG. a

That a function may be sometimes increasing and sometimes decreasing is shown by the graph (Fig. c) of

(1) $\quad y = 2x^3 - 9x^2 + 12x - 3$.

As we move along the curve from left to right the curve rises until we reach the point A, falls from A to B, and rises to the right of B. Hence

(a) *from $x = -\infty$ to $x = 1$ the function is increasing;*

(b) *from $x = 1$ to $x = 2$ the function is decreasing;*

(c) *from $x = 2$ to $x = +\infty$ the function is increasing.*

FIG. b

At any point, such as C, where the function is increasing, the tangent makes an acute angle with the x-axis. The slope is positive. On the other hand, at a point, such as D, where the function is decreasing, the tangent makes an obtuse angle with the x-axis, and the slope is negative. Hence the following criterion:

A function is increasing when its derivative is positive, and decreasing when its derivative is negative.

FIG. c

For example, differentiating (1) above, we have

(2) $\quad \dfrac{dy}{dx} = f'(x) = 6x^2 - 18x + 12 = 6(x-1)(x-2)$.

When $x < 1$, $f'(x)$ is positive, and $f(x)$ is increasing.
When $1 < x < 2$, $f'(x)$ is negative, and $f(x)$ is decreasing.
When $x > 2$, $f'(x)$ is positive, and $f(x)$ is increasing.

These results agree with the conclusions arrived at above from the graph.

52 DIFFERENTIAL CALCULUS

46. Maximum and minimum values of a function; definitions. A *maximum* value of a function is one that is *greater* than any value immediately preceding or following.

A *minimum* value of a function is one that is *less* than any value immediately preceding or following.

For example, in Fig. c, Art. 45, it is clear that the function has a maximum value MA $(= y = 2)$ when $x = 1$, and a minimum value NB $(= y = 1)$ when $x = 2$.

The student should observe that a maximum value is not necessarily the greatest possible value of a function nor a minimum value the least. For in Fig. c it is seen that the function $(= y)$ will have values to the right of B that are greater than the maximum MA, and values to the left of A that are less than the minimum NB.

If $f(x)$ is an increasing function of x when x is slightly less than a, and a decreasing function of x when x is slightly greater than a, that is, if $f'(x)$ changes sign from $+$ to $-$ as x increases through a, then $f(x)$ has a maximum value when $x = a$. Therefore, if continuous, $f'(x)$ must vanish when $x = a$.

Thus, in the above example (Fig. c), at C, $f'(x)$ is positive; at A, $f'(x) = 0$; at D, $f'(x)$ is negative.

On the other hand, if $f(x)$ is a decreasing function when x is slightly less than a, and an increasing function when x is slightly greater than a, that is, if $f'(x)$ changes sign from $-$ to $+$ as x increases through a, then $f(x)$ has a minimum value when $x = a$. Therefore, if continuous, $f'(x)$ must vanish when $x = a$.

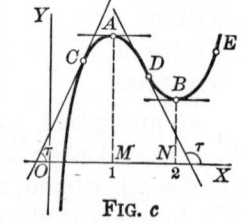

FIG. c

Thus, in Fig. c, at D, $f'(x)$ is negative; at B, $f'(x) = 0$; at E, $f'(x)$ is positive.

We may then state the conditions in general for maximum and minimum values of $f(x)$.

$f(x)$ is a **maximum** if $f'(x) = 0$ and $f'(x)$ changes sign from $+$ to $-$.

$f(x)$ is a **minimum** if $f'(x) = 0$ and $f'(x)$ changes sign from $-$ to $+$.

The values of the variable satisfying the equation $f'(x) = 0$ are called **critical values**; thus, from (2), Art. 45, $x = 1$ and $x = 2$ are the critical values of the variable for the function whose graph is shown in Fig. c. The critical values determine *turning points* where the tangent is parallel to OX.

To determine the sign of the first derivative at points near a particular turning point, substitute in it, first, a value of the variable slightly less than the corresponding critical value, and then one

VARIOUS APPLICATIONS OF THE DERIVATIVE

slightly greater. If the first sign is + and the second −, then the function has a maximum value for the critical value considered.

If the first sign is − and the second +, then the function has a minimum value.

If the sign is the same in both cases, then the function has neither a maximum nor a minimum value for the critical value considered. For example, take the above function in (1), Art. 45,

(1) $$y = f(x) = 2x^3 - 9x^2 + 12x - 3.$$

Then, as we have seen,

(2) $$f'(x) = 6(x-1)(x-2).$$

Setting $f'(x) = 0$, we find the critical values $x = 1$, $x = 2$. Let us first test $x = 1$. We consider values of x near this critical value to be substituted in the right-hand member of (2), and observe the signs of the factors. (Compare Art. 45.)

When $x < 1$, $f'(x) = (-)(-) = +$.
When $x > 1$, $f'(x) = (+)(-) = -$.

Hence $f(x)$ has a maximum value when $x = 1$. By the table, this value is $y = f(1) = 2$.

x	y
1	2
2	1

Next, test $x = 2$. Proceed as before, taking values of x now near the critical value 2.

When $x < 2$, $f'(x) = (+)(-) = -$.
When $x > 2$, $f'(x) = (+)(+) = +$.

Hence $f(x)$ has a minimum value when $x = 2$. By the table above, this value is $y = f(2) = 1$.

We shall now summarize our results in a *working rule*.

47. First method for examining a function for maximum and minimum values. Working rule.

FIRST STEP. *Find the first derivative of the function.*

SECOND STEP. *Set the first derivative equal to zero and solve the resulting equation for real roots. These roots are the critical values of the variable.*

THIRD STEP. *Considering one critical value at a time, test the first derivative, first for a value a trifle less* and then for a value a trifle greater than the critical value. If the sign of the derivative is first + and then −, the function has a maximum value for that particular critical value of the variable; but if the reverse is true, then it has a minimum value. If the sign does not change, the function has neither.*

---
* In this connection the term "little less," or "trifle less," means any value between the next smaller root (critical value) and the one under consideration; and the term "little greater," or "trifle greater," means any value between the root under consideration and the next larger one.

DIFFERENTIAL CALCULUS

In the Third Step, it is often convenient to resolve $f'(x)$ into factors, as in Art. 46.

ILLUSTRATIVE EXAMPLE 1. In the first problem worked out in Art. 44 we showed by means of the graph of the function

$$A = x\sqrt{100 - x^2}$$

that the rectangle of maximum area inscribed in a circle of radius 5 in. contained 50 sq. in. This may now be proved analytically as follows by applying the above rule.

Solution. $\qquad f(x) = x\sqrt{100 - x^2}.$

First Step. $\qquad f'(x) = \dfrac{100 - 2x^2}{\sqrt{100 - x^2}}.$

Second Step. Setting $f'(x) = 0$, we have

$$x = 5\sqrt{2} = 7.07,$$

which is the critical value. Only the positive sign of the radical is taken, since, from the nature of the problem, the negative sign has no meaning.

Third Step. When $x < 5\sqrt{2}$, then $2x^2 < 100$, and $f'(x)$ is $+$.
When $x > 5\sqrt{2}$, then $2x^2 > 100$, and $f'(x)$ is $-$.
Since the sign of the first derivative changes from $+$ to $-$, the function has a maximum value $f(5\sqrt{2}) = 5\sqrt{2} \cdot 5\sqrt{2} = 50$. *Ans.*

ILLUSTRATIVE EXAMPLE 2. Examine the function $(x-1)^2(x+1)^3$ for maximum and minimum values.

Solution. $f(x) = (x-1)^2(x+1)^3.$
First Step. $f'(x) = 2(x-1)(x+1)^3 + 3(x-1)^2(x+1)^2 = (x-1)(x+1)^2(5x-1).$
Second Step. $(x-1)(x+1)^2(5x-1) = 0.$

Hence $x = 1, -1, \frac{1}{5}$, are the critical values.
Third Step. $\qquad f'(x) = 5(x-1)(x+1)^2(x-\frac{1}{5}).$

Examine first the critical value $x = 1$ (C in figure).

When $x < 1$, $f'(x) = 5(-)(+)^2(+) = -$.
When $x > 1$, $f'(x) = 5(+)(+)^2(+) = +$.

Therefore, when $x = 1$ the function has a minimum value $f(1) = 0$ (= ordinate of C).
Examine now the critical value $x = \frac{1}{5}$ (B in figure).

When $x < \frac{1}{5}$, $f'(x) = 5(-)(+)^2(-) = +$.
When $x > \frac{1}{5}$, $f'(x) = 5(-)(+)^2(+) = -$.

Therefore, when $x = \frac{1}{5}$ the function has a maximum value $f(\frac{1}{5}) = 1.11$ (= ordinate of B).
Examine lastly the critical value $x = -1$ (A in figure).

When $x < -1$, $f'(x) = 5(-)(-)^2(-) = +$.
When $x > -1$, $f'(x) = 5(-)(+)^2(-) = +$.

Therefore, when $x = -1$ the function has neither a maximum nor a minimum value.

VARIOUS APPLICATIONS OF THE DERIVATIVE

48. Maximum or minimum values when $f'(x)$ becomes infinite and $f(x)$ is continuous. Consider the graph in the figure. At B, or G,

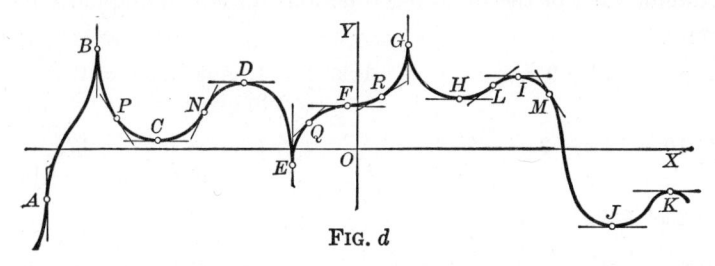

FIG. d

$f(x)$ is continuous and has a maximum value, but $f'(x)$ becomes infinite, since the tangent line at B is parallel to the y-axis. At E, $f(x)$ has a minimum value, and $f'(x)$ again becomes infinite. In our discussion of all possible maximum and minimum values of $f(x)$, we must therefore include as *critical values* also those values of x for which $f'(x)$ becomes infinite, or, what is the same thing, values of x satisfying

$$\text{(1)} \qquad \frac{1}{f'(x)} = 0.$$

The Second Step of the rule of the preceding section must then be modified as indicated by (1). The other steps are unchanged.

In Fig. d above, note that $f'(x)$ also becomes infinite at A, but the function is neither a maximum nor a minimum at A.

ILLUSTRATIVE EXAMPLE. Examine the function $a - b(x - c)^{\frac{2}{3}}$ for maxima and minima.

Solution.
$$f(x) = a - b(x - c)^{\frac{2}{3}}.$$
$$f'(x) = -\frac{2b}{3(x-c)^{\frac{1}{3}}}.$$
$$\frac{1}{f'(x)} = -\frac{3(x-c)^{\frac{1}{3}}}{2b}.$$

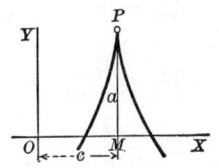

Since $x = c$ is a critical value for which $\frac{1}{f'(x)} = 0$ (and $f'(x) = \infty$), but for which $f(x)$ is not infinite, let us test the function for maximum and minimum values when $x = c$.

When $x < c$, $f'(x) = +$.
When $x > c$, $f'(x) = -$.

Hence, when $x = c = OM$, the function has a maximum value $f(c) = a = MP$.

PROBLEMS

Examine each of the following functions for maximum and minimum values.

1. $x^3 - 6x^2 + 9x$.   Ans. $x = 1$, gives max. $= 4$.
$x = 3$, gives min. $= 0$.

2. $10 + 12x - 3x^2 - 2x^3$. $x = 1$, gives max. $= 17$.
$x = -2$, gives min. $= -10$.

3. $2x^3 + 3x^2 + 12x - 4$.   No max. or min.

4. $x^3 + 2x^2 - 15x - 20$.

5. $2x^2 - x^4$. $x = 0$, gives min. $= 0$.
$x = \pm 1$, gives max. $= 1$.

6. $x^4 - 4x$. $x = 1$, gives min. $= -3$.

7. $x^4 - x^2 + 1$.

8. $3x^4 - 4x^3 - 12x^2$. $x = -1$, gives min. $= -5$.
$x = 0$, gives max. $= 0$.
$x = 2$, gives min. $= -32$.

9. $x^5 - 5x^4$. $x = 0$, gives max. $= 0$.
$x = 4$, gives min. $= -256$.

10. $3x^5 - 20x^3$.

11. $x^2 + \dfrac{2a^3}{x}$. $x = a$, gives min. $= 3a^2$.

12. $2x - \dfrac{a^3}{x^2}$.

13. $x^2 + \dfrac{a^4}{x^2}$. $x = \pm a$, gives min. $= 2a^2$.

14. $\dfrac{ax}{x^2 + a^2}$. $x = -a$, gives min. $= -\frac{1}{2}$.
$x = a$, gives max. $= \frac{1}{2}$.

15. $\dfrac{x^2}{x + a}$.

16. $\dfrac{x^2}{x^2 + a^2}$.

17. $\dfrac{x^2 + 2a^2}{x^2 + a^2}$.

18. $(2 + x)^2(1 - x)^2$.

19. $(2 + x)^2(1 - x)^3$.

20. $b + c(x - a)^{\frac{2}{3}}$. $x = a$, gives min. $= b$.

21. $a - b(x - c)^{\frac{1}{3}}$.   No max. or min.

VARIOUS APPLICATIONS OF THE DERIVATIVE 57

22. $(2+x)^{\frac{1}{3}}(1-x)^{\frac{2}{3}}$.   Ans. $x = 1$, gives min. $= 0$.
$x = -1$, gives max. $= \sqrt[3]{4} = 1.6$.

23. $x(a+x)^2(a-x)^3$. $x = -a$, gives max. $= 0$.
$x = -\frac{1}{2}a$, gives min. $= -\frac{27}{64}a^6$.
$x = \frac{1}{3}a$, gives max. $= \frac{128}{729}a^6$.
$x = a$, gives neither.

24. $(2x-a)^{\frac{1}{3}}(x-a)^{\frac{2}{3}}$. $x = \frac{2}{3}a$, gives max. $= \frac{1}{3}a$.
$x = a$, gives min. $= 0$.
$x = \frac{1}{2}a$, gives neither.

25. $\dfrac{x+2}{x^2+2x+4}$. $x = 0$, gives max. $= \frac{1}{2}$.
$x = -4$, gives min. $= -\frac{1}{6}$.

26. $\dfrac{x^2+x+4}{x+1}$. $x = -3$, gives max. $= -5$.
$x = 1$, gives min. $= 3$.

27. $\dfrac{x^2+x+4}{x^2+2x+4}$. $x = -2$, gives max. $= \frac{3}{2}$.
$x = 2$, gives min. $= \frac{5}{6}$.

28. $\dfrac{(x-a)(b-x)}{x^2}$. $x = \dfrac{2ab}{a+b}$, gives max. $= \dfrac{(b-a)^2}{4ab}$.

29. $\dfrac{a^2}{x} + \dfrac{b^2}{a-x}$. $x = \dfrac{a^2}{a+b}$, gives min. $= \dfrac{(a+b)^2}{a}$.
$x = \dfrac{a^2}{a-b}$, gives max. $= \dfrac{(a-b)^2}{a}$.

30. $\dfrac{(a-x)^3}{a-2x}$. $x = \dfrac{a}{4}$, gives min. $= \frac{27}{32}a^2$.

31. $\dfrac{x^2+x-1}{x^2-x+1}$.

49. Maximum and minimum values. Applied problems. In many problems we must first construct, from the given conditions, the function whose maximum and minimum values are required, as was done in the two examples worked out in Art. 44. This sometimes offers considerable difficulty. No rule applicable in all cases can be given, but in many problems we may be guided by the following

General directions

(a) *Set up the function whose maximum or minimum value is involved in the problem.*

(b) *If the resulting expression contains more than one variable, the conditions of the problem will furnish enough relations between the variables so that all may be expressed in terms of a single one.*

(c) *To the resulting function of a single variable apply the above rule (p. 53) for finding maximum and minimum values.*

(d) *In practical problems it is usually easy to tell which critical value will give a maximum and which a minimum value, so it is not always necessary to apply the third step.*

(e) *Draw the graph of the function in order to check the work.*

The work of finding maximum and minimum values may frequently be simplified by the aid of the following principles, which follow at once from our discussion of the subject.

(a) *The maximum and minimum values of a continuous function must occur alternately.*

(b) *When c is a positive constant, cf(x) is a maximum or a minimum for such values of x, and such only, as make f(x) a maximum or a minimum.*

Hence, in determining the critical values of x and testing for maxima and minima, any constant factor may be omitted.

When c is negative, cf(x) is a maximum when f(x) is a minimum, and conversely.

(c) *If c is a constant, f(x) and c + f(x) have maximum and minimum values for the same values of x.*

Hence a constant term may be omitted when finding critical values of x and testing.

PROBLEMS

1. It is desired to make an open-top box of greatest possible volume from a square piece of tin whose side is a, by cutting equal squares out of the corners and then folding up the tin to form the sides. What should be the length of a side of the squares cut out?

Solution. Let x = side of small square = depth of box; then $a - 2x$ = side of square forming bottom of box, and volume is $V = (a - 2x)^2 x$, which is the function to be made a maximum by varying x. Applying the rule, p. 53,

First Step. $\dfrac{dV}{dx} = (a - 2x)^2 - 4x(a - 2x) = a^2 - 8ax + 12x^2.$

Second Step. Solving $a^2 - 8ax + 12x^2 = 0$ gives critical values $x = \dfrac{a}{2}$ and $\dfrac{a}{6}$.

It is evident from the figure that $x = \dfrac{a}{2}$ must give a minimum, for then all the tin would be cut away, leaving no material out of which to make a box. By the usual test, $x = \dfrac{a}{6}$ is found to give a maximum volume $\dfrac{2 a^3}{27}$. Hence the side of the square to be cut out is one sixth of the side of the given square.

VARIOUS APPLICATIONS OF THE DERIVATIVE 59

The drawing of the graph of the function in this and the following problems is left to the student.

2. Assuming that the strength of a beam with rectangular cross section varies directly as the breadth and as the square of the depth, what are the dimensions of the strongest beam that can be sawed out of a round log whose diameter is d?

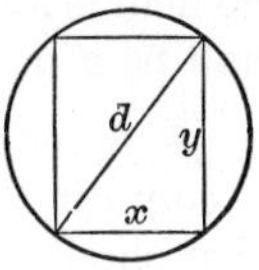

Solution. If $x =$ breadth and $y =$ depth, then the beam will have maximum strength when the function xy^2 is a maximum. From the figure, $y^2 = d^2 - x^2$; hence we should test the function
$$f(x) = x(d^2 - x^2).$$

First Step. $\quad f'(x) = -2x^2 + d^2 - x^2 = d^2 - 3x^2.$

Second Step. $d^2 - 3x^2 = 0.$ $\therefore x = \dfrac{d}{\sqrt{3}} =$ critical value which gives a maximum.

Therefore, if the beam is cut so that

$$\text{Depth} = \sqrt{\tfrac{2}{3}} \text{ of diameter of log,}$$

and $\qquad\qquad\text{Breadth} = \sqrt{\tfrac{1}{3}} \text{ of diameter of log,}$

the beam will have maximum strength.

3. What is the width of the rectangle of maximum area that can be inscribed in a given segment OAA' of a parabola?

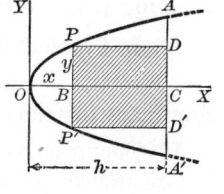

HINT. If $OC = h$, $BC = h - x$ and $PP' = 2y$; therefore the area of rectangle $PDD'P'$ is
$$2(h - x)y.$$
But since P lies on the parabola $y^2 = 2px$, the function to be tested is
$$f(x) = 2(h - x)\sqrt{2px}.$$

Ans. Width $= \tfrac{2}{3} h$.

4. Find the altitude of the cone of maximum volume that can be inscribed in a sphere of radius r.

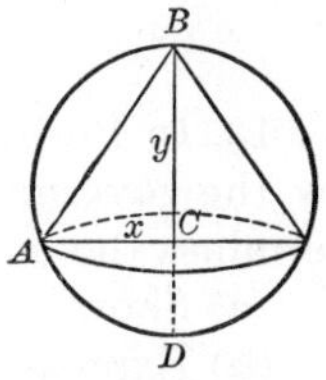

HINT. Volume of cone $= \tfrac{1}{3} \pi x^2 y$. But
$$x^2 = BC \times CD = y(2r - y);$$
therefore the function to be tested is
$$f(y) = \frac{\pi}{3} y^2(2r - y).$$

Ans. Altitude of cone $= \tfrac{4}{3} r$.

5. Find the altitude of the cylinder of maximum volume that can be inscribed in a given right cone.

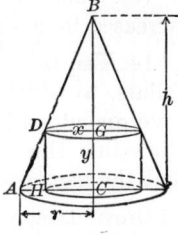

HINT. Let $AC = r$ and $BC = h$. Volume of cylinder $= \pi x^2 y$. But from similar triangles ABC and DBG,
$$r : x = h : h - y. \therefore x = \frac{r(h - y)}{h}.$$
Hence the function to be tested is
$$f(y) = \frac{r^2}{h^2} y(h - y)^2.$$

Ans. Altitude $= \tfrac{1}{3} h$.

6. If three sides of a trapezoid are each 10 in. long, how long must the fourth side be if the area is a maximum? *Ans.* 20 in.

7. It is required to inclose a rectangular field by a fence, and then to divide it into two lots by a fence parallel to one of the sides. If the area of the field is given, find the ratio of the sides so that the total length of fence shall be a minimum. *Ans.* 2/3.

8. A rectangular garden is to be laid out along a neighbor's lot and is to contain 432 sq. rd. If the neighbor pays for half the dividing fence, what should be the dimensions of the garden so that the cost to the owner of inclosing it may be a minimum? *Ans.* 18 rd. × 24 rd.

9. A radio manufacturer finds that he can sell x instruments per week at p dollars each, where $5x = 375 - 3p$. The cost of production is $(500 + 15x + \frac{1}{5}x^2)$ dollars. Show that the maximum profit is obtained when the production is about 30 instruments per week.

10. In Problem 9 suppose the relation between x and p is
$$x = 100 - 20\sqrt{\frac{p}{5}}.$$
Show that the manufacturer should produce only about 25 instruments per week for maximum profit.

11. In Problem 9 suppose the relation between x and p is
$$x^2 = 2500 - 20p.$$
How many instruments should be produced each week for maximum profit?

12. The total cost of producing x articles per week is $(ax^2 + bx + c)$ dollars and the price (p dollars) at which each can be sold is $p = \beta - \alpha x^2$. Show that the output for maximum profit is
$$x = \frac{\sqrt{a^2 + 3\alpha(\beta - b)} - a}{3\alpha}.$$

13. In Problem 9 suppose a tax of t dollars per instrument is imposed by the government. The manufacturer adds the tax to his cost and determines the output and price under the new conditions.

(a) Show that the price increases by a little less than half the tax.

(b) Express the receipts from the tax in terms of t and determine the tax for maximum return.

(c) When the tax determined in (b) is imposed, show that the price is increased by about 33 per cent.

14. The total cost of producing x articles per week is $(ax^2 + bx + c)$ dollars, to which is added a tax of t dollars per article imposed by the government, and the price (p dollars) at which each can be sold is $p = \beta - \alpha x$. Show that the tax brings in the maximum return when $t = \frac{1}{2}(\beta - b)$ and that the increase in price is always less than the tax.

NOTE. In applications in economics a, b, c, α, β are positive numbers.

VARIOUS APPLICATIONS OF THE DERIVATIVE

15. A steel plant is capable of producing x tons per day of a low-grade steel and y tons per day of a high-grade steel, where $y = \dfrac{40 - 5x}{10 - x}$. If the fixed market price of low-grade steel is half that of high-grade steel, show that about $5\frac{1}{2}$ tons of low-grade steel are produced per day for maximum receipts.

16. A telephone company finds there is a net profit of $15 per instrument if an exchange has 1000 subscribers or less. If there are over 1000 subscribers, the profits per instrument decrease 1¢ for each subscriber above that number. How many subscribers would give the maximum net profit? *Ans.* 1250.

17. The cost of manufacturing a given article is p dollars and the number which can be sold varies inversely as the nth power of the selling price. Find the selling price which will yield the greatest total net profit.
Ans. $\dfrac{np}{n-1}$.

18. What should be the diameter of a tin can holding 1 qt. (58 cu. in.) and requiring the least amount of tin (a) if the can is open at the top? (b) if the can has a cover?
Ans. (a) $\sqrt[3]{\dfrac{464}{\pi}} = 5.29$ in.; (b) $\sqrt[3]{\dfrac{232}{\pi}} = 4.20$ in.

19. The lateral surface of a right circular cylinder is 4π sq. ft. From the cylinder is cut a hemisphere whose diameter equals the diameter of the cylinder. Find the dimensions of the cylinder if the remaining volume is a maximum or minimum. Determine whether it is a maximum or a minimum. *Ans.* Radius $= 1$ ft., altitude $= 2$ ft.; maximum.

20. Find the area of the largest rectangle with sides parallel to the coördinate axes which can be inscribed in the figure bounded by the two parabolas $3y = 12 - x^2$ and $6y = x^2 - 12$. *Ans.* 16.

21. Two vertices of a rectangle are on the x-axis. The other two vertices are on the lines whose equations are $y = 2x$ and $3x + y = 30$. For what value of y will the area of the rectangle be a maximum? *Ans.* $y = 6$.

22. One base of an isosceles trapezoid is a diameter of a circle of radius a, and the ends of the other base lie on the circumference of the circle. Find the length of the other base if the area is a maximum. *Ans.* a.

23. A rectangle is inscribed in a parabolic segment with one side of the rectangle along the base of the segment. Show that the ratio of the area of the largest rectangle to the area of the segment is $\dfrac{1}{\sqrt{3}}$.

24. The strength of a rectangular beam varies as the product of the breadth and the square of the depth. Find the dimensions of the strongest beam that can be cut from a log whose cross section is an ellipse of semiaxes a and b. *Ans.* Breadth $= 2b\sqrt{\tfrac{1}{3}}$; depth $= 2a\sqrt{\tfrac{2}{3}}$.

DIFFERENTIAL CALCULUS

25. The stiffness of a rectangular beam varies as the product of the breadth and the cube of the depth. Find the dimensions of the stiffest beam that can be cut from a cylindrical log whose radius is a.

Ans. $a \times a\sqrt{3}$.

26. The equation of the path of a ball is $y = mx - \dfrac{(m^2+1)x^2}{800}$, where the origin is taken at the point from which the ball is thrown and m is the slope of the curve at the origin. For what value of m will the ball strike (a) at the greatest distance along the same horizontal level? (b) at the greatest height on a vertical wall 300 ft. away? Ans. (a) 1; (b) $\frac{4}{3}$.

27. A window of perimeter p ft. is in the form of a rectangle surmounted by an isosceles right triangle. Show that the window will admit the most light when the sides of the rectangle are equal to the sides of the right triangle.

28. The sum of the surfaces of a sphere and a cube being given, show that the sum of their volumes will be least when the diameter of the sphere is equal to the edge of the cube. When will the sum of the volumes be greatest?

29. Find the dimensions of the largest rectangle which can be inscribed in the ellipse $\dfrac{x^2}{a^2} + \dfrac{y^2}{b^2} = 1$. Ans. $a\sqrt{2} \times b\sqrt{2}$.

30. Find the area of the largest rectangle which can be drawn with its base on the x-axis and with two vertices on the witch whose equation is $y = \dfrac{8\,a^3}{x^2 + 4\,a^2}$. (See figure in Chapter XXVI.) Ans. $4\,a^2$.

31. Find the ratio of the area of the smallest ellipse that can be circumscribed about a rectangle to the area of the rectangle. The area of an ellipse is πab, where a and b are the semiaxes. Ans. $\frac{1}{2}\pi$.

32. The two lower vertices of an isosceles trapezoid are the points $(-6, 0)$ and $(6, 0)$. The two upper vertices lie on the curve $x^2 + 4\,y = 36$. Find the area of the largest trapezoid which can be drawn in this way.

Ans. 64.

33. The distance between the centers of two spheres of radii a and b, respectively, is c. Find from what point P on the line of centers AB the greatest amount of spherical surface is visible. (The area of the curved surface of a zone of height h is $2\,\pi r h$, where r is the radius of the sphere.)

Ans. $\dfrac{c a^{\frac{3}{2}}}{a^{\frac{3}{2}} + b^{\frac{3}{2}}}$ units from A.

34. Find the dimensions of the largest rectangular parallelepiped with a square base which can be cut from a solid sphere of radius r.

Ans. $h = \frac{2}{3} r\sqrt{3}$.

VARIOUS APPLICATIONS OF THE DERIVATIVE

35. Given a sphere of radius 6 in. Calculate the altitude of each of the following solids:
  (a) inscribed right circular cylinder of maximum volume;
  (b) inscribed right circular cylinder of maximum total surface;
  (c) circumscribed right cone of minimum volume.
  Ans. (a) $4\sqrt{3}$ in.; (b) 6.31 in.; (c) 24 in.

36. Prove that a conical tent of a given capacity will require the least amount of canvas when the height is $\sqrt{2}$ times the radius of the base. Show that when the canvas is laid out flat it will be a circle with a sector of 152° 9′ cut out. How much canvas would be required for a tent 10 ft. high? *Ans.* 272 sq. ft.

37. Given a point on the axis of the parabola $y^2 = 2\,px$ at a distance a from the vertex; find the abscissa of the point on the curve nearest to it.
 Ans. $x = a - p$.

38. Find the point on the curve $2y = x^2$ which is nearest to the point (4, 1). *Ans.* (2, 2).

39. If PQ is the longest or shortest line segment which can be drawn from $P(a, b)$ to the curve $y = f(x)$, prove that PQ is perpendicular to the tangent to the curve at Q.

40. A formula for the efficiency of a screw is $E = \dfrac{h(1 - h \tan \theta)}{h + \tan \theta}$, where θ is the angle of friction and h is the pitch of the screw. Find h for maximum efficiency. *Ans.* $h = \sec \theta - \tan \theta$.

41. The distance between two sources of heat A and B, with intensities a and b respectively, is l. The total intensity of heat at a point P between A and B is given by the formula

$$I = \frac{a}{x^2} + \frac{b}{(l - x)^2},$$

where x is the distance of P from A. For what position of P will the temperature be lowest?
 Ans. $x = \dfrac{a^{\frac{1}{3}} l}{a^{\frac{1}{3}} + b^{\frac{1}{3}}}$.

42. The lower base of an isosceles trapezoid is the major axis of an ellipse; the ends of the upper base are points on the ellipse. Show that the maximum trapezoid of this type has the length of its upper base half that of the lower.

43. An isosceles triangle with vertex at $(0, b)$ is to be inscribed in the ellipse $b^2x^2 + a^2y^2 = a^2b^2$. Find the equation of the base if the area of the triangle is a maximum. *Ans.* $2y + b = 0$.

44. Find the base and altitude of the isosceles triangle of minimum area which circumscribes the ellipse $b^2x^2 + a^2y^2 = a^2b^2$, and whose base is parallel to the x-axis. *Ans.* Altitude $= 3b$, base $= 2a\sqrt{3}$.

45. Let $P(a, b)$ be a point in the first quadrant of a set of rectangular axes. Draw a line through P cutting the positive ends of the axes at A and B. Calculate the intercepts of this line on OX and OY in each of the following cases.

   (a) when the area OAB is a minimum;
   (b) when the length AB is a minimum;
   (c) when the sum of the intercepts is a minimum;
   (d) when the perpendicular distance from O to AB is a maximum.

Ans. (a) $2a$, $2b$; (b) $a + a^{\frac{1}{3}}b^{\frac{2}{3}}$, $b + a^{\frac{2}{3}}b^{\frac{1}{3}}$; (c) $a + \sqrt{ab}$, $b + \sqrt{ab}$; (d) $\dfrac{a^2 + b^2}{a}$, $\dfrac{a^2 + b^2}{b}$.

50. Derivative as the rate of change. In Art. 23 the functional relation

(1) $$y = x^2$$

gave as the ratio of corresponding increments

(2) $$\frac{\Delta y}{\Delta x} = 2x + \Delta x.$$

When $x = 4$ and $\Delta x = 0.5$, equation (2) becomes

(3) $$\frac{\Delta y}{\Delta x} = 8.5.$$

Then, we say, the average rate of change of y with respect to x equals 8.5 when x increases from $x = 4$ to $x = 4.5$.

In general, the ratio

(A) $\dfrac{\Delta y}{\Delta x} =$ average rate of change of y with respect to x when x changes from x to $x + \Delta x$.

Constant rate of change. When

(4) $$y = ax + b,$$

we have $\dfrac{\Delta y}{\Delta x} = a.$

That is, the average rate of change of y with respect to x equals a, the slope of the straight line (4), and is constant. In this case, and in this case only, the change in y (Δy), when x increases from any value x to $x + \Delta x$, equals the rate of change a times Δx.

Instantaneous rate of change. If the interval from x to $x + \Delta x$ decreases and $\Delta x \to 0$, then the average rate of change of y with respect to x in this interval becomes at the limit the *instantaneous rate of change of y with respect to x.* Hence, by Art. 24,

(B) $\dfrac{dy}{dx} =$ instantaneous rate of change of y with respect to x for a definite value of x.

For example, from (1) above,

(5) $$\frac{dy}{dx} = 2x.$$

When $x = 4$, the instantaneous rate of change of y is 8 units per unit change in x. The word "instantaneous" is often dropped in (B).

Geometric interpretation. Let the graph of

(6) $$y = f(x)$$

be drawn, as in the figure. When x increases from OM to ON, then y increases from MP to NQ. The average rate of change of y with respect to x equals the slope of the secant line PQ. The instantaneous rate when $x = OM$ equals the slope of the tangent line PT.

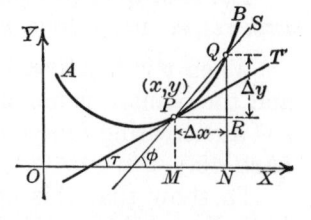

Hence the instantaneous rate of change of y at $P(x, y)$ is equal to the constant rate of change of y along the tangent line at P.

When $x = x_0$, the instantaneous rate of change of y, or $f(x)$, in (6) is $f'(x_0)$. If x now increases from x_0 to $x_0 + \Delta x$, the *exact* change in y is *not* equal to $f'(x_0)\Delta x$, unless $f'(x)$ is constant, as in (4). We shall see later, however, that this product is equal to Δy, nearly, when Δx is sufficiently small.

51. Velocity in rectilinear motion. Important applications arise when the independent variable in a rate is the time. The rate is then called a **time-rate**. Velocity in rectilinear motion affords a simple example.

Consider the motion of a point P on the straight line AB. Let s be the distance measured from some fixed point, as O, to any position of P, and let t be the corresponding elapsed time. To each value of t corresponds a position of P and therefore a distance (or space) s. Hence s will be a function of t, and we may write

$$s = f(t).$$

Now let t take on an increment Δt; then s takes on an increment Δs, and

(1) $$\frac{\Delta s}{\Delta t} = \text{the average velocity}$$

of P when the point moves from P to P', during the time interval Δt. If P moves with uniform motion (constant velocity), the above ratio will have the same value for every interval of time and is the velocity at any instant.

DIFFERENTIAL CALCULUS

For the general case of any kind of motion, uniform or not, we define *the velocity (time-rate of change of s) at any instant as the limit of the average velocity* as Δt approaches zero as a limit; that is,

(C) $$v = \frac{ds}{dt}.$$

The velocity at any instant is the derivative of the distance (= space) with respect to the time, or the time-rate of change of the distance.

When v is positive, the distance s is an increasing function of t, and the point P is moving in the direction AB. When v is negative, s is a decreasing function of t, and P is moving in the direction BA. (Art. 45.)

To show that this definition agrees with the conception we already have of velocity, let us find the velocity of a falling body at the end of two seconds.

By experiment it has been found that a body falling freely from rest in a vacuum near the earth's surface follows approximately the law

(2) $$s = 16.1\, t^2,$$

where $s =$ distance of fall in feet, $t =$ time in seconds. Apply the General Rule (Art. 27) to (2).

FIRST STEP. $s + \Delta s = 16.1(t + \Delta t)^2 = 16.1\, t^2 + 32.2\, t \cdot \Delta t + 16.1(\Delta t)^2.$

SECOND STEP. $\Delta s = 32.2\, t \cdot \Delta t + 16.1(\Delta t)^2.$

THIRD STEP. $\dfrac{\Delta s}{\Delta t} = 32.2\, t + 16.1\, \Delta t =$ average velocity throughout the time interval Δt.

Placing $t = 2$,

(3) $\dfrac{\Delta s}{\Delta t} = 64.4 + 16.1\, \Delta t =$ average velocity throughout the time interval Δt after two seconds of falling.

Our notion of velocity tells at once that (3) does not give us the actual velocity *at the end of two seconds*; for even if we take Δt very small, say $\frac{1}{100}$ or $\frac{1}{1000}$ of a second, (3) still gives only the *average velocity* during the corresponding small interval of time. But what we do mean by the velocity at the end of two seconds is *the limit of the average velocity when Δt diminishes toward zero*; that is, the velocity at the end of two seconds is, from (3), 64.4 feet per second. Thus even the everyday notion of velocity which we get from experience involves the idea of a limit, or, in our notation,

$$v = \lim_{\Delta t \to 0} \left(\frac{\Delta s}{\Delta t}\right) = 64.4 \text{ ft. per second.}$$

VARIOUS APPLICATIONS OF THE DERIVATIVE

52. Related rates. In many problems several variables are involved each of which is a function of the time. Relations between the variables are established by the conditions of the problem. The relations between their time-rates of change are then found by differentiation.

As a guide in solving rate problems use the following *rule*.

FIRST STEP. *Draw a figure illustrating the problem. Denote by x, y, z, etc. the quantities which vary with the time.*

SECOND STEP. *Obtain a relation between the variables involved which will hold true at any instant.*

THIRD STEP. *Differentiate with respect to the time.*

FOURTH STEP. *Make a list of the given and required quantities.*

FIFTH STEP. *Substitute the known quantities in the result found by differentiating (third step), and solve for the unknown.*

PROBLEMS

1. A man is walking at the rate of 5 mi. per hour toward the foot of a tower 60 ft. high. At what rate is he approaching the top when he is 80 ft. from the foot of the tower?

Solution. Apply the above rule.

First Step. Draw the figure. Let $x =$ distance of the man from the foot, and $y =$ his distance from the top, of the tower at any instant.

Second Step. Since we have a right triangle,
$$y^2 = x^2 + 3600.$$

Third Step. Differentiating, we get
$$2y \frac{dy}{dt} = 2x \frac{dx}{dt}, \text{ or}$$

(1) $$\frac{dy}{dt} = \frac{x}{y} \frac{dx}{dt}.$$

This means that at *any* instant whatever
$$(\text{Rate of change of } y) = \left(\frac{x}{y}\right) \text{ times (rate of change of } x).$$

Fourth Step. $x = 80$, $\frac{dx}{dt} = -5$ mi. an hour
$\phantom{Fourth Step.\ x = 80,\ \frac{dx}{dt}} = -5 \times 5280$ ft. an hour.
$y = \sqrt{x^2 + 3600}$ $\frac{dy}{dt} = ?$
$ = 100.$

Fifth Step. Substituting in (1),
$$\frac{dy}{dt} = -\frac{80}{100} \times 5 \times 5280 \text{ ft. per hour}$$
$$\phantom{\frac{dy}{dt}} = -4 \text{ mi. per hour. } Ans.$$

2. A point moves on the parabola $6y = x^2$ in such a way that when $x = 6$ the abscissa is increasing at the rate of 2 ft. per second. At what rate is the ordinate increasing at that instant?

Solution. *First Step.* Plot the parabola.

Second Step. $6y = x^2$.

Third Step. $6\dfrac{dy}{dt} = 2x\dfrac{dx}{dt}$, or

(2) $\qquad \dfrac{dy}{dt} = \dfrac{x}{3} \cdot \dfrac{dx}{dt}.$

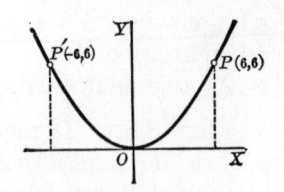

This means that at *any* point on the parabola

$$(\text{Rate of change of ordinate}) = \left(\dfrac{x}{3}\right) \text{ times (rate of change of abscissa).}$$

Fourth Step. $\quad x = 6. \qquad \dfrac{dx}{dt} = 2$ ft. per second.

$$y = \dfrac{x^2}{6} = 6. \qquad \dfrac{dy}{dt} = ?$$

Fifth Step. Substituting in (2),

$$\dfrac{dy}{dt} = \dfrac{6}{3} \times 2 = 4 \text{ ft. per second. } Ans.$$

From the first result we note that at the point $P(6, 6)$ the ordinate changes twice as rapidly as the abscissa.

If we consider the point $P'(-6, 6)$ instead, the result is $\dfrac{dy}{dt} = -4$ ft. per second, the minus sign indicating that the ordinate is decreasing as the abscissa increases.

3. A circular plate of metal expands by heat so that its radius increases at the rate of 0.01 in. per second. At what rate is the surface increasing when the radius is 2 in.?

Solution. Let $x =$ radius and $y =$ area of plate. Then

$$y = \pi x^2.$$

(3) $\qquad \dfrac{dy}{dt} = 2\pi x \dfrac{dx}{dt}.$

That is, at any instant the area of the plate is increasing in square inches $2\pi x$ times as fast as the radius is increasing in linear inches.

$$x = 2, \quad \dfrac{dx}{dt} = 0.01, \quad \dfrac{dy}{dt} = ?$$

Substituting in (3),

$$\dfrac{dy}{dt} = 2\pi \times 2 \times 0.01 = 0.04\,\pi \text{ sq. in. per second. } Ans.$$

4. An arc light is hung 12 ft. directly above a straight horizontal walk on which a boy 5 ft. tall is walking. How fast is the boy's shadow

VARIOUS APPLICATIONS OF THE DERIVATIVE

lengthening when he is walking away from the light at the rate of 168 ft. per minute?

Solution. Let x = distance of the boy from a point directly under the light L, and y = length of boy's shadow. From the figure,

$$y : y + x = 5 : 12,$$

or $\qquad y = \tfrac{5}{7} x.$

Differentiating, $\qquad \dfrac{dy}{dt} = \dfrac{5}{7} \dfrac{dx}{dt};$

that is, the shadow is lengthening $\tfrac{5}{7}$ as fast as the boy is walking, or 120 ft. per minute.

5. A point moves along the parabola $y^2 = 12\,x$ in such a way that its abscissa increases uniformly at the rate of 2 in. per second. At what point do the abscissa and ordinate increase at the same rate? *Ans.* (3, 6).

6. Find the values of x for which the rate of change of

$$x^3 - 12\,x^2 + 45\,x - 13$$

is zero. *Ans.* 3 and 5.

7. A barge whose deck is 12 ft. below the level of a dock is drawn up to it by means of a cable attached to a ring in the floor of the dock, the cable being hauled in by a windlass on deck at the rate of 8 ft. per minute. How fast is the barge moving towards the dock when 16 ft. away?
Ans. 10 ft. per minute.

8. A boat is fastened to a rope which is wound about a windlass 20 ft. above the level at which the rope is attached to the boat. The boat is drifting away at the rate of 8 ft. per second. How fast is it unwinding the rope when 30 ft. from the point directly under the windlass?
Ans. 6.66 ft. per second.

9. One end of a ladder 50 ft. long is leaning against a perpendicular wall standing on a horizontal plane. Suppose the foot of the ladder to be pulled away from the wall at the rate of 3 ft. per minute. (a) How fast is the top of the ladder descending when its foot is 14 ft. from the wall? (b) When will the top and bottom of the ladder move at the same rate? (c) When is the top of the ladder descending at the rate of 4 ft. per minute?
Ans. (a) $\tfrac{7}{8}$ ft. per minute; (b) when $25\sqrt{2}$ ft. from the wall; (c) when 40 ft. from the wall.

10. One ship was sailing south at the rate of 6 mi. per hour; another east at the rate of 8 mi. per hour. At 4 P.M. the second crossed the track of the first where the first was 2 hr. before. (a) How was the distance between the ships changing at 3 P.M.? (b) How at 5 P.M.? (c) When was the distance between them not changing?
Ans. (a) Decreasing 2.8 mi. per hour; (b) increasing 8.73 mi. per hour; (c) 3.17 P.M.

DIFFERENTIAL CALCULUS

11. The side of an equilateral triangle is a in. long, and is increasing at the rate of k in. per hour. How fast is the area increasing?

Ans. $\frac{1}{2} ak\sqrt{3}$ sq. in. per hour.

12. The edges of a regular tetrahedron are 10 in. long and are increasing at the rate of 0.1 in. per minute. Find the rate of increase of the volume.

13. If at a certain instant the two dimensions of a rectangle are a and b and these dimensions are changing at the rates m, n respectively, show that the area is changing at the rate $an + bm$.

14. At a certain instant the three dimensions of a rectangular parallelepiped are 6 in., 8 in., 10 in., and these are increasing at the respective rates of 0.2 in. per second, 0.3 in. per second, 0.1 in. per second. How fast is the volume increasing?

15. The period (P sec.) of a complete oscillation of a pendulum of length l in. is given by the formula $P = 0.324\sqrt{l}$. Find the rate of change of the period with respect to the length when $l = 9$ in. By means of this result approximate the change in P due to a change in l from 9 to 9.2 in.

Ans. 0.054 sec. per inch; 0.0108 sec.

16. The diameter and altitude of a right circular cylinder are found at a certain instant to be 10 in. and 20 in. respectively. If the diameter is increasing at the rate of 1 in. per minute, what change in the altitude will keep the volume constant? *Ans.* Decreasing 4 in. per minute.

17. The radius of the base of a certain cone is increasing at the rate of 3 in. per minute and the altitude is decreasing at the rate of 4 in. per minute. Find the rate of change of the total surface of the cone when the radius is 7 in. and the altitude is 24 in.

Ans. Increasing $96\,\pi$ sq. in. per minute.

18. A cylinder of radius r and altitude h has a hemisphere of radius r attached to each end. If r is increasing at the rate of $\frac{1}{2}$ in. per minute, find the rate at which h must change to keep the volume of the solid fixed at the instant when r is 10 in. and h is 20 in.

19. A conical funnel is 12 in. across the top and 10 in. deep. A liquid is flowing in at the rate of 10 cu. in./sec., and out at the rate of 1 cu. in./sec. How fast is the surface of the liquid rising when it is 6 in. deep?

20. A gas holder contains 1000 cu. ft. of gas at a pressure of 5 lb. per square inch. If the pressure is decreasing at the rate of 0.05 lb. per square inch per hour, find the rate of increase of the volume. (Assume Boyle's Law: $pv = c$.) *Ans.* 10 cu. ft. per hour.

21. The adiabatic law for the expansion of air is $PV^{1.4} = C$. If at a given time the volume is observed to be 10 cu. ft. and the pressure is 50 lb. per square inch, at what rate is the pressure changing if the volume is decreasing 1 cu. ft. per second?

Ans. Increasing 7 lb. per square inch per second.

VARIOUS APPLICATIONS OF THE DERIVATIVE

22. If $y = 4x - x^3$ and x is increasing steadily at the rate of $\frac{1}{3}$ unit per second, find how fast the slope of the graph is changing at the instant when $x = 2$.   *Ans.* Decreasing 4 units per second.

23. A rod 10 ft. long moves so that its ends A and B remain constantly on the x-axis and y-axis, respectively. If A is 8 ft. from the origin and is moving away at the rate of 2 ft. per second,
 (a) at what rate is the end B coming down?
 (b) at what rate is the area formed by AB and the axes changing?
 (c) If P is the midpoint of AB, at what rate is OP changing?
   Ans. (a) $\frac{8}{3}$ ft. per second.
   (b) Decreasing $\frac{14}{3}$ sq. ft. per second.
   (c) OP is constant.

24. If r denotes the radius of a sphere, S the surface, and V the volume, prove the relation $\dfrac{dV}{dt} = \dfrac{r}{2}\dfrac{dS}{dt}$.

25. A railroad track crosses a highway at an angle of 60°. A locomotive is 500 ft. from the intersection and moving away from it at the rate of 60 mi. per hour. An automobile is 500 ft. from the intersection and moving toward it at the rate of 30 mi. per hour. What is the rate of change of the distance between them?
   Ans. Increasing 15 mi. per hour or $15\sqrt{3}$ mi. per hour.

26. A horizontal trough 10 ft. long has a vertical section in the shape of an isosceles right triangle. If water is poured into it at the rate of 8 cu. ft. per minute, at what rate is the surface of the water rising when the water is 2 ft. deep?   *Ans.* $\frac{1}{5}$ ft. per minute.

27. In Problem 26, at what rate must water be poured into the trough to make the level rise $\frac{1}{6}$ ft. per minute when the water is 3 ft. deep?

28. A horizontal trough 12 ft. long has a vertical cross section in the shape of a trapezoid, the bottom being 3 ft. wide and the sides inclined to the vertical at an angle whose sine is $\frac{4}{5}$. Water is being poured into it at the rate of 10 cu. ft. per minute. How fast is the water level rising when the water is 2 ft. deep?

29. In Problem 28, at what rate is water being drawn from the trough if the level is falling 0.1 ft. per minute when the water is 3 ft. deep?

30. The x-intercept of the tangent line to the positive branch of the hyperbola $xy = 4$ is increasing 3 units per second. Let the y-intercept be OB. Find the velocity of B at the end of 5 sec., the x-intercept starting at the origin.   *Ans.* $-\frac{16}{75}$ unit per second.

31. A point P moves along the parabola $y^2 = x$ so that its abscissa increases at the constant rate of k units per second. The projection of P on the x-axis is M. At what rate is the area of triangle OMP changing when P is at the point where $x = a$?   *Ans.* $\frac{3}{4}k\sqrt{a}$ units per second.

ADDITIONAL PROBLEMS

1. Rectangles inscribed in the area bounded by the parabola $y^2 = 16\,x$ and its latus rectum and such that one side always lies along the latus rectum serve as the bases of rectangular parallelepipeds whose altitudes are always the same as the side parallel to the x-axis. Find the volume of the largest parallelepiped. *Ans.* $\frac{4096}{125}\sqrt{5} = 73.27$.

2. An ellipse symmetrical with respect to the coördinate axes passes through the fixed point (h, k). Find the equation of the ellipse if its area is a minimum. *Ans.* $k^2x^2 + h^2y^2 = 2\,h^2k^2$.

3. The curve $x^3 - 3\,xy + y^3 = 0$ has a loop in the first quadrant symmetric with respect to the line $y = x$. Isosceles triangles having a common vertex at the origin and bases along the line $x + y = a$ are inscribed in this loop. Find the value of a if the area of the triangle is a maximum.
Ans. $\frac{1}{2}(1 + \sqrt{13}) = 2.303$.

4. At a point P in the first quadrant on the curve $y = 7 - x^2$ a tangent is drawn, meeting the coördinate axes at A and B. Find the position of P which makes AB a minimum. *Ans.* Ordinate $= \frac{42}{8}$.

5. The cost of erecting an office building is $50,000 for the first story, $52,500 for the second, $55,000 for the third, and so on. Other expenses (lot, plans, basement, etc.) are $350,000. The net annual income is $5000 for each story. How many stories will give the greatest rate of interest on the investment? *Ans.* 17.

6. For a certain article the increase in the number of pounds consumed is proportional to the decrease in the tax on each pound. If the consumption is m lb. when there is no tax and n lb. when the tax is t dollars per pound, find the tax which should be imposed on each pound to bring in the most revenue.

7. A triangle ABC is formed by a chord BC of the parabola $y = kx^2$ and the tangents AB and AC at each extremity of the chord. If BC remains perpendicular to the axis of the parabola and approaches the vertex at the rate of 2 units per second, find the rate at which the area of the triangle is changing when the chord BC is 4 units above the vertex.

8. A vertical cylindrical tank of radius 10 in. has a hole of radius 1 in. in its base. The velocity with which the water contained runs out of the tank is given by the formula $v^2 = 2\,gh$, where h is the depth of the water and g is the acceleration of gravity. How rapidly is the velocity changing?
Ans. Decreasing $\frac{1}{100}\,g$ ft. per second per second.

9. A light is 20 ft. from a wall and 10 ft. above the center of a path which is perpendicular to the wall. A man 6 ft. tall is walking on the path toward the wall at the rate of 2 ft. per second. When he is 4 ft. from the wall, how fast is the shadow of his head moving up the wall?
Ans. $\frac{5}{8}$ ft. per second.

CHAPTER VI

SUCCESSIVE DIFFERENTIATION AND APPLICATIONS

53. Definition of successive derivatives. We have seen that the derivative of a function of x is in general also a function of x. This new function may also be differentiable, in which case the derivative of the *first derivative* is called the *second derivative* of the original function. Similarly, the derivative of the second derivative is called the *third derivative*; and so on to the nth *derivative*. Thus, if

$$y = 3x^4,$$

$$\frac{dy}{dx} = 12x^3,$$

$$\frac{d}{dx}\left(\frac{dy}{dx}\right) = 36x^2,$$

$$\frac{d}{dx}\left[\frac{d}{dx}\left(\frac{dy}{dx}\right)\right] = 72x. \text{ Etc.}$$

Notation. The symbols for the successive derivatives are usually abbreviated as follows.

$$\frac{d}{dx}\left(\frac{dy}{dx}\right) = \frac{d^2y}{dx^2}, \quad \frac{d}{dx}\left(\frac{d^2y}{dx^2}\right) = \frac{d^3y}{dx^3}, \quad \text{etc.}$$

If $y = f(x)$, the successive derivatives are also denoted by

$$\frac{dy}{dx} = y' = f'(x); \quad \frac{d^2y}{dx^2} = y'' = f''(x);$$

$$\frac{d^3y}{dx^3} = y''' = f'''(x); \quad \cdots; \quad \frac{d^ny}{dx^n} = y^{(n)} = f^{(n)}(x).$$

In the example given above, the notation $y = 3x^4$, $y' = 12x^3$, $y'' = 36x^2$, $y''' = 72x$, $y^{\text{iv}} = 72$ is most convenient.

54. Successive differentiation of implicit functions. To illustrate the process we shall find $\dfrac{d^2y}{dx^2}$ from the equation of the hyperbola

(1) $$b^2x^2 - a^2y^2 = a^2b^2.$$

Differentiating with respect to x (Art. 41),

$$2b^2x - 2a^2y \frac{dy}{dx} = 0,$$

or

(2) $$\frac{dy}{dx} = \frac{b^2x}{a^2y}.$$

Differentiating again, remembering that y is a function of x,

$$\frac{d^2y}{dx^2} = \frac{a^2yb^2 - b^2xa^2 \frac{dy}{dx}}{a^4y^2}.$$

Substituting for $\frac{dy}{dx}$ its value from (2),

$$\frac{d^2y}{dx^2} = \frac{a^2b^2y - a^2b^2x\left(\frac{b^2x}{a^2y}\right)}{a^4y^2} = -\frac{b^2(b^2x^2 - a^2y^2)}{a^4y^3}.$$

But, from the given equation, $b^2x^2 - a^2y^2 = a^2b^2$.

$$\therefore \frac{d^2y}{dx^2} = -\frac{b^4}{a^2y^3}.$$

PROBLEMS

Prove each of the following differentiations.

1. $y = 3x^4 - 2x^3 + 6x.$ $\quad \frac{d^2y}{dx^2} = 36x^2 - 12x.$

2. $s = \sqrt{a + bt}.$ $\quad \frac{d^3s}{dt^3} = \frac{3b^3}{8(a+bt)^{\frac{5}{2}}}.$

3. $y = \frac{a + bx}{a - bx}.$ $\quad \frac{d^2y}{dx^2} = \frac{4ab^2}{(a-bx)^3}.$

4. $u = \sqrt{a^2 + v^2}.$ $\quad \frac{d^2u}{dv^2} = \frac{a^2}{(a^2+v^2)^{\frac{3}{2}}}.$

5. $y = \frac{x^2}{a + x}.$ $\quad \frac{d^2y}{dx^2} = \frac{2a^2}{(a+x)^3}.$

6. $s = \frac{t}{\sqrt{2t+1}}.$ $\quad \frac{d^2s}{dt^2} = \frac{-(t+2)}{(2t+1)^{\frac{5}{2}}}.$

7. $f(x) = \frac{x^3 - 2x^2}{1 - x}.$ $\quad f^{\text{iv}}(x) = \frac{-\lfloor 4}{(1-x)^5}.$

8. $y = \frac{2}{x + 1}.$ $\quad \frac{d^n y}{dx^n} = \frac{2(-1)^n \lfloor n}{(x+1)^{n+1}}.$

SUCCESSIVE DIFFERENTIATION AND APPLICATIONS

9. $x^2 + y^2 = r^2$. $\quad\dfrac{d^2y}{dx^2} = -\dfrac{r^2}{y^3}.$

10. $y^2 = 4\,ax$. $\quad\dfrac{d^2y}{dx^2} = -\dfrac{4\,a^2}{y^3}.$

11. $b^2x^2 + a^2y^2 = a^2b^2$. $\quad\dfrac{d^2y}{dx^2} = -\dfrac{b^4}{a^2y^3};\ \dfrac{d^3y}{dx^3} = -\dfrac{3\,b^6x}{a^4y^5}.$

12. $ax^2 + 2\,hxy + by^2 = 1$. $\quad\dfrac{d^2y}{dx^2} = \dfrac{h^2 - ab}{(hx + by)^3}.$

13. $x^3 + y^3 = 1$. $\quad\dfrac{d^2y}{dx^2} = -\dfrac{2\,x}{y^5}.$

14. $x^4 + 2\,x^2y^2 = a^4$. $\quad\dfrac{d^2y}{dx^2} = \dfrac{2\,y^4 - x^2y^2 - x^4}{x^2y^3}.$

In Problems 15–25 find the values of y' and y'' for the given values of the variables.

15. $y = \sqrt{ax} + \dfrac{a^2}{\sqrt{ax}};\ x = a.$ $\qquad$ Ans. $y' = 0,\ y'' = \dfrac{1}{2\,a}.$

16. $y = \sqrt{25 - 3\,x};\ x = 3.$ $\qquad\qquad y' = -\tfrac{3}{8},\ y'' = -\tfrac{9}{256}.$

17. $y = x\sqrt{x^2 + 9};\ x = 4.$ $\qquad\qquad y' = \tfrac{41}{5},\ y'' = \tfrac{236}{125}.$

18. $x^2 - 4\,y^2 = 9;\ x = 5,\ y = 2.$ $\qquad y' = \tfrac{5}{8},\ y'' = -\tfrac{9}{128}.$

19. $x^2 + 4\,xy + y^2 + 3 = 0;\ x = 2,\ y = -1.$ $\qquad y' = 0,\ y'' = -\tfrac{1}{3}.$

20. $y = (3 - x^2)^4;\ x = 1.$

21. $y = \sqrt{1 + 2\,x};\ x = 4.$

22. $y = \sqrt[3]{x^2 + 4};\ x = 2.$ $\qquad$ 24. $y^2 + 2\,xy = 16;\ x = 3,\ y = 2.$

23. $y = x\sqrt{3\,x - 2};\ x = 2.$ $\qquad$ 25. $x^3 - xy^2 + y^3 = 8;\ x = 2,\ y = 2.$

In each of the following problems find $\dfrac{d^2y}{dx^2}.$

26. $y = x^3 - \dfrac{3}{x}.$ $\qquad\qquad$ 29. $y = x\sqrt{a^2 - x^2}.$

27. $y = \dfrac{x^2}{x^2 + a^2}.$ $\qquad\qquad$ 30. $y^2 - 4\,xy = 16.$

$\qquad\qquad\qquad\qquad\qquad\qquad$ 31. $x^3 - 3\,axy + y^3 = b^3.$

28. $y = \sqrt[3]{2 - 3\,x}.$

55. Direction of bending of a curve. When the point $P(x, y)$ traces a curve, the slope of the tangent line at P varies. When the tangent line is below the curve (Fig. *a*), the arc is concave upward; when above the curve (Fig. *b*), the arc is concave downward. In Fig. *a* the slope increases when

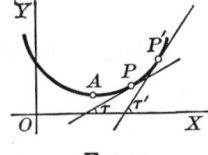

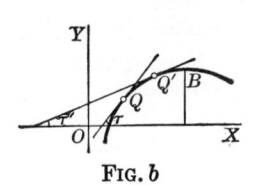

P describes the arc AP'. Hence $f'(x)$ is an increasing function of x. On the other hand, in Fig. b, when P describes the arc QB, the slope decreases, and $f'(x)$ is a decreasing function. In the first case, therefore, $f''(x)$ is positive; in the second case, negative (Art. 45). Hence we have the following criterion for determining the direction of bending at a point.

The graph of $y = f(x)$ is concave upward if the second derivative of y with respect to x is positive, and concave downward if this derivative is negative.

56. Second method for testing for maximum and minimum values. At A in Fig. a of the preceding section, the arc is concave upward, and the ordinate has a minimum value. That is, $f'(x) = 0$ and $f''(x)$ is positive. At B in Fig. b, $f'(x) = 0$ and $f''(x)$ is negative.

We may then state the sufficient conditions for maximum and minimum values of $f(x)$ for critical values of the variable as follows.

$f(x)$ is a maximum if $f'(x) = 0$ and $f''(x) =$ a negative number.

$f(x)$ is a minimum if $f'(x) = 0$ and $f''(x) =$ a positive number.

Following is the corresponding **working rule** for applying this test for maximum and minimum values.

FIRST STEP. *Find the first derivative of the function.*

SECOND STEP. *Set the first derivative equal to zero and solve the resulting equation for real roots in order to find the critical values of the variable.*

THIRD STEP. *Find the second derivative.*

FOURTH STEP. *Substitute each critical value for the variable in the second derivative. If the result is negative, then the function is a maximum for that critical value; if the result is positive, the function is a minimum.*

When $f''(x) = 0$ or does not exist, the above process fails, although there may even then be a maximum or a minimum; in that case the first method given in Art. 47 holds, being fundamental. Usually the second method does apply, and when the process of finding the second derivative is not too long or tedious, it is generally the shortest method.

ILLUSTRATIVE EXAMPLE 1. Let us now apply the above rule to test analytically the function
$$M = x^2 + \frac{432}{x}$$
found in the example worked out on page 49.

Solution. $$f(x) = x^2 + \frac{432}{x}.$$

First Step. $$f'(x) = 2x - \frac{432}{x^2}.$$

SUCCESSIVE DIFFERENTIATION AND APPLICATIONS

Second Step. $\quad 2x - \dfrac{432}{x^2} = 0,$

$\quad\quad x = 6,$ critical value.

Third Step. $\quad f''(x) = 2 + \dfrac{864}{x^3}.$

Fourth Step. $\quad f''(6) = +.$
Hence $\quad f(6) = 108,$ minimum value.

Illustrative Example 2. Examine $x^3 - 3x^2 - 9x + 5$ for maxima and minima. Use the second method.

Solution. $\quad f(x) = x^3 - 3x^2 - 9x + 5.$
First Step. $\quad f'(x) = 3x^2 - 6x - 9.$
Second Step. $\quad 3x^2 - 6x - 9 = 0;$
hence the critical values are $\quad x = -1$ and $3.$
Third Step. $\quad f''(x) = 6x - 6.$
Fourth Step. $f''(-1) = -12.$
$\quad\quad \therefore f(-1) = 10 =$ (ordinate of A) $=$ maximum value.
$f''(3) = +12. \ \therefore f(3) = -22$ (ordinate of B) $=$ minimum value.

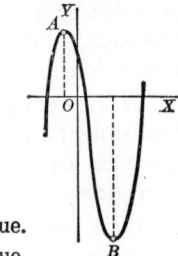

PROBLEMS

Examine each of the following functions for maxima and minima.

1. $x^3 + 3x^2 - 2.$
   Ans. $x = -2,$ gives max. $= 2.$
   $x = 0,$ gives min. $= -2.$

2. $x^3 - 3x + 4.$
   $x = -1,$ gives max. $= 6.$
   $x = 1,$ gives min. $= 2.$

3. $2x^3 - 3ax^2 + a^3. \quad (a > 0)$
   $x = 0,$ gives max. $= a^3.$
   $x = a,$ gives min. $= 0.$

4. $2 + 12x + 3x^2 - 2x^3.$
   $x = 2,$ gives max. $= 22.$
   $x = -1,$ gives min. $= -5.$

5. $3x - 2x^2 - \dfrac{4x^3}{3}.$
   $x = \tfrac{1}{2},$ gives max. $= \tfrac{5}{6}.$
   $x = -\tfrac{3}{2},$ gives min. $= -\tfrac{9}{2}.$

6. $3x^4 - 4x^3 - 12x^2 + 2.$
   $x = 0,$ gives max. $= 2.$
   $x = -1,$ gives min. $= -3.$
   $x = 2,$ gives min. $= -30.$

7. $x^4 - 4x^2 + 4.$
   $x = 0,$ gives max. $= 4.$
   $x = \pm\sqrt{2},$ gives min. $= 0.$

8. $\dfrac{ax}{x^2 + a^2}.$
   $x = a,$ gives max. $= \tfrac{1}{2}.$
   $x = -a,$ gives min. $= -\tfrac{1}{2}.$

9. $x^3 + 9x^2 + 27x + 9.$

10. $12x + 9x^2 - 4x^3.$

11. $x^2(x-4)^2.$

12. $x^2 + \dfrac{x^3}{3} - \dfrac{x^4}{4}.$

13. $x^2 - \dfrac{a^4}{x^2}.$

14. A rectangular box with a square base and an open top is to be made. Find the volume of the largest box that can be made from 1200 sq. ft. of material. *Ans.* 4000 cu. ft.

15. A water tank is to be constructed with a square base and an open top, and is to hold 125 cu. yd. If the cost of the sides is $2 a square yard, and of the bottom $4 a square yard, what are the dimensions when the cost is a minimum? *Ans.* A cube of edge 5 yd.

16. A rectangular flower bed is to contain 800 sq. ft. It is to be surrounded by a walk which is 3 ft. wide along the sides and 6 ft. wide across the ends. If the total area of the bed and walk is a minimum, what are the dimensions of the flower bed? *Ans.* 20 ft. × 40 ft.

17. A rectangular field to contain a given area is to be fenced off along the bank of a straight river. If no fence is needed along the river, show that the least amount of fencing will be required if the field is twice as long as it is wide.

18. A trough is to be made of a long rectangular piece of tin by bending up two edges so as to give a rectangular cross section. If the width of the piece is 14 in., how deep should the trough be made in order that its carrying capacity may be a maximum? *Ans.* 3.5 in.

19. A window composed of a rectangle surmounted by an equilateral triangle is 15 ft. in perimeter. Find its dimensions if it admits the maximum amount of light. *Ans.* Rectangle is 3.51 ft. wide and 2.23 ft. high.

20. A solid wooden sphere weighs w lb. What is the weight of the heaviest right circular cylinder which can be cut from the sphere?

Ans. $\dfrac{w}{\sqrt{3}}$ lb.

21. The slant height of a right circular cone is a given constant a. Find the altitude if the volume is a maximum. *Ans.* $\dfrac{a}{\sqrt{3}}$

22. An oil can consists of a cylinder surmounted by a cone with altitude $= \tfrac{2}{3}$ diameter. Show that for a given capacity the least material is required if the altitude of the cylinder is equal to the altitude of the cone.

23. Given the parabola $y^2 = 8x$ and the point $P(6, 0)$ on the axis, find the coördinates of the points on the parabola nearest to P. *Ans.* $(2, \pm 4)$.

24. A given isosceles triangle has a base of 20 ft. and an altitude of 8 ft. What are the dimensions of the maximum inscribed parallelogram, one side coinciding with the base of the triangle, if the acute angle of the parallelogram is arc tan $\tfrac{4}{3}$? *Ans.* 5 ft. × 10 ft.

25. A miner wishes to dig a tunnel from a point A to a point B 200 ft. below and 600 ft. to the east of A. Below the level of A it is bed rock and above A is soft earth. If the cost of tunneling through earth is $5 and through rock is $13 per linear foot, find the minimum cost of a tunnel.

Ans. $5400.

26. An oil can consists of a cylinder surmounted by a cone with $h = cr$, where h is the altitude of the cone and r is the radius of its base. If c is a constant, show that the least material (M) required for given capacity (V) is given by

$$M^3 = 9\pi V^2(3 - 2c + 3\sqrt{1 + c^2}).$$

If c is a variable parameter, show that the minimum value of M is given by

$$M^3 = 9\pi V^2(3 + \sqrt{5}).$$

57. Points of inflection. A *point of inflection* (or *inflectional point*) on a curve separates arcs having opposite directions of bending (see Art. 55).

In the figure, B is a point of inflection. When the tracing point on a curve passes through such a point, the second derivative will change sign, and if continuous must vanish at the point. Hence we must have

(1)   At points of inflection, $f''(x) = 0$.

Solving the equation resulting from (1) gives the abscissas of the points of inflection. To determine the direction of bending in the vicinity of a point of inflection, test $f''(x)$ for values of x first a trifle less and then a trifle greater than the abscissa at that point.

If $f''(x)$ changes sign, we have a point of inflection, and the signs obtained determine if the curve is concave upward or concave downward in the neighborhood.

The student should observe that near a point where the curve is concave upward (as at A) the curve lies above the tangent, and at a point where the curve is concave downward (as at C) the curve lies below the tangent. At a point of inflection (as at B) the tangent evidently crosses the curve.

Following is a rule for finding points of inflection of the curve whose equation is $y = f(x)$. This rule includes also directions for examining the direction of bending.

FIRST STEP. *Find $f''(x)$.*

SECOND STEP. *Set $f''(x) = 0$, and solve the resulting equation for real roots.*

THIRD STEP. *Test $f''(x)$ for values of x first a trifle less and then a trifle greater than each root found in the second step. If $f''(x)$ changes sign, we have a point of inflection.*

When $f''(x) = +$, the curve is concave upward ⌣.*
When $f''(x) = -$, the curve is concave downward ⌢.

Before the Third Step it is sometimes convenient to factor $f''(x)$.
It is assumed that $f'(x)$ and $f''(x)$ are continuous. The solution of Problem 2, below, shows how to discuss a case where $f'(x)$ and $f''(x)$ are both infinite.

PROBLEMS

Examine the following curves for points of inflection and direction of bending.

1. $y = 3x^4 - 4x^3 + 1$.

Solution. $\quad\quad f(x) = 3x^4 - 4x^3 + 1$.
First Step. $\quad\quad f''(x) = 36x^2 - 24x$.
Second Step. $\quad 36x^2 - 24x = 0$.
$\quad\quad\quad\therefore x = \frac{2}{3}$ and $x = 0$ are the roots.
Third Step. $\quad\quad f''(x) = 36x(x - \frac{2}{3})$.

When $x < 0$, $f''(x) = +$.
When $\frac{2}{3} > x > 0$, $f''(x) = -$.

Therefore the curve is concave upward to the left and concave downward to the right of $x = 0$ (A in figure).

When $0 < x < \frac{2}{3}$, $f''(x) = -$.
When $x > \frac{2}{3}$, $f''(x) = +$.

Therefore the curve is concave downward to the left and concave upward to the right of $x = \frac{2}{3}$ (B in figure).

Hence the points $A(0, 1)$ and $B(\frac{2}{3}, \frac{11}{27})$ are points of inflection.

The curve is evidently concave upward everywhere to the left of A, concave downward between $A(0, 1)$ and $B(\frac{2}{3}, \frac{11}{27})$, and concave upward everywhere to the right of B.

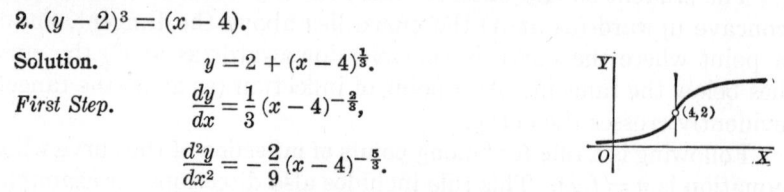

2. $(y - 2)^3 = (x - 4)$.

Solution. $\quad\quad y = 2 + (x - 4)^{\frac{1}{3}}$.

First Step. $\quad\quad \dfrac{dy}{dx} = \dfrac{1}{3}(x - 4)^{-\frac{2}{3}}$,

$\quad\quad\quad\quad \dfrac{d^2y}{dx^2} = -\dfrac{2}{9}(x - 4)^{-\frac{5}{3}}$.

Second Step. When $x = 4$, both first and second derivatives become infinite.

Third Step. $\quad\quad$ When $x < 4$, $\dfrac{d^2y}{dx^2} = +$.

$\quad\quad\quad\quad$ When $x > 4$, $\dfrac{d^2y}{dx^2} = -$.

---
* This may easily be remembered if we say that a vessel shaped like the curve where it is concave upward will hold (+) water, and where it is concave downward will spill (−) water.

We may therefore conclude that the tangent at (4, 2) is perpendicular to the x-axis, that to the left of (4, 2) the curve is concave upward, and to the right of (4, 2) it is concave downward. Therefore (4, 2) is a point of inflection.

3. $y = x^2$.   *Ans.* Concave upward everywhere.

4. $y = 5 - 2x - x^2$.   Concave downward everywhere.

5. $y = x^3$.   Concave downward to the left and concave upward to the right of (0, 0).

6. $y = x^4$.   Concave upward everywhere.

7. $y = 2x^3 - 3x^2 - 36x + 25$.   Concave downward to the left and concave upward to the right of $x = \frac{1}{2}$.

8. $y = 24x^2 - x^4$.   9. $y = x + \frac{1}{x}$.   10. $y = x^2 + \frac{1}{x}$.

58. Curve-tracing. The elementary method of tracing (or plotting) a curve whose equation is given in rectangular coördinates, and one with which the student is already familiar, is to solve its equation for y (or x), assume arbitrary values of x (or y), calculate the corresponding values of y (or x), plot the respective points, and draw a smooth curve through them, the result being an approximation to the required curve. This process is laborious at best, and in case the equation of the curve is of a degree higher than the second, it may not be possible to solve the equation for y or x. The general form of a curve is usually all that is desired, and the calculus furnishes us with powerful methods for determining the shape of a curve with very little computation.

The first derivative gives us the slope of the curve at any point; the second derivative determines the intervals within which the curve is concave upward or concave downward, and the points of inflection separate these intervals; the maximum points are the high points, and the minimum points are the low points on the curve. As a guide in his work the student may follow the

Rule for tracing curves, using rectangular coördinates

FIRST STEP. *Find the first derivative; place it equal to zero; solve to find the abscissas of the maximum and minimum points. Test these values.*

SECOND STEP. *Find the second derivative; place it equal to zero; solve to find the abscissas of the points of inflection. Test these values.*

THIRD STEP. *Calculate the corresponding ordinates of the points whose abscissas were found in the first two steps. Calculate as many* more points as may be necessary to give a good idea of the shape of the curve. Make a table such as is shown in the problem worked out below.

FOURTH STEP. Plot the points determined and sketch in the curve to correspond with the results shown in the table.

If the calculated values of the ordinates are large, it is best to reduce the scale on the y-axis so that the general shape of the curve will be shown within the limits of the paper used. Coördinate plotting paper should be employed. Results should be tabulated as in the problems solved. In this table the values of x should follow one another, increasing *algebraically*.

PROBLEMS

Trace the following curves, making use of the above rule. Also find the equations of the tangent and normal at each point of inflection.

1. $y = x^3 - 9x^2 + 24x - 7$.

Solution. Use the above rule.

First Step. $\qquad y' = 3x^2 - 18x + 24,$
$$3x^2 - 18x + 24 = 0,$$
$$x = 2, 4.$$

Second Step. $\qquad y'' = 6x - 18,$
$$6x - 18 = 0,$$

Third Step. $\qquad x = 3.$

x	y	y'	y''	Remarks	Direction of Curve
0	−7	+	−		
2	13	0	−	max.	} concave down
3	11	−	0	pt. of infl.	
4	9	0	+	min.	} concave up
6	29	+	+		

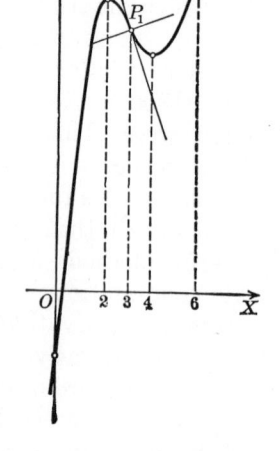

Fourth Step. Plotting the points and sketching in the curve, we get the figure shown.

To find the equations of the tangent and normal to the curve at the point of inflection $P_1(3, 11)$, use formulas (1), (2), Art. 43. This gives $3x + y = 20$ for the tangent and $3y - x = 30$ for the normal.

2. $3y = x^3 - 3x^2 - 9x + 11$.

Ans. Max. $(-1, \frac{16}{3})$; min. $(3, -\frac{16}{3})$; point of inflection, $(1, 0)$; tangent, $4x + y - 4 = 0$; normal, $x - 4y - 1 = 0$.

3. $6y = 12 - 24x - 15x^2 - 2x^3$.

Ans. Max. $(-1, \frac{23}{6})$; min. $(-4, -\frac{2}{3})$; point of inflection, $(-\frac{5}{2}, \frac{19}{12})$.

SUCCESSIVE DIFFERENTIATION AND APPLICATIONS

4. $y = x^4 - 8x^2$.
   Ans. Max. $(0, 0)$; min. $(\pm 2, -16)$; points of inflection, $(\pm \frac{2}{3}\sqrt{3}, -\frac{80}{9})$.

5. $y = 5x - x^5$.
   Ans. Max. $(1, 4)$; min. $(-1, -4)$; point of inflection, $(0, 0)$.

6. $y = \dfrac{6x}{x^2 + 3}$.
   Ans. Max. $(\sqrt{3}, \sqrt{3})$; min. $(-\sqrt{3}, -\sqrt{3})$; points of inflection, $(-3, -\frac{3}{2})$, $(0, 0)$, $(3, \frac{3}{2})$.

7. $y = x^3 + 6x^2$.
8. $y = 4 + 3x - x^3$.
9. $3y = 4x^3 - 18x^2 + 15x$.
10. $y = (x - a)^3 + b$.
11. $12y = (x-1)^4 - 24(x-1)^2$.
12. $y = x^2(9 - x^2)$.
13. $y = 2x^5 - 5x^2$.
14. $y = 3x^5 - 5x^3$.
15. $y = x^5 - 5x^4$.
16. $y = x(x^2 - 4)^2$.
17. $ay = x^2 + \dfrac{a^4}{x^2}$.
18. $ay = x^2 + \dfrac{2a^3}{x}$.
19. $a^2 y = x^3 + \dfrac{a^4}{x}$.
20. $a^2 y = x^3 + \dfrac{a^5}{x^2}$.
21. $y = \dfrac{8a^3}{x^2 + 4a^2}$.
22. $y = \dfrac{x}{(x + a)^2}$.
23. $x^2 y = (x^2 + 1)^2$.
24. $x^3 y + 16y - x^3 = 0$.

59. Acceleration in rectilinear motion. In Art. 51 velocity in rectilinear motion was defined as the time-rate of change of the distance. We now *define acceleration as the time-rate of change of the velocity.* That is,

(A) $$\text{Acceleration} = a = \frac{dv}{dt}.$$

From (C), Art. 51, we obtain also, since $v = \dfrac{ds}{dt}$,

(B) $$a = \frac{d^2 s}{dt^2}.$$

Referring to Arts. 45, 47, and 56, we have the following criteria which apply to a definite instant $t = t_0$:

If $a > 0$, v is increasing (algebraically).
If $a < 0$, v is decreasing (algebraically).
If $a > 0$ and $v = 0$, s has a minimum value.
If $a < 0$ and $v = 0$, s has a maximum value.
If $a = 0$ and changes sign from $+$ to $-$ (from $-$ to $+$) when t passes through t_0, then v has a maximum (a minimum) value when $t = t_0$.

DIFFERENTIAL CALCULUS

In *uniformly accelerated* rectilinear motion, a is constant. Thus in the case of a body falling freely under the action of gravity only, $a = 32.2$ ft. per second per second. Namely, from (2), Art. 51,

$$s = 16.1\, t^2, \quad v = \frac{ds}{dt} = 32.2\, t, \quad a = \frac{dv}{dt} = 32.2.$$

PROBLEMS

1. By experiment it has been found that a body falling freely from rest in a vacuum near the earth's surface follows approximately the law $s = 16.1\, t^2$, where $s =$ space (height) in feet, $t =$ time in seconds. Find the velocity and acceleration (a) at any instant; (b) at end of the first second; (c) at end of the fifth second.

Solution. (1) $s = 16.1\, t^2$.

(a) Differentiating, $\quad\dfrac{ds}{dt} = 32.2\, t,$

or, from (C), Art. 51, (2) $v = 32.2\, t$ ft. per second.

Differentiating again, $\quad\dfrac{dv}{dt} = 32.2,$

or, from (A) above, (3) $a = 32.2$ ft. per (sec.)^2,

which tells us that the acceleration of a falling body is constant; in other words, the velocity increases 32.2 ft. per second every second it keeps on falling.

(b) To find v and a at the end of the first second, substitute $t = 1$ in (2) and (3).
Then $\quad v = 32.2$ ft. per second, $\quad a = 32.2$ ft. per (sec.)^2.

(c) To find v and a at the end of the fifth second, substitute $t = 5$ in (2) and (3).
Then $\quad v = 161$ ft. per second, $\quad a = 32.2$ ft. per (sec.)^2.

Given the following equations of rectilinear motion; find the position, velocity, and acceleration at the instant indicated.

2. $s = 4\, t^2 - 6\, t;\ t = 2.\qquad$ Ans. $s = 4,\ v = 10,\ a = 8.$

3. $s = 120\, t - 16\, t^2;\ t = 4.\qquad s = 224,\ v = -8,\ a = -32.$

4. $x = 32\, t - 8\, t^2;\ t = 2.\qquad x = 32,\ v = 0,\ a = -16.$

5. $y = 6\, t^2 - 2\, t^3;\ t = 1.\qquad y = 4,\ v = 6,\ a = 0.$

6. $s = \dfrac{t}{t+1};\ t = 2.\qquad s = \tfrac{2}{3},\ v = \tfrac{1}{9},\ a = -\tfrac{2}{27}.$

7. $x = 16\, t^2 - 20\, t + 4;\ t = 2.$

8. $y = 100 - 4\, t - 8\, t^2;\ t = 3.$

9. $s = \sqrt{5\, t} + \dfrac{10}{\sqrt{5\, t}};\ t = 5.$

10. $s = \sqrt[3]{3\, t + 2};\ t = 2.$

SUCCESSIVE DIFFERENTIATION AND APPLICATIONS

Given the following equations of rectilinear motion; find the position and acceleration when the particle first comes to rest.

11. $s = 16\,t^2 - 64\,t + 64$.   Ans. $s = 0,\ a = 32$.

12. $s = 80\,t - 16\,t^2$.

13. $s = 3\,c^2 t - t^3$.

14. $s = t + \dfrac{32}{(t+1)^2}$.

15. A ball thrown directly upward moves according to the law
$$s = 80\,t - 16\,t^2.$$
Find (a) its position and velocity after 2 sec. and after 3 sec.; (b) how high it will rise; (c) how far it will move in the fourth second.

16. If the equation of a rectilinear motion is $s = \sqrt{t+1}$, show that the acceleration is negative and proportional to the cube of the velocity.

17. The height (s ft.) reached in t sec. by a body projected vertically upward with a velocity of v_1 ft. per second is given by the formula $s = v_1 t - \tfrac{1}{2} g t^2$. Find a formula for the greatest height reached by the body.

18. In the preceding problem suppose $v_1 = 160$, $g = 32$. Find (a) the velocity at the end of 4 sec. and at the end of 6 sec.; (b) the distance moved during the fourth second and during the sixth second.

19. The velocity of a point moving along a straight line is given by $v^2 = a + \dfrac{2\,b}{s}$, where a and b are constants. Show that the acceleration is $-b/s^2$.

20. A car makes a trip in 10 min. and moves according to the law $s = 250\,t^2 - \tfrac{5}{4}t^4$, where t is measured in minutes and s in feet. (a) How far does the car go? (b) What is its maximum speed? (c) How far has the car moved when its maximum speed is reached?

    Ans. (a) 12,500 ft.; (b) 1924 ft. per minute; (c) 6944 ft.

ADDITIONAL PROBLEMS

1. Trace the curve $(4 - 2x + x^2)y = 2x - x^2$, and find the equations of the tangent and normal at each point of inflection.

Ans. Max. $(1, \tfrac{1}{3})$. Point of inflection $(0, 0)$: tangent, $x - 2y = 0$; normal, $2x + y = 0$. Point of inflection $(2, 0)$: tangent, $x + 2y - 2 = 0$; normal, $2x - y - 4 = 0$.

2. A certain curve (the tractrix) is such that the length of every tangent from its point of contact $P(x, y)$ to its intersection A with the x-axis is the constant c $(AP = c)$. Show that

(a) $\dfrac{dy}{dx} = \dfrac{\pm y}{\sqrt{c^2 - y^2}}$;   (b) $\dfrac{d^2 y}{dx^2} = \dfrac{c^2 y}{(c^2 - y^2)^2}$.

3. Determine k so that the normals at the points of inflection of the curve $y = k(x^2 - 3)^2$ will pass through the origin.   Ans. $k = \dfrac{1}{4\sqrt{2}}$.

CHAPTER VII

DIFFERENTIATION OF TRANSCENDENTAL FUNCTIONS. APPLICATIONS

We consider now functions such as

$$\sin 2x, \quad 3^x, \quad \log(1+x^2),$$

called *transcendental functions*, as distinguished from the algebraic functions hitherto discussed.

60. Formulas for derivatives; second list. The following formulas, grouped here for convenience of reference, will be derived in this chapter, and, with the formulas of Art. 29, comprise all formulas for derivatives used in this book.

X $\qquad \dfrac{d}{dx}(\ln v) = \dfrac{\frac{dv}{dx}}{v} = \dfrac{1}{v}\dfrac{dv}{dx}.$ $\qquad$ ($\ln v = \log_e v$)

X a $\qquad \dfrac{d}{dx}(\log v) = \dfrac{\log e}{v}\dfrac{dv}{dx}.$

XI $\qquad \dfrac{d}{dx}(a^v) = a^v \ln a \dfrac{dv}{dx}.$

XI a $\qquad \dfrac{d}{dx}(e^v) = e^v \dfrac{dv}{dx}.$

XII $\qquad \dfrac{d}{dx}(u^v) = vu^{v-1}\dfrac{du}{dx} + \ln u \cdot u^v \dfrac{dv}{dx}.$

XIII $\qquad \dfrac{d}{dx}(\sin v) = \cos v \dfrac{dv}{dx}.$

XIV $\qquad \dfrac{d}{dx}(\cos v) = -\sin v \dfrac{dv}{dx}.$

XV $\qquad \dfrac{d}{dx}(\tan v) = \sec^2 v \dfrac{dv}{dx}.$

XVI $\qquad \dfrac{d}{dx}(\operatorname{ctn} v) = -\csc^2 v \dfrac{dv}{dx}.$

XVII $\qquad \dfrac{d}{dx}(\sec v) = \sec v \tan v \dfrac{dv}{dx}.$

XVIII $\quad \dfrac{d}{dx}(\csc v) = -\csc v \ctn v \dfrac{dv}{dx}.$

XIX $\quad \dfrac{d}{dx}(\text{vers } v) = \sin v \dfrac{dv}{dx}.$

XX $\quad \dfrac{d}{dx}(\text{arc sin } v) = \dfrac{\dfrac{dv}{dx}}{\sqrt{1-v^2}}.$

XXI $\quad \dfrac{d}{dx}(\text{arc cos } v) = -\dfrac{\dfrac{dv}{dx}}{\sqrt{1-v^2}}.$

XXII $\quad \dfrac{d}{dx}(\text{arc tan } v) = \dfrac{\dfrac{dv}{dx}}{1+v^2}.$

XXIII $\quad \dfrac{d}{dx}(\text{arc ctn } v) = -\dfrac{\dfrac{dv}{dx}}{1+v^2}.$

XXIV $\quad \dfrac{d}{dx}(\text{arc sec } v) = \dfrac{\dfrac{dv}{dx}}{v\sqrt{v^2-1}}.$

XXV $\quad \dfrac{d}{dx}(\text{arc csc } v) = -\dfrac{\dfrac{dv}{dx}}{v\sqrt{v^2-1}}.$

XXVI $\quad \dfrac{d}{dx}(\text{arc vers } v) = \dfrac{\dfrac{dv}{dx}}{\sqrt{2v-v^2}}.$

61. The number *e*. Natural logarithms. One of the most important limits in the calculus is

(1) $$\lim_{x \to 0} (1+x)^{\frac{1}{x}} = 2.71828\cdots.$$

This limit is denoted by *e*. To prove rigorously that such a limit *e* exists is beyond the scope of this book. For the present we shall content ourselves with plotting the locus of the equation

(2) $$y = (1+x)^{\frac{1}{x}}$$

and showing graphically that as $x \to 0$ the function $(1+x)^{\frac{1}{x}}\ (=y)$ takes on values in the near neighborhood of $2.718\cdots$ and therefore $e = 2.718\cdots$ approximately.

As $x \to 0$ from the left, y decreases and approaches e as a limit. As $x \to 0$ from the right, y increases and also approaches e as a limit.

x	y	x	y
10	1.2710		
5	1.4310		
2	1.7320		
1	2.0000		
.5	2.2500	$-.5$	4.0000
.1	2.5937	$-.1$	2.8680
.01	2.7048	$-.01$	2.7320
.001	2.7169	$-.001$	2.7195

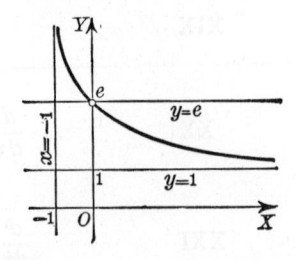

The fact expressed in (1) is used in Art. 63.

As $x \to +\infty$, y approaches 1 as a limit; and as $x \to -1$ from the right, y increases without limit. The lines $y = 1$ and $x = -1$ are asymptotes.

In Chapter XX we shall show how to calculate the value of e to any number of decimal places.

Natural, or *Napierian*, *logarithms* are those which have the number e for base. These logarithms play a very important rôle in mathematics. To distinguish between natural logarithms and common logarithms when the base is not explicitly stated, the following notation will be used.

Natural logarithm of v (base e) = $\ln v$.

Common logarithm of v (base 10) = $\log v$.

By definition, the natural logarithm of a number N is the exponent x in the equation

(3) $\qquad e^x = N$; that is, $x = \ln N$.

If $x = 0$, $N = 1$, and $\ln 1 = 0$. If $x = 1$, $N = e$, and $\ln e = 1$. If $x \to -\infty$, then $N \to 0$, and we write $\ln 0 = -\infty$.

The student is familiar with the use of tables of common logarithms, where the base is 10. The **common logarithm** of a number N is the exponent y in the equation

(4) $\qquad 10^y = N$, or $y = \log N$.

Let us find the relation between $\ln N$ and $\log N$.

In (3), take logarithms of both members to the base 10. Then we have, from (2), p. 1,

(5) $\qquad x \log e = \log N$.

Solving for x, which equals $\ln N$, by (3), we get the desired relation

(A) $\qquad \ln N = \dfrac{\log N}{\log e}.$

That is, *we obtain the natural logarithm of any number by dividing its common logarithm by log e.*

Equation (*A*) may be written

(6) $$\log N = \log e \cdot \ln N.$$

Hence the *common logarithm of a number is obtained by multiplying its natural logarithm by log e.* This multiplier is called the **modulus** (= *M*) of common logarithms.

By tables, $\log e = 0.4343$, and $\dfrac{1}{\log e} = 2.303$.

Equation (*A*) may now be written

(7) $$\ln N = 2.303 \log N.$$

Tables of natural logarithms should be at hand.

62. Exponential and logarithmic functions. The function of x defined by

(1) $$y = e^x \qquad (e = 2.718 \cdots)$$

is called an *exponential function*. Its graph is shown in the figure. The function is an increasing function for all values of x, as we shall see later, and it is everywhere continuous.

From (1), we have, by definition,

(2) $$x = \ln y.$$

The functions e^x and $\ln y$ are inverse functions (Art. 39). Interchanging x and y in (2), we have

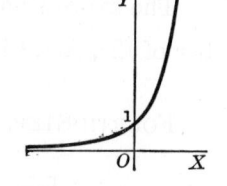

(3) $$y = \ln x,$$

in which y is now a *logarithmic function* of x. The graph is shown in the figure. The function is not defined for negative values of x, nor for $x = 0$. It is an increasing function for all values of $x > 0$, and is everywhere continuous. That is (Art. 17), for any value a of x greater than zero

(4) $$\lim_{x \to a} \ln x = \ln a.$$

When $x \to 0$, then $y \to -\infty$, as remarked above. The y-axis is an asymptote in the graph.

The functions a^x and $\log_a x$ ($a > 0$) have the same properties as e^x and $\ln x$ and graphs similar to the above.

63. Differentiation of a logarithm.

Let $$y = \ln v. \qquad (v > 0)$$

Differentiating by the General Rule (Art. 27), considering v as the independent variable, we have

First Step. $\quad y + \Delta y = \ln (v + \Delta v).$

Second Step. $\Delta y = \ln (v + \Delta v) - \ln v$
$$= \ln \left(\frac{v + \Delta v}{v}\right) = \ln \left(1 + \frac{\Delta v}{v}\right). \qquad \text{By (2), p. 1}$$

Third Step. $\quad \dfrac{\Delta y}{\Delta v} = \dfrac{1}{\Delta v} \ln \left(1 + \dfrac{\Delta v}{v}\right).$

We cannot evaluate the limit of the right-hand member as it stands by Art. 16, since the denominator Δv approaches zero as a limit. But we can rewrite the equation as follows:

$$\frac{\Delta y}{\Delta v} = \frac{1}{v} \cdot \frac{v}{\Delta v} \ln \left(1 + \frac{\Delta v}{v}\right)$$

$$\left[\text{Multiplying by } \frac{v}{v}.\right]$$

$$= \frac{1}{v} \ln \left(1 + \frac{\Delta v}{v}\right)^{\frac{v}{\Delta v}}. \qquad \text{By (2), p. 1}$$

The expression following ln is in the form of the right-hand member of (2), Art. 61, with $x = \dfrac{\Delta v}{v}$.

Fourth Step. $\quad \dfrac{dy}{dv} = \dfrac{1}{v} \ln e = \dfrac{1}{v}.$

$$\left[\begin{array}{l}\text{When } \Delta v \to 0, \dfrac{\Delta v}{v} \to 0. \text{ Hence } \lim_{\Delta v \to 0} \left(1 + \dfrac{\Delta v}{v}\right)^{\frac{v}{\Delta v}} = e, \text{ by} \\ \text{(1), Art. 61. Using (4), Art. 62, we have the result.}\end{array}\right]$$

Since v is a function of x and it is required to differentiate $\ln v$ with respect to x, we must use formula **(A)**, Art. 38, for differentiating a *function of a function*, namely,

$$\frac{dy}{dx} = \frac{dy}{dv} \cdot \frac{dv}{dx}.$$

Substituting the value of $\dfrac{dy}{dv}$ from the result of the Fourth Step, we get

X $$\frac{d}{dx}(\ln v) = \frac{\frac{dv}{dx}}{v} = \frac{1}{v}\frac{dv}{dx}.$$

TRANSCENDENTAL FUNCTIONS

The derivative of the natural logarithm of a function is equal to the derivative of the function divided by the function (or the reciprocal of the function times its derivative).

Since $\log v = \log e \ln v$, we have at once (IV, Art. 29)

X a $$\frac{d}{dx}(\log v) = \frac{\log e}{v}\frac{dv}{dx}.$$

64. Differentiation of the exponential function

Let $$y = a^v. \qquad (a > 0)$$

Taking the logarithm of both sides to the base e, we get

$$\ln y = v \ln a,$$

or $$v = \frac{\ln y}{\ln a} = \frac{1}{\ln a} \cdot \ln y.$$

Differentiating with respect to y by formula X,

$$\frac{dv}{dy} = \frac{1}{\ln a} \cdot \frac{1}{y}.$$

From (C), Art. 39, relating to *inverse functions*, we get

$$\frac{dy}{dv} = \ln a \cdot y,$$

or

(1) $$\frac{dy}{dv} = \ln a \cdot a^v.$$

Since v is a function of x and it is required to differentiate a^v with respect to x, we use formula (A), Art. 38. Thus we get

$$\frac{dy}{dx} = \ln a \cdot a^v \cdot \frac{dv}{dx}.$$

XI $$\therefore \frac{d}{dx}(a^v) = \ln a \cdot a^v \cdot \frac{dv}{dx}.$$

When $a = e$, $\ln a = \ln e = 1$, and XI becomes

XI a $$\frac{d}{dx}(e^v) = e^v \frac{dv}{dx}.$$

The derivative of a constant with a variable exponent is equal to the product of the natural logarithm of the constant, the constant with the variable exponent, and the derivative of the exponent.

65. Differentiation of the general exponential function. Proof of the Power Rule

Let
$$y = u^v. \qquad (u > 0)$$

Taking the logarithm of both sides to the base e,
$$\ln y = v \ln u,$$

or
$$y = e^{v \ln u}. \qquad \text{By (3), Art. 61}$$

Differentiating by formula **XI** a,

$$\frac{dy}{dx} = e^{v \ln u} \frac{d}{dx}(v \ln u)$$

$$= e^{v \ln u}\left(\frac{v}{u}\frac{du}{dx} + \ln u \frac{dv}{dx}\right) \qquad \text{by V and X}$$

$$= u^v\left(\frac{v}{u}\frac{du}{dx} + \ln u \frac{dv}{dx}\right).$$

XII $\qquad \therefore \dfrac{d}{dx}(u^v) = v u^{v-1}\dfrac{du}{dx} + \ln u \cdot u^v \dfrac{dv}{dx}.$

The derivative of a function with a variable exponent is equal to the sum of the two results obtained by first differentiating by **VI**, *regarding the exponent as constant, and again differentiating by* **XI**, *regarding the function as constant.*

Let $v = n$, any constant; then **XII** reduces to
$$\frac{d}{dx}(u^n) = n u^{n-1}\frac{du}{dx}.$$

Thus we have shown that the Power Rule **VI** holds true for any value of the constant n.

ILLUSTRATIVE EXAMPLE 1. Differentiate $y = \ln(x^2 + a)$.

Solution. $\qquad \dfrac{dy}{dx} = \dfrac{\dfrac{d}{dx}(x^2 + a)}{x^2 + a} \qquad$ by **X**

$[v = x^2 + a.]$

$\qquad\qquad\quad = \dfrac{2x}{x^2 + a}.$ Ans.

ILLUSTRATIVE EXAMPLE 2. Differentiate $y = \log \dfrac{2x}{1 + x^2}$.

Solution. By (2), p. 1, we may write this
$$y = \log 2x - \log(1 + x^2).$$

Then $\qquad \dfrac{dy}{dx} = \dfrac{\log e}{2x}\dfrac{d}{dx} 2x - \dfrac{\log e}{1 + x^2}\dfrac{d}{dx}(1 + x^2) \qquad$ by **III** and **X** a

$\qquad = \log e\left(\dfrac{1}{x} - \dfrac{2x}{1 + x^2}\right) = \log e\,\dfrac{1 - x^2}{x(1 + x^2)}.$ Ans.

ILLUSTRATIVE EXAMPLE 3. Differentiate $y = a^{3x^2}$.

Solution. $\dfrac{dy}{dx} = \ln a \cdot a^{3x^2} \dfrac{d}{dx}(3x^2)$ by XI

$= 6x \ln a \cdot a^{3x^2}$. Ans.

ILLUSTRATIVE EXAMPLE 4. Differentiate $y = be^{c^2 + x^2}$.

Solution. $\dfrac{dy}{dx} = b \dfrac{d}{dx}(e^{c^2 + x^2})$ by IV

$= be^{c^2 + x^2} \dfrac{d}{dx}(c^2 + x^2)$ by XI a

$= 2bxe^{c^2 + x^2}$. Ans.

ILLUSTRATIVE EXAMPLE 5. Differentiate $y = x^{e^x}$.

Solution. $\dfrac{dy}{dx} = e^x x^{e^x - 1} \dfrac{d}{dx}(x) + x^{e^x} \ln x \dfrac{d}{dx}(e^x)$ by XII

$= e^x x^{e^x - 1} + x^{e^x} \ln x \cdot e^x$

$= e^x x^{e^x} \left(\dfrac{1}{x} + \ln x \right)$. Ans.

66. Logarithmic differentiation. Instead of applying **X** and **X** a at once in differentiating logarithmic functions, we may sometimes simplify the work by first making use of one of the formulas of (2) on page 1. It is important that these formulas should be used whenever this is possible.

ILLUSTRATIVE EXAMPLE 1. Differentiate $y = \ln \sqrt{1 - x^2}$.

Solution. By using (2), p. 1, we may write this in a form free from radicals, as follows:

$y = \tfrac{1}{2} \ln (1 - x^2)$.

Then $\dfrac{dy}{dx} = \dfrac{1}{2} \dfrac{\dfrac{d}{dx}(1 - x^2)}{1 - x^2}$ by X

$= \dfrac{1}{2} \cdot \dfrac{-2x}{1 - x^2} = \dfrac{x}{x^2 - 1}$. Ans.

ILLUSTRATIVE EXAMPLE 2. Differentiate $y = \ln \sqrt{\dfrac{1 + x^2}{1 - x^2}}$.

Solution. Simplifying by means of (2), p. 1,

$y = \tfrac{1}{2} [\ln (1 + x^2) - \ln (1 - x^2)]$.

Then $\dfrac{dy}{dx} = \dfrac{1}{2} \left[\dfrac{\dfrac{d}{dx}(1 + x^2)}{1 + x^2} - \dfrac{\dfrac{d}{dx}(1 - x^2)}{1 - x^2} \right]$ by III and X

$= \dfrac{x}{1 + x^2} + \dfrac{x}{1 - x^2} = \dfrac{2x}{1 - x^4}$. Ans.

In differentiating an exponential function, especially a variable with a variable exponent, the best plan is first to take the natural logarithm of the function and then to differentiate. Thus Illustrative Example 5, Art. 65, is solved more elegantly as follows:

ILLUSTRATIVE EXAMPLE 3. Differentiate $y = x^{e^x}$.

Solution. Taking the natural logarithm of both sides,
$$\ln y = e^x \ln x. \qquad \text{By (2), p. 1}$$

Now differentiate both sides with respect to x.

$$\frac{\frac{dy}{dx}}{y} = e^x \frac{d}{dx}(\ln x) + \ln x \frac{d}{dx}(e^x) \qquad \text{by X and V}$$

$$= e^x \cdot \frac{1}{x} + \ln x \cdot e^x, \qquad \text{by X and XI } a$$

or
$$\frac{dy}{dx} = e^x \cdot y \left(\frac{1}{x} + \ln x\right)$$

$$= e^x x^{e^x} \left(\frac{1}{x} + \ln x\right). \quad Ans.$$

ILLUSTRATIVE EXAMPLE 4. Differentiate $y = (4x^2 - 7)^{2+\sqrt{x^2-5}}$.

Solution. Taking the natural logarithm of both sides,
$$\ln y = (2 + \sqrt{x^2 - 5}) \ln (4x^2 - 7).$$

Differentiating both sides with respect to x,
$$\frac{1}{y}\frac{dy}{dx} = (2 + \sqrt{x^2 - 5}) \frac{8x}{4x^2 - 7} + \ln (4x^2 - 7) \cdot \frac{x}{\sqrt{x^2 - 5}}.$$

$$\frac{dy}{dx} = x(4x^2 - 7)^{2+\sqrt{x^2-5}} \left[\frac{8(2 + \sqrt{x^2 - 5})}{4x^2 - 7} + \frac{\ln (4x^2 - 7)}{\sqrt{x^2 - 5}}\right]. \quad Ans.$$

In the case of a function consisting of a number of factors it is sometimes convenient to take the natural logarithm and simplify by (2), p. 1, before differentiating. Thus,

ILLUSTRATIVE EXAMPLE 5. Differentiate $y = \sqrt{\frac{(x-1)(x-2)}{(x-3)(x-4)}}$.

Solution. Taking the natural logarithm of both sides,
$$\ln y = \tfrac{1}{2}[\ln (x-1) + \ln (x-2) - \ln (x-3) - \ln (x-4)].$$

Differentiating both sides with respect to x,

$$\frac{1}{y}\frac{dy}{dx} = \frac{1}{2}\left[\frac{1}{x-1} + \frac{1}{x-2} - \frac{1}{x-3} - \frac{1}{x-4}\right]$$

$$= -\frac{2x^2 - 10x + 11}{(x-1)(x-2)(x-3)(x-4)},$$

or
$$\frac{dy}{dx} = -\frac{2x^2 - 10x + 11}{(x-1)^{\frac{1}{2}}(x-2)^{\frac{1}{2}}(x-3)^{\frac{3}{2}}(x-4)^{\frac{3}{2}}}. \quad Ans.$$

TRANSCENDENTAL FUNCTIONS

PROBLEMS

Differentiate each of the following functions.

1. $y = \ln(ax+b)$.   Ans. $\dfrac{dy}{dx} = \dfrac{a}{ax+b}$.

2. $y = \ln(ax^2+b)$. $\dfrac{dy}{dx} = \dfrac{2ax}{ax^2+b}$.

3. $y = \ln(ax+b)^2$. $\dfrac{dy}{dx} = \dfrac{2a}{ax+b}$.

4. $y = \ln ax^n$. $\dfrac{dy}{dx} = \dfrac{n}{x}$.

5. $y = \ln x^3$. $\dfrac{dy}{dx} = \dfrac{3}{x}$.

6. $y = \ln^3 x \ [= (\ln x)^3]$. $\dfrac{dy}{dx} = \dfrac{3 \ln^2 x}{x}$.

7. $y = \ln(2x^3 - 3x^2 + 4)$. $\dfrac{dy}{dx} = \dfrac{6x(x-1)}{2x^3 - 3x^2 + 4}$.

8. $y = \log \dfrac{2}{x}$. $\dfrac{dy}{dx} = -\dfrac{\log e}{x}$.

9. $y = \ln \dfrac{x^2}{1+x^2}$. $\dfrac{dy}{dx} = \dfrac{2}{x(1+x^2)}$.

10. $y = \ln \sqrt{9 - 2x^2}$. $\dfrac{dy}{dx} = \dfrac{-2x}{9 - 2x^2}$.

11. $y = \ln(ax\sqrt{a+x})$. $\dfrac{dy}{dx} = \dfrac{2a+3x}{2x(a+x)}$.

12. $f(x) = x \ln x$. $f'(x) = 1 + \ln x$.

13. $f(x) = \ln(x + \sqrt{1+x^2})$. $f'(x) = \dfrac{1}{\sqrt{1+x^2}}$.

14. $s = \ln \sqrt{\dfrac{a+bt}{a-bt}}$. $\dfrac{ds}{dt} = \dfrac{ab}{a^2 - b^2 t^2}$.

15. $f(x) = x^2 \ln x^2$. $f'(x) = 2x(1 + 2 \ln x)$.

16. $y = e^{nx}$. $\dfrac{dy}{dx} = ne^{nx}$.

17. $y = 10^{nx}$. $\dfrac{dy}{dx} = n \, 10^{nx} \ln 10$.

18. $y = e^{x^2}$. $\dfrac{dy}{dx} = 2xe^{x^2}$.

19. $y = \dfrac{2}{e^x}$. $\dfrac{dy}{dx} = -\dfrac{2}{e^x}$.

20. $s = e^{\sqrt{t}}$. $\dfrac{ds}{dt} = \dfrac{e^{\sqrt{t}}}{2\sqrt{t}}$.

21. $z = b^{2y}$. $\dfrac{dz}{dy} = 2 b^{2y} \ln b$.

96 DIFFERENTIAL CALCULUS

22. $u = se^s$.   Ans. $\dfrac{du}{ds} = e^s(s+1)$.

23. $v = \dfrac{e^u}{u}$. $\dfrac{dv}{du} = \dfrac{e^u(u-1)}{u^2}$.

24. $y = \dfrac{\ln x}{x}$. $\dfrac{dy}{dx} = \dfrac{1-\ln x}{x^2}$.

25. $y = \ln(x^2 e^x)$. $\dfrac{dy}{dx} = \dfrac{2}{x} + 1$.

26. $y = \dfrac{e^x - 1}{e^x + 1}$. $\dfrac{dy}{dx} = \dfrac{2e^x}{(e^x+1)^2}$.

27. $y = x^2 e^{-x}$. $\dfrac{dy}{dx} = e^{-x}(2x - x^2)$.

28. $y = \dfrac{a}{2}\left(e^{\frac{x}{a}} - e^{-\frac{x}{a}}\right)$. $\dfrac{dy}{dx} = \dfrac{1}{2}\left(e^{\frac{x}{a}} + e^{-\frac{x}{a}}\right)$.

29. $y = \dfrac{e^x - e^{-x}}{e^x + e^{-x}}$. $\dfrac{dy}{dx} = \dfrac{4}{(e^x + e^{-x})^2}$.

30. $s = \dfrac{\ln t^2}{t^2}$. $\dfrac{ds}{dt} = \dfrac{2 - 4\ln t}{t^3}$.

31. $f(x) = \ln \dfrac{\sqrt{x^2+1} - x}{\sqrt{x^2+1} + x}$. $f'(x) = \dfrac{-2}{\sqrt{x^2+1}}$.

HINT. First rationalize the denominator.

32. $y = x^x$.   Ans. $y' = x^x(1 + \ln x)$.

33. $y = x^{\sqrt{x}}$. $y' = \dfrac{x^{\sqrt{x}}(2 + \ln x)}{2\sqrt{x}}$.

34. $s = \left(\dfrac{a}{t}\right)^t$. $\dfrac{ds}{dt} = \left(\dfrac{a}{t}\right)^t \left(\ln \dfrac{a}{t} - 1\right)$.

35. $y = \dfrac{x\sqrt[3]{3x+a}}{\sqrt{2x+b}}$. $\dfrac{dy}{dx} = y\left[\dfrac{1}{x} + \dfrac{1}{3x+a} - \dfrac{1}{2x+b}\right]$.

36. $y = \dfrac{\sqrt{4+x^2}}{x\sqrt{4-x^2}}$. $\dfrac{dy}{dx} = y\left[\dfrac{x}{4+x^2} - \dfrac{1}{x} + \dfrac{x}{4-x^2}\right]$.

37. $y = x^n(a+bx)^m$. $\dfrac{dy}{dx} = y\left[\dfrac{n}{x} + \dfrac{mb}{a+bx}\right]$.

In Problems 38–47 find the value of $\dfrac{dy}{dx}$ for the given value of x.

38. $y = \ln(x^2+2)$; $x = 4$.   Ans. $y' = \frac{4}{9}$.

39. $y = \log(4x-3)$; $x = 2$. $y' = 0.3474$.

40. $y = x\ln\sqrt{x+3}$; $x = 6$. $y' = 1.4319$.

41. $y = xe^{-2x}$; $x = \frac{1}{2}$. $y' = 0$.

42. $y = \dfrac{\ln x^2}{x}$; $x = 4$. $y' = -0.0483$.

43. $y = \dfrac{e^{\frac{x}{2}}}{x+1}$; $x = 1$.

44. $y = \log \sqrt{25 - 4x}$; $x = 5$.

45. $y = 10^{\sqrt{x}}$; $x = 4$.

46. $y = \left(\dfrac{3}{x}\right)^x$; $x = 3$.

47. $y = \dfrac{x^3 \sqrt{x^2 + 9}}{\sqrt[3]{20 - 3x}}$; $x = 4$.

Find $\dfrac{d^2y}{dx^2}$ for each of the following functions.

48. $y = \ln cx$.

49. $y = e^{ax}$.

50. $y = x \ln x$.

51. $y = e^{x^2}$.

52. $y = \ln \dfrac{x-a}{x+a}$.

53. $y = \dfrac{e^x}{x^2}$.

Differentiate each of the following functions.

54. $\ln \dfrac{\sqrt{a^2 - x^2}}{x}$.

55. $\dfrac{\ln \sqrt{a^2 - x^2}}{x}$.

56. $\log \sqrt{\dfrac{x^2 + a^2}{x + a}}$.

57. $\ln \dfrac{t}{\sqrt{2t + 3}}$.

58. $e^{\sqrt{x}} \ln \sqrt{x}$.

59. $10^t \log t$.

60. $(ae)^{nx}$.

61. $2^s s^2$.

62. $\left(\dfrac{x}{a}\right)^{\sqrt{x}}$.

67. The function sin x. The graph of

(1) $$y = \sin x$$

is shown in the figure. Any value of x is assumed to be the measure of an angle in radians (Art. 2).

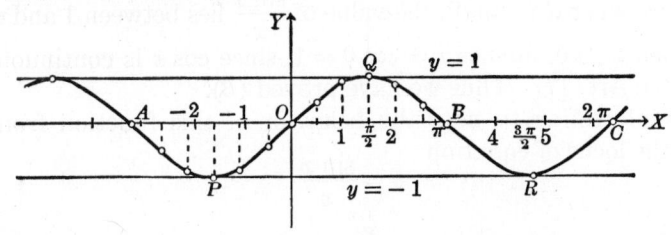

Thus for $x = 1$, $y = \sin (1 \text{ radian}) = \sin 57° 18' = 0.841$. The function $\sin x$ is defined and is continuous for all values of x. It is important to note that $\sin x$ is a *periodic function* with the period 2π. For
$$\sin (x + 2\pi) = \sin x.$$

That is, when the value of x is increased by a period, the value of y is repeated.

The property of periodicity has the following interpretation in the graph on page 97: *The portion of the curve for values of x from 0 to 2 π (arc OQBRC in the figure) may be displaced parallel to OX either to the right or left a distance equal to any multiple of the period 2 π, and it will be part of the locus in its new position.*

68. Theorem. Before differentiating sin x (Art. 69) it is necessary to prove that

$$(B) \qquad \lim_{x \to 0} \frac{\sin x}{x} = 1.$$

This limit cannot be found by Art. 16. We proceed by geometry and trigonometry.

Let O be the center of a circle whose radius is unity. Let x = angle AOM measured in radians. Since the radius is unity, arc $AM = x$, also.

Lay off arc $AM' =$ arc AM and draw MT and $M'T$ tangent to the circle at M and M' respectively. From geometry,

$$MM' < \text{arc } MAM' < MT + M'T.$$

Or, by trigonometry,

$$2 \sin x < 2x < 2 \tan x.$$

Dividing through by $2 \sin x$, we get

$$1 < \frac{x}{\sin x} < \frac{1}{\cos x}.$$

Replacing each term by its reciprocal and reversing the inequality signs, we have

$$1 > \frac{\sin x}{x} > \cos x.$$

Therefore when x is small, the value of $\frac{\sin x}{x}$ lies between 1 and $\cos x$. But when $x \to 0$, lim $\cos x = \cos 0 = 1$, since $\cos x$ is continuous for $x = 0$ (see Art. 17). Thus we have proved **(B)**.

It is interesting to note the behavior of this function from its graph, the locus of equation

$$y = \frac{\sin x}{x}.$$

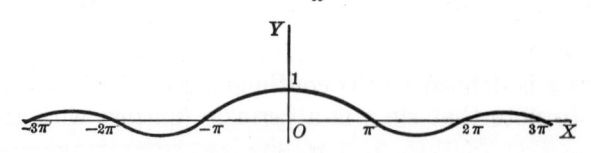

The function is not defined for $x = 0$. Let us, however, assign the value 1 to it for $x = 0$. Then the function is defined and **is** continuous for all values of x (see Art. 17).

69. Differentiation of sin v

Let $$y = \sin v.$$
By the General Rule, Art. 27, considering v as the independent variable, we have

FIRST STEP. $\quad y + \Delta y = \sin(v + \Delta v).$
SECOND STEP. $\quad \Delta y = \sin(v + \Delta v) - \sin v.$

The right-hand member must be transformed in order to evaluate the limit in the Fourth Step. To this end, use the formula from (6), p. 3,
$$\sin A - \sin B = 2\cos\tfrac{1}{2}(A+B)\sin\tfrac{1}{2}(A-B),$$
setting $\quad A = v + \Delta v, \quad B = v.$
Then $\quad \tfrac{1}{2}(A+B) = v + \tfrac{1}{2}\Delta v, \quad \tfrac{1}{2}(A-B) = \tfrac{1}{2}\Delta v.$
Substituting,
$$\sin(v + \Delta v) - \sin v = 2\cos(v + \tfrac{1}{2}\Delta v)\sin\tfrac{1}{2}\Delta v.$$
Hence $\quad \Delta y = 2\cos\left(v + \dfrac{\Delta v}{2}\right)\sin\dfrac{\Delta v}{2}.$

THIRD STEP. $\quad \dfrac{\Delta y}{\Delta v} = \cos\left(v + \dfrac{\Delta v}{2}\right)\dfrac{\sin\dfrac{\Delta v}{2}}{\dfrac{\Delta v}{2}}.$

FOURTH STEP. $\quad \dfrac{dy}{dv} = \cos v.$

$\left[\text{Since }\lim\limits_{\Delta v \to 0}\left(\dfrac{\sin\frac{\Delta v}{2}}{\frac{\Delta v}{2}}\right) = 1, \text{ by Art. 68, and }\lim\limits_{\Delta v \to 0}\cos\left(v + \tfrac{\Delta v}{2}\right) = \cos v.\right]$

Substituting this value of $\dfrac{dy}{dv}$ in (A), Art. 38, we get
$$\frac{dy}{dx} = \cos v\,\frac{dv}{dx}.$$

XIII $\quad \therefore \dfrac{d}{dx}(\sin v) = \cos v\,\dfrac{dv}{dx}.$

The statement of the corresponding rules will now be left to the student.

70. The other trigonometric functions.
The function $\cos x$ is defined and is continuous for any value of x. It is periodic, with the period 2π. The graph of
$$y = \cos x$$
is obtained from the graph of Art. 67 of $\sin x$ by taking the line $x = \tfrac{1}{2}\pi$ as the y-axis.

The graph of $y = \tan x$
(see figure) shows that the function $\tan x$ is discontinuous for an infinite number of values of the independent variable x; namely, when $x = (n + \tfrac{1}{2})\pi$, where n denotes any positive or negative integer.

In fact, when $x \to \tfrac{1}{2}\pi$, $\tan x$ becomes infinite. But from the relation $\tan (\pi + x) = \tan x$, we see that the function has the period π, and the values $x = (n + \tfrac{1}{2})\pi$ differ from $\tfrac{1}{2}\pi$ by a multiple of the period.

The function $\operatorname{ctn} x$ has the period π. It is defined and is continuous for all values of x except $x = n\pi$, n being any integer as before. For these values $\operatorname{ctn} x$ becomes infinite. Finally, $\sec x$ and $\csc x$ are periodic, each with the period 2π. The former is discontinuous only when $x = (n + \tfrac{1}{2})\pi$, the latter only when $x = n\pi$. The values of x for which these functions become infinite determine vertical asymptotes in the graphs.

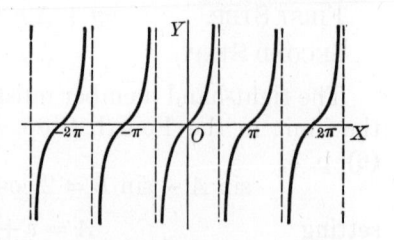

71. Differentiation of cos v

Let $y = \cos v.$

By (3), p. 2, this may be written

$$y = \sin\left(\frac{\pi}{2} - v\right).$$

Differentiating by formula XIII,

$$\frac{dy}{dx} = \cos\left(\frac{\pi}{2} - v\right)\frac{d}{dx}\left(\frac{\pi}{2} - v\right)$$

$$= \cos\left(\frac{\pi}{2} - v\right)\left(-\frac{dv}{dx}\right)$$

$$= -\sin v \frac{dv}{dx}.$$

$\left[\text{Since } \cos\left(\tfrac{\pi}{2} - v\right) = \sin v, \text{ by (3), p. 2.}\right]$

XIV $\qquad \therefore \dfrac{d}{dx}(\cos v) = -\sin v \dfrac{dv}{dx}.$

72. Proofs of formulas XV–XIX. These formulas are readily derived by expressing the function concerned in terms of other functions whose derivatives have been found, and differentiating.

Proof of XV. Let $y = \tan v.$

By (2), p. 2, this may be written

$$y = \frac{\sin v}{\cos v}.$$

TRANSCENDENTAL FUNCTIONS

Differentiating by formula VII,

$$\frac{dy}{dx} = \frac{\cos v \frac{d}{dx}(\sin v) - \sin v \frac{d}{dx}(\cos v)}{\cos^2 v}$$

$$= \frac{\cos^2 v \frac{dv}{dx} + \sin^2 v \frac{dv}{dx}}{\cos^2 v}$$

$$= \frac{\frac{dv}{dx}}{\cos^2 v} = \sec^2 v \frac{dv}{dx}. \qquad \text{Using (2), p. 2}$$

XV $\qquad \therefore \frac{d}{dx}(\tan v) = \sec^2 v \frac{dv}{dx}.$

To prove XVI–XIX, differentiate the form as given for each of the functions below.

XVI. $\operatorname{ctn} v = \dfrac{1}{\tan v}.$ **XVII.** $\sec v = \dfrac{1}{\cos v}.$ **XVIII.** $\csc v = \dfrac{1}{\sin v}.$

XIX. $\qquad\qquad$ versine $v = \operatorname{vers} v = 1 - \cos v.$

The details are left as exercises.

73. Comments. In the derivation of formulas I–XIX it was necessary to apply the General Rule, Art. 27, only for the following.

III $\qquad \dfrac{d}{dx}(u + v - w) = \dfrac{du}{dx} + \dfrac{dv}{dx} - \dfrac{dw}{dx}.$ Algebraic sum.

V $\qquad \dfrac{d}{dx}(uv) = u\dfrac{dv}{dx} + v\dfrac{du}{dx}.$ Product.

VII $\qquad \dfrac{d}{dx}\left(\dfrac{u}{v}\right) = \dfrac{v\dfrac{du}{dx} - u\dfrac{dv}{dx}}{v^2}.$ Quotient.

VIII $\qquad \dfrac{dy}{dx} = \dfrac{dy}{dv} \cdot \dfrac{dv}{dx}.$ Function of a function.

IX $\qquad \dfrac{dy}{dx} = \dfrac{1}{\dfrac{dx}{dy}}.$ Inverse functions.

X $\qquad \dfrac{d}{dx}(\ln v) = \dfrac{\dfrac{dv}{dx}}{v}.$ Logarithm.

XIII $\qquad \dfrac{d}{dx}(\sin v) = \cos v \dfrac{dv}{dx}.$ Sine.

Not only do all the other formulas deduced depend on these, but all we shall deduce hereafter depend on them as well. Hence we see

102 DIFFERENTIAL CALCULUS

that the derivation of the fundamental formulas for differentiation involves the calculation of only two limits of any difficulty, namely,

$$\lim_{v \to 0} \frac{\sin v}{v} = 1 \qquad \text{by Art. 68}$$

and
$$\lim_{v \to 0} (1+v)^{\frac{1}{v}} = e. \qquad \text{By Art. 61}$$

PROBLEMS

Differentiate the following functions.

1. $y = \sin ax^2$.

Solution. $\quad \dfrac{dy}{dx} = \cos ax^2 \dfrac{d}{dx}(ax^2) \qquad$ by XIII

$[v = ax^2.]$

$= 2ax \cos ax^2$. Ans.

2. $y = \tan \sqrt{1-x}$.

Solution. $\quad \dfrac{dy}{dx} = \sec^2 \sqrt{1-x} \dfrac{d}{dx}(1-x)^{\frac{1}{2}} \qquad$ by XV

$[v = \sqrt{1-x}.]$

$= \sec^2 \sqrt{1-x} \cdot \tfrac{1}{2}(1-x)^{-\frac{1}{2}}(-1)$

$= -\dfrac{\sec^2 \sqrt{1-x}}{2\sqrt{1-x}}$. Ans.

3. $y = \cos^3 x$.

Solution. This may also be written

$y = (\cos x)^3$.

$\dfrac{dy}{dx} = 3(\cos x)^2 \dfrac{d}{dx}(\cos x) \qquad$ by VI

$[v = \cos x \text{ and } n = 3.]$

$= 3 \cos^2 x (-\sin x) \qquad$ by XIV

$= -3 \sin x \cos^2 x$. Ans.

4. $y = \sin nx \sin^n x$.

Solution. $\quad \dfrac{dy}{dx} = \sin nx \dfrac{d}{dx}(\sin x)^n + \sin^n x \dfrac{d}{dx}(\sin nx) \qquad$ by V

$[u = \sin nx \text{ and } v = \sin^n x.]$

$= \sin nx \cdot n (\sin x)^{n-1} \dfrac{d}{dx}(\sin x)$

$\qquad + \sin^n x \cos nx \dfrac{d}{dx}(nx) \qquad$ by VI and XIII

$= n \sin nx \cdot \sin^{n-1} x \cos x + n \sin^n x \cos nx$

$= n \sin^{n-1} x (\sin nx \cos x + \cos nx \sin x)$

$= n \sin^{n-1} x \sin(n+1)x$. Ans.

TRANSCENDENTAL FUNCTIONS

5. $y = \sin ax.$ Ans $y' = a \cos ax.$

6. $y = 3 \cos 2x.$ $y' = -6 \sin 2x.$

7. $s = \tan 3t.$ $s' = 3 \sec^2 3t.$

8. $u = 2 \operatorname{ctn} \dfrac{v}{2}.$ $\dfrac{du}{dv} = -\csc^2 \dfrac{v}{2}.$

9. $y = \sec 4x.$ $y' = 4 \sec 4x \tan 4x.$

10. $\rho = a \csc b\theta.$ $\rho' = -ab \csc b\theta \operatorname{ctn} b\theta.$

11. $y = \tfrac{1}{2} \sin^2 x.$ $y' = \sin x \cos x.$

12. $s = \sqrt{\cos 2t}.$ $\dfrac{ds}{dt} = \dfrac{-\sin 2t}{\sqrt{\cos 2t}}.$

13. $\rho = \sqrt[3]{\tan 3\theta}.$ $\dfrac{d\rho}{d\theta} = \dfrac{\sec^2 3\theta}{(\tan 3\theta)^{\frac{2}{3}}}.$

14. $y = \dfrac{4}{\sqrt{\sec x}}.$ $\dfrac{dy}{dx} = \dfrac{-2 \tan x}{\sqrt{\sec x}}.$

15. $y = x \cos x.$ $y' = \cos x - x \sin x.$

16. $f(\theta) = \tan \theta - \theta.$ $f'(\theta) = \tan^2 \theta.$

17. $\rho = \dfrac{\sin \theta}{\theta}.$ $\dfrac{d\rho}{d\theta} = \dfrac{\theta \cos \theta - \sin \theta}{\theta^2}.$

18. $y = \sin 2x \cos x.$ $y' = 2 \cos 2x \cos x - \sin 2x \sin x.$

19. $y = \ln \sin ax.$ $y' = a \operatorname{ctn} ax.$

20. $y = \ln \sqrt{\cos 2x}.$ $y' = -\tan 2x.$

21. $y = e^{ax} \sin bx.$ $y' = e^{ax}(a \sin bx + b \cos bx).$

22. $s = e^{-t} \cos 2t.$ $s' = -e^{-t}(2 \sin 2t + \cos 2t).$

23. $y = \ln \tan \dfrac{x}{2}.$ $y' = \tfrac{1}{2} \operatorname{ctn} \dfrac{x}{2} \sec^2 \dfrac{x}{2}.$

24. $y = \ln \sqrt{\dfrac{1+\sin x}{1-\sin x}}.$ $y' = \sec x.$

25. $f(\theta) = \sin(\theta + a)\cos(\theta - a).$ $f'(\theta) = \cos 2\theta.$

26. $f(x) = \sin^2(\pi - x).$ $f'(x) = -2 \sin(\pi - x)\cos(\pi - x).$

27. $\rho = \tfrac{1}{3}\tan^3 \theta - \tan \theta + \theta.$ $\rho' = \tan^4 \theta.$

28. $y = x^{\sin x}.$ $\dfrac{dy}{dx} = x^{\sin x}\left(\dfrac{\sin x}{x} + \cos x \ln x\right).$

29. $y = (\cos x)^x.$ $y' = y(\ln \cos x - x \tan x).$

Find the second derivative of each of the following functions.

30. $y = \sin kx$. Ans. $\dfrac{d^2y}{dx^2} = -k^2 \sin kx$.

31. $\rho = \frac{1}{4} \cos 2\theta$. $\dfrac{d^2\rho}{d\theta^2} = -\cos 2\theta$.

32. $u = \tan v$. $\dfrac{d^2u}{dv^2} = 2 \sec^2 v \tan v$.

33. $y = x \cos x$. $\dfrac{d^2y}{dx^2} = -2 \sin x - x \cos x$.

34. $y = \dfrac{\sin x}{x}$. $\dfrac{d^2y}{dx^2} = \dfrac{2 \sin x - 2x \cos x - x^2 \sin x}{x^3}$.

35. $s = e^t \cos t$. $\dfrac{d^2s}{dt^2} = -2 e^t \sin t$.

36. $s = e^{-t} \sin 2t$. $\dfrac{d^2s}{dt^2} = -e^{-t}(3 \sin 2t + 4 \cos 2t)$.

37. $y = e^{ax} \sin bx$. $\dfrac{d^2y}{dx^2} = e^{ax}[(a^2 - b^2) \sin bx + 2ab \cos bx]$.

Find $\dfrac{dy}{dx}$ from each of the following equations.

38. $y = \cos(x - y)$. Ans. $\dfrac{dy}{dx} = \dfrac{\sin(x-y)}{\sin(x-y) - 1}$.

39. $e^y = \sin(x + y)$. $\dfrac{dy}{dx} = \dfrac{\cos(x+y)}{e^y - \cos(x+y)}$.

40. $\cos y = \ln(x + y)$. $\dfrac{dy}{dx} = \dfrac{-1}{1 + (x+y) \sin y}$.

In Problems 41–50 find the value of $\dfrac{dy}{dx}$ for the given value of x (in radians).

41. $y = x - \cos x$; $x = 1$. Ans. $y' = 1.841$.

42. $y = x \sin \dfrac{x}{2}$; $x = 2$. $y' = 1.381$.

43. $y = \ln \cos x$; $x = 0.5$. $y' = -0.546$.

44. $y = \dfrac{e^x}{x}$; $x = -0.5$. $y' = -3.639$.

45. $y = \sin x \cos 2x$; $x = 1$. $y' = -1.754$.

46. $y = \ln \sqrt{\tan x}$; $x = \frac{1}{4}\pi$. $y' = 1$.

47. $y = e^x \sin x$; $x = 2$. $y' = 3.643$.

48. $y = 10 e^{-x} \cos \pi x$; $x = 1$. $y' = 3.679$.

49. $y = 5 e^{\frac{x}{2}} \sin \dfrac{\pi x}{2}$; $x = 2$. $y' = -21.35$.

50. $y = 10 e^{-\frac{x}{10}} \sin 3x$; $x = 1$. $y' = -27.00$.

74. Inverse trigonometric functions.

From the equation

(1) $$y = \sin x,$$

we may read "x is the measure of an angle in radians whose sine equals y." For a central angle in a circle with radius unity, x equals also the intercepted arc (see Art. 2). The statement in quotation marks is then abbreviated thus

(2) $$x = \text{arc sin } y,$$

read "x equals an arc whose sine is y." Interchanging x and y in (2), we obtain

(3) $$y = \text{arc sin } x,$$

and arc sin x is called the *inverse sine function* of x. It is defined for any value of x numerically less than or equal to 1. From (1) and (2), it appears that sin x and arc sin y are inverse functions (Art. 39).

Equation (3) is often written $y = \sin^{-1} x$, read "the inverse sine of x." This notation is inconvenient, for the reason that $\sin^{-1} x$, as thus written, might be read as sin x with the exponent -1.

Consider the value of y determined by $x = \frac{1}{2}$ in (3), that is, by

(4) $$y = \text{arc sin } \tfrac{1}{2}.$$

One value of y satisfying (4) is $y = \frac{1}{6}\pi$, since $\sin \frac{1}{6}\pi = \sin 30° = \frac{1}{2}$. A second value is $y = \frac{5}{6}\pi$, since $\sin \frac{5}{6}\pi = \sin 150° = \frac{1}{2}$. To each of these solutions any multiple of 2π may be added or subtracted. Hence *the number of values of y satisfying* (4) *is without limit*. The function arc sin x is then said to be "multiple-valued."

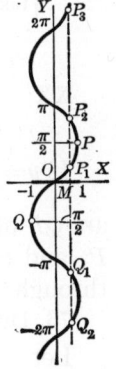

The graph of arc sin x (see figure) shows this property well. When $x = OM$, $y = MP_1$, MP_2, MP_3, $\cdots$, MQ_1, MQ_2, $\cdots$.

For most purposes in the calculus it is allowable and advisable to select *one* of the many values of y. We select, then, the value between $-\frac{1}{2}\pi$ and $\frac{1}{2}\pi$, that is, the *smallest numerical value*. Thus, for example,

(5) $\text{arc sin } \tfrac{1}{2} = \tfrac{1}{6}\pi$, arc sin $0 = 0$, arc sin $(-1) = -\tfrac{1}{2}\pi$.

The function arc sin x is now *single-valued*, and if

(6) $$y = \text{arc sin } x, \text{ then } -\tfrac{1}{2}\pi \leq y \leq \tfrac{1}{2}\pi.$$

In the graph we confine ourselves to the arc QOP.

In the same manner each of the other inverse trigonometric functions may be made single-valued. Thus, for arc cos x, if

(7) $\qquad y = \text{arc cos } x, \quad \text{then} \quad 0 \leq y \leq \pi.$

As examples,

arc cos $\tfrac{1}{2} = \tfrac{1}{3}\pi$, arc cos $(-\tfrac{1}{2}) = \tfrac{2}{3}\pi$, arc cos $(-1) = \pi$.

From (6) and (7) we now have the identical relation

(8) $\qquad \text{arc sin } x + \text{arc cos } x = \tfrac{1}{2}\pi.$

In the graph of arc cos x (see figure), we confine ourselves to the arc QP_1P.

Definitions establishing a single value for each of the other inverse trigonometric functions are given below.

75. Differentiation of arc sin v

Let $\qquad y = \text{arc sin } v;\qquad (-\tfrac{1}{2}\pi \leq y \leq \tfrac{1}{2}\pi)$

then $\qquad v = \sin y.$

Differentiating with respect to y by XIII,

$$\frac{dv}{dy} = \cos y :$$

therefore $\qquad \dfrac{dy}{dv} = \dfrac{1}{\cos y}.\qquad$ By (**C**), Art. 39

Since v is a function of x, this may be substituted in (**A**), Art. 38, giving

$$\frac{dy}{dx} = \frac{1}{\cos y} \cdot \frac{dv}{dx} = \frac{1}{\sqrt{1-v^2}} \frac{dv}{dx}.$$

$\left[\cos y = \sqrt{1 - \sin^2 y} = \sqrt{1 - v^2}, \text{ the positive sign of the radical being taken, since } \cos y \text{ is positive for all values of } y \text{ between } -\dfrac{\pi}{2} \text{ and } \dfrac{\pi}{2} \text{ inclusive.}\right]$

XX $\qquad \therefore \dfrac{d}{dx}(\text{arc sin } v) = \dfrac{\dfrac{dv}{dx}}{\sqrt{1-v^2}}.$

If $y = \text{arc sin } x$, $y' = \dfrac{dy}{dx} = \dfrac{1}{\sqrt{1-x^2}}$. The graph is the arc QP of the figure. The slope becomes infinite at Q and P, and equals 1 at O. The function increases ($y' > 0$) throughout the interval $x = -1$ to $x = 1$.

76. Differentiation of arc cos v

Let $\qquad y = \text{arc cos } v;\qquad (0 \leq y \leq \pi)$

then $\qquad v = \cos y.$

TRANSCENDENTAL FUNCTIONS

Differentiating with respect to y by **XIV**,
$$\frac{dv}{dy} = -\sin y;$$

therefore
$$\frac{dy}{dv} = -\frac{1}{\sin y}. \qquad \text{By } (C), \text{ Art. 39}$$

But since v is a function of x, this may be substituted in (A), Art. 38, giving
$$\frac{dy}{dx} = -\frac{1}{\sin y} \cdot \frac{dv}{dx} = -\frac{1}{\sqrt{1-v^2}} \frac{dv}{dx}.$$

$\left[\begin{array}{l}\sin y = \sqrt{1-\cos^2 y} = \sqrt{1-v^2}, \text{ the plus sign of the radical being taken,} \\ \text{since } \sin y \text{ is positive for all values of } y \text{ between } 0 \text{ and } \pi \text{ inclusive.}\end{array}\right]$

XXI $\qquad \therefore \dfrac{d}{dx}(\text{arc cos } v) = -\dfrac{\dfrac{dv}{dx}}{\sqrt{1-v^2}}.$

If $y = \text{arc cos } x$, then $y' = -\dfrac{1}{\sqrt{1-x^2}}$. When x increases from -1 to $+1$ (arc PQ of the first figure on page 106), y decreases from π to 0 ($y' < 0$).

77. Differentiation of arc tan v. Let

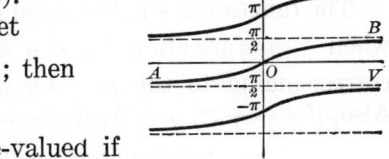

(1) $\qquad y = \text{arc tan } v;$ then

(2) $\qquad v = \tan y.$

The function (1) becomes single-valued if we choose the *least numerical value* of y, that is, a value between $-\frac{1}{2}\pi$ and $\frac{1}{2}\pi$, corresponding to arc AB of the figure. Also, when $v \to -\infty$, $y \to -\frac{1}{2}\pi$; when $v \to +\infty$, $y \to \frac{1}{2}\pi$. Or, symbolically,

(3) $\qquad \text{arc tan }(+\infty) = \frac{1}{2}\pi, \quad \text{arc tan }(-\infty) = -\frac{1}{2}\pi.$

Differentiating (2) with respect to y by **XV**,
$$\frac{dv}{dy} = \sec^2 y;$$

and
$$\frac{dy}{dv} = \frac{1}{\sec^2 y}. \qquad \text{By } (C), \text{ Art. 39}$$

Since v is a function of x, this may be substituted in (A), Art. 38, giving
$$\frac{dy}{dx} = \frac{1}{\sec^2 y} \cdot \frac{dv}{dx} = \frac{1}{1+v^2} \frac{dv}{dx}.$$

$[\sec^2 y = 1 + \tan^2 y = 1 + v^2.]$

XXII $\qquad \therefore \dfrac{d}{dx}(\text{arc tan } v) = \dfrac{\dfrac{dv}{dx}}{1+v^2}.$

If $y = \arc\tan x$, then $y' = \dfrac{1}{1+x^2}$ and the function is an increasing function for all values of x.

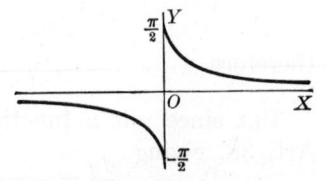

The function $\arc\tan \dfrac{1}{x}$ furnishes a good example of a discontinuous function. Confining ourselves to one branch of the graph of

$$y = \arc\tan \dfrac{1}{x},$$

we see that as x approaches zero from the left, y approaches $-\tfrac{1}{2}\pi$ as a limit, and as x approaches zero from the right, y approaches $+\tfrac{1}{2}\pi$ as a limit. Hence the function is discontinuous when $x = 0$ (Art. 17). Its value for $x = 0$ can be assigned at pleasure.

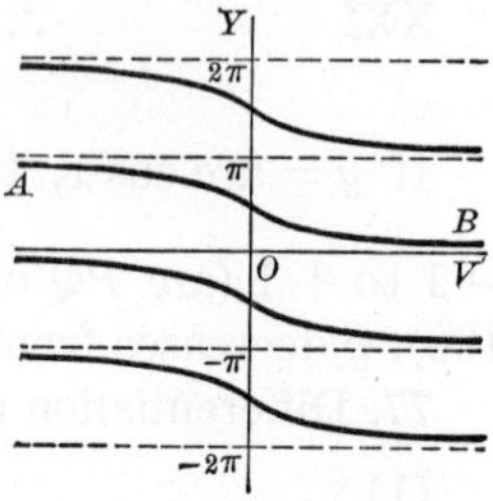

78. Differentiation of arc ctn v. Following the method of the last article, we get

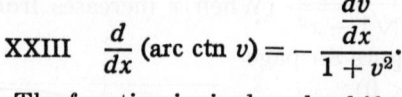

$$\text{XXIII} \quad \dfrac{d}{dx}(\arc\ctn v) = -\dfrac{\dfrac{dv}{dx}}{1+v^2}.$$

The function is single-valued if, when $y = \arc\ctn v$, $0 < y < \pi$, corresponding to the arc AB of the figure. Also, if $v \to +\infty$, $y \to 0$; if $v \to -\infty$, $y \to \pi$. That is, symbolically,

$$\arc\ctn(+\infty) = 0\,; \quad \arc\ctn(-\infty) = \pi.$$

79. Differentiation of arc sec v and arc csc v. Let

(1) $\qquad\qquad\qquad y = \arc\sec v.$

This function is defined for all values of v except those lying between -1 and $+1$. To make the function single-valued (see figure),

when v is positive, choose y between 0 and $\tfrac{1}{2}\pi$ (arc AB);

when v is negative choose y between $-\pi$ and $-\tfrac{1}{2}\pi$ (arc CD).

Also, if $v \to +\infty$, $y \to \tfrac{1}{2}\pi$;

if $\qquad v \to -\infty$, $y \to -\tfrac{1}{2}\pi$.

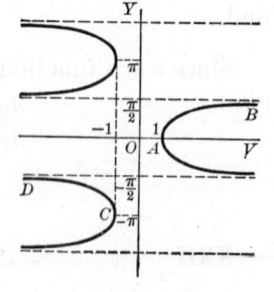

Solving (1), $v = \sec y$.

Differentiating with respect to y by XVII,

$$\dfrac{dv}{dy} = \sec y \tan y\,;$$

therefore $\dfrac{dy}{dv} = \dfrac{1}{\sec y \tan y}.$ By (C), Art. 39

TRANSCENDENTAL FUNCTIONS

Since v is a function of x, this may be substituted in (A), Art. 38, giving

$$\frac{dy}{dx} = \frac{1}{\sec y \tan y} \cdot \frac{dv}{dx} = \frac{1}{v\sqrt{v^2-1}} \frac{dv}{dx}.$$

$\left[\sec y = v, \text{ and } \tan y = \sqrt{\sec^2 y - 1} = \sqrt{v^2 - 1}, \text{ the plus sign of the radical being taken,} \right.$
$\left. \text{since } \tan y \text{ is positive for all values of } y \text{ between 0 and } \frac{\pi}{2} \text{ and between } -\pi \text{ and } -\frac{\pi}{2}. \right]$

XXIV $\quad \therefore \dfrac{d}{dx}(\text{arc sec } v) = \dfrac{\dfrac{dv}{dx}}{v\sqrt{v^2-1}}.$

Differentiation of arc csc v. Let

$$y = \text{arc csc } v;$$

then
$$v = \csc y.$$

Differentiating with respect to y by **XVIII** and following the method of the last section, we get

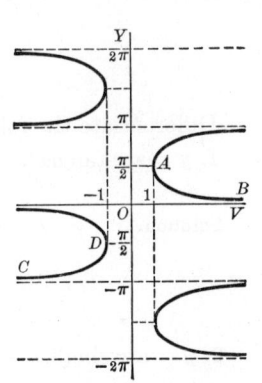

XXV $\quad \dfrac{d}{dx}(\text{arc csc } v) = - \dfrac{\dfrac{dv}{dx}}{v\sqrt{v^2-1}}.$

The function $y = \text{arc csc } v$ is defined for all values of v except those lying between -1 and $+1$, and is many-valued. To make the function single-valued (see figure above),

when v is positive, choose y between 0 and $\frac{1}{2}\pi$ (arc AB):

when v is negative, choose y between $-\pi$ and $-\frac{1}{2}\pi$ (arc CD).

80. Differentiation of arc vers v

Let $\qquad y = \text{arc vers } v;$ *

then $\qquad v = \text{vers } y.$

Differentiating with respect to y by **XIX**,

$$\frac{dv}{dy} = \sin y;$$

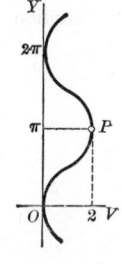

therefore $\qquad \dfrac{dy}{dv} = \dfrac{1}{\sin y}.$ $\qquad$ By (C), Art. 39

---

* Defined only for values of v between 0 and 2 inclusive. and many-valued. To make the function single-valued, y is taken as the smallest positive arc whose versed sine is v; that is, y lies between 0 and π inclusive. Hence we confine ourselves to arc OP of the graph.

110   DIFFERENTIAL CALCULUS

Since v is a function of x, this may be substituted in (A), Art. 38, giving

$$\frac{dy}{dx} = \frac{1}{\sin y} \cdot \frac{dv}{dx} = \frac{1}{\sqrt{2v - v^2}} \frac{dv}{dx}.$$

$\left[\sin y = \sqrt{1 - \cos^2 y} = \sqrt{1 - (1 - \text{vers } y)^2} = \sqrt{2v - v^2}, \text{ the plus sign of the radical being taken, since } \sin y \text{ is positive for all values of } y \text{ between } 0 \text{ and } \pi \text{ inclusive.}\right]$

XXVI $\qquad \therefore \dfrac{d}{dx}(\text{arc vers } v) = \dfrac{\dfrac{dv}{dx}}{\sqrt{2v - v^2}}.$

PROBLEMS

Differentiate the following functions.

1. $y = \text{arc tan } ax^2.$

Solution. $\qquad \dfrac{dy}{dx} = \dfrac{\dfrac{d}{dx}(ax^2)}{1 + (ax^2)^2} \qquad$ by XXII

$[v = ax^2.]$

$= \dfrac{2ax}{1 + a^2x^4}.$ Ans.

2. $y = \text{arc sin }(3x - 4x^3).$

Solution. $\qquad \dfrac{dy}{dx} = \dfrac{\dfrac{d}{dx}(3x - 4x^3)}{\sqrt{1 - (3x - 4x^3)^2}} \qquad$ by XX

$[v = 3x - 4x^3.]$

$= \dfrac{3 - 12x^2}{\sqrt{1 - 9x^2 + 24x^4 - 16x^6}} = \dfrac{3}{\sqrt{1 - x^2}}.$ Ans.

3. $y = \text{arc sec } \dfrac{x^2 + 1}{x^2 - 1}.$

Solution. $\qquad \dfrac{dy}{dx} = \dfrac{\dfrac{d}{dx}\left(\dfrac{x^2+1}{x^2-1}\right)}{\dfrac{x^2+1}{x^2-1}\sqrt{\left(\dfrac{x^2+1}{x^2-1}\right)^2 - 1}} \qquad$ by XXIV

$\left[v = \dfrac{x^2+1}{x^2-1}.\right]$

$= \dfrac{\dfrac{(x^2-1)2x - (x^2+1)2x}{(x^2-1)^2}}{\dfrac{x^2+1}{x^2-1} \cdot \dfrac{2x}{x^2-1}} = -\dfrac{2}{x^2+1}.$ Ans.

4. $y = \text{arc cos } \dfrac{x}{a}.$ $\qquad$ Ans. $\dfrac{dy}{dx} = -\dfrac{1}{\sqrt{a^2 - x^2}}.$

5. $y = \text{arc sec } \dfrac{x}{a}.$ $\qquad\qquad\qquad\quad \dfrac{dy}{dx} = \dfrac{a}{x\sqrt{x^2 - a^2}}.$

TRANSCENDENTAL FUNCTIONS

6. $y = \operatorname{arc\,ctn} \dfrac{x}{a}$. Ans. $\dfrac{dy}{dx} = \dfrac{-a}{a^2 + x^2}$.

7. $y = \operatorname{arc\,sec} \dfrac{1}{x}$. $\dfrac{dy}{dx} = \dfrac{-1}{\sqrt{1-x^2}}$.

8. $y = \operatorname{arc\,csc} 2x$. $\dfrac{dy}{dx} = \dfrac{-1}{x\sqrt{4x^2-1}}$.

9. $y = \operatorname{arc\,sin} \sqrt{x}$. $\dfrac{dy}{dx} = \dfrac{1}{2\sqrt{x-x^2}}$.

10. $\theta = \operatorname{arc\,vers} \rho^2$. $\dfrac{d\theta}{d\rho} = \dfrac{2}{\sqrt{2-\rho^2}}$.

11. $y = x \operatorname{arc\,sin} 2x$. $\dfrac{dy}{dx} = \operatorname{arc\,sin} 2x + \dfrac{2x}{\sqrt{1-4x^2}}$.

12. $y = x^2 \operatorname{arc\,cos} x$. $\dfrac{dy}{dx} = 2x \operatorname{arc\,cos} x - \dfrac{x^2}{\sqrt{1-x^2}}$.

13. $f(u) = u\sqrt{a^2-u^2} + a^2 \operatorname{arc\,sin} \dfrac{u}{a}$. $f'(u) = 2\sqrt{a^2-u^2}$.

14. $f(x) = \sqrt{a^2-x^2} + a \operatorname{arc\,sin} \dfrac{x}{a}$. $f'(x) = \sqrt{\dfrac{a-x}{a+x}}$.

15. $v = a^2 \operatorname{arc\,sin} \dfrac{u}{a} - u\sqrt{a^2-u^2}$. $\dfrac{dv}{du} = \dfrac{2u^2}{\sqrt{a^2-u^2}}$.

16. $v = \dfrac{u}{\sqrt{a^2-u^2}} - \operatorname{arc\,sin} \dfrac{u}{a}$. $\dfrac{dv}{du} = \dfrac{u^2}{(a^2-u^2)^{\frac{3}{2}}}$.

17. $v = \operatorname{arc\,sin} \dfrac{u}{a} + \dfrac{\sqrt{a^2-u^2}}{u}$. $\dfrac{dv}{du} = -\dfrac{\sqrt{a^2-u^2}}{u^2}$.

18. $v = a \operatorname{arc\,cos}\left(1 - \dfrac{u}{a}\right) + \sqrt{2au - u^2}$. $\dfrac{dv}{du} = \dfrac{\sqrt{2au-u^2}}{u}$.

19. $\phi = \operatorname{arc\,tan} \dfrac{a+r}{1-ar}$. $\dfrac{d\phi}{dr} = \dfrac{1}{1+r^2}$.

20. $x = r \operatorname{arc\,vers} \dfrac{y}{r} - \sqrt{2ry - y^2}$. $\dfrac{dx}{dy} = \dfrac{y}{\sqrt{2ry-y^2}}$.

21. $y = \tfrac{1}{3}x^3 \operatorname{arc\,tan} x + \tfrac{1}{6} \ln(x^2+1) - \tfrac{1}{6}x^2$. $\dfrac{dy}{dx} = x^2 \operatorname{arc\,tan} x$.

In Problems 22–27 find the value of $\dfrac{dy}{dx}$ for the given value of x.

22. $y = x \operatorname{arc\,sin} x$; $x = \tfrac{1}{2}$. Ans. $y' = 1.101$.

23. $y = x \operatorname{arc\,cos} x$; $x = -\tfrac{1}{2}$. $y' = 2.671$.

24. $y = \dfrac{\operatorname{arc\,tan} x}{x}$; $x = 1$. $y' = -0.285$.

25. $y = \sqrt{x} \operatorname{arc\,ctn} \dfrac{x}{4}$; $x = 4$. $y' = -0.054$.

26. $y = \dfrac{\operatorname{arc\,sec} 2x}{\sqrt{x}}$; $x = 1$. $y' = 0.053$.

27. $y = x^2 \operatorname{arc\,csc} \sqrt{x}$; $x = 2$. $y' = 2.142$.

Differentiate each of the following functions.

28. $\arcsin \sqrt{x}$.

29. $\arctan \dfrac{2}{x}$.

30. $x \arccos \dfrac{x}{2}$.

31. $\dfrac{\operatorname{arc\,ctn} 2x}{x}$.

32. $\operatorname{arc\,vers}(1-x)$.

33. $\operatorname{arc\,sec} \sqrt{x}$.

34. $e^x \arccos x$.

35. $\ln \arctan x$.

36. $\sqrt{\arcsin 2x}$.

37. $\dfrac{\arccos \sqrt{x}}{\sqrt{x}}$.

PROBLEMS

Sketch the following curves, and find the slope at each point where the curve crosses the axes of coördinates.

1. $y = \ln x$.   Ans. At $(1, 0)$, $m = 1$.
2. $y = \log x$.   At $(1, 0)$, $m = 0.434$.
3. $y = \ln(4-x)$.   At $(3, 0)$, $m = -1$; at $x = 0$, $m = -\frac{1}{4}$.
4. $y = \ln \sqrt{4-x^2}$.
5. Show that if $y = \frac{1}{2} a\left(e^{\frac{x}{a}} + e^{-\frac{x}{a}}\right)$, then $y'' = \dfrac{y}{a^2}$.

Find the angles of intersection of each of the following pairs of curves.

6. $y = \ln(x+1)$, $y = \ln(7-2x)$.   Ans. 127° 53′.
7. $y = \ln(x+3)$, $y = \ln(5-x^2)$.
8. $y = \sin x$, $y = \cos x$.   109° 28′.
9. $y = \tan x$, $y = \operatorname{ctn} x$.   53° 8′.
10. $y = \cos x$, $y = \sin 2x$.

Find the maximum, minimum, and inflectional points on the following curves and draw their graphs.

11. $y = x \ln x$.   Ans. Min. $\left(\dfrac{1}{e}, -\dfrac{1}{e}\right)$.
12. $y = \dfrac{x}{\ln x}$.   Min. (e, e); inflectional point, $(e^2, \frac{1}{2}e^2)$.
13. $y = \ln(8x - x^2)$.   Max. $(4, \ln 16)$.
14. $y = xe^x$.   Min. $\left(-1, -\dfrac{1}{e}\right)$; inflectional point, $\left(-2, -\dfrac{2}{e^2}\right)$.
15. $y = x^2 e^{-x}$.

TRANSCENDENTAL FUNCTIONS

16. A submarine telegraph cable consists of a core of copper wires with a covering made of nonconducting material. If x denotes the ratio of the radius of the core to the thickness of the covering, it is known that the speed of signaling varies as $x^2 \ln \frac{1}{x}$. Show that the greatest speed is attained when $x = \frac{1}{\sqrt{e}}$.

17. What is the minimum value of $y = ae^{kx} + be^{-kx}$? Ans. $2\sqrt{ab}$.

18. Find the maximum point and the points of inflection of the graph of $y = e^{-x^2}$, and draw the curve.
Ans. Max. $(0, 1)$; points of inflection, $\left(\pm \frac{1}{\sqrt{2}}, \frac{1}{\sqrt{e}}\right)$.

19. Show that the maximum rectangle with one side on the x-axis which can be inscribed under the curve in Problem 18 has two of its vertices at the points of inflection.

Find the maximum, minimum, and inflectional points for the range indicated, and sketch the following curves.

20. $y = \frac{1}{2} x - \sin x$; (0 to 2π).
Ans. Min. $(\frac{1}{3}\pi, -0.3424)$; max. $(\frac{5}{3}\pi, 3.4840)$; inflectional points, $(0, 0)$, $(\pi, \frac{1}{2}\pi)$, $(2\pi, \pi)$.

21. $y = 2x - \tan x$; (0 to π).
Ans. Max. $(\frac{1}{4}\pi, 0.571)$; min. $(\frac{3}{4}\pi, 5.712)$; inflectional points, $(0, 0)$, $(\pi, 2\pi)$.

22. $y = \tan x - 4x$; (0 to π).
Ans. Min. $(\frac{1}{3}\pi, -2.457)$; max. $(\frac{2}{3}\pi, -10.11)$; inflectional points, $(0, 0)$, $(\pi, -4\pi)$.

23. $y = 3 \sin x - 4 \cos x$; (0 to 2π).
Ans. Max. $(2.498, 5)$; min. $(5.640, -5)$; inflectional points, $(0.927, 0)$, $(4.069, 0)$.

24. $y = x + \cos 2x$; (0 to π).

25. $y = \sin \pi x - \cos \pi x$; (0 to 2).

26. $y = \frac{1}{2} x + \sin 2x$; (0 to π).

27. $y = x - 2 \cos 2x$; (0 to π).

28. $y = \frac{1}{2} \pi x + \sin \pi x$; (0 to 2).

29. Show that the maximum value of $y = a \sin x + b \cos x$ is $\sqrt{a^2 + b^2}$.

30. The base of an isosceles triangle is 8 ft. If the altitude is 4 ft. and if it is increasing at the rate of 3 in. per minute, at what rate is the vertex angle changing?
Ans. Decreasing $3° 35'$ per minute.

31. Find the maximum and minimum points for the range indicated and sketch the following curves.

(a) $y = 10 \, e^{-x} \sin x$; (0 to 2π).

 Ans. Max. ($\tfrac{1}{4}\pi$, 3.224); min. ($\tfrac{5}{4}\pi$, -0.139).

(b) $y = 10 \, e^{-x} \cos x$; (0 to 2π). (c) $y = 10 \, e^{-\tfrac{1}{2}x} \sin 2x$; (0 to π).

32. Find the dimensions of the cylinder of maximum volume which can be inscribed in a sphere of radius 6 in. (Use the angle θ subtended by the radius of the base of the inscribed cylinder as a parameter. Then $r = 6 \sin \theta$, $h = 12 \cos \theta$.)

33. Solve Problem 32 if the convex surface of the cylinder is to be a maximum, using the same parameter.

34. A body of weight W is dragged along a horizontal plane by means of a force P whose line of action makes an angle x with the plane. The magnitude of the force is given by the equation

$$P = \frac{mW}{m \sin x + \cos x},$$

where m denotes the coefficient of friction. Show that the pull is least when $\tan x = m$.

35. If a projectile is fired from O so as to strike an inclined plane which makes a constant angle α with the horizontal at O, the range is given by the formula

$$R = \frac{2 v^2 \cos \theta \sin (\theta - \alpha)}{g \cos^2 \alpha},$$

where v and g are constants and θ is the angle of elevation. Calculate the value of θ giving the maximum range up the plane. Ans. $\theta = \tfrac{1}{4}\pi + \tfrac{1}{2}\alpha$.

36. For a square-threaded screw with pitch θ and angle of friction ϕ the efficiency is given by the formula

$$E = \frac{\tan \theta}{\tan (\theta + \phi) + f},$$

where f is a constant. Find the value of θ for maximum efficiency when ϕ is a known constant angle.

ADDITIONAL PROBLEMS

1. The curves $y = x \ln x$ and $y = x \ln (1 - x)$ intersect at the origin and at another point A. Find the angle of intersection at A. Ans. $103° \, 30'$.

2. Sketch the following curves on the same axes and find their angle of intersection.

$$y = \ln \left(\frac{x^3}{8} - 1\right), \quad y = \ln \left(3x - \frac{x^2}{4} - 1\right). \quad \text{Ans. } 32° \, 28'.$$

3. The line AB is tangent to the curve whose equation is $y = e^x + 1$ at A and crosses the x-axis at B. Find the coördinates of A if the length of AB is a minimum. Ans. (0, 2).

CHAPTER VIII

APPLICATIONS TO PARAMETRIC EQUATIONS, POLAR EQUATIONS, AND ROOTS

81. Parametric equations of a curve. Slope. The coördinates x and y of a point on a curve are often expressed as functions of a third variable, or *parameter*, t, in the form

(1) $$\begin{cases} x = f(t), \\ y = \phi(t). \end{cases}$$

Each value of t gives a value of x and a value of y and determines a point on the curve. Equations (1) are called *parametric equations* of the curve. If we eliminate t from equations (1), we obtain the *rectangular equation* of the curve. For example,

(2) $$\begin{cases} x = r \cos t, \\ y = r \sin t \end{cases}$$

are parametric equations of the circle in the figure, t being the parameter. For if we eliminate t by squaring and adding the results, we have

$$x^2 + y^2 = r^2(\cos^2 t + \sin^2 t) = r^2,$$

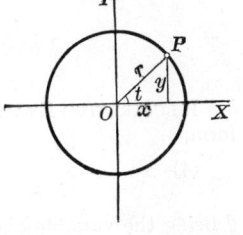

the rectangular equation of the circle. It is evident that if t varies from 0 to 2π, the point $P(x, y)$ will describe a complete circumference.

Since, from (1), y is a function of t, and t is a function (inverse) of x, we have

$$\frac{dy}{dx} = \frac{dy}{dt}\frac{dt}{dx} \qquad \text{by (A), Art. 38}$$

$$= \frac{dy}{dt} \cdot \frac{1}{\frac{dx}{dt}}; \qquad \text{by (C), Art. 39}$$

that is,

(A) $$\frac{dy}{dx} = \frac{\frac{dy}{dt}}{\frac{dx}{dt}} = \frac{\phi'(t)}{f'(t)} = \text{slope at } P(x, y).$$

By this formula we may find the slope of a curve whose parametric equations are given.

DIFFERENTIAL CALCULUS

ILLUSTRATIVE EXAMPLE 1. Find the equations of the tangent and normal, and the lengths of the subtangent and subnormal to the ellipse*

(3) $$\begin{cases} x = a \cos \phi, \\ y = b \sin \phi \end{cases}$$

at the point where $\phi = 45°$.

Solution. The parameter being ϕ, $\dfrac{dx}{d\phi} = -a \sin \phi$, $\dfrac{dy}{d\phi} = b \cos \phi$.

Substituting in (A), $\dfrac{dy}{dx} = -\dfrac{b \cos \phi}{a \sin \phi} = -\dfrac{b}{a} \operatorname{ctn} \phi = $ slope at any point $= m$.

Substituting $\phi = 45°$ in the given equations (3), we get $x_1 = \tfrac{1}{2} a\sqrt{2}$, $y_1 = \tfrac{1}{2} b\sqrt{2}$ as the point of contact, and the slope m becomes

$$m_1 = -\frac{b}{a} \operatorname{ctn} 45° = -\frac{b}{a}.$$

Substituting in (1) and (2), Art. 43, and reducing, we get

$$bx + ay = \sqrt{2}\, ab = \text{equation of the tangent,}$$
$$\sqrt{2}(ax - by) = a^2 - b^2 = \text{equation of the normal.}$$

Substituting in (3) and (4), Art. 43,

$$\tfrac{1}{2} b\sqrt{2} \left(-\frac{a}{b}\right) = -\tfrac{1}{2} a\sqrt{2} = \text{length of subtangent,}$$

$$\tfrac{1}{2} b\sqrt{2} \left(-\frac{b}{a}\right) = -\frac{b^2 \sqrt{2}}{2a} = \text{length of subnormal.}$$

ILLUSTRATIVE EXAMPLE 2. Given the equations of the cycloid† in parametric form,

(4) $$\begin{cases} x = a(\theta - \sin \theta), \\ y = a(1 - \cos \theta), \end{cases}$$

θ being the variable parameter; find the lengths of the subtangent, subnormal, and normal at the point (x_1, y_1) where $\theta = \theta_1$.

---

* As in the figure, draw the major and minor auxiliary circles of the ellipse. Through two points B and C on the same radius draw BA parallel to OY and DP parallel to OX. These lines will intersect in a point $P(x, y)$ on the ellipse, because

$$x = OA = OB \cos \phi = a \cos \phi$$

and $$y = AP = OD = OC \sin \phi = b \sin \phi,$$

or $$\frac{x}{a} = \cos \phi \quad \text{and} \quad \frac{y}{b} = \sin \phi.$$

Now squaring and adding, we get

$$\frac{x^2}{a^2} + \frac{y^2}{b^2} = \cos^2 \phi + \sin^2 \phi = 1,$$

the rectangular equation of the ellipse. ϕ is sometimes called the eccentric angle of the point P of the ellipse.

† The path described by a point on the circumference of a circle which rolls without sliding on a fixed straight line is called a cycloid. Let the radius of the rolling circle be a, P the tracing point, and M the point of contact with the fixed line OX, which is called the

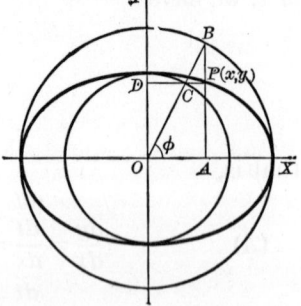

PARAMETRIC AND POLAR EQUATIONS

Solution. Differentiating, $\frac{dx}{d\theta} = a(1 - \cos\theta)$, $\frac{dy}{d\theta} = a\sin\theta$.

Substituting in (A), Art. 81,

$$\frac{dy}{dx} = \frac{\sin\theta}{1 - \cos\theta} = m = \text{slope at any point.}$$

When $\theta = \theta_1$, $y = y_1 = a(1 - \cos\theta_1)$, $m = m_1 = \frac{\sin\theta_1}{1 - \cos\theta_1}$.

Following Art. 43, we find (see figure at foot of this page)

$$TN = \text{subtangent} = \frac{a(1 - \cos\theta_1)^2}{\sin\theta_1}; \quad NM = \text{subnormal} = a\sin\theta_1.$$

$MP = \text{length of normal} = a\sqrt{2(1 - \cos\theta_1)} = 2a\sin\tfrac{1}{2}\theta_1$. By (5), p. 3.

In the figure, $PA = a\sin\theta_1$ (if $\theta = \theta_1$) = the subnormal NM as above. Hence the construction for the normal PM and tangent PB is as indicated.

Horizontal and vertical tangents. From (A), and referring to Art. 42, we see that the values of the parameter t for the points of contact of these tangent lines are determined thus:

Horizontal tangents: solve $\frac{dy}{dt} = 0$ for t.

Vertical tangents: solve $\frac{dx}{dt} = 0$ for t.

ILLUSTRATIVE EXAMPLE 3. Find the points of contact of the horizontal and vertical tangents to the cardioid (see figure)

(5) $\quad \begin{cases} x = a\cos\theta - \tfrac{1}{2}a\cos 2\theta - \tfrac{1}{2}a, \\ y = a\sin\theta - \tfrac{1}{2}a\sin 2\theta. \end{cases}$

Solution. $\frac{dx}{d\theta} = a(-\sin\theta + \sin 2\theta)$; $\frac{dy}{d\theta} = a(\cos\theta - \cos 2\theta)$.

Horizontal tangents. Then $\cos\theta - \cos 2\theta = 0$. Substituting (using (5), p. 3) $\cos 2\theta = 2\cos^2\theta - 1$, and solving, we get $\theta = 0, 120°, 240°$.

Vertical tangents. Then $-\sin\theta + \sin 2\theta = 0$. Substituting (using (5), p. 3) $\sin 2\theta = 2\sin\theta\cos\theta$, and solving, $\theta = 0, 60°, 180°, 300°$.

base. If arc PM equals OM in length, then P will touch at O if the circle is rolled to the left. We have, denoting the angle PCM by θ,

$$x = ON = OM - NM = a\theta - a\sin\theta = a(\theta - \sin\theta),$$
$$y = NP = MC - AC = a - a\cos\theta = a(1 - \cos\theta),$$

the parametric equations of the cycloid, the angle θ through which the radius of the rolling circle turns being the parameter. $OD = 2\pi a$ is called the base of one arch of the cycloid, and the point V is called the vertex. Eliminating θ, we get the rectangular equation

$$x = a\arccos\left(\frac{a-y}{a}\right) - \sqrt{2ay - y^2}.$$

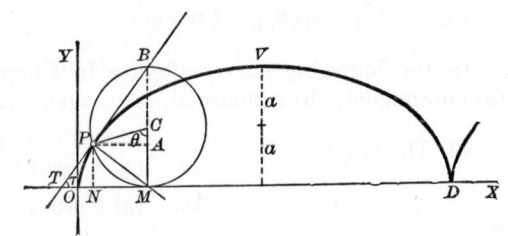

The common root $\theta = 0$ should be rejected. For both numerator and denominator in (A) become zero, and the slope is indeterminate (see Art. 12). From (5), $x = y = 0$ when $\theta = 0$. The point O is called a cusp.

Substituting the other values in (5), the results are as follows:

Horizontal tangents: points of contact $(-\tfrac{3}{4} a, \pm \tfrac{3}{4} a\sqrt{3})$.
Vertical tangents: points of contact $(\tfrac{1}{4} a, \pm \tfrac{1}{4} a\sqrt{3})$, $(-2a, 0)$.
Two vertical tangents coincide, forming a "double tangent" line.
These results agree with the figure.

PROBLEMS

Find the equations of the tangent and normal, and the lengths of the subtangent and subnormal, to each of the following curves at the point indicated.

	Tangent	Normal	Subt.	Subn.
1. $x = t^2, y = 2t+1$; $t = 1$. Ans.	$x - y + 2 = 0$,	$x + y - 4 = 0$,	3,	3.
2. $x = t^3, y = 3t$; $t = -1$.	$x - y - 2 = 0$,	$x + y + 4 = 0$,	-3,	-3.
3. $x = 3t, y = \dfrac{2}{t}$; $t = 2$.	$x + 6y - 12 = 0$,	$6x - y - 35 = 0$,	-6,	$-\tfrac{1}{6}$.
4. $x = e^t, y = 3 e^{-t}$; $t = 0$.	$3x + y - 6 = 0$,	$x - 3y + 8 = 0$,	-1,	-9.

5. $x = \cos 2\theta, y = \sin\theta$; $\theta = \tfrac{1}{6}\pi$.
6. $x = t^2, y = 2 - t$; $t = 1$.
7. $3x = t^3, 2y = t^2$; $t = 2$.
8. $x = 6t - t^2, y = 2t + 3$; $t = 0$.
9. $x = t^2, y = t^3 + 3t$; $t = 1$.
10. $x = \dfrac{1}{t}, y = 2t$; $t = -1$.
11. $x = \tan\theta, y = \operatorname{ctn}\theta$; $\theta = \tfrac{1}{4}\pi$.
12. $x = -3 e^{-t}, y = 2 e^t$; $t = 0$.
13. $x = 3\cos\alpha, y = 5\sin\alpha$; $\alpha = \tfrac{1}{4}\pi$.
14. $x = \sin 2\theta, y = \cos\theta$; $\theta = \tfrac{1}{3}\pi$.
15. $x = \ln(t-2), 3y = t$; $t = 3$.

In each of the following problems plot the curves and find the points of contact of the horizontal and vertical tangents.

16. $x = 3t - t^3, y = t + 1$. Ans. Horizontal tangents, none; vertical tangents, $(2, 2), (-2, 0)$.
17. $x = 3 - 4\sin\theta, y = 4 + 3\cos\theta$.
Ans. Horizontal tangents, $(3, 1), (3, 7)$; vertical tangents, $(7, 4), (-1, 4)$.
18. $x = t^2 - 2t, y = t^3 - 12t$.
19. $x = h + r\cos\theta, y = k + r\sin\theta$.
20. $x = \sin 2t, y = \sin t$.
21. $x = \cos^4\theta, y = \sin^4\theta$.

In the following curves (figures in Chapter XXVI) find lengths of (a) subtangent, (b) subnormal, (c) tangent, (d) normal, at any point.

22. The curve $\begin{cases} x = a(\cos t + t\sin t), \\ y = a(\sin t - t\cos t). \end{cases}$

Ans. (a) $y \operatorname{ctn} t$, (b) $y \tan t$, (c) $\dfrac{y}{\sin t}$, (d) $\dfrac{y}{\cos t}$.

PARAMETRIC AND POLAR EQUATIONS

23. The hypocycloid (astroid) $\begin{cases} x = 4a\cos^3 t, \\ y = 4a\sin^3 t. \end{cases}$ (Figure, p. 533)

Ans. (a) $-y\operatorname{ctn} t$, (b) $-y\tan t$, (c) $\dfrac{y}{\sin t}$, (d) $\dfrac{y}{\cos t}$.

24. The circle $\begin{cases} x = r\cos t, \\ y = r\sin t. \end{cases}$

25. The cardioid $\begin{cases} x = a(2\cos t - \cos 2t), \\ y = a(2\sin t - \sin 2t). \end{cases}$ (Figure, p. 117)

26. The folium $\begin{cases} x = \dfrac{3t}{1+t^3}, \\ y = \dfrac{3t^2}{1+t^3}. \end{cases}$ (Figure, p. 533)

27. The hyperbolic spiral $\begin{cases} x = \dfrac{a}{t}\cos t, \\ y = \dfrac{a}{t}\sin t. \end{cases}$ (Figure, p. 534)

82. Parametric equations. Second derivative. Using y' as symbol for the first derivative of y with respect to x, then **(A)**, Art. 81, will give y' as a function of t,

(1) $\qquad\qquad y' = h(t).$

To find the second derivative y'', use this formula **(A)** again, replacing y by y'. Then we have

(B) $\qquad\qquad y'' = \dfrac{dy'}{dx} = \dfrac{\dfrac{dy'}{dt}}{\dfrac{dx}{dt}} = \dfrac{h'(t)}{f'(t)},$

if $x = f(t)$, as in (1), Art. 81.

ILLUSTRATIVE EXAMPLE. Find y'' for the cycloid (see Illustrative Example 2, Art. 81)

$$\begin{cases} x = a(\theta - \sin\theta), \\ y = a(1 - \cos\theta). \end{cases}$$

Solution. We found $y' = \dfrac{\sin\theta}{1-\cos\theta}$, and $\dfrac{dx}{d\theta} = a(1-\cos\theta)$.

Also, differentiating,

$$\dfrac{dy'}{d\theta} = \dfrac{(1-\cos\theta)\cos\theta - \sin^2\theta}{(1-\cos\theta)^2} = \dfrac{\cos\theta - 1}{(1-\cos\theta)^2} = -\dfrac{1}{(1-\cos\theta)}.$$

Substituting in **(B)**,

$$y'' = -\dfrac{1}{a(1-\cos\theta)^2}. \quad Ans.$$

Note that y'' is negative, and the curve therefore is concave downward, as in the figure for the cycloid, p. 117.

PROBLEMS

1. In each of the following examples find $\dfrac{dy}{dx}$ and $\dfrac{d^2y}{dx^2}$ in terms of t.

(a) $x = t - 1,\ y = t^2 + 1$.   Ans. $\dfrac{dy}{dx} = 2t,\ \dfrac{d^2y}{dx^2} = 2$.

(b) $x = \dfrac{t^2}{2},\ y = 1 - t$. $\dfrac{dy}{dx} = -\dfrac{1}{t},\ \dfrac{d^2y}{dx^2} = \dfrac{1}{t^3}$.

(c) $x = 2t,\ y = \dfrac{t^3}{3}$.

(d) $x = \dfrac{t^3}{6},\ y = \dfrac{t^2}{2}$.

(e) $x = a \cos t,\ y = b \sin t$.
(f) $x = 2(1 - \sin t),\ y = 4 \cos t$.
(g) $x = \sin t,\ y = \sin 2t$.
(h) $x = \cos 2t,\ y = \sin t$.

2. Show that the curve $x = \sec\theta,\ y = \tan\theta$ has no point of inflection.

3. In each of the following examples plot the curve and find the maximum, minimum, and inflectional points:

(a) $x = 2a \operatorname{ctn}\theta,\ y = 2a \sin^2\theta$.
  Ans. Max. $(0, 2a)$; points of inflection, $\left(\pm\dfrac{2a}{\sqrt{3}},\ \dfrac{3a}{2}\right)$.

(b) $x = \tan t,\ y = \sin t \cos t$.
  Ans. Max. $(1, \tfrac{1}{2})$; min. $(-1, -\tfrac{1}{2})$;
  points of inflection, $\left(-\sqrt{3},\ -\dfrac{\sqrt{3}}{4}\right)$, $(0, 0)$, $\left(\sqrt{3},\ \dfrac{\sqrt{3}}{4}\right)$.

83. Curvilinear motion. Velocity. When the parameter t in the parametric equations (1), Art. 81, is the time, and the functions $f(t)$ and $\phi(t)$ are continuous, if t varies continuously the point $P(x, y)$ will trace the curve or path. We then have a *curvilinear motion*, and

(1) $\qquad x = f(t),\quad y = \phi(t)$

are called the **equations of motion**.

The velocity v of the moving point $P(x, y)$ at any instant is determined by its horizontal and vertical components.

The horizontal component v_x is equal to the velocity along OX of the projection M of P, and is therefore the time rate of change of x. Hence, from (**C**), Art. 51, when s is replaced by x, we get

(**C**) $\qquad v_x = \dfrac{dx}{dt}$.

In the same way the vertical component v_y, or time rate of change of y, is

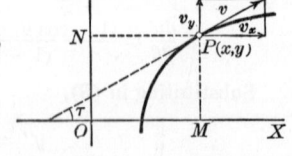

(**D**) $\qquad v_y = \dfrac{dy}{dt}$.

Lay off the vectors v_x and v_y from P as in the figure, complete the rectangle, and draw the diagonal from P. This is the required vector velocity v. From the figure, its magnitude and direction are given by the formulas

(E) $$v^2 = v_x^2 + v_y^2, \quad \tan \tau = \frac{v_y}{v_x} = \frac{\dfrac{dy}{dt}}{\dfrac{dx}{dt}}.$$

Comparing with (A), Art. 81, we see that $\tan \tau$ equals the slope of the path at P. Therefore the direction of v lies along the tangent line at P. The magnitude of the vector velocity is called the *speed*.

84. Curvilinear motion. Component accelerations. In treatises on mechanics it is shown that in curvilinear motion the vector acceleration α is not, like the vector velocity, directed along the tangent, but toward the concave side of the path of motion. It may be resolved into a tangential component, α_t, and a normal component, α_n, where

$$a_t = \frac{dv}{dt}; \quad a_n = \frac{v^2}{R}.$$

(R is the radius of curvature. See Art. 105.)

The acceleration may also be resolved into components parallel to the axes of coördinates. Following the same plan used in Art. 83 for component velocities, we define the *component accelerations* parallel to OX and OY,

(F) $$a_x = \frac{dv_x}{dt}; \quad a_y = \frac{dv_y}{dt}.$$

Also, if a rectangle is constructed with vertex P and sides α_x and α_y, then α is the diagonal from P. Hence

(G) $$a = \sqrt{(a_x)^2 + (a_y)^2},$$

which gives the magnitude (always positive) of the vector acceleration at any instant.

In Problem 1 below we make use of the equations of motion of a projectile, which illustrate very well this and the preceding article.

PROBLEMS

1. Neglecting the resistance of the air, the equations of motion for a projectile are

$$x = v_1 \cos \phi \cdot t, \quad y = v_1 \sin \phi \cdot t - 16.1\, t^2;$$

where v_1 = initial velocity, ϕ = angle of projection with horizon, and t = time of flight in seconds, x and y being measured in feet. Find the component velocities, component accelerations, velocity, and acceleration (a) at any instant; (b) at the end of the first second, having given $v_1 = 100$ ft. per second, $\phi = 30°$.

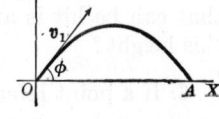

Find (c) the direction of motion at the end of the first second; (d) the rectangular equation of the path.

Solution. From (C) and (D),

(a) $\quad v_x = v_1 \cos \phi; \quad v_y = v_1 \sin \phi - 32.2\, t.$

Also, from (E), $\quad v = \sqrt{v_1^2 - 64.4\, tv_1 \sin \phi + 1036.8\, t^2}.$

From (F) and (G), $\quad \alpha_x = 0; \; \alpha_y = -32.2; \; \alpha = 32.2,$ direction downward.

(b) Substituting $t = 1,\, v_1 = 100,\, \phi = 30°$ in these results, we get

$v_x = 86.6$ ft. per sec. $\qquad \alpha_x = 0.$

$v_y = 17.8$ ft. per sec. $\qquad \alpha_y = -32.2$ ft. per (sec.)^2.

v (speed) $= 88.4$ ft. per sec. $\qquad \alpha = 32.2$ ft. per (sec.)^2.

(c) $\tau = \arctan \dfrac{v_y}{v_x} = \arctan \dfrac{17.8}{86.6} = 11° 37' =$ angle of direction of motion with the horizontal.

(d) When $v_1 = 100,\, \phi = 30°$, the equations of motion become

$$x = 50\, t\sqrt{3}, \quad y = 50\, t - 16.1\, t^2.$$

Eliminating t, the result is $y = \dfrac{x}{\sqrt{3}} - \dfrac{0.161}{75} x^2$, a parabola.

2. Show that the rectangular equation of the path of the projectile in Problem 1 is

$$y = x \tan \phi - \frac{16.1}{v_1^2}(1 + \tan^2 \phi)x^2.$$

3. If a projectile be given an initial velocity of 160 ft. per second in a direction inclined 45° with the horizontal, find (a) the component velocities at the end of the second and fourth seconds; (b) the velocity and direction of motion at the same instants.

Ans. (a) When $t = 2$, $v_x = 113.1$ ft. per sec., $v_y = 48.7$ ft. per sec.,
when $t = 4$, $v_x = 113.1$ ft. per sec., $v_y = -15.7$ ft. per sec.;
(b) when $t = 2$, $v = 123.1$ ft. per sec., $\tau = 23° 18'$,
when $t = 4$, $v = 114.2$ ft. per sec., $\tau = 172° 6'$.

4. With the data as in Problem 3 find the greatest height reached by the projectile. If the projectile strikes the ground at the same horizontal level from which it started, find the time of flight and the angle of impact.

5. A projectile with an initial velocity of 160 ft. per second is hurled at a vertical wall 480 ft. away. Show that the highest point on this wall that can be hit is at a height above the x-axis of 253 ft. What is ϕ for this height? $\qquad$ Ans. $\phi = 59°$.

6. If a point referred to rectangular coördinates moves so that

$$x = a \cos t + b \quad \text{and} \quad y = a \sin t + c,$$

show that its velocity has a constant magnitude.

PARAMETRIC AND POLAR EQUATIONS 123

7. If the path of a moving point is the sine curve
$$\begin{cases} x = at, \\ y = b \sin at, \end{cases}$$
show (a) that the x-component of the velocity is constant; (b) that the acceleration of the point at any instant is proportional to its distance from the x-axis.

8. Given the equations of motion $x = t^2$, $y = (t-1)^2$. (a) Find the equation of the path in rectangular coördinates. (b) Draw the path with the velocity and acceleration vectors for $t = \frac{1}{2}$, $t = 1$, $t = 2$. (c) For what values of the time is the speed a minimum? (d) Where is the point when its speed is 10 ft. per second?
  Ans. (a) Parabola, $x^{\frac{1}{2}} + y^{\frac{1}{2}} = 1$; (c) $t = \frac{1}{2}$; (d) (16, 9).

9. In uniform motion (speed constant) in a circle, show that the acceleration at any point P is constant in magnitude and directed along the radius from P toward the center of the circle.

10. The equations of a curvilinear motion are $x = 2 \cos 2t$, $y = 3 \cos t$. (a) Show that the moving point oscillates on an arc of the parabola $4y^2 - 9x - 18 = 0$. Draw the path. (b) Draw the acceleration vectors at the points where $v = 0$. (c) Draw the velocity vector at the point where the speed is a maximum.

Given the following equations of curvilinear motion, find at the given instant v_x, v_y, v; α_x, α_y, α; position of point (coördinates); direction of motion. Also find the equation of the path in rectangular coördinates.

11. $x = t^2$, $y = 2t$; $t = 2$.
12. $x = 2t$, $y = t^3$; $t = 1$.
13. $x = t^3$, $y = t^2$; $t = 2$.
14. $x = 3t$, $y = t^2 - 3$; $t = 3$.
15. $x = 2 - t$, $y = 1 + t^2$; $t = 0$.
16. $x = a \cos t$, $y = a \sin t$; $t = \frac{3}{4}\pi$.
17. $x = 4 \sin t$, $y = 2 \cos t$; $t = \frac{1}{2}\pi$.
18. $x = \sin 2t$, $y = 2 \cos t$; $t = \frac{1}{2}\pi$.
19. $x = 2 \sin t$; $y = \cos 2t$; $t = \frac{1}{2}\pi$.
20. $x = \tan t$; $y = \operatorname{ctn} t$; $t = \frac{1}{4}\pi$.

85. Polar coördinates. Angle between the radius vector and the tangent line. Let the equation of a curve in polar coördinates ρ, θ, be

(1) $$\rho = f(\theta).$$

We proceed to prove the

Theorem. *If ψ is the angle between the radius vector OP and the tangent line at P, then*

(H) $$\tan \psi = \frac{\rho}{\rho'}$$

where $\rho' = \dfrac{d\rho}{d\theta}$.

Proof. Through P and a point $Q(\rho + \Delta\rho, \theta + \Delta\theta)$ on the curve near P draw the secant line AB. Draw PR perpendicular to OQ.

Then (see figure) $OQ = \rho + \Delta\rho$, angle $POQ = \Delta\theta$, $PR = \rho \sin \Delta\theta$, and $OR = \rho \cos \Delta\theta$. Also,

(2) $$\tan PQR = \frac{PR}{RQ} = \frac{PR}{OQ - OR} = \frac{\rho \sin \Delta\theta}{\rho + \Delta\rho - \rho \cos \Delta\theta}.$$

Denote by ψ the angle between the radius vector OP and the tangent line PT. If we now let $\Delta\theta$ approach zero as a limit, then

(a) the point Q will approach P;
(b) the secant AB will turn about P and approach the tangent line PT as a limiting position; and
(c) the angle PQR will approach ψ as a limit.

Hence

(3) $$\tan \psi = \lim_{\Delta\theta \to 0} \frac{\rho \sin \Delta\theta}{\rho + \Delta\rho - \rho \cos \Delta\theta}.$$

To get this fraction in a form so that the theorems of Art. 16 will apply, we transform it as shown in the following equations:

$$\frac{\rho \sin \Delta\theta}{\rho(1 - \cos \Delta\theta) + \Delta\rho} = \frac{\rho \sin \Delta\theta}{2\rho \sin^2 \dfrac{\Delta\theta}{2} + \Delta\rho}.$$

$\left[\text{Since from (5), p. 3. } \rho - \rho \cos \Delta\theta = \rho(1 - \cos \Delta\theta) = 2\rho \sin^2 \dfrac{\Delta\theta}{2}.\right]$

$$= \frac{\rho \cdot \dfrac{\sin \Delta\theta}{\Delta\theta}}{\rho \sin \dfrac{\Delta\theta}{2} \cdot \dfrac{\sin \dfrac{\Delta\theta}{2}}{\dfrac{\Delta\theta}{2}} + \dfrac{\Delta\rho}{\Delta\theta}}.$$

[Dividing both numerator and denominator by $\Delta\theta$ and factoring.]

PARAMETRIC AND POLAR EQUATIONS 125

When $\Delta\theta \to 0$, then, by Art. 68,

$$\lim \frac{\sin \Delta\theta}{\Delta\theta} = 1, \quad \text{and} \quad \lim \frac{\sin \frac{\Delta\theta}{2}}{\frac{\Delta\theta}{2}} = 1.$$

Also, $\quad \lim \sin \frac{\Delta\theta}{2} = 0, \quad \lim \frac{\Delta\rho}{\Delta\theta} = \frac{d\rho}{d\theta} = \rho'.$

Hence the limits of numerator and denominator are, respectively, ρ and ρ'. Thus (*H*) is proved.

To find the slope ($\tan \tau$ in the figure), proceed as follows. Take rectangular axes OX, OY, as usual. Then for $P(x, y)$ we have

(4) $\qquad x = \rho \cos \theta, \quad y = \rho \sin \theta.$

Using (1), these equations become parametric equations of the curve, θ being the parameter. The slope is found by (*A*). Thus, from (4),

$$\frac{dx}{d\theta} = \rho' \cos \theta - \rho \sin \theta, \quad \frac{dy}{d\theta} = \rho' \sin \theta + \rho \cos \theta,$$

(*I*) $\qquad$ Slope of tangent $= \tan \tau = \dfrac{\rho' \sin \theta + \rho \cos \theta}{\rho' \cos \theta - \rho \sin \theta}.$

Formula (*I*) is easily verified for the figure on page 124. For, from the triangle OPT, $\tau = \theta + \psi$. Then $\tan \tau = \tan(\theta + \psi) = \dfrac{\tan \theta + \tan \psi}{1 - \tan \theta \tan \psi}.$ Substituting $\tan \theta = \dfrac{\sin \theta}{\cos \theta}$, $\tan \psi = \dfrac{\rho}{\rho'}$, and reducing, we have (*I*).

ILLUSTRATIVE EXAMPLE 1. Find $\tan \psi$ and the slope for the cardioid $\rho = a(1 - \cos \theta)$.

Solution. $\dfrac{d\rho}{d\theta} = \rho' = a \sin \theta.$ Substituting in (*H*) and (*I*),

$$\tan \psi = \frac{\rho}{\rho'} = \frac{a(1 - \cos \theta)}{a \sin \theta} = \frac{2a \sin^2 \frac{1}{2}\theta}{2a \sin \frac{1}{2}\theta \cos \frac{1}{2}\theta}$$

$= \tan \frac{1}{2}\theta.$ $\qquad$ ((5), p. 3)

$$\tan \tau = \frac{a \sin^2 \theta + a(1 - \cos \theta) \cos \theta}{a \sin \theta \cos \theta - a(1 - \cos \theta) \sin \theta}$$

$= \dfrac{\cos \theta - \cos 2\theta}{\sin 2\theta - \sin \theta} = \tan \frac{3}{2}\theta.$ ((5), (6), p. 3)

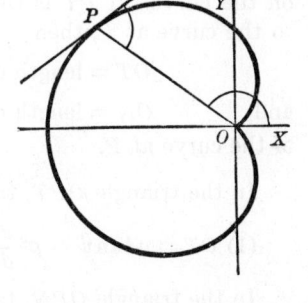

At P in the figure, $\psi = $ angle $OPT = \frac{1}{2}\theta = \frac{1}{2}$ angle XOP. If the tangent line PT is produced to cross the axis OX, forming with it the angle τ, we have angle $XOP = 180° - $ angle $OPT + \tau.$
Therefore $\tau = \frac{3}{2}\theta - 180°$, and $\tan \tau = \tan \frac{3}{2}\theta$, as above ((3), p. 3).

NOTE. Formula (*H*) has been derived for the figure on page 124. In each problem, the relations between the angles ψ, τ, and θ should be determined by examining the signs of their trigonometric functions and drawing a figure.

126   DIFFERENTIAL CALCULUS

To find the angle of intersection ϕ of two curves C and C' whose equations are given in polar coördinates, we may proceed as follows:

Angle TPT' = angle OPT' − angle OPT,

or $\quad \phi = \psi' - \psi$. Hence

(J) $\quad \tan \phi = \dfrac{\tan \psi' - \tan \psi}{1 + \tan \psi' \tan \psi},$

where $\tan \psi'$ and $\tan \psi$ are calculated by (H) from the equations of the curves and evaluated for the point of intersection.

ILLUSTRATIVE EXAMPLE 2. Find the angle of intersection of the curves $\rho = a \sin 2\theta$, $\rho = a \cos 2\theta$.

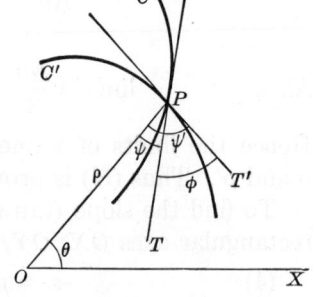

Solution. Solving the two equations simultaneously, we get, at the point of intersection,
$$\tan 2\theta = 1, \ 2\theta = 45°, \ \theta = 22\tfrac{1}{2}°.$$

From the first curve, using (H),
$$\tan \psi' = \tfrac{1}{2} \tan 2\theta = \tfrac{1}{2}, \text{ for } \theta = 22\tfrac{1}{2}°.$$

From the second curve,
$$\tan \psi = -\tfrac{1}{2} \cot 2\theta = -\tfrac{1}{2}, \text{ for } \theta = 22\tfrac{1}{2}°.$$

Substituting in (J),
$$\tan \phi = \dfrac{\tfrac{1}{2} + \tfrac{1}{2}}{1 - \tfrac{1}{4}} = \tfrac{4}{3}. \ \therefore \phi = \text{arc tan } \tfrac{4}{3}. \text{ Ans.}$$

The curves are shown in Chapter XXVI.

86. Lengths of polar subtangent and polar subnormal. Draw a line NT through the origin perpendicular to the radius vector of the point P on the curve. If PT is the tangent and PN the normal to the curve at P, then

$\qquad OT$ = length of polar subtangent,

and $\qquad ON$ = length of polar subnormal,

of the curve at P.

In the triangle OPT, $\tan \psi = \dfrac{OT}{\rho}$. Therefore

(1) $\quad OT = \rho \tan \psi = \rho^2 \dfrac{d\theta}{d\rho} =$ length of polar subtangent.*

In the triangle OPN, $\tan \psi = \dfrac{\rho}{ON}$. Therefore

(2) $\quad ON = \dfrac{\rho}{\tan \psi} = \dfrac{d\rho}{d\theta} =$ length of polar subnormal.

* When θ increases with ρ, $\dfrac{d\theta}{d\rho}$ is positive and ψ is an acute angle, as in the above figure. Then the subtangent OT is positive and is measured to the right of an observer placed at O and looking along OP. When $\dfrac{d\theta}{d\rho}$ is negative, the subtangent is negative and is measured to the left of the observer.

PARAMETRIC AND POLAR EQUATIONS

The *length of the polar tangent* ($= PT$) and the *length of the polar normal* ($= PN$) may be found from the figure, each being the hypotenuse of a right triangle.

ILLUSTRATIVE EXAMPLE. Find the lengths of the polar subtangent and polar subnormal to the lemniscate $\rho^2 = a^2 \cos 2\theta$ (figure in Chapter XXVI).

Solution. Differentiating the equation of the curve, regarding ρ as an implicit function of θ,

$$2\rho \frac{d\rho}{d\theta} = -2 a^2 \sin 2\theta, \quad \text{or} \quad \frac{d\rho}{d\theta} = -\frac{a^2 \sin 2\theta}{\rho}.$$

Substituting in (1) and (2), we get

$$\text{Length of polar subtangent} = -\frac{\rho^3}{a^2 \sin 2\theta},$$

$$\text{Length of polar subnormal} = -\frac{a^2 \sin 2\theta}{\rho}.$$

If we wish to express the results in terms of θ, find ρ in terms of θ from the given equation and substitute. Thus, in the above, $\rho = \pm a\sqrt{\cos 2\theta}$; therefore the length of the polar subtangent $= \pm a \operatorname{ctn} 2\theta \sqrt{\cos 2\theta}$.

PROBLEMS

1. In the circle $\rho = a \sin \theta$, find ψ and τ in terms of θ.

    Ans. $\psi = \theta, \tau = 2\theta$.

2. In the parabola $\rho = a \sec^2 \frac{\theta}{2}$, show that $\tau + \psi = \pi$.

3. Show that ψ is constant in the logarithmic spiral $\rho = e^{a\theta}$. Since the tangent makes a constant angle with the radius vector, this curve is also called the equiangular spiral. (Figure, p. 534)

4. Show that $\tan \psi = \theta$ in the spiral of Archimedes, $\rho = a\theta$. Find values of ψ when $\theta = 2\pi$ and 4π. (Figure, p. 534)

    Ans. $\psi = 80° 57'$ and $85° 27'$.

Find the slopes of the following curves at the points designated.

5. $\rho = a(1 - \cos \theta); \; \theta = \frac{\pi}{2}$.   Ans. -1.
6. $\rho = a \sec^2 \theta; \; \rho = 2a$.   3.
7. $\rho = a \sin 4\theta$; origin. $0, 1, \infty, -1$.
8. $\rho^2 = a^2 \sin 4\theta$; origin. $0, 1, \infty, -1$.
9. $\rho = a \sin 3\theta$; origin. $0, \sqrt{3}, -\sqrt{3}$.
10. $\rho = a \cos 3\theta$; origin.
11. $\rho = a \cos 2\theta$; origin.
12. $\rho = a \sin 2\theta; \; \theta = \frac{\pi}{4}$.
13. $\rho = a \sin 3\theta; \; \theta = \frac{\pi}{6}$.
14. $\rho = a\theta; \; \theta = \frac{\pi}{2}$.
15. $\rho\theta = a; \; \theta = \frac{\pi}{2}$.
16. $\rho = e^\theta; \; \theta = 0$.

Find the angle of intersection between the following pairs of curves.

17. $\rho \cos \theta = 2a$, $\rho = 5a \sin \theta$. *Ans.* arc tan $\frac{3}{4}$.
18. $\rho = a \sin \theta$, $\rho = a \sin 2\theta$.
   Ans. At origin, $0°$; at two other points, arc tan $3\sqrt{3}$.
19. $\rho \sin \theta = 2a$, $\rho = a \sec^2 \frac{\theta}{2}$. *Ans.* $45°$.
20. $\rho = 4 \cos \theta$, $\rho = 4(1 - \cos \theta)$. $60°$.
21. $\rho = 6 \cos \theta$, $\rho = 2(1 + \cos \theta)$. $30°$.
22. $\rho = \sin \theta$, $\rho = \cos 2\theta$. $0°$ and arc tan $\frac{3\sqrt{3}}{5}$.
23. $\rho^2 \sin 2\theta = 4$, $\rho^2 = 16 \sin 2\theta$. $60°$.
24. $\rho = a(1 + \cos \theta)$, $\rho = b(1 - \cos \theta)$.
25. $\rho = \sin 2\theta$, $\rho = \cos 2\theta + 1$.
26. $\rho^2 \sin 2\theta = 8$, $\rho = 2 \sec \theta$.

Show that the following pairs of curves intersect at right angles.

27. $\rho = 2 \sin \theta$, $\rho = 2 \cos \theta$.
28. $\rho = a\theta$, $\rho\theta = a$.
29. $\rho = a(1 + \cos \theta)$, $\rho = a(1 - \cos \theta)$.
30. $\rho^2 \sin 2\theta = a^2$, $\rho^2 \cos 2\theta = b^2$.
31. $\rho = a \sec^2 \frac{\theta}{2}$, $\rho = b \csc^2 \frac{\theta}{2}$.

32. Find the lengths of the polar subtangent, subnormal, tangent, and normal of the spiral of Archimedes, $\rho = a\theta$.

Ans. Subtangent $= \frac{\rho^2}{a}$, tangent $= \frac{\rho}{a}\sqrt{a^2 + \rho^2}$,

subnormal $= a$, normal $= \sqrt{a^2 + \rho^2}$.

The student should note the fact that the subnormal is constant.

33. Find the lengths of the polar subtangent, subnormal, tangent, and normal in the logarithmic spiral $\rho = a^\theta$.

Ans. Subtangent $= \frac{\rho}{\ln a}$, tangent $= \rho\sqrt{1 + \frac{1}{\ln^2 a}}$,

subnormal $= \rho \ln a$, normal $= \rho\sqrt{1 + \ln^2 a}$.

34. Show that the reciprocal spiral $\rho\theta = a$ has a constant polar subtangent.

87. Real roots of equations. Graphical methods. A value of x which satisfies the equation

(1) $\qquad\qquad\qquad f(x) = 0$

is called a *root* of the equation (or a *root of* $f(x)$). A root of (1) may be a real number or an imaginary (complex) number. Methods of determining real roots approximately will now be developed.

PARAMETRIC AND POLAR EQUATIONS

Location and number of the roots.

FIRST METHOD. If the graph of $f(x)$, that is, the locus of

(2) $$y = f(x)$$

is constructed, following Art. 58, the *intercepts on the x-axis are the roots*. From the figure, therefore, we know at once the number of roots and their approximate values.

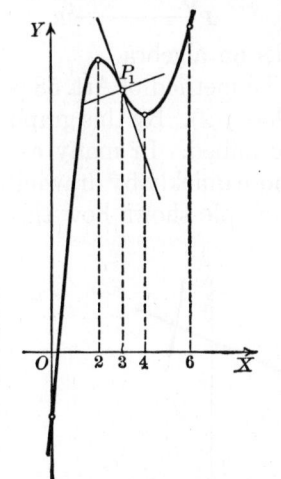

ILLUSTRATIVE EXAMPLE. Locate all real roots of

(3) $x^3 - 9x^2 + 24x - 7 = 0$.

x	$f(x)$
0	-7
1	9

Solution. The graph has been constructed in Art. 58, Problem 1. It crosses the axis of x between 0 and 1. Hence there is one real root between these values, and there are no other real roots.

The table gives the values of $f(0)$ and $f(1)$, showing a change of sign.

The table of values of x and y used in plotting the graph may locate a root exactly, namely, if $y = 0$ for some value of x. If not, the values of y for two successive values $x = a$, $x = b$ may have opposite signs. The corresponding points

x	y
a	$f(a)$
x_0	$f(x_0) = 0$
b	$f(b)$

$P(a, f(a))$, $Q(b, f(b))$ are, therefore, on opposite sides of the x-axis, and the graph of (2) joining these points will cross this axis. That is, a root x_0 will lie between a and b.

An exact statement of the principle involved here is as follows.

If a continuous function $f(x)$ changes sign in an interval $a < x < b$ and if its derivative does not change sign, then the equation $f(x) = 0$ has one root, and only one, between a and b.

Location of a root by trial depends upon this principle. If a and b are not far apart, a further approximation can be found by *interpolation*. This amounts to determining the intercept on the x-axis of the chord PQ. That is, the portion of the graph joining P and Q is replaced, as an approximation, by the chord.

ILLUSTRATIVE EXAMPLE (CONTINUED). The root between 0 and 1 may be located by calculation more closely between 0.3 and 0.4. See table. Let $0.3 + z$ be this root. Then, by interpolation (proportion),

x	$f(x) = y$
0.4	1.224
$0.3 + z$(root)	0
0.3	-0.583
Diff. 0.1	1.807

$$\frac{z}{0.1} = \frac{.583}{1.807}, \quad z = .032.$$

Hence $x = 0.332$ is a second approximation. This is the intercept on the x-axis of the line that joins the points $Q(0.4, 1.224)$ and $P(0.3, -0.583)$, which lie

130 DIFFERENTIAL CALCULUS

on the graph of (3). In the figure, $MP = -0.583$, $NQ = 1.224$, drawn to scale. The abscissas of M and N are 0.3 and 0.4 respectively. Also, $MC = z$, and the homologous sides of the similar triangles MPC and PQR give the above proportion.

For an algebraic equation, of which (3) is an example, Horner's method is best adapted to calculating a numerical root to any desired degree of accuracy, as explained in textbooks on algebra.

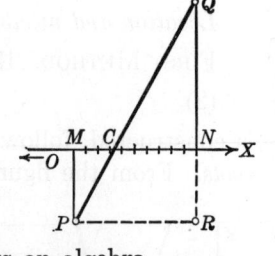

88. Second method for locating real roots. The method of Art. 58 is well adapted to constructing quickly the graph of $f(x)$. By this graph the roots are located and their number determined. In many examples, however, the same result is attained more quickly by drawing certain intersecting curves. The following example shows how this is done.

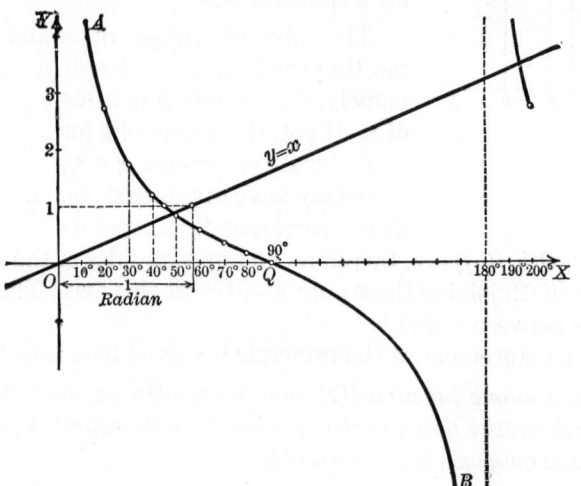

ILLUSTRATIVE EXAMPLE. Determine the number of real roots (x in radians) of the equation

(1) $\operatorname{ctn} x - x = 0$,

and locate the smallest root.

Solution. Transpose and write (1) thus:

(2) $\operatorname{ctn} x = x$.

If we draw the curves

(3) $y = \operatorname{ctn} x$ and $y = x$

on the same axes, *the abscissas of the points of intersection will be roots of* (1). For, obviously, eliminating y from (3) gives equation (1), from which the values of x of the points of intersection are to be obtained.

	$y = \operatorname{ctn} x$	
x (degrees)	x (radians)	y
0	0	∞
10	.175	5.67
20	.349	2.75
30	.524	1.73
40	.698	1.19
45	.785	1.000
50	.873	.839
60	1.047	.577
70	1.222	.364
80	1.396	.176
90	1.571	.0

PARAMETRIC AND POLAR EQUATIONS 131

In plotting it is well to lay off carefully both scales (degrees and radians) on OX.
Number of solutions. The curve $y = \operatorname{ctn} x$ consists of an infinite number of branches
congruent to AQB of the figure (see
Art. 70). The line $y = x$ will obviously
cross each branch. Hence the equation
(1) has an infinite number of solutions.

Using tables of natural cotangents
and radian equivalents of degrees, we
may locate the smallest root more closely
as shown in the table. By interpolation
we find $x = 0.860$. *Ans.*

x (degrees)	x (radians)	ctn x	ctn $x - x$
50	0.873 root	0.839	-0.034
49	0.855	0.869	$+0.014$
Diff.	0.018		-0.048

The Second Method may be described as follows.

Transpose certain selected terms of $f(x) = 0$ so that it becomes

(4) $\qquad f_1(x) = f_2(x).$

Plot the curves

(5) $\qquad y = f_1(x), \quad y = f_2(x)$

on the same axes, choosing suitable scales (not necessarily the same on
both axes).

The number of points of intersection of these curves equals the number of real roots of $f(x) = 0$, and the abscissas of these points are the roots.

The terms selected in (4) can often be chosen so that one or both
of the curves in (5) are standard curves.

For example, to locate the real roots of
$$x^3 + 4x - 5 = 0,$$
write the equation $\qquad x^3 = 5 - 4x.$

The curves in (5) are now the standard curves
$$y = x^3, \quad y = 5 - 4x,$$
a *cubical parabola* and a straight line.

As a second example, consider
$$2 \sin 2x + 1 - x^2 = 0.$$
Write this in the form $\quad \sin 2x = \tfrac{1}{2}(x^2 - 1).$

Then the curves in (5) are the standard curve
$$y = \sin 2x$$
and the parabola $\qquad y = \tfrac{1}{2}(x^2 - 1).$

89. Newton's method. Having located a root, Newton's method
affords a procedure to calculate its approximate value.

The figure shows two points
$$P(a, f(a)), \quad Q(b, f(b))$$

on the graph of $f(x)$ on opposite sides of the x-axis. Let PT be the tangent line at P (Fig. a). The intercept a' of this line on the

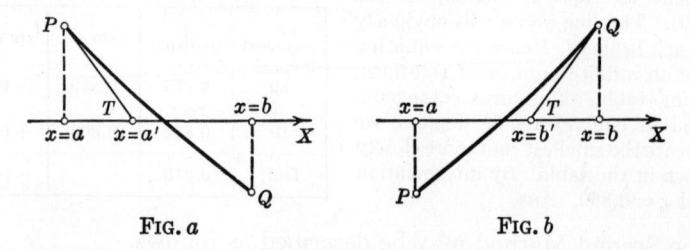

FIG. a   FIG. b

x-axis is, obviously, an approximate value of the intercept of the graph and hence of the corresponding root of $f(x) = 0$. Newton's method determines the x-intercept of PT.

We find this intercept a' as follows. The coördinates of P are $x_1 = a$, $y_1 = f(a)$. The slope of PT is $m_1 = f'(a)$. Hence the equation of PT is ((1), Art. 43)

(1) $$y - f(a) = f'(a)(x - a)$$

Putting $y = 0$ and solving for $x (= a')$ gives Newton's *formula for approximation*

(K) $$a' = a - \frac{f(a)}{f'(a)}.$$

Having found a' by (K), we may substitute a' for a in the right-hand member, obtaining
$$a'' = a' - \frac{f(a')}{f'(a')}$$
as a second approximation. The process might be continued, giving a sequence of values
$$a, a', a'', a''', \cdots$$
approaching the exact root.

Or the tangent may be drawn at Q (Fig. b). Then replacing a in (K) by b, we obtain b', and from b' we obtain b'' etc., giving values
$$b', b'', b''', \cdots$$
approaching the exact root.

ILLUSTRATIVE EXAMPLE. Find the smallest root of
$$\operatorname{ctn} x - x = 0$$
by Newton's method.

Solution. Here $f(x) = \operatorname{ctn} x - x$,
$$f'(x) = -\csc^2 x - 1 = -2 - \operatorname{ctn}^2 x.$$

PARAMETRIC AND POLAR EQUATIONS

By the illustrative example of Art. 88, we take $a = 0.855$. Then, by the table in Art. 88,
$$f(a) = 0.014.$$
Also, $\quad f'(a) = -2 - (0.869)^2 = -2.76.$

Hence, by (K), $a' = 0.855 + \dfrac{0.014}{2.76} = 0.860.$ *Ans.*

If we used $b = 0.873$ in (K), then
$$b' = 0.873 - \frac{0.034}{2.704} = 0.861.$$

By interpolation we found $x = 0.860$. The above results are valid to three places of decimals.

From the figures on page 132 we observe that the graph crosses the x-axis between the tangent PT and the chord PQ. *Hence the exact root lies between the value found by Newton's method and that found by interpolation.* This statement is, however, subject to the reservation that $f''(x) = 0$ has no root between a and b, that is, that *there is no point of inflection on the arc PQ.*

PROBLEMS

Determine graphically the number and approximate location of the real roots of each of the following equations. Calculate each root to two decimals.

1. $x^3 + 2x - 8 = 0.$ *Ans.* 1.67.
2. $x^3 - 4x + 2 = 0.$ $-2.21, 0.54, 1.67.$
3. $x^3 - 8x - 5 = 0.$ $-2.44, -0.66, 3.10.$
4. $x^3 - 3x - 1 = 0.$ $-1.53, -0.35, 1.88.$
5. $x^3 - 3x^2 + 3 = 0.$ $-0.88, 1.35, 2.53.$
6. $x^3 + 3x^2 - 10 = 0.$ $1.49.$
7. $x^3 - 3x^2 - 4x + 7 = 0.$ $-1.71, 1.14, 3.57.$
8. $x^3 + 2x^2 - 5x - 8 = 0.$ $-2.76, -1.36, 2.12.$
9. $2x^3 - 14x^2 + 2x + 5 = 0.$ $-0.51, 0.71, 6.80.$
10. $x^4 + 8x - 12 = 0.$ $-2.36, 1.22.$
11. $x^4 - 4x^3 - 6x^2 + 20x + 9 = 0.$ $-2.16, -0.41, 2.41, 4.16.$
12. $x^4 + 4x^3 - 6x^2 - 20x - 23 = 0.$ $-4.60, 2.60.$

Determine graphically the number of real roots of each of the following equations. Calculate the smallest root (different from zero), using both interpolation and Newton's formula.

13. $\cos x + x = 0.$ *Ans.* One root; $x = -0.739.$
14. $\tan x - x = 0.$ Infinite number of roots.
15. $\cos 2x - x = 0.$ One root; $x = 0.515.$

DIFFERENTIAL CALCULUS

16. $3 \sin x - x = 0$.   *Ans.* Three roots; $x = 2.279$.

17. $2 \sin x - x^2 = 0$.   Two roots; $x = 1.404$.

18. $\cos x - 2 x^2 = 0$.   Two roots; $x = 0.635$.

19. $\operatorname{ctn} x + x^2 = 0$.   Infinite number of roots; $x = 3.032$.

20. $2 \sin 2x - x = 0$.   Three roots; $x = 1.237$.

21. $\sin x + x - 1 = 0$.   One root; $x = 0.511$.

22. $\cos x + x - 1 = 0$.   One root; $x = 0$.

23. $e^{-x} - \cos x = 0$.   Infinite number of roots; $x = 1.29$.

24. $\tan x - \log x = 0$.   Infinite number of roots; $x = 3.65$.

25. $e^x + x - 3 = 0$.   One root; $x = 0.792$.

26. $\sin 3x - \cos 2x = 0$.   Infinite number of roots; $x = 0.314$.

27. $2 \sin \tfrac{1}{2} x - \cos 2x = 0$.   Infinite number of roots; $x = 0.517$.

28. $\tan x - 2 e^x = 0$.   Infinite number of roots; $x = 1.44$.

29. The inner radius (r) and outer radius (R) in inches of the hollow steel driving shaft of a steamer transmitting H horse power at a speed of N revolutions per minute satisfy the relation $R^4 - r^4 = \dfrac{33\,HR}{N\pi^2}$. If $H = 2500$, $N = 160$, $r = 6$, find R.

30. A cylindrical shell with a hemispherical end has a diameter d in., and contains V cu. in. The length of the cylindrical part is h in. Show that $d^3 + 3\,hd^2 = \dfrac{12\,V}{\pi}$. Given $h = 20$, $V = 800$, find d.   *Ans.* $d = 6.77$.

31. Determine graphically the number of real roots of the following equation and calculate the largest root.

     $x + 1 - 4 \sin x = 0$.   *Ans.* Three roots; $x = 2.210$.

32. Find, to three decimals, the maximum value of each of the following functions for x between 0 and π.

     (a) $x \sin x$.   *Ans.* 1.820.   (b) $x^2 \cos x$.   *Ans.* 0.548.

33. If V cu. ft. is the volume of 1 lb. of superheated steam at a temperature $T°$ F. and pressure P lb. per square inch,

$$V = 0.6490\,\frac{T}{P} - \frac{22.58}{P^{\frac{3}{4}}}.$$

Given $V = 2.8$, $T = 420°$, find P.

34. The chord c of an arc s in a circle of radius r is given approximately by the formula

$$c = s - \frac{s^3}{24\,r^2}.$$

If $r = 4$ ft., $c = 5.60$ ft., find s.   *Ans.* $s = 6.23$ ft.

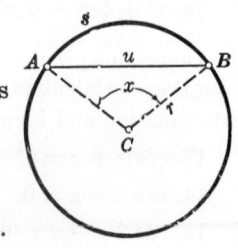

35. The area u of a circular segment whose arc s subtends the central angle x (in radians) is $u = \frac{1}{2} r^2 (x - \sin x)$. Find the value of x if $r = 8$ in. and $u = 64$ sq. in. *Ans.* $x = 2.554$ radians.

36. The volume V of a spherical segment of one base of height $CD = h$ is
$$V = \pi (rh^2 - \tfrac{1}{3} h^3).$$
Find h if $r = 4$ ft., $V = 150$ cu. ft.
Ans. $h = 4.32$ ft.

37. The volume V of a spherical shell of radius R and thickness t is
$$V = 4\pi t(R^2 - Rt + \tfrac{1}{3} t^2).$$
Derive this result. If $R = 4$ ft. and V is one half the volume of a solid sphere of equal radius, find t. *Ans.* $t = 0.827$ ft.

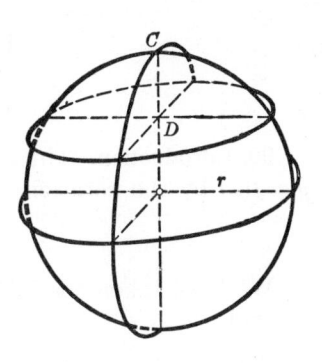

38. A solid wooden sphere of specific gravity S and diameter d sinks in water to a depth h. Let $x = \dfrac{h}{d}$ and show that $2x^3 - 3x^2 + S = 0$. (See Problem 36.) Find x for a maple ball for which $S = 0.786$.
Ans. 0.702.

39. Find the smallest positive value of θ for which the curves $\rho = \cos \theta$ and $\rho = e^{-\theta}$ intersect. Find the angle of intersection at this point.
Ans. $\theta = 1.29$ radians; $29°$.

ADDITIONAL PROBLEMS

1. Find the angle of intersection of the curves $\rho = 2 \cos \theta$ and $\rho = e^\theta$ at the point of intersection farthest from the origin.
Ans. Point of intersection is $\theta = 0.54$ radian; $75° 56'$.

2. Show that the curve $\rho = a \sin^4 \tfrac{1}{4} \theta$ cuts itself at right angles.

3. Any radius vector of the cardioid $\rho = a(1 + \cos \theta)$ is OP. From the center C of the circle $\rho = a \cos \theta$ a radius of the circle CQ is drawn parallel to OP and in the same direction. Prove that PQ is normal to the cardioid.

4. A square, one of whose diagonals lies along the polar axis, is circumscribed about the cardioid $\rho = a(1 - \cos \theta)$. Show that its area is $\tfrac{27}{16}(2 + \sqrt{3})a^2$.

5. The path of a particle is the ellipse $\rho = \dfrac{ep}{1 - e \cos \theta}$. The particle moves so that the radius vector ρ describes area at a constant time rate. Find the ratio of the velocities of the particle at the ends of the major axis.
Ans. $\dfrac{1-e}{1+e}$.

CHAPTER IX

DIFFERENTIALS

90. Introduction. Thus far we have represented the derivative of $y = f(x)$ by the notation

$$\frac{dy}{dx} = f'(x).$$

We have taken special pains to impress on the student that the symbol

$$\frac{dy}{dx}$$

was to be considered not as an ordinary fraction with dy as numerator and dx as denominator, but as a single symbol denoting the limit of the quotient

$$\frac{\Delta y}{\Delta x}$$

as Δx approaches zero as a limit.

Problems occur where it is important to give meanings to dx and dy separately, and this is especially useful in applications of the integral calculus. How this may be done is explained in what follows.

91. Definitions. If $f'(x)$ is the derivative of $f(x)$ for a particular value of x, and Δx is an arbitrarily chosen increment of x, then the **differential** of $f(x)$, denoted by the symbol $df(x)$, is defined by the equation

(A) $$df(x) = f'(x)\Delta x = \frac{dy}{dx}\Delta x.$$

If now $f(x) = x$, then $f'(x) = 1$, and (A) reduces to

$$dx = \Delta x.$$

Thus, when x is the independent variable, the **differential of** $x (= dx)$ is identical with Δx. Hence, if $y = f(x)$, (A) may in general be written in the form

(B) $$dy = f'(x)dx^* = \frac{dy}{dx}dx.$$

* On account of the position which the derivative $f'(x)$ here occupies, it is sometimes called the differential coefficient.

The differential of a function equals its derivative multiplied by the differential of the independent variable.

Let us illustrate what this means geometrically.

Draw the curve $y = f(x)$.
Let $f'(x)$ be the value of the derivative at P.
Take $dx = PQ$, then

$$dy = f'(x)dx = \tan \tau \cdot PQ = \frac{QT}{PQ} \cdot PQ = QT.$$

Therefore dy, or $df(x)$, is the increment ($= QT$) of the ordinate of the tangent corresponding to dx.

This gives the following interpretation of the derivative as a fraction.

If an arbitrarily chosen increment of the independent variable x for a point $P(x, y)$ on the curve $y = f(x)$ be denoted by dx, then in the derivative

$$\frac{dy}{dx} = f'(x) = \tan \tau,$$

dy denotes the corresponding increment of the ordinate of the tangent line at P.

The student should note especially that the differential ($= dy$) and the increment ($= \Delta y$) of the function corresponding to the same value of $dx (= \Delta x)$ are not in general equal. For, in the figure, $dy = QT$, but $\Delta y = QP'$.

92. Approximation of increments by means of differentials. From Art. 91 it is clear that $\Delta y (= QP'$ in the figure) and $dy (= QT)$ are approximately equal when $dx (= PQ)$ is small. When only an *approximate value* of the increment of a function is desired, it is usually easier to calculate the value of the corresponding differential and use this value.

ILLUSTRATIVE EXAMPLE 1. Find the volume approximately of a spherical shell of outside diameter 10 in. and thickness $\frac{1}{16}$ in.

Solution. The volume V of a sphere of diameter x is

(1) $$V = \tfrac{1}{6} \pi x^3.$$

Obviously, the *exact* volume of the shell is the difference ΔV between the volumes of two solid spheres with diameters 10 in. and $9\frac{7}{8}$ in. respectively. Since only an approximate value of ΔV is required, we find dV. From (1) and (*B*),

$$dV = \tfrac{1}{2} \pi x^2 \, dx, \quad \text{since} \quad \frac{dV}{dx} = \tfrac{1}{2} \pi x^2.$$

Substituting $x = 10$, $dx = -\frac{1}{8}$, we obtain $dV = 19.63$ cu. in., approximately, neglecting the sign, which merely means that V decreases as x decreases. The exact value is $\Delta V = 19.4$ cu. in. Note that the approximation is close, for dx is *relatively small*, that is, small as compared with x ($= 10$). The method would be worthless otherwise.

ILLUSTRATIVE EXAMPLE 2. Calculate $\tan 46°$, approximately, using differentials, given $\tan 45° = 1$, $\sec 45° = \sqrt{2}$, $1° = 0.01745$ radians.

Solution. Let $y = \tan x$. Then, by (B),

(1) $$dy = \sec^2 x\, dx.$$

When x changes to $x + dx$, y will change to $y + dy$, approximately. In (1), substitute $x = \frac{1}{4}\pi$ (45°), $dx = 0.0175$. Then $dy = 0.0350$. Since $y = \tan 45° = 1$, $y + dy = 1.0350 = \tan 46°$, approximately. *Ans.*
(Four-place tables give $\tan 46° = 1.0355$.)

93. Small errors. A second application of differentials is afforded when small errors in calculation are to be determined.

ILLUSTRATIVE EXAMPLE 1. The diameter of a circle is found by measurement to be 5.2 in., with a maximum error of 0.05 in. Find the approximate maximum error in the area when calculated by the formula

(1) $$A = \tfrac{1}{4}\pi x^2.$$ ($x =$ diameter)

Solution. Obviously, the exact maximum error in A will be the change (ΔA) in its value found by (1) when x changes from 5.2 in. to 5.25 in. The approximate error is the corresponding value of dA. Hence

$$dA = \tfrac{1}{2}\pi x\, dx = \tfrac{1}{2}\pi \times 5.2 \times 0.05 = 0.41 \text{ sq. in. } Ans.$$

Relative and percentage errors. If du is the error in u, then the ratio

(2) $$\frac{du}{u} = \text{the relative error;}$$

(3) $$100\,\frac{du}{u} = \text{the percentage error.}$$

The relative error may be found directly by logarithmic differentiation (Art. 66).

ILLUSTRATIVE EXAMPLE 2. Find the relative and percentage errors in the preceding example.

Solution. Taking natural logarithms in (1),

$$\log A = \log \tfrac{1}{4}\pi + 2\log x.$$

Differentiating, $\quad \dfrac{1}{A}\dfrac{dA}{dx} = \dfrac{2}{x}$, and $\dfrac{dA}{A} = \dfrac{2\,dx}{x}$.

Substituting $x = 5.2$, $dx = 0.05$, we find

Relative error in $A = 0.0192$; percentage error $= 1\tfrac{92}{100}\%$. *Ans.*

The errors in calculation considered here are due to small errors in the data upon which the calculation is based. The latter may arise from lack of precision in the measurements or from other causes.

DIFFERENTIALS

PROBLEMS

1. If A is the area of a square of side x, find dA. Draw a figure showing the square, dA, and ΔA. *Ans.* $dA = 2\,x\,dx$.

2. Find an approximate formula for the area of a circular ring of radius r and width dr. What is the exact formula?
Ans. $dA = 2\,\pi r\,dr$; $\Delta A = \pi(2\,r + \Delta r)\Delta r$.

3. What is the approximate error in the volume and surface of a cube of edge 6 in. if an error of 0.02 in. is made in measuring the edge?
Ans. Volume, ± 2.16 cu. in.; surface, ± 1.44 sq. in.

4. The formulas for the surface and volume of a sphere are $S = 4\,\pi r^2$ and $V = \tfrac{4}{3}\pi r^3$. If the radius is found to be 3 in. by measuring, (a) what is the approximate maximum error in S and V if measurements are accurate to 0.01 in.? (b) what is the maximum percentage error in each case?
Ans. (a) S, $0.24\,\pi$ sq. in.; V, $0.36\,\pi$ cu. in.;
(b) S, $\tfrac{2}{3}\%$; V, 1%.

5. Show by means of differentials that

$$\frac{1}{x + dx} = \frac{1}{x} - \frac{dx}{x^2} \text{ (approximately).}$$

6. Find an approximate formula for the volume of a thin cylindrical shell with open ends if the radius is r, the length h, and the thickness t.
Ans. $2\,\pi rht$.

7. A box is to be constructed in the form of a cube to hold 1000 cu. ft. How accurately must the inner edge be made so that the volume will be correct to within 3 cu. ft.? *Ans.* Error ≤ 0.01 ft.

8. If $y = x^{\frac{2}{3}}$ and the possible error in measuring x is 0.9 when $x = 27$, what is the possible error in the value of y? Use this result to obtain approximate values of $(27.9)^{\frac{2}{3}}$ and $(26.1)^{\frac{2}{3}}$. *Ans.* 0.2; 9.2; 8.8.

Use differentials to find an approximate value of each of the following expressions.

9. $\sqrt{66}$.   11. $\sqrt[3]{120}$.   13. $\tfrac{1}{96}$.   15. $\sqrt[5]{35}$.

10. $\sqrt{98}$.   12. $\sqrt[3]{1010}$.   14. $\dfrac{1}{\sqrt{51}}$.   16. $\sqrt[4]{15}$.

17. If $\ln 10 = 2.303$, approximate $\ln 10.2$ by means of differentials.
Ans. 2.323.

18. If $e^2 = 7.39$, approximate $e^{2.1}$ by means of differentials. *Ans.* 8.13.

19. Given $\sin 60° = 0.86603$, $\cos 60° = 0.5$, and $1° = 0.01745$ radians, use differentials to compute the values of each of the following functions to four decimals: (a) $\sin 62°$; (b) $\cos 61°$; (c) $\sin 59°$; (d) $\cos 58°$.
Ans. (a) 0.8835; (b) 0.4849; (c) 0.8573; (d) 0.5302.

20. The time of one vibration of a pendulum is given by the formula $t^2 = \dfrac{\pi^2 l}{g}$, where t is measured in seconds, $g = 32.2$, and l, the length of the pendulum, is measured in feet. Find (a) the length of a pendulum vibrating once a second; (b) the change in t if the pendulum in (a) is lengthened 0.01 ft.; (c) how much a clock with this error would lose or gain in a day.

Ans. (a) 3.26 ft.; (b) 0.00153 sec.; (c) − 2 min. 12 sec.

21. How exactly must the diameter of a circle be measured in order that the area shall be correct to within 1 per cent? *Ans.* Error $\leq \tfrac{1}{2}\%$.

22. Show that the relative error in the volume of a sphere, due to an error in measuring the diameter, is three times the relative error in the radius.

23. Show that the relative error in the nth power of a number is n times the relative error in the number.

24. Show that the relative error in the nth root of a number is $\dfrac{1}{n}$ times the relative error in the number.

25. When a cubical block of metal is heated, each edge increases $\tfrac{1}{10}$ per cent per degree increase in temperature. Show that the surface increases $\tfrac{2}{10}$ per cent per degree, and that the volume increases $\tfrac{3}{10}$ per cent per degree.

94. Formulas for finding the differentials of functions. Since the differential of a function is its derivative multiplied by the differential of the independent variable, it follows at once that the formulas for finding differentials are the same as those for finding derivatives given in Arts. 29 and 60, if we multiply each one by dx.

This gives

I	$d(c) = 0.$
II	$d(x) = dx.$
III	$d(u + v - w) = du + dv - dw.$
IV	$d(cv) = c\, dv.$
V	$d(uv) = u\, dv + v\, du.$
VI	$d(v^n) = nv^{n-1} dv.$
VI a	$d(x^n) = nx^{n-1} dx.$
VII	$d\left(\dfrac{u}{v}\right) = \dfrac{v\, du - u\, dv}{v^2}.$
VII a	$d\left(\dfrac{u}{c}\right) = \dfrac{du}{c}.$
X	$d(\ln v) = \dfrac{dv}{v}.$
XI	$d(a^v) = a^v \ln a\, dv.$
XI a	$d(e^v) = e^v\, dv.$
XII	$d(u^v) = vu^{v-1} du + \ln u \cdot u^v\, dv.$
XIII	$d(\sin v) = \cos v\, dv.$

DIFFERENTIALS

XIV $d(\cos v) = -\sin v\, dv.$
XV $d(\tan v) = \sec^2 v\, dv.$ Etc.
XX $d(\arcsin v) = \dfrac{dv}{\sqrt{1-v^2}}.$ Etc.

The term "differentiation" also includes the operation of finding differentials.

In finding differentials the easiest way is to find the derivative as usual, and then multiply the result by dx.

ILLUSTRATIVE EXAMPLE 1. Find the differential of
$$y = \frac{x+3}{x^2+3}.$$
Solution. $dy = d\left(\dfrac{x+3}{x^2+3}\right) = \dfrac{(x^2+3)d(x+3) - (x+3)d(x^2+3)}{(x^2+3)^2}$
$= \dfrac{(x^2+3)dx - (x+3)2x\,dx}{(x^2+3)^2} = \dfrac{(3 - 6x - x^2)dx}{(x^2+3)^2}.$ Ans.

ILLUSTRATIVE EXAMPLE 2. Find dy from
$$b^2 x^2 - a^2 y^2 = a^2 b^2.$$
Solution. $2 b^2 x\, dx - 2 a^2 y\, dy = 0.$
$$\therefore dy = \frac{b^2 x}{a^2 y} dx.\ Ans.$$

ILLUSTRATIVE EXAMPLE 3. Find $d\rho$ from
$$\rho^2 = a^2 \cos 2\theta.$$
Solution. $2 \rho\, d\rho = -a^2 \sin 2\theta \cdot 2\, d\theta.$
$$\therefore d\rho = -\frac{a^2 \sin 2\theta}{\rho} d\theta.\ Ans.$$

ILLUSTRATIVE EXAMPLE 4. Find $d[\arcsin(3t - 4t^3)]$.
Solution. $d[\arcsin(3t - 4t^3)] = \dfrac{d(3t - 4t^3)}{\sqrt{1 - (3t - 4t^3)^2}} = \dfrac{3\,dt}{\sqrt{1 - t^2}}.$ Ans.

PROBLEMS

Find each of the following differentials.

1. $y = x^3 - 3x.$ Ans. $dy = 3(x^2 - 1)dx.$
2. $y = \dfrac{x}{a} + \dfrac{a}{x}.$ $dy = \left(\dfrac{1}{a} - \dfrac{a}{x^2}\right)dx.$
3. $y = \sqrt{ax + b}.$ $dy = \dfrac{a\,dx}{2\sqrt{ax+b}}.$
4. $y = x\sqrt{a^2 - x^2}.$ $dy = \dfrac{(a^2 - 2x^2)dx}{\sqrt{a^2 - x^2}}.$
5. $s = ae^{bt}.$ $ds = abe^{bt}\,dt.$
6. $u = \ln cv.$ $du = \dfrac{dv}{v}.$
7. $\rho = \sin a\theta.$ $d\rho = a\cos a\theta\, d\theta.$

8. $y = \ln \sin x$. Ans. $dy = \ctn x\, dx$.
9. $\rho = \theta \cos \theta$. $d\rho = (\cos \theta - \theta \sin \theta) d\theta$.
10. $s = e^t \cos \pi t$. $ds = e^t(\cos \pi t - \pi \sin \pi t) dt$.

Find the differential of each of the following functions.

11. $y = \sqrt{\dfrac{x}{a}} - \sqrt{\dfrac{a}{x}}$. 15. $\rho = 2 \sin \dfrac{\theta}{2}$.

12. $u = \sqrt{e^v + 1}$. 16. $s = e^{-at} \sin bt$.

13. $y = \dfrac{x}{\sqrt{a^2 - x^2}}$. 17. $\rho = \sqrt{\ctn \theta}$.

14. $y = \sqrt{\dfrac{a-x}{a+x}}$. 18. $y = \ln \sqrt[3]{\dfrac{6x-5}{4-3x}}$.

19. If $x^2 + y^2 = a^2$, show that $dy = -\dfrac{x\, dx}{y}$.

Find dy in terms of x, y, and dx from each of the following equations.

20. $2x^2 + 3xy + 4y^2 = 20$. Ans. $dy = -\dfrac{(4x + 3y)dx}{3x + 8y}$.

21. $x^3 + 6xy^2 + 2y^3 = 10$. 24. $x^{\frac{2}{3}} + y^{\frac{2}{3}} = a^{\frac{2}{3}}$.

22. $x + 4\sqrt{xy} + 2y = a$. 25. $x - y = e^{x+y}$.

23. $\sqrt{x} + \sqrt{y} = \sqrt{a}$. 26. $\sin(x-y) = \cos(x+y)$.

27. The legs of a right triangle are found by measurement to be 14.5 ft. and 21.4 ft. respectively. The maximum error in each measurement is ± 0.1 ft. Find the maximum error in degrees in calculating the angle opposite the smaller side by using the formula for the tangent of that angle.

95. Differential of the arc in rectangular coördinates. Let s be the length of the arc AP measured from a fixed point A on the curve. Denote the increment of s ($=$ arc PQ) by Δs. The following proof depends on the assumption that, as Q approaches P,

$$\lim \left(\frac{\text{chord } PQ}{\text{arc } PQ} \right) = 1.$$

From the figure,

(1) $(\text{Chord } PQ)^2 = (\Delta x)^2 + (\Delta y)^2$.

Multiplying and dividing by $(\Delta s)^2$ in the left-hand member and dividing both members by $(\Delta x)^2$, we get

(2) $\left(\dfrac{\text{Chord } PQ}{\Delta s} \right)^2 \left(\dfrac{\Delta s}{\Delta x} \right)^2 = 1 + \left(\dfrac{\Delta y}{\Delta x} \right)^2$.

DIFFERENTIALS

Now let Q approach P as a limiting position; then $\Delta x \to 0$ and we have

(3) $$\left(\frac{ds}{dx}\right)^2 = 1 + \left(\frac{dy}{dx}\right)^2.$$

Multiplying both members by dx^2, we get the result

(C) $$ds^2 = dx^2 + dy^2.$$

Or, if we extract the square root in (3) and multiply both members by dx,

(D) $$ds = \left(1 + \left(\frac{dy}{dx}\right)^2\right)^{\frac{1}{2}} dx.$$

From (C), we may readily show also that

(E) $$ds = \left(1 + \left(\frac{dx}{dy}\right)^2\right)^{\frac{1}{2}} dy.$$

All these forms are useful.

From (D), since

$$1 + \left(\frac{dy}{dx}\right)^2 = 1 + \tan^2 \tau = \sec^2 \tau,$$

we obtain $ds = \sec \tau\, dx$, assuming the angle τ to be acute. Hence we may easily prove

(F) $$\frac{dx}{ds} = \cos \tau, \quad \frac{dy}{ds} = \sin \tau.$$

$$\left[\frac{dy}{ds} = \frac{dy}{dx} \cdot \frac{dx}{ds} = \tan \tau \cos \tau = \sin \tau.\right]$$

For later reference, we add the formulas, setting $y' = \frac{dy}{dx}$,

(G) $$\cos \tau = \frac{1}{(1+y'^2)^{\frac{1}{2}}}, \quad \sin \tau = \frac{y'}{(1+y'^2)^{\frac{1}{2}}}.$$

If the angle τ is obtuse ($y' < 0$), a negative sign must be placed before the denominators in (G) and before $\cos \tau$ in (F).

In the accompanying figure, $PQ = \Delta x = dx$, PT is tangent at P, and τ is acute. Angle PQT is a right angle.

Therefore $QT = \tan \tau\, dx = dy$. By Art. 91

Then $PT = \sqrt{dx^2 + dy^2} = ds$. By (C)

The figure will help in memorizing the relations above.

The assumption made at the beginning of this article is proved in Art. 99.

96. Differential of the arc in polar coördinates. From the relations

(1) $$x = \rho \cos \theta, \quad y = \rho \sin \theta$$

between the rectangular and polar coördinates of a point, we obtain, by V, XIII, XIV, of Art. 94,

(2) $\quad dx = \cos \theta \, d\rho - \rho \sin \theta \, d\theta, \quad dy = \sin \theta \, d\rho + \rho \cos \theta \, d\theta.$

Substituting in (*C*), Art. 95, reducing, and extracting the square root, we obtain the result

(*H*) $$ds = \sqrt{d\rho^2 + \rho^2 d\theta^2}.$$

This may be written

(*I*) $$ds = \left[\rho^2 + \left(\frac{d\rho}{d\theta}\right)^2\right]^{\frac{1}{2}} d\theta.$$

The figure is drawn so that the angle ψ between the radius vector OP and the tangent line PT is acute (Art. 85). Also ρ, $\Delta\theta$, and $\Delta\rho$ ($= OP' - OP$) are positive. Take ρ for the independent variable. Then $\Delta\rho = d\rho$. In the right triangle PQT, take $PQ = d\rho$. Then $QT = \tan \psi \, d\rho$.

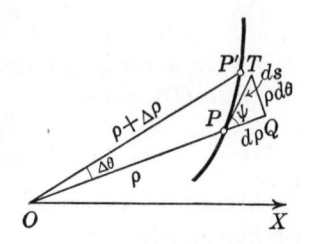

But $\qquad \tan \psi = \rho \dfrac{d\theta}{d\rho}.$ $\qquad\qquad$ By (*H*), Art. 85

Therefore $\qquad QT = \rho \dfrac{d\theta}{d\rho} d\rho = \rho \, d\theta;$ $\qquad\qquad$ by (*B*)

hence $\qquad PT = \sqrt{d\rho^2 + \rho^2 \, d\theta^2} = ds.$ $\qquad\qquad$ By (*H*)

ILLUSTRATIVE EXAMPLE 1. Find the differential of the arc of the circle $x^2 + y^2 = r^2$.

Solution. Differentiating, $\qquad \dfrac{dy}{dx} = -\dfrac{x}{y}.$

To find ds in terms of x we substitute in (*D*), giving

$$ds = \left(1 + \frac{x^2}{y^2}\right)^{\frac{1}{2}} dx = \left(\frac{y^2 + x^2}{y^2}\right)^{\frac{1}{2}} dx = \left(\frac{r^2}{y^2}\right)^{\frac{1}{2}} dx = \frac{r \, dx}{\sqrt{r^2 - x^2}}.$$

To find ds in terms of y we substitute in (*E*), giving

$$ds = \left(1 + \frac{y^2}{x^2}\right)^{\frac{1}{2}} dy = \left(\frac{x^2 + y^2}{x^2}\right)^{\frac{1}{2}} dy = \left(\frac{r^2}{x^2}\right)^{\frac{1}{2}} dy = \frac{r \, dy}{\sqrt{r^2 - y^2}}.$$

ILLUSTRATIVE EXAMPLE 2. Find the differential of the arc of the cycloid $x = a(\theta - \sin \theta)$, $y = a(1 - \cos \theta)$, in terms of θ and $d\theta$. (See Illustrative Example 2, Art. 81.)

Solution. Differentiating,

$$dx = a(1 - \cos \theta) d\theta, \quad dy = a \sin \theta \, d\theta.$$

DIFFERENTIALS

Substituting in (C),
$$ds^2 = a^2(1 - \cos\theta)^2 d\theta^2 + a^2 \sin^2\theta\, d\theta^2 = 2a^2(1 - \cos\theta)d\theta^2.$$
From (5), Art. 2, $1 - \cos\theta = 2\sin^2 \tfrac{1}{2}\theta$. Hence $ds = 2a\sin\tfrac{1}{2}\theta\, d\theta$. *Ans.*

ILLUSTRATIVE EXAMPLE 3. Find the differential of the arc of the cardioid $\rho = a(1 - \cos\theta)$ in terms of θ.

Solution. Differentiating, $\dfrac{d\rho}{d\theta} = a\sin\theta$.

Substituting in (I) gives
$$ds = [a^2(1-\cos\theta)^2 + a^2\sin^2\theta]^{\tfrac{1}{2}} d\theta = a(2 - 2\cos\theta)^{\tfrac{1}{2}} d\theta = a\left(4\sin^2\tfrac{\theta}{2}\right)^{\tfrac{1}{2}} d\theta = 2a\sin\tfrac{\theta}{2} d\theta.$$

PROBLEMS

For each of the following curves find ds in terms of x and dx.

1. $2y = x^2$.   *Ans.* $ds = \sqrt{1 + x^2}\, dx$.

2. $y^2 = 2\,px$. $ds = \sqrt{\dfrac{2x + p}{2x}}\, dx$.

3. $b^2 x^2 + a^2 y^2 = a^2 b^2$. $ds = \sqrt{\dfrac{a^4 - (a^2 - b^2)x^2}{a^2(a^2 - x^2)}}\, dx$.

4. $6\,xy = x^4 + 3$. $ds = \dfrac{(x^4 + 1)dx}{2x^2}$.

5. $y = \ln\sec x$. $ds = \sec x\, dx$.

6. $a^2 y = x^3$.

7. $ay^2 = x^3$.

8. $\sqrt{x} + \sqrt{y} = \sqrt{a}$.

9. $2y = e^x + e^{-x}$.

10. $y = \sin x$.

11. $y = \cos^2 x$.

For each of the following curves find ds in terms of y and dy.

12. $y^2 = 2\,px$.   *Ans.* $ds = \dfrac{\sqrt{y^2 + p^2}\, dy}{p}$.

13. $ay^2 = x^3$. $ds = \dfrac{\sqrt{9 y^{\tfrac{2}{3}} + 4 a^{\tfrac{2}{3}}}\, dy}{3 y^{\tfrac{1}{3}}}$.

14. $x^{\tfrac{2}{3}} + y^{\tfrac{2}{3}} = a^{\tfrac{2}{3}}$. $ds = \sqrt[3]{\dfrac{a}{y}}\, dy$.

15. $a^2 y = x^3$.

16. $y^2 - 2x - 3y = 0$.   17. $2xy^2 - y^2 - 4 = 0$.

For each of the following curves find ds, $\sin\tau$, and $\cos\tau$ in terms of t and dt.

18. $x = 2t + 3,\ y = t^2 - 2$.   20. $x = a\sin t,\ y = a\cos t$.

19. $x = 3t^2,\ y = 2t^3$.   21. $x = 4\cos t,\ y = 3\sin t$.

For each of the following curves find ds in terms of θ and $d\theta$.

22. $\rho = a \cos \theta$.   Ans. $ds = a\, d\theta$.

23. $\rho = 5 \cos \theta + 12 \sin \theta$. $ds = 13\, d\theta$.

24. $\rho = 1 - \sin \theta$. $ds = \sqrt{2 - 2 \sin \theta}\, d\theta$.

25. $\rho = 3 \sin \theta - 4 \cos \theta$.

26. $\rho = 1 + \cos \theta$.

27. $\rho = \sec^2 \dfrac{\theta}{2}$.

28. $\rho = 2 - \cos \theta$.

29. $\rho = 2 + 3 \sin \theta$.

30. $\rho = a \cos n\theta$.

31. $\rho = 4 \sin^3 \dfrac{\theta}{3}$.

32. $\rho = \dfrac{4}{1 + \cos \theta}$.

33. $\rho = \dfrac{4}{3 - \cos \theta}$.

34. $\rho = \dfrac{4}{1 - 3 \cos \theta}$.

97. Velocity as the time-rate of change of arc. In the discussion of curvilinear motion in Art. 83, the velocity, or, more correctly, the speed v was given by (*E*),

(1) $$v^2 = v_x{}^2 + v_y{}^2.$$

By (*C*) and (*D*) in Art. 83, $v_x = \dfrac{dx}{dt}, \quad v_y = \dfrac{dy}{dt}.$

Substituting in (1), using differentials and (*C*), Art. 95, the result is

(2) $$v^2 = \dfrac{dx^2 + dy^2}{dt^2} = \dfrac{ds^2}{dt^2}.$$

Extracting the square root, taking the positive sign, we have

$$v = \dfrac{ds}{dt}.$$

Hence, *in curvilinear motion the speed of the moving point is the time-rate of change of the length of arc of the path.*

This statement should be compared with the definition of velocity in rectilinear motion as the time-rate of change of distance (Art. 51).

98. Differentials as infinitesimals. In applied mathematics differentials are often treated as infinitesimals (Art. 20), that is, as variables approaching zero as a limit. Conversely, relations between infinitesimals are frequently established in which these are replaced by differentials. The "principle of replacement" involved here is very useful.

DIFFERENTIALS

If x is the independent variable, we have seen that $\Delta x = dx$, and thus Δx *may be replaced by dx in any equation.* If $\Delta x \to 0$, so will also $dx \to 0$. On the other hand, Δy and dy are not in general equal. But, when x has a fixed value and Δx ($= dx$) is an infinitesimal, so also is Δy, and, from **(B)**, Art. 91, dy as well. Furthermore it is easy to prove the relation

(1) $$\lim_{\Delta x \to 0} \frac{\Delta y}{dy} = 1.$$

Proof. Since $\lim_{\Delta x \to 0} \frac{\Delta y}{\Delta x} = f'(x),$

we may write $\frac{\Delta y}{\Delta x} = f'(x) + i$, if $\lim_{\Delta x \to 0} i = 0.$

Clearing of fractions, and using **(B)**,

$$\Delta y = dy + i\, \Delta x.$$

Dividing both members by Δy, and transposing, the result is

$$\frac{dy}{\Delta y} = 1 - i\frac{\Delta x}{\Delta y}.$$

Hence $\lim_{\Delta x \to 0} \frac{dy}{\Delta y} = 1,$ or also $\lim_{\Delta x \to 0} \frac{\Delta y}{dy} = 1.$ Q. E. D.

We now state, without proof, the

Replacement Theorem. *In problems involving only the ratios of infinitesimals which simultaneously approach zero, an infinitesimal may be replaced by a second infinitesimal so related to it that the limit of their ratio is unity.*

From the above theorem, Δy may be replaced by dy, and, in general, any increment by the corresponding differential.

In an equation which is *homogeneous* in infinitesimals the above theorem is simple in application.

ILLUSTRATIVE EXAMPLE 1. By (5), p. 3, if $x = \frac{1}{2} i$, $1 - \cos i = 2 \sin^2 \frac{1}{2} i$. Let i be an infinitesimal. Then, by **(B)**, Art. 68, $\sin i$ may be replaced by i, $\sin^2 \frac{1}{2} i$ by $\frac{1}{4} i^2$, and therefore $1 - \cos i$ by $\frac{1}{2} i^2$. Also $\tan i$ ($= \sin i / \cos i$) may be replaced by i.

ILLUSTRATIVE EXAMPLE 2. In (1), Art. 95, all quantities are infinitesimals ultimately, since $\Delta x \to 0$. The equation is homogeneous, each term being of the second degree. By the theorem, we may replace the infinitesimals as follows:

Chord PQ by arc $PQ = \Delta s$, and Δs by ds; Δy by dy; and Δx by dx.

Then (1) becomes $ds^2 = dx^2 + dy^2$, that is, **(C)**.

99. Order of infinitesimals. Differentials of higher order.

Let i and j be infinitesimals which simultaneously approach zero, and let

$$\lim \frac{j}{i} = L.$$

If L is not zero, i and j are said *to be of the same order*.
If $L = 0$, j is said *to be of higher order than* i.
If L becomes infinite, j is said *to be of lower order than* i.
Let $L = 1$. Then $j - i$ is of higher order than i.

$$\left[\lim \left(\frac{j-i}{i}\right) = \lim \left(\frac{j}{i} - 1\right) = \lim \frac{j}{i} - 1 = 0.\right]$$

The converse also is true. In this case ($L = 1$), j is said *to differ from i by an infinitesimal of higher order*.

For example, dy and Δx are of the same order if $f'(x)$ neither vanishes nor becomes infinite. Then Δy and Δx are of the same order, but $\Delta y - dy$ is of higher order than Δx. For this reason dy is called the "principal part of Δy." Obviously powers of an infinitesimal i are of higher order than i.

ILLUSTRATIVE EXAMPLE. Prove the assumption of Art. 95,

$$\lim \left(\frac{\text{chord } PQ}{\text{arc } PQ}\right) = 1.$$

Solution. In the figure we have, by geometry,

$$\text{chord } PQ < \text{arc } PQ < PT + TQ.$$

Therefore, by division,

$$1 < \frac{\text{arc } PQ}{\text{chord } PQ} < \frac{PT}{\text{chord } PQ} + \frac{TQ}{\text{chord } PQ}.$$

Now $\text{chord } PQ = \sec \phi \, \Delta x$, $PT = \sec \tau \, \Delta x$, $TQ = \Delta y - dy$,

and hence $\dfrac{PT}{\text{chord } PQ} = \dfrac{\sec \tau}{\sec \phi}$, $\dfrac{TQ}{\text{chord } PQ} = \cos \phi \, \dfrac{\Delta y - dy}{\Delta x}$.

Then $\lim\limits_{\Delta x \to 0}\left(\dfrac{PT}{\text{chord } PQ}\right) = 1$, $\lim\limits_{\Delta x \to 0}\left(\dfrac{TQ}{\text{chord } PQ}\right) = 0$. $\therefore \lim\limits_{\Delta x \to 0}\left(\dfrac{\text{arc } PQ}{\text{chord } PQ}\right) = 1$.

Differentials of higher order. Let $y = f(x)$. The equation

$$d^2y = f''(x) \, \Delta x^2 = y'' \, \Delta x^2$$

defines the *second differential* of y. If y'' neither vanishes nor becomes infinite, d^2y is of the same order as Δx^2 and therefore of higher order than dy. In a similar manner $d^3y, \cdots, d^ny$ may be defined.

PROBLEM

In triangle ABC the sides a, b, c are infinitesimals which simultaneously approach zero, and c is of higher order than b. Prove $\lim \dfrac{a}{b} = 1$.

CHAPTER X

CURVATURE. RADIUS AND CIRCLE OF CURVATURE

100. Curvature. In Art. 55 the direction of bending of a curve was discussed. The shape of a curve at a point (its flatness or sharpness) depends upon the *rate of change of direction*. This rate is called the *curvature at the point* and is denoted by K. Let us find the mathematical expression for K.

In the figure, P' is a second point on a curve near P. When the point of contact of the tangent line describes the arc $PP'(=\Delta s)$, the tangent line turns through the angle $\Delta \tau$. That is, $\Delta \tau$ is the change in the inclination of the tangent line. We now set down the following definitions.

$$\frac{\Delta \tau}{\Delta s} = \text{average curvature of the arc } PP'.$$

The *curvature at P $(= K)$ is the limiting value of the average curvature when P' approaches P as a limiting position*, that is

(A) $\qquad K = \lim\limits_{\Delta s \to 0} \frac{\Delta \tau}{\Delta s} = \frac{d\tau}{ds} = \text{curvature at } P.$

In formal terms the curvature is the *rate of change of the inclination with respect to the arc* (compare Art. 50).

Since the angle $\Delta \tau$ is measured in radians and the length of arc Δs in units of length, it follows that *the unit of curvature at a point is one radian per unit of length*.

101. Curvature of a circle

Theorem. *The curvature of a circle at any point equals the reciprocal of the radius, and is therefore the same at all points.*

Proof. In the figure the angle $\Delta \tau$ between the tangent lines at P and P' equals the central angle PCP' between the radii CP and CP'. Hence

$$\frac{\Delta \tau}{\Delta s} = \frac{\text{angle } PCP'}{\Delta s} = \frac{\frac{\Delta s}{R}}{\Delta s} = \frac{1}{R},$$

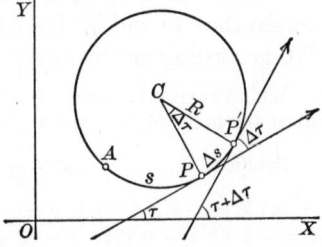

149

since the angle PCP' is *measured in radians*. That is, the average curvature of the arc PP' is equal to a constant. Letting $\Delta s \to 0$, we have the result stated in the theorem.

From the standpoint of curvature, the circle is the simplest curve, since a circle bends at a uniform rate. Obviously, the curvature of a straight line is everywhere zero.

102. Formulas for curvature; rectangular coördinates

Theorem. *When the equation of a curve is given in rectangular coördinates, then*

$$(B) \qquad K = \frac{y''}{(1+y'^2)^{\frac{3}{2}}},$$

where y' and y'' are, respectively, the first and second derivatives of y with respect to x.

Proof. Since $\tau = \arctan y'$, $\qquad \left(y' = \dfrac{dy}{dx}\right)$

differentiating, we have

(1) $\qquad \dfrac{d\tau}{dx} = \dfrac{y''}{1+y'^2}.$ $\qquad$ By **XXII**, Art. 60

But

(2) $\qquad \dfrac{ds}{dx} = (1+y'^2)^{\frac{1}{2}}.$ $\qquad$ By (3), Art. 95

Dividing (1) by (2) gives **(B)**. $\hfill$ Q.E.D.

Exercise. If y is the independent variable, show that

$$(C) \qquad K = \frac{-x''}{(1+x'^2)^{\frac{3}{2}}}$$

where x' and x'' are, respectively, the first and second derivatives of x with respect to y.

Formula **(C)** can be used as an alternative formula in cases where differentiation with respect to y is simpler. Also, **(B)** fails when y' becomes infinite, that is, when the tangent at P is vertical. Then in **(C)**

$$x' = 0 \quad \text{and} \quad K = -x''.$$

Sign of K. Choosing the positive sign in the denominator of **(B)**, we see that K and y'' have like signs. That is, K is positive or negative according as the curve is concave upward or concave downward.

Illustrative Example 1. Find the curvature of the parabola $y^2 = 4x$ (a) at the point $(1, 2)$; (b) at the vertex.

Solution. $\qquad y' = \dfrac{2}{y}, \quad y'' = \dfrac{d}{dx}\left(\dfrac{2}{y}\right) = -\dfrac{2y'}{y^2}.$

(a) When $x = 1$ and $y = 2$, then $y' = 1$, $y'' = -\frac{1}{2}$. Substituting in **(B)**, $K = -\frac{1}{8}\sqrt{2} = -0.177$. Hence at $(1, 2)$ the curve is concave downward and the inclination of the tangent is changing at the rate 0.177 radian per unit arc. Since 0.177 radian $= 10°\ 7'$, the angle between the tangent lines at $P(1, 2)$ and at a point Q such that arc $PQ = 0.1$ unit is approximately $1°$.

(b) At the vertex $(0, 0)$, y' becomes infinite. Hence use (C).

$$x' = \frac{1}{2} y, \quad x'' = \frac{1}{2} \frac{dy}{dy} = \frac{1}{2}. \quad K = -\frac{1}{2}. \quad Ans.$$

ILLUSTRATIVE EXAMPLE 2. Find K for the cycloid (see Art. 81)

$$x = a(\theta - \sin \theta), \quad y = a(1 - \cos \theta).$$

Solution. In Illustrative Example 2, Art. 81, we found

$$y' = \frac{\sin \theta}{1 - \cos \theta}.$$

Hence
$$1 + y'^2 = \frac{2}{1 - \cos \theta}.$$

Also, in the Illustrative Example, Art. 82, it was shown that

$$y'' = -\frac{1}{a(1 - \cos \theta)^2}.$$

Substituting in (B),

$$K = -\frac{1}{2\,a\sqrt{2 - 2 \cos \theta}} = -\frac{1}{4\,a \sin \frac{1}{2}\theta}. \quad Ans.$$

103. Special formula for parametric equations. From equation (A), Art. 81, we have, by differentiation,

(1) $$\frac{dy'}{dt} = \frac{\dfrac{dx}{dt}\dfrac{d^2y}{dt^2} - \dfrac{dy}{dt}\dfrac{d^2x}{dt^2}}{\left(\dfrac{dx}{dt}\right)^2}.$$

Whence, using (B), Art. 82, and substituting in (B), Art. 102, and reducing, we obtain

(D) $$K = \frac{x'y'' - y'x''}{(x'^2 + y'^2)^{\frac{3}{2}}},$$

where the accents indicate derivatives with respect to t; that is,

$$x' = \frac{dx}{dt}, \quad x'' = \frac{d^2x}{dt^2}, \quad y' = \frac{dy}{dt}, \quad y'' = \frac{d^2y}{dt^2}.$$

Formula (D) is convenient, but it is often better to proceed as in Illustrative Example 2, Art. 102, finding y' as in Art. 81, y'' as in Art. 82, and substituting directly in (B).

104. Formula for curvature; polar coördinates

Theorem. *When the equation of a curve is given in polar coördinates,*

(E) $$K = \frac{\rho^2 + 2\,\rho'^2 - \rho\rho''}{(\rho^2 + \rho'^2)^{\frac{3}{2}}},$$

where ρ' and ρ'' are, respectively, the first and second derivatives *of* ρ with respect to θ.

Proof. By **(I)**, Art. 85, $\tau = \theta + \psi$.

Hence

(1) $$\frac{d\tau}{d\theta} = 1 + \frac{d\psi}{d\theta}.$$

Also, by **(H)**, Art. 85, $\psi = \arctan \dfrac{\rho}{\rho'}.$

Hence
$$\frac{d\psi}{d\theta} = \frac{\rho'^2 - \rho\rho''}{\rho'^2 + \rho^2}.$$

Then, by (1),

(2) $$\frac{d\tau}{d\theta} = \frac{\rho^2 + 2\rho'^2 - \rho\rho''}{\rho^2 + \rho'^2}.$$

From **(I)**, Art. 96,

(3) $$\frac{ds}{d\theta} = (\rho^2 + \rho'^2)^{\frac{1}{2}}.$$

Dividing (2) by (3) gives **(E)**.   Q.E.D.

ILLUSTRATIVE EXAMPLE. Find the curvature of the logarithmic spiral $\rho = e^{a\theta}$ at any point.

Solution. $\dfrac{d\rho}{d\theta} = \rho' = ae^{a\theta} = a\rho;\quad \dfrac{d^2\rho}{d\theta^2} = \rho'' = a^2 e^{a\theta} = a^2\rho.$

Substituting in **(E)**, $\quad K = \dfrac{1}{\rho\sqrt{1+a^2}}.\quad$ *Ans.*

105. Radius of curvature. The radius of curvature R at a point on a curve equals the reciprocal of the curvature at that point. Hence, from **(B)**,

(F) $$R = \frac{1}{K} = \frac{(1+y'^2)^{\frac{3}{2}}}{y''}.$$

ILLUSTRATIVE EXAMPLE. Find the radius of curvature at any point of the catenary $y = \dfrac{a}{2}\left(e^{\frac{x}{a}} + e^{-\frac{x}{a}}\right)$ (figure in Chapter XXVI).

Solution. $y' = \dfrac{1}{2}\left(e^{\frac{x}{a}} - e^{-\frac{x}{a}}\right);\quad y'' = \dfrac{1}{2a}\left(e^{\frac{x}{a}} + e^{-\frac{x}{a}}\right) = \dfrac{y}{a^2}.$

$1 + y'^2 = 1 + \dfrac{1}{4}\left(e^{\frac{x}{a}} - e^{-\frac{x}{a}}\right)^2 = \dfrac{1}{4}\left(e^{\frac{x}{a}} + e^{-\frac{x}{a}}\right)^2 = \dfrac{y^2}{a^2}.\quad \therefore R = \dfrac{y^2}{a}.\quad$ *Ans.*

106. Railroad or transition curves. In laying out the curves on a railroad it will not do, on account of the high speed of trains, to pass abruptly from a straight stretch of track to a circular curve. In order to make the change of curvature gradual, engineers make use of transition curves to connect the straight part of a track with a

RADIUS AND CIRCLE OF CURVATURE

circular track. This curve should have zero curvature at its point of junction with the straight track and the curvature of the circular track where it joins the latter. Arcs of cubical parabolas are generally employed as transition curves.

ILLUSTRATIVE EXAMPLE. The transition curve on a railway track has the shape of an arc of the cubical parabola $y = \frac{1}{3} x^3$. At what rate is a car on this track changing its direction (1 mi. = unit of length) when it is passing through (a) the point (3, 9)? (b) the point $(2, \frac{8}{3})$? (c) the point $(1, \frac{1}{3})$?

Solution. $\quad \dfrac{dy}{dx} = x^2, \quad \dfrac{d^2y}{dx^2} = 2\,x.$

Substituting in (B), $\quad K = \dfrac{2\,x}{(1+x^4)^{\frac{3}{2}}}.$

(a) At (3, 9), $K = \dfrac{6}{(82)^{\frac{3}{2}}}$ radian per mile $= 28'$ per mile. *Ans.*

(b) At $(2, \frac{8}{3})$, $K = \dfrac{4}{(17)^{\frac{3}{2}}}$ radian per mile $= 3°\,16'$ per mile. *Ans.*

(c) At $(1, \frac{1}{3})$, $K = \dfrac{2}{(2)^{\frac{3}{2}}} = \dfrac{1}{\sqrt{2}}$ radian per mile $= 40°\,30'$ per mile. *Ans.*

107. Circle of curvature. Consider any point P on the curve C. The tangent line drawn to the curve at P has the same slope as the curve itself at P (Art. 42). In an analogous manner we may construct for each point of the curve a tangent circle whose curvature is the same as the curvature of the curve itself at that point. To do this, proceed as follows. Draw the normal to the curve at P on the concave side of the curve. Lay off on this normal the distance $Pc =$ radius of curvature ($= R$) at P. With c as a center draw the circle passing through P. The curvature of this circle is then

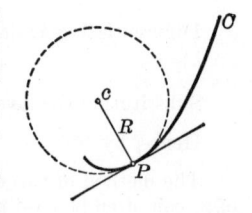

$$K = \frac{1}{R},$$

which also equals the curvature of the curve itself at P. The circle so constructed is called the *circle of curvature* for the point P on the curve.

In general, the circle of curvature of a curve at a point will cross the curve at that point. This is illustrated in the above figure. (Compare with the tangent line at a point of inflection (Art. 57).)

Just as the tangent at P shows the direction of the curve at P, so the circle of curvature at P aids us very materially in forming a geometric concept of the curvature of the curve at P, the rate of change of direction of the curve and of the circle being the same at P.

DIFFERENTIAL CALCULUS

In a subsequent section (Art. 114) the circle of curvature will be defined as the limiting position of a secant circle, a definition analogous to that of the tangent given in Art. 28.

ILLUSTRATIVE EXAMPLE. Find the radius of curvature at the point (3, 4) on the equilateral hyperbola $xy = 12$, and draw the corresponding circle of curvature.

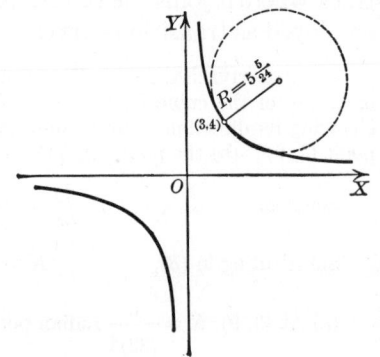

Solution. $\dfrac{dy}{dx} = -\dfrac{y}{x}$, $\dfrac{d^2y}{dx^2} = \dfrac{2\,y}{x^2}$.

For (3, 4), $\dfrac{dy}{dx} = -\dfrac{4}{3}$, $\dfrac{d^2y}{dx^2} = \dfrac{8}{9}$.

$\therefore R = \dfrac{[1 + \tfrac{16}{9}]^{\tfrac{3}{2}}}{\tfrac{8}{9}} = \dfrac{125}{24} = 5\tfrac{5}{24}.$

The circle of curvature crosses the curve at two points.

ILLUSTRATIVE EXAMPLE 2. Find R at (2, 1) for the hyperbola
$$x^2 + 4\,xy - 2\,y^2 = 10.$$

Solution. Differentiating, regarding y as an implicit function of x, we get
$$x + 2\,y + 2\,xy' - 2\,yy' = 0.$$
Differentiating again, regarding y and y' as implicit functions of x, we get
$$1 + 4\,y' - 2\,y'^2 + 2(x - y)y'' = 0.$$
Substituting the given values $x = 2$, $y = 1$, we find $y' = -2$, $y'' = \tfrac{15}{2}$.

Hence, by (F), $\qquad R = \tfrac{2}{3}\sqrt{5}$. Ans.

The method of this example (namely, regarding y and y' as implicit functions of x) can often be used to advantage when only the numerical values of y' and y'' are required, and not general expressions for them in terms of x and y.

PROBLEMS

Find the radius of curvature for each of the following curves at the point indicated. Draw the curve and the corresponding circle of curvature.

1. $2\,y = x^2$; (0, 0).   Ans. $R = 1$.
2. $6\,y = x^3$; (2, $\tfrac{4}{3}$). $R = \tfrac{5}{2}\sqrt{5}$.
3. $y^2 = x^3$; (1, 1). $R = \tfrac{13}{6}\sqrt{13}$.
4. $y = \sin x$; ($\tfrac{1}{2}\,\pi$, 1). $R = 1$.
5. $y = e^x$; (0, 1). $R = 2\sqrt{2}$.
6. $x^2 - 4\,y^2 = 9$; (5, 2).   8. $y = 2 \sin 2\,x$; ($\tfrac{1}{4}\,\pi$, 2).
7. $y^2 = x^3 + 8$; (1, 3).   9. $y = \tan x$; ($\tfrac{1}{4}\,\pi$, 1).

RADIUS AND CIRCLE OF CURVATURE

Calculate the radius of curvature at any point (x_1, y_1) on each of the following curves.

10. $y = x^3$. $\qquad$ Ans. $R = \dfrac{(1 + 9 x_1^4)^{\frac{3}{2}}}{6 x_1}$.

11. $y^2 = 2 px$.

12. $b^2 x^2 - a^2 y^2 = a^2 b^2$. $\qquad R = \dfrac{(b^4 x_1^2 + a^4 y_1^2)^{\frac{3}{2}}}{a^4 b^4}$.

13. $b^2 x^2 + a^2 y^2 = a^2 b^2$.

14. $x^{\frac{1}{2}} + y^{\frac{1}{2}} = a^{\frac{1}{2}}$. $\qquad R = \dfrac{2(x_1 + y_1)^{\frac{3}{2}}}{a^{\frac{1}{2}}}$.

15. $x^{\frac{2}{3}} + y^{\frac{2}{3}} = a^{\frac{2}{3}}$. $\qquad R = 3(a x_1 y_1)^{\frac{1}{3}}$.

16. $x = r \text{ arc vers } \dfrac{y}{r} - \sqrt{2 ry - y^2}$. $\qquad R = 2\sqrt{2 r y_1}$.

17. $y = \ln \sec x$. $\qquad R = \sec x_1$.

18. If the point of contact of the tangent line at $(2, 4)$ to the parabola $y^2 = 8 x$ moves along the curve a distance $\Delta s = 0.1$, through what angle, approximately, will the tangent line turn? (Use differentials.) *3 minutes*

19. The inclination of the curve $27 y = x^3$ at the point $A(3, 1)$ is $45°$. Use differentials to find approximately the inclination of the curve at the point B on the curve such that the distance along the curve from A to B is $\Delta s = 0.2$ units.

Calculate the radius of curvature at any point (ρ_1, θ_1) on each of the following curves.

20. The circle $\rho = a \sin \theta$. $\qquad$ Ans. $R = \frac{1}{2} a$.

21. The spiral of Archimedes $\rho = a\theta$. (Fig., p. 534) $\qquad R = \dfrac{(\rho_1^2 + a^2)^{\frac{3}{2}}}{\rho_1^2 + 2 a^2}$.

22. The cardioid $\rho = a(1 - \cos \theta)$. (Fig., p. 533) $\qquad R = \frac{2}{3} \sqrt{2 a \rho_1}$.

23. The lemniscate $\rho^2 = a^2 \cos 2\theta$. (Fig., p. 532) $\qquad R = \dfrac{a^2}{3 \rho_1}$.

24. The parabola $\rho = a \sec^2 \frac{1}{2} \theta$. (Fig., p. 537) $\qquad R = 2 a \sec^3 \frac{1}{2} \theta_1$.

25. The curve $\rho = a \sin^3 \frac{1}{3} \theta$. $\qquad R = \frac{3}{4} a \sin^2 \frac{1}{3} \theta_1$.

26. The trisectrix $\rho = 2 a \cos \theta - a$. $\qquad R = \dfrac{a(5 - 4 \cos \theta_1)^{\frac{3}{2}}}{9 - 6 \cos \theta_1}$.

27. The equilateral hyperbola $\rho^2 \cos 2\theta = a^2$. $\qquad R = \dfrac{\rho_1^3}{a^2}$.

28. The conic $\rho = \dfrac{a(1 - e^2)}{1 - e \cos \theta}$. $\qquad R = \dfrac{a(1 - e^2)(1 - 2 e \cos \theta_1 + e^2)^{\frac{3}{2}}}{(1 - e \cos \theta_1)^3}$.

Find the radius of curvature for each of the following curves at the point indicated. Draw the curve and the corresponding circle of curvature.

29. $x = 2t$, $y = t^2 - 1$; $t = 1$.   Ans. $R = 4\sqrt{2}$.

30. $x = 3t^2$, $y = 3t - t^3$; $t = 1$. $R = 6$.

31. $x = 2e^t$, $y = 2e^{-t}$; $t = 0$. $R = 2\sqrt{2}$.

32. $x = a \cos t$, $y = a \sin t$; $t = t_1$. $R = a$.

33. $x = 2t$, $y = \dfrac{4}{t}$; $t = 1$.   36. $x = 2 \sin t$, $y = \cos 2t$; $t = \tfrac{1}{6}\pi$.

34. $x = t^2 + 1$, $y = t^3 - 1$; $t = 1$.   37. $x = \tan t$, $y = \operatorname{ctn} t$; $t = \tfrac{1}{4}\pi$.

35. $x = 4 \cos t$, $y = 2 \sin t$; $y = 1$.   38. $x = t - \sin t$, $y = 1 - \cos t$; $t = \pi$.

39. Find the radius of curvature at any point $(t = t_1)$ on the hypocycloid $x = a \cos^3 t$, $y = a \sin^3 t$.   Ans. $R = 3 a \sin t_1 \cos t_1$.

40. Find the radius of curvature at any point $(t = t_1)$ on the involute of the circle
$$x = a(\cos t + t \sin t),$$
$$y = a(\sin t - t \cos t).$$
     Ans. $R = at_1$.

41. Find the point on the curve $y = e^x$ where the curvature is a maximum.   Ans. $x = -0.347$.

42. Find the points on the curve $3y = x^3 - 2x$ where the curvature is a maximum.   Ans. $x = \pm 0.931$.

43. Show that the radius of curvature becomes infinite at a point of inflection.

44. Given the curve $y = 3x - x^3$.

(a) Find the radius of curvature at the maximum point of the curve and draw the corresponding circle of curvature.

(b) Prove that the maximum point of the curve is not the point of maximum curvature.

(c) Find to the nearest hundredth of a unit the abscissa of the point of maximum curvature.   Ans. $x = 1.01$.

45. Find the radius of curvature at each maximum and minimum point on the curve $y = x^4 - 2x^2$. Draw the curve and the circles of curvature. Find the points on the curve where the radius of curvature is a minimum.

46. Show that at a point of minimum radius of curvature on the curve $y = f(x)$ we have
$$3\left(\frac{dy}{dx}\right)\left(\frac{d^2y}{dx^2}\right)^2 = \frac{d^3y}{dx^3}\left[1 + \left(\frac{dy}{dx}\right)^2\right].$$

47. Show that the curvature of the cubical parabola $3a^2y = x^3$ increases from zero to a maximum value when x increases from zero to $\tfrac{1}{5}a\sqrt[4]{125}$. Find the minimum value of the radius of curvature.
     Ans. $0.983\,a$.

RADIUS AND CIRCLE OF CURVATURE

108. Center of curvature. The tangent line at $P(x, y)$ has the property that x, y, and y' have the same values at P for the tangent line and the curve. The circle of curvature at P has a similar property; namely, x, y, y', and y'' have the same values at P for the circle of curvature and the curve.

DEFINITION. *The center of curvature (α, β) for a point $P(x, y)$ on a curve is the center of the circle of curvature.*

Theorem. *The coördinates (α, β) of the center of curvature for $P(x, y)$ are*

(G) $$\alpha = x - \frac{y'(1+y'^2)}{y''}, \quad \beta = y + \frac{(1+y'^2)}{y''}.$$

Proof. The equation of the circle of curvature is

(1) $$(x-\alpha)^2 + (y-\beta)^2 = R^2,$$

where R is given by **(F)**. Differentiating (1),

(2) $$y' = -\frac{x-\alpha}{y-\beta}, \quad y'' = -\frac{R^2}{(y-\beta)^3}.$$

From the second of these equations, after substituting the value of R from **(F)**, we obtain

(3) $$(y-\beta)^3 = -\frac{(1+y'^2)^3}{y''^3}. \quad \therefore y - \beta = -\frac{1+y'^2}{y''}.$$

From the first of equations (2), we get, using (3),

(4) $$x - \alpha = -y'(y-\beta) = \frac{y'(1+y'^2)}{y''}.$$

Solving in (3) for β, in (4) for α, we have **(G)**. Q.E.D.

EXERCISE 1. Work out **(G)** directly from the accompanying figure, using **(G)**, Art. 95. ($\alpha = x - R \sin \tau$, $\beta = y + R \cos \tau$, etc.)

EXERCISE 2. If x' and x'' are, respectively, the first and second derivatives of x with respect to y, derive **(G)** in the form

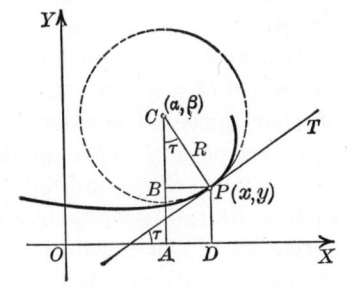

(H) $$\alpha = x + \frac{1+x'^2}{x''}, \quad \beta = y - \frac{x'(1+x'^2)}{x''}.$$

Formulas **(H)** may be used when y' becomes infinite, or if differentiation with respect to y is simpler.

158   DIFFERENTIAL CALCULUS

ILLUSTRATIVE EXAMPLE. Find the coördinates of the center of curvature of the parabola $y^2 = 4\,px$ corresponding (a) to any point on the curve; (b) to the vertex.

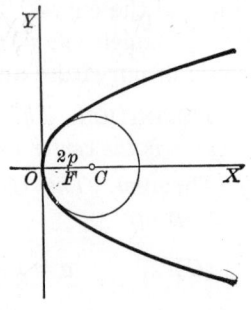

Solution. Use (H). Then $x' = \dfrac{y}{2\,p}$, $x'' = \dfrac{1}{2\,p}$.

Hence $\alpha = x + \dfrac{y^2 + 4\,p^2}{2\,p} = 3\,x + 2\,p$,

$$\beta = y - \dfrac{y(y^2 + 4\,p^2)}{4\,p^2} = -\dfrac{y^3}{4\,p^2}.$$

Therefore (a) $\left(3\,x + 2\,p,\ -\dfrac{y^3}{4\,p^2}\right)$ is the center of curvature corresponding to any point on the curve.

(b) $(2\,p, 0)$ is the center of curvature corresponding to the vertex $(0, 0)$.

From Art. 57 we know that at a point of inflection (as Q in the next figure)

$$\frac{d^2y}{dx^2} = 0.$$

Therefore, by (B), Art. 102, the curvature $K = 0$; and from (F), Art. 105, and (G), Art. 108, we see that in general α, β, and R increase without limit as the second derivative approaches zero, unless the tangent line is vertical. That is, if we suppose P with its tangent to move along the curve to P', at the point of inflection Q the curvature is zero, the rotation of the tangent is momentarily arrested, and as the direction of rotation changes, the center of curvature moves out indefinitely and the radius of curvature becomes infinite.

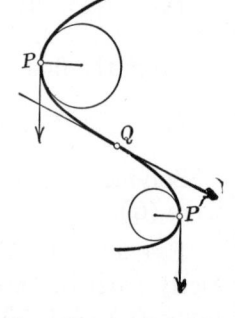

109. Evolutes. The locus of the centers of curvature of a given curve is called the *evolute* of that curve. Consider the circle of curvature at a point P on a curve. If P moves along the curve, we may suppose the corresponding circle of curvature to roll along the curve with it, its radius varying so as to be always equal to the radius of curvature of the curve at the point P. The curve CC_7 described by the center of the circle is the evolute of PP_7.

Formulas (G) and (H), Art. 108, give the coördinates of any point (α, β) on the evolute expressed in terms of the coördinates of the corresponding point (x, y) of the given curve. But y is a function of x; therefore these formulas give us at once *the parametric equations of the evolute in terms of the parameter* x.

RADIUS AND CIRCLE OF CURVATURE

To find the ordinary rectangular equation of the evolute we eliminate x and y between the two expressions and the equation of the given curve. No general process of elimination can be given that will apply in all cases, the method to be adopted depending on the form of the given equation. In a large number of cases, however, the student can find the rectangular equation of the evolute by taking the following steps.

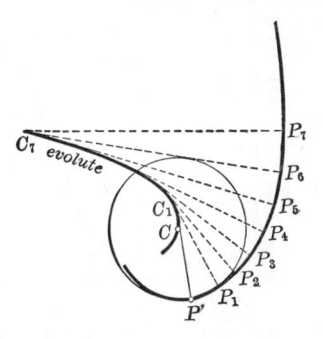

General directions for finding the equation of the evolute in rectangular coördinates.

FIRST STEP. *Find α and β from (G) or (H), Art. 108.*

SECOND STEP. *Solve the two resulting equations for x and y in terms of α and β.*

THIRD STEP. *Substitute these values of x and y in the given equation and reduce. This gives a relation between the variables α and β which is the equation of the evolute.*

ILLUSTRATIVE EXAMPLE 1. Find the equation of the evolute of the parabola $y^2 = 4\,px$.

Solution. $\quad \dfrac{dy}{dx} = \dfrac{2\,p}{y}, \quad \dfrac{d^2y}{dx^2} = -\dfrac{4\,p^2}{y^3}.$

First Step. $\quad \alpha = 3\,x + 2\,p, \quad \beta = -\dfrac{y^3}{4\,p^2}.$

Second Step. $\quad x = \dfrac{\alpha - 2\,p}{3}, \quad y = -(4\,p^2\beta)^{\frac{1}{3}}.$

Third Step. $(4\,p^2\beta)^{\frac{2}{3}} = 4\,p\left(\dfrac{\alpha - 2\,p}{3}\right),$

or $\quad p\beta^2 = \tfrac{4}{27}(\alpha - 2\,p)^3.$

Remembering that α denotes the abscissa and β the ordinate of a rectangular system of coördinates, we see that the evolute of the parabola AOB is the semicubical parabola $DC'E$, the centers of curvature for O. P, P_1, P_2 being at C', C, C_1, C_2 respectively.

160 DIFFERENTIAL CALCULUS

ILLUSTRATIVE EXAMPLE 2. Find the equation of the evolute of the ellipse $b^2x^2 + a^2y^2 = a^2b^2$.

Solution. $\dfrac{dy}{dx} = -\dfrac{b^2x}{a^2y},\ \dfrac{d^2y}{dx^2} = -\dfrac{b^4}{a^2y^3}$.

First Step. $\alpha = \dfrac{(a^2 - b^2)x^3}{a^4},$

$\beta = -\dfrac{(a^2 - b^2)y^3}{b^4}.$

Second Step. $x = \left(\dfrac{a^4\alpha}{a^2 - b^2}\right)^{\frac{1}{3}},$

$y = -\left(\dfrac{b^4\beta}{a^2 - b^2}\right)^{\frac{1}{3}}.$

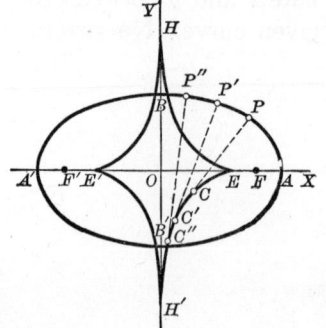

Third Step. $(a\alpha)^{\frac{2}{3}} + (b\beta)^{\frac{2}{3}} = (a^2 - b^2)^{\frac{2}{3}}$, the equation of the evolute $EHE'H'$ of the ellipse $ABA'B'$. E, E', H', H are the centers of curvature corresponding to the points A, A', B, B', on the curve, and C, C', C'' correspond to the points P, P', P''.

ILLUSTRATIVE EXAMPLE 3. The parametric equations of a curve are

(1) $\qquad x = \dfrac{t^2 + 1}{4},\quad y = \dfrac{t^3}{6}.$

Find the equation of the evolute in parametric form, plot the curve and the evolute, find the radius of curvature at the point where $t = 1$, and draw the corresponding circle of curvature.

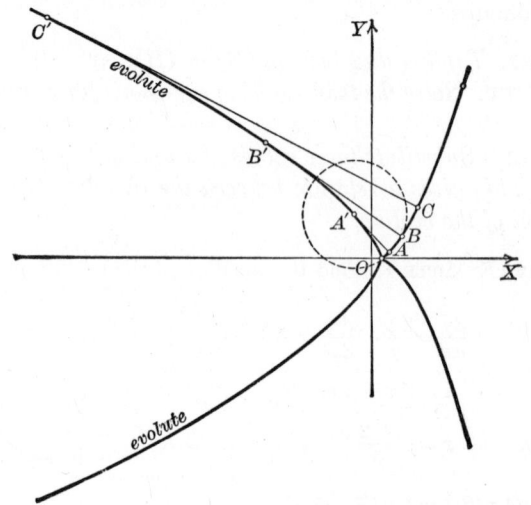

Solution. $\qquad \dfrac{dx}{dt} = \dfrac{t}{2},\ \dfrac{dy}{dt} = \dfrac{1}{2}t^2.\ \therefore y' = t.$ By (A), Art. 81

$\dfrac{dy'}{dt} = 1.\ \therefore y'' = \dfrac{2}{t}.$ By (B), Art. 82

Substituting in (G) and reducing gives

(2) $\qquad \alpha = \dfrac{1 - t^2 - 2\,t^4}{4},\quad \beta = \dfrac{4\,t^3 + 3\,t}{6},$

RADIUS AND CIRCLE OF CURVATURE

the parametric equations of the evolute. Assuming values of the parameter t, we calculate x, y from (1) and α, β from (2), and tabulate the results.

Now plot the curve and its evolute. The point $(\tfrac{1}{4}, 0)$ is common to the given curve and its evolute. The given curve (semicubical parabola) lies entirely to the right and the evolute entirely to the left of $x = \tfrac{1}{4}$.

The circle of curvature at $A(\tfrac{1}{2}, \tfrac{1}{6})$, where $t = 1$, will have its center at $A'(-\tfrac{1}{2}, \tfrac{7}{6})$ on the evolute and radius $= AA'$. To verify our work find the radius of curvature at A. From (F), Art. 105, we get

t	x	y	α	β
-3	$\tfrac{5}{2}$	$-\tfrac{9}{2}$		
-2	$\tfrac{5}{4}$	$-\tfrac{4}{3}$	$-\tfrac{35}{4}$	$-\tfrac{19}{3}$
$-\tfrac{3}{2}$	$\tfrac{13}{16}$	$-\tfrac{9}{16}$	$-\tfrac{91}{32}$	-3
-1	$\tfrac{1}{2}$	$-\tfrac{1}{6}$	$-\tfrac{1}{2}$	$-\tfrac{7}{6}$
0	$\tfrac{1}{4}$	0	$\tfrac{1}{4}$	0
1	$\tfrac{1}{2}$	$\tfrac{1}{6}$	$-\tfrac{1}{2}$	$\tfrac{7}{6}$
$\tfrac{3}{2}$	$\tfrac{13}{16}$	$\tfrac{9}{16}$	$-\tfrac{91}{32}$	3
2	$\tfrac{5}{4}$	$\tfrac{4}{3}$	$-\tfrac{35}{4}$	$\tfrac{19}{3}$
3	$\tfrac{5}{2}$	$\tfrac{9}{2}$		

$$R = \frac{t(1+t^2)^{\frac{3}{2}}}{2} = \sqrt{2} \text{ when } t = 1.$$

This should equal the distance

$$AA' = \sqrt{(\tfrac{1}{2}+\tfrac{1}{2})^2 + (\tfrac{1}{6}-\tfrac{7}{6})^2} = \sqrt{2}. \qquad \text{By (1), Art. 3}$$

ILLUSTRATIVE EXAMPLE 4. Find the parametric equations of the evolute of the cycloid

(3) $\qquad \begin{cases} x = a(t - \sin t), \\ y = a(1 - \cos t). \end{cases}$

Solution. As in the Illustrative Example of Art. 82, we get

$$\frac{dy}{dx} = \frac{\sin t}{1 - \cos t}, \quad \frac{d^2y}{dx^2} = -\frac{1}{a(1 - \cos t)^2}.$$

Substituting these results in formulas (G), Art. 108, we get

(4) $\qquad \begin{cases} \alpha = a(t + \sin t), \\ \beta = -a(1 - \cos t). \end{cases}$ Ans.

NOTE. If we eliminate t between equations (4), there results the rectangular equation of the evolute $OO'Q^v$ referred to the axes $O'\alpha$ and $O'\beta$. The coördinates of O with respect to these axes are $(-\pi a, -2a)$. Let us transform equations (4) to the new set of axes OX and OY. Then

$\alpha = x - \pi a, \quad \beta = y - 2a.$
Also, let $t = t' - \pi.$

Substituting in (4) and reducing, the equations of the evolute become

(5) $\qquad \begin{cases} x = a(t' - \sin t'), \\ y = a(1 - \cos t'). \end{cases}$

Since (5) and (3) are identical in form, we have:

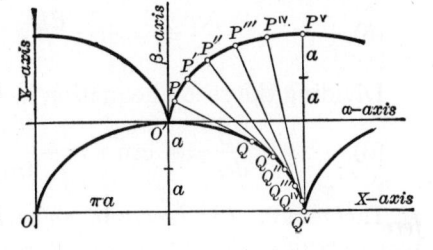

The evolute of a cycloid is itself a cycloid whose generating circle equals that of the given cycloid.

162 DIFFERENTIAL CALCULUS

110. Properties of the evolute. The evolute has two interesting properties.

Theorem 1. *The normal at $P(x, y)$ to the given curve is tangent to the evolute at the center of curvature $C(\alpha, \beta)$ for P.* (See figures in the preceding article.)

Proof. From the figure,

(1) $\qquad \alpha = x - R \sin \tau,$
$\qquad \qquad \beta = y + R \cos \tau.$

The line PC lies along the normal at P, and the

(2) Slope of $PC = \dfrac{y - \beta}{x - \alpha} = -\dfrac{1}{\tan \tau}$
$\qquad$ = slope of normal at P.

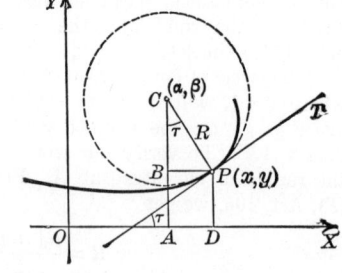

We show now that the slope of the evolute equals the slope of PC. Note that
$$\text{Slope of evolute} = \frac{d\beta}{d\alpha},$$
since α and β are the rectangular coördinates of any point on the evolute.

Let us choose as independent variable the length of arc on the given curve; then $x, y, R, \tau, \alpha, \beta$ are functions of s. Differentiating (1) with respect to s gives

(3) $\qquad \dfrac{d\alpha}{ds} = \dfrac{dx}{ds} - R \cos \tau \dfrac{d\tau}{ds} - \sin \tau \dfrac{dR}{ds},$

(4) $\qquad \dfrac{d\beta}{ds} = \dfrac{dy}{ds} - R \sin \tau \dfrac{d\tau}{ds} + \cos \tau \dfrac{dR}{ds}.$

But $\dfrac{dx}{ds} = \cos \tau$, $\dfrac{dy}{ds} = \sin \tau$, from Art. 95; and $\dfrac{d\tau}{ds} = \dfrac{1}{R}$.

Substituting in (3) and (4), and reducing, we obtain

(5) $\qquad \dfrac{d\alpha}{ds} = -\sin \tau \dfrac{dR}{ds}, \quad \dfrac{d\beta}{ds} = \cos \tau \dfrac{dR}{ds}.$

Dividing the second equation in (5) by the first gives

(6) $\qquad \dfrac{d\beta}{d\alpha} = -\operatorname{ctn} \tau = -\dfrac{1}{\tan \tau} = \text{slope of } PC. \qquad$ **Q.E.D.**

Theorem 2. *The length of an arc of the evolute is equal to the difference between the radii of curvature of the given curve which are tangent to this arc at its extremities, provided that along the arc of the given curve R increases or decreases.*

Proof. Squaring equations (5) and adding, we get

(7) $$\left(\frac{d\alpha}{ds}\right)^2+\left(\frac{d\beta}{ds}\right)^2=\left(\frac{dR}{ds}\right)^2.$$

But if $s' =$ length of arc of the evolute,
$$ds'^2 = d\alpha^2 + d\beta^2,$$
by **(C)**, Art. 95, if $s = s'$, $x = \alpha$, $y = \beta$. Hence (7) asserts that

(8) $$\left(\frac{ds'}{ds}\right)^2=\left(\frac{dR}{ds}\right)^2, \quad \text{or} \quad \frac{ds'}{ds} = \pm \frac{dR}{ds}.$$

Confining ourselves to an arc on the given curve for which the right-hand member does not change sign, we may write

(9) $$\frac{ds'}{dR} = +1 \quad \text{or} \quad \frac{ds'}{dR} = -1.$$

That is, *the rate of change of the arc of the evolute with respect to R is $+1$ or -1.* Hence, by Art. 50, corresponding increments of s' and R are numerically equal. That is,

(10) $$s' - s'_0 = \pm (R - R_0),$$

or (first figure, p. 159) Arc $CC_1 = \pm (P_1 C_1 - PC)$.

Thus the theorem is proved.

In Illustrative Example 4, Art.109, we observe that at O', $R = 0$; at P^v, $R = 4a$. Hence arc $O'QQ^v = 4a$.

The length of one arch of the cycloid (as $OO'Q^v$) is eight times the length of the radius of the generating circle.

111. Involutes and their mechanical construction. Let a flexible ruler be bent in the form of the curve $C_1 C_9$, the evolute of the curve $P_1 P_9$, and suppose a string of length R_9, with one end fastened at C_9, to be stretched along the ruler (or curve). It is clear from the results of the last article that when the string is unwound and kept taut, the free end will describe the curve $P_1 P_9$. Hence the name *evolute*.

The curve $P_1 P_9$ is said to be an *involute* of $C_1 C_9$. Obviously any point on the string will describe an involute, so that a given curve has an infinite number of involutes but only **one evolute**.

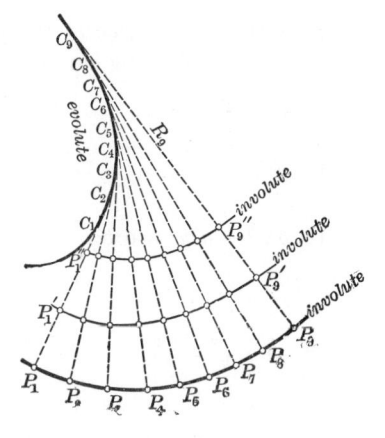

164   DIFFERENTIAL CALCULUS

The involutes P_1P_9, $P_1'P_9'$, $P_1''P_9''$ are called *parallel curves* since the distance between any two of them measured along their common normals is constant.

The student should observe how the parabola and the ellipse on pages 159, 160 may be constructed in this way from their evolutes.

PROBLEMS

Find the radius and center of curvature for each of the following curves at the given point. Check your results by proving (a) that the center of curvature lies on the normal to the curve at the given point, and (b) that the distance from the given point to the center of curvature is equal to the radius of curvature.

1. $2\,py = x^2$;   (0, 0).   *Ans.* (0, p).
2. $x^2 + 4\,y^2 = 25$;   (3, 2).   ($\frac{81}{100}$, $-\frac{96}{25}$).
3. $x^3 - y^3 = 19$;   (3, 2).   ($\frac{519}{76}$, $\frac{17}{57}$).
4. $xy = 6$;   (2, 3).   ($\frac{63}{12}$, $\frac{31}{6}$).
5. $y = e^x$;   (0, 1).   (-2, 3).
6. $y = \cos x$;   (0, 1).   (0, 0).
7. $y = \ln x$;   (1, 0).   (3, -2).
8. $y = 2 \sin 2\,x$;   ($\tfrac{1}{4}\pi$, 2).   ($\tfrac{1}{4}\pi$, $\tfrac{15}{8}$).
9. $(x + 6)^3 + xy^2 = 0$;   (-3, 3).   (-13, 8).
10. $2\,y = x^2 - 4$;   (0, -2).
11. $xy = x^2 + 2$;   (2, 3).
12. $y = \sin \pi x$;   ($\tfrac{1}{2}$, 1).
13. $y = \tfrac{1}{2}\tan 2\,x$;   ($\tfrac{1}{8}\pi$, $\tfrac{1}{2}$).

Find the coördinates of the center of curvature at any point (x, y) of each of the following curves.

14. $y^2 = 2\,px$.   *Ans.* $\alpha = \dfrac{3\,y^2 + 2\,p^2}{2\,p}$, $\beta = -\dfrac{y^3}{p^2}$.

15. $y^3 = a^2x$. $\alpha = \dfrac{a^4 + 15\,y^4}{6\,a^2y}$, $\beta = \dfrac{a^4y - 9\,y^5}{2\,a^4}$.

16. $b^2x^2 - a^2y^2 = a^2b^2$. $\alpha = \dfrac{(a^2 + b^2)x^3}{a^4}$,

    $\beta = -\dfrac{(a^2 + b^2)y^3}{b^4}$.

17. $x^{\frac{2}{3}} + y^{\frac{2}{3}} = a^{\frac{2}{3}}$. $\alpha = x + 3\,x^{\frac{1}{3}}y^{\frac{2}{3}}$,

    $\beta = y + 3\,x^{\frac{2}{3}}y^{\frac{1}{3}}$.

RADIUS AND CIRCLE OF CURVATURE

18. Find the radii and centers of curvature for the curve $xy = 4$ at the points (1, 4) and (2, 2). Draw the arc of the evolute between these centers. What is its length?

Ans. At (1, 4), $R_1 = \tfrac{17}{8}\sqrt{17}$, $\alpha = \tfrac{19}{2}$, $\beta = \tfrac{49}{8}$;
at (2, 2), $R_2 = 2\sqrt{2}$, $\alpha = 4$, $\beta = 4$;
$R_1 - R_2 = 5.933$.

Find the parametric equations of the evolute of each of the following curves in terms of the parameter t. Draw the curve and its evolute, and draw at least one circle of curvature.

19. $x = 3\,t^2$, $y = 3\,t - t^3$. Ans. $\alpha = \tfrac{3}{2}(1 + 2\,t^2 - t^4)$, $\beta = -\,4\,t^3$.
20. $x = 3\,t$, $y = t^2 - 6$. $\alpha = -\,\tfrac{4}{3}t^3$, $\beta = 3\,t^2 - \tfrac{3}{2}$.
21. $x = 6 - t^2$, $y = 2\,t$. $\alpha = 4 - 3\,t^2$, $\beta = -\,2\,t^3$.
22. $x = 2\,t$, $y = t^2 - 2$. $\alpha = -\,2\,t^3$, $\beta = 3\,t^2$.
23. $x = 4\,t$, $y = 3 + t^2$. $\alpha = -\,t^3$, $\beta = 11 + 3\,t^2$.
24. $x = 9 - t^2$, $y = 2\,t$. $\alpha = 7 - 3\,t^2$, $\beta = -\,2\,t^3$.
25. $x = 2\,t$, $y = \dfrac{3}{t}$. $\alpha = \dfrac{12\,t^4 + 9}{4\,t^3}$, $\beta = \dfrac{27 + 4\,t^4}{6\,t}$.

26. $x = a \cos t$, $y = b \sin t$. $\alpha = \dfrac{(a^2 - b^2)}{a} \cos^3 t$,
$\beta = \dfrac{(b^2 - a^2)}{b} \sin^3 t$.

27. $x = a \cos^3 t$, $\alpha = a \cos^3 t + 3\,a \cos t \sin^2 t$,
   $y = a \sin^3 t$. $\beta = 3\,a \cos^2 t \sin t + a \sin^3 t$.
28. $x = a(\cos t + t \sin t)$, $\alpha = a \cos t$, $\beta = a \sin t$.
    $y = a(\sin t - t \cos t)$.
29. $x = 4 - t^2$, $y = 2\,t$.
30. $x = 2\,t$, $y = 16 - t^2$.
31. $x = t^2$, $y = \tfrac{1}{6} t^3$.
32. $x = 1 - \cos t$, $y = t - \sin t$.
33. $x = \cos^4 t$, $y = \sin^4 t$.
34. $x = a \sec t$, $y = b \tan t$.
35. $x = \cos t$, $y = t$.
36. $x = 6 \sin t$, $y = 3 \cos t$.
37. $x = 3 \csc t$, $y = 4 \ctn t$.
38. $x = a(t + \sin t)$.
    $y = a(1 - \cos t)$.
39. $x = 2 \cos t + \cos 2\,t$.
    $y = 2 \sin t + \sin 2\,t$.

40. Show that in the parabola $x^{\frac{1}{2}} + y^{\frac{1}{2}} = a^{\frac{1}{2}}$ we have the relation $\alpha + \beta = 3(x + y)$.

41. Given the equation of the equilateral hyperbola $2xy = a^2$, show that
$$\alpha + \beta = \frac{(y+x)^3}{a^2}, \quad \alpha - \beta = \frac{(y-x)^3}{a^2}.$$

From this derive the equation of the evolute
$$(\alpha + \beta)^{\frac{2}{3}} - (\alpha - \beta)^{\frac{2}{3}} = 2 a^{\frac{2}{3}}.$$

112. Transformation of derivatives. Some of the formulas derived above independently can be deduced from others by formulas which establish relations between derivatives. Two cases will be presented here.

Interchange of dependent and independent variables.

NOTATION. Let $y' = \dfrac{dy}{dx}, \quad y'' = \dfrac{dy'}{dx} = \dfrac{d^2y}{dx^2}, \quad$ etc.,

$x' = \dfrac{dx}{dy}, \quad x'' = \dfrac{dx'}{dy} = \dfrac{d^2x}{dy^2}, \quad$ etc.

By IX, Art. 29,

(*I*) $\qquad\qquad y' = \dfrac{1}{x'}.$

Now $\qquad\qquad y'' = \dfrac{dy'}{dx} = \dfrac{\dfrac{dy'}{dy}}{x'}.$

Using (*I*), we get $\qquad \dfrac{dy'}{dy} = -\dfrac{x''}{x'^2}.$

(*J*) $\qquad\qquad \therefore y'' = -\dfrac{x''}{x'^3}.$

Again, $\qquad\qquad y''' = \dfrac{dy''}{dx} = \dfrac{\dfrac{dy''}{dy}}{x'}.$

Using (*J*), $\qquad \dfrac{dy''}{dy} = -\dfrac{x'x''' - 3x''^2}{x'^4}.$

(*K*) $\qquad\qquad \therefore y''' = -\dfrac{x'x''' - 3x''^2}{x'^5}.$

And so on for higher derivatives. By these formulas equations in $y', y'', y''',$ etc. can be transformed into equations in $x', x'', x''',$ etc.

RADIUS AND CIRCLE OF CURVATURE

ILLUSTRATIVE EXAMPLE. Transform (*B*), Art. 102, into (*C*) in that article.

Solution. Using (*I*) and (*J*) above,

$$K = \frac{y''}{(1+y'^2)^{\frac{3}{2}}} = \frac{-\dfrac{x''}{x'^3}}{\left(1+\dfrac{1}{x'^2}\right)^{\frac{3}{2}}} = -\frac{x''}{(x'^2+1)^{\frac{3}{2}}}. \quad Ans.$$

Transformation from rectangular to polar coördinates. Given the relations

(1) $\qquad x = \rho \cos \theta, \quad y = \rho \sin \theta$

between the rectangular and polar coördinates of a point. If the polar equation of a curve is $\rho = f(\theta)$, then equations (1) are parametric equations for that curve, θ being the parameter.

NOTATION. The independent variable is θ, and $x', x'', y', y'', \rho', \rho''$ denote successive derivatives of these variables with respect to θ.

Differentiating (1),

(2) $x' = -\rho \sin \theta + \rho' \cos \theta, \qquad y' = \rho \cos \theta + \rho' \sin \theta;$

(3) $x'' = -2\rho' \sin \theta + (\rho'' - \rho) \cos \theta, \quad y'' = 2\rho' \cos \theta + (\rho'' - \rho) \sin \theta.$

By formulas (1), (2), (3), equations in x, y, x', y', x'', y'' may be transformed into equations in $\rho, \theta, \rho', \rho''$.

ILLUSTRATIVE EXAMPLE. Derive (*E*), Art. 104, directly from (*D*), Art. 103.

Solution. Taking numerator and denominator in (*D*) separately, substituting from (2), (3), and reducing, we obtain the results

$$x'y'' - y'x'' = \rho^2 + 2\rho'^2 - \rho\rho''; \quad x'^2 + y'^2 = \rho^2 + \rho'^2.$$

Putting these values in (*D*) gives (*E*).

PROBLEMS

In Problems 1–5 interchange the dependent and independent variables

1. $x\dfrac{d^2y}{dx^2} + y\dfrac{dy}{dx} = 0.$ Ans. $x\dfrac{d^2x}{dy^2} - y\left(\dfrac{dx}{dy}\right)^2 = 0.$

2. $\dfrac{dy}{dx} + \left(\dfrac{dy}{dx}\right)^3 + (y-2)\dfrac{d^2y}{dx^2} = 0.$ $1 + \left(\dfrac{dx}{dy}\right)^2 - (y-2)\dfrac{d^2x}{dy^2} = 0.$

3. $(y-4)\left(\dfrac{dy}{dx}\right)^3 + \dfrac{dy}{dx} - \dfrac{d^2y}{dx^2} = 0.$ $\dfrac{d^2x}{dy^2} + \left(\dfrac{dx}{dy}\right)^2 + y - 4 = 0.$

4. $xy\dfrac{d^3y}{dx^3} + y^2\left(\dfrac{dy}{dx}\right)\left(\dfrac{d^2y}{dx^2}\right) = \left(\dfrac{dy}{dx}\right)^4.$

5. $\left(\dfrac{d^2y}{dx^2}\right)\left(\dfrac{d^3y}{dx^3}\right) = y\left(\dfrac{dy}{dx}\right)^4.$

DIFFERENTIAL CALCULUS

6. Transform $\dfrac{x\dfrac{dy}{dx} - y}{\sqrt{1 + \left(\dfrac{dy}{dx}\right)^2}}$ by assuming $x = \rho \cos \theta$, $y = \rho \sin \theta$.

Ans. $\dfrac{\rho^2}{\sqrt{\rho^2 + \left(\dfrac{d\rho}{d\theta}\right)^2}}$.

7. Transform the equation $\dfrac{d^2y}{dx^2} - \dfrac{x}{1-x^2}\dfrac{dy}{dx} + \dfrac{y}{1-x^2} = 0$ by assuming $x = \cos t$.

Ans. $\dfrac{d^2y}{dt^2} + y = 0$.

8. Transform the equation $x^2\dfrac{d^2y}{dx^2} + 2x\dfrac{dy}{dx} + \dfrac{a^2}{x^2}y = 0$ by assuming $x = \dfrac{1}{t}$.

Ans. $\dfrac{d^2y}{dt^2} + a^2y = 0$.

ADDITIONAL PROBLEMS

1. Given the curve $x = 3\cos t + \cos 3t$, $y = 3\sin t - \sin 3t$. Find the parametric equations of the evolute. Find the center of curvature for $t = 0$ and show that it coincides with the corresponding point on the given curve. Ans. $\alpha = 6\cos t - 2\cos 3t$, $\beta = 6\sin t + 2\sin 3t$.

2. If R is the radius of curvature at any point of the ellipse $b^2x^2 + a^2y^2 = a^2b^2$ and D the perpendicular distance from the origin to the tangent drawn at this point, prove that $RD^3 = a^2b^2$.

3. Find the equations of the evolute of the parabola $y^2 = 4x$, using x as a parameter. Find the points of the parabola for which the corresponding centers of curvature are also points of the parabola. Hence find the length of the part of the evolute inside the parabola.

Ans. $(2, \pm 2\sqrt{2})$; $4(\sqrt{27} - 1)$.

4. (a) At every point (x, y) of a certain curve, its slope is equal to $\dfrac{x(1+y)}{\sqrt{5-x^2}}$ and the curve passes through the point $(2, 0)$. Verify that the equation of the curve is $\ln(1+y) = 1 - \sqrt{5-x^2}$.

(b) Find the curvature of the curve at this point and draw a small portion of the curve near it. Ans. $K = \tfrac{9}{25}\sqrt{5}$.

(c) Draw the circle of curvature at this point. Ans. $\alpha = \tfrac{8}{9}$, $\beta = \tfrac{5}{9}$.

5. The slope of the tangent to a certain curve C at any point P is given by $\dfrac{dy}{dx} = \dfrac{s}{a}$, where s is the length of arc (measured from some fixed point) and a is a constant. The center of curvature of C at P is P'. Denote the radius of curvature of C at P by R and the radius of curvature of the evolute of C at P' by R'. Prove

(a) $R = \dfrac{s^2 + a^2}{a}$; (b) $R' = \dfrac{2s(s^2+a^2)}{a^2}$.

CHAPTER XI

THEOREM OF MEAN VALUE AND ITS APPLICATIONS

113. Rolle's Theorem. A theorem which lies at the foundation of the theoretical development of the calculus will now be explained.

Let $y = f(x)$ be a single-valued function of x, continuous throughout the interval $[a, b]$ (Art. 7) and vanishing at its extremities ($f(a) = 0$, $f(b) = 0$). Suppose also that $f(x)$ has a derivative $f'(x)$ at each interior point ($a < x < b$) of the interval. The function will then be represented graphically by a continuous curve as in the figure. Geometric intuition shows us at once that for *at least one value of x between a and b* the tangent is parallel to the x-axis (as at P); that is, the slope is zero. This illustrates

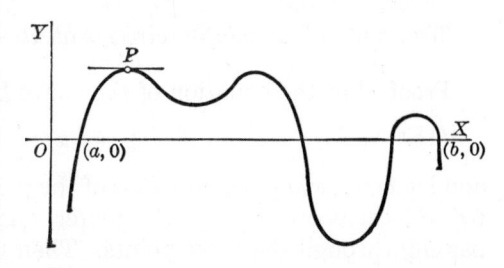

Rolle's Theorem. *If $f(x)$ is continuous throughout the interval $[a, b]$ and vanishes at its extremities, and if it has a derivative $f'(x)$ at every interior point of the interval, then $f'(x)$ must vanish for at least one value of x between a and b.*

The proof is simple. For $f(x)$ must be positive or negative in some parts of the interval if it does not vanish at all points. But in this special case the theorem is obviously true. Suppose, then, that $f(x)$ is positive in a part of the interval. Then $f(x)$ will have a maximum value at some point within the interval. Similarly, if $f(x)$ is negative, it will have a minimum value. But if $f(X)$ is a maximum or minimum ($a < X < b$), then $f'(X) = 0$. Otherwise, $f(x)$ would increase or decrease as x passes through X (Art. 51).

The figure illustrates a case in which Rolle's Theorem does not hold; $f(x)$ is continuous throughout the interval $[a, b]$. $f'(x)$, however, does not exist for $x = c$, but becomes infinite. At no point of the graph is the tangent parallel to the x-axis.

We give first two applications of Rolle's Theorem to geometry.

114. Osculating circle. If a circle be drawn through three neighboring points P_0, P_1, P_2 on a curve, and if P_1 and P_2 be made to approach P_0 along the curve as a limiting position, then this circle will in general approach in magnitude and position a limiting circle called the *osculating circle of the curve at the point P_0*.

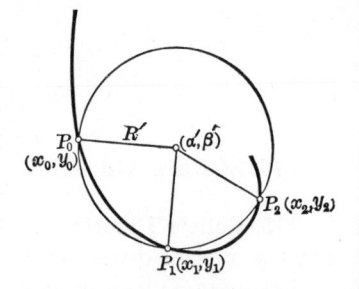

Theorem. *The osculating circle is identical with the circle of curvature.*

Proof. Let the equation of the curve be

(1) $$y = f(x);$$

and let x_0, x_1, x_2 be the abscissas of the points P_0, P_1, P_2 respectively, (α', β') the coördinates of the center, and R' the radius of the circle passing through the three points. Then the equation of the circle is

$$(x - \alpha')^2 + (y - \beta')^2 = R'^2;$$

and since the coördinates of the points P_0, P_1, P_2 must satisfy this equation, we have

(2) $$\begin{cases} (x_0 - \alpha')^2 + (y_0 - \beta')^2 - R'^2 = 0, \\ (x_1 - \alpha')^2 + (y_1 - \beta')^2 - R'^2 = 0, \\ (x_2 - \alpha')^2 + (y_2 - \beta')^2 - R'^2 = 0. \end{cases}$$

Now consider the *function of x* defined by

$$F(x) = (x - \alpha')^2 + (y - \beta')^2 - R'^2,$$

in which y is defined by (1).

Then from equations (2) we get

$$F(x_0) = 0, \quad F(x_1) = 0, \quad F(x_2) = 0.$$

Hence, by Rolle's Theorem (Art. 113), $F'(x)$ must vanish for at least two values of x, one lying between x_0 and x_1, say x', and the other lying between x_1 and x_2, say x''; that is,

$$F'(x') = 0, \quad F'(x'') = 0.$$

Again, for the same reason, $F''(x)$ must vanish for some value of x between x' and x'', say x_3; hence

$$F''(x_3) = 0.$$

THEOREM OF MEAN VALUE AND ITS APPLICATIONS

Therefore the elements α', β', R' of the circle passing through the points P_0, P_1, P_2 must satisfy the three equations
$$F(x_0) = 0, \quad F'(x') = 0, \quad F''(x_3) = 0.$$

Now let the points P_1 and P_2 approach P_0 as a limiting position; then x_1, x_2, x', x'', x_3 will all approach x_0 as a limit, and the elements α, β, R of the osculating circle are therefore determined by the three equations
$$F(x_0) = 0, \quad F'(x_0) = 0, \quad F''(x_0) = 0;$$
or, dropping the subscripts, by

(3) $\quad (x - \alpha)^2 + (y - \beta)^2 = R^2,$

(4) $\quad (x - \alpha) + (y - \beta)y' = 0,$ differentiating (3).

(5) $\quad 1 + y'^2 + (y - \beta)y'' = 0,$ differentiating (4).

Solving (4) and (5) for $x - \alpha$ and $y - \beta$, we get ($y'' \neq 0$),

(6) $\quad x - \alpha = \dfrac{y'(1 + y'^2)}{y''}, \quad y - \beta = -\dfrac{1 + y'^2}{y''}.$

Solving (6) for α and β, the result is identical with (G), Art. 108. Substituting from (6) in (3), and solving for R, the result is (F), Art. 105. Hence the osculating circle is identical with the circle of curvature.

In Art. 28 the tangent line at P was defined as the limiting position of a secant line drawn through P and a neighboring point Q on the curve. We now see that the circle of curvature at P may be defined as the limiting position of a circle drawn through P and two other points Q, R on the curve.

115. Limiting point of intersection of consecutive normals

Theorem. *The center of curvature C for a point P on a curve is the limiting position of the intersection of the normal to the curve at P with a neighboring normal.*

Proof. Let the equation of a curve be

(1) $\quad y = f(x).$

The equations of the normals to the curve at two neighboring points P_0 and P_1 are
$$(x_0 - x) + (y_0 - y)f'(x_0) = 0,$$
$$(x_1 - x) + (y_1 - y)f'(x_1) = 0.$$

If the normals intersect at $C'(\alpha', \beta')$, the coördinates of this point must satisfy both equations, giving

(2) $\quad \begin{cases} (x_0 - \alpha') + (y_0 - \beta')f'(x_0) = 0, \\ (x_1 - \alpha') + (y_1 - \beta')f'(x_1) = 0. \end{cases}$

Now consider the *function of x* defined by
$$\phi(x) = (x - \alpha') + (y - \beta')y',$$
in which y is defined by (1).
Then equations (2) show that
$$\phi(x_0) = 0, \quad \phi(x_1) = 0.$$
But then, by Rolle's Theorem (Art. 113), $\phi'(x)$ must vanish for some value of x between x_0 and x_1, say x'. Therefore α' and β' are determined by the two equations
$$\phi(x_0) = 0, \quad \phi'(x') = 0.$$
If now P_1 approaches P_0 as a limiting position, then x' approaches x_0, giving
$$\phi(x_0) = 0, \quad \phi'(x_0) = 0;$$
and $C'(\alpha', \beta')$ will approach as a limiting position a point $C(\alpha, \beta)$ on the normal at P_0. Dropping the subscripts and accents, the last two equations are
$$(x - \alpha) + (y - \beta)y' = 0,$$
$$1 + y'^2 + (y - \beta)y'' = 0.$$
Solving for α and β, the results are identical with **(G)**, Art. 108. Q.E.D.

116. Theorems of Mean Value (Laws of the Mean). For later applications we need the

Theorem. *If $f(x)$ and $F(x)$ and their first derivatives are continuous throughout the interval $[a, b]$, and if, moreover, $F'(x)$ does not vanish within the interval, then for some value $x = x_1$ between a and b,*

(A) $$\frac{f(b) - f(a)}{F(b) - F(a)} = \frac{f'(x_1)}{F'(x_1)}. \qquad (a < x_1 < b)$$

Proof. Form the function

(1) $$\phi(x) \equiv \frac{f(b) - f(a)}{F(b) - F(a)} [F(x) - F(a)] - [f(x) - f(a)].$$

Evidently $\phi(a) = \phi(b) = 0$, and Rolle's Theorem, Art. 113, may be applied. Differentiating,

(2) $$\phi'(x) = \frac{f(b) - f(a)}{F(b) - F(a)} F'(x) - f'(x).$$

This must vanish for a value $x = x_1$ between a and b.

(3) $$\therefore \frac{f(b) - f(a)}{F(b) - F(a)} F'(x_1) - f'(x_1) = 0.$$

THEOREM OF MEAN VALUE AND ITS APPLICATIONS

Dividing through by $F'(x_1)$ (remembering that $F'(x_1)$ does not vanish), and transposing, the result is (A). Q.E.D.

If $F(x) = x$, (A) becomes

(B) $\quad\dfrac{f(b) - f(a)}{b - a} = f'(x_1).\quad\quad (a < x_1 < b)$

In this form the theorem has a simple geometric interpretation. In the figure the curve is the graph of $f(x)$. Also,

$OC = a, \quad CA = f(a),$
$OD = b, \quad DB = f(b).$

Hence

$\dfrac{f(b) - f(a)}{b - a} = $ slope of chord AB.

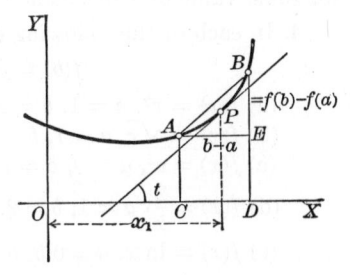

Now $f'(x_1)$ in (B) is the slope of the curve at a point on the arc AB, and (B) states that the slope at this point equals the slope of AB. Hence *there is at least one point on the arc AB at which the tangent line is parallel to the chord AB.*

The student should draw curves (as the first curve in Art. 113) to show that there may be more than one such point in the interval, and curves to illustrate, on the other hand, that the theorem may not be true if $f(x)$ becomes discontinuous for any value of x between a and b, or if $f'(x)$ becomes discontinuous (as in the second figure of Art. 113).

Clearing (B) of fractions, we may also write the theorem in the form

(C) $\quad\quad f(b) = f(a) + (b - a)f'(x_1).$

Let $b = a + \Delta a$; then $b - a = \Delta a$, and since x_1 is a number lying between a and b, we may write

$$x_1 = a + \theta \cdot \Delta a,$$

where θ is a positive proper fraction. Substituting in (C), we get *another form of the Theorem of Mean Value,*

(D) $\quad\quad f(a + \Delta a) - f(a) = \Delta a f'(a + \theta \cdot \Delta a). \quad (0 < \theta < 1)$

PROBLEMS

1. Verify Rolle's Theorem by finding the values of x for which $f(x)$ and $f'(x)$ vanish in each of the following cases.

(a) $f(x) = x^3 - 3x$.
(b) $f(x) = 6x^2 - x^3$.
(c) $f(x) = a + bx + cx^2$.
(d) $f(x) = \sin x$.

(e) $f(x) = \sin \pi x - \cos \pi x$.
(f) $f(x) = \tan x - x$.
(g) $f(x) = x \ln x$.
(h) $f(x) = xe^x$.

2. Given $f(x) = \tan x$. Then $f(0) = 0$ and $f(\pi) = 0$. Does Rolle's Theorem justify the conclusion that $f'(x)$ vanishes for some value of x between 0 and π? Explain your answer.

3. Given $(y + 1)^3 = x^2$. Then $y = 0$ when $x = -1$ and $y = 0$ when $x = +1$. Does Rolle's Theorem justify the conclusion that y' vanishes for some value of x between -1 and $+1$? Explain your answer.

4. In each of the following cases find x_1 such that
$$f(b) = f(a) + (b - a)f'(x_1).$$

(a) $f(x) = x^2$, $a = 1$, $b = 2$.   Ans. $x_1 = 1.5$.
(b) $f(x) = \sqrt{x}$, $a = 1$, $b = 4$. $x_1 = 2.25$.
(c) $f(x) = e^x$, $a = 0$, $b = 1$. $x_1 = \ln(e - 1) = 0.54$.
(d) $f(x) = \dfrac{2}{x}$, $a = 1$, $b = 2$.
(e) $f(x) = \ln x$, $a = 0.5$, $b = 1.5$.
(f) $f(x) = \sin \dfrac{\pi x}{2}$, $a = 0$, $b = 1$.

5. Given $f(x) = \dfrac{1}{x}$, $a = -1$, $b = 1$. For what value of x_1, if any, will $f(b) = f(a) + (b - a)f'(x_1)$?

6. Given $f(x) = x^{\frac{2}{3}}$, $a = -1$, $b = 1$. For what value of x_1, if any, will $f(b) = f(a) + (b - a)f'(x_1)$?

117. Indeterminate forms. When, for a particular value of the independent variable, a function takes on one of the forms

$$\frac{0}{0}, \quad \frac{\infty}{\infty}, \quad 0 \times \infty, \quad \infty - \infty, \quad 0^0, \quad \infty^0, \quad 1^\infty,$$

it is said to be *indeterminate*, and the function is *not* defined for that value of the independent variable by the given analytical expression. For example, suppose we have

$$y = \frac{f(x)}{F(x)},$$

where for some value of the variable, as $x = a$,

$$f(a) = 0, \quad F(a) = 0.$$

For this value of x our function is *not* defined and we may therefore assign to it any value we please. It is evident from what has gone before (Case II, Art. 17) that it is desirable to assign to the function a value that will make it continuous when $x = a$ whenever it is possible to do so.

118. Evaluation of a function taking on an indeterminate form. If the function $f(x)$ assumes an indeterminate form when $x = a$, then if

$$\lim_{x \to a} f(x)$$

THEOREM OF MEAN VALUE AND ITS APPLICATIONS

exists and is finite, we assign this value to the function for $x = a$, which now becomes continuous for $x = a$ (Art. 17).

The limiting value can sometimes be found after simple transformations, as the following examples show.

ILLUSTRATIVE EXAMPLE 1. Given $f(x) = \dfrac{x^2 - 4}{x - 2}$. Prove $\lim\limits_{x \to 2} f(x) = 4$.

Solution. $f(2)$ is indeterminate. But, dividing numerator by denominator, $f(x) = x + 2$. Then $\lim\limits_{x \to 2} (x + 2) = 4$.

ILLUSTRATIVE EXAMPLE 2. Given $f(x) = \sec x - \tan x$. Prove $\lim\limits_{x \to \frac{1}{2}\pi} f(x) = 0$.

Solution. $f(x)$ is indeterminate ($\infty - \infty$). Transform as follows:

$$\sec x - \tan x = \frac{1 - \sin x}{\cos x} = \frac{1 - \sin x}{\cos x} \cdot \frac{1 + \sin x}{1 + \sin x} = \frac{\cos x}{1 + \sin x}.$$

Hence the limit is 0.

See also Art. 18. General methods for evaluating the indeterminate forms of Art. 117 depend upon the calculus.

119. Evaluation of the indeterminate form $\dfrac{0}{0}$. Given a function of the form $\dfrac{f(x)}{F(x)}$ such that $f(a) = 0$ and $F(a) = 0$. The function is indeterminate when $x = a$. It is then required to find

$$\lim\limits_{x \to a} \frac{f(x)}{F(x)}.$$

We shall prove the equation

(E) $$\lim\limits_{x \to a} \frac{f(x)}{F(x)} = \lim\limits_{x \to a} \frac{f'(x)}{F'(x)}.$$

Proof. Referring to (A), Art. 116, and setting $b = x$, remembering that $f(a) = F(a) = 0$, we have

(1) $$\frac{f(x)}{F(x)} = \frac{f'(x_1)}{F'(x_1)}. \qquad (a < x_1 < x)$$

If $x \to a$, so also $x_1 \to a$. Hence, if the right-hand member of (1) approaches a limit when $x_1 \to a$, then the left-hand member will approach the same limit. Thus (E) is proved.

From (E), if $f'(a)$ and $F'(a)$ are not both zero, we shall have

(2) $$\lim\limits_{x \to a} \frac{f(x)}{F(x)} = \frac{f'(a)}{F'(a)}.$$

Rule for evaluating the indeterminate form $\dfrac{0}{0}$. *Differentiate the numerator for a new numerator and the denominator for a new denominator. The value of this new fraction for the assigned value of the variable will be the limiting value of the original fraction.*

176　DIFFERENTIAL CALCULUS

In case it happens that $f'(a) = 0$ and $F'(a) = 0$, that is, the first derivatives also vanish for $x = a$, then (E) can be applied to the ratio

$$\frac{f'(x)}{F'(x)},$$

and the rule will give us $\lim\limits_{x \to a} \dfrac{f(x)}{F(x)} = \dfrac{f''(a)}{F''(a)}.$

It may be necessary to repeat the process several times.

The student is warned against the very careless but common mistake of differentiating the whole expression as a fraction by VII.

If $a = \infty$, the substitution $x = \dfrac{1}{z}$ reduces the problem to the evaluation of the limit for $z = 0$.

Thus $\lim\limits_{x \to \infty} \dfrac{f(x)}{F(x)} = \lim\limits_{z \to 0} \dfrac{-f'\left(\dfrac{1}{z}\right)\dfrac{1}{z^2}}{-F'\left(\dfrac{1}{z}\right)\dfrac{1}{z^2}} = \lim\limits_{z \to 0} \dfrac{f'\left(\dfrac{1}{z}\right)}{F'\left(\dfrac{1}{z}\right)} = \lim\limits_{x \to \infty} \dfrac{f'(x)}{F'(x)}.$

Therefore the rule holds in this case also.

ILLUSTRATIVE EXAMPLE 1. Prove $\lim\limits_{x \to 0} \dfrac{\sin nx}{x} = n.$

Solution. Let $f(x) = \sin nx$, $F(x) = x$. Then $f(0) = 0$, $F(0) = 0$. Therefore, by (E),

$$\lim\limits_{x \to 0} \frac{f(x)}{F(x)} = \lim\limits_{x \to 0} \frac{f'(x)}{F'(x)} = \lim\limits_{x \to 0} \frac{n \cos nx}{1} = n. \qquad \text{Q.E.D.}$$

ILLUSTRATIVE EXAMPLE 2. Prove $\lim\limits_{x \to 1} \dfrac{x^3 - 3x + 2}{x^3 - x^2 - x + 1} = \dfrac{3}{2}.$

Solution. Let $f(x) = x^3 - 3x + 2$, $F(x) = x^3 - x^2 - x + 1$. Then $f(1) = 0$, $F(1) = 0$. Therefore, by (E),

$$\lim\limits_{x \to 1} \frac{f(x)}{F(x)} = \lim\limits_{x \to 1} \frac{f'(x)}{F'(x)} = \lim\limits_{x \to 1} \frac{3x^2 - 3}{3x^2 - 2x - 1} = \frac{0}{0}. \quad \therefore \text{ indeterminate.}$$

$$= \lim\limits_{x \to 1} \frac{f''(x)}{F''(x)} = \lim\limits_{x \to 1} \frac{6x}{6x - 2} = \frac{3}{2}. \qquad \text{Q.E.D.}$$

ILLUSTRATIVE EXAMPLE 3. Prove $\lim\limits_{x \to 0} \dfrac{e^x - e^{-x} - 2x}{x - \sin x} = 2.$

Solution. Let $f(x) = e^x - e^{-x} - 2x$, $F(x) = x - \sin x$. Then $f(0) = 0$, $F(0) = 0$. Therefore, by (E),

$$\lim\limits_{x \to 0} \frac{f(x)}{F(x)} = \lim\limits_{x \to 0} \frac{f'(x)}{F'(x)} = \lim\limits_{x \to 0} \frac{e^x + e^{-x} - 2}{1 - \cos x} = \frac{0}{0}. \quad \therefore \text{ indeterminate.}$$

$$= \lim\limits_{x \to 0} \frac{f''(x)}{F''(x)} = \lim\limits_{x \to 0} \frac{e^x - e^{-x}}{\sin x} = \frac{0}{0}. \quad \therefore \text{ indeterminate.}$$

$$= \lim\limits_{x \to 0} \frac{f'''(x)}{F'''(x)} = \lim\limits_{x \to 0} \frac{e^x + e^{-x}}{\cos x} = 2. \qquad \text{Q.E.D.}$$

THEOREM OF MEAN VALUE AND ITS APPLICATIONS

PROBLEMS

Evaluate each of the following indeterminate forms by differentiation.*

1. $\lim\limits_{x \to 4} \dfrac{x^2 - 16}{x^2 + x - 20}$. Ans. $\frac{8}{9}$.

2. $\lim\limits_{x \to a} \dfrac{x - a}{x^n - a^n}$. $\dfrac{1}{na^{n-1}}$.

3. $\lim\limits_{x \to 1} \dfrac{\ln x}{x - 1}$. 1.

4. $\lim\limits_{x \to 0} \dfrac{e^x - e^{-x}}{\sin x}$. 2.

5. $\lim\limits_{x \to 0} \dfrac{\tan x - x}{x - \sin x}$. 2.

6. $\lim\limits_{x \to \frac{\pi}{2}} \dfrac{\ln \sin x}{(\pi - 2x)^2}$. $-\frac{1}{8}$.

7. $\lim\limits_{x \to 0} \dfrac{a^x - b^x}{x}$. $\ln \dfrac{a}{b}$.

8. $\lim\limits_{\theta \to 0} \dfrac{\theta - \arcsin \theta}{\sin^3 \theta}$. $-\frac{1}{6}$.

9. $\lim\limits_{x \to \phi} \dfrac{\sin x - \sin \phi}{x - \phi}$. $\cos \phi$.

10. $\lim\limits_{y \to 0} \dfrac{e^y + \sin y - 1}{\ln(1 + y)}$. 2.

11. $\lim\limits_{\phi \to \frac{\pi}{4}} \dfrac{\sec^2 \phi - 2 \tan \phi}{1 + \cos 4\phi}$. $\frac{1}{2}$.

12. $\lim\limits_{r \to a} \dfrac{r^3 - ar^2 - a^2 r + a^3}{r^2 - a^2}$.

13. $\lim\limits_{x \to 3} \dfrac{\sqrt{3x} - \sqrt{12 - x}}{2x - 3\sqrt{19 - 5x}}$.

14. $\lim\limits_{x \to 2} \dfrac{\sqrt{16x - x^4} - 2\sqrt[3]{4x}}{2 - \sqrt[4]{2x^3}}$.

15. $\lim\limits_{\theta \to 0} \dfrac{\tan \theta + \sec \theta - 1}{\tan \theta - \sec \theta + 1}$.

16. $\lim\limits_{x \to 0} \dfrac{x - \sin x}{x^3}$.

17. $\lim\limits_{x \to 0} \dfrac{\tan x - \sin x}{\sin^3 x}$.

18. Given a circle with center at O, radius r, and a tangent line AT. In the figure, AM equals arc AP, and B is the intersection of the line through M and P and the line through A and O. Find the limiting position of B as P approaches A as a limiting position. Ans. $OB = 2r$.

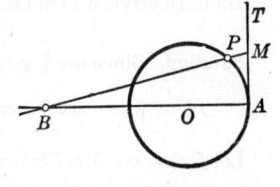

* After differentiating, the student should in every case reduce the resulting expression to its simplest possible form before substituting the value of the variable.

178   DIFFERENTIAL CALCULUS

120. Evaluation of the indeterminate form $\frac{\infty}{\infty}$. In order to find

$$\lim_{x \to a} \frac{f(x)}{F(x)}$$

when both $f(x)$ and $F(x)$ become infinite when $x \to a$, we follow the same rule as that given in Art. 119 for evaluating the indeterminate form $\frac{0}{0}$. Hence

Rule for evaluating the indeterminate form $\frac{\infty}{\infty}$. *Differentiate the numerator for a new numerator and the denominator for a new denominator. The value of this new fraction for the assigned value of the variable will be the limiting value of the original fraction.*

A rigorous proof of this rule is beyond the scope of this book.

ILLUSTRATIVE EXAMPLE. Prove $\lim\limits_{x \to 0} \dfrac{\ln x}{\csc x} = 0$.

Solution. Let $f(x) = \ln x$, $F(x) = \csc x$. Then $f(0) = -\infty$, $F(0) = \infty$. Hence, by the rule,

$$\lim_{x \to 0} \frac{f(x)}{F(x)} = \lim_{x \to 0} \frac{f'(x)}{F'(x)} = \lim_{x \to 0} \frac{\frac{1}{x}}{-\csc x \, \ctn x} = \lim_{x \to 0} \frac{-\sin^2 x}{x \cos x} = \frac{0}{0}.$$

Then, by (*E*), $\quad \lim\limits_{x \to 0} \dfrac{-\sin^2 x}{x \cos x} = \lim\limits_{x \to 0} \dfrac{-2 \sin x \cos x}{\cos x - x \sin x} = 0.$ Q.E.D.

121. Evaluation of the indeterminate form $0 \cdot \infty$. If a function $f(x) \cdot \phi(x)$ takes on the indeterminate form $0 \cdot \infty$ for $x = a$, we write the given function

$$f(x) \cdot \phi(x) = \frac{f(x)}{\frac{1}{\phi(x)}} \left(\text{or} = \frac{\phi(x)}{\frac{1}{f(x)}} \right)$$

so as to cause it to take on one of the forms $\frac{0}{0}$ or $\frac{\infty}{\infty}$, thus bringing it under Art. 119 or Art. 120.

As shown, the product $f(x) \cdot \phi(x)$ may be rewritten in either of the two forms given. As a rule, one of these forms is better than the other, and the choice will depend upon the example.

ILLUSTRATIVE EXAMPLE. Prove $\lim\limits_{x \to \frac{1}{2}\pi} (\sec 3x \cos 5x) = -\frac{5}{3}$.

Solution. Since $\sec \frac{3}{2} \pi = \infty$, $\cos \frac{5}{2} \pi = 0$, we write

$$\sec 3x \cos 5x = \frac{1}{\cos 3x} \cdot \cos 5x = \frac{\cos 5x}{\cos 3x}.$$

Let $f(x) = \cos 5x$, $F(x) = \cos 3x$. Then $f(\frac{1}{2}\pi) = 0$, $F(\frac{1}{2}\pi) = 0$. Hence, by (*E*),

$$\lim_{x \to \frac{1}{2}\pi} \frac{f(x)}{F(x)} = \lim_{x \to \frac{1}{2}\pi} \frac{f'(x)}{F'(x)} = \lim_{x \to \frac{1}{2}\pi} \frac{-5 \sin 5x}{-3 \sin 3x} = -\frac{5}{3}. \qquad \text{Q.E.D.}$$

THEOREM OF MEAN VALUE AND ITS APPLICATIONS

122. Evaluation of the indeterminate form $\infty - \infty$. It is possible in general to transform the expression into a fraction which will assume either the form $\dfrac{0}{0}$ or $\dfrac{\infty}{\infty}$.

ILLUSTRATIVE EXAMPLE. Prove $\lim\limits_{x \to \frac{1}{2}\pi} (\sec x - \tan x) = 0$.

Solution. We have $\sec \frac{1}{2}\pi - \tan \frac{1}{2}\pi = \infty - \infty$. $\therefore$ indeterminate.

By (2), p. 2, $\sec x - \tan x = \dfrac{1}{\cos x} - \dfrac{\sin x}{\cos x} = \dfrac{1 - \sin x}{\cos x}$.

Let $f(x) = 1 - \sin x$, $F(x) = \cos x$. Then $f(\frac{1}{2}\pi) = 0$, $F(\frac{1}{2}\pi) = 0$. Hence, by (E),

$$\lim_{x \to \frac{1}{2}\pi} \frac{f(x)}{F(x)} = \lim_{x \to \frac{1}{2}\pi} \frac{f'(x)}{F'(x)} = \lim_{x \to \frac{1}{2}\pi} \frac{-\cos x}{-\sin x} = 0. \qquad \text{Q.E.D.}$$

PROBLEMS

Evaluate each of the following indeterminate forms.

1. $\lim\limits_{x \to \infty} \dfrac{\ln x}{x^n}$.   Ans. 0.

2. $\lim\limits_{x \to 0} \dfrac{\ctn x}{\ctn 2x}$.   2.

3. $\lim\limits_{\theta \to \frac{\pi}{2}} \dfrac{\tan 3\theta}{\tan \theta}$. $\frac{1}{3}$.

4. $\lim\limits_{x \to \infty} \dfrac{x^3}{e^x}$.   0.

5. $\lim\limits_{x \to \infty} \dfrac{e^x}{\ln x}$. ∞.

6. $\lim\limits_{x \to 0} \dfrac{\ctn x}{\ln x}$. $-\infty$.

7. $\lim\limits_{x \to 0} \dfrac{\ln \sin 2x}{\ln \sin x}$.   1.

8. $\lim\limits_{x \to 0} x \ln \sin x$.   0.

9. $\lim\limits_{\phi \to 0} \dfrac{\pi}{\phi} \tan \dfrac{\pi\phi}{2}$. $\frac{1}{2}\pi^2$.

10. $\lim\limits_{x \to \infty} x \sin \dfrac{a}{x}$. a.

11. $\lim\limits_{x \to \frac{\pi}{2}} (\pi - 2x)\tan x$.   2.

12. $\lim\limits_{\theta \to \frac{\pi}{4}} (1 - \tan \theta) \sec 2\theta$.   Ans. 1.

13. $\lim\limits_{x \to 1} \left[\dfrac{2}{x^2 - 1} - \dfrac{1}{x - 1}\right]$. $-\frac{1}{2}$.

14. $\lim\limits_{x \to 1} \left[\dfrac{1}{\ln x} - \dfrac{x}{\ln x}\right]$. -1.

15. $\lim\limits_{\phi \to 0} \left[\dfrac{2}{\sin^2 \phi} - \dfrac{1}{1 - \cos \phi}\right]$. $\frac{1}{2}$.

16. $\lim\limits_{y \to 1} \left[\dfrac{y}{y - 1} - \dfrac{1}{\ln y}\right]$. $\frac{1}{2}$.

17. $\lim\limits_{x \to 0} \left[\dfrac{1}{\sin^2 x} - \dfrac{1}{x^2}\right]$. $\frac{1}{3}$.

18. $\lim\limits_{x \to \infty} \dfrac{x + \ln x}{x \ln x}$.

19. $\lim\limits_{\theta \to 0} \theta \csc 2\theta$.

20. $\lim\limits_{x \to 0} \dfrac{\ctn 2x}{\ctn 3x}$.

21. $\lim\limits_{\phi \to a} (a^2 - \phi^2) \tan \dfrac{\pi\phi}{2a}$.

22. $\lim\limits_{\theta \to \frac{\pi}{2}} (\sec 5\theta - \tan \theta)$.

DIFFERENTIAL CALCULUS

23. $\lim\limits_{x \to 0} \left[\dfrac{\pi}{4x} - \dfrac{\pi}{2x(e^{\pi x} + 1)} \right]$.

24. $\lim\limits_{x \to 0} \left[\dfrac{1}{x^2} - \dfrac{1}{x \tan x} \right]$.

25. $\lim\limits_{x \to \frac{\pi}{2}} \left[x \tan x - \dfrac{\pi}{2} \sec x \right]$.

26. $\lim\limits_{x \to 2} \dfrac{x^2 - 4}{x^2} \tan \dfrac{\pi x}{4}$.

27. $\lim\limits_{x \to 0} \left[\dfrac{1}{\ln(1+x)} - \dfrac{1}{x} \right]$.

28. $\lim\limits_{x \to 0} \left[\dfrac{1}{\sin^3 x} - \dfrac{1}{x^3} \right]$.

123. Evaluation of the indeterminate forms 0^0, 1^∞, ∞^0. Given a function of the form
$$f(x)^{\phi(x)}.$$

In order that the function shall take on one of the above three forms, we must have, for a certain value of x,

$$f(x) = 0, \quad \phi(x) = 0, \quad \text{giving } 0^0;$$
or $\quad f(x) = 1, \quad \phi(x) = \infty, \quad \text{giving } 1^\infty;$
or $\quad f(x) = \infty, \quad \phi(x) = 0, \quad \text{giving } \infty^0.$

Let $\qquad y = f(x)^{\phi(x)}.$

Taking the natural logarithm of both sides,
$$\ln y = \phi(x) \ln f(x).$$

In any of the above cases the natural logarithm of y (the function) will take on the indeterminate form
$$0 \cdot \infty.$$

Evaluating this by the process illustrated in Art. 121 gives the limit of the logarithm of the function. This being equal to the logarithm of the limit of the function, the limit of the function is known. For if limit $\ln y = a$, then $\lim y = e^a$.

Illustrative Example 1. Prove $\lim\limits_{x \to 0} x^x = 1$.

Solution. The function assumes the indeterminate form 0^0 when $x = 0$.

Let $\qquad\qquad y = x^x;$
then $\qquad\qquad \ln y = x \ln x = 0 \cdot -\infty, \qquad\qquad$ when $x = 0.$

By Art. 121, $\qquad \ln y = \dfrac{\ln x}{\dfrac{1}{x}} = \dfrac{-\infty}{\infty}, \qquad\qquad$ when $x = 0.$

By Art. 120, $\qquad \lim\limits_{x \to 0} \dfrac{\ln x}{\dfrac{1}{x}} = \lim\limits_{x \to 0} \dfrac{\dfrac{1}{x}}{-\dfrac{1}{x^2}} = 0.$

Therefore $\quad \lim\limits_{x \to 0} \ln y = 0,$ and $\lim\limits_{x \to 0} y = \lim\limits_{x \to 0} x^x = e^0 = 1.$ Q.E.D.

THEOREM OF MEAN VALUE AND ITS APPLICATIONS

ILLUSTRATIVE EXAMPLE 2. Prove $\lim\limits_{x \to 1} (2-x)^{\tan \frac{1}{2}\pi x} = e^{\frac{2}{\pi}}$.

Solution. The function assumes the indeterminate form 1^∞ when $x = 1$.

Let $\qquad y = (2-x)^{\tan \frac{1}{2}\pi x}$;

then $\qquad \ln y = \tan \frac{1}{2}\pi x \ln (2-x) = \infty \cdot 0,$ when $x = 1$.

By Art. 121, $\quad \ln y = \dfrac{\ln(2-x)}{\operatorname{ctn} \frac{1}{2}\pi x} = \dfrac{0}{0},$ when $x = 1$.

By Art. 119, $\quad \lim\limits_{x \to 1} \dfrac{\ln(2-x)}{\operatorname{ctn} \frac{1}{2}\pi x} = \lim\limits_{x \to 1} \dfrac{-\dfrac{1}{2-x}}{-\frac{1}{2}\pi \csc^2 \frac{1}{2}\pi x} = \dfrac{2}{\pi}.$

Therefore $\lim\limits_{x \to 1} \ln y = \dfrac{2}{\pi},$ and $\lim\limits_{x \to 1} y = \lim\limits_{x \to 1} (2-x)^{\tan \frac{1}{2}\pi x} = e^{\frac{2}{\pi}}.$ Q.E.D.

ILLUSTRATIVE EXAMPLE 3. Prove $\lim\limits_{x \to 0} (\operatorname{ctn} x)^{\sin x} = 1$.

Solution. The function assumes the indeterminate form ∞^0 when $x = 0$.

Let $\qquad y = (\operatorname{ctn} x)^{\sin x}$;

then $\qquad \ln y = \sin x \ln \operatorname{ctn} x = 0 \cdot \infty,$ when $x = 0$.

By Art. 121, $\quad \ln y = \dfrac{\ln \operatorname{ctn} x}{\csc x} = \dfrac{\infty}{\infty},$ when $x = 0$.

By Art. 120, $\lim\limits_{x \to 0} \dfrac{\ln \operatorname{ctn} x}{\csc x} = \lim\limits_{x \to 0} \dfrac{\dfrac{-\csc^2 x}{\operatorname{ctn} x}}{-\csc x \operatorname{ctn} x} = \lim\limits_{x \to 0} \dfrac{\sin x}{\cos^2 x} = 0.$

Therefore $\lim\limits_{x \to 0} \ln y = 0,$ and $\lim\limits_{x \to 0} y = \lim\limits_{x \to 0} (\operatorname{ctn} x)^{\sin x} = e^0 = 1.$ Q.E.D.

PROBLEMS

Evaluate each of the following indeterminate forms.

1. $\lim\limits_{x \to \frac{\pi}{2}} (\sin x)^{\tan x}$.   *Ans.* 1.   8. $\lim\limits_{x \to \infty} \left(\cos \dfrac{2}{x}\right)^x$.

2. $\lim\limits_{x \to \infty} \left(\dfrac{2}{x}+1\right)^x$. e^2.   9. $\lim\limits_{x \to \infty} \left(\cos \dfrac{2}{x}\right)^{x^2}$.

3. $\lim\limits_{x \to 1} x^{\frac{1}{1-x}}$. $\dfrac{1}{e}$.   10. $\lim\limits_{x \to \infty} \left(\cos \dfrac{2}{x}\right)^{x^3}$.

4. $\lim\limits_{y \to \infty} \left(1+\dfrac{a}{y}\right)^y$. e^a.   11. $\lim\limits_{x \to 0} (e^{2x}+2x)^{\frac{1}{4x}}$.

5. $\lim\limits_{x \to 0} (1+\sin x)^{\operatorname{ctn} x}$. e.   12. $\lim\limits_{x \to 0} (x+1)^{\operatorname{ctn} x}$.

6. $\lim\limits_{x \to 0} (e^x+x)^{\frac{1}{x}}$. e^2.   13. $\lim\limits_{x \to 0} \left(\dfrac{1}{x}\right)^{\sin x}$.

7. $\lim\limits_{t \to 0} (1+nt)^{\frac{1}{t}}$. e^n.   14. $\lim\limits_{x \to 0} (1+x)^{\ln x}$.

124. The Extended Theorem of Mean Value. Let the constant R be defined by the equation

(1) $\quad f(b) - f(a) - (b-a)f'(a) - \frac{1}{2}(b-a)^2 R = 0.$

Let $F(x)$ be a function formed by replacing b by x in the left-hand member of (1); that is,

(2) $\quad F(x) = f(x) - f(a) - (x-a)f'(a) - \frac{1}{2}(x-a)^2 R.$

From (1), $F(b) = 0$; and from (2), $F(a) = 0$; therefore, by Rolle's Theorem (Art. 113), at least one value of x between a and b, say x_1, will cause $F'(x)$ to vanish. Hence, since

$$F'(x) = f'(x) - f'(a) - (x-a)R,$$

we get $\quad F'(x_1) = f'(x_1) - f'(a) - (x_1 - a)R = 0.$

Since $F'(x_1) = 0$ and $F'(a) = 0$, it is evident that $F'(x)$ also satisfies the conditions of Rolle's Theorem, so that *its derivative*, namely $F''(x)$, must vanish for at least one value of x between a and x_1, say x_2, and therefore x_2 also lies between a and b. But

$$F''(x) = f''(x) - R; \text{ therefore } F''(x_2) = f''(x_2) - R = 0,$$

and $\quad R = f''(x_2).$

Substituting this result in (1), we get

(F) $\quad f(b) = f(a) + (b-a)f'(a) + \dfrac{1}{\underline{|2}}(b-a)^2 f''(x_2). \quad (a < x_2 < b)$

By continuing this process we get the general result,

(G) $\quad f(b) = f(a) + \dfrac{(b-a)}{\underline{|1}} f'(a) + \dfrac{(b-a)^2}{\underline{|2}} f''(a)$

$\quad\quad + \dfrac{(b-a)^3}{\underline{|3}} f'''(a) + \cdots + \dfrac{(b-a)^{n-1}}{\underline{|n-1}} f^{(n-1)}(a)$

$\quad\quad + \dfrac{(b-a)^n}{\underline{|n}} f^{(n)}(x_1). \quad\quad\quad\quad (a < x_1 < b)$

Equation **(G)** is called the *Extended Theorem of Mean Value*, or the *Extended Law of the Mean*.

125. Maxima and minima treated analytically. By making use of Art. 116 and the results of the last section we can now give a general discussion of *maxima and minima of functions of a single independent variable*.

Given the function $f(x)$. Let h be a positive number as small as we please; then the definitions given in Art. 46 may be stated as follows.

THEOREM OF MEAN VALUE AND ITS APPLICATIONS 183

If, for all values of x different from a in the interval $[a-h, a+h]$,

(1) $\qquad f(x) - f(a) =$ a negative number,

then $f(x)$ is said to be a *maximum when $x = a$*.
If, on the other hand,

(2) $\qquad f(x) - f(a) =$ a positive number,

then $f(x)$ is said to be a *minimum when $x = a$*.
We begin with an analytical proof of the criterion on page 51.

A function is increasing when the derivative is positive, and decreasing when the derivative is negative.

For, let $y = f(x)$. When Δx is small, numerically, $\dfrac{\Delta y}{\Delta x}$ and the derivative $f'(x)$ will agree in sign (Art. 24). Let $f'(x) > 0$. Then, when Δx is positive, so is Δy, and when Δx is negative, so is Δy. Therefore $f(x)$ is increasing. A similar proof holds when the derivative is negative.

The truth of the following statement is now easily deduced.

If $f(a)$ is a maximum or minimum value of $f(x)$, then $f'(a) = 0$.

For, if $f'(a) \neq 0$, $f(x)$ would increase or decrease as x increases through a. But then $f(a)$ is neither a maximum nor a minimum.

We seek general sufficient conditions for maxima and minima. Consider the following cases.

I. Let $f'(a) = 0$, and $f''(a) \neq 0$.

From (*F*), Art. 124, replacing b by x and transposing $f(a)$,

(3) $\qquad f(x) - f(a) = \dfrac{(x-a)^2}{\lfloor 2} f''(x_2). \qquad (a < x_2 < x)$

Since $f''(a) \neq 0$, and $f''(x)$ is assumed as continuous, we may choose our interval $[a - h, a + h]$ so small that $f''(x)$ will have the same sign as $f''(a)$. Also, $(x - a)^2$ does not change sign. Therefore the second member of (3) will not change sign, and the difference

$$f(x) - f(a)$$

will have the same sign for all values of x in the interval $[a - h, a + h]$, and, moreover, *this sign will be the same as the sign of $f''(a)$*. It therefore follows from our definitions (1) and (2) that

(4) *$f(a)$ is a maximum if $f'(a) = 0$ and $f''(a) =$ a negative number;*

(5) *$f(a)$ is a minimum if $f'(a) = 0$ and $f''(a) =$ a positive number.*

These conditions are the same as those in Art. 56.

184 DIFFERENTIAL CALCULUS

II. Let $f'(a) = f''(a) = 0$, and $f'''(a) \neq 0$.

From (**G**), Art. 124, putting $n = 3$, replacing b by x, and transposing $f(a)$,

(6) $$f(x) - f(a) = \frac{1}{\underline{|3}} (x - a)^3 f'''(x_3). \qquad (a < x_3 < x)$$

As before, $f'''(x)$ will have the same sign as $f'''(a)$. But $(x - a)^3$ changes its sign from $-$ to $+$ as x increases through a. Therefore the difference
$$f(x) - f(a)$$
must change sign, and $f(a)$ is neither a maximum nor a minimum.

III. Let $f'(a) = f''(a) = \cdots = f^{(n-1)}(a) = 0$, and $f^{(n)}(a) \neq 0$.

By continuing the process as illustrated in I and II, it is seen that if the first of the derivatives of $f(x)$ which does not vanish for $x = a$ is of *even* order ($= n$), then

(**H**) $f(a)$ is a maximum if $f^{(n)}(a) =$ a negative number;
(**I**) $f(a)$ is a minimum if $f^{(n)}(a) =$ a positive number.*

If the first derivative of $f(x)$ which does not vanish for $x = a$ is of *odd order*, then $f(a)$ will be neither a maximum nor a minimum.

ILLUSTRATIVE EXAMPLE 1. Examine $x^3 - 9x^2 + 24x - 7$ for maximum and minimum values.

Solution. $\quad f(x) = x^3 - 9x^2 + 24x - 7.$
$\quad\quad\quad\quad\quad f'(x) = 3x^2 - 18x + 24.$
Solving, $\quad 3x^2 - 18x + 24 = 0$

gives the critical values $x = 2$ and $x = 4$. $\therefore f'(2) = 0$, and $f'(4) = 0$.

Differentiating again, $\quad f''(x) = 6x - 18.$
Since $f''(2) = -6$, we know, from (**H**), that $f(2) = 13$ is a maximum.
Since $f''(4) = +6$, we know, from (**I**), that $f(4) = 9$ is a minimum.

ILLUSTRATIVE EXAMPLE 2. Examine $e^x + 2 \cos x + e^{-x}$ for maximum and minimum values.

Solution. $\quad f(x) = e^x + 2 \cos x + e^{-x},$
$\quad\quad\quad\quad\quad f'(x) = e^x - 2 \sin x - e^{-x} = 0,$ for $x = 0,$†
$\quad\quad\quad\quad\quad f''(x) = e^x - 2 \cos x + e^{-x} = 0,$ for $x = 0,$
$\quad\quad\quad\quad\quad f'''(x) = e^x + 2 \sin x - e^{-x} = 0,$ for $x = 0,$
$\quad\quad\quad\quad\quad f^{iv}(x) = e^x + 2 \cos x + e^{-x} = 4,$ for $x = 0.$

Hence, from (**I**), $f(0) = 4$ is a minimum.

---
* As in Art. 46, a critical value $x = a$ is found by placing the first derivative equal to zero and solving the resulting equation for real roots.

† $x = 0$ is the only root of the equation $e^x - 2 \sin x - e^{-x} = 0$.

THEOREM OF MEAN VALUE AND ITS APPLICATIONS

PROBLEMS

Examine each of the following functions for maximum and minimum values, using the method of the last section.

1. $x^4 - 4x^3 + 5$.   Ans. $x = 0$, gives neither,
$x = 3$, gives min. $= -22$.

2. $x^3 + 3x^2 + 3x$. $x = -1$, gives neither.

3. $x^3(x-2)^2$. $x = 0$, gives neither,
$x = \frac{6}{5}$, gives max. $= 1.11$,
$x = 2$, gives min. $= 0$.

4. $x(x-1)^2(x+1)^3$.

5. $\sin^3 x$, $(-\pi < x < \pi)$.   Ans. $x = -\frac{\pi}{2}$, gives min. $= -1$,
$x = 0$, gives neither,
$x = \frac{\pi}{2}$, gives max. $= 1$.

6. $x^4 e^x$. $x = -4$, gives max. $= 4.69$,
$x = 0$, gives min. $= 0$.

7. $e^{-x} \cos^2 x$, $(0 < x < \pi)$. $x = \frac{\pi}{2}$, gives min. $= 0$,
$x = 2.68$, gives max. $= 0.055$.

8. Investigate $4x^5 - 15x^4 + 20x^3 - 10x^2$ at $x = 1$.

9. Investigate $x^2 \sin x$ at $x = 0$. Show that the function has an extreme value where $2 \tan x + x = 0$, whence $x = 2.29$ gives max. $= 3.96$.

10. Investigate $x \sin^2 x$ at $x = 0$. Show that the function has an extreme value where $\tan x + 2x = 0$, whence $x = 1.84$ gives max. $= 1.71$.

11. Show that if the first derivative of $f(x)$ which does not vanish for $x = a$ is of odd order ($= n$), then $f(x)$ is an increasing or decreasing function when $x = a$, according as $f^{(n)}(a)$ is positive or negative.

ADDITIONAL PROBLEMS

1. If $y = e^x + e^{-x}$, find dx in terms of y and dy.   Ans. $dx = \dfrac{\pm \, dy}{\sqrt{y^2 - 4}}$.

2. Prove that

$$\frac{d}{dx} \ln \left(3x + 2 + \sqrt{9x^2 + 12x}\right) = \frac{3}{\sqrt{9x^2 + 12x}}.$$

3. Prove that

$$\frac{d}{dx}\left[\frac{x}{4}(x^2+1)^{\frac{3}{2}} - \frac{x}{8}\sqrt{x^2+1} - \frac{1}{8}\ln\left(x + \sqrt{x^2+1}\right)\right] = x^2\sqrt{x^2+1}.$$

4. Show that the curve $x = t^3 + 2t^2$, $y = 3t^2 + 4t$ has no point of inflection.

5. Prove that the points of intersection of the curves $2\,y = x \sin x$ and $y = \cos x$ are points of inflection of the first curve. Sketch both curves on the same axes.

6. Given the damped harmonic motion
$$s = ae^{-bt} \sin ct,$$
where a, b, and c are positive constants; prove that the successive values of t for which $v = 0$ form an arithmetic progression and that the corresponding values of s form a decreasing geometric progression.

7. The abscissa of a point P moving on the parabola $y = ax^2$ is increasing at the rate of one unit per second. Let O denote the origin and let T be the intersection of the x-axis with the tangent to the parabola at P. Show that the rate of increase of the arc length OP is numerically equal to the ratio $\dfrac{TP}{OT}$.

8. Let MP be the ordinate at any point P on the catenary (see page 532). The line MA is drawn perpendicular to the tangent at P. Prove that the length of MA is constant and equal to a.

9. The curve $x^2y + 12\,y = 144$ has one maximum point and two points of inflection. Find the area of the triangle formed by the tangents to the curve at these three points. *Ans.* 1.

10. Given $\ln 6 = 1.792$ and $\ln 7 = 1.946$. Calculate $\ln 6.15$ first by interpolation and second by differentials. Show graphically that the true value lies between the two approximations.

11. Given the ellipse $b^2x^2 + a^2y^2 = a^2b^2$, find the length of the shortest tangent intercepted between the coördinate axes. *Ans.* $a + b$.

12. Given the area in the first quadrant bounded by the curves $y^2 = x$ and $y^2 = x^3$. A rectangle with sides parallel respectively to the axes is drawn within the limits of this area. The width of the rectangle (measured horizontally) is $\tfrac{1}{3}$ and one of the diagonals has an extremity on each curve. Find the area of the rectangle of maximum area which can be constructed in this way. *Ans.* 0.019.

13. Rectangles are drawn with one side along the x-axis, a second side along the line $x = \tfrac{1}{2}$, and one vertex on the curve $y = e^{-x^2}$. Find the area of the largest of these rectangles. *Ans.* $e^{-0.25} = 0.7788$.

14. Find the maximum and minimum values of y, if
$$y = ae^{\frac{x}{a}} - 3\,x - 2\,ae^{-\frac{x}{a}}.$$
Ans. Max. $= -a$; min. $= a(1 - 3 \ln 2)$.

15. Given $x^2 + 3\,xy + 2\,y^2 - 5\,x - 6\,y + 5 = 0$, find the maximum and minimum values of y. *Ans.* Max. $= 1$; min. $= 5$.

INTEGRAL CALCULUS

CHAPTER XII

INTEGRATION; RULES FOR INTEGRATING STANDARD ELEMENTARY FORMS

126. Integration. The student is already familiar with the mutually inverse operations of addition and subtraction, multiplication and division, raising to a power and extracting roots. In the examples which follow, the second members of one column are, respectively, the inverse of the second members of the other column.

$y = x^2 + 1,$ $x = \pm \sqrt{y-1}\,;$
$y = a^x,$ $x = \log_a y\,;$
$y = \sin x,$ $x = \arcsin y.$

From the differential calculus we have learned how to calculate the derivative $f'(x)$ of a given function $f(x)$, an operation indicated by

$$\frac{d}{dx} f(x) = f'(x),$$

or, if we are using differentials, by

$$df(x) = f'(x)dx.$$

The problems of the integral calculus depend on the *inverse operation*, namely:

To find a function $f(x)$ whose derivative

(1) $\qquad\qquad f'(x) = \phi(x)$

is given.

Or, since it is customary to use differentials in the integral calculus, we may write

(2) $\qquad\qquad df(x) = f'(x)dx = \phi(x)dx$

and state the problem as follows:

Having given the differential of a function, to find the function itself.

INTEGRAL CALCULUS

The function $f(x)$ thus found is called an *integral* of the given differential expression, the process of finding it is called *integration*, and the operation is indicated by writing the *integral sign* * $\int$ in front of the given differential expression; thus

(3) $$\int f'(x)dx = f(x),$$

read *the integral of $f'(x)dx$ equals $f(x)$*. The differential dx indicates that x is *the variable of integration*. For example,

(a) If $f(x) = x^3$, then $f'(x)dx = 3\,x^2\,dx$, and
$$\int 3\,x^2\,dx = x^3.$$

(b) If $f(x) = \sin x$, then $f'(x)dx = \cos x\,dx$, and
$$\int \cos x\,dx = \sin x.$$

(c) If $f(x) = \arctan x$, then $f'(x)dx = \dfrac{dx}{1+x^2}$, and
$$\int \frac{dx}{1+x^2} = \arctan x.$$

Let us now emphasize what is apparent from the preceding explanations, namely, that

Differentiation and integration are inverse operations.

Differentiating (3) gives

(4) $$d\int f'(x)dx = f'(x)dx.$$

Substituting the value of $f'(x)dx\ [= df(x)]$ from (2) in (3), we get

(5) $$\int df(x) = f(x).$$

Therefore, considered as symbols of operation, $\dfrac{d}{dx}$ and $\int \cdots dx$ are *inverse to each other*; or, if we are using differentials, d and $\int$ are inverse to each other.

When d is followed by $\int$ they annul each other, as in (4), but when $\int$ is followed by d, as in (5), that will not in general be the case. The reason for this will appear at once from the definition of the constant of integration given in the next section.

---
* Historically this sign is a distorted S, the initial letter of the word *sum*. See Art. 155.

INTEGRATION

127. Constant of integration. Indefinite integral. From the preceding section it follows that

since $d(x^3) = 3\,x^2 dx$, we have $\int 3\,x^2 dx = x^3$;

since $d(x^3 + 2) = 3\,x^2 dx$, we have $\int 3\,x^2 dx = x^3 + 2$;

since $d(x^3 - 7) = 3\,x^2 dx$, we have $\int 3\,x^2 dx = x^3 - 7$.

In fact, since $d(x^3 + C) = 3\,x^2 dx$, where C is any arbitrary constant, we have

$$\int 3\,x^2 dx = x^3 + C.$$

A constant C arising in this way is called a *constant of integration*, a number independent of the *variable of integration*. Since we can give C as many values as we please, it follows that if a given differential expression has one integral, it has infinitely many differing only by constants. Hence

$$\int f'(x) dx = f(x) + C;$$

and since C is unknown and *indefinite*, the expression

$$f(x) + C$$

is called the *indefinite integral of $f'(x) dx$*.

It is evident that if $\phi(x)$ is a function the derivative of which is $f(x)$, then $\phi(x) + C$, where C is any constant whatever, is likewise a function the derivative of which is $f(x)$. Hence the

Theorem. *If two functions differ by a constant, they have the same derivative.*

It is, however, not obvious that if $\phi(x)$ is a function the derivative of which is $f(x)$, then *all* functions having the same derivative $f(x)$ are of the form

$$\phi(x) + C,$$

where C is any constant. In other words, there remains to be proved the

Converse theorem. *If two functions have the same derivative, their difference is a constant.*

Proof. Let $\phi(x)$ and $\psi(x)$ be two functions having the same derivative $f(x)$. Place

$$F(x) = \phi(x) - \psi(x)\,; \text{ then, by hypothesis,}$$

(1) $\qquad F'(x) = \dfrac{d}{dx}\,[\phi(x) - \psi(x)] = f(x) - f(x) = 0.$

190 INTEGRAL CALCULUS

But from the Theorem of Mean Value (*D*), Art. 116, we have

$$F(x + \Delta x) - F(x) = \Delta x \, F'(x + \theta \cdot \Delta x). \quad (0 < \theta < 1)$$

$$\therefore F(x + \Delta x) - F(x) = 0,$$

[Since by (1) the derivative of $F(x)$ is zero for all values of x.]

and $$F(x + \Delta x) = F(x).$$

This means that the function

$$F(x) = \phi(x) - \psi(x)$$

does not change in value at all when x takes on the increment Δx, that is, $\phi(x)$ and $\psi(x)$ differ only by a constant.

In any given case the value of C can be found when we know the value of the integral for some value of the variable, and this will be illustrated by numerous examples in the next chapter. For the present we shall content ourselves with first learning how to find the indefinite integrals of given differential expressions. In what follows we shall assume that *every continuous function has an indefinite integral*, a statement the rigorous proof of which is beyond the scope of this book. For all elementary functions, however, the truth of the statement will appear in the chapters which follow.

In all cases of indefinite integration the test to be applied in verifying the results is that *the differential of the integral must be equal to the given differential expression*.

128. Rules for integrating standard elementary forms. The differential calculus furnished us with a General Rule for differentiation (Art. 27). The integral calculus gives us no corresponding general rule that can be readily applied in practice for performing the inverse operation of integration.* Each case requires special treatment, and we arrive at the integral of a given differential expression through our previous knowledge of the known results of differentiation. That is, we must be able to answer the question, *What function, when differentiated, will yield the given differential expression?*

Integration, then, is essentially a tentative process, and to expedite the work, tables of known integrals are formed called *standard forms*. To effect any integration we compare the given differential expression with these forms, and if it is found to be identical with one of them, the integral is known. If it is not identical with one of them, we strive to reduce it to one of the standard forms by various methods, many of which employ artifices which can be suggested by practice

* Even though the integral of a given differential expression may be known to exist, yet it may not be possible for us actually to find it in terms of known functions.

INTEGRATION

only. Accordingly a large portion of our text will be devoted to the explanation of methods for integrating those functions which frequently appear in the process of solving practical problems.

From any result of differentiation may always be derived a formula for integration.

The following two rules are useful in reducing differential expressions to standard forms.

(a) *The integral of any algebraic sum of differential expressions equals the same algebraic sum of the integrals of these expressions taken separately.*

Proof. Differentiating the expression

$$\int du + \int dv - \int dw,$$

u, v, w being functions of a single variable, we get

$$du + dv - dw. \qquad \text{By III, Art. 94}$$

(1) $\qquad \therefore \int (du + dv - dw) = \int du + \int dv - \int dw.$

(b) *A constant factor may be written either before or after the integral sign.*

Proof. Differentiating the expression

$$a \int dv$$

gives $\qquad a\, dv. \qquad \text{By IV, Art. 94}$

(2) $\qquad \therefore \int a\, dv = a \int dv.$

On account of their importance we shall write the above two rules as formulas at the head of the following list of "Standard Elementary Forms."

Standard Elementary Forms

(1) $\quad \int (du + dv - dw) = \int du + \int dv - \int dw.$

(2) $\quad \int a\, dv = a \int dv.$

(3) $\quad \int dx = x + C.$

(4) $\quad \int v^n dv = \dfrac{v^{n+1}}{n+1} + C. \qquad (n \neq -1)$

(5) $\displaystyle\int \frac{dv}{v} = \ln v + C$
$= \ln v + \ln c = \ln cv.$
[Placing $C = \ln c$.]

(6) $\displaystyle\int a^v \, dv = \frac{a^v}{\ln a} + C.$

(7) $\displaystyle\int e^v \, dv = e^v + C.$

(8) $\displaystyle\int \sin v \, dv = -\cos v + C.$

(9) $\displaystyle\int \cos v \, dv = \sin v + C.$

(10) $\displaystyle\int \sec^2 v \, dv = \tan v + C.$

(11) $\displaystyle\int \csc^2 v \, dv = -\ctn v + C.$

(12) $\displaystyle\int \sec v \tan v \, dv = \sec v + C.$

(13) $\displaystyle\int \csc v \ctn v \, dv = -\csc v + C.$

(14) $\displaystyle\int \tan v \, dv = -\ln \cos v + C = \ln \sec v + C.$

(15) $\displaystyle\int \ctn v \, dv = \ln \sin v + C.$

(16) $\displaystyle\int \sec v \, dv = \ln(\sec v + \tan v) + C.$

(17) $\displaystyle\int \csc v \, dv = \ln(\csc v - \ctn v) + C.$

(18) $\displaystyle\int \frac{dv}{v^2 + a^2} = \frac{1}{a} \arctan \frac{v}{a} + C.$

(19) $\displaystyle\int \frac{dv}{v^2 - a^2} = \frac{1}{2a} \ln \frac{v-a}{v+a} + C.$ \qquad $(v^2 > a^2)$

(19 a) $\displaystyle\int \frac{dv}{a^2 - v^2} = \frac{1}{2a} \ln \frac{a+v}{a-v} + C.$ \qquad $(v^2 < a^2)$

(20) $\displaystyle\int \frac{dv}{\sqrt{a^2 - v^2}} = \arcsin \frac{v}{a} + C.$

(21) $\displaystyle\int \frac{dv}{\sqrt{v^2 \pm a^2}} = \ln(v + \sqrt{v^2 \pm a^2}) + C.$

(22) $\displaystyle\int \sqrt{a^2 - v^2}\, dv = \frac{v}{2}\sqrt{a^2 - v^2} + \frac{a^2}{2}\arcsin\frac{v}{a} + C.$

(23) $\displaystyle\int \sqrt{v^2 \pm a^2}\, dv = \frac{v}{2}\sqrt{v^2 \pm a^2} \pm \frac{a^2}{2}\ln(v + \sqrt{v^2 \pm a^2}) + C.$

129. Formulas (3), (4), (5). These are easily proved.

Proof of (3). Since $d(x + C) = dx,$ II, Art. 94

we get
$$\int dx = x + C.$$

Proof of (4). Since
$$d\left(\frac{v^{n+1}}{n+1} + C\right) = v^n dv, \qquad \text{VI, Art. 94}$$

we get
$$\int v^n dv = \frac{v^{n+1}}{n+1} + C.$$

This holds true for all values of n except $n = -1$. For when $n = -1$, (4) involves division by zero.

The case when $n = -1$ comes under (5).

Proof of (5). Since
$$d(\ln v + C) = \frac{dv}{v}, \qquad \text{X, Art. 94}$$

we get
$$\int \frac{dv}{v} = \ln v + C.$$

The results we get from (5) may be put in more compact form if we denote the constant of integration by $\log_e c$. Thus

$$\int \frac{dv}{v} = \ln v + \ln c = \ln cv.$$

Formula (5) states that *if the expression under the integral sign is a fraction whose numerator is the differential of the denominator, then the integral is the natural logarithm of the denominator.*

ILLUSTRATIVE EXAMPLES*

Work out the following integrations:

1. $\displaystyle\int x^6 dx = \frac{x^{6+1}}{6+1} + C = \frac{x^7}{7} + C$, by (4), where $v = x$ and $n = 6$.

2. $\displaystyle\int \sqrt{x}\, dx = \int x^{\frac{1}{2}} dx = \frac{x^{\frac{3}{2}}}{\frac{3}{2}} + C = \frac{2}{3} x^{\frac{3}{2}} + C$, by (4), where $v = x$ and $n = \frac{1}{2}$.

---
* When learning to integrate, the student should have oral drill in integrating simple functions.

3. $\int \dfrac{dx}{x^3} = \int x^{-3} dx = \dfrac{x^{-2}}{-2} + C = -\dfrac{1}{2\,x^2} + C,$ by **(4)**, where $v = x$ and $n = -3.$

4. $\int ax^5 dx = a \int x^5 dx = \dfrac{ax^6}{6} + C.$ By **(2)** and **(4)**

5. $\int (2\,x^3 - 5\,x^2 - 3\,x + 4)dx$

$\qquad = \int 2\,x^3 dx - \int 5\,x^2 dx - \int 3\,x\,dx + \int 4\,dx$ by **(1)**

$\qquad = 2\int x^3 dx - 5\int x^2 dx - 3\int x\,dx + 4\int dx$ by **(2)**

$\qquad = \dfrac{x^4}{2} - \dfrac{5\,x^3}{3} - \dfrac{3\,x^2}{2} + 4\,x + C.$

NOTE. Although each separate integration requires an arbitrary constant, we write down only a single constant denoting their algebraic sum.

6. $\int \left(\dfrac{2\,a}{\sqrt{x}} - \dfrac{b}{x^2} + 3\,c\sqrt[3]{x^2} \right) dx$

$\qquad = \int 2\,ax^{-\frac{1}{2}} dx - \int bx^{-2} dx + \int 3\,cx^{\frac{2}{3}} dx$ by **(1)**

$\qquad = 2\,a\int x^{-\frac{1}{2}} dx - b\int x^{-2} dx + 3\,c\int x^{\frac{2}{3}} dx$ by **(2)**

$\qquad = 2\,a \cdot \dfrac{x^{\frac{1}{2}}}{\frac{1}{2}} - b \cdot \dfrac{x^{-1}}{-1} + 3\,c \cdot \dfrac{x^{\frac{5}{3}}}{\frac{5}{3}} + C$ by **(4)**

$\qquad = 4\,a\sqrt{x} + \dfrac{b}{x} + \dfrac{9}{5} cx^{\frac{5}{3}} + C.$

7. $\int (a^{\frac{2}{3}} - x^{\frac{2}{3}})^3 dx = a^2 x + \dfrac{9}{7} a^{\frac{2}{3}} x^{\frac{7}{3}} - \dfrac{9}{5} a^{\frac{4}{3}} x^{\frac{5}{3}} - \dfrac{x^3}{3} + C.$

HINT. First expand.

8. $\int (a^2 + b^2 x^2)^{\frac{1}{2}} x\,dx = \dfrac{(a^2 + b^2 x^2)^{\frac{3}{2}}}{3\,b^2} + C.$

Solution. This may be brought into the form **(4)**. For insert the factor $2\,b^2$ *after* the integral sign before $x\,dx$, and its reciprocal *before* the integral sign. Those operations balance each other by **(2)**.

Comparison with **(4)**.
$v = a^2 + b^2 x^2, \quad n = \tfrac{1}{2},$
$dv = 2\,b^2 x\,dx.$

$\int (a^2 + b^2 x^2)^{\frac{1}{2}} x\,dx = \dfrac{1}{2\,b^2} \int (a^2 + b^2 x^2)^{\frac{1}{2}} (2\,b^2 x\,dx) \left[= \dfrac{1}{2\,b^2} \int v^{\frac{1}{2}} dv = \dfrac{v^{\frac{3}{2}}}{3\,b^2} + C, \text{ by } \mathbf{(4)} \right]$

$\qquad = \dfrac{(a^2 + b^2 x^2)^{\frac{3}{2}}}{3\,b^2} + C.$

NOTE. The student is warned against transferring any function of the variable from one side of the integral sign to the other, since that would change the value of the integral.

9. $\int \dfrac{3\,ax\,dx}{b^2 + c^2 x^2} = \dfrac{3\,a}{2\,c^2} \ln (b^2 + c^2 x^2) + C.$

Solution. $\qquad\qquad \int \dfrac{3\,ax\,dx}{b^2 + c^2 x^2} = 3\,a \int \dfrac{x\,dx}{b^2 + c^2 x^2}.$ By **(2)**

INTEGRATION

This resembles (5). If we insert the factor $2c^2$ after the integral sign and its reciprocal before it, the value of the expression will not be changed.

Comparison with (5).
$v = b^2 + c^2x^2$, $dv = 2c^2x\,dx$.

Hence $3a\int \dfrac{x\,dx}{b^2+c^2x^2} = \dfrac{3a}{2c^2}\int \dfrac{2c^2x\,dx}{b^2+c^2x^2}\left[=\dfrac{3a}{2c^2}\int \dfrac{dv}{v} = \dfrac{3a}{2c^2}\ln v + C, \text{ by (5)}\right]$

$= \dfrac{3a}{2c^2}\ln(b^2+c^2x^2) + C.$

10. $\int \dfrac{x^3\,dx}{x+1} = x - \dfrac{x^2}{2} + \dfrac{x^3}{3} - \ln(x+1) + C.$

Solution. First divide the numerator by the denominator. Then

$$\dfrac{x^3}{x+1} = x^2 - x + 1 - \dfrac{1}{x+1}.$$

Substituting in the integral, using (1), and integrating gives the answer.

11. $\int \dfrac{2x-1}{2x+3}\,dx = x - \ln(2x+3)^2 + C.$

Solution. Dividing, $\dfrac{2x-1}{2x+3} = 1 - \dfrac{4}{2x+3}$. Substitute and use (1) etc.

The function to be integrated is called the *integrand*. Thus in Illustrative Example 1, p. 193, the integrand is x^6.

PROBLEMS

Verify the following integrations.

1. $\int x^4\,dx = \dfrac{x^5}{5} + C.$

2. $\int \dfrac{dx}{x^2} = -\dfrac{1}{x} + C.$

3. $\int x^{\frac{2}{3}}\,dx = \dfrac{3x^{\frac{5}{3}}}{5} + C.$

4. $\int \dfrac{dx}{\sqrt{x}} = 2\sqrt{x} + C.$

5. $\int \dfrac{dx}{\sqrt[3]{x}} = \dfrac{3x^{\frac{2}{3}}}{2} + C.$

6. $\int 3ay^2\,dy = ay^3 + C.$

7. $\int \dfrac{2\,dt}{t^2} = -\dfrac{2}{t} + C.$

8. $\int \sqrt{ax}\,dx = \dfrac{2x\sqrt{ax}}{3} + C.$

9. $\int \dfrac{dx}{\sqrt{2x}} = \sqrt{2x} + C.$

10. $\int \sqrt[3]{3t}\,dt = \dfrac{(3t)^{\frac{4}{3}}}{4} + C.$

11. $\int (x^{\frac{3}{2}} - 2x^{\frac{2}{3}} + 5\sqrt{x} - 3)\,dx = \dfrac{2x^{\frac{5}{2}}}{5} - \dfrac{6x^{\frac{5}{3}}}{5} + \dfrac{10x^{\frac{3}{2}}}{3} - 3x + C.$

12. $\int \dfrac{4x^2 - 2\sqrt{x}}{x}\,dx = 2x^2 - 4\sqrt{x} + C.$

13. $\int \left(\dfrac{x^2}{2} - \dfrac{2}{x^2}\right)dx = \dfrac{x^3}{6} + \dfrac{2}{x} + C.$

14. $\int \sqrt{x}(3x-2)\,dx = \dfrac{6x^{\frac{5}{2}}}{5} - \dfrac{4x^{\frac{3}{2}}}{3} + C.$

INTEGRAL CALCULUS

15. $\int \dfrac{x^3 - 6x + 5}{x} dx = \dfrac{x^3}{3} - 6x + 5 \ln x + C.$

16. $\int \sqrt{a + bx}\, dx = \dfrac{2(a + bx)^{\frac{3}{2}}}{3b} + C.$

17. $\int \dfrac{dy}{\sqrt{a - by}} = -\dfrac{2\sqrt{a - by}}{b} + C.$

18. $\int (a + bt)^2\, dt = \dfrac{(a + bt)^3}{3b} + C.$

19. $\int x(2 + x^2)^2\, dx = \dfrac{(2 + x^2)^3}{6} + C.$

20. $\int y(a - by^2)\, dy = -\dfrac{(a - by^2)^2}{4b} + C.$

21. $\int t\sqrt{2t^2 + 3}\, dt = \dfrac{(2t^2 + 3)^{\frac{3}{2}}}{6} + C.$

22. $\int x(2x + 1)^2\, dx = x^4 + \dfrac{4x^3}{3} + \dfrac{x^2}{2} + C.$

23. $\int \dfrac{4x^2\, dx}{\sqrt{x^3 + 8}} = \dfrac{8\sqrt{x^3 + 8}}{3} + C.$

24. $\int \dfrac{6z\, dz}{(5 - 3z^2)^2} = \dfrac{1}{5 - 3z^2} + C.$

25. $\int (\sqrt{a} - \sqrt{x})^2\, dx = ax - \dfrac{4x\sqrt{ax}}{3} + \dfrac{x^2}{2} + C.$

26. $\int \dfrac{(\sqrt{a} - \sqrt{x})^2\, dx}{\sqrt{x}} = -\dfrac{2(\sqrt{a} - \sqrt{x})^3}{3} + C.$

27. $\int \sqrt{x}(\sqrt{a} - \sqrt{x})^2\, dx = \dfrac{2\, ax^{\frac{3}{2}}}{3} - x^2\sqrt{a} + \dfrac{2\, x^{\frac{5}{2}}}{5} + C.$

28. $\int \dfrac{t^3\, dt}{\sqrt{a^4 + t^4}} = \dfrac{\sqrt{a^4 + t^4}}{2} + C.$

29. $\int \dfrac{dy}{(a + by)^3} = -\dfrac{1}{2b(a + by)^2} + C.$

30. $\int \dfrac{x\, dx}{(a + bx^2)^3} = -\dfrac{1}{4b(a + bx^2)^2} + C.$

31. $\int \dfrac{t^2\, dt}{(a + bt^3)^2} = -\dfrac{1}{3b(a + bt^3)} + C.$

32. $\int z(a + bz^3)^2\, dz = \dfrac{a^2 z^2}{2} + \dfrac{2\, abz^5}{5} + \dfrac{b^2 z^8}{8} + C.$

33. $\int x^{n-1}\sqrt{a + bx^n}\, dx = \dfrac{2(a + bx^n)^{\frac{3}{2}}}{3nb} + C.$

34. $\int \dfrac{(2x + 3)\, dx}{\sqrt{x^2 + 3x}} = 2\sqrt{x^2 + 3x} + C.$

35. $\int \dfrac{(x^2 + 1)\, dx}{\sqrt{x^3 + 3x}} = \dfrac{2\sqrt{x^3 + 3x}}{3} + C.$

INTEGRATION

36. $\int \dfrac{(2 + \ln x)dx}{x} = \dfrac{(2 + \ln x)^2}{2} + C.$

37. $\int \sin^2 x \cos x \, dx = \int (\sin x)^2 \cos x \, dx = \dfrac{(\sin x)^3}{3} + C = \dfrac{\sin^3 x}{3} + C.$

HINT. Use (4), making $v = \sin x$, $dv = \cos x \, dx$, $n = 2$.

38. $\int \sin ax \cos ax \, dx = \dfrac{\sin^2 ax}{2a} + C.$

39. $\int \sin 2x \cos^2 2x \, dx = -\dfrac{\cos^3 2x}{6} + C.$

40. $\int \tan \dfrac{x}{2} \sec^2 \dfrac{x}{2} dx = \tan^2 \dfrac{x}{2} + C.$

41. $\int \dfrac{\cos ax \, dx}{\sqrt{b + \sin ax}} = \dfrac{2\sqrt{b + \sin ax}}{a} + C.$

42. $\int \left(\dfrac{\sec x}{1 + \tan x}\right)^2 dx = -\dfrac{1}{1 + \tan x} + C.$

43. $\int \dfrac{dx}{2 + 3x} = \dfrac{\ln(2 + 3x)}{3} + C.$

44. $\int \dfrac{x^2 \, dx}{2 + x^3} = \dfrac{\ln(2 + x^3)}{3} + C.$

45. $\int \dfrac{t \, dt}{a + bt^2} = \dfrac{\ln(a + bt^2)}{2b} + C.$

46. $\int \dfrac{(2x + 3)dx}{x^2 + 3x} = \ln(x^2 + 3x) + C.$

47. $\int \dfrac{(y + 2)dy}{y^2 + 4y} = \dfrac{\ln(y^2 + 4y)}{2} + C.$

48. $\int \dfrac{e^\theta \, d\theta}{a + be^\theta} = \dfrac{\ln(a + be^\theta)}{b} + C.$

49. $\int \dfrac{\sin x \, dx}{1 - \cos x} = \ln(1 - \cos x) + C.$

50. $\int \dfrac{\sec^2 y \, dy}{a + b \tan y} = \dfrac{1}{b} \ln(a + b \tan y) + C.$

51. $\int \dfrac{(2x + 3)dx}{x + 2} = 2x - \ln(x + 2) + C.$

52. $\int \dfrac{(x^2 + 2)dx}{x + 1} = \dfrac{x^2}{2} - x + 3 \ln(x + 1) + C$

53. $\int \dfrac{(x + 4)dx}{2x + 3} = \dfrac{x}{2} + \dfrac{5 \ln(2x + 3)}{4} + C.$

54. $\int \dfrac{e^{2s} \, ds}{e^{2s} + 1} = \tfrac{1}{2} \ln(e^{2s} + 1) + C.$

55. $\int \dfrac{ae^\theta + b}{ae^\theta - b} d\theta = 2 \ln(ae^\theta - b) - \theta + C.$

Work out each of the following integrals and verify your results by differentiation.

56. $\int \dfrac{2x\,dx}{\sqrt[3]{6-5x^2}}$.

Solution. $\int \dfrac{2x\,dx}{\sqrt[3]{6-5x^2}} = -\dfrac{1}{5}\int (6-5x^2)^{-\frac{1}{3}}(-10x\,dx) = -\dfrac{3}{10}(6-5x^2)^{\frac{2}{3}} + C.$

Verification. $d\{-\tfrac{3}{10}(6-5x^2)^{\frac{2}{3}} + C\} = -\tfrac{3}{10}\cdot\tfrac{2}{3}(6-5x^2)^{-\frac{1}{3}}(-10x)dx$
$$= \dfrac{2x\,dx}{\sqrt[3]{6-5x^2}}.$$

57. $\int (x^3 + 3x^2)dx.$

58. $\int \dfrac{(x^2-4)dx}{x^4}.$

59. $\int \left(\dfrac{\sqrt{5}\,x}{5} + \dfrac{5}{\sqrt{5}\,x}\right)dx.$

60. $\int \sqrt[3]{by^2}\,dy.$

61. $\int \dfrac{dt}{t\sqrt{2\,t}}.$

62. $\int \sqrt[3]{2-3x}\,dx.$

63. $\int \dfrac{\sin 2\theta\,d\theta}{\sqrt{\cos 2\theta}}.$

64. $\int \dfrac{e^x\,dx}{\sqrt{e^x-5}}.$

65. $\int \dfrac{2\,dx}{\sqrt{3+2x}}.$

66. $\int \dfrac{3\,dx}{2+3x}.$

67. $\int \dfrac{x\,dx}{\sqrt{1-2x^2}}.$

68. $\int \dfrac{t\,dt}{3t^2+4}.$

69. $\int \left(\sqrt{x} - \dfrac{1}{\sqrt{x}}\right)^2 dx.$

70. $\int \left(y^2 - \dfrac{1}{y^2}\right)^3 dy.$

71. $\int \dfrac{\sin a\theta\,d\theta}{\cos a\theta + b}.$

72. $\int \dfrac{\csc^2\phi\,d\phi}{\sqrt{2\ctn\phi+3}}.$

73. $\int \dfrac{(2x+5)dx}{x^2+5x+6}.$

74. $\int \dfrac{(2x+7)dx}{x+3}.$

75. $\int \dfrac{(x^2+2)dx}{x+2}.$

76. $\int \dfrac{(x^3+3x)dx}{x^2+1}.$

77. $\int \dfrac{(4x+3)dx}{\sqrt[3]{1+3x+2x^2}}.$

78. $\int \dfrac{(e^t+2)dt}{e^t+2t}.$

79. $\int \dfrac{(e^x+\sin x)dx}{\sqrt{e^x-\cos x}}.$

80. $\int \dfrac{\sec 2\theta\tan 2\theta\,d\theta}{3\sec 2\theta - 2}.$

81. $\int \dfrac{\sec^2 2t\,dt}{\sqrt{5+3\tan 2t}}.$

130. Proofs of (6) and (7). These follow at once from the corresponding formulas for differentiation, XI and XI a, Art. 94.

ILLUSTRATIVE EXAMPLE. Prove $\int ba^{2x}\,dx = \dfrac{ba^{2x}}{2\ln a} + C.$

Solution. $\int ba^{2x}\,dx = b\int a^{2x}\,dx.$ By (2)

This resembles (6). Let $v = 2x$; then $dv = 2\,dx$. If we then insert the factor 2 before dx and the factor $\tfrac{1}{2}$ before the integral sign, we have

$b\int a^{2x}\,dx = \dfrac{b}{2}\int a^{2x}\,2\,dx = \dfrac{b}{2}\int a^{2x}\,d(2x) \left[= \dfrac{b}{2}\int a^v\,dv = \dfrac{b}{2}\dfrac{a^v}{\ln a}\right] = \dfrac{b}{2}\cdot\dfrac{a^{2x}}{\ln a} + C.$ By (6)

INTEGRATION

PROBLEMS

Work out the following integrals.

1. $\int 6\, e^{3x}\, dx = 2\, e^{3x} + C.$

2. $\int e^{\frac{x}{n}}\, dx = n e^{\frac{x}{n}} + C.$

3. $\int \dfrac{dx}{e^x} = -\dfrac{1}{e^x} + C.$

4. $\int 10^x\, dx = \dfrac{10^x}{\ln 10} + C.$

5. $\int a^{ny}\, dy = \dfrac{a^{ny}}{n \ln a} + C.$

6. $\int \dfrac{e^{\sqrt{x}}\, dx}{\sqrt{x}} = 2 e^{\sqrt{x}} + C.$

7. $\int \left(e^{\frac{x}{a}} + e^{-\frac{x}{a}}\right) dx = a\left(e^{\frac{x}{a}} - e^{-\frac{x}{a}}\right) + C.$

8. $\int \left(e^{\frac{x}{a}} - e^{-\frac{x}{a}}\right)^2 dx = \dfrac{a}{2}\left(e^{\frac{2x}{a}} - e^{-\frac{2x}{a}}\right) - 2x + C.$

9. $\int x e^{x^2}\, dx = \tfrac{1}{2} e^{x^2} + C.$

10. $\int e^{\sin x} \cos x\, dx = e^{\sin x} + C.$

11. $\int e^{\tan \theta} \sec^2 \theta\, d\theta = e^{\tan \theta} + C.$

12. $\int \sqrt{e^t}\, dt = 2\sqrt{e^t} + C.$

13. $\int a^x e^x\, dx = \dfrac{a^x e^x}{1 + \ln a} + C.$

14. $\int a^{2x}\, dx = \dfrac{a^{2x}}{2 \ln a} + C.$

15. $\int (e^{5x} + a^{5x}) dx = \dfrac{1}{5}\left(e^{5x} + \dfrac{a^{5x}}{\ln a}\right) + C.$

Work out each of the following integrals and verify your results by differentiation.

16. $\int 5\, e^{ax}\, dx.$

17. $\int \dfrac{3\, dx}{e^x}.$

18. $\int \dfrac{4\, dt}{\sqrt{e^t}}.$

19. $\int c^{ax}\, dx.$

20. $\int \dfrac{dx}{4^{2x}}.$

21. $\int x^2 e^{x^3}\, dx.$

22. $\int \left(\dfrac{e^x + 4}{e^x}\right) dx.$

23. $\int \dfrac{e^x\, dx}{e^x - 2}.$

24. $\int x(e^{x^2} + 2)\, dx.$

25. $\int \dfrac{e^{\sqrt{x}} - 3}{\sqrt{x}}\, dx.$

26. $\int t\, 2^{t^2}\, dt.$

27. $\int \dfrac{a\, d\theta}{b^{3\theta}}.$

28. $\int 6\, x e^{-x^2}\, dx.$

29. $\int (e^{2x})^2\, dx.$

30. $\int e^{\cos 2x} \sin 2x\, dx.$

31. $\int \dfrac{x^2\, dx}{e^{x^3}}.$

INTEGRAL CALCULUS

131. Proofs of (8)–(17). Formulas (8)–(13) follow at once from the corresponding formulas for differentiation, XIII, etc., Art. 94.

Proof of (14).
$$\int \tan v \, dv = \int \frac{\sin v \, dv}{\cos v}$$
$$= -\int \frac{-\sin v \, dv}{\cos v}$$
$$= -\int \frac{d(\cos v)}{\cos v}$$
$$= -\ln \cos v + C \qquad \text{by (5)}$$
$$= \ln \sec v + C.$$

$\left[\text{Since } -\ln \cos v = -\ln \frac{1}{\sec v} = -\ln 1 + \ln \sec v = \ln \sec v. \right]$

Proof of (15).
$$\int \ctn v \, dv = \int \frac{\cos v \, dv}{\sin v} = \int \frac{d(\sin v)}{\sin v}$$
$$= \ln \sin v + C. \qquad \text{By (5)}$$

Proof of (16). Since $\sec v = \sec v \dfrac{\sec v + \tan v}{\sec v + \tan v}$

$$= \frac{\sec v \tan v + \sec^2 v}{\sec v + \tan v},$$

$$\int \sec v \, dv = \int \frac{\sec v \tan v + \sec^2 v}{\sec v + \tan v} dv$$
$$= \int \frac{d(\sec v + \tan v)}{\sec v + \tan v}$$
$$= \ln (\sec v + \tan v) + C. \qquad \text{By (5)}$$

Proof of (17). Since $\csc v = \csc v \dfrac{\csc v - \ctn v}{\csc v - \ctn v}$

$$= \frac{-\csc v \ctn v + \csc^2 v}{\csc v - \ctn v},$$

$$\int \csc v \, dv = \int \frac{-\csc v \ctn v + \csc^2 v}{\csc v - \ctn v} dv$$
$$= \int \frac{d(\csc v - \ctn v)}{\csc v - \ctn v}$$
$$= \ln (\csc v - \ctn v) + C. \qquad \text{By (5)}$$

An alternative form of (17) is

$$\int \csc v \, dv = \ln \tan \tfrac{1}{2} v + C \text{ (see Problem 4, p. 201).}$$

INTEGRATION

ILLUSTRATIVE EXAMPLE 1. Prove the following integration.
$$\int \sin 2\,ax\,dx = -\frac{\cos 2\,ax}{2\,a} + C.$$

Solution. This resembles (8). For let $v = 2\,ax$; then $dv = 2\,a\,dx$. If we now insert the factor $2\,a$ before dx and the factor $\frac{1}{2\,a}$ before the integral sign, we get

$$\int \sin 2\,ax\,dx = \frac{1}{2\,a}\int \sin 2\,ax \cdot 2\,a\,dx \left[= \frac{1}{2\,a}\int \sin v\,dv = -\frac{1}{2\,a}\cos v + C,\text{ by (8)}\right]$$
$$= \frac{1}{2\,a}\cdot -\cos 2\,ax + C = -\frac{\cos 2\,ax}{2\,a} + C.$$

ILLUSTRATIVE EXAMPLE 2. Prove the following integration.
$$\int (\tan 2\,s - 1)^2\,ds = \tfrac{1}{2}\tan 2\,s + \ln \cos 2\,s + C.$$

Solution. $(\tan 2\,s - 1)^2 = \tan^2 2\,s - 2\tan 2\,s + 1.$
$\tan^2 2\,s = \sec^2 2\,s - 1.$ By (2), Art. 2
Hence, substituting,
$$\int (\tan 2\,s - 1)^2\,ds = \int (\sec^2 2\,s - 2\tan 2\,s)\,ds = \int \sec^2 2\,s\,ds - 2\int \tan 2\,s\,ds.$$

Now let $v = 2\,s$. Then $dv = 2\,ds$. Using (10) and (14), the steps are as follows.

$$\int \sec^2 2\,s\,ds = \tfrac{1}{2}\int \sec^2 2\,s\,d(2\,s)\left[= \tfrac{1}{2}\int \sec^2 v\,dv = \tfrac{1}{2}\tan v \right] = \tfrac{1}{2}\tan 2\,s.$$
$$\int \tan 2\,s\,ds = \tfrac{1}{2}\int \tan 2\,s\,d(2\,s)\left[= \tfrac{1}{2}\int \tan v\,dv = -\tfrac{1}{2}\ln \cos v \right] = -\tfrac{1}{2}\ln \cos 2\,s.$$

PROBLEMS

Verify the following integrations.

1. $\int \cos mx\,dx = \dfrac{1}{m}\sin mx + C.$

2. $\int \tan bx\,dx = \dfrac{1}{b}\ln \sec bx + C.$

3. $\int \sec ax\,dx = \dfrac{1}{a}\ln (\sec ax + \tan ax) + C.$

4. $\int \csc v\,dv = \ln \tan \tfrac{1}{2}v + C.$

5. $\int \sec 3\,t \tan 3\,t\,dt = \tfrac{1}{3}\sec 3\,t + C.$

6. $\int \csc ay \operatorname{ctn} ay\,dy = -\dfrac{1}{a}\csc ay + C.$

7. $\int \csc^2 3\,x\,dx = -\tfrac{1}{3}\operatorname{ctn} 3\,x + C.$

8. $\int \operatorname{ctn}\dfrac{x}{2}\,dx = 2\ln \sin \dfrac{x}{2} + C.$

9. $\int x^2 \sec^2 x^3\,dx = \tfrac{1}{3}\tan x^3 + C.$

10. $\int \dfrac{dx}{\sin^2 x} = -\operatorname{ctn} x + C.$

INTEGRAL CALCULUS

11. $\int \dfrac{ds}{\cos^2 s} = \tan s + C.$

12. $\int (\tan \theta + \operatorname{ctn} \theta)^2 \, d\theta = \tan \theta - \operatorname{ctn} \theta + C.$

13. $\int (\sec \phi - \tan \phi)^2 \, d\phi = 2(\tan \phi - \sec \phi) - \phi + C.$

14. $\int \dfrac{dx}{1 + \cos x} = -\operatorname{ctn} x + \csc x + C.$

HINT. Multiply both numerator and denominator by $1 - \cos x$ and reduce before integrating.

15. $\int \dfrac{dx}{1 + \sin x} = \tan x - \sec x + C.$

16. $\int \dfrac{\sin s \, ds}{1 + \cos s} = -\ln(1 + \cos s) + C.$

17. $\int \dfrac{\sec^2 x \, dx}{1 + \tan x} = \ln(1 + \tan x) + C.$

18. $\int x \cos x^2 \, dx = \tfrac{1}{2} \sin x^2 + C.$

19. $\int (x + \sin 2x) dx = \tfrac{1}{2}(x^2 - \cos 2x) + C.$

20. $\int \dfrac{\sin x \, dx}{\sqrt{4 - \cos x}} = 2\sqrt{4 - \cos x} + C.$

21. $\int \dfrac{(1 + \cos x) dx}{x + \sin x} = \ln(x + \sin x) + C.$

22. $\int \dfrac{\sec^2 \theta \, d\theta}{\sqrt{1 + 2\tan \theta}} = \sqrt{1 + 2\tan \theta} + C.$

Work out each of the following integrals and verify your results by differentiation.

23. $\int \sin \dfrac{2x}{3} \, dx.$

24. $\int \cos (b + ax) dx.$

25. $\int \csc^2 (a - bx) dx.$

26. $\int \sec \dfrac{\theta}{2} \tan \dfrac{\theta}{2} \, d\theta.$

27. $\int \csc \dfrac{a\phi}{b} \operatorname{ctn} \dfrac{a\phi}{b} \, d\phi.$

28. $\int e^x \operatorname{ctn} e^x \, dx.$

29. $\int \sec^2 2ax \, dx.$

30. $\int \tan \dfrac{x}{3} \, dx.$

31. $\int \dfrac{dt}{\tan 5t}.$

32. $\int \dfrac{d\theta}{\sin^2 4\theta}.$

33. $\int \dfrac{dy}{\operatorname{ctn} 7y}.$

34. $\int \dfrac{\sin \sqrt{x} \, dx}{\sqrt{x}}.$

INTEGRATION

35. $\int \dfrac{dt}{\sin^2 3\, t}$.

36. $\int \dfrac{d\phi}{\cos 4\, \phi}$.

37. $\int \dfrac{a\, dx}{\cos^2 bx}$.

38. $\int \left(\sec 2\, \theta - \csc \dfrac{\theta}{2} \right) d\theta$.

39. $\int (\tan \phi + \sec \phi)^2\, d\phi$.

40. $\int \left(\tan 4\, s - \ctn \dfrac{s}{4} \right) ds$.

41. $\int (\ctn x - 1)^2\, dx$.

42. $\int (\sec t - 1)^2\, dt$.

43. $\int (1 - \csc y)^2\, dy$.

44. $\int \dfrac{dx}{1 - \cos x}$.

45. $\int \dfrac{dx}{1 - \sin x}$.

46. $\int \dfrac{\sin 2\, x\, dx}{3 + \cos 2\, x}$.

47. $\int \dfrac{\cos t\, dt}{\sqrt{a + b \sin t}}$.

48. $\int \dfrac{\csc \theta \ctn \theta\, d\theta}{5 - 4 \csc \theta}$.

49. $\int \dfrac{\csc^2 x\, dx}{\sqrt[3]{3 - \ctn x}}$.

50. $\int \dfrac{\sqrt{5 + 2 \tan x}\, dx}{\cos^2 x}$.

132. Proofs of (18)–(21). Formulas (18) and (20) follow easily from the corresponding formulas for differentiation.

Proof of (18). Since

$$d\left(\dfrac{1}{a} \arctan \dfrac{v}{a} + C \right) = \dfrac{1}{a} \dfrac{d\left(\dfrac{v}{a}\right)}{1 + \left(\dfrac{v}{a}\right)^2} = \dfrac{dv}{v^2 + a^2}, \text{ by XXII, Art. 60}$$

we get

$$\int \dfrac{dv}{v^2 + a^2} = \dfrac{1}{a} \arctan \dfrac{v}{a} + C.$$

Proofs of (19) and (19 a). We prove (19) first. By algebra, we have

$$\dfrac{1}{v - a} - \dfrac{1}{v + a} = \dfrac{2\, a}{v^2 - a^2}.$$

Hence

$$\dfrac{1}{v^2 - a^2} = \dfrac{1}{2\, a}\left[\dfrac{1}{v - a} - \dfrac{1}{v + a} \right].$$

Then

$$\int \dfrac{dv}{v^2 - a^2} = \dfrac{1}{2\, a} \int \dfrac{dv}{v - a} - \dfrac{1}{2\, a} \int \dfrac{dv}{v + a} \qquad \text{by (1)}$$

$$= \dfrac{1}{2\, a} \ln (v - a) - \dfrac{1}{2\, a} \ln (v + a) \qquad \text{by (5)}$$

$$= \dfrac{1}{2\, a} \ln \dfrac{v - a}{v + a} + C. \qquad \text{By (2), Art. 1}$$

To prove **(19 a)**, by algebra,

$$\dfrac{1}{a + v} + \dfrac{1}{a - v} = \dfrac{2\, a}{a^2 - v^2}.$$

204 INTEGRAL CALCULUS

The rest of the proof proceeds as above.

NOTE. The integrals in (19) and (19 a) satisfy the relation

$$\int \frac{dv}{v^2 - a^2} = -\int \frac{dv}{a^2 - v^2}.$$

Hence either formula can be applied in any given case. Later we shall see that one form must be chosen in many numerical examples.

Proof of (20). Since

$$d\left(\arcsin \frac{v}{a} + C\right) = \frac{d\left(\frac{v}{a}\right)}{\sqrt{1 - \left(\frac{v}{a}\right)^2}} = \frac{dv}{\sqrt{a^2 - v^2}}, \text{ by XX, Art. 94}$$

we get

$$\int \frac{dv}{\sqrt{a^2 - v^2}} = \arcsin \frac{v}{a} + C.$$

Proof of (21). Assume $v = a \tan z$, where z is a new variable; differentiating, $dv = a \sec^2 z \, dz$. Hence, by substitution,

$$\int \frac{dv}{\sqrt{v^2 + a^2}} = \int \frac{a \sec^2 z \, dz}{\sqrt{a^2 \tan^2 z + a^2}} = \int \frac{\sec^2 z \, dz}{\sqrt{\tan^2 z + 1}}$$

$$= \int \sec z \, dz = \ln (\sec z + \tan z) + c \qquad \text{by (16)}$$

$$= \ln (\tan z + \sqrt{\tan^2 z + 1}) + c. \qquad \text{By (2), Art. 2}$$

But $\tan z = \frac{v}{a}$; hence,

$$\int \frac{dv}{\sqrt{v^2 + a^2}} = \ln \left(\frac{v}{a} + \sqrt{\frac{v^2}{a^2} + 1}\right) + c$$

$$= \ln \frac{v + \sqrt{v^2 + a^2}}{a} + c$$

$$= \ln (v + \sqrt{v^2 + a^2}) - \ln a + c.$$

Placing $C = -\ln a + c$, we get

$$\int \frac{dv}{\sqrt{v^2 + a^2}} = \ln (v + \sqrt{v^2 + a^2}) + C.$$

In the same manner, by assuming $v = a \sec z$, $dv = a \sec z \tan z \, dz$, we get

$$\int \frac{dv}{\sqrt{v^2 - a^2}} = \int \frac{a \sec z \tan z \, dz}{\sqrt{a^2 \sec^2 z - a^2}} = \int \sec z \, dz$$

$$= \ln (\sec z + \tan z) + c \qquad \text{by (16)}$$

$$= \ln (\sec z + \sqrt{\sec^2 z - 1}) + c \qquad \text{by (2), Art. 2}$$

$$= \ln \left(\frac{v}{a} + \sqrt{\frac{v^2}{a^2} - 1}\right) + c = \ln (v + \sqrt{v^2 - a^2}) + C.$$

INTEGRATION

ILLUSTRATIVE EXAMPLE. Work out the following integration.

$$\int \frac{dx}{4x^2 + 9} = \frac{1}{6} \arctan \frac{2x}{3} + C.$$

Solution. This resembles (18). For, let $v^2 = 4x^2$ and $a^2 = 9$; then $v = 2x$, $dv = 2\,dx$, and $a = 3$. Hence if we multiply the numerator by 2 and divide in front of the integral sign by 2, we get

$$\int \frac{dx}{4x^2 + 9} = \frac{1}{2} \int \frac{2\,dx}{(2x)^2 + (3)^2} \left[= \frac{1}{2} \int \frac{dv}{v^2 + a^2} = \frac{1}{2a} \arctan \frac{v}{a} + C. \text{ By (18)} \right]$$

$$= \frac{1}{6} \arctan \frac{2x}{3} + C.$$

PROBLEMS

Work out the following integrals.

1. $\displaystyle\int \frac{dx}{x^2 + 9} = \frac{1}{3} \arctan \frac{x}{3} + C.$

2. $\displaystyle\int \frac{dx}{x^2 - 4} = \frac{1}{4} \ln \left(\frac{x-2}{x+2}\right) + C.$

3. $\displaystyle\int \frac{dy}{\sqrt{25 - y^2}} = \arcsin \frac{y}{5} + C.$

4. $\displaystyle\int \frac{ds}{\sqrt{s^2 - 16}} = \ln (s + \sqrt{s^2 - 16}) + C.$

5. $\displaystyle\int \frac{dx}{9x^2 - 4} = \frac{1}{12} \ln \left(\frac{3x-2}{3x+2}\right) + C.$

6. $\displaystyle\int \frac{dx}{\sqrt{16 - 9x^2}} = \frac{1}{3} \arcsin \frac{3x}{4} + C.$

7. $\displaystyle\int \frac{dx}{9x^2 - 1} = \frac{1}{6} \ln \left(\frac{3x-1}{3x+1}\right) + C.$

8. $\displaystyle\int \frac{dt}{4 - 9t^2} = \frac{1}{12} \ln \left(\frac{2+3t}{2-3t}\right) + C.$

9. $\displaystyle\int \frac{e^x\,dx}{1 + e^{2x}} = \arctan e^x + C.$

10. $\displaystyle\int \frac{\cos\theta\,d\theta}{4 - \sin^2\theta} = \frac{1}{4} \ln \left(\frac{2+\sin\theta}{2-\sin\theta}\right) + C.$

11. $\displaystyle\int \frac{b\,dx}{a^2 x^2 - c^2} = \frac{b}{2ac} \ln \left(\frac{ax-c}{ax+c}\right) + C.$

12. $\displaystyle\int \frac{5x\,dx}{\sqrt{1 - x^4}} = \frac{5}{2} \arcsin x^2 + C.$

13. $\displaystyle\int \frac{ax\,dx}{x^4 + b^4} = \frac{a}{2b^2} \arctan \frac{x^2}{b^2} + C.$

14. $\displaystyle\int \frac{dt}{(t-2)^2 + 9} = \frac{1}{3} \arctan \left(\frac{t-2}{3}\right) + C.$

15. $\int \dfrac{dy}{\sqrt{1 + a^2 y^2}} = \dfrac{1}{a} \ln \left(ay + \sqrt{1 + a^2 y^2}\right) + C.$

16. $\int \dfrac{du}{\sqrt{4 - (u + 3)^2}} = \arcsin \left(\dfrac{u + 3}{2}\right) + C.$

Work out each of the following integrals and verify your results by differentiation.

17. $\int \dfrac{dx}{\sqrt{9 - 16 x^2}}.$

18. $\int \dfrac{dy}{\sqrt{9 y^2 + 4}}.$

19. $\int \dfrac{dt}{4 t^2 + 25}.$

20. $\int \dfrac{dx}{25 x^2 - 4}.$

21. $\int \dfrac{7 \, dx}{3 + 7 x^2}.$

22. $\int \dfrac{3 \, dy}{9 y^2 - 16}.$

23. $\int \dfrac{ds}{\sqrt{4 s^2 + 5}}.$

24. $\int \dfrac{t \, dt}{\sqrt{t^4 - 4}}.$

25. $\int \dfrac{x \, dx}{\sqrt{5 x^2 + 3}}.$

26. $\int \dfrac{2 e^x \, dx}{\sqrt{1 - e^{2x}}}.$

27. $\int \dfrac{6 t \, dt}{8 - 3 t^2}.$

28. $\int \dfrac{\sin \theta \, d\theta}{\sqrt{4 + \cos^2 \theta}}.$

29. $\int \dfrac{dx}{m^2 + (x + n)^2}.$

30. $\int \dfrac{du}{4 - (2 u - 1)^2}.$

31. $\int \dfrac{7 x^2 \, dx}{5 - x^6}.$

The standard formulas (18)–(21) involve quadratic expressions ($v^2 \pm a^2$, $a^2 - v^2$) with two terms only. If an integral involves a quadratic expression containing three terms, the latter may be reduced to one with two terms by completing the square, as shown in the following examples.

ILLUSTRATIVE EXAMPLE 1. Verify the following:

$$\int \dfrac{dx}{x^2 + 2 x + 5} = \dfrac{1}{2} \arctan \dfrac{x + 1}{2} + C.$$

Solution. $x^2 + 2 x + 5 = x^2 + 2 x + 1 + 4 = (x + 1)^2 + 4.$

$$\therefore \int \dfrac{dx}{x^2 + 2 x + 5} = \int \dfrac{dx}{(x + 1)^2 + 2^2}.$$

This is in the form (18). For let $v = x + 1$, and $a = 2$. Then $dv = dx$. Hence the above becomes

$$\int \dfrac{dv}{v^2 + a^2} = \dfrac{1}{a} \arctan \dfrac{v}{a} + C = \dfrac{1}{2} \arctan \dfrac{x + 1}{2} + C.$$

ILLUSTRATIVE EXAMPLE 2. $\int \dfrac{2 \, dx}{\sqrt{2 + x - x^2}} = 2 \arcsin \dfrac{2 x - 1}{3} + C.$

Solution. This is in the form (20), *since the coefficient of x^2 is negative*.

Now $2 + x - x^2 = 2 - (x^2 - x + \tfrac{1}{4}) + \tfrac{1}{4} = \tfrac{9}{4} - (x - \tfrac{1}{2})^2.$

Let $v = x - \tfrac{1}{2}$, $a = \tfrac{3}{2}$. Then $dv = dx.$

$\int \dfrac{2 \, dx}{\sqrt{2 + x - x^2}} = 2 \int \dfrac{dx}{\sqrt{\tfrac{9}{4} - (x - \tfrac{1}{2})^2}} = 2 \int \dfrac{dv}{\sqrt{a^2 - v^2}} = 2 \arcsin \dfrac{v}{a} + C,$ by (20)

$= 2 \arcsin \dfrac{2 x - 1}{3} + C.$

INTEGRATION 207

ILLUSTRATIVE EXAMPLE 3. $\int \dfrac{dx}{3\,x^2 + 4\,x - 7} = \dfrac{1}{10} \ln \dfrac{3\,x - 3}{3\,x + 7} + C.$

Solution. $3\,x^2 + 4\,x - 7 = 3(x^2 + \tfrac{4}{3}\,x - \tfrac{7}{3}) = 3(x^2 + \tfrac{4}{3}\,x + \tfrac{4}{9} - \tfrac{25}{9})$
$\qquad\qquad\qquad\qquad = 3[(x + \tfrac{2}{3})^2 - \tfrac{25}{9}].$

$\therefore \int \dfrac{dx}{3\,x^2 + 4\,x - 7} = \int \dfrac{dx}{3[(x + \tfrac{2}{3})^2 - \tfrac{25}{9}]} = \dfrac{1}{3}\int \dfrac{dv}{v^2 - a^2},$

by form (19), if $v = x + \tfrac{2}{3}$, $a = \tfrac{5}{3}$, since also $dv = dx$. Then we have

$\dfrac{1}{3}\int \dfrac{dv}{v^2 - a^2} = \dfrac{1}{6\,a}\ln\dfrac{v - a}{v + a} + C = \dfrac{1}{10}\ln\dfrac{x + \tfrac{2}{3} - \tfrac{5}{3}}{x + \tfrac{2}{3} + \tfrac{5}{3}} + C,$ etc.

PROBLEMS

Work out the following integrals.

1. $\displaystyle\int \dfrac{dx}{x^2 + 4\,x + 3} = \dfrac{1}{2}\ln\left(\dfrac{x + 1}{x + 3}\right) + C.$

2. $\displaystyle\int \dfrac{dx}{2\,x - x^2 - 10} = -\dfrac{1}{3}\arctan\left(\dfrac{x - 1}{3}\right) + C.$

3. $\displaystyle\int \dfrac{3\,dx}{x^2 - 8\,x + 25} = \arctan\left(\dfrac{x - 4}{3}\right) + C.$

4. $\displaystyle\int \dfrac{dx}{\sqrt{3\,x - x^2 - 2}} = \arcsin(2\,x - 3) + C.$

5. $\displaystyle\int \dfrac{dv}{v^2 - 6\,v + 5} = \dfrac{1}{4}\ln\left(\dfrac{v - 5}{v - 1}\right) + C.$

6. $\displaystyle\int \dfrac{dx}{2\,x^2 - 2\,x + 1} = \arctan(2\,x - 1) + C.$

7. $\displaystyle\int \dfrac{dx}{\sqrt{15 + 2\,x - x^2}} = \arcsin\left(\dfrac{x - 1}{4}\right) + C.$

8. $\displaystyle\int \dfrac{dx}{x^2 + 2\,x} = \dfrac{1}{2}\ln\left(\dfrac{x}{x + 2}\right) + C.$

9. $\displaystyle\int \dfrac{dx}{4\,x - x^2} = \dfrac{1}{4}\ln\left(\dfrac{x}{x - 4}\right) + C.$

10. $\displaystyle\int \dfrac{dx}{\sqrt{2\,x - x^2}} = \arcsin(x - 1) + C.$

11. $\displaystyle\int \dfrac{ds}{\sqrt{2\,as + s^2}} = \ln(s + a + \sqrt{2\,as + s^2}) + C.$

12. $\displaystyle\int \dfrac{dy}{y^2 + 3\,y + 1} = \dfrac{1}{\sqrt{5}}\ln\left(\dfrac{2\,y + 3 - \sqrt{5}}{2\,y + 3 + \sqrt{5}}\right) + C.$

13. $\displaystyle\int \dfrac{dx}{1 + x + x^2} = \dfrac{2}{\sqrt{3}}\arctan\left(\dfrac{2\,x + 1}{\sqrt{3}}\right) + C.$

14. $\int \dfrac{dx}{\sqrt{1+x+x^2}} = \ln\left(x+\dfrac{1}{2}+\sqrt{1+x+x^2}\right) + C.$

15. $\int \dfrac{dx}{4x^2+4x+5} = \dfrac{1}{4}\arctan\left(\dfrac{2x+1}{2}\right) + C.$

16. $\int \dfrac{dx}{3x^2-2x+4} = \dfrac{1}{\sqrt{11}}\arctan\left(\dfrac{3x-1}{\sqrt{11}}\right) + C.$

17. $\int \dfrac{dx}{\sqrt{2-3x-4x^2}} = \dfrac{1}{2}\arcsin\left(\dfrac{8x+3}{\sqrt{41}}\right) + C.$

Work out each of the following integrals and verify your results by differentiation.

18. $\int \dfrac{dx}{x^2+2x+10}.$

19. $\int \dfrac{dx}{x^2+2x-3}.$

20. $\int \dfrac{dy}{3-2y-y^2}.$

21. $\int \dfrac{3\,du}{\sqrt{5-4u-u^2}}.$

22. $\int \dfrac{5\,dx}{\sqrt{x^2+2x+5}}.$

23. $\int \dfrac{dx}{\sqrt{x^2+4x+3}}.$

24. $\int \dfrac{dx}{\sqrt{x^2+2x}}.$

25. $\int \dfrac{dt}{\sqrt{3t-2t^2}}.$

26. $\int \dfrac{dx}{x^2-4x+5}.$

27. $\int \dfrac{dx}{2+2x-x^2}.$

28. $\int \dfrac{dr}{r^2-2r-3}.$

29. $\int \dfrac{4\,dx}{\sqrt{x^2-4x+13}}.$

30. $\int \dfrac{dz}{\sqrt{3+2z-z^2}}.$

31. $\int \dfrac{dv}{\sqrt{v^2-8v+15}}.$

32. $\int \dfrac{x\,dx}{x^4-x^2-1}.$

33. $\int \dfrac{dt}{\sqrt{1-t-2t^2}}.$

34. $\int \dfrac{dx}{3x^2+4x+1}.$

35. $\int \dfrac{dw}{2w^2+2w+1}.$

36. $\int \dfrac{x^2\,dx}{9x^6-3x^3-1}.$

37. $\int \dfrac{dt}{15+4t-t^2}.$

38. $\int \dfrac{dx}{\sqrt{9x^2+12x+8}}.$

39. $\int \dfrac{dx}{\sqrt{4x^2-12x+7}}.$

When the integrand is a fraction of which the numerator is an expression of the first degree while the denominator is an expression of the second degree or the square root of such an expression, the integral can be reduced to standard forms, as shown in the following examples.

INTEGRATION

ILLUSTRATIVE EXAMPLE 1. Prove the following:

$$\int \frac{3x-1}{\sqrt{4x^2+9}}\, dx = \frac{3}{4}\sqrt{4x^2+9} - \frac{1}{2}\ln(2x+\sqrt{4x^2+9}) + C.$$

Solution. Multiply through by dx and apply (1).

$$\int \frac{3x-1}{\sqrt{4x^2+9}}\, dx = \int \frac{3x\, dx}{\sqrt{4x^2+9}} - \int \frac{dx}{\sqrt{4x^2+9}}.$$

Then by (4) and (21) we obtain the answer.

ILLUSTRATIVE EXAMPLE 2.

$$\int \frac{2x-3}{3x^2+4x-7}\, dx = \frac{1}{3}\ln\left(x^2+\frac{4}{3}x-\frac{7}{3}\right) - \frac{13}{30}\ln\frac{3x-3}{3x+7} + C.$$

Solution. $3x^2 + 4x - 7 = 3[(x+\frac{2}{3})^2 - \frac{25}{9}]$, by Illustrative Example 3, p. 207. Let $v = x + \frac{2}{3}$. Then $x = v - \frac{2}{3}$, and $dx = dv$. Substituting,

$$\int \frac{2x-3}{3x^2+4x-7}\, dx = \int \frac{2(v-\frac{2}{3})-3}{3(v^2-\frac{25}{9})}\, dv = \frac{1}{9}\int \frac{6v-13}{v^2-\frac{25}{9}}\, dv$$
$$= \frac{1}{3}\int \frac{2v\, dv}{v^2-\frac{25}{9}} - \frac{13}{9}\int \frac{dv}{v^2-\frac{25}{9}}.$$

Using (5) and (19), and substituting back $v = x + \frac{2}{3}$, we have the above result.

PROBLEMS

Work out the following integrals.

1. $\displaystyle \int \frac{(1+2x)dx}{1+x^2} = \arctan x + \ln(1+x^2) + C.$

2. $\displaystyle \int \frac{(2x+1)dx}{\sqrt{x^2-1}} = 2\sqrt{x^2-1} + \ln(x+\sqrt{x^2-1}) + C.$

3. $\displaystyle \int \frac{(x-1)dx}{\sqrt{1-x^2}} = -\sqrt{1-x^2} - \arcsin x + C.$

4. $\displaystyle \int \frac{(3x-1)dx}{x^2+9} = \frac{3}{2}\ln(x^2+9) - \frac{1}{3}\arctan\frac{x}{3} + C.$

5. $\displaystyle \int \frac{(3s-2)ds}{\sqrt{9-s^2}} = -3\sqrt{9-s^2} - 2\arcsin\frac{s}{3} + C.$

6. $\displaystyle \int \frac{(x+3)dx}{\sqrt{x^2+4}} = \sqrt{x^2+4} + 3\ln(x+\sqrt{x^2+4}) + C.$

7. $\displaystyle \int \frac{(2x-5)dx}{3x^2-2} = \frac{1}{3}\ln(3x^2-2) - \frac{5\sqrt{6}}{12}\ln\left(\frac{3x-\sqrt{6}}{3x+\sqrt{6}}\right) + C.$

8. $\displaystyle \int \frac{(5t-1)dt}{\sqrt{3t^2-9}} = \frac{5}{3}\sqrt{3t^2-9} - \frac{\sqrt{3}}{3}\ln(t\sqrt{3}+\sqrt{3t^2-9}) + C.$

9. $\displaystyle \int \frac{(x+3)dx}{6x-x^2} = -\frac{1}{2}\ln(6x-x^2) - \ln\left(\frac{x-6}{x}\right) + C.$

10. $\displaystyle \int \frac{(2x+5)dx}{x^2+2x+5} = \ln(x^2+2x+5) + \frac{3}{2}\arctan\left(\frac{x+1}{2}\right) + C.$

210 INTEGRAL CALCULUS

11. $\int \dfrac{(1-x)dx}{4x^2-4x-3} = -\dfrac{1}{8}\ln(4x^2-4x-3) + \dfrac{1}{16}\ln\left(\dfrac{2x-3}{2x+1}\right) + C.$

12. $\int \dfrac{(3x-2)dx}{1-6x-9x^2} = -\dfrac{1}{6}\ln(1-6x-9x^2)$
$\qquad + \dfrac{\sqrt{2}}{4}\ln\left(\dfrac{3x+1-\sqrt{2}}{3x+1+\sqrt{2}}\right) + C.$

13. $\int \dfrac{(x+3)dx}{\sqrt{x^2+2x}} = \sqrt{x^2+2x} + 2\ln(x+1+\sqrt{x^2+2x}) + C.$

14. $\int \dfrac{(x+2)dx}{\sqrt{4x-x^2}} = -\sqrt{4x-x^2} + 4\arcsin\left(\dfrac{x-2}{2}\right) + C.$

15. $\int \dfrac{x\,dx}{\sqrt{27+6x-x^2}} = -\sqrt{27+6x-x^2} + 3\arcsin\left(\dfrac{x-3}{6}\right) + C.$

16. $\int \dfrac{(3x+2)dx}{\sqrt{19-5x+x^2}} = 3\sqrt{19-5x+x^2}$
$\qquad + \tfrac{19}{2}\ln(x-\tfrac{5}{2}+\sqrt{19-5x+x^2}) + C.$

17. $\int \dfrac{(3x-2)dx}{\sqrt{4x^2-4x+5}} = \dfrac{3}{4}\sqrt{4x^2-4x+5}$
$\qquad - \tfrac{1}{4}\ln(2x-1+\sqrt{4x^2-4x+5}) + C.$

18. $\int \dfrac{(8x-3)dx}{\sqrt{12x-4x^2-5}} = -2\sqrt{12x-4x^2-5}$
$\qquad + \dfrac{9}{2}\arcsin\left(\dfrac{2x-3}{2}\right) + C.$

Work out each of the following integrals and verify your results by differentiation.

19. $\int \dfrac{(4x+3)dx}{x^2+1}.$

20. $\int \dfrac{(3x-4)dx}{x^2-1}.$

21. $\int \dfrac{(3-x)dx}{4-3x^2}.$

22. $\int \dfrac{(2x+3)dx}{\sqrt{2-3x^2}}.$

23. $\int \dfrac{(4x-1)dx}{\sqrt{3+5x^2}}.$

24. $\int \dfrac{(3x-5)dx}{x^2+4x}.$

25. $\int \dfrac{(4x+5)dx}{\sqrt{3x-x^2}}.$

26. $\int \dfrac{(x+2)dx}{x^2-6x+5}.$

27. $\int \dfrac{(3-4x)dx}{\sqrt{3x-x^2-2}}.$

28. $\int \dfrac{(5x+2)dx}{\sqrt{x^2+2x+5}}.$

29. $\int \dfrac{(1-x)dx}{\sqrt{x^2+4x+3}}.$

30. $\int \dfrac{(8-3x)dx}{x^2+x+1}.$

31. $\int \dfrac{(x+4)dx}{\sqrt{x^2+x+1}}.$

32. $\int \dfrac{(2x+7)dx}{2x^2+2x+1}.$

33. $\int \dfrac{(3x+8)dx}{9x^2-3x-1}.$

34. $\int \dfrac{(6-x)dx}{\sqrt{4x^2-12x+7}}.$

INTEGRATION

133. Proofs of (22) and (23). To prove (22), substitute
$$v = a \sin z.$$
Then
$$dv = a \cos z \, dz,$$
and
$$\sqrt{a^2 - v^2} = \sqrt{a^2 - a^2 \sin^2 z} = a \cos z.$$
Hence
$$\int \sqrt{a^2 - v^2} \, dv = a^2 \int \cos^2 z \, dz = \frac{a^2}{2} \int (\cos 2z + 1) dz \text{ by (5), Art. 2}$$
$$= \frac{a^2}{4} \sin 2z + \frac{a^2}{2} z + C.$$

To obtain the result in terms of v, we have, from above,
$$z = \arcsin \frac{v}{a}, \quad \text{and} \quad \sin 2z = 2 \sin z \cos z = 2 \frac{v}{a} \frac{\sqrt{a^2 - v^2}}{a}.$$
Substituting, we obtain (22).

Proof of (23). By substituting $v = a \tan z$, we show (see Art. 132) readily that

(1) $\quad \int \sqrt{v^2 + a^2} \, dv = \int a \sec z \cdot a \sec^2 z \, dz = a^2 \int \sec^3 z \, dz.$

In a later section it is proved that

(2) $\quad \int \sec^3 z \, dz = \frac{1}{2} \sec z \tan z + \frac{1}{2} \ln (\sec z + \tan z) + C.$

Since $\tan z = \frac{v}{a}$, and $\sec z = \frac{\sqrt{v^2 + a^2}}{a}$, we derive, from (1) and (2),

(3) $\quad \int \sqrt{v^2 + a^2} \, dv = \frac{v}{2} \sqrt{v^2 + a^2} + \frac{a^2}{2} \ln (v + \sqrt{v^2 + a^2}) + C',$

where $C' = C - \frac{a^2}{2} \ln a$. Hence (23) is proved when the positive sign holds.

By substituting $v = a \sec z$, we obtain (see Art. 132)

(4) $\quad \int \sqrt{v^2 - a^2} \, dv = \int a \tan z \cdot a \sec z \tan z \, dz = a^2 \int \tan^2 z \sec z \, dz$
$$= a^2 \int \sec^3 z \, dz - a^2 \int \sec z \, dz.$$

Comparing (4) with (2), we have

(5) $\quad \int \sqrt{v^2 - a^2} \, dv = \frac{a^2}{2} \sec z \tan z - \frac{a^2}{2} \ln (\sec z + \tan z) + C.$

But $\sec z = \frac{v}{a}$, and hence $\tan z = \frac{\sqrt{v^2 - a^2}}{a}$. Substituting in (5), we obtain (23) for the negative sign before a^2.

212　　　　　　　INTEGRAL CALCULUS

ILLUSTRATIVE EXAMPLE 1. Prove the following:
$$\int \sqrt{4-9x^2}\,dx = \frac{x}{2}\sqrt{4-9x^2} + \frac{2}{3}\arcsin\frac{3x}{2} + C.$$
Solution. Compare with (22), and let $a^2 = 4$, $v = 3x$. Then $dv = 3\,dx$. Hence
$$\int \sqrt{4-9x^2}\,dx = \frac{1}{3}\int \sqrt{4-9x^2}\,3\,dx = \frac{1}{3}\int \sqrt{a^2-v^2}\,dv.$$
Using (22), and setting $v = 3x$, $a^2 = 4$, we have the answer.

ILLUSTRATIVE EXAMPLE 2.
$$\int \sqrt{3x^2+4x-7}\,dx$$
$$= \tfrac{1}{6}(3x+2)\sqrt{3x^2+4x-7} - \frac{25\sqrt{3}}{18}\ln\left(3x+2+\sqrt{9x^2+12x-21}\right) + C.$$
Solution. By Illustrative Example 3, p. 207,
$$3x^2+4x-7 = 3\left[(x+\tfrac{2}{3})^2 - \tfrac{25}{9}\right] = 3(v^2-a^2)$$
if $v = x + \tfrac{2}{3}$, $a = \tfrac{5}{3}$. Then $dv = dx$.
$$\therefore \int \sqrt{3x^2+4x-7}\,dx = \sqrt{3}\int \sqrt{v^2-a^2}\,dv.$$
Using (23), and setting $v = x + \tfrac{2}{3}$, $a = \tfrac{5}{3}$, we obtain the answer.

PROBLEMS

Verify the following integrations:

1. $\int \sqrt{1-4x^2}\,dx = \frac{x}{2}\sqrt{1-4x^2} + \frac{1}{4}\arcsin 2x + C.$

2. $\int \sqrt{1+9x^2}\,dx = \frac{x}{2}\sqrt{1+9x^2} + \frac{1}{6}\ln(3x+\sqrt{1+9x^2}) + C.$

3. $\int \sqrt{\frac{x^2}{4}-1}\,dx = \frac{x}{4}\sqrt{x^2-4} - \ln(x+\sqrt{x^2-4}) + C.$

4. $\int \sqrt{25-9x^2}\,dx = \frac{x}{2}\sqrt{25-9x^2} + \frac{25}{6}\arcsin\frac{3x}{5} + C.$

5. $\int \sqrt{4x^2+9}\,dx = \frac{x}{2}\sqrt{4x^2+9} + \frac{9}{4}\ln(2x+\sqrt{4x^2+9}) + C.$

6. $\int \sqrt{5-3x^2}\,dx = \frac{x}{2}\sqrt{5-3x^2} + \frac{5}{2\sqrt{3}}\arcsin x\sqrt{\frac{3}{5}} + C.$

7. $\int \sqrt{3-2x-x^2}\,dx = \frac{x+1}{2}\sqrt{3-2x-x^2} + 2\arcsin\frac{x+1}{2} + C.$

8. $\int \sqrt{5-2x+x^2}\,dx = \frac{x-1}{2}\sqrt{5-2x+x^2}$
$$+ 2\ln(x-1+\sqrt{5-2x+x^2}) + C.$$

9. $\int \sqrt{2x-x^2}\,dx = \frac{x-1}{2}\sqrt{2x-x^2} + \frac{1}{2}\arcsin(x-1) + C.$

10. $\int \sqrt{10-4x+4x^2}\,dx = \frac{2x-1}{4}\sqrt{10-4x+4x^2}$
$$+ \tfrac{9}{4}\ln(2x-1+\sqrt{10-4x+4x^2}) + C.$$

INTEGRATION

Work out each of the following integrals and verify your results by differentiation.

11. $\int \sqrt{16 - 9x^2}\, dx.$

12. $\int \sqrt{4 + 25x^2}\, dx.$

13. $\int \sqrt{9x^2 - 1}\, dx.$

14. $\int \sqrt{8 - 3x^2}\, dx.$

15. $\int \sqrt{5 + 2x^2}\, dx.$

16. $\int \sqrt{5 - 4x - x^2}\, dx.$

17. $\int \sqrt{5 + 2x + x^2}\, dx.$

18. $\int \sqrt{x^2 - 8x + 7}\, dx.$

19. $\int \sqrt{4 - 2x - x^2}\, dx.$

20. $\int \sqrt{x^2 - 2x + 8}\, dx.$

134. Trigonometric differentials. We shall now consider some trigonometric differentials of frequent occurrence which may be readily integrated by transformation into standard forms by means of simple trigonometric reductions.

Example I. *To find* $\int \sin^m u \cos^n u\, du.$

When either m or n is a positive odd integer, no matter what the other may be, this integration may be performed by means of simple transformations and formula (4),

$$\int v^n\, dv = \frac{v^{n+1}}{n+1} + C.$$

For example, if m is odd, we write

$$\sin^m u = \sin^{m-1} u \sin u.$$

Then, since $m - 1$ is even, the first term of the right-hand member will be a power of $\sin^2 u$, and can be expressed in powers of $\cos^2 u$ by substituting

$$\sin^2 u = 1 - \cos^2 u.$$

Then the integral takes the form

(1) $\qquad \int (\text{sum of terms involving } \cos u) \sin u\, du.$

Since $\sin u\, du = -d(\cos u)$, each term to be integrated is of the form $v^n\, dv$ if $v = \cos u$.

Similarly, if n is odd, write $\cos^n u = \cos^{n-1} u \cos u$, and use the substitution $\cos^2 u = 1 - \sin^2 u.$ Then the integral becomes

(2) $\qquad \int (\text{sum of terms involving } \sin u) \cos u\, du.$

214 INTEGRAL CALCULUS

ILLUSTRATIVE EXAMPLE 1. Find $\int \sin^2 x \cos^5 x \, dx$.

Solution. $\int \sin^2 x \cos^5 x \, dx = \int \sin^2 x \cos^4 x \cos x \, dx$

$= \int \sin^2 x (1 - \sin^2 x)^2 \cos x \, dx$ by (2), Art. 2

$= \int (\sin^2 x - 2 \sin^4 x + \sin^6 x) \cos x \, dx$

$= \int (\sin x)^2 \cos x \, dx - 2 \int (\sin x)^4 \cos x \, dx + \int (\sin x)^6 \cos x \, dx$

$= \dfrac{\sin^3 x}{3} - \dfrac{2 \sin^5 x}{5} + \dfrac{\sin^7 x}{7} + C.$ By (4)

Here $v = \sin x$, $dv = \cos x \, dx$, and $n = 2, 4,$ and 6 respectively.

ILLUSTRATIVE EXAMPLE 2. Prove $\int \sin^3 \tfrac{1}{2} x \, dx = \tfrac{2}{3} \cos^3 \tfrac{1}{2} x - 2 \cos \tfrac{1}{2} x + C.$

Solution. Let $\tfrac{1}{2} x = u$. Then $x = 2u$, $dx = 2 \, du$. Substituting,

(3) $\qquad \int \sin^3 \tfrac{1}{2} x \, dx = 2 \int \sin^3 u \, du.$

Now $\int \sin^3 u \, du = \int \sin^2 u \cdot \sin u \, du = \int (1 - \cos^2 u) \sin u \, du$

$= \int \sin u \, du - \int \cos^2 u \sin u \, du = - \cos u + \tfrac{1}{3} \cos^3 u + C.$

Using this result in the right-hand member of (3) and substituting back $u = \tfrac{1}{2} x$ we have the answer.

PROBLEMS

Work out the following integrals.

1. $\int \sin^3 x \, dx = \tfrac{1}{3} \cos^3 x - \cos x + C.$

2. $\int \sin^2 \theta \cos \theta \, d\theta = \tfrac{1}{3} \sin^3 \theta + C.$

3. $\int \cos^2 \phi \sin \phi \, d\phi = - \tfrac{1}{3} \cos^3 \phi + C.$

4. $\int \sin^3 6x \cos 6x \, dx = \tfrac{1}{24} \sin^4 6x + C.$

5. $\int \cos^3 2\theta \sin 2\theta \, d\theta = - \tfrac{1}{8} \cos^4 2\theta + C.$

6. $\int \dfrac{\cos^3 x}{\sin^4 x} \, dx = \csc x - \tfrac{1}{3} \csc^3 x + C.$

7. $\int \dfrac{\sin^3 \phi}{\cos^2 \phi} \, d\phi = \sec \phi + \cos \phi + C.$

8. $\int \cos^4 x \sin^3 x \, dx = - \tfrac{1}{5} \cos^5 x + \tfrac{1}{7} \cos^7 x + C.$

9. $\int \sin^5 x \, dx = - \cos x + \tfrac{2}{3} \cos^3 x - \tfrac{1}{5} \cos^5 x + C.$

10. $\int \cos^5 x \, dx = \sin x - \frac{2}{3} \sin^3 x + \frac{1}{5} \sin^5 x + C.$

11. $\int \frac{\sin^5 y}{\sqrt{\cos y}} \, dy = -2\sqrt{\cos y} \left(1 - \frac{2}{5} \cos^2 y + \frac{1}{9} \cos^4 y\right) + C.$

12. $\int \frac{\cos^5 t}{\sqrt[3]{\sin t}} \, dt = \frac{3}{2} \sin^{\frac{2}{3}} t \left(1 - \frac{1}{2} \sin^2 t + \frac{1}{7} \sin^4 t\right) + C.$

Work out each of the following integrals and verify your results by differentiation.

13. $\int \sin^3 2\theta \, d\theta.$

14. $\int \cos^3 \frac{\theta}{2} \, d\theta.$

15. $\int \sin 2x \cos 2x \, dx.$

16. $\int \sin^3 t \cos^3 t \, dt.$

17. $\int \cos^3 \frac{\phi}{2} \sin^2 \frac{\phi}{2} \, d\phi.$

18. $\int \sin^3 mt \cos^2 mt \, dt.$

19. $\int \sin^5 nx \, dx.$

20. $\int \cos^3 (a + bt) dt.$

21. $\int \frac{\operatorname{ctn} \theta}{\sqrt{\sin \theta}} \, d\theta.$

22. $\int \frac{\sin^3 2x}{\sqrt[3]{\cos 2x}} \, dx.$

Example II. To find $\int \tan^n u \, du$ or $\int \operatorname{ctn}^n u \, du.$

These forms can be readily integrated, when n is an integer, on somewhat the same plan as in the previous examples.

The method consists in writing, as the first step,

$$\tan^n u = \tan^{n-2} u \tan^2 u = \tan^{n-2} u (\sec^2 u - 1);$$

or $\quad \operatorname{ctn}^n u = \operatorname{ctn}^{n-2} u \operatorname{ctn}^2 u = \operatorname{ctn}^{n-2} u (\csc^2 u - 1).$ By (2), Art. 2

The examples illustrate the subsequent steps.

ILLUSTRATIVE EXAMPLE 1. Find $\int \tan^4 x \, dx.$

Solution. $\int \tan^4 x \, dx = \int \tan^2 x (\sec^2 x - 1) dx$

$\qquad = \int \tan^2 x \sec^2 x \, dx - \int \tan^2 x \, dx$

$\qquad = \int (\tan x)^2 \, d(\tan x) - \int (\sec^2 x - 1) dx$

$\qquad = \frac{\tan^3 x}{3} - \tan x + x + C. \qquad$ By (4) and (10)

ILLUSTRATIVE EXAMPLE 2. Prove

$$\int \operatorname{ctn}^3 2x \, dx = -\frac{1}{4} \operatorname{ctn}^2 2x - \frac{1}{2} \ln \sin 2x + C.$$

Solution. Let $2x = u$. Then $x = \frac{1}{2} u$, $dx = \frac{1}{2} du$. Substituting,

(4) $\qquad \int \operatorname{ctn}^3 2x \, dx = \frac{1}{2} \int \operatorname{ctn}^3 u \, du.$

Now
$$\int \operatorname{ctn}^3 u\, du = \int \operatorname{ctn} u \cdot \operatorname{ctn}^2 u\, du$$
$$= \int \operatorname{ctn} u(\csc^2 u - 1)du$$
$$= \int \operatorname{ctn} u \csc^2 u\, du - \int \operatorname{ctn} u\, du$$
$$= -\tfrac{1}{2} \operatorname{ctn}^2 u - \ln \sin u + C. \quad \text{By (4) and (15)}$$

Using this result in the right-hand member of (4) and substituting back $u = 2x$, we have the answer.

Example III. *To find* $\int \sec^n u\, du$ *or* $\int \csc^n u\, du.$

These can be easily integrated when n is a positive even integer. The first step is to write

$$\sec^n u = \sec^{n-2} u \sec^2 u = (\tan^2 u + 1)^{\frac{n-2}{2}} \sec^2 u;$$

or $\csc^n u = \csc^{n-2} u \csc^2 u = (\operatorname{ctn}^2 u + 1)^{\frac{n-2}{2}} \csc^2 u.$ By (2), Art. 2

The example shows the subsequent steps.

ILLUSTRATIVE EXAMPLE 3. Prove $\int \sec^4 \tfrac{1}{2} x\, dx = \tfrac{2}{3} \tan^3 \tfrac{1}{2} x + 2 \tan \tfrac{1}{2} x + C.$

Solution. Let $\tfrac{1}{2} x = u$. Then $x = 2u$, $dx = 2\, du$. Substituting,

(5) $\qquad \int \sec^4 \tfrac{1}{2} x\, dx = 2 \int \sec^4 u\, du.$

Now $\qquad \int \sec^4 u\, du = \int \sec^2 u \cdot \sec^2 u\, du$
$$= \int (\tan^2 u + 1) \sec^2 u\, du \qquad \text{by (2), Art. 2}$$
$$= \int \tan^2 u \sec^2 u\, du + \int \sec^2 u\, du$$
$$= \tfrac{1}{3} \tan^3 u + \tan u + C. \qquad \text{By (4) and (10)}$$

Substituting back in the right-hand member of (5) and putting $u = \tfrac{1}{2} x$ gives the answer.

EXERCISE. Set $\sec^2 u = 1 + \tan^2 u$ in the right-hand member of (5), square, and follow Illustrative Example 1 above.

Example IV. *To find* $\int \tan^m u \sec^n u\, du$ *or* $\int \operatorname{ctn}^m u \csc^n u\, du.$

When n is a positive even integer we proceed as in Example III.

ILLUSTRATIVE EXAMPLE 4. Find $\int \tan^6 x \sec^4 x\, dx.$

Solution. $\int \tan^6 x \sec^4 x\, dx = \int \tan^6 x (\tan^2 x + 1) \sec^2 x\, dx \qquad$ by (2), Art. 2
$$= \int (\tan x)^8 \sec^2 x\, dx + \int \tan^6 x \sec^2 x\, dx$$
$$= \frac{\tan^9 x}{9} + \frac{\tan^7 x}{7} + C. \qquad \text{By (4)}$$

Here $v = \tan x$, $dv = \sec^2 x\, dx$, etc.

INTEGRATION 217

When m is odd we may proceed as in the following example.

ILLUSTRATIVE EXAMPLE 5. Find $\int \tan^5 x \sec^3 x \, dx$.

Solution. $\int \tan^5 x \sec^3 x \, dx = \int \tan^4 x \sec^2 x \sec x \tan x \, dx$

$$= \int (\sec^2 x - 1)^2 \sec^2 x \sec x \tan x \, dx \quad \text{by (2), Art. 2}$$

$$= \int (\sec^6 x - 2 \sec^4 x + \sec^2 x) \sec x \tan x \, dx$$

$$= \frac{\sec^7 x}{7} - \frac{2 \sec^5 x}{5} + \frac{\sec^3 x}{3} + C. \quad \text{By (4)}$$

Here $v = \sec x$, $dv = \sec x \tan x \, dx$, etc.

The methods used in the above examples are obviously limited in their application. For example, they fail in the following case.

$$\int \sec^3 u \, du = \int \sec u \sec^2 u \, du$$

$$= \int \sec u \tan^2 u \, du + \ln (\sec u + \tan u).$$

For we cannot proceed further by the elementary standard forms. Later other methods will be developed of more general use.

PROBLEMS

Work out the following integrals.

1. $\int \tan^3 x \, dx = \frac{1}{2} \tan^2 x + \ln \cos x + C.$

2. $\int \ctn^3 \frac{x}{3} \, dx = -\frac{3}{2} \ctn^2 \frac{x}{3} - 3 \ln \sin \frac{x}{3} + C.$

3. $\int \ctn^3 2x \csc 2x \, dx = \frac{1}{2} \csc 2x - \frac{1}{6} \csc^3 2x + C.$

4. $\int \csc^4 \frac{x}{4} \, dx = -\frac{4}{3} \ctn^3 \frac{x}{4} - 4 \ctn \frac{x}{4} + C.$

5. $\int \tan^5 3\theta \, d\theta = \frac{1}{12} \tan^4 3\theta - \frac{1}{6} \tan^2 3\theta + \frac{1}{3} \ln \sec 3\theta + C.$

6. $\int \frac{\sin^2 \phi \, d\phi}{\cos^4 \phi} = \frac{1}{3} \tan^3 \phi + C.$

7. $\int \frac{dx}{\sin^2 2x \cos^4 2x} = \tan 2x + \frac{1}{6} \tan^3 2x - \frac{1}{2} \ctn 2x + C.$

8. $\int \frac{\cos^4 x \, dx}{\sin^6 x} = -\frac{1}{5} \ctn^5 x + C.$

9. $\int \frac{\sin^{\frac{3}{2}} x \, dx}{\cos^{\frac{11}{2}} x} = \frac{2}{5} \tan^{\frac{5}{2}} x + \frac{2}{9} \tan^{\frac{9}{2}} x + C.$

10. $\int \tan^3 \alpha \sec^{\frac{5}{2}} \alpha \, d\alpha = \frac{2}{9} \sec^{\frac{9}{2}} \alpha - \frac{2}{5} \sec^{\frac{5}{2}} \alpha + C.$

11. $\int \left(\frac{\sec ax}{\tan ax}\right)^4 dx = -\frac{1}{a}\left(\operatorname{ctn} ax + \frac{1}{3}\operatorname{ctn}^3 ax\right) + C.$

12. $\int (\operatorname{ctn}^2 2\theta + \operatorname{ctn}^4 2\theta) d\theta = -\frac{1}{6}\operatorname{ctn}^3 2\theta + C.$

13. $\int (\tan bt - \operatorname{ctn} bt)^3 \, dt = \frac{1}{2b}[\tan^2 bt + \operatorname{ctn}^2 bt] + \frac{4}{b}\ln \sin 2bt + C.$

Work out each of the following integrals and verify your results by differentiation.

14. $\int \operatorname{ctn}^5 ax \, dx.$

15. $\int \sec^6 \theta \, d\theta.$

16. $\int \csc^6 \frac{x}{2} \, dx.$

17. $\int \frac{\sec^4 t \, dt}{\tan^3 t}.$

18. $\int \frac{\sec^4 x \, dx}{\sqrt{\tan x}}.$

19. $\int \left(\frac{\csc ax}{\operatorname{ctn} ax}\right)^4 dx.$

20. $\int \tan^3 \frac{x}{3} \sec^3 \frac{x}{3} \, dx.$

21. $\int \frac{dx}{\sin^4 3x \cos^2 3x}.$

22. $\int \left(\frac{\csc bx}{\tan bx}\right)^2 dx.$

23. $\int \left(\frac{\tan \phi}{\operatorname{ctn} \phi}\right)^3 d\phi.$

24. $\int \left(\frac{\tan at}{\cos at}\right)^4 dt.$

25. $\int \frac{\tan^3 x \, dx}{\sqrt{\sec x}}.$

26. $\int \tan^n x \sec^4 x \, dx.$

27. $\int \frac{\tan^5 2\theta \, d\theta}{\sec^3 2\theta}.$

Example V. *To find* $\int \sin^m u \cos^n u \, du$ *by means of multiple angles.*

When either m or n is a positive odd integer, the shortest method is that shown in Example I, p. 213. When m and n are both positive even integers, the given differential expression may be transformed by suitable trigonometric substitutions into an expression involving sines and cosines of multiple angles, and then integrated. For this purpose we employ the following formulas.

$$\sin u \cos u = \tfrac{1}{2} \sin 2u, \qquad \text{by (5), Art. 2}$$
$$\sin^2 u = \tfrac{1}{2} - \tfrac{1}{2}\cos 2u, \qquad \text{by (5), Art. 2}$$
$$\cos^2 u = \tfrac{1}{2} + \tfrac{1}{2}\cos 2u. \qquad \text{by (5), Art. 2}$$

ILLUSTRATIVE EXAMPLE 1. Find $\int \cos^2 u \, du.$

Solution. $\int \cos^2 u \, du = \int (\tfrac{1}{2} + \tfrac{1}{2}\cos 2u) du$

$= \dfrac{1}{2}\int du + \dfrac{1}{2}\int \cos 2u \, du = \dfrac{u}{2} + \dfrac{1}{4}\sin 2u + C.$

INTEGRATION

ILLUSTRATIVE EXAMPLE 2. Find $\int \sin^2 x \cos^2 x \, dx$.

Solution. $\int \sin^2 x \cos^2 x \, dx = \tfrac{1}{4} \int \sin^2 2x \, dx$ by (5), Art. 2

$\qquad = \tfrac{1}{4} \int (\tfrac{1}{2} - \tfrac{1}{2} \cos 4x) \, dx$ by (5), Art. 2

$\qquad = \dfrac{x}{8} - \dfrac{1}{32} \sin 4x + C.$

ILLUSTRATIVE EXAMPLE 3. Find $\int \sin^4 x \cos^2 x \, dx$.

Solution. $\int \sin^4 x \cos^2 x \, dx = \int (\sin x \cos x)^2 \sin^2 x \, dx$

$\qquad = \int \tfrac{1}{4} \sin^2 2x (\tfrac{1}{2} - \tfrac{1}{2} \cos 2x) \, dx$

$\qquad = \tfrac{1}{8} \int \sin^2 2x \, dx - \tfrac{1}{8} \int \sin^2 2x \cos 2x \, dx$

$\qquad = \tfrac{1}{8} \int (\tfrac{1}{2} - \tfrac{1}{2} \cos 4x) \, dx - \tfrac{1}{8} \int \sin^2 2x \cos 2x \, dx$

$\qquad = \dfrac{x}{16} - \dfrac{\sin 4x}{64} - \dfrac{\sin^3 2x}{48} + C.$

Example VI. To find $\int \sin mx \cos nx \, dx, \int \sin mx \sin nx \, dx,$ or $\int \cos mx \cos nx \, dx,$ when $m \neq n$.

By (6), Art. 2, $\sin mx \cos nx = \tfrac{1}{2} \sin (m+n)x + \tfrac{1}{2} \sin (m-n)x.$

$\therefore \int \sin mx \cos nx \, dx = \tfrac{1}{2} \int \sin(m+n)x \, dx + \tfrac{1}{2} \int \sin(m-n)x \, dx$

$\qquad = -\dfrac{\cos (m+n)x}{2(m+n)} - \dfrac{\cos (m-n)x}{2(m-n)} + C.$

Similarly, we find

$\int \sin mx \sin nx \, dx = -\dfrac{\sin (m+n)x}{2(m+n)} + \dfrac{\sin (m-n)x}{2(m-n)} + C,$

$\int \cos mx \cos nx \, dx = \dfrac{\sin (m+n)x}{2(m+n)} + \dfrac{\sin (m-n)x}{2(m-n)} + C.$

PROBLEMS

Work out the following integrals.

1. $\int \sin^2 x \, dx = \dfrac{x}{2} - \dfrac{\sin 2x}{4} + C.$

2. $\int \sin^4 x \, dx = \dfrac{3x}{8} - \dfrac{\sin 2x}{4} + \dfrac{\sin 4x}{32} + C.$

3. $\int \cos^4 x \, dx = \dfrac{3x}{8} + \dfrac{\sin 2x}{4} + \dfrac{\sin 4x}{32} + C.$

4. $\int \sin^6 x \, dx = \dfrac{5x}{16} - \dfrac{\sin 2x}{4} + \dfrac{\sin^3 2x}{48} + \dfrac{3 \sin 4x}{64} + C.$

5. $\int \cos^6 x \, dx = \dfrac{5x}{16} + \dfrac{\sin 2x}{4} - \dfrac{\sin^3 2x}{48} + \dfrac{3 \sin 4x}{64} + C.$

6. $\int \sin^2 ax \, dx = \dfrac{x}{2} - \dfrac{\sin 2ax}{4a} + C.$

7. $\int \sin^2 \dfrac{x}{2} \cos^2 \dfrac{x}{2} \, dx = \dfrac{x}{8} - \dfrac{\sin 2x}{16} + C.$

8. $\int \sin^4 ax \, dx = \dfrac{3x}{8} - \dfrac{\sin 2ax}{4a} + \dfrac{\sin 4ax}{32a} + C.$

9. $\int \sin^2 2x \cos^4 2x \, dx = \dfrac{x}{16} + \dfrac{\sin^3 4x}{96} - \dfrac{\sin 8x}{128} + C.$

10. $\int (2 - \sin \theta)^2 \, d\theta = \dfrac{9\theta}{2} + 4 \cos \theta - \dfrac{\sin 2\theta}{4} + C.$

11. $\int (\sin^2 \phi + \cos \phi)^2 \, d\phi = \dfrac{7\phi}{8} + \dfrac{2 \sin^3 \phi}{3} + \dfrac{\sin 4\phi}{32} + C.$

12. $\int \sin 2x \cos 4x \, dx = \dfrac{\cos 2x}{4} - \dfrac{\cos 6x}{12} + C.$

13. $\int \sin 3x \sin 2x \, dx = \dfrac{\sin x}{2} - \dfrac{\sin 5x}{10} + C.$

14. $\int \cos 4x \cos 3x \, dx = \dfrac{\sin x}{2} + \dfrac{\sin 7x}{14} + C.$

Work out each of the following integrals and verify your results by differentiation.

15. $\int \cos^2 ax \, dx.$

16. $\int \cos^4 ax \, dx.$

17. $\int \sin^2 ax \cos^2 ax \, dx.$

18. $\int \sin^4 \dfrac{\theta}{2} \cos^2 \dfrac{\theta}{2} \, d\theta.$

19. $\int \sin^4 2\alpha \cos^4 2\alpha \, d\alpha.$

20. $\int \sin^2 x \cos^6 x \, dx.$

21. $\int (1 + \cos x)^3 \, dx.$

22. $\int (\sqrt{\sin 2\theta} - \cos 2\theta)^2 \, d\theta.$

23. $\int (\sqrt{\cos \theta} - 2 \sin \theta)^2 \, d\theta.$

24. $\int (\sin 2x - \sin 3x)^2 \, dx.$

25. $\int (\sin x + \cos 2x)^2 \, dx.$

26. $\int (\cos x + 2 \cos 2x)^2 \, dx.$

INTEGRATION

135. Integration of expressions containing $\sqrt{a^2 - u^2}$ or $\sqrt{u^2 \pm a^2}$ by trigonometric substitution. In many cases the shortest method of integrating such expressions is to change the variable as follows.

When $\sqrt{a^2 - u^2}$ occurs, let $u = a \sin z$.
When $\sqrt{a^2 + u^2}$ occurs, let $u = a \tan z$.
When $\sqrt{u^2 - a^2}$ occurs, let $u = a \sec z$.

These substitutions were used in Arts. 132–133. By them the radical sign is in each case eliminated. For

(1) $\sqrt{a^2 - a^2 \sin^2 z} = a\sqrt{1 - \sin^2 z} = a \cos z$;
(2) $\sqrt{a^2 + a^2 \tan^2 z} = a\sqrt{1 + \tan^2 z} = a \sec z$;
(3) $\sqrt{a^2 \sec^2 z - a^2} = a\sqrt{\sec^2 z - 1} = a \tan z$.

ILLUSTRATIVE EXAMPLE 1. Find $\int \dfrac{du}{(a^2 - u^2)^{\frac{3}{2}}}$.

Solution. Let $u = a \sin z$; then $du = a \cos z\, dz$, and, using (1),

$$\int \frac{du}{(a^2 - u^2)^{\frac{3}{2}}} = \int \frac{a \cos z\, dz}{a^3 \cos^3 z} = \frac{1}{a^2} \int \frac{dz}{\cos^2 z} = \frac{1}{a^2} \int \sec^2 z\, dz$$

$$= \frac{\tan z}{a^2} + C = \frac{u}{a^2\sqrt{a^2 - u^2}} + C.$$

For, since $\sin z = \dfrac{u}{a}$, draw a right triangle and mark the sides as in the figure. Then $\tan z = \dfrac{u}{\sqrt{a^2 - u^2}}$.

ILLUSTRATIVE EXAMPLE 2. Prove $\int \dfrac{dx}{x\sqrt{4x^2 + 9}} = \dfrac{1}{3} \ln \dfrac{\sqrt{4x^2 + 9} - 3}{2x} + C.$

Solution. Here $\sqrt{4x^2 + 9} = \sqrt{u^2 + a^2}$ if $u = 2x$, $a = 3$.
Hence let $2x = u$; then $x = \frac{1}{2} u$, $dx = \frac{1}{2} du$. Substituting,

(4) $\int \dfrac{dx}{x\sqrt{4x^2 + 9}} = \int \dfrac{\frac{1}{2} du}{\frac{1}{2} u\sqrt{u^2 + a^2}} = \int \dfrac{du}{u\sqrt{u^2 + a^2}}.$

Let $u = a \tan z$. Then $du = a \sec^2 z\, dz$, and, using (2),

$$\int \frac{du}{u\sqrt{u^2 + a^2}} = \int \frac{a \sec^2 z\, dz}{a \tan z \cdot a \sec z} = \frac{1}{a}\int \frac{\sec z\, dz}{\tan z} = \frac{1}{a}\int \frac{dz}{\sin z}$$

$$= \frac{1}{a}\int \csc z\, dz = \frac{1}{a} \ln (\csc z - \operatorname{ctn} z) + C.$$

Since $\tan z = \dfrac{u}{a}$, draw a right triangle and mark the sides as in the figure. Then

$$\csc z = \frac{\sqrt{u^2 + a^2}}{u}, \quad \cot z = \frac{a}{u}.$$

Hence $\int \dfrac{du}{u\sqrt{u^2 + a^2}} = \dfrac{1}{a} \ln \dfrac{\sqrt{u^2 + a^2} - a}{u} + C.$

Substituting back in (4), and setting, as above, $u = 2x$, $a = 3$, we have the answer.

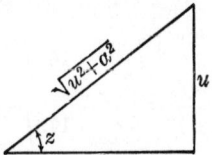

PROBLEMS

Work out the following integrals.

1. $\displaystyle\int \frac{dx}{(x^2+2)^{\frac{3}{2}}} = \frac{x}{2\sqrt{x^2+2}} + C.$

2. $\displaystyle\int \frac{x^2\,dx}{\sqrt{x^2-6}} = \frac{x}{2}\sqrt{x^2-6} + 3\ln\left(x+\sqrt{x^2-6}\right) + C.$

3. $\displaystyle\int \frac{dx}{(5-x^2)^{\frac{3}{2}}} = \frac{x}{5\sqrt{5-x^2}} + C.$

4. $\displaystyle\int \frac{t^2\,dt}{\sqrt{4-t^2}} = -\frac{t}{2}\sqrt{4-t^2} + 2\arcsin\frac{t}{2} + C.$

5. $\displaystyle\int \frac{x^2\,dx}{(x^2+8)^{\frac{3}{2}}} = -\frac{x}{\sqrt{x^2+8}} + \ln\left(x+\sqrt{x^2+8}\right) + C.$

6. $\displaystyle\int \frac{u^2\,du}{(9-u^2)^{\frac{3}{2}}} = \frac{u}{\sqrt{9-u^2}} - \arcsin\frac{u}{3} + C.$

7. $\displaystyle\int \frac{dx}{x\sqrt{x^2+4}} = \frac{1}{2}\ln\left(\frac{x}{2+\sqrt{x^2+4}}\right) + C.$

8. $\displaystyle\int \frac{dx}{x\sqrt{25-x^2}} = \frac{1}{5}\ln\left(\frac{x}{5+\sqrt{25-x^2}}\right) + C.$

9. $\displaystyle\int \frac{dy}{y^2\sqrt{y^2-7}} = \frac{\sqrt{y^2-7}}{7y} + C.$

10. $\displaystyle\int \frac{dx}{x^2\sqrt{5-x^2}} = -\frac{\sqrt{5-x^2}}{5x} + C.$

11. $\displaystyle\int \frac{dx}{x^3\sqrt{x^2-9}} = \frac{\sqrt{x^2-9}}{18x^2} + \frac{1}{54}\operatorname{arc\,sec}\frac{x}{3} + C.$

12. $\displaystyle\int \frac{\sqrt{16-t^2}\,dt}{t^2} = -\frac{\sqrt{16-t^2}}{t} - \arcsin\frac{t}{4} + C.$

Work out each of the following integrals and verify your results by differentiation.

13. $\displaystyle\int \frac{\sqrt{x^2+16}\,dx}{x}.$

14. $\displaystyle\int \frac{\sqrt{y^2-9}\,dy}{y}.$

15. $\displaystyle\int \frac{dx}{x^3\sqrt{4-x^2}}.$

16. $\displaystyle\int \frac{\sqrt{x^2+9}\,dx}{x^2}.$

17. $\displaystyle\int \frac{\sqrt{100-u^2}\,du}{u}.$

18. $\displaystyle\int \frac{dx}{x^3\sqrt{x^2+1}}.$

19. $\displaystyle\int \frac{dv}{(v^2-3)^{\frac{3}{2}}}.$

20. $\displaystyle\int \frac{x^2\,dx}{\sqrt{x^2+5}}.$

21. $\displaystyle\int \frac{dx}{x^4\sqrt{x^2-5}}.$

22. $\displaystyle\int \frac{\sqrt{x^2+9}\,dx}{x^6}.$

INTEGRATION

136. Integration by parts. If u and v are functions of a single independent variable, we have, from the formula for the differentiation of a product (V, Art. 94),
$$d(uv) = u\,dv + v\,du,$$
or, transposing, $\quad u\,dv = d(uv) - v\,du.$

Integrating this, we get the inverse formula,

(A) $$\int u\,dv = uv - \int v\,du,$$

called the **formula for integration by parts**. This formula makes the integration of $u\,dv$, which we may not be able to integrate directly, depend on the integration of dv and $v\,du$, which may be in such form as to be readily integrable. This method of *integration by parts* is one of the most useful in the integral calculus.

To apply this formula in any given case the given differential must be separated into two factors, namely, u and dv. No general directions can be given for choosing these factors, except that

(a) *dx is always a part of dv;*
(b) *it must be possible to integrate dv; and*
(c) *when the expression to be integrated is the product of two functions, it is usually best to choose the most complicated-looking one that it is possible to integrate as part of dv.*

The following examples will show in detail how the formula is applied:

ILLUSTRATIVE EXAMPLE 1. Find $\int x \cos x\,dx$.

Solution. Let $\quad u = x \quad$ and $\quad dv = \cos x\,dx;$

then $\quad du = dx \quad$ and $\quad v = \int \cos x\,dx = \sin x.$

Substituting in (A),
$$\int \overbrace{x}^{u} \overbrace{\cos x\,dx}^{dv} = \overbrace{x}^{u} \overbrace{\sin x}^{v} - \int \overbrace{\sin x}^{v} \overbrace{dx}^{du} = x \sin x + \cos x + C.$$

ILLUSTRATIVE EXAMPLE 2. Find $\int x \ln x\,dx$.

Solution. Let $\quad u = \ln x \quad$ and $\quad dv = x\,dx;$

then $\quad du = \dfrac{dx}{x} \quad$ and $\quad v = \int x\,dx = \dfrac{x^2}{2}.$

Substituting in (A),
$$\int x \ln x\,dx = \ln x \cdot \frac{x^2}{2} - \int \frac{x^2}{2} \cdot \frac{dx}{x}$$
$$= \frac{x^2}{2} \ln x - \frac{x^2}{4} + C.$$

INTEGRAL CALCULUS

ILLUSTRATIVE EXAMPLE 3. Find $\int xe^{ax}\,dx$.

Solution. Let $u = e^{ax}$ and $dv = x\,dx$;

then $du = e^{ax} \cdot a\,dx$ and $v = \int x\,dx = \dfrac{x^2}{2}$.

Substituting in (A),

$$\int xe^{ax}\,dx = e^{ax} \cdot \frac{x^2}{2} - \int \frac{x^2}{2} e^{ax} a\,dx$$

$$= \frac{x^2 e^{ax}}{2} - \frac{a}{2}\int x^2 e^{ax}\,dx.$$

But $x^2 e^{ax}\,dx$ is not as simple to integrate as $xe^{ax}\,dx$, which fact indicates that we did not choose our factors suitably. Instead, let

$u = x$ and $dv = e^{ax}\,dx$;

then $du = dx$ and $v = \int e^{ax}\,dx = \dfrac{e^{ax}}{a}$.

Substituting in (A),

$$\int xe^{ax}\,dx = x \cdot \frac{e^{ax}}{a} - \int \frac{e^{ax}}{a}\,dx$$

$$= \frac{xe^{ax}}{a} - \frac{e^{ax}}{a^2} + C = \frac{e^{ax}}{a}\left(x - \frac{1}{a}\right) + C.$$

It may be necessary to apply the formula for integration by parts more than once, as in the following example.

ILLUSTRATIVE EXAMPLE 4. Find $\int x^2 e^{ax}\,dx$.

Solution. Let $u = x^2$ and $dv = e^{ax}\,dx$;

then $du = 2x\,dx$ and $v = \int e^{ax}\,dx = \dfrac{e^{ax}}{a}$.

Substituting in (A),

$$\int x^2 e^{ax}\,dx = x^2 \cdot \frac{e^{ax}}{a} - \int \frac{e^{ax}}{a} \cdot 2x\,dx$$

(1) $$= \frac{x^2 e^{ax}}{a} - \frac{2}{a}\int xe^{ax}\,dx.$$

The integral in the last term may be found by applying formula (A) again, which gives

$$\int xe^{ax}\,dx = \frac{e^{ax}}{a}\left(x - \frac{1}{a}\right) + C.$$

Substituting this result in (1), we get

$$\int x^2 e^{ax}\,dx = \frac{x^2 e^{ax}}{a} - \frac{2}{a^2}e^{ax}\left(x - \frac{1}{a}\right) + C = \frac{e^{ax}}{a}\left(x^2 - \frac{2x}{a} + \frac{2}{a^2}\right) + C.$$

ILLUSTRATIVE EXAMPLE 5. Prove

$$\int \sec^3 z\,dz = \tfrac{1}{2}\sec z \tan z + \tfrac{1}{2}\ln(\sec z + \tan z) + C.$$

Solution. Let $u = \sec z$ and $dv = \sec^2 z\,dz$;

then $du = \sec z \tan z\,dz$ and $v = \tan z$.

INTEGRATION

Substituting in (A),

$$\int \sec^3 z \, dz = \sec z \tan z - \int \sec z \tan^2 z \, dz.$$

In the new integral, substitute $\tan^2 z = \sec^2 z - 1$. Then we get

$$\int \sec^3 z \, dz = \sec z \tan z - \int \sec^3 z \, dz + \ln(\sec z + \tan z) + C.$$

Transposing the integral in the right-hand member and dividing by 2, we have the required result.

ILLUSTRATIVE EXAMPLE 6. Prove

$$\int e^{ax} \sin nx \, dx = \frac{e^{ax}(a \sin nx - n \cos nx)}{a^2 + n^2} + C.$$

Solution. Let $\quad u = e^{ax} \quad$ and $\quad dv = \sin nx \, dx;$
then $\quad\quad\quad\quad du = ae^{ax} dx \quad$ and $\quad v = -\dfrac{\cos nx}{n}.$

Substituting in formula (A), the result is

(2) $\quad \displaystyle\int e^{ax} \sin nx \, dx = -\frac{e^{ax} \cos nx}{n} + \frac{a}{n} \int e^{ax} \cos nx \, dx.$

Integrate the new integral by parts.

Let $\quad u = e^{ax} \quad$ and $\quad dv = \cos nx \, dx;$
then $\quad\quad\quad du = ae^{ax} dx \quad$ and $\quad v = \dfrac{\sin nx}{n}.$

Hence, by (A),

(3) $\quad \displaystyle\int e^{ax} \cos nx \, dx = \frac{e^{ax} \sin nx}{n} - \frac{a}{n} \int e^{ax} \sin nx \, dx.$

Substituting in (2), we obtain

(4) $\quad \displaystyle\int e^{ax} \sin nx \, dx = \frac{e^{ax}}{n^2}\left(a \sin nx - n \cos nx\right) - \frac{a^2}{n^2} \int e^{ax} \sin nx \, dx.$

The two integrals in (4) are the same. Transposing the one in the right-hand member and solving, the result is as above.

Among the most important applications of the method of integration by parts is the integration of

(a) *differentials involving products,*
(b) *differentials involving logarithms,*
(c) *differentials involving inverse circular functions.*

PROBLEMS

Work out the following integrals.

1. $\displaystyle\int x \sin x \, dx = \sin x - x \cos x + C.$

2. $\displaystyle\int \ln x \, dx = x(\ln x - 1) + C.$

3. $\int x \sin \frac{x}{2} dx = 4 \sin \frac{x}{2} - 2 x \cos \frac{x}{2} + C.$

4. $\int x \cos nx \, dx = \frac{\cos nx}{n^2} + \frac{x \sin nx}{n} + C.$

5. $\int u \sec^2 u \, du = u \tan u + \ln \cos u + C.$

6. $\int v \sin^2 3 v \, dv = \frac{1}{4} v^2 - \frac{1}{12} v \sin 6 v - \frac{1}{72} \cos 6 v + C.$

7. $\int y^2 \sin ny \, dy = \frac{2 \cos ny}{n^3} + \frac{2 y \sin ny}{n^2} - \frac{y^2 \cos ny}{n} + C.$

8. $\int x a^x \, dx = a^x \left[\frac{x}{\ln a} - \frac{1}{\ln^2 a} \right] + C.$

9. $\int x^n \ln x \, dx = \frac{x^{n+1}}{n+1} \left(\ln x - \frac{1}{n+1} \right) + C.$

10. $\int \arcsin x \, dx = x \arcsin x + \sqrt{1 - x^2} + C.$

11. $\int \arctan x \, dx = x \arctan x - \frac{1}{2} \ln (1 + x^2) + C.$

12. $\int \operatorname{arcctn} y \, dy = y \operatorname{arcctn} y + \frac{1}{2} \ln (1 + y^2) + C.$

13. $\int \arccos 2 x \, dx = x \arccos 2 x - \frac{1}{2} \sqrt{1 - 4 x^2} + C.$

14. $\int \operatorname{arcsec} y \, dy = y \operatorname{arcsec} y - \ln (y + \sqrt{y^2 - 1}) + C.$

15. $\int \operatorname{arccsc} \frac{t}{2} dt = t \operatorname{arccsc} \frac{t}{2} + 2 \ln (t + \sqrt{t^2 - 4}) + C.$

16. $\int x \arctan x \, dx = \frac{x^2 + 1}{2} \arctan x - \frac{x}{2} + C.$

17. $\int \arctan \sqrt{x} \, dx = (x + 1) \arctan \sqrt{x} - \sqrt{x} + C.$

18. $\int x^2 e^{-x} dx = - e^{-x}(2 + 2 x + x^2) + C.$

19. $\int e^\theta \cos \theta \, d\theta = \frac{e^\theta}{2} (\sin \theta + \cos \theta) + C.$

20. $\int \frac{\ln x \, dx}{(x+1)^2} = \frac{x}{x+1} \ln x - \ln (x+1) + C.$

21. $\int x^2 \arcsin x \, dx = \frac{x^3}{3} \arcsin x + \frac{x^2+2}{9} \sqrt{1-x^2} + C.$

22. $\int \frac{\ln (x+1) dx}{\sqrt{x+1}} = 2\sqrt{x+1} \,[\ln (x+1) - 2] + C.$

23. $\int \frac{x e^x \, dx}{(1+x)^2} = \frac{e^x}{1+x} + C.$

24. $\int e^{-t} \cos \pi t \, dt = \frac{e^{-t}(\pi \sin \pi t - \cos \pi t)}{\pi^2 + 1} + C.$

INTEGRATION

Work out each of the following integrals and verify your results by differentiation.

25. $\int x \sec^2 \dfrac{x}{2}\, dx.$

26. $\int x \cos^2 2x\, dx.$

27. $\int x^2 \cos x\, dx.$

28. $\int \operatorname{arc\,sin} mx\, dx.$

29. $\int \operatorname{arc\,ctn} \dfrac{x}{2}\, dx.$

30. $\int \operatorname{arc\,cos} \dfrac{1}{x}\, dx.$

31. $\int \operatorname{arc\,sec} \dfrac{1}{y}\, dy.$

32. $\int \operatorname{arc\,csc} nt\, dt.$

33. $\int \operatorname{arc\,sin} \sqrt{\dfrac{x}{2}}\, dx.$

34. $\int x^3 \operatorname{arc\,sin} x\, dx.$

35. $\int \dfrac{x \operatorname{arc\,sin} x\, dx}{\sqrt{1-x^2}}.$

36. $\int \dfrac{\operatorname{arc\,tan}\sqrt{x}\, dx}{x^2}.$

37. $\int x^3 \operatorname{arc\,tan} x\, dx.$

38. $\int (e^x + 2x)^2\, dx.$

39. $\int (2^x + x^2)^2\, dx.$

40. $\int e^{-\theta} \cos \dfrac{\theta}{2}\, d\theta.$

41. $\int e^{\frac{t}{5}} \sin \pi t\, dt.$

42. $\int e^{3x} \cos \dfrac{x}{3}\, dx.$

43. $\int e^{-\frac{t}{2}} \cos 2t\, dt.$

44. $\int e^{\frac{t}{4}} \cos \pi t\, dt.$

45. $\int e^{-\frac{t}{4}} \sin \dfrac{\pi t}{4}\, dt.$

46. $\int \csc^3 \theta\, d\theta.$

137. Comments. Integration is, on the whole, a more difficult operation than differentiation. In fact, so simple an integral (in appearance) as

$$\int \sqrt{x} \sin x\, dx$$

cannot be worked out; that is, there is *no elementary function* whose derivative is $\sqrt{x} \sin x$. To assist in the technique of integration, elaborate tables of integrals have been prepared. A short table is given in Chapter XXVII in this book. The use of this table is explained below in Art. 176. At this point let it suffice to remark that the methods thus far presented are adequate for many problems. Other methods will be developed in later chapters.

MISCELLANEOUS PROBLEMS

Work out each of the following integrals and verify your results by differentiation.

1. $\int \dfrac{3\,x\,dx}{\sqrt{5-2\,x^2}}$.

2. $\int \dfrac{3\,x\,dx}{5-2\,x^2}$.

3. $\int \dfrac{(ax+b)dx}{\sqrt{c^2-x^2}}$.

4. $\int x \cos 2\,x\,dx$.

5. $\int \dfrac{(4\,x+3)dx}{x^2+4\,x+8}$.

6. $\int \dfrac{(4\,x+3)dx}{\sqrt{x^2+4\,x+8}}$.

7. $\int \dfrac{dx}{(a^2-x^2)^{\frac{3}{2}}}$.

8. $\int \dfrac{dx}{x^2-6\,x+9}$.

9. $\int \dfrac{dx}{x^2-6\,x+8}$.

10. $\int \dfrac{dx}{x^2-6\,x+10}$.

11. $\int (e^{2x}+2\,e^{-x})^2\,dx$.

12. $\int (e^{2x}-2\,x)^2\,dx$.

13. $\int \dfrac{dx}{e^x-4\,e^{-x}}$.

14. $\int \sin^2 ax \cos ax\,dx$.

15. $\int \sin^2 ax \cos^2 ax\,dx$.

16. $\int \ln(1-\sqrt{x})dx$.

17. $\int (2 \tan 2\,\theta - \operatorname{ctn}\theta)^2\,d\theta$.

18. $\int \dfrac{4\,x\,dx}{1-4\,x^4}$.

19. $\int \dfrac{x^3\,dx}{\sqrt{x^2+1}}$.

20. $\int \dfrac{x^3\,dx}{\sqrt{x^2-1}}$.

21. $\int \dfrac{x^3\,dx}{\sqrt{1-x^2}}$.

22. $\int \dfrac{x^3\,dx}{x-1}$.

23. $\int \dfrac{4\,x\,dx}{\sqrt{1-4\,x^4}}$.

24. $\int e^{2t} \cos 3\,t\,dt$.

25. $\int \sin^5 \dfrac{\theta}{4}\,d\theta$.

26. $\int \sin^4 \dfrac{\theta}{5}\,d\theta$.

27. $\int \dfrac{(t-\csc^2 2\,t)dt}{t^2+\operatorname{ctn} 2\,t}$.

28. $\int \sqrt{\dfrac{\arcsin x}{1-x^2}}\,dx$.

29. $\int \dfrac{5\,dx}{x^2-x+1}$.

30. $\int \dfrac{5\,dx}{\sqrt{x^2-x+1}}$.

31. $\int x^3 \arctan \dfrac{x}{2}\,dx$.

32. $\int (e^x + \sin x)^2\,dx$.

33. $\int (x - \cos x)^2\,dx$.

34. $\int (1 + \tan x)^3\,dx$.

35. $\int \dfrac{\sin\theta\,d\theta}{(1-\cos\theta)^3}$.

36. $\int \dfrac{(1+\sin t)^3\,dt}{\cos t}$.

37. $\int e^{-t} \sin 2\,t\,dt$.

38. $\int \sin 2\,\theta \cos 3\,\theta\,d\theta$.

39. $\int \sin\phi \sin 4\,\phi\,d\phi$.

40. $\int \cos\alpha \cos 2\,\alpha\,d\alpha$.

CHAPTER XIII

CONSTANT OF INTEGRATION

138. Determination of the constant of integration by means of initial conditions. As was pointed out on page 190, the constant of integration may be found in any given case when we know the value of the integral for some value of the variable. In fact, it is necessary, in order to be able to determine the constant of integration, to have some data given in addition to the differential expression to be integrated. Let us illustrate this by means of an example.

ILLUSTRATIVE EXAMPLE. Find a function whose first derivative is $3\ x^2 - 2\ x + 5$, and which shall have the value 12 when $x = 1$.

Solution. $(3\ x^2 - 2\ x + 5)dx$ is the differential expression to be integrated.

Thus $$\int (3\ x^2 - 2\ x + 5)dx = x^3 - x^2 + 5\ x + C,$$

where C is the constant of integration. From the conditions of our problem this result must equal 12 when $x = 1$; that is,

$$12 = 1 - 1 + 5 + C, \quad \text{or} \quad C = 7.$$

Hence $x^3 - x^2 + 5\ x + 7$ is the required function.

139. Geometrical signification of the constant of integration. We shall illustrate this by means of examples.

ILLUSTRATIVE EXAMPLE 1. Determine the equation of the curve at every point of which the tangent line has the slope $2\ x$.

Solution. Since the slope of the tangent to a curve at any point is $\dfrac{dy}{dx}$, we have, by hypothesis,

$$\frac{dy}{dx} = 2\ x,$$

or $\qquad dy = 2\ x\ dx.$

Integrating, $\qquad y = 2 \int x\ dx,$ or

(1) $\qquad y = x^2 + C,$

where C is the constant of integration. Now if we give to C a series of values, say 6, 0, -3, (1) yields the equations

$$y = x^2 + 6, \quad y = x^2, \quad y = x^2 - 3,$$

whose loci are parabolas with axes coinciding with the y-axis and having 6, 0, -3 respectively as intercepts on the y-axis.

All the parabolas (1) have the same value of $\dfrac{dy}{dx}$; that is, they have the same direction (or slope) for the same value of x. It will also be noticed that the difference

229

INTEGRAL CALCULUS

in the lengths of their ordinates remains the same for all values of x. Hence all the parabolas can be obtained by moving any one of them vertically up or down, the value of C in this case not affecting the slope of the curve.

If in the above example we impose the additional condition that the curve shall pass through the point (1, 4), then the coördinates of this point must satisfy (1), giving
$$4 = 1 + C, \quad \text{or} \quad C = 3.$$
Hence the particular curve required is the parabola $y = x^2 + 3$.

ILLUSTRATIVE EXAMPLE 2. Determine the equation of a curve such that the slope of the tangent line to the curve at any point is the ratio of the abscissa to the ordinate with sign changed.

Solution. The condition of the problem is expressed by the equation
$$\frac{dy}{dx} = -\frac{x}{y},$$
or, separating the variables,
$$y\,dy = -x\,dx.$$
Integrating, $\quad \dfrac{y^2}{2} = -\dfrac{x^2}{2} + C,$

or $\quad x^2 + y^2 = 2\,C.$

This, we see, represents a family of concentric circles with their centers at the origin. If, in addition, we impose the condition that the curve must pass through the point (3, 4), then $\quad 9 + 16 = 2\,C.$

Hence the particular curve required is the circle $x^2 + y^2 = 25$.

PROBLEMS

The following expressions have been obtained by differentiating certain functions. Find the function in each case for the given values of the variable and the function.

	Derivative of function	Value of variable	Corresponding value of function	Answer
1.	$x - 3$	2	9	$\frac{1}{2}x^2 - 3x + 13$.
2.	$3 + x - 5x^2$	6	-20	$304 + 3x + \frac{1}{2}x^2 - \frac{5}{3}x^3$.
3.	$y^3 - b^2 y$	2	0	$\frac{1}{4}y^4 - \frac{1}{2}b^2y^2 + 2b^2 - 4$
4.	$\sin\theta + \cos\theta$	$\frac{1}{2}\pi$	2	$\sin\theta - \cos\theta + 1$.
5.	$\dfrac{1}{t} - \dfrac{1}{2 - t}$	1	0	$\ln(2t - t^2)$.
6.	$\sec^2\phi + \tan\phi$	0	5	$\tan\phi + \ln\sec\phi + 5$.
7.	$\dfrac{1}{x^2 + a^2}$	a	$\dfrac{\pi}{2a}$	$\dfrac{1}{a}\arctan\dfrac{x}{a} + \dfrac{\pi}{4a}$.
8.	$bx^3 + ax + 4$	b	10	
9.	$\sqrt{t} + \dfrac{1}{\sqrt{t}}$	4	0	
10.	$\operatorname{ctn}\theta - \csc^2\theta$	$\frac{1}{2}\pi$	3	
11.	$3\,te^{2t^2}$	0	4	

CONSTANT OF INTEGRATION

Find the equation of the family of curves such that the slope of the tangent at any point is as follows.

12. m.   *Ans.* Straight lines, $y = mx + C$.

13. x.   Parabolas, $y = \frac{1}{2}x^2 + C$.

14. $\dfrac{1}{y}$.   Parabolas, $\frac{1}{2}y^2 = x + C$.

15. $\dfrac{x^2}{y}$.   Semicubical parabolas, $\frac{1}{2}y^2 = \frac{1}{3}x^3 + C$.

16. $\dfrac{x}{y^2}$.   Semicubical parabolas, $\frac{1}{3}y^3 = \frac{1}{2}x^2 + C$.

17. $3x^2$.   Cubical parabolas, $y = x^3 + C$.

18. $\dfrac{1}{y^2}$.   Cubical parabolas, $\frac{1}{3}y^3 = x + C$.

19. $\dfrac{x}{y}$.   Equilateral hyperbolas, $y^2 - x^2 = C$.

20. $-\dfrac{y}{x}$.   Equilateral hyperbolas, $xy = C$.

21. $\dfrac{b^2 x}{a^2 y}$.   Hyperbolas, $b^2 x^2 - a^2 y^2 = C$.

22. $-\dfrac{b^2 x}{a^2 y}$.   Ellipses, $b^2 x^2 + a^2 y^2 = C$.

23. $\dfrac{1+x}{1-y}$.   Circles, $x^2 + y^2 + 2x - 2y = C$.

In each of the following examples find the equation of the curve whose slope at any point is the given function of the coördinates and which passes through the assigned particular point.

24. x; $(1, 1)$.   *Ans.* $2y = x^2 + 1$.

25. $4y$; $(1, 1)$. $\ln y = 4x - 4$.

26. $2xy$; $(3, 1)$. $\ln y = x^2 - 9$.

27. $-xy$; $(0, 2)$. $y = 2e^{-\frac{x^2}{2}}$.

28. $\dfrac{x+1}{y+1}$; $(0, 1)$. $(y+1)^2 = (x+1)^2 + 3$.

29. $\dfrac{h-x}{y-k}$; $(0, 0)$. $x^2 + y^2 - 2hx - 2ky = 0$.

30. $\dfrac{y}{x^2}$; $(1, 1)$. $x \ln y = x - 1$.

31. $y\sqrt{x}$; $(4, 1)$. $3 \ln y = 2(x\sqrt{x} - 8)$.

32. $\dfrac{4xy}{4x^2 - 15}$; $(2, 1)$. $4x^2 - y^2 = 15$.

33. $\dfrac{y^2}{x}$; (1, 4).

34. $x\sqrt{y}$; (1, 9).

35. $\dfrac{x-3}{1-y}$; (3, 0).

36. $\dfrac{xy}{x^2+4}$; (1, 2).

37. $\dfrac{4-x}{2y-3}$; (4, 2).

38. $\sqrt{\dfrac{2+x}{3+y}}$; (2, 6).

39. $\sqrt{\dfrac{y-1}{x-2}}$; (3, 5).

40. $x\cos^2 y$; (4, $\tfrac{1}{4}\pi$).

41. Given $dy = (2x+1)dx$, $y = 7$ when $x = 1$. Find the value of y when $x = 3$. *Ans.* 17.

42. Given $dA = \sqrt{2px}\,dx$, $A = \dfrac{p^2}{3}$ when $x = \dfrac{p}{2}$. Find the value of A when $x = 2p$. *Ans.* $\tfrac{8}{3}p^2$.

43. Given $dy = x\sqrt{100-x^2}\,dx$, $y = 0$ when $x = 0$. Find the value of y when $x = 8$. *Ans.* $\tfrac{184}{3}$.

44. Given $d\rho = \cos 2\theta\,d\theta$, $\rho = 6$ when $\theta = \tfrac{1}{2}\pi$. Find the value of ρ when $\theta = \tfrac{3}{4}\pi$.

45. Given $ds = t\sqrt{4t+1}\,dt$, $s = 0$ when $t = 0$. Find the value of s when $t = 2$.

46. At every point of a certain curve $y'' = x$. Find the equation of the curve if it passes through the point (3, 0) and has the slope $\tfrac{7}{2}$ at that point.
Ans. $6y = x^3 - 6x - 9$.

47. At every point of a certain curve $y'' = \dfrac{12}{x^3}$. Find the equation of the curve if it passes through the point (1, 0) and is tangent to the line $6x + y = 6$ at that point. *Ans.* $xy + 6x = 6$.

48. Find the equation of the curve at every point of which $y'' = \dfrac{3}{\sqrt{x+3}}$ and which passes through the point (1, 1) with an inclination of 45°.

49. Find the equation of the curve at every point of which $y'' = \dfrac{1}{x}$ and which passes through the point (1, 0) with an inclination of 135°.

50. Find the equation of the curve whose subnormal is constant and equal to $2a$. *Ans.* $y^2 = 4ax + C$, a parabola.

HINT. From (4), Art. 43, subnormal $= y\dfrac{dy}{dx}$.

51. Find the curve whose subtangent is constant and equal to a (see (3), Art. 43). *Ans.* $a \ln y = x + C$.

52. Find the curve whose subnormal equals the abscissa of the point of contact. *Ans.* $y^2 - x^2 = 2C$, an equilateral hyperbola.

53. Find the curve whose normal is constant $(= R)$, assuming that $y = R$ when $x = 0$. Ans. $x^2 + y^2 = R^2$, a circle.

HINT. From Art. 43, length of normal $= y\sqrt{1 + \left(\dfrac{dy}{dx}\right)^2}$, or
$$dx = \pm (R^2 - y^2)^{-\frac{1}{2}} y \, dy.$$

54. Determine the curves in which the length of the subnormal is proportional to the square of the ordinate. Ans. $y = Ce^{kx}$.

55. Find the equation of the curve in which the angle between the radius vector and the tangent is one half the vectorial angle.
Ans. $\rho = c(1 - \cos\theta)$.

56. Find the curves in which the angle between the radius vector and the tangent at any point is n times the vectorial angle. Ans. $\rho^n = c \sin n\theta$.

140. Physical signification of the constant of integration. The following examples will illustrate what is meant.

ILLUSTRATIVE EXAMPLE 1. Find the laws governing the motion of a point which moves in a straight line with constant acceleration.

Solution. Since the acceleration $\left[= \dfrac{dv}{dt}, \text{ from } (A), \text{ Art. 59}\right]$ is constant, say f, we have
$$\frac{dv}{dt} = f,$$

or $\qquad dv = f \, dt$. Integrating,

(1) $\qquad v = ft + C.$

To determine C, suppose that the *initial* velocity is v_0; that is, let $v = v_0$ when $t = 0$.

These values substituted in (1) give

$\qquad v_0 = 0 + C, \text{ or } C = v_0.$

Hence (1) becomes

(2) $\qquad v = ft + v_0.$

Initial Conditions		
t	v	s
0	v_0	s_0

Since $v = \dfrac{ds}{dt}$ ((C), Art. 51), we get from (2)
$$\frac{ds}{dt} = ft + v_0,$$

or $\qquad ds = ft \, dt + v_0 \, dt.$ Integrating,

(3) $\qquad s = \tfrac{1}{2} ft^2 + v_0 t + C.$

To determine C, suppose that the *initial* distance is s_0; that is, let $s = s_0$ when $t = 0$. These values substituted in (3) give

$\qquad s_0 = 0 + 0 + C, \text{ or } C = s_0.$

Hence (3) becomes

(4) $\qquad s = \tfrac{1}{2} ft^2 + v_0 t + s_0.$

By substituting the values $f = g$, $v_0 = 0$, $s_0 = 0$, $s = h$, in (2) and (4), we get the laws of motion of a body falling from rest in a vacuum, namely,

$$v = gt, \quad \text{and} \quad h = \tfrac{1}{2} gt^2.$$

Eliminating t between these equations gives $v = \sqrt{2\,gh}$.

ILLUSTRATIVE EXAMPLE 2. Discuss the motion of a projectile having an initial velocity v_0 inclined at an angle α with the horizontal, the resistance of the air being neglected.

Solution. Assume the XOY-plane as the plane of motion, OX as horizontal, and OY as vertical, and let the projectile be thrown from the origin.

Suppose the projectile to be acted upon by gravity alone. Then the acceleration in the horizontal direction will be zero and in the vertical direction $-g$. Hence, from (*F*), Art. 84,

$$\frac{dv_x}{dt} = 0, \quad \text{and} \quad \frac{dv_y}{dt} = -g.$$

Integrating, $\quad v_x = C_1 \quad \text{and} \quad v_y = -gt + C_2.$

But $\quad v_0 \cos \alpha = $ initial velocity in the horizontal direction,

and $\quad v_0 \sin \alpha = $ initial velocity in the vertical direction.

Hence $\quad C_1 = v_0 \cos \alpha \quad \text{and} \quad C_2 = v_0 \sin \alpha$, giving

(5) $\quad v_x = v_0 \cos \alpha \quad \text{and} \quad v_y = -gt + v_0 \sin \alpha.$

But from (*C*) and (*D*), Art. 83, $v_x = \dfrac{dx}{dt}$ and $v_y = \dfrac{dy}{dt}$; therefore (5) gives

$$\frac{dx}{dt} = v_0 \cos \alpha \quad \text{and} \quad \frac{dy}{dt} = -gt + v_0 \sin \alpha,$$

or $\quad dx = v_0 \cos \alpha\, dt \quad \text{and} \quad dy = -gt\, dt + v_0 \sin \alpha\, dt.$

Integrating, we get

(6) $\quad x = v_0 \cos \alpha \cdot t + C_3 \quad \text{and} \quad y = -\tfrac{1}{2} gt^2 + v_0 \sin \alpha \cdot t + C_4.$

To determine C_3 and C_4 we observe that when $t = 0$, $x = 0$ and $y = 0$. Substituting these values in (6) gives

$$C_3 = 0 \quad \text{and} \quad C_4 = 0.$$

Hence

(7) $\quad x = v_0 \cos \alpha \cdot t$, and

(8) $\quad y = -\tfrac{1}{2} gt^2 + v_0 \sin \alpha \cdot t.$

Eliminating t between (7) and (8), we obtain

(9) $$y = x \tan \alpha - \frac{gx^2}{2\,v_0^2 \cos^2 \alpha},$$

which is the equation of the *trajectory* and shows that the projectile will move in a parabola.

PROBLEMS

In the following problems the relation between v and t is given. Find the relation between s and t if $s = 2$ when $t = 1$.

1. $v = a + bt.\quad$ Ans. $s = a(t-1) + \tfrac{1}{2} b(t^2 - 1) + 2.$

2. $v = \sqrt{t-1}.$

3. $v = t^2 + \dfrac{1}{t^2}.$

In the following problems the expression for the acceleration is given. Find the relation between v and t if $v = 2$ when $t = 3$.

4. $4 - t^2$. Ans. $v = 4t - \frac{1}{3}t^3 - 1$.   5. $\sqrt{t} + 3$.   6. $\frac{1}{t^2} - t$.

In the following problems the expression for the acceleration is given. Find the relation between s and t if $s = 0$, $v = 20$ when $t = 0$.

7. -32. Ans. $s = 20t - 16t^2$.   8. $4 - t$.   9. $-16\cos 2t$.

10. With what velocity will a stone strike the ground if dropped from the top of a building 120 ft. high? ($g = 32$.)   Ans. 87.64 ft. per second.

11. With what velocity will the stone of Problem 10 strike the ground if thrown downward with a speed of 20 ft. per second? if thrown upward with a speed of 20 ft. per second?   Ans. 89.89 ft. per second.

12. A stone dropped from a balloon which was rising at the rate of 15 ft. per second reached the ground in 8 sec. How high was the balloon when the stone was dropped?   Ans. 904 ft.

13. In Problem 12, if the balloon had been falling at the rate of 15 ft. per second, how long would the stone have taken to reach the ground?
Ans. $7\frac{1}{16}$ sec.

14. A train leaving a railroad station has an acceleration of $0.5 + 0.02\,t$ ft. per second per second. Find how far it will move in 20 sec. Ans. 126.7 ft.

15. A particle sliding on a certain inclined plane is subject to an acceleration downward of 4 ft. per second per second. If it is started upward from the bottom of the plane with a velocity of 6 ft. per second, find the distance moved after t sec. How far will it go before sliding backward?   Ans. 4.5 ft.

16. If the inclined plane in Problem 15 is 20 ft. long, find the necessary initial speed in order that the particle may just reach the top of the plane.
Ans. $4\sqrt{10}$ ft. per second.

17. A ball thrown upward from the ground reaches a height of 80 ft. in 1 sec. Find how high the ball will go.

18. A projectile with an initial velocity of 160 ft. per second is fired at a vertical wall 480 ft. from the point of projection.

(a) If $\alpha = 45°$, find the height of the point struck on the wall.
Ans. 192 ft.

(b) Find α so that the projectile will strike the base of the wall.
Ans. 18° or 72°.

(c) Find α so that the projectile will strike 80 ft. above the base.
Ans. 29° or 70°.

(d) Find α for the maximum height on the wall and this height.
Ans. 59°; 256 ft.

19. If the acceleration of a particle is given by $-ks$, where k is a constant, and if $v = v_0$ when $s = s_0$, show that

$$v^2 - v_0^2 + k(s^2 - s_0^2) = 0.$$

20. The acceleration of a body below the surface of the earth is directed toward the center of the earth, and its magnitude varies directly as the distance of the body from the center. If a body were dropped into a hole 1000 mi. deep, with what speed would it reach the bottom?

Ans. $3\frac{1}{4}$ mi. per second, approximately.

ADDITIONAL PROBLEMS

1. The temperature of a liquid in a room of temperature 20° is observed to be 70°, and after 5 min. to be 60°. Assuming the rate of cooling to be proportional to the difference of the temperatures of the liquid and the room, find the temperature of the liquid 30 min. after the first observation. *Ans.* 33.1°.

2. Find the equation of the curve whose polar subtangent is n times the length of the corresponding radius vector and which passes through the point $(a, 0)$.

Ans. $\rho = ae^{\frac{\theta}{n}}$.

3. Find the equation of the curve whose polar subnormal is n times the length of the corresponding radius vector, and which passes through the point $(a, 0)$. *Ans.* $\rho = ae^{n\theta}$.

4. A particle moves in the xy-plane so that the components of velocity parallel to the x-axis and the y-axis are ky and kx, respectively. Prove that the path is an equilateral hyperbola.

5. A particle projected from the top of a tower at an angle of 45° above the horizontal plane strikes the ground in 5 sec. at a horizontal distance from the foot of the tower equal to its height. Find the height of the tower $(g = 32)$. *Ans.* 200 ft.

6. A particle starts from the origin of coördinates and in t sec. its x-component of velocity is $t^2 - 4$ and its y-component is $4t$.

(a) Find the position of the particle after t sec.

Ans. $x = \frac{1}{3}t^3 - 4t$, $y = 2t^2$.

(b) Find the distance traversed along the path. *Ans.* $s = \frac{1}{3}t^3 + 4t$.

(c) Find the equation of the path. *Ans.* $72x^2 = y^3 - 48y^2 + 576y$.

7. Find an equation of a curve for which the length of the tangent (Art. 43) is constant $(= c)$.

SUGGESTION. Choose the minus sign in Problem 2 (a), p. 85, and assume $y = c$ when $x = 0$.

Ans. $x = c \ln\left(\dfrac{c + \sqrt{c^2 - y^2}}{y}\right) - \sqrt{c^2 - y^2}$.

8. Find the equation of the curve for which (Art. 96) $a^2 \, ds = \rho^3 \, d\theta$, and which passes through the point $(a, 0)$. *Ans.* $\rho^2 = a^2 \sec 2\theta$.

CHAPTER XIV

THE DEFINITE INTEGRAL

141. Differential of the area under a curve. Consider the continuous function $\phi(x)$, and let

$$y = \phi(x)$$

be the equation of the curve AB. Let CD be a fixed and MP a variable ordinate, and let u be the measure of the area $CMPD$. When x takes on a small increment Δx, u takes on an increment Δu ($=$ area $MNQP$). Completing the rectangles $MNRP$ and $MNQS$, we see that

Area $MNRP <$ area $MNQP <$ area $MNQS$,

or $\quad MP \cdot \Delta x < \Delta u < NQ \cdot \Delta x$;

and, dividing by Δx,

$$MP < \frac{\Delta u}{\Delta x} < NQ.*$$

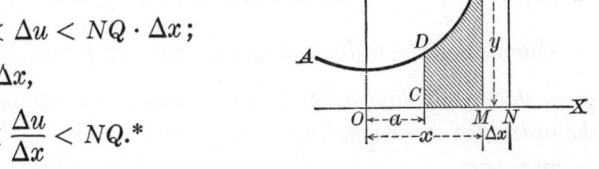

Now let Δx approach zero as a limit; then since MP remains fixed and NQ approaches MP as a limit (since y is a continuous function of x), we get

$$\frac{du}{dx} = y (= MP),$$

or, using differentials, $\quad du = y\, dx$.

Theorem. *The differential of the area bounded by any curve, the x-axis, a fixed ordinate, and a variable ordinate is equal to the product of the variable ordinate and the differential of the corresponding abscissa.*

142. The definite integral. It follows from the theorem in the last article that if the curve AB is the locus of

$$y = \phi(x),$$

then $\quad du = y\, dx$, or

(1) $\quad du = \phi(x)dx$,

---

* In this figure MP is less than NQ; if MP happens to be greater than NQ, simply reverse the inequality signs.

where du is the differential of the area between the curve, the x-axis, and two ordinates. Integrating, we get

$$u = \int \phi(x) \, dx.$$

Denote $\int \phi(x) \, dx$ by $f(x) + C$.

(2) $\therefore u = f(x) + C.$

We determine C by observing that $u = 0$ when $x = a$. Substituting these values in (2), we get

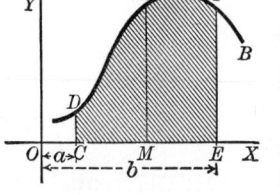

$$0 = f(a) + C,$$

and hence $C = -f(a).$

Then (2) becomes

(3) $u = f(x) - f(a).$

The required area $CEFD$ is the value of u in (3) when $x = b$. Hence we have

(A) Area $CEFD = f(b) - f(a).$

Theorem. *The difference of the values of $\int y \, dx$ for $x = a$ and $x = b$ gives the area bounded by the curve whose ordinate is y, the x-axis, and the ordinates corresponding to $x = a$ and $x = b$.*

This difference is represented by the symbol*

(4) $$\int_a^b y \, dx \quad \text{or} \quad \int_a^b \phi(x) \, dx,$$

and is read "the integral from a to b of $y \, dx$." The operation is called *integration between limits*, a being the *lower* and b the *upper* limit.†

Since (4) always has a *definite* value, it is called a *definite integral.* For, if

$$\int \phi(x) \, dx = f(x) + C,$$

then $\int_a^b \phi(x) \, dx = \Big[f(x) + C\Big]_a^b = [f(b) + C] - [f(a) + C],$

or $\int_a^b \phi(x) \, dx = f(b) - f(a),$

the *constant of integration* having disappeared.

---

* This notation is due to Joseph Fourier (1768–1830).

† The word "limit" in this connection means merely the value of the variable at one end of its range (end value), and should not be confused with the meaning of the word in the **Theory of Limits.**

THE DEFINITE INTEGRAL

We may accordingly define the symbol

$$\int_a^b \phi(x)\,dx \quad \text{or} \quad \int_a^b y\,dx$$

as the numerical measure of the area bounded by the curve $y = \phi(x)$,* the x-axis, and the ordinates of the curve at $x = a$ and $x = b$. This definition presupposes that these lines bound an area; that is, the curve does not rise or fall to infinity and does not cross the x-axis, and both a and b are finite.

143. Calculation of a definite integral. The process may be summarized as follows:

FIRST STEP. *Integrate the given differential expression.*

SECOND STEP. *Substitute in this indefinite integral first the upper limit and then the lower limit for the variable, and subtract the last result from the first.*

It is not necessary to bring in the constant of integration, since it always disappears in subtracting.

ILLUSTRATIVE EXAMPLE 1. Find $\int_1^4 x^2\,dx$.

Solution. $\int_1^4 x^2\,dx = \left[\dfrac{x^3}{3}\right]_1^4 = \dfrac{64}{3} - \dfrac{1}{3} = 21$. *Ans.*

ILLUSTRATIVE EXAMPLE 2. Find $\int_0^\pi \sin x\,dx$.

Solution. $\int_0^\pi \sin x\,dx = \left[-\cos x\right]_0^\pi = \left[-(-1)\right] - \left[-1\right] = 2$. *Ans.*

ILLUSTRATIVE EXAMPLE 3. Prove $\int_0^a \dfrac{dx}{a^2 + x^2} = \dfrac{\pi}{4a}$.

Solution. $\int_0^a \dfrac{dx}{a^2 + x^2} = \left[\dfrac{1}{a}\arctan\dfrac{x}{a}\right]_0^a = \dfrac{1}{a}\arctan 1 - \dfrac{1}{a}\arctan 0 = \dfrac{\pi}{4a}$.

ILLUSTRATIVE EXAMPLE 4. Prove $\int_{-1}^0 \dfrac{dx}{4x^2 - 9} = -\dfrac{1}{12}\ln 5 = -0.134$.

Solution. Comparing with (19) or (19 a), $v = 2x$, $a = 3$, $dv = 2\,dx$.

To decide between the use of (19) or (19 a), consider the limits. The values of x increase from -1 to 0.

Then $v \,(= 2x)$ increases from -2 to 0.

Hence $v^2 \leqq 4$. But $a^2 = 9$. Therefore $v^2 < a^2$, and (19 a) must be used. Thus

(1) $\quad \int_{-1}^0 \dfrac{dx}{4x^2 - 9} = -\int_{-1}^0 \dfrac{dx}{9 - 4x^2} = -\dfrac{1}{12}\left[\ln\dfrac{3 + 2x}{3 - 2x}\right]_{-1}^0$. By (19 a)

Evaluating in (1) gives the answer. The result is negative because the curve and the bounding ordinates lie below the x-axis.

* $\phi(x)$ is continuous and single-valued throughout the interval [a, b].

144. Change in limits corresponding to change in variable. When integrating by the substitution of a new variable it is sometimes rather troublesome to translate the result back into the original variable. When integrating between limits, however, we may avoid the process of restoring the original variable by changing the limits to correspond with the new variable. This process will now be illustrated by an example.

ILLUSTRATIVE EXAMPLE. Calculate $\int_0^{16} \frac{x^{\frac{1}{4}} dx}{1 + x^{\frac{1}{2}}}$.

Solution. Assume $x = z^4$.
Then $dx = 4 z^3 dz$, $x^{\frac{1}{2}} = z^2$, $x^{\frac{1}{4}} = z$. Also, to change the limits we observe that when

$$x = 0, \quad z = 0,$$
and when
$$x = 16, \quad z = 2.$$

$$\therefore \int_0^{16} \frac{x^{\frac{1}{4}} dx}{1 + x^{\frac{1}{2}}} = \int_0^2 \frac{z \cdot 4 z^3 dz}{1 + z^2} = 4 \int_0^2 \left(z^2 - 1 + \frac{1}{1+z^2}\right) dz$$

$$= 4 \int_0^2 z^2 dz - 4 \int_0^2 dz + 4 \int_0^2 \frac{dz}{1+z^2} = \left[\frac{4 z^3}{3} - 4 z + 4 \arctan z\right]_0^2$$

$$= \tfrac{8}{3} + 4 \arctan 2. \quad Ans.$$

The relation between the old and the new variable should be such that to each value of one within the limits of integration there is always one, and only one, finite value of the other. When one is given as a many-valued function of the other, care must be taken to choose the right values.

PROBLEMS

1. Prove that $\int_a^b f(x) dx = -\int_b^a f(x) dx$.

Work out the following integrals.

2. $\int_0^a (a^2 x - x^3) dx = \frac{a^4}{4}$.

3. $\int_1^e \frac{dx}{x} = 1$.

4. $\int_0^1 \frac{dx}{\sqrt{3 - 2x}} = \sqrt{3} - 1$.

5. $\int_2^3 \frac{2 t \, dt}{1 + t^2} = \ln 2$.

6. $\int_0^2 \frac{x^3 \, dx}{x + 1} = \tfrac{8}{3} - \ln 3$.

7. $\int_0^r \frac{r \, dx}{\sqrt{r^2 - x^2}} = \frac{\pi r}{2}$.

8. $\int_0^a (\sqrt{a} - \sqrt{x})^2 \, dx = \frac{a^2}{6}$.

9. $\int_0^4 \frac{x^2 \, dx}{x + 1} = 5.6094$.

10. $\int_0^1 \frac{dx}{e^{3x}} = 0.3167$.

11. $\int_0^{\frac{\pi}{2}} \cos \phi \, d\phi = 1$.

12. $\int_0^\pi \sqrt{2 + 2 \cos \theta} \, d\theta = 4$.

13. $\int_0^{\frac{\pi}{2}} \sin^3 x \cos^3 x \, dx = \tfrac{1}{12}$.

14. $\int_0^{\frac{\pi}{4}} \sec^4 \theta \, d\theta = \tfrac{4}{3}$.

THE DEFINITE INTEGRAL

Find the value of each of the following definite integrals.

15. $\int_0^4 \dfrac{dx}{\sqrt{9-2x}}.$

16. $\int_0^3 \dfrac{t\,dt}{\sqrt{t^2+16}}.$

17. $\int_0^2 \dfrac{y\,dy}{\sqrt{25-4y^2}}.$

18. $\int_0^a \sqrt{a^2-x^2}\,dx.$

19. $\int_1^5 \dfrac{dx}{\sqrt{2x-1}}.$

20. $\int_0^1 xe^{-x^2}\,dx.$

21. $\int_0^{\frac{\pi}{2}} \cos^2\theta\,d\theta.$

22. $\int_0^\pi \sin^2\dfrac{\theta}{2}\cos\dfrac{\theta}{2}\,d\theta.$

23. $\int_{-1}^2 \dfrac{x^2\,dx}{x+2}.$

24. $\int_0^1 \dfrac{x^2\,dx}{x^2+1}.$

145. Calculation of areas. On page 238 it was shown that the area between a curve, the x-axis, and the ordinates $x=a$ and $x=b$ is given by the formula

(B) $\qquad\text{Area} = \displaystyle\int_a^b y\,dx,$

where the value of y in terms of x is substituted from the equation of the given curve.

ILLUSTRATIVE EXAMPLE 1. Find the area bounded by the parabola $y=x^2$, the x-axis, and the ordinates $x=2$ and $x=4$.

Solution. Substituting in the formula,

$\text{Area } ABDC = \displaystyle\int_2^4 x^2\,dx = \left[\dfrac{x^3}{3}\right]_2^4$

$= \tfrac{64}{3} - \tfrac{8}{3} = \tfrac{56}{3} = 18\tfrac{2}{3}.$ Ans.

ILLUSTRATIVE EXAMPLE 2. Find the area bounded by the circle $x^2+y^2=25$, the x-axis, and the ordinates $x=-3$, $x=4$.

Solution. Solving, $y=\sqrt{25-x^2}$. Hence

$\text{Area} = \displaystyle\int_{-3}^4 \sqrt{25-x^2}\,dx$

$= \left[\dfrac{x}{2}\sqrt{25-x^2} + \dfrac{25}{2}\arcsin\dfrac{x}{5}\right]_{-3}^4 \qquad$ by (22)

$= 6 + \tfrac{25}{2}\arcsin\tfrac{4}{5} + 6 - \tfrac{25}{2}\arcsin(-\tfrac{3}{5}) = 31.6.$ Ans.

The answer should be compared with the area of the semicircle, which is $\tfrac{1}{2}(25\,\pi) = 39.3$.

ILLUSTRATIVE EXAMPLE 3. *Area under a parabola whose axis is parallel to the y-axis.* In the figure (p. 242) the point P' on the parabolic arc PP'' is chosen so that $AO=OB$. The ordinates of P, P', P'' are, respectively, y, y', y''. Prove that the area between the parabola, the x-axis, and the ordinates of P and P'' equals $\tfrac{1}{3}h(y+4y'+y'')$ if $2h$ is the distance apart of the ordinates of P and P''.

242　INTEGRAL CALCULUS

Solution. Take the y-axis along the ordinate of P', as in the figure. Then $AB = 2h$. The equation of a parabola with axis parallel to the y-axis is, by **(7)**, p. 4, $(x-h)^2 = 2p(y-k)$. If this is solved for y, the result takes the form

(1) $\qquad y = ax^2 + 2bx + c.$

The required area $APP''B \;(= u)$ is, by **(B)**,

(2) $\qquad u = \int_{-h}^{h} (ax^2 + 2bx + c)dx = \tfrac{2}{3} ah^3 + 2 ch.$

By (1), if $x = -h,\; y = AP = ah^2 - 2bh + c$;
$\phantom{\text{By (1), }}$if $x = 0, y' = OP' = c$;
$\phantom{\text{By (1), }}$if $x = h, y'' = BP'' = ah^2 + 2bh + c.$

Therefore $\tfrac{1}{3} h(y + 4y' + y'') = \tfrac{2}{3} ah^3 + 2 ch = u.$ Q.E.D.

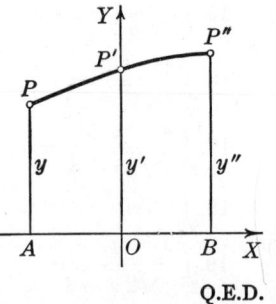

146. Area when the equations of the curve are given in parametric form.* Let the equations of the curve be given in the parametric form

$$x = f(t), \quad y = \phi(t).$$

We then have $y = \phi(t)$, and $dx = f'(t)dt$. Hence

(1) $\qquad \text{Area} = \int_a^b y\, dx = \int_{t_1}^{t_2} \phi(t) f'(t)\, dt,$

where $\quad t = t_1$ when $x = a,\;$ and $\;t = t_2$ when $x = b.$

ILLUSTRATIVE EXAMPLE. Find the area of the ellipse whose parametric equations (Art. 81) are

$$x = a \cos \phi, \quad y = b \sin \phi.$$

Solution. Here $\quad y = b \sin \phi,$
and $\qquad\qquad dx = -a \sin \phi\, d\phi.$
When $\qquad x = 0,\; \phi = \tfrac{1}{2}\pi$;
and when $\qquad x = a,\; \phi = 0.$

Substituting these in (1), above, we get

$$\frac{\text{Area}}{4} = \int_0^a y\, dx = -\int_{\tfrac{1}{2}\pi}^{0} ab \sin^2 \phi\, d\phi = \frac{\pi ab}{4}.$$

Hence the entire area equals $\pi ab.$ Ans.

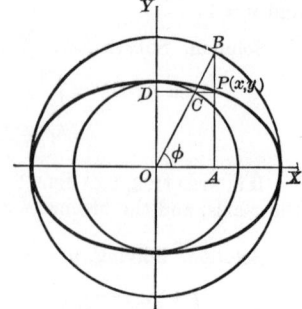

PROBLEMS

1. Find by integration the area of the triangle bounded by the line $y = 2x$, the x-axis, and the ordinate $x = 4$. Verify your result by finding the area as half the product of the base and altitude.

2. Find by integration the area of the trapezoid bounded by the line $x + y = 10$, the x-axis, and the ordinates $x = 1$ and $x = 8$. Verify your result by finding the area as half the product of the sum of the parallel sides and the altitude.

---
* For a rigorous proof of this substitution the student is referred to more advanced treatises on the calculus.

THE DEFINITE INTEGRAL

Find the area bounded by the given curve, the x-axis, and the given ordinates.

3. $y = x^3$; $x = 0$, $x = 4$. *Ans.* 64.
4. $y = 9 - x^2$; $x = 0$, $x = 3$. 18.
5. $y = x^3 + 3x^2 + 2x$; $x = -3$, $x = 3$. 59.
6. $y = x^2 + x + 1$; $x = 2$, $x = 3$. $9\frac{5}{6}$.
7. $xy = k^2$; $x = a$, $x = b$. $k^2 \ln\left(\dfrac{b}{a}\right)$.
8. $y = 2x + \dfrac{1}{x^2}$; $x = 1$, $x = 4$. $15\frac{3}{4}$.
9. $y = \dfrac{10}{\sqrt{x+4}}$; $x = 0$, $x = 5$. 20.
10. $ay = x\sqrt{a^2 - x^2}$; $x = 0$, $x = a$. $\frac{1}{3} a^2$.
11. $y^2 + 4x = 0$; $x = -1$, $x = 0$.
12. $y^2 = 4x + 16$; $x = -2$, $x = 0$.
13. $y = x^2 + 4x$; $x = -4$, $x = -2$.
14. $y = 4x - x^2$; $x = 1$, $x = 3$.
15. $y^2 = 9 - x$; $x = 0$, $x = 8$.
16. $2y^2 = x^3$; $x = 0$, $x = 2$.

Find the area bounded by the given curve, the y-axis, and the given lines.

17. $y^2 = 4x$; $y = 0$, $y = 4$. *Ans.* $5\frac{1}{3}$.
18. $y = 4 - x^2$; $y = 0$, $y = 3$. $4\frac{2}{3}$.
19. $x = 9y - y^3$; $y = 0$, $y = 3$.
20. $xy = 8$; $y = 1$, $y = 4$.
21. $y^3 = a^2 x$; $y = 0$, $y = a$.
22. $ay^2 = x^3$; $y = 0$, $y = a$.

Sketch each of the following curves and find the area of one arch.

23. $y = 2 \cos x$. *Ans.* 4.
24. $y = 2 \sin \frac{1}{2} \pi x$. $\dfrac{8}{\pi}$.
25. $y = \cos 2x$. 1.
26. $y = \sin \frac{1}{2} x$. 4.

27. Find the area bounded by the coördinate axes and the parabola $\sqrt{x} + \sqrt{y} = \sqrt{a}$.

28. Prove that the area of any segment of a parabola cut off by a chord perpendicular to the axis of the parabola is two thirds of the circumscribing rectangle.

29. P and Q are any two points on an equilateral hyperbola $xy = k$. Show that the area bounded by the arc PQ, the ordinates of P and Q, and the x-axis is equal to the area bounded by PQ, the abscissas of P and Q, and the y-axis.

30. Find the area bounded by the catenary $y = \frac{1}{2} a \left(e^{\frac{x}{a}} + e^{-\frac{x}{a}}\right)$, the x-axis, and the lines $x = a$ and $x = -a$. *Ans.* $a^2 \left(e - \dfrac{1}{e}\right)$.

31. Find the area included between the two parabolas $y^2 = 2\,px$ and $x^2 = 2\,py$. *Ans.* $\frac{4}{3}p^2$.

32. Find the area included between the two parabolas $y^2 = ax$ and $x^2 = by$. *Ans.* $\frac{1}{3}ab$.

33. Find the area inclosed by the loop of the curve whose equation is $4\,y^2 = x^2(4-x)$. *Ans.* $\frac{128}{15}$.

34. Find the area bounded by the curve whose equation is $y^2 = x^2(x^2 - 1)$ and by the line $x = 2$. *Ans.* $2\sqrt{3}$.

35. Find the area inclosed by the loop of the curve whose equation is $y^2 = x^2(9 - x)$. *Ans.* $\frac{648}{5}$.

36. Find the area bounded by the curve whose equation is $y^2 = x^3 - x^2$ and by the line $x = 2$. *Ans.* $\frac{32}{15}$.

37. Find the area inclosed by the loop of the curve whose equation is $y^2 = x(x - 2)^2$. *Ans.* $\frac{32}{15}\sqrt{2}$.

38. Find the area inclosed by the loop of the curve whose equation is $4\,y^2 = x^4(4 - x)$. *Ans.* $\frac{2048}{105}$.

39. Find the area bounded by the hyperbola $x^2 - y^2 = a^2$ and the line $x = 2\,a$. *Ans.* $a^2[2\sqrt{3} - \ln(2 + \sqrt{3})]$.

40. Find the area bounded by the hyperbola $x^2 - 4\,y^2 = 4$ and the line $x = 6$.

41. Find the area bounded by one arch of the cycloid $x = a(\theta - \sin\theta)$, $y = a(1 - \cos\theta)$, and the x-axis. *Ans.* $3\,\pi a^2$.

42. Find the area of the cardioid
$$x = a(2\cos t - \cos 2t),$$
$$y = a(2\sin t - \sin 2t). \quad \textit{Ans. } 6\,\pi a^2.$$

43. The locus in the figure is called the "companion to the cycloid." Its equations are
$$x = a\theta,$$
$$y = a(1 - \cos\theta).$$
Find the area of one arch. *Ans.* $2\,\pi a^2$.

44. Find the area of the hypocycloid
$$\begin{cases} x = a\cos^3\theta, \\ y = a\sin^3\theta, \end{cases}$$

θ being the parameter. *Ans.* $\dfrac{3\,\pi a^2}{8}$; that is, three eighths of the area of the circumscribing circle.

147. Geometrical representation of an integral. In the preceding articles the definite integral appeared as an area. This does not necessarily mean that every definite integral is an area, for the physical interpretation of the result depends on the nature of the quantities represented by the abscissa and the ordinate. Thus, if x and y are considered as simply the coördinates of a point, then the integral in (B), Art. 145, is indeed an area. But suppose the ordinate represents the speed of a moving point, and the corresponding abscissa the time at which the point has that speed; then the graph is the speed curve of the motion, and the area under it and between any two ordinates will represent the distance passed through in the corresponding interval of time. That is, the *number* which denotes the area equals the number which denotes the distance (or value of the integral). Similarly, a definite integral standing for volume, surface, mass, force, etc. may be represented geometrically by an area.

148. Approximate integration. Trapezoidal rule. We now prove two rules for evaluating

(1) $$\int_a^b f(x)dx,$$

approximately. These rules are useful when the integration in (1) is difficult, or impossible in terms of elementary functions.

The *exact* numerical value of (1) is the measure of the area bounded by the curve

(2) $\quad y = f(x),$

the x-axis, and the ordinates $x=a$, $x=b$. This area may be evaluated approximately by adding together trapezoids, as follows.

Divide the segment $b-a$ on OX into n equal parts, each of length Δx. Let the successive abscissas of the points of division be $x_0(=a), x_1, x_2, \cdots, x_n(=b)$. At these points erect the corresponding ordinates of the curve (2). Let these be

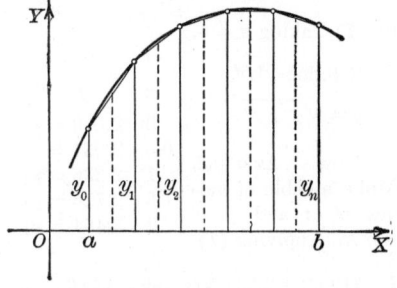

$$y_0 = f(x_0), \quad y_1 = f(x_1), \quad y_2 = f(x_2), \quad \cdots, \quad y_n = f(x_n).$$

Join the extremities of consecutive ordinates by straight lines (chords) forming trapezoids. Then, the area of a trapezoid being one half the product of the sum of the parallel sides multiplied by the altitude, we get

$\frac{1}{2}(y_0 + y_1)\Delta x =$ area of first trapezoid,

$\frac{1}{2}(y_1 + y_2)\Delta x =$ area of second trapezoid,

$\cdots \cdots \cdots \cdots$

$\frac{1}{2}(y_{n-1} + y_n)\Delta x =$ area of nth trapezoid.

INTEGRAL CALCULUS

Adding, we get the **trapezoidal rule**,

(T) Area $= (\tfrac{1}{2} y_0 + y_1 + y_2 + \cdots + y_{n-1} + \tfrac{1}{2} y_n)\Delta x.$

It is clear that the greater the number of intervals (that is, the smaller Δx is), the closer will the sum of the areas of the trapezoids approach the area under the curve.

ILLUSTRATIVE EXAMPLE 1. Calculate $\int_1^{12} x^2 \, dx$ by the trapezoidal rule, dividing $x = 1$ to $x = 12$ into eleven intervals.

Solution. Here $\dfrac{b-a}{n} = \dfrac{12-1}{11} = 1 = \Delta x$. The area in question is under the curve $y = x^2$. Substituting the abscissas $x = 1, 2, 3, \cdots, 12$ in this equation, we get the ordinates $y = 1, 4, 9, \cdots, 144$. Hence, from (T),

Area $= (\tfrac{1}{2} + 4 + 9 + 16 + 25 + 36 + 49 + 64 + 81 + 100 + 121 + \tfrac{1}{2} \cdot 144) \cdot 1 = 577\tfrac{1}{2}.$

By integration $\int_1^{12} x^2 \, dx = \left[\dfrac{x^3}{3}\right]_1^{12} = 575\tfrac{2}{3}$. Hence, in this example, the trapezoidal rule is in error by less than one third of 1 per cent.

ILLUSTRATIVE EXAMPLE 2. Find the approximate value of

$$I = \int_0^2 \sqrt{4 + x^3} \, dx$$

by (T), taking $n = 4$.

Solution. Let

$y = \sqrt{4 + x^3}.$

Now $\Delta x = 0.5$. Make a table of values of x and y as shown. Applying (T),

x	y
0	$2.000 = y_0$
0.5	$2.031 = y_1$
1	$2.236 = y_2$
1.5	$2.716 = y_3$
2	$3.464 = y_4$

$I = (1.000 + 2.031 + 2.236 + 2.716 + 1.732) \times 0.5 = 4.858.$ *Ans.*

If we take $n = 10$, we obtain $I = 4.826$, a closer approximation.

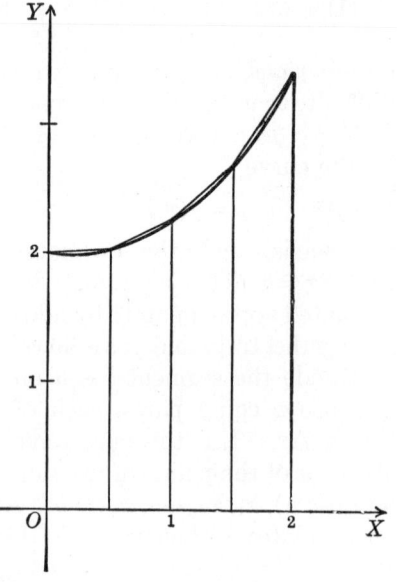

PROBLEMS

Compute the approximate values of the following integrals by the trapezoidal rule, using the values of n indicated. Check your results by performing the integrations.

1. $\int_3^{10} \dfrac{dx}{x}$; $n = 7.$

2. $\int_0^5 x\sqrt{25 - x^2} \, dx$; $n = 10.$

3. $\int_4^8 \sqrt{64 - x^2} \, dx$; $n = 8.$

4. $\int_0^3 \sqrt{16 + x^2} \, dx$; $n = 6.$

Compute the approximate values of the following integrals by the trapezoidal rule, using the values of n indicated.

5. $\int_0^4 \dfrac{dx}{\sqrt{4+x^3}}$; $n=4$.   Ans. 1.227.   10. $\int_{-2}^3 \sqrt{20+x^4}\,dx$; $n=5$.

6. $\int_0^2 \sqrt{1+x^3}\,dx$; $n=4$.   3.283.   11. $\int_0^2 x^2\sqrt{16-x^4}\,dx$; $n=4$.

7. $\int_0^{10} \sqrt[3]{125-x^2}\,dx$; $n=5$.   44.17.   12. $\int_1^6 \sqrt[3]{x^2+3\,x}\,dx$; $n=5$.

8. $\int_1^5 \sqrt{126-x^3}\,dx$; $n=4$.   34.78.   13. $\int_1^4 \dfrac{x\,dx}{\sqrt[3]{10+x^3}}$; $n=6$.

9. $\int_2^8 \dfrac{x\,dx}{\sqrt[3]{4+x^2}}$; $n=6$.   9.47.   14. $\int_2^4 \dfrac{x^2\,dx}{\sqrt[3]{10+x^2}}$; $n=4$.

149. Simpson's rule (parabolic rule). Instead of connecting the extremities of successive ordinates by chords and forming trapezoids, we can get a still closer approximation to the area by joining them with arcs of parabolas and summing up the areas under these arcs. A parabola with a vertical axis may be passed through any three points on a curve, and a series of such arcs will fit the curve more closely than the broken line of chords. In fact, the equation of such a parabola is of the form (1) in Illustrative Example 3, Art. 145, and the values of the constants a, b, c may be determined so that this parabola shall pass through three given points. In the present investigation, however, this is not necessary.

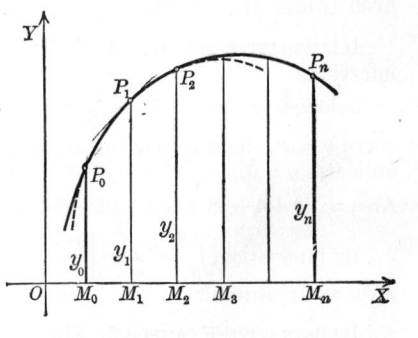

We now divide the interval from $x=a=OM_0$ to $x=b=OM_n$ into an *even* number ($=n$) of parts, each equal to Δx. Through each successive set of three points P_0, P_1, P_2; P_2, P_3, P_4; etc., are drawn arcs of parabolas with vertical axes. The ordinates of these points are y_0, y_1, y_2, $\cdots$, y_n, as indicated in the figure. The area $M_0P_0P_2\cdots P_nM_n$ is thus replaced by a set of "double parabolic strips" such as $M_0P_0P_1P_2M_2$, whose upper boundary is in every case a parabolic arc (1) of Illustrative Example 3, Art. 145. The area of each of these double strips is found by using the formula

$$u = \tfrac{1}{3} h(y + 4\,y' + y'')$$

of this example.

For the first one, $h=\Delta x$, $y=y_0$, $y'=y_1$, $y''=y_2$. Hence

248 INTEGRAL CALCULUS

Area of first parabolic strip $M_0P_0P_1P_2M_2 = \dfrac{\Delta x}{3}(y_0 + 4y_1 + y_2)$.

Similarly, Second (double) strip $= \dfrac{\Delta x}{3}(y_2 + 4y_3 + y_4)$,

Third (double) strip $= \dfrac{\Delta x}{3}(y_4 + 4y_5 + y_6)$,

. . . .

Last (double) strip $= \dfrac{\Delta x}{3}(y_{n-2} + 4y_{n-1} + y_n)$.

Adding, we get Simpson's rule (n being even),

(S) Area $= \dfrac{\Delta x}{3}(y_0 + 4y_1 + 2y_2 + 4y_3 + 2y_4 + \cdots + y_n)$.

As in the case of the trapezoidal rule, the greater the number of parts into which M_0M_n is divided, the closer will the result be to the area under the curve.

ILLUSTRATIVE EXAMPLE 1. Calculate $\int_0^{10} x^3 \, dx$ by Simpson's rule, taking ten intervals.

Solution. Here $\dfrac{b-a}{n} = \dfrac{10-0}{10} = 1 = \Delta x$. The area in question is under the curve $y = x^3$. Substituting the abscissas $x = 0, 1, 2, \cdots, 10$ in $y = x^3$, we get the ordinates $y = 0, 1, 8, 27, \cdots, 1000$. Hence, from (S),

Area $= \tfrac{1}{3}(0 + 4 + 16 + 108 + 128 + 500 + 432 + 1372 + 1024 + 2916 + 1000) = 2500$.

By integration, $\int_0^{10} x^3 \, dx = \left[\dfrac{x^4}{4}\right]_0^{10} = 2500$, so that in this example Simpson's rule gives an exact result.

ILLUSTRATIVE EXAMPLE 2. Find the approximate value of

$$I = \int_0^2 \sqrt{4 + x^3} \, dx$$

by (S), taking $n = 4$.

Solution. The table of values is given in Illustrative Example 2 of the preceding article. Hence

$$I = (2.000 + 8.124 + 4.472 + 10.864 + 3.464) \times \dfrac{0.5}{3} = 4.821.$$

Compare this result with that given by (T) when $n = 10$, namely 4.826.
In this case formula (S) gives a better approximation than (T) when $n = 4$.

PROBLEMS

Compute the approximate values of the following integrals by Simpson's rule, using the values of n indicated. Check your results by performing the integrations.

1. $\int_3^6 \dfrac{x \, dx}{4 + x^2}$; $n = 6$.

2. $\int_0^4 x\sqrt{25 - x^2} \, dx$; $n = 4$.

3. $\int_2^8 \sqrt{64 - x^2} \, dx$; $n = 6$.

4. $\int_4^7 \sqrt{16 + x^2} \, dx$; $n = 6$.

Compute the approximate values of the following integrals by Simpson's rule, using the values of n indicated.

5. $\displaystyle\int_0^4 \frac{dx}{\sqrt{4+x^3}}$; $n=4$. 	Ans. 1.236.

6. $\displaystyle\int_0^2 \sqrt{1+x^3}\, dx$; $n=4$. 	3.239.

7. $\displaystyle\int_1^5 \sqrt{126-x^3}\, dx$; $n=4$. 	35.68.

8. $\displaystyle\int_2^8 \frac{x\, dx}{\sqrt[3]{4+x^2}}$; $n=6$. 	9.49.

9. $\displaystyle\int_1^5 \sqrt[3]{6+x^2}\, dx$; $n=4$. 	11. $\displaystyle\int_1^5 \sqrt[3]{x^3-x}\, dx$; $n=4$.

10. $\displaystyle\int_2^5 \frac{x^3\, dx}{\sqrt{1+x^3}}$; $n=6$. 	12. $\displaystyle\int_2^4 \frac{x\, dx}{\sqrt{5+x^3}}$; $n=4$.

Calculate the approximate values of the following integrals by both the trapezoidal and Simpson's rules. If the indefinite integral can be found, calculate also the exact value of the integral.

13. $\displaystyle\int_2^4 \sqrt{16-x^2}\, dx$; $n=4$. 	18. $\displaystyle\int_0^2 e^{-\frac{x^2}{2}}\, dx$; $n=4$.

14. $\displaystyle\int_2^4 x\sqrt{16-x^2}\, dx$; $n=4$. 	19. $\displaystyle\int_0^{\frac{\pi}{2}} \frac{10\, d\theta}{\sqrt{2+\sin^2\theta}}$; $n=6$.

15. $\displaystyle\int_3^7 \frac{x\, dx}{\sqrt{64-x^2}}$; $n=4$. 	20. $\displaystyle\int_0^{\frac{\pi}{2}} \sqrt{2-\cos^2\theta}\, d\theta$; $n=6$.

16. $\displaystyle\int_3^7 \frac{dx}{\sqrt{64-x^2}}$; $n=4$. 	21. $\displaystyle\int_0^2 \frac{10\, d\theta}{\sqrt{1+\cos^2\frac{1}{4}\pi\theta}}$; $n=4$.

17. $\displaystyle\int_2^8 \frac{x\, dx}{\sqrt{3+x^3}}$; $n=6$. 	22. $\displaystyle\int_0^{\frac{1}{2}} \sqrt{4-3\sin^2\pi\theta}\, d\theta$; $n=8$.

150. Interchange of limits. Since

$$\int_a^b \phi(x)dx = f(b) - f(a),$$

and 	$\displaystyle\int_b^a \phi(x)dx = f(a) - f(b) = -[f(b) - f(a)],$

we have 	$\displaystyle\int_a^b \phi(x)dx = -\int_b^a \phi(x)dx.$

Theorem. *Interchanging the limits is equivalent to changing the sign of the definite integral.*

250   INTEGRAL CALCULUS

151. Decomposition of the interval of integration of the definite integral. Since

$$\int_a^{x_1} \phi(x)dx = f(x_1) - f(a), \qquad (a < x_1 < b)$$

and

$$\int_{x_1}^b \phi(x)dx = f(b) - f(x_1),$$

we get, by addition,

$$\int_a^{x_1} \phi(x)dx + \int_{x_1}^b \phi(x)dx = f(b) - f(a).$$

But

$$\int_a^b \phi(x)dx = f(b) - f(a);$$

therefore, by comparing the last two expressions, we obtain

(C) $$\int_a^b \phi(x)dx = \int_a^{x_1} \phi(x)dx + \int_{x_1}^b \phi(x)dx.$$

Interpreting this theorem geometrically, as in Art. 142, we see that the integral on the left-hand side represents the whole area $CEFD$, the first integral on the right-hand side the area $CMPD$, and the second integral on the right-hand side the area $MEFP$. The truth of the theorem is therefore obvious.

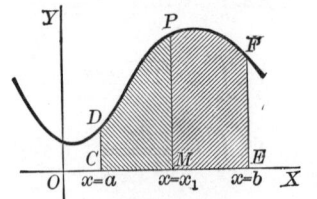

Evidently the definite integral may be decomposed into any number of separate definite integrals in this way.

152. The definite integral a function of its limits

From

$$\int_a^b \phi(x)dx = f(b) - f(a)$$

we see that the definite integral is a function of its limits. Thus $\int_a^b \phi(z)dz$ has precisely the same value as $\int_a^b \phi(x)dx$.

Theorem. *A definite integral is a function of its limits.*

153. Improper integrals. Infinite limits. So far the limits of the integral have been assumed as finite. Even in elementary work, however, it is sometimes desirable to remove this restriction and to consider integrals with infinite limits. This is possible in certain cases by making use of the following *definitions*.

THE DEFINITE INTEGRAL

When the upper limit is infinite,

$$\int_a^{+\infty} \phi(x)dx = \lim_{b\to +\infty} \int_a^b \phi(x)dx,$$

and when the lower limit is infinite,

$$\int_{-\infty}^b \phi(x)dx = \lim_{a\to -\infty} \int_a^b \phi(x)dx,$$

provided the limits exist.

ILLUSTRATIVE EXAMPLE 1. Find $\int_1^{+\infty} \frac{dx}{x^2}$.

Solution. $\int_1^{\infty} \frac{dx}{x^2} = \lim_{b\to +\infty} \int_1^b \frac{dx}{x^2} = \lim_{b\to +\infty}\left[-\frac{1}{x}\right]_1^b = \lim_{b\to +\infty}\left[-\frac{1}{b}+1\right] = 1.$ *Ans.*

ILLUSTRATIVE EXAMPLE 2. Find $\int_0^{+\infty} \frac{8\,a^3\,dx}{x^2+4\,a^2}$.

Solution. $\int_0^{+\infty} \frac{8\,a^3\,dx}{x^2+4\,a^2} = \lim_{b\to +\infty} \int_0^b \frac{8\,a^3\,dx}{x^2+4\,a^2} = \lim_{b\to +\infty}\left[4\,a^2\arctan\frac{x}{2\,a}\right]_0^b$

$= \lim_{b\to +\infty}\left[4\,a^2\arctan\frac{b}{2\,a}\right] = 4\,a^2\cdot\frac{\pi}{2} = 2\,\pi a^2.$ *Ans.*

Let us interpret this result geometrically. The graph of our function is the witch, the locus of

$$y = \frac{8\,a^3}{x^2+4\,a^2}.$$

Area $OPQb = \int_0^b \frac{8\,a^3\,dx}{x^2+4\,a^2} = 4\,a^2\arctan\frac{b}{2\,a}.$

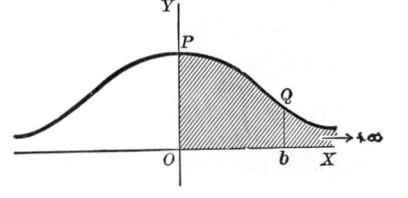

Then as the ordinate bQ moves indefinitely to the right, the area $OPQb$ approaches a finite limit $2\,\pi a^2$.

ILLUSTRATIVE EXAMPLE 3. Find $\int_1^{+\infty} \frac{dx}{x}$.

Solution. $\int_1^{+\infty} \frac{dx}{x} = \lim_{b\to +\infty} \int_1^b \frac{dx}{x} = \lim_{b\to +\infty}(\ln b).$

The limit of $\ln b$ as b increases without limit does not exist; hence the integral has in this case no meaning.

154. Improper integrals. When $y = \phi(x)$ is discontinuous. Let us now consider cases where the function to be integrated is discontinuous for isolated values of the variable lying within the limits of integration.

Consider first the case where the function to be integrated is continuous for all values of x between the limits a and b except $x = a$.

If $a < b$ and ϵ is positive, we use the *definition*

(1) $$\int_a^b \phi(x)dx = \lim_{\epsilon \to 0} \int_{a+\epsilon}^b \phi(x)dx.$$

Likewise, when $\phi(x)$ is continuous except at $x = b$, we *define*

(2) $$\int_a^b \phi(x)dx = \lim_{\epsilon \to 0} \int_a^{b-\epsilon} \phi(x)dx,$$

provided the limits exist.

ILLUSTRATIVE EXAMPLE 1. Find $\int_0^a \dfrac{dx}{\sqrt{a^2 - x^2}}$.

Solution. Here $\dfrac{1}{\sqrt{a^2 - x^2}}$ becomes infinite for $x = a$. Therefore, by (2),

$$\int_0^a \frac{dx}{\sqrt{a^2-x^2}} = \lim_{\epsilon \to 0} \int_0^{a-\epsilon} \frac{dx}{\sqrt{a^2-x^2}} = \lim_{\epsilon \to 0} \left[\arcsin \frac{x}{a}\right]_0^{a-\epsilon}$$
$$= \lim_{\epsilon \to 0} \left[\arcsin\left(1 - \frac{\epsilon}{a}\right)\right] = \arcsin 1 = \frac{\pi}{2}. \text{ Ans.}$$

ILLUSTRATIVE EXAMPLE 2. Find $\int_0^1 \dfrac{dx}{x^2}$.

Solution. Here $\dfrac{1}{x^2}$ becomes infinite for $x = 0$. Therefore, by (1),

$$\int_0^1 \frac{dx}{x^2} = \lim_{\epsilon \to 0} \int_\epsilon^1 \frac{dx}{x^2} = \lim_{\epsilon \to 0} \left(\frac{1}{\epsilon} - 1\right).$$

In this case there is no limit, and therefore the integral does not exist.

If c lies between a and b, and $\phi(x)$ is continuous except at $x = c$, then, ϵ and ϵ' being positive numbers, the integral between a and b is *defined by*

(3) $$\int_a^b \phi(x)dx = \lim_{\epsilon \to 0} \int_a^{c-\epsilon} \phi(x)dx + \lim_{\epsilon' \to 0} \int_{c+\epsilon'}^b \phi(x)dx,$$

provided each separate limit exists.

ILLUSTRATIVE EXAMPLE 3. Find $\int_0^{3a} \dfrac{2x\,dx}{(x^2 - a^2)^{\frac{2}{3}}}$.

Solution. Here the function to be integrated becomes discontinuous for $x = a$, that is, for a value of x between the limits of integration 0 and $3a$. Hence the above definition (3) must be employed. Thus

$$\int_0^{3a} \frac{2x\,dx}{(x^2-a^2)^{\frac{2}{3}}} = \lim_{\epsilon \to 0} \int_0^{a-\epsilon} \frac{2x\,dx}{(x^2-a^2)^{\frac{2}{3}}} + \lim_{\epsilon' \to 0} \int_{a+\epsilon'}^{3a} \frac{2x\,dx}{(x^2-a^2)^{\frac{2}{3}}}$$
$$= \lim_{\epsilon \to 0} \left[3(x^2-a^2)^{\frac{1}{3}}\right]_0^{a-\epsilon} + \lim_{\epsilon' \to 0} \left[3(x^2-a^2)^{\frac{1}{3}}\right]_{a+\epsilon'}^{3a}$$
$$= \lim_{\epsilon \to 0}\left[3\sqrt[3]{(a-\epsilon)^2 - a^2} + 3a^{\frac{2}{3}}\right] + \lim_{\epsilon' \to 0}\left[3\sqrt[3]{8a^2} - 3\sqrt[3]{(a+\epsilon')^2 - a^2}\right]$$
$$= 3a^{\frac{2}{3}} + 6a^{\frac{2}{3}} = 9a^{\frac{2}{3}}. \text{ Ans.}$$

To interpret this geometrically, let us plot the graph, that is, the locus, of

$$y = \frac{2x}{(x^2-a^2)^{\frac{2}{3}}}$$

and note that $x = a$ is an asymptote.

Area $OPE = \int_0^{a-\epsilon} \dfrac{2x\,dx}{(x^2-a^2)^{\frac{2}{3}}}$
$= 3\sqrt[3]{(a-\epsilon)^2 - a^2} + 3a^{\frac{2}{3}}.$

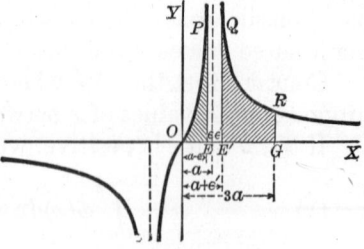

THE DEFINITE INTEGRAL

Then as PE moves to the right toward the asymptote, that is, as ϵ approaches zero, the area OPE approaches $3\,a^{\frac{2}{3}}$ as a limit.

Similarly,

$$\text{Area } E'QRG = \int_{a+\epsilon'}^{3a} \frac{2\,x\,dx}{(x^2 - a^2)^{\frac{2}{3}}} = 3\sqrt[3]{8\,a^2} - 3\sqrt[3]{(a+\epsilon')^2 - a^2}$$

approaches $6\,a^{\frac{2}{3}}$ as a limit as QE' moves to the left toward the asymptote, that is, as ϵ' approaches zero. Adding these results, we get $9\,a^{\frac{2}{3}}$.

ILLUSTRATIVE EXAMPLE 4. Find $\int_0^{2a} \frac{dx}{(x-a)^2}$.

Solution. This function also becomes infinite between the limits of integration. Hence, by (3),

$$\int_0^{2a} \frac{dx}{(x-a)^2} = \lim_{\epsilon \to 0} \int_0^{a-\epsilon} \frac{dx}{(x-a)^2} + \lim_{\epsilon' \to 0} \int_{a+\epsilon'}^{2a} \frac{dx}{(x-a)^2}$$

$$= \lim_{\epsilon \to 0}\left[-\frac{1}{x-a}\right]_0^{a-\epsilon} + \lim_{\epsilon' \to 0}\left[-\frac{1}{x-a}\right]_{a+\epsilon'}^{2a}$$

$$= \lim_{\epsilon \to 0}\left(\frac{1}{\epsilon} - \frac{1}{a}\right) + \lim_{\epsilon' \to 0}\left(-\frac{1}{a} + \frac{1}{\epsilon'}\right).$$

In this case the limits do not exist and the integral has no meaning.

If we plot the graph of this function the condition of things appears very much the same as in the last example. We see, however, that the limits do not exist, and therein lies the difference.

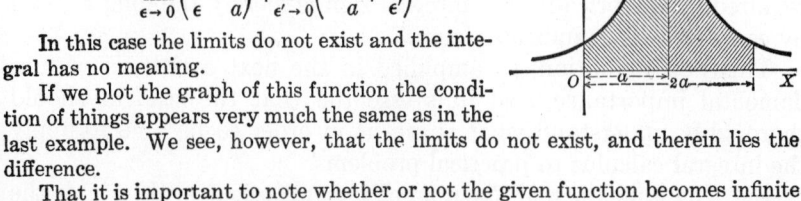

That it is important to note whether or not the given function becomes infinite within the limits of integration will appear at once if we apply our integration formula without any investigation. Thus

$$\int_0^{2a} \frac{dx}{(x-a)^2} = \left[-\frac{1}{x-a}\right]_0^{2a} = -\frac{2}{a},$$

a result which is absurd in view of the above discussion.

PROBLEMS

Work out each of the following integrals.

1. $\int_0^{+\infty} \frac{dx}{a^2 + b^2 x^2} = \frac{\pi}{2\,ab}$.

2. $\int_0^{+\infty} x e^{-x^2} dx = \frac{1}{2}$.

3. $\int_1^5 \frac{x\,dx}{\sqrt{5-x}} = \frac{44}{3}$.

4. $\int_0^a \frac{x^2 dx}{\sqrt{a^2 - x^2}} = \frac{\pi a^2}{4}$.

5. $\int_1^2 \frac{dx}{x^2\sqrt{4 - x^2}} = \frac{\sqrt{3}}{4}$.

6. $\int_a^{2a} \frac{x^2 dx}{\sqrt{x^2 - a^2}} = 2.39\,a^2$.

Show that the following integrals do not exist.

7. $\int_0^{+\infty} \frac{dx}{\sqrt{2\,x+1}}$. 8. $\int_1^2 \frac{dx}{(x-1)^{\frac{3}{2}}}$. 9. $\int_a^{+\infty} \frac{x\,dx}{\sqrt{x^2 - a^2}}$. 10. $\int_0^1 \frac{dx}{(2\,x-1)^{\frac{4}{3}}}$.

CHAPTER XV

INTEGRATION A PROCESS OF SUMMATION

155. Introduction. Thus far we have defined integration as the *inverse of differentiation*. In a great many of the applications of the integral calculus, however, it is preferable to define integration as a *process of summation*. In fact, the integral calculus was invented in the attempt to calculate the area bounded by curves, by supposing the given area to be divided into an "infinite number of infinitesimal parts called *elements*, the sum of all these elements being the area required." Historically, the integral sign is merely the long S, used by early writers to indicate "sum."

This new definition, as amplified in the next article, is of fundamental importance, and it is essential that the student should thoroughly understand what is meant in order to be able to apply the integral calculus to practical problems.

156. The Fundamental Theorem of integral calculus. If $\phi(x)$ is the derivative of $f(x)$, then it has been shown in Art. 142 that the value of the definite integral

(1) $$\int_a^b \phi(x)\,dx = f(b) - f(a)$$

gives the area bounded by the curve $y = \phi(x)$, the x-axis, and the ordinates erected at $x = a$ and $x = b$.

Now let us make the following construction in connection with this area. Divide the interval from $x = a$ to $x = b$ into any number n of equal subintervals, erect ordinates at these points of division, and complete rectangles by drawing horizontal lines through the extremities of the ordinates, as in the figure. It is clear that the sum of the areas of these n rectangles (the shaded area) is an approximate value for the area in question. It is further evident that the *limit* of the sum of the areas of these rectangles when their number n is indefinitely increased will *equal* the area under the curve.

INTEGRATION A PROCESS OF SUMMATION

Let us now carry through the following more general construction. Divide the interval into n subintervals, *not necessarily equal*, and erect ordinates at the points of division. Choose a point within each subdivision *in any manner*, erect ordinates at these points, and through their extremities draw horizontal lines to form rectangles, as in the figure. Then, as before, the sum of the areas of these n rectangles (the shaded area) equals approximately the area under the curve; and the *limit of this sum* as n increases without limit, and each subinterval approaches zero as a limit, is precisely the area under the curve. These considerations show that the definite integral (1) may be regarded as the *limit of a sum*. Let us now formulate this result.

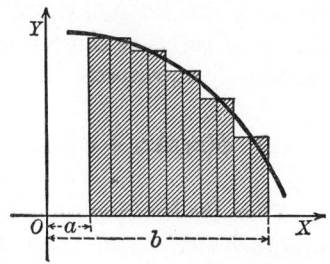

(a) Denote the lengths of the successive subintervals by

$$\Delta x_1, \quad \Delta x_2, \quad \Delta x_3, \quad \cdots, \quad \Delta x_n.$$

(b) Denote the abscissas of the points chosen in the subintervals by

$$x_1, \quad x_2, \quad x_3, \quad \cdots, \quad x_n.$$

Then the ordinates of the curve at these points are

$$\phi(x_1), \quad \phi(x_2), \quad \phi(x_3), \quad \cdots, \quad \phi(x_n).$$

(c) The areas of the successive rectangles are obviously

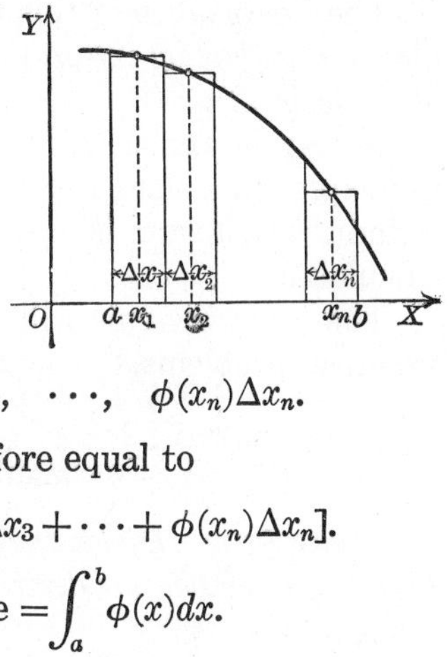

$$\phi(x_1)\Delta x_1, \quad \phi(x_2)\Delta x_2, \quad \phi(x_3)\Delta x_3, \quad \cdots, \quad \phi(x_n)\Delta x_n.$$

(d) The area under the curve is therefore equal to

$$\lim_{n \to \infty} [\phi(x_1)\Delta x_1 + \phi(x_2)\Delta x_2 + \phi(x_3)\Delta x_3 + \cdots + \phi(x_n)\Delta x_n].$$

But from (1) the area under the curve $= \int_a^b \phi(x)dx$. Therefore our discussion gives

(A) $\quad \int_a^b \phi(x)dx = \lim_{n \to \infty} [\phi(x_1)\Delta x_1 + \phi(x_2)\Delta x_2 + \cdots + \phi(x_n)\Delta x_n].$

This equation has been derived by making use of the notion of area. Intuition has aided us in establishing the result. Let us now regard (A) *simply as a theorem in analysis*, which may then be stated as follows:

Fundamental Theorem of the Integral Calculus

Let $\phi(x)$ be continuous for the interval $x = a$ to $x = b$. Let this interval be divided into n subintervals whose lengths are $\Delta x_1, \Delta x_2, \cdots, \Delta x_n$, and points be chosen, one in each subinterval, their abscissas being $x_1, x_2, \cdots, x_n$ respectively. Consider the sum

(2) $\quad \phi(x_1)\Delta x_1 + \phi(x_2)\Delta x_2 + \cdots + \phi(x_n)\Delta x_n = \sum_{i=1}^{n} \phi(x_i)\Delta x_i.$

Then the limiting value of this sum when n increases without limit, and each subinterval approaches zero as a limit, equals the value of the definite integral

$$\int_a^b \phi(x)dx.$$

Equation (A) may be abbreviated as follows.

(3) $\quad \int_a^b \phi(x)dx = \lim_{n \to \infty} \sum_{i=1}^{n} \phi(x_i)\Delta x_i.$

The importance of this theorem results from the fact that *we are able to calculate by integration a magnitude which is the limit of a sum of the form* (2).

It may be remarked that each term in the sum (2) is a *differential* expression, since the lengths $\Delta x_1, \Delta x_2, \cdots, \Delta x_n$ approach zero as a limit. Each term is also called an **element** of the magnitude to be calculated.

The following rule will be of service in applying this theorem to practical problems.

Fundamental Theorem. Rule

FIRST STEP. *Divide the required magnitude into similar parts such that it is clear that the desired result will be found by taking the limit of a sum of such parts.*

SECOND STEP. *Find expressions for the magnitudes of these parts such that their sum will be of the form* (2).

THIRD STEP. *Having chosen the proper limits $x = a$ and $x = b$, we apply the Fundamental Theorem*

$$\lim_{n \to \infty} \sum_{i=1}^{n} \phi(x_i)\Delta x_i = \int_a^b \phi(x)dx$$

and integrate.

INTEGRATION A PROCESS OF SUMMATION

157. Analytical proof of the Fundamental Theorem. As in the last article, divide the interval from $x = a$ to $x = b$ into any number n of subintervals, not necessarily equal, and denote the abscissas of these points of division by $b_1, b_2, \cdots, b_{n-1}$, and the lengths of the subintervals by $\Delta x_1, \Delta x_2, \cdots, \Delta x_n$. Now, however, we let $x'_1, x'_2, \cdots, x'_n$ denote abscissas, one in each interval, determined by the Theorem of Mean Value (Art. 116), erect ordinates at these points, and through their extremities draw horizontal lines to form rectangles, as in the figure. Note that here $\phi(x)$ takes the place of $f'(x)$. Applying (B), Art. 116 to the first interval ($a = a$, $b = b_1$, and x'_1 lies between a and b_1), we have

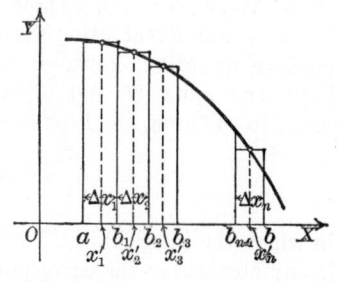

$$\frac{f(b_1) - f(a)}{b_1 - a} = \phi(x'_1),$$

or, since $b_1 - a = \Delta x_1$,

$$f(b_1) - f(a) = \phi(x'_1)\Delta x_1.$$

Also, $f(b_2) - f(b_1) = \phi(x'_2)\Delta x_2$, for the second interval,
$f(b_3) - f(b_2) = \phi(x'_3)\Delta x_3$, for the third interval,

$\cdots \cdots \cdots \cdots \cdots \cdots \cdots \cdots \cdots$

$f(b) - f(b_{n-1}) = \phi(x'_n)\Delta x_n$, for the nth interval.

Adding these, we get

(1) $\quad f(b) - f(a) = \phi(x'_1)\Delta x_1 + \phi(x'_2)\Delta x_2 + \cdots + \phi(x'_n)\Delta x_n.$

But $\quad \phi(x'_1) \cdot \Delta x_1 =$ area of the first rectangle,

$\phi(x'_2) \cdot \Delta x_2 =$ area of the second rectangle, etc.

Hence the sum on the right-hand side of (1) equals the sum of the areas of the rectangles. But from (1), Art. 156, the left-hand side of (1) equals the area between the curve $y = \phi(x)$, the x-axis, and the ordinates at $x = a$ and $x = b$. Then the sum

(2) $\quad \sum_{i=1}^{n} \phi(x'_i) \Delta x_i$

equals this area. And while the corresponding sum

(3) $\quad \sum_{i=1}^{n} \phi(x_i) \Delta x_i \qquad$ (where x_i is *any* abscissa of the subinterval Δx_i)

(formed as in last article) does not also give the area, nevertheless we may show that the two sums (2) and (3) approach equality when n increases without limit and each subinterval approaches zero as a limit. For the difference $\phi(x'_i) - \phi(x_i)$ does not exceed in numerical value the difference of the greatest and smallest ordinates in Δx_i. And, furthermore, it is always possible* to make all these differences less in numerical

---

* That such is the case is shown in advanced works on the calculus.

value than any assignable positive number ϵ, however small, by continuing the process of subdivision far enough, that is, by choosing n sufficiently large. Hence for such a choice of n the difference of the sums (2) and (3) is less in numerical value than $\epsilon(b - a)$, that is, less than any assignable positive quantity, however small. Accordingly, as n increases without limit, the sums (2) and (3) approach equality, and since (2) is always equal to the area, the fundamental result follows that

$$\int_a^b \phi(x)dx = \lim_{n \to \infty} \sum_{i=1}^n \phi(x_i)\Delta x_i,$$

in which the interval $[a, b]$ is subdivided in any manner whatever and x_i is any abscissa in the corresponding subinterval.

158. Areas of plane curves; rectangular coördinates. As already explained, the area between a curve, the x-axis, and the ordinates at $x = a$ and $x = b$ is given by the formula

(B) $\quad$ Area $= \int_a^b y\, dx,$

the value of y in terms of x being substituted from the equation of the curve.

Equation (B) is readily memorized by observing that the element of the area is a rectangle (as CR) of base dx and altitude y. The required area $ABQP$ is the limit of the sum of all such rectangles (strips) between the ordinates AP and BQ.

Let us now apply the Fundamental Theorem (Art. 156) to the calculation of the area bounded by the curve $x = \phi(y)$ (AB in figure), the y-axis, and the horizontal lines $y = c$ and $y = d$.

FIRST STEP. Construct the n rectangles as in the figure. The required area is clearly the limit of the sum of the areas of these rectangles as their number increases without limit and the altitude of each one approaches zero as a limit.

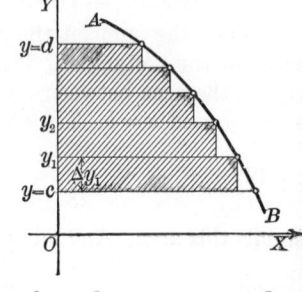

SECOND STEP. Denote the altitudes by Δy_1, Δy_2, etc. Take a point in each interval at the upper extremity and denote their ordinates by y_1, y_2, etc. Then the bases are $\phi(y_1)$, $\phi(y_2)$, etc., and the sum of the areas of the rectangles is

$$\phi(y_1)\Delta y_1 + \phi(y_2)\Delta y_2 + \cdots + \phi(y_n)\Delta y_n = \sum_{i=1}^n \phi(y_i)\Delta y_i.$$

INTEGRATION A PROCESS OF SUMMATION 259

THIRD STEP. Applying the Fundamental Theorem gives

$$\lim_{n\to\infty} \sum_{i=1}^{n} \phi(y_i)\Delta y_i = \int_c^d \phi(y)dy.$$

Hence the area between a curve, the y-axis, and the horizontal lines $y = c$ and $y = d$ is given by the formula

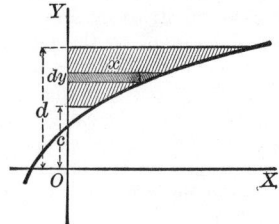

(C) $\quad$ Area $= \int_c^d x\,dy,$

the value of x in terms of y being substituted from the equation of the curve. Formula (C) is remembered as indicating the limit of the sum of all horizontal strips (rectangles) within the required area, x and dy being, respectively, the base and altitude of any strip. The element of the area is one of these rectangles.

Meaning of the negative sign before an area. In formula (B), a is less than b. Since we now interpret the right-hand member as the limit of the sum of n terms resulting from $y_i \Delta x_i$ by letting $i = 1, 2, 3, \cdots, n$, then, if y is *negative*, each term of this sum will be negative, and (B) will give the area with a negative sign prefixed. This means that the area lies below the x-axis.

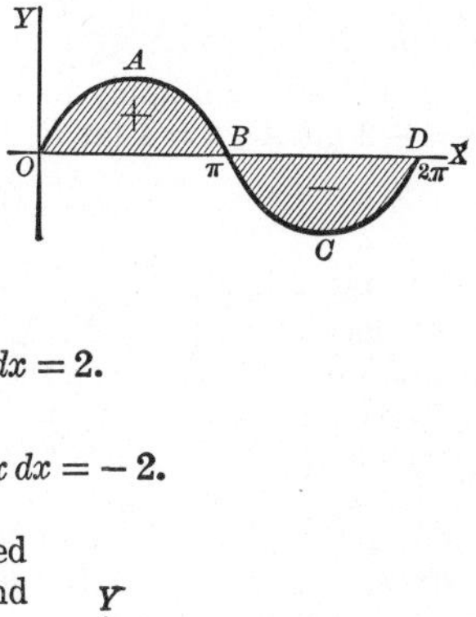

ILLUSTRATIVE EXAMPLE 1. Find the area of one arch of the sine curve $y = \sin x$.

Solution. Placing $y = 0$ and solving for x, we find
$$x = 0,\ \pi,\ 2\pi,\ \text{etc.}$$

Substituting in (B),

$$\text{Area } OAB = \int_a^b y\,dx = \int_0^\pi \sin x\,dx = 2.$$

Also, $\quad$ Area $BCD = \int_a^b y\,dx = \int_\pi^{2\pi} \sin x\,dx = -2.$

ILLUSTRATIVE EXAMPLE 2. Find the area bounded by the semicubical parabola $ay^2 = x^3$, the y-axis, and the lines $y = a$ and $y = 2a$.

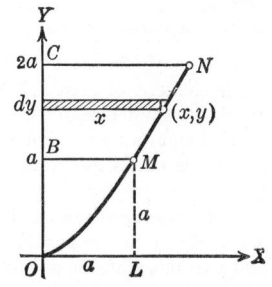

Solution. By (C) above, and the figure, the element of area $= x\,dy = a^{\frac{1}{3}}y^{\frac{2}{3}}\,dy$, substituting the value of x from the equation of the curve MN. Hence

$$\text{Area } BMNC = \int_a^{2a} a^{\frac{1}{3}}y^{\frac{2}{3}}\,dy$$
$$= \tfrac{3}{5}a^2(\sqrt[3]{32} - 1) = 1.304\ a^2.\ \text{Ans.}$$

Note that $a^2 =$ area $OLMB$.

In the area given by (*B*) one boundary is the *x*-axis. In (*C*) one boundary is the *y*-axis. Consider now the area bounded by two curves.

ILLUSTRATIVE EXAMPLE 3. Find the area bounded by the parabola $y^2 = 2x$ and the straight line $x - y = 4$.

Solution. The curves intersect at $A(2, -2)$, $B(8, 4)$. Divide the area into horizontal strips by a system of equidistant lines parallel to OX drawn from the parabola AOB to the line AB. Let their common distance apart be dy. Consider the strip in the figure whose upper side has the extremities (x_1, y), (x_2, y). From these points drop perpendiculars on the lower side. Then a rectangle is formed and its area is given by

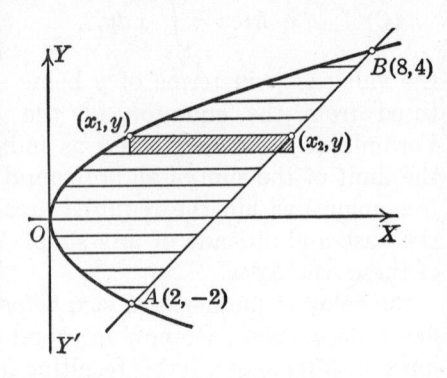

(1) $\quad dA = (x_2 - x_1)dy. \quad (x_2 > x_1)$

This is the element of area. For the required area is obviously the limit of the sum of all such rectangles. That is, by the Fundamental Theorem,

(2) $\quad \text{Area} = \int_c^d (x_2 - x_1)dy,$

in which x_2 and x_1 are functions of y determined from the equations of the bounding curves. Thus, in this example, from $x - y = 4$ we find $x = x_2 = 4 + y$; from $y^2 = 2x$ we find $x = x_1 = \frac{1}{2}y^2$. Then we have, by (1),

(3) $\quad dA = (4 + y - \frac{1}{2}y^2)dy.$

This formula will apply to the rectangle formed from any strip. The limits are $c = -2$ (at A), $d = 4$ (at B). Hence

$$\text{Area} = \int_{-2}^{4}(4 + y - \frac{1}{2}y^2)dy = 18. \text{ Ans.}$$

In this example the area can be divided also into strips by a system of equidistant lines parallel to OY. Let Δx be their common distance apart. The *upper* end of each line will lie on the parabola OB. But the *lower* end will lie on the parabola OA when drawn to the left of A, but on the line AB when drawn to the right of A. If (x, y_2) is the upper extremity, and (x, y_1) the lower, the rectangle whose area is

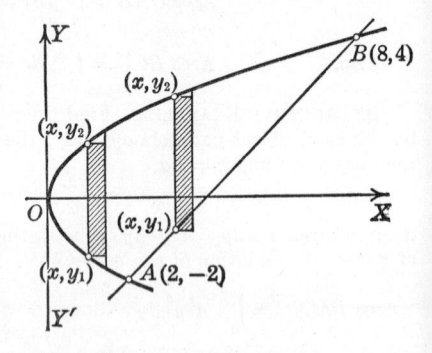

(4) $\quad dA = (y_2 - y_1)dx \quad (y_2 > y_1)$

is the element of area. But in this example it is not possible to find by (4) a *single* formula to represent the area of every one of the rectangles. For while $y_2 = \sqrt{2x}$, we have $y_1 = -\sqrt{2x}$ or $y_1 = x - 4$, according as the lower vertex of dA is on the parabola or on AB. Thus from (4), we have two forms of dA, and two integrations are necessary.

INTEGRATION A PROCESS OF SUMMATION

In any problem, therefore, the strips should be constructed so that only a *single* formula for the element of area is necessary. Formula (4) is used when these strips are constructed by drawing lines parallel to the y-axis.

In the Fundamental Theorem some or all of the elements $\phi(x_i) \Delta x_i$ may be negative. Hence the limit of their sum (the definite integral) may be zero or negative. For example, if $\phi(x) = \sin x$, $a = 0$, $b = 2\pi$, the definite integral (3), Art. 156, is zero. The interpretation of this result using areas appears from Illustrative Example 1 above.

PROBLEMS

1. Find the area bounded by the hyperbola $xy = a^2$, the x-axis, and the ordinates $x = a$ and $x = 2a$. *Ans.* $a^2 \ln 2$.

2. Find the area bounded by the curve $y = \ln x$, the x-axis, and the line $x = 10$. *Ans.* 14.026.

3. Find the area bounded by the curve $y = xe^x$, the x-axis, and the line $x = 4$. *Ans.* 164.8.

4. Find the area bounded by the parabola $\sqrt{x} + \sqrt{y} = \sqrt{a}$, and the coördinate axes. *Ans.* $\frac{1}{6} a^2$.

5. Find the entire area of the hypocycloid $x^{\frac{2}{3}} + y^{\frac{2}{3}} = a^{\frac{2}{3}}$. *Ans.* $\frac{3}{8}\pi a^2$.

Find the areas bounded by the following curves. In each case draw the figure, showing the element of area.

6. $y^2 = 6x$, $x^2 = 6y$. *Ans.* 12.   10. $y^2 = 2x$, $x^2 + y^2 = 4x$.

7. $y^2 = 4x$, $x^2 = 6y$.   8.   11. $y = 6x - x^2$, $y = x$.

8. $y^2 = 4x$, $2x - y = 4$.   9.   12. $y = x^3 - 3x$, $y = x$.

9. $y = 4 - x^2$, $y = 4 - 4x$. $10\frac{2}{3}$.   13. $y^2 = 4x$, $x = 12 + 2y - y^2$.

14. Find the area bounded by the parabola $y = 6 + 4x - x^2$ and the chord joining $(-2, -6)$ and $(4, 6)$. *Ans.* 36.

15. Find the area bounded by the semicubical parabola $y^3 = x^2$ and the chord joining $(-1, 1)$ and $(8, 4)$. *Ans.* 2.7.

16. Find an expression for the area bounded by the equilateral hyperbola $x^2 - y^2 = a^2$, the x-axis, and a line drawn from the origin to any point (x, y) on the curve.   *Ans.* $\dfrac{a^2}{2} \ln\left(\dfrac{x+y}{a}\right)$.

17. Find the area bounded by the curve $y = x(1 \pm \sqrt{x})$ and the line $x = 4$. *Ans.* $12\frac{8}{5}$.

18. Find the area bounded by the curve $x^2 y = x^2 - 1$ and the lines $y = 1$, $x = 1$, and $x = 4$. *Ans.* $\frac{3}{4}$.

19. Find the area bounded by the curve $y = x^3 - 9x^2 + 24x - 7$, the y-axis, and the line $y = 29$. *Ans.* 108.

262　INTEGRAL CALCULUS

A square is formed by the coördinate axes and parallel lines through the point (1, 1). Calculate the ratio of the larger to the smaller of the two areas into which it is divided by each of the following curves.

20. $y = x^2$.　　　Ans. 2.　　　26. $y = \sin \dfrac{\pi x}{2}$.　　Ans. $\dfrac{2}{\pi - 2}$.

21. $y = x^3$.　　　3.

22. $y = x^4$.　　　4.　　　27. $y = xe^{x-1}$.

23. $y^3 = x^2$.　　　$\tfrac{3}{2}$.

24. $\sqrt{x} + \sqrt{y} = 1$.　　　5.　　　28. $y = \tan \dfrac{\pi x}{4}$.

25. $x^{\frac{2}{3}} + y^{\frac{2}{3}} = 1$.　　$\dfrac{32 - 3\pi}{3\pi}$.　　29. $x^{\frac{1}{3}} + y^{\frac{1}{3}} = 1$.

For each of the following curves calculate the area in the first quadrant lying under the arc which extends from the y-axis to the first intercept on the x-axis.

30. $x + y + y^2 = 2$.　　Ans. $1\tfrac{1}{6}$.　　34. $y = e^{\frac{x}{2}} \cos 2x$.

31. $y = x^3 - 8x^2 + 15x$.　　$15\tfrac{3}{4}$.　　35. $y = 4 e^{-\frac{x}{2}} \cos \tfrac{1}{2} \pi x$.

32. $y = e^x \sin x$.　　　12.07.　　36. $y = \sin (x + 1)$.

33. $y^2 = (4 - x)^3$.

159. Areas of plane curves; polar coördinates. Let it be required to find the area bounded by a curve and two of its radii vectores.

Assume the equation of the curve to be
$$\rho = f(\theta),$$
and let OP_1 and OD be the two radii. Denote by α and β the angles which the radii make with the polar axis. Apply the Fundamental Theorem, Art. 156.

FIRST STEP. The required area is clearly the limit of the sum of circular sectors constructed as in the figure.

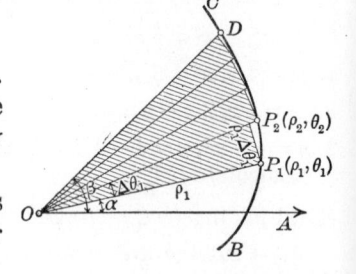

SECOND STEP. Let the central angles of the successive sectors be $\Delta\theta_1$, $\Delta\theta_2$, etc., and their radii ρ_1, ρ_2, etc. Then the sum of the areas of the sectors is

$$\tfrac{1}{2} \rho_1^2 \Delta\theta_1 + \tfrac{1}{2} \rho_2^2 \Delta\theta_2 + \cdots + \tfrac{1}{2} \rho_n^2 \Delta\theta_n = \sum_{i=1}^{n} \tfrac{1}{2} \rho_i^2 \Delta\theta_i.$$

For the area of a circular sector = $\tfrac{1}{2}$ radius × arc. Hence the area of the first sector = $\tfrac{1}{2} \rho_1 \cdot \rho_1 \Delta\theta_1 = \tfrac{1}{2} \rho_1^2 \Delta\theta_1$, etc.

THIRD STEP. Applying the Fundamental Theorem,

$$\lim_{n \to \infty} \sum_{i=1}^{n} \tfrac{1}{2} \rho_i^2 \Delta\theta_i = \int_{\alpha}^{\beta} \tfrac{1}{2} \rho^2 \, d\theta.$$

INTEGRATION A PROCESS OF SUMMATION

Hence the area swept over by the radius vector of the curve in moving from the position OP_1 to the position OD is given by the formula

(D) $$\text{Area} = \tfrac{1}{2} \int_\alpha^\beta \rho^2 \, d\theta,$$

the value of ρ in terms of θ being substituted from the equation of the curve.

The element of area for (D) is a circular sector with radius ρ and central angle $d\theta$. Hence its area is $\tfrac{1}{2} \rho^2 \, d\theta$.

ILLUSTRATIVE EXAMPLE. Find the entire area of the lemniscate $\rho^2 = a^2 \cos 2\theta$.

Solution. Since the figure is symmetrical with respect to both OX and OY, the whole area $= 4$ times the area of OAB.

Since $\rho = 0$ when $\theta = \tfrac{\pi}{4}$, we see that if θ varies from 0 to $\tfrac{\pi}{4}$, the radius vector OP sweeps over the area OAB. Hence, substituting in (D)

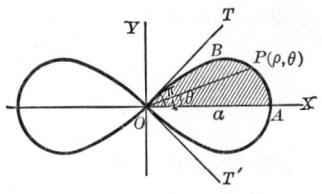

$$\text{Entire area} = 4 \times \text{area } OAB = 4 \cdot \tfrac{1}{2} \int_\alpha^\beta \rho^2 \, d\theta = 2 a^2 \int_0^{\frac{\pi}{4}} \cos 2\theta \, d\theta = a^2;$$

that is, the area of both loops equals the area of a square constructed on OA as one side.

PROBLEMS

1. Find the area bounded by the circle $\rho = a \cos \theta$ and the lines $\theta = 0$ and $\theta = 60°$. *Ans.* $0.37 \, a^2$.

2. Find the entire area of the curve $\rho = a \sin 2\theta$. *Ans.* $\tfrac{1}{2} \pi a^2$.

Calculate the area inclosed by each of the following curves.

3. $\rho^2 = 4 \sin 2\theta$. *Ans.* 4.
4. $\rho = a \cos 3\theta$. $\tfrac{1}{4} \pi a^2$.
5. $\rho = a(1 - \cos \theta)$. $\tfrac{3}{2} \pi a^2$.
6. $\rho = 2 - \cos \theta$. $\tfrac{9}{2} \pi$.
7. $\rho = \sin^2 \tfrac{\theta}{2}$. $\tfrac{3}{8} \pi$.
8. $\rho = \tfrac{1}{2} + \cos 2\theta$. $\tfrac{3}{4} \pi$.
9. $\rho = 2 + \sin 3\theta$. $\tfrac{9}{2} \pi$.

10. $\rho = 3 + \cos 3\theta$.
11. $\rho = a \cos \theta + b \sin \theta$.
12. $\rho = 2 \cos^2 \tfrac{\theta}{2}$.
13. $\rho = a \sin n\theta$.
14. $\rho = \cos 3\theta - \cos \theta$.
15. $\rho = \cos 3\theta - 2 \cos \theta$.

16. Find the area bounded by the parabola $\rho(1 + \cos \theta) = a$ and the lines $\theta = 0$ and $\theta = 120°$. *Ans.* $0.866 \, a^3$.

17. Find the area bounded by the hyperbola $\rho^2 \cos 2\theta = a^2$ and the lines $\theta = 0$ and $\theta = 30°$. *Ans.* $0.33 \, a^2$.

18. Prove that the area generated by the radius vector of the spiral $\rho = e^\theta$ equals one fourth of the area of the square described on the radius vector.

19. Find the area of that part of the parabola $\rho = a \sec^2 \dfrac{\theta}{2}$ which is intercepted between the curve and the latus rectum.

Ans. $\tfrac{8}{3} a^2$.

20. Show that the area bounded by any two radii vectors of the hyperbolic spiral $\rho\theta = a$ is proportional to the difference between the lengths of these radii.

21. Find the area of the ellipse $\rho^2 = \dfrac{a^2 b^2}{a^2 \sin^2 \theta + b^2 \cos^2 \theta}$. *Ans.* πab.

22. Find the entire area of the curve $\rho = a(\sin 2\theta + \cos 2\theta)$.

Ans. πa^2.

23. Find the area below OX within the curve $\rho = a \sin^3 \dfrac{\theta}{3}$.

Ans. $\tfrac{1}{64}(10\pi + 27\sqrt{3})a^2$.

24. Find the area bounded by $\rho^2 = a^2 \sin 4\theta$. *Ans.* a^2.

Find the area bounded by the following curves and the given lines.

25. $\rho = \tan \theta$; $\theta = 0$, $\theta = \tfrac{1}{4}\pi$. **27.** $\rho = \sec \theta + \tan \theta$; $\theta = 0$, $\theta = \tfrac{1}{4}\pi$.
26. $\rho = e^{\frac{1}{2}\theta}$; $\theta = \tfrac{1}{4}\pi$, $\theta = \tfrac{1}{2}\pi$. **28.** $\rho = a \sin \theta + b \cos \theta$; $\theta = 0$, $\theta = \tfrac{1}{2}\pi$.

Calculate the area which the curves in each of the following pairs have in common.

29. $\rho = 3 \cos \theta$, $\rho = 1 + \cos \theta$. *Ans.* $\tfrac{5}{4}\pi$.

30. $\rho = 1 + \cos \theta$, $\rho = 1$. $\tfrac{5}{4}\pi - 2$.

31. $\rho = 1 - \cos \theta$, $\rho = \sin \theta$. $\tfrac{1}{2}\pi - 1$.

32. $\rho^2 = 2 \cos 2\theta$, $\rho = 1$. $\tfrac{1}{3}\pi + 2 - \sqrt{3}$.

33. $\rho^2 = \cos 2\theta$, $\rho^2 = \sin 2\theta$. $1 - \tfrac{1}{2}\sqrt{2}$.

34. $\rho = \sqrt{6} \cos \theta$, $\rho^2 = 9 \cos 2\theta$. $\tfrac{1}{2}(\pi + 9 - 3\sqrt{3})$.

35. $\rho = \sqrt{2} \sin \theta$, $\rho^2 = \cos 2\theta$. $\tfrac{1}{6}(\pi + 3 - 3\sqrt{3})$.

36. $\rho = \sqrt{2} \cos \theta$, $\rho^2 = \sqrt{3} \sin 2\theta$.

37. Find the area which is inside the circle $3\rho = \sqrt{3} \cos \theta$ and inside the loop of the curve $\rho = \cos 2\theta$ from $\theta = -\dfrac{\pi}{4}$ to $\theta = \dfrac{\pi}{4}$.

38. $3\rho = \sqrt{6} \sin 2\theta$, $\rho^2 = \cos 2\theta$.

39. Find the area of the inside loop of the trisectrix $\rho = a(1 - 2 \cos \theta)$. For figure, see limaçon, Chapter XXVI. *Ans.* $\tfrac{1}{2}a^2(2\pi - 3\sqrt{3})$.

INTEGRATION A PROCESS OF SUMMATION

160. Volumes of solids of revolution. Let V denote the volume of the solid generated by revolving the plane surface $ABCD$ about the x-axis, the equation of the plane curve DC being

$$y = f(x).$$

FIRST STEP. Divide the segment AB into n parts of lengths $\Delta x_1, \Delta x_2, \cdots, \Delta x_n$, and pass a plane perpendicular to the x-axis through each point of division. These planes will divide the solid into n circular plates. If rectangles are constructed with bases Δx_1, $\Delta x_2, \cdots, \Delta x_n$ within the area $ABCD$, then each rectangle generates a cylinder of revolution when area $ABCD$ is revolved. Thus a cylinder is formed corresponding to each of the circular plates. (In the figure $n = 4$ and two cylinders are shown.) The limit of the sum of these n cylinders ($n \to \infty$) is the required volume.

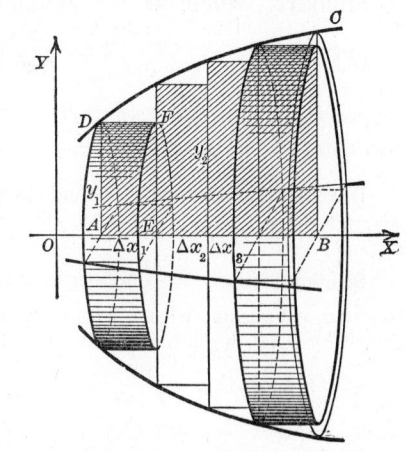

SECOND STEP. Let the ordinates of the curve DC at the points of division on the x-axis be $y_1, y_2, \cdots, y_n$. Then the volume of the cylinder generated by the rectangle $AEFD$ will be $\pi y_1^2 \, \Delta x_1$, and the sum of the volumes of all such cylinders is

$$\pi y_1^2 \, \Delta x_1 + \pi y_2^2 \, \Delta x_2 + \cdots + \pi y_n^2 \, \Delta x_n = \sum_{i=1}^{n} \pi y_i^2 \, \Delta x_i.$$

THIRD STEP. Applying the Fundamental Theorem (using limits $OA = a$ and $OB = b$),

$$\lim_{n \to \infty} \sum_{i=1}^{n} \pi y_i^2 \, \Delta x_i = \int_a^b \pi y^2 \, dx.$$

Hence the volume generated by revolving about the x-axis the area bounded by the curve, the x-axis, and the ordinates $x = a$ and $x = b$ is given by the formula

$$(E) \qquad V_x = \pi \int_a^b y^2 \, dx,$$

where the value of y in terms of x must be substituted from the equation of the given curve.

This formula is easily remembered if we consider a thin slice or disk of the solid between two planes perpendicular to the axis of revolution and regard this circular plate as, approximately, a cylinder of altitude dx with a base of area πy^2 and hence of volume $\pi y^2\, dx$. This cylinder is the element of volume.

Similarly, when OY is the axis of revolution, we use the formula

(F) $$V_y = \pi \int_c^v x^2\, dy,$$

where the value of x in terms of y must be substituted from the equation of the given curve.

ILLUSTRATIVE EXAMPLE 1. Find the volume generated by revolving the ellipse $\dfrac{x^2}{a^2} + \dfrac{y^2}{b^2} = 1$ about the x-axis.

Solution. Since $y^2 = \dfrac{b^2}{a^2}(a^2 - x^2)$, and the required volume is twice the volume generated by OAB, we get, substituting in (E),

$$\frac{V_x}{2} = \pi \int_0^a y^2\, dx = \pi \int_0^a \frac{b^2}{a^2}(a^2 - x^2)dx$$
$$= \frac{2\,\pi ab^2}{3}.$$
$$\therefore V_x = \frac{4\,\pi ab^2}{3}.$$

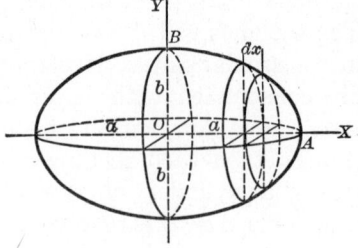

To check this result, let $b = a$. Then $V_x = \dfrac{4\,\pi a^3}{3}$, the volume of a sphere, which is only a special case of the ellipsoid. When the ellipse is revolved about its major axis, the solid generated is called a prolate spheroid; when about its minor axis, an oblate spheroid.

ILLUSTRATIVE EXAMPLE 2. The area bounded by the semicubical parabola

(1) $$ay^2 = x^3,$$

the y-axis, and the line AB $(y = a)$ is revolved about AB. Find the volume of the solid of revolution generated.

Solution. In the figure, $OPAB$ is the area revolved. Divide the segment AB into n equal parts each of length Δx. In the figure, NM is one of these parts. The rectangle $NMPQ$ when revolved about AB generates a cylinder of revolution, whose volume is an element of the required volume. Hence

Element of volume $= \pi r^2 h = \pi(a - y)^2\, \Delta x$,
since $\qquad r = PM = RM - RP = a - y$
and $\qquad h = NM = \Delta x$.

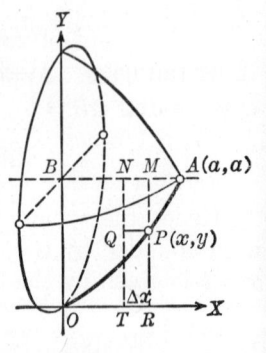

Then, by the Fundamental Theorem,

(2) Volume of solid $= V = \pi \displaystyle\int_0^a (a - y)^2\, dx = \pi \int_0^a (a^2 - 2\,ay + y^2)dx,$

for the limits are $x = 0$ and $x = AB = a$. Substituting for y its value given by (1), the answer is $V = 0.45\,\pi a^3$. *Ans.*

This should be compared with the volume of the cone of revolution with altitude $AB\ (= a)$ and base of radius $OB\ (= a)$. Volume of cone $= \frac{1}{3}\,\pi a^3$.

If the equations of the curve CD in the figure on page 265 are given in parametric form
$$x = f(t), \quad y = \phi(t),$$
then in (E) substitute $y = \phi(t)$, $dx = f'(t)dt$, and change the limits to t_1 and t_2, if
$$t = t_1 \text{ when } x = a, \quad t = t_2 \text{ when } x = b.$$

Volume of a hollow solid of revolution. When a plane area is revolved about an axis not crossing the area, a hollow solid of revolution is formed. Consider the solid obtained by revolving about the x-axis the area $ACBDA$ of the figure. Pass through the solid a system of equidistant planes perpendicular to the axis of revolution OX. Let Δx be their common distance apart. Then the solid is divided into hollow circular plates each of thickness Δx. If one of the planes dividing the solid passes through M, the hollow circular plate with one base in this plane is approximately a hollow circular cylinder whose inner and outer radii are respectively $MP_1 (= y_1)$ and $MP_2 (= y_2)$. Its altitude is Δx. Hence its volume is $\pi(y_2^2 - y_1^2)\Delta x$. Let there be n hollow cylinders, where $b - a = n \cdot \Delta x$. The limit of the sum of these n hollow cylinders when $n \to \infty$ is the volume of the hollow solid of revolution. Hence

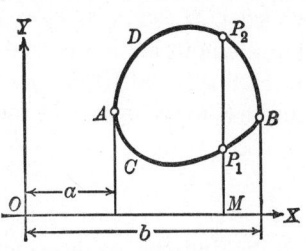

(3) $$V_x = \pi \int_a^b (y_2^2 - y_1^2)dx. \qquad (y_2 > y_1)$$

The element of volume in (3) is a hollow circular cylinder with inner radius y_1, outer radius y_2, and altitude Δx. The radii y_1 and y_2 are functions of x $(= OM)$ found from the equations of the curves (or the equation of the curve) bounding the area revolved.

ILLUSTRATIVE EXAMPLE 3. Find the volume of the ring solid (*anchor ring* or *torus*) obtained by revolving a circle of radius a about an external axis in its plane b units from its center $(b > a)$.

Solution. Let the equation of the circle be
$$x^2 + (y - b)^2 = a^2,$$
and let the x-axis be the axis of revolution.

Solving for y, we have
$$y_2 = b + \sqrt{a^2 - x^2}, \quad y_1 = b - \sqrt{a^2 - x^2}.$$
$$\therefore dV = \pi(y_2^2 - y_1^2)\Delta x = 4\pi b\sqrt{a^2 - x^2}\,\Delta x.$$

By (3), $$V_x = 4\pi b \int_{-a}^{a} \sqrt{a^2 - x^2}\, dx = 2\pi^2 a^2 b. \quad Ans.$$

A solid of revolution may be divided into cylindrical shells by passing through it a system of circular cylinders whose common axis is the axis of revolution. If the area $ACBD$ of the figure be revolved about the y-axis, it may be shown that

(4) $\quad V_y = 2\pi \int_a^b (y_2 - y_1) x \, dx,$

where $OM = x$, $MP_1 = y_1$, $MP_2 = y_2$. The element dV is now a cylindrical shell of radius r, altitude $y_2 - y_1$, and thickness Δx. Illustrative Example 3 may be solved by (4).

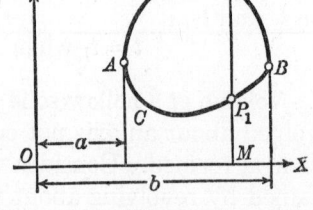

PROBLEMS

1. Find the volume of the sphere generated by revolving the circle $x^2 + y^2 = r^2$ about a diameter. *Ans.* $\frac{4}{3}\pi r^3$.

2. Find by integration the volume of the truncated cone generated by revolving the area bounded by $y = 6 - x$, $y = 0$, $x = 0$, $x = 4$ about OX. Verify geometrically.

3. Find the volume of the paraboloid of revolution whose surface is generated by revolving about its axis the arc of the parabola $y^2 = 2\,px$ between the origin and the point (x_1, y_1).

Ans. $\pi p x_1^2 = \frac{1}{2}\pi y_1^2 x_1$; that is, one half of the volume of the circumscribing cylinder.

4. Find the volume of the figure generated by revolving the arc in Problem 3 about OY.

Ans. $\frac{1}{5}\pi x_1^2 y_1$; that is, one fifth of the cylinder of altitude y_1 and radius of base x_1.

Find the volume generated by revolving about OX the areas bounded by the following loci.

5. $y = x^3$, $y = 0$, $x = 2$. *Ans.* $\frac{128}{7}\pi$.

6. $ay^2 = x^3$, $y = 0$, $x = a$. $\quad \frac{1}{4}\pi a^3$.

7. The parabola $\sqrt{x} + \sqrt{y} = \sqrt{a}$, $x = 0$, $y = 0$. $\quad \frac{1}{15}\pi a^3$.

8. The hypocycloid $x^{\frac{2}{3}} + y^{\frac{2}{3}} = a^{\frac{2}{3}}$. $\quad \frac{32}{105}\pi a^3$.

9. One arch of $y = \sin x$. $\quad \frac{1}{2}\pi^2$.

10. One arch of $y = \cos 2x$. $\quad \frac{1}{4}\pi^2$.

11. $y = e^{-x}$, $y = 0$, $x = 0$, $x = 5$. $\quad \frac{1}{2}\pi(1 - e^{-10})$.

12. $9x^2 + 16y^2 = 144$.   Ans. 48π.
13. $y = xe^x$, $y = 0$, $x = 1$. $\frac{1}{4}\pi(e^2 - 1)$.
14. The witch $(x^2 + 4a^2)y = 8a^3$, $y = 0$. $4\pi^2 a^3$.
15. $\left(\dfrac{x}{a}\right)^2 + \left(\dfrac{y}{b}\right)^{\frac{2}{3}} = 1$. $\frac{32}{35}\pi ab^2$.
16. $y^2(2a - x) = x^3$, $y = 0$, $x = a$. $0.2115\,\pi a^3$.
17. $y = x^2 - 6x$, $y = 0$.
18. $y^2 = (2-x)^3$, $y = 0$, $x = 0$, $x = 1$.
19. $y^2(4 + x^2) = 1$, $y = 0$, $x = 0$, $x = \infty$.
20. $(x-1)y = 2$, $y = 0$, $x = 2$, $x = 5$.

Find the volume generated by revolving about OY the areas bounded by the following loci.

21. $y = x^3$, $y = 0$, $x = 2$.   Ans. $\frac{64}{5}\pi$.
22. $2y^2 = x^3$, $y = 0$, $x = 2$. $\frac{32}{7}\pi$.
23. $y = e^x$, $y = 0$, $x = 0$. 2π.
24. $9x^2 + 16y^2 = 144$. 64π.
25. $\left(\dfrac{x}{a}\right)^2 + \left(\dfrac{y}{b}\right)^{\frac{2}{3}} = 1$. $\frac{4}{5}\pi a^2 b$.
26. $y^2 = 9 - x$, $x = 0$.
27. $x^2 = 16 - y$, $y = 0$.
28. $y^2 = ax$, $y = 0$, $x = a$.

29. The equation of the curve OA in the figure is $y^2 = x^3$. Find the volume generated when the area

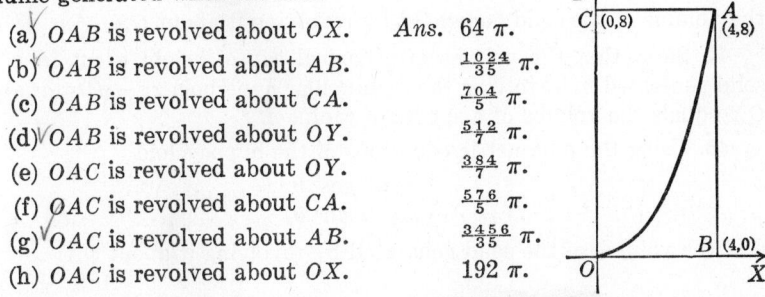

(a) OAB is revolved about OX.   Ans. 64π.
(b) OAB is revolved about AB. $\frac{1024}{35}\pi$.
(c) OAB is revolved about CA. $\frac{704}{5}\pi$.
(d) OAB is revolved about OY. $\frac{512}{7}\pi$.
(e) OAC is revolved about OY. $\frac{384}{7}\pi$.
(f) OAC is revolved about CA. $\frac{576}{5}\pi$.
(g) OAC is revolved about AB. $\frac{3456}{35}\pi$.
(h) OAC is revolved about OX. 192π.

Figure: $C(0,8)$, $A(4,8)$, $B(4,0)$, with curve OA shown.

30. Find the volume of the oblate spheroid generated by revolving the area bounded by the ellipse $b^2 x^2 + a^2 y^2 = a^2 b^2$ about the y-axis.   Ans. $\frac{4}{3}\pi a^2 b$.

31. A segment of one base of thickness h is cut from a sphere of radius r. Show by integration that its volume is $\dfrac{\pi h^2 (3r - h)}{3}$.

Calculate the volume generated by revolving about each of the following lines the area which it cuts from the corresponding curve.

32. $y = 3$; $y = 4x - x^2$. Ans. $\frac{16}{15}\pi$.

33. $x = 4$; $y^2 = x^3$. $\frac{2048}{35}\pi$.

34. $y = -4$; $y = 4 + 6x - 2x^2$. $\frac{1250}{3}\pi$.

35. $y = x$; $y = x^2$. $\frac{1}{60}\pi\sqrt{2}$.

36. $y = x$; $y = 3x - x^2$. $\frac{8}{15}\pi\sqrt{2}$.

37. $4y = 4x + 33$; $y = 9 - x^2$. $\frac{8}{15}\pi\sqrt{2}$.

38. $x + y = 1$; $\sqrt{x} + \sqrt{y} = 1$. $\frac{1}{15}\pi\sqrt{2}$.

39. $x + y = 7$; $xy = 6$.

40. Find the volume generated by revolving one arch of the cycloid
$$x = r \text{ arc vers } \frac{y}{r} - \sqrt{2ry - y^2}$$
about OX, its base.

HINT. Substitute $dx = \dfrac{y\,dy}{\sqrt{2ry - y^2}}$, and limits $y = 0$, $y = 2r$, in (E), Art. 160.

Ans. $5\pi^2 r^3$.

41. Find the volume generated by revolving the catenary $y = \dfrac{a}{2}\left(e^{\frac{x}{a}} + e^{-\frac{x}{a}}\right)$ about the x-axis from $x = 0$ to $x = b$.

Ans. $\dfrac{\pi a^3}{8}\left(e^{\frac{2b}{a}} - e^{-\frac{2b}{a}}\right) + \dfrac{\pi a^2 h}{2}$.

42. Find the volume of the solid generated by revolving the cissoid $y^2 = \dfrac{x^3}{2a - x}$ about its asymptote $x = 2a$. Ans. $2\pi^2 a^3$.

43. Given the slope of the tangent to the tractrix $\dfrac{dy}{dx} = -\dfrac{y}{\sqrt{a^2 - y^2}}$, find the volume of the solid generated by revolving it about OX. Ans. $\frac{2}{3}\pi a^3$.

44. Show that the volume of a conical cap of height a cut from the solid generated by revolving the rectangular hyperbola $x^2 - y^2 = a^2$ about OX equals the volume of a sphere of radius a.

45. Using the parametric equations of the hypocycloid
$$\begin{cases} x = a\cos^3\theta, \\ y = a\sin^3\theta, \end{cases}$$
find the volume of the solid generated by revolving it about OX.

Ans. $\frac{32}{105}\pi a^3$.

46. Find the volume generated by revolving one arch of the cycloid
$$\begin{cases} x = a(\theta - \sin\theta), \\ y = a(1 - \cos\theta) \end{cases}$$
about its base OX. Ans. $5\pi^2 a^3$.

Show that if the arch be revolved about OY, the volume generated is $6\pi^3 a^3$.

47. Find the volume generated if the area bounded by the curve $y = \sec \frac{1}{2} \pi x$, the x-axis, and the lines $x = \pm \frac{1}{2}$ is revolved about the x-axis.

Ans. 4.

48. The area under the curve $y = e^x \sin x$ from $x = 0$ to $x = \pi$ is revolved about the x-axis. Find the volume generated.

49. Given the curve $x = t^2$, $y = 4t - t^3$. Find (a) the area of the loop and (b) the volume generated by the area inside the loop when revolved about the x-axis. *Ans.* (a) $\frac{256}{15}$; (b) 67.02.

50. Revolve the area bounded by the two parabolas $y^2 = 4x$ and $y^2 = 5 - x$ about each axis and calculate the respective volumes.

Ans. OX: 10π; OY: $17\frac{6}{3}\pi$.

51. Revolve about the polar axis the part of the cardioid $\rho = 4 + 4\cos\theta$ between the lines $\theta = 0$ and $\theta = \dfrac{\pi}{2}$ and compute the volume.

Ans. 160π.

161. Length of a curve. By the *length of a straight line* we commonly mean the number of times we can superpose upon it another straight line employed as a unit of length, as when the carpenter measures the length of a board by making end-to-end applications of his foot rule.

Since it is impossible to make a straight line coincide with an arc of a curve, we cannot measure curves in the same manner as we measure straight lines. We proceed, then, as follows.

Divide the curve (as AB) into any number of parts in any manner whatever (as at C, D, E) and connect the adjacent points of division, forming chords (as AC, CD, DE, EB).

The length of the curve is defined as the limit of the sum of the chords as the number of points of division increases without limit in such a way that at the same time each chord separately approaches zero as a limit.

Since this limit will also be the measure of the length of some straight line, the finding of the length of a curve is also called "the rectification of the curve."

The student has already made use of this definition for the length of a curve in his geometry. Thus the circumference of a circle is defined as the limit of the perimeter of the inscribed (or circumscribed) regular polygon when the number of sides increases without limit.

The method of the next article for finding the length of a plane curve is based on the above definition, and the student should note very carefully how it is applied.

162. Lengths of plane curves; rectangular coördinates. We shall now proceed to express, in analytical form, the definition of the last article, making use of the Fundamental Theorem.

Given the curve
$$y = f(x)$$
and the points $P'(a, c)$, $Q(b, d)$ on it; to find the length of the arc $P'Q$.

FIRST STEP. Take any number n of points on the curve between P' and Q and draw the chords joining the adjacent points, as in the figure. The required length of arc $P'Q$ is evidently the limit of the sum of the lengths of such chords.

SECOND STEP. Consider any one of these chords, $P'P''$ for example, and let the coördinates of P' and P'' be

$$P'(x', y') \quad \text{and} \quad P''(x' + \Delta x', y' + \Delta y').$$

Then, as in Art. 95,

$$P'P'' = \sqrt{(\Delta x')^2 + (\Delta y')^2},$$

or
$$P'P'' = \left[1 + \left(\frac{\Delta y'}{\Delta x'}\right)^2\right]^{\frac{1}{2}} \Delta x'.$$

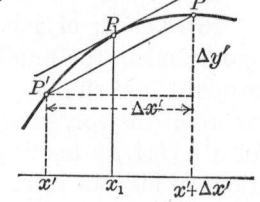

[Dividing inside the radical by $(\Delta x')^2$ and multiplying outside by $\Delta x'$.]

But from the Theorem of Mean Value (Art. 116) (if $\Delta y'$ is denoted by $f(b) - f(a)$ and $\Delta x'$ by $b - a$), we get

$$\frac{\Delta y'}{\Delta x'} = f'(x_1), \qquad (x' < x_1 < x' + \Delta x')$$

x_1 being the abscissa of a point P_1 on the curve *between P' and P''* at which the tangent is parallel to the chord.

Substituting, $P'P'' = [1 + f'(x_1)^2]^{\frac{1}{2}} \Delta x' = $ length of first chord.

Similarly, $\quad P''P''' = [1 + f'(x_2)^2]^{\frac{1}{2}} \Delta x'' = $ length of second chord.

$$\cdot \quad \cdot \quad \cdot \quad \cdot \quad \cdot \quad \cdot \quad \cdot \quad \cdot$$

$$P^{(n)}Q = [1 + f'(x_n)^2]^{\frac{1}{2}} \Delta x^{(n)} = \text{length of } n\text{th chord}.$$

The length of the inscribed broken line joining P' and Q (sum of the chords) is then the sum of these expressions, namely,

$$[1 + f'(x_1)^2]^{\frac{1}{2}} \Delta x' + [1 + f'(x_2)^2]^{\frac{1}{2}} \Delta x'' + \cdots + [1 + f'(x_n)^2]^{\frac{1}{2}} \Delta x^{(n)}$$
$$= \sum_{i=1}^{n} [1 + f'(x_i)^2]^{\frac{1}{2}} \Delta x^{(i)}.$$

INTEGRATION A PROCESS OF SUMMATION 273

Third Step. Applying the Fundamental Theorem,

$$\lim_{n \to \infty} \sum_{i=1}^{n} [1 + f'(x_i)^2]^{\frac{1}{2}} \Delta x^{(i)} = \int_a^b [1 + f'(x)^2]^{\frac{1}{2}} dx.$$

Hence, denoting the length of arc $P'Q$ by s, we have the formula for the length of the arc

$$s = \int_a^b [1 + f'(x)^2]^{\frac{1}{2}} dx, \text{ or}$$

(G) $$s = \int_a^b [1 + y'^2]^{\frac{1}{2}} dx,$$

where $y' = \dfrac{dy}{dx}$ must be found in terms of x from the equation of the given curve.

Sometimes it is more convenient to use y as the independent variable. To derive a formula to cover this case, we know from Art. 39 that

$$\frac{dy}{dx} = \frac{1}{\dfrac{dx}{dy}}; \text{ hence } dx = x' \, dy.$$

Substituting these values in (G), and noting that the corresponding y limits are c and d, we get the formula for the length of the arc,

(H) $$s = \int_c^d [x'^2 + 1]^{\frac{1}{2}} dy,$$

where $x' = \dfrac{dx}{dy}$ must be found in terms of y from the equation of the given curve.

Formula (G) may be derived in another way. In Art. 95, formula (D),

(1) $$ds = (1 + y'^2)^{\frac{1}{2}} dx$$

gives the differential of the arc of a curve. If we proceed from (1) as in Art. 142, we obtain (G). Also, (H) follows from (E) in Art. 95. Finally, if the curve is defined by parametric equations

(2) $$x = f(t), \quad y = \phi(t),$$

it is convenient to use

(3) $$s = \int (dx^2 + dy^2)^{\frac{1}{2}} = \int_{t_1}^{t_2} [f'^2(t) + \phi'^2(t)]^{\frac{1}{2}} dt,$$

since, from (2), $dx = f'(t)dt$, $dy = \phi'(t)dt$.

ILLUSTRATIVE EXAMPLE 1. Find the length of the circumference of the circle $x^2 + y^2 = r^2$.

Solution. Differentiating, $\dfrac{dy}{dx} = -\dfrac{x}{y}$.

Substituting in (G),

$$\text{Arc } BA = \int_0^r \left[1 + \frac{x^2}{y^2}\right]^{\frac{1}{2}} dx$$

$$= \int_0^r \left[\frac{y^2 + x^2}{y^2}\right]^{\frac{1}{2}} dx = \int_0^r \left[\frac{r^2}{r^2 - x^2}\right]^{\frac{1}{2}} dx.$$

[Substituting $y^2 = r^2 - x^2$ from the equation of the circle in order to get everything in terms of x.]

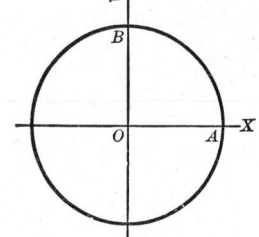

$$\therefore \text{arc } BA = r \int_0^r \frac{dx}{\sqrt{r^2 - x^2}} = \frac{\pi r}{2}. \quad \text{(See Illustrative Example 1, Art. 154.)}$$

Hence the total length equals $2\pi r$. *Ans.*

ILLUSTRATIVE EXAMPLE 2. Find the length of arc of one arch of the cycloid

$$x = a(\theta - \sin \theta), \quad y = a(1 - \cos \theta).$$

See Illustrative Example 2, Art. 81.

Solution. $\quad dx = a(1 - \cos \theta)d\theta, \quad dy = a \sin \theta \, d\theta.$

Then $\quad dx^2 + dy^2 = 2\,a^2(1 - \cos \theta)d\theta^2 = 4\,a^2 \sin^2 \tfrac{1}{2}\,\theta\,d\theta^2.$ By (5), Art. 2

Using (3), $\quad s = \int_0^{2\pi} 2\,a \sin \tfrac{1}{2}\,\theta\,d\theta = 8\,a.$ *Ans.*

The limits are the values of θ at O and D (see figure, Illustrative Example 2, Art. 81), that is, $\theta = 0$ and $\theta = 2\pi$.

ILLUSTRATIVE EXAMPLE 3. Find the length of the arc of the curve $25\,y^2 = x^5$ from $x = 0$ to $x = 2$.

Solution. Differentiating, $y' = \tfrac{1}{2}\,x^{\frac{3}{2}}$. Hence, by (G),

(4) $$s = \int_0^2 (1 + \tfrac{1}{4}\,x^3)^{\frac{1}{2}} dx = \tfrac{1}{2} \int_0^2 (4 + x^3)^{\frac{1}{2}} dx.$$

The integral in (4) was evaluated approximately in Illustrative Example 2, Art. 148, by the trapezoidal rule, and also in Illustrative Example 2, Art. 149, by Simpson's rule. Taking the latter value, $s = \tfrac{1}{2}(4.821) = 2.41$ *linear units. Ans.*

163. Lengths of plane curves; polar coördinates. From (*I*), Art. 96, by proceeding as in Art. 142, we get the formula for the length of the arc,

(*I*) $$s = \int_\alpha^\beta \left[\rho^2 + \left(\frac{d\rho}{d\theta}\right)^2\right]^{\frac{1}{2}} d\theta,$$

where ρ and $\dfrac{d\rho}{d\theta}$ in terms of θ must be substituted from the equation of the given curve.

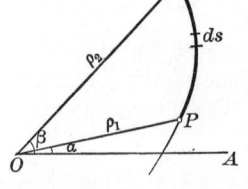

In case it is more convenient to use ρ as the independent variable, and the equation is in the form

$$\theta = \phi(\rho),$$

then $\qquad\qquad d\theta = \phi'(\rho)d\rho = \dfrac{d\theta}{d\rho}\,d\rho.$

Substituting this in $[\rho^2 d\theta^2 + d\rho^2]^{\frac{1}{2}}$ gives

$$\left[\rho^2 \left(\frac{d\theta}{d\rho}\right)^2 + 1\right]^{\frac{1}{2}} d\rho.$$

Hence if ρ_1 and ρ_2 are the corresponding limits of the independent variable ρ, we get the formula for the length of the arc,

(J) $$s = \int_{\rho_1}^{\rho_2} \left[\rho^2 \left(\frac{d\theta}{d\rho}\right)^2 + 1\right]^{\frac{1}{2}} d\rho,$$

where $\frac{d\theta}{d\rho}$ in terms of ρ must be substituted from the equation of the given curve.

ILLUSTRATIVE EXAMPLE. Find the perimeter of the cardioid $\rho = a(1 + \cos\theta)$

Solution. Here $\frac{d\rho}{d\theta} = -a \sin\theta$.

If we let θ vary from 0 to π, the point P will generate one half of the curve. Substituting in (I),

$$\frac{s}{2} = \int_0^\pi [a^2(1+\cos\theta)^2 + a^2 \sin^2\theta]^{\frac{1}{2}} d\theta$$

$$= a \int_0^\pi (2 + 2\cos\theta)^{\frac{1}{2}} d\theta = 2a \int_0^\pi \cos\frac{\theta}{2} d\theta = 4a.$$

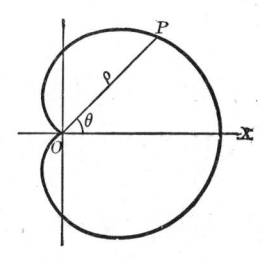

∴ $s = 8a$. Ans.

PROBLEMS

1. Find the length of the curve whose equation is $y^3 = x^2$ between the points (0, 0) and (8, 4). *Ans.* 9.07.

2. Find the length of the arc of the semicubical parabola $ay^2 = x^3$ from the origin to the ordinate $x = 5a$. *Ans.* $\frac{335\,a}{27}$.

3. Find the length of the curve whose equation is $y = \frac{x^3}{6} + \frac{1}{2x}$ from the point where $x = 1$ to the point where $x = 3$. *Ans.* $\frac{14}{3}$.

4. Find the length of the arc of the parabola $y^2 = 2\,px$ from the vertex to one extremity of the latus rectum. *Ans.* $\frac{p\sqrt{2}}{2} + \frac{p}{2} \ln(1 + \sqrt{2})$.

5. Find the length of arc of the curve $y^2 = x^3$ from the point where $x = 0$ to the point where $x = \frac{5}{9}$. *Ans.* $\frac{19}{27}$.

6. Find the length of arc of the parabola $6y = x^2$ from the origin to the point $(4, \frac{8}{3})$. *Ans.* 4.98.

7. Approximate by Simpson's rule the length of arc of the curve $3y = x^3$ from the origin to the point $(1, \frac{1}{3})$. *Ans.* 1.09.

8. Find the length of arc of the curve $y = \ln \sec x$ from the origin to the point $\left(\dfrac{\pi}{3}, \ln 2\right)$. *Ans.* $\ln(2 + \sqrt{3})$.

9. Find the length of arc of the hyperbola $x^2 - y^2 = 9$ from $(3, 0)$ to $(5, 4)$. [Use Simpson's rule.] *Ans.* 4.56.

10. Find the length of the arch of the parabola $y = 4x - x^2$ which lies above the x-axis. *Ans.* 9.29.

11. Find the entire length of the hypocycloid $x^{\frac{2}{3}} + y^{\frac{2}{3}} = a^{\frac{2}{3}}$. *Ans.* $6a$.

12. Rectify the catenary $y = \dfrac{a}{2}(e^{\frac{x}{a}} + e^{-\frac{x}{a}})$ from $x = 0$ to the point (x, y).

Ans. $\dfrac{a}{2}(e^{\frac{x}{a}} - e^{-\frac{x}{a}})$.

13. Find the length of one complete arch of the cycloid

$$x = r \text{ arc vers } \dfrac{y}{r} - \sqrt{2ry - y^2}. \qquad \textit{Ans. } 8r.$$

HINT. Use (H), Art. 162. Here $\dfrac{dx}{dy} = \dfrac{y}{\sqrt{2ry - y^2}}$.

14. Rectify the curve $9ay^2 = x(x - 3a)^2$ from $x = 0$ to $x = 3a$.

Ans. $2a\sqrt{3}$.

15. Find the length in one quadrant of the curve $\left(\dfrac{x}{a}\right)^{\frac{2}{3}} + \left(\dfrac{y}{b}\right)^{\frac{2}{3}} = 1$.

Ans. $\dfrac{a^2 + ab + b^2}{a + b}$.

16. Find the length between $x = a$ and $x = b$ of the curve $e^y = \dfrac{e^x + 1}{e^x - 1}$.

Ans. $\ln \dfrac{e^{2b} - 1}{e^{2a} - 1} + a - b$.

17. The equations of the involute of a circle are

$$\begin{cases} x = a(\cos\theta + \theta\sin\theta), \\ y = a(\sin\theta - \theta\cos\theta). \end{cases}$$

Find the length of the arc from $\theta = 0$ to $\theta = \theta_1$. *Ans.* $\tfrac{1}{2} a\theta_1^2$.

18. Find the length of arc of curve $\begin{cases} x = e^\theta \sin\theta \\ y = e^\theta \cos\theta \end{cases}$ from $\theta = 0$ to $\theta = \dfrac{\pi}{2}$.

Ans. $\sqrt{2}(e^{\frac{\pi}{2}} - 1)$.

Find the length of arc of each of the following curves.

19. $y = \ln(1 - x^2)$ from $x = 0$ to $x = \tfrac{1}{2}$.

20. $y = \dfrac{x^2}{4} - \dfrac{1}{2}\ln x$ from $x = 1$ to $x = 2$.

21. $y = \ln \csc x$ from $x = \dfrac{\pi}{6}$ to $x = \dfrac{\pi}{2}$.

22. $3x^2 = y^3$ from $y = 1$ to $y = 20$.

23. One arch of the curve $y = \sin x$.

INTEGRATION A PROCESS OF SUMMATION 277

24. Find the length of the spiral of Archimedes, $\rho = a\theta$, from the origin to the end of the first revolution.

$$\text{Ans. } \pi a \sqrt{1 + 4\pi^2} + \frac{a}{2} \ln(2\pi + \sqrt{1 + 4\pi^2}).$$

25. Rectify the spiral $\rho = e^{a\theta}$ from the origin to the point (ρ, θ).

HINT. Use (J).

$$\text{Ans. } \frac{\rho}{a}\sqrt{a^2 + 1}.$$

26. Find the length of the curve $\rho = a \sec^2 \frac{\theta}{2}$ from $\theta = 0$ to $\theta = \frac{\pi}{2}$.

$$\text{Ans. } [\sqrt{2} + \ln(\sqrt{2} + 1)]a.$$

27. Find the length of arc of the parabola $\rho = \dfrac{2}{1 + \cos\theta}$ from $\theta = 0$ to $\theta = \dfrac{\pi}{2}$.

$$\text{Ans. } \sqrt{2} + \ln(\sqrt{2} + 1).$$

28. Find the length of the hyperbolic spiral $\rho\theta = a$ from (ρ_1, θ_1) to (ρ_2, θ_2).

$$\text{Ans. } \sqrt{a^2 + \rho_1^2} - \sqrt{a^2 + \rho_2^2} + a \ln \frac{\rho_1(a + \sqrt{a^2 + \rho_2^2})}{\rho_2(a + \sqrt{a^2 + \rho_1^2})}.$$

29. Show that the entire length of the curve $\rho = a \sin^3 \dfrac{\theta}{3}$ is $\dfrac{3\pi a}{2}$. Show that OA, AB, BC (see figure) are in arithmetical progression.

30. Find the length of arc of the cissoid $\rho = 2 a \tan\theta \sin\theta$ from $\theta = 0$ to $\theta = \dfrac{\pi}{4}$.

31. Approximate the perimeter of one leaf of the curve $\rho = \sin 2\theta$.

164. Areas of surfaces of revolution. A surface of revolution is generated by revolving the arc CD of the curve

$$y = f(x)$$

about the axis of X.

It is desired to measure the area of this surface by making use of the Fundamental Theorem.

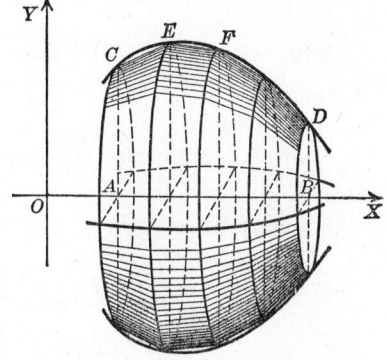

FIRST STEP. As before, divide the interval AB into subintervals Δx_1, Δx_2, etc., and erect ordinates at the points of division. Draw the chords CE, EF, etc. of the curve. When the curve is revolved, each chord generates the lateral surface of a frustum of a cone of revolution. The area of the required surface of revolution is defined as the limit of the sum of the lateral areas of these frustums.

SECOND STEP. For the sake of clearness let us draw the first frustum on a larger scale. Let M be the middle point of the chord CE. Then

(1) $\quad$ Lateral area $= 2\,\pi NM \cdot CE$.*

In order to apply the Fundamental Theorem it is necessary to express this product as a function of the abscissa of some point in the interval Δx_1. As in Art. 162, we get, using the Theorem of Mean Value, the length of the chord

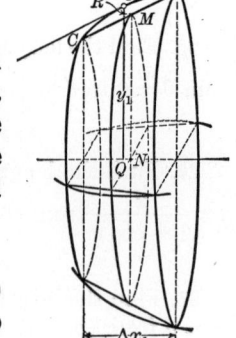

(2) $\quad CE = [1 + f'(x_1)^2]^{\frac{1}{2}}\,\Delta x_1,$

where x_1 is the abscissa of the point $P_1(x_1, y_1)$ on the arc CE where the tangent is parallel to the chord CE. Let the horizontal line through M intersect QP_1 (the ordinate of P_1) at R, and denote RP_1 by ϵ_1.†
Then

(3) $\quad NM = y_1 - \epsilon_1.$

Substituting (2) and (3) in (1), we get

$2\,\pi(y_1 - \epsilon_1)[1 + f'(x_1)^2]^{\frac{1}{2}}\,\Delta x_1 =$ lateral area of first frustum.

Similarly,

$2\,\pi(y_2 - \epsilon_2)[1 + f'(x_2)^2]^{\frac{1}{2}}\Delta x_2 =$ lateral area of second frustum,

. . . .

$2\,\pi(y_n - \epsilon_n)[1 + f'(x_n)^2]^{\frac{1}{2}}\Delta x_n =$ lateral area of last frustum.

Hence

$\sum\limits_{i=1}^{n} 2\,\pi(y_i - \epsilon_i)[1 + f'(x_i)^2]^{\frac{1}{2}}\,\Delta x_i =$ sum of lateral areas of frustums.

This may be written

(4) $\quad \sum\limits_{i=1}^{n} 2\,\pi y_i[1 + f'(x_i)^2]^{\frac{1}{2}}\Delta x_i - 2\,\pi \sum\limits_{i=1}^{n} \epsilon_i[1 + f'(x_i)^2]^{\frac{1}{2}}\,\Delta x_i.$

* The lateral area of the frustum of a cone of revolution is equal to the circumference of the middle section multiplied by the slant height.

† The student will observe that as Δx_1 approaches zero as a limit, ϵ_1 also approaches the limit zero.

THIRD STEP. Applying the Fundamental Theorem to the first sum (using the limits $OA = a$ and $OB = b$), we get

$$\lim_{n\to\infty} \sum_{i=1}^{n} 2\pi y_i [1 + f'(x_i)^2]^{\frac{1}{2}} \Delta x_i = \int_a^b 2\pi y [1 + f'(x)^2]^{\frac{1}{2}} dx.$$

The limit of the second sum of (4) when $n \to \infty$ is zero.* Hence the area of the surface of revolution generated by revolving the arc CD about OX is given by the formula

(K) $$S_x = 2\pi \int_a^b y \left[1 + \left(\frac{dy}{dx}\right)^2\right]^{\frac{1}{2}} dx,$$

where S_x denotes the required area. Or we may write the formula in the form

(L) $$S = 2\pi \int_a^b y \, ds.$$

Similarly, when OY is the axis of revolution we use the formula

(M) $$S_y = 2\pi \int_c^d x \, ds.$$

In (L) and (M) ds will have one of the three forms (C), (D), (E), of Art. 95, namely,

$$ds = \left[1 + \left(\frac{dy}{dx}\right)^2\right]^{\frac{1}{2}} dx = \left[1 + \left(\frac{dx}{dy}\right)^2\right]^{\frac{1}{2}} dy = (dx^2 + dy^2)^{\frac{1}{2}},$$

depending upon the choice of the independent variable. The last form must be used when the given curve is defined by parametric equations. In using (L) or (M), calculate ds first.

The formula (L) is easily remembered if we consider a narrow band of the surface included between two planes perpendicular to the axis of revolution, and regard it approximately as the convex surface of a frustum of a cone of revolution of slant height ds, with a middle section whose circumference equals $2\pi y$, and hence of area $2\pi y \, ds$.

ILLUSTRATIVE EXAMPLE 1. The arc of the cubical parabola
(5) $$a^2 y = x^3$$
between $x = 0$ and $x = a$ is revolved about OX. Find the area of the surface of revolution generated.

---

* This is easily seen as follows. Denote the second sum by S_n. If ϵ equals the largest of the positive numbers $|\epsilon_1|, |\epsilon_2|, \cdots, |\epsilon_n|$, then

$$S_n \leq \epsilon \sum_{i=1}^{n} [1 + f'(x_i)^2]^{\frac{1}{2}} \Delta x_i.$$

The sum on the right is, by Art. 162, equal to the sum of the chords CE, EF, etc. Let this sum be l_n. Then $S_n \leq \epsilon l_n$. Since $\lim_{n\to\infty} \epsilon = 0$, S_n is an infinitesimal, and therefore $\lim_{n\to\infty} S_n = 0$.

Solution. From (5), $y' = \dfrac{3\,x^2}{a^2}$. Hence

$$ds = (1 + y'^2)^{\frac{1}{2}}\,dx = \dfrac{1}{a^2}(a^4 + 9\,x^4)^{\frac{1}{2}}\,dx.$$

Then the element of area $= 2\,\pi y\,ds = \dfrac{2\,\pi}{a^4}(a^4 + 9\,x^4)^{\frac{1}{2}} x^3\,dx.$

Therefore, by (L),

$$S_x = \dfrac{2\,\pi}{a^4}\int_0^a (a^4 + 9\,x^4)^{\frac{1}{2}} x^3\,dx = \dfrac{\pi}{27\,a^4}\Big[(a^4 + 9\,x^4)^{\frac{3}{2}}\Big]_0^a$$
$$= \dfrac{\pi}{27}(10\sqrt{10} - 1)a^2 = 3.6\,a^2. \quad Ans.$$

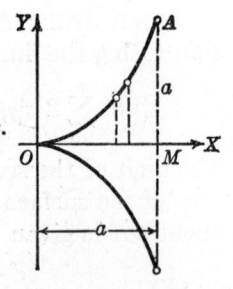

ILLUSTRATIVE EXAMPLE 2. Find the area of the ellipsoid of revolution generated by revolving the ellipse whose parametric equations are (see (3), Art. 81) $x = a\cos\phi$, $y = b\sin\phi$ about OX.

Solution. We have
$$dx = -a\sin\phi\,d\phi, \quad dy = b\cos\phi\,d\phi,$$
and
$$ds = (dx^2 + dy^2)^{\frac{1}{2}} = (a^2\sin^2\phi + b^2\cos^2\phi)^{\frac{1}{2}}\,d\phi.$$

Hence the element of area $= 2\,\pi y\,ds = 2\,\pi b(a^2\sin^2\phi + b^2\cos^2\phi)^{\frac{1}{2}}\sin\phi\,d\phi.$

(6) $\quad \therefore \tfrac{1}{2}S_x = 2\,\pi b \int_0^{\frac{\pi}{2}} (a^2\sin^2\phi + b^2\cos^2\phi)^{\frac{1}{2}}\sin\phi\,d\phi.$

To integrate, let $u = \cos\phi$. Then $du = -\sin\phi\,d\phi$. Also,
$$a^2\sin^2\phi + b^2\cos^2\phi = a^2(1 - \cos^2\phi) + b^2\cos^2\phi = a^2 - (a^2 - b^2)u^2.$$

Hence, using the new limits $u = 1$, $u = 0$, and interchanging the u limits (Art. 150), the result is

$$\tfrac{1}{2}S_x = 2\,\pi b\int_0^1 [a^2 - (a^2 - b^2)u^2]^{\frac{1}{2}}\,du. \qquad (a > b)$$

Working this out by (22), we get

$$S_x = 2\,\pi b^2 + \dfrac{2\,\pi ab}{e}\,\text{arc sin } e, \text{ where } e = \text{eccentricity} = \dfrac{\sqrt{a^2 - b^2}}{a}. \quad Ans.$$

ILLUSTRATIVE EXAMPLE 3. Find the area of the surface of revolution generated by revolving the hypocycloid $x^{\frac{2}{3}} + y^{\frac{2}{3}} = a^{\frac{2}{3}}$ about the x-axis.

Solution. Here $\dfrac{dy}{dx} = -\dfrac{y^{\frac{1}{3}}}{x^{\frac{1}{3}}}$, $y = (a^{\frac{2}{3}} - x^{\frac{2}{3}})^{\frac{3}{2}}$;

$$ds = \left(1 + \dfrac{y^{\frac{2}{3}}}{x^{\frac{2}{3}}}\right)^{\frac{1}{2}} dx = \left(\dfrac{x^{\frac{2}{3}} + y^{\frac{2}{3}}}{x^{\frac{2}{3}}}\right)^{\frac{1}{2}} = \dfrac{a^{\frac{1}{3}}}{x^{\frac{1}{3}}}\,dx.$$

Substituting in (L), noting that the arc BA generates only one half of the surface, we get

$$\dfrac{S_x}{2} = 2\,\pi a^{\frac{1}{3}}\int_0^a (a^{\frac{2}{3}} - x^{\frac{2}{3}})^{\frac{3}{2}} x^{-\frac{1}{3}}\,dx.$$

This is an improper integral, since the function to be integrated is discontinuous (becomes infinite) when $x = 0$. Using the definition (1), Art. 154, the result is $\dfrac{S_x}{2} = \dfrac{6\,\pi a^2}{5}. \quad \therefore S_x = \dfrac{12\,\pi a^2}{5}.$

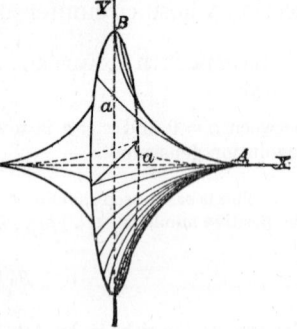

INTEGRATION A PROCESS OF SUMMATION

PROBLEMS

1. Find by integration the area of the surface of the sphere generated by revolving the circle $x^2 + y^2 = r^2$ about a diameter. *Ans.* $4\pi r^2$.

2. Find by integration the area of the surface of the cone generated by revolving about OX the line joining the origin to the point (a, b).
$$Ans.\ \pi b\sqrt{a^2 + b^2}.$$

3. Find by integration the area of the surface of the cone generated by revolving the line $y = 2x$ from $x = 0$ to $x = 2$ (a) about OX; (b) about OY. Verify your results geometrically.

4. Find by integration the lateral area of the frustum of a cone generated by revolving about OX the line $2y = x - 4$ from $x = 0$ to $x = 5$. Verify your result geometrically.

5. Find the area of the surface generated by revolving about OY the arc of the parabola $y = x^2$ from $y = 0$ to $y = 2$. *Ans.* $\tfrac{13}{3}\pi$.

6. Find the area of the surface generated by revolving about OX the arc of the parabola $y = x^2$ from $(0, 0)$ to $(2, 4)$.

7. Find the area of the surface generated by revolving about OX the arc of the parabola $y^2 = 4 - x$ which lies in the first quadrant. *Ans.* 36.18.

8. Find the area of the surface generated by revolving about OX the arc of the parabola $y^2 = 2px$ from $x = 0$ to $x = 4p$. *Ans.* $\tfrac{52}{3}\pi p^2$.

9. Find the area of the surface generated by revolving about OY the arc of $y = x^3$ from $(0, 0)$ to $(2, 8)$.

Find the area of the surface generated by revolving each of the following curves about OX.

10. $9y = x^3$, from $x = 0$ to $x = 2$. *Ans.* $\tfrac{98}{81}\pi$.

11. $y^2 = 9x$, from $x = 0$ to $x = 4$. 49π.

12. $y^2 = 24 - 4x$, from $x = 3$ to $x = 6$. $\tfrac{56}{3}\pi$.

13. $6y = x^2$, from $x = 0$ to $x = 4$. $\dfrac{(820 - 81\ln 3)\pi}{72}$.

14. $y = e^{-x}$, from $x = 0$ to $x = \infty$. $\pi[\sqrt{2} + \ln(1 + \sqrt{2})]$.

15. The loop of $9\,ay^2 = x(3a - x)^2$. $3\pi a^2$.

16. $6a^2xy = x^4 + 3a^4$, from $x = a$ to $x = 2a$. $\tfrac{47}{16}\pi a^2$.

17. One loop of $8a^2y^2 = a^2x^2 - x^4$. $\tfrac{1}{4}\pi a^2$.

18. $y^2 + 4x = 2\ln y$, from $y = 1$ to $y = 2$. $\tfrac{10}{3}\pi$.

19. The cycloid $\begin{cases} x = a(\theta - \sin\theta), \\ y = a(1 - \cos\theta). \end{cases}$ *Ans.* $\tfrac{64}{3}\pi a^2$.

20. The cardioid $\begin{cases} x = a(2\cos\theta - \cos 2\theta), \\ y = a(2\sin\theta - \sin 2\theta). \end{cases}$ $\tfrac{128}{5}\pi a^2$.

21. $y^2 = 4x$, from $x = 0$ to $x = 3$.
22. $x^2 + y^2 = 4$, from $x = 1$ to $x = 2$.
23. $x^2 + 4y^2 = 36$.
24. $9x^2 + 4y^2 = 36$.

Find the area of the surface generated by revolving each of the following curves about OY.

25. $x = y^3$, from $y = 0$ to $y = 3$.   Ans. $\frac{1}{27}\pi[(730)^{\frac{3}{2}} - 1]$.
26. $y = x^3$, from $y = 0$ to $y = 3$.
27. $6a^2xy = x^4 + 3a^4$, from $x = a$ to $x = 3a$. $(20 + \ln 3)\pi a^2$.
28. $4y = x^2 - 2\ln x$, from $x = 1$ to $x = 4$. 24π.
29. $2y = x\sqrt{x^2 - 1} + \ln(x - \sqrt{x^2 - 1})$, from $x = 2$ to $x = 5$.
    Ans. 78π.
30. $y^2 = x^3$, from $x = 0$ to $x = 8$.   713.
31. $4y = x^2$, from $y = 0$ to $y = 4$.   33. $4x^2 + y^2 = 64$.
32. $x^2 + 4y^2 = 16$.   34. $9x = y^3$, from $y = 0$ to $y = 3$.

Find the area of the surface generated by revolving each of the following curves.

                           About OX   *About OY*

35. The ellipse $\frac{x^2}{a^2} + \frac{y^2}{b^2} = 1$. $2\pi a^2 + \frac{\pi b^2}{e}\ln\frac{1+e}{1-e}$.

HINT. e = eccentricity of ellipse
$$= \frac{\sqrt{a^2 - b^2}}{a}.$$

36. The catenary $y = \frac{a}{2}\left(e^{\frac{x}{a}} + e^{-\frac{x}{a}}\right)$,

from $x = 0$ to $x = a$. $\frac{\pi a^2}{4}(e^2 + 4 - e^{-2})$. $2\pi a^2(1 - e^{-1})$.
(Figure, p. 532)

37. $x^4 + 3 = 6xy$, from $x = 1$ to $x = 2$. $\frac{47}{16}\pi$. $\pi(\frac{15}{4} + \ln 2)$.
38. $\begin{cases} x = e^\theta \sin\theta, \\ y = e^\theta \cos\theta, \end{cases}$ from $\theta = 0$ to $\frac{\pi}{2}$. $\frac{2\sqrt{2}\,\pi}{5}(e^\pi - 2)$. $\frac{2\sqrt{2}\,\pi}{5}(2e^\pi + 1)$.
39. $3x^2 + 4y^2 = 3a^2$. $\left(\frac{3}{2} + \frac{\pi}{\sqrt{3}}\right)\pi a^2$. $(4 + 3\ln 3)\frac{\pi a^2}{2}$.

40. The slope of the tractrix at any point of the curve in the first quadrant is given by $\frac{dy}{dx} = \frac{-y}{\sqrt{c^2 - y^2}}$. Show that the surface generated by revolving about OX the arc joining the points (x_1, y_1) and (x_2, y_2) on the tractrix is $2\pi c(y_1 - y_2)$. (Figure, p. 537)

41. The area in the first quadrant bounded by the curves whose equations are $y = x^3$ and $y = 4x$ is revolved about OX. Find the total surface of the solid generated.   Ans. 410.3.

INTEGRATION A PROCESS OF SUMMATION 283

42. The area bounded by the y-axis and the curves whose equations are $x^2 = 4y$ and $x - 2y + 4 = 0$ is revolved about OY. Find the total surface of the solid generated. *Ans.* 141.5.

43. Find the surface generated by revolving about OX the arc of the curve whose equation is $y = \dfrac{x^3}{6} + \dfrac{1}{2x}$ from $x = 1$ to $x = 3$. *Ans.* $\dfrac{208\,\pi}{9}$.

44. Find the entire surface of the solid generated by revolving about OX the area bounded by the two parabolas $y^2 = 4x$ and $y^2 = x + 3$.

Ans. $\tfrac{1}{6}\pi(17\sqrt{17} + 32\sqrt{2} - 17) = 51.53$.

45. Find the area of the surface generated by revolving about OX one arch of the curve $y = \sin x$. *Ans.* 14.42.

165. Solids with known parallel cross sections. In Art. 160 we discussed the volume of a solid of revolution, such as is shown in the accompanying figure. All cross sections in planes perpendicular to the x-axis are circles. If $OM = x$, $MC = y$, then

(1) Area cross section
$$ACBD = \pi y^2 = \pi[\phi(x)]^2,$$

if $y = \phi(x)$ is the equation of the generating curve OCG. *Hence the area of the cross section in any plane perpendicular to OX is a function of its perpendicular distance* $(= x)$ *from the point O.*

We shall now discuss the calculation of volumes of solids that are not solids of revolution when it is possible to express the area of any plane section of the solid which is perpendicular to a fixed line (as OX) as a function of its distance from a fixed point (as O).

Divide the solid into n slices by equidistant sections perpendicular to OX, each of thickness Δx.

Let FDE be one face of such a slice, and let $ON = x$. Then, by hypothesis,

(2) $\qquad$ Area $FDE = A(x)$.

The volume of this slice is equal, approximately, to

(3) $\qquad$ Area $FDE \times \Delta x = A(x)\Delta x$ (base $\times$ altitude)

Then $\sum_{i=1}^{n} A(x_i)\Delta x_i =$ sum of volumes of all such prisms. It is evident that the required volume is the limit of this sum; hence, by the Fundamental Theorem,

$$\lim_{n \to \infty} \sum_{i=1}^{n} A(x_i)\Delta x_i = \int A(x)dx,$$

and we have the formula

(N) $$V = \int A(x)dx,$$

where $A(x)$ is defined in (2).

The element of volume is a prism (in some cases a cylinder) whose altitude is dx and whose base has the area $A(x)$. That is,

$$dV = A(x)dx.$$

ILLUSTRATIVE EXAMPLE 1. The base of a solid is a circle of radius r. All sections perpendicular to a fixed diameter of the base are squares. Find the volume of the solid.

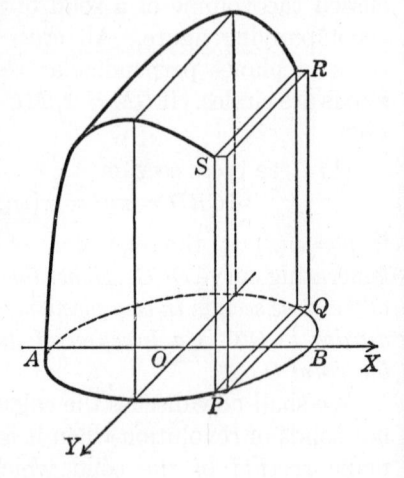

Solution. Take the circle $x^2 + y^2 = r^2$ in the XY-plane as base, and OX as the fixed diameter. Then the section $PQRS$ perpendicular to OX is a square of area $4y^2$, if $PQ = 2y$. (In the figure, the portion of the solid on the right of the section $PQRS$ is omitted.)

Hence $A(x) = 4y^2 = 4(r^2 - x^2)$, and, by (N),

Volume $= 4\int_{-r}^{r} (r^2 - x^2)dx = \tfrac{16}{3} r^3$. Ans.

ILLUSTRATIVE EXAMPLE 2. Find the volume of a right conoid with circular base, the radius of the base being r and the altitude a.

Solution. Placing the conoid as shown in the figure, consider a section PQR perpendicular to OX. This section is an isosceles triangle; and, since

$$RM = \sqrt{2rx - x^2}$$

(found by solving $x^2 + y^2 = 2rx$, the equation of the circle $ORAQ$, for y) and

$$MP = a,$$

the area of the section is

$$a\sqrt{2rx - x^2} = A(x).$$

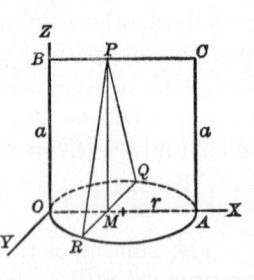

Substituting in (N),

$$V = a \int_0^{2r} \sqrt{2rx - x^2}\, dx = \frac{\pi r^2 a}{2}. \text{ Ans.}$$

This is one half the volume of the cylinder of the same base and altitude.

INTEGRATION A PROCESS OF SUMMATION

ILLUSTRATIVE EXAMPLE 3. Calculate the volume of the ellipsoid

$$\frac{x^2}{a^2} + \frac{y^2}{b^2} + \frac{z^2}{c^2} = 1$$

by a single integration.

Solution. Consider a section of the ellipsoid perpendicular to OX, as $ABCD$, with semiaxes b' and c'. The equation of the ellipse $HEJG$ in the XOY-plane is

$$\frac{x^2}{a^2} + \frac{y^2}{b^2} = 1.$$

Solving this for y $(= b')$ in terms of x $(= OM)$ gives

$$b' = \frac{b}{a}\sqrt{a^2 - x^2}.$$

Similarly, from the equation of the ellipse $EFGI$ in the XOZ-plane we get

$$c' = \frac{c}{a}\sqrt{a^2 - x^2}.$$

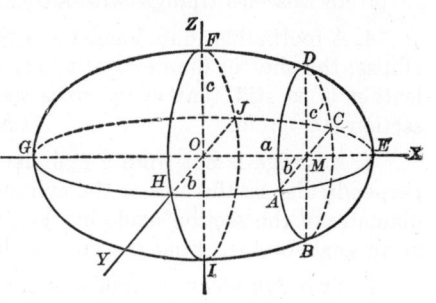

Hence the area of the ellipse (section) $ABCD$ is

$$\pi b'c' = \frac{\pi bc}{a^2}(a^2 - x^2) = A(x).$$

Substituting in (N),

$$V = \frac{\pi bc}{a^2} \int_{-a}^{+a} (a^2 - x^2) dx = \frac{4}{3}\pi abc. \text{ Ans.}$$

PROBLEMS

1. A solid has a circular base of radius r. The line AB is a diameter of the base. Find the volume of the solid if every plane section perpendicular to AB is

(a) an equilateral triangle;   Ans. $\frac{4}{3} r^3 \sqrt{3}$.
(b) an isosceles right triangle with its hypotenuse in the plane of the base;   Ans. $\frac{4}{3} r^3$.
(c) an isosceles right triangle with one leg in the plane of the base;   Ans. $\frac{8}{3} r^3$.
(d) an isosceles triangle with its altitude equal to 20 in.;   Ans. $10 \pi r^2$.
(e) an isosceles triangle with its altitude equal to its base. Ans. $\frac{8}{3} r^3$.

2. A solid has a base in the form of an ellipse with major axis 20 in. long and minor axis 10 in. long. Find the volume of the solid if every section perpendicular to the major axis is

(a) a square;   Ans. 1,333 cu. in.
(b) an equilateral triangle;   577.3 cu. in.
(c) an isosceles triangle with altitude 10 in.   785.4 cu. in.

3. The base of a solid is a segment of a parabola cut off by a chord perpendicular to its axis. The chord has a length of 16 in. and is distant 8 in. from the vertex of the parabola. Find the volume of the solid if every section perpendicular to the axis of the base is
(a) a square; *Ans.* 1024 cu. in.
(b) an equilateral triangle; 443.4 cu. in.
(c) an isosceles triangle with altitude 10 in. 426.7 cu. in.

4. A football is 16 in. long, and a plane section containing a seam is an ellipse, the shorter diameter of which is 8 in. Find the volume (a) if the leather is so stiff that every cross section is a square; (b) if the cross section is a circle. *Ans.* (a) $341\frac{1}{3}$ cu. in.; (b) 535.9 cu. in.

5. A wedge is cut from a cylinder of radius 5 in. by two planes, one perpendicular to the axis of the cylinder and the other passing through a diameter of the section made by the first plane and inclined to this plane at an angle of 45°. Find the volume of the wedge. *Ans.* $2\frac{50}{3}$ cu. in.

6. Two cylinders of equal radius r have their axes meeting at right angles. Find the volume of the common part. *Ans.* $\frac{16}{3} r^3$.

7. Find the volume of the solid generated by revolving about the x-axis the area bounded by the curves whose equations are
(a) $y = 6x - x^2$, $y = x$. *Ans.* $\frac{625}{3} \pi$.
(b) $y^2 = 4x$, $2x - y = 4$. $\frac{64}{3} \pi$.

8. A variable equilateral triangle moves with its plane perpendicular to the x-axis and the ends of its base on the points on the curves $y^2 = 16 ax$ and $y^2 = 4 ax$, respectively, above the x-axis. Find the volume generated by the triangle as it moves from the origin to the points whose abscissa is a. *Ans.* $\frac{1}{2}\sqrt{3}\, a^3$.

9. A rectangle moves from a fixed point, one side being always equal to the distance from this point, and the other equal to the square of this distance. What is the volume generated while the rectangle moves a distance of 2 ft.? *Ans.* 4 cu. ft.

10. On the double ordinates of the ellipse $\dfrac{x^2}{a^2} + \dfrac{y^2}{b^2} = 1$, isosceles triangles of vertical angle 90° are described in planes perpendicular to that of the ellipse. Find the volume of the solid generated by supposing such a variable triangle moving from one extremity to the other of the major axis of the ellipse. *Ans.* $\frac{4}{3} ab^2$.

Calculate the volumes bounded by the following quadric surfaces and the given planes.

11. $z = x^2 + 4y^2$; $z = 1$. *Ans.* $\frac{1}{4} \pi$.
12. $4x^2 + 9z^2 + y = 0$; $y + 1 = 0$. $\frac{1}{12} \pi$.
13. $x^2 + 4y^2 = 1 + z^2$; $z + 1 = 0$; $z - 1 = 0$. $\frac{4}{3} \pi$.
14. $25 y^2 + 9 z^2 = 1 + x^2$; $x = 0$; $x = 2$. $\frac{14}{45} \pi$.

INTEGRATION A PROCESS OF SUMMATION 287

15. $x^2 + 4y^2 + 9z^2 = 1$. $\qquad$ Ans. $\frac{2}{9}\pi$.

16. $z^2 = x^2 + 9y^2$; $z + 1 = 0$. $\qquad$ $\frac{1}{9}\pi$.

17. Given the parabola $z = 4 - x^2$ in the XZ-plane and the circle $x^2 + y^2 = 4$ in the XY-plane. From each point on the parabola lying above the circle two lines are drawn parallel to the YZ-plane to meet the circle. Calculate the volume of the wedge-shaped solid thus formed.

$\qquad$ Ans. 6π.

18. Find the volume of the solid bounded by the hyperboloid of one sheet $\dfrac{z^2}{c^2} = \dfrac{x^2}{a^2} - \dfrac{y^2}{b^2} + 1$ and the planes $x = 0$, $x = a$. $\qquad$ Ans. $\frac{4}{3}\pi abc$.

19. A solid is bounded by one nappe of the hyperboloid of two sheets $\dfrac{z^2}{c^2} = \dfrac{x^2}{a^2} - \dfrac{y^2}{b^2} - 1$ and the plane $x = 2a$. Find the volume. $\qquad$ Ans. $\frac{4}{3}\pi abc$.

20. Find the volume of the solid bounded by the surface

$$\frac{x^4}{a^4} + \frac{y^2}{b^2} + \frac{z^2}{c^2} = 1.$$ $\qquad$ Ans. $\frac{8}{5}\pi abc$.

ADDITIONAL PROBLEMS

1. Find the area of the loop of the curve

$$y^2 = (x+4)(x^2 - x + 2y - 4).$$ $\qquad$ Ans. $\frac{256}{15}$.

2. A point moves along a parabola in such a way that the radius joining it to the focus generates area at a constant rate. If the point moves from the vertex to one end of the latus rectum in 1 sec., what will be its position at the end of the next 8 sec.?

$\qquad$ Ans. Distance from focus $= \frac{5}{2}$ latus rectum.

3. Find the perimeter of the figure bounded by the line $y = 1$ and the curve $4y = e^{2x} + e^{-2x}$. $\qquad$ Ans. $\sqrt{3} + \ln(2 + \sqrt{3}) = 3.05$.

4. The arc OP of the curve $xy = x - y$ joins the origin to the point $P(x_1, y_1)$, and bounds with the x-axis and the line $x = x_1$ an area A. The same arc bounds with the y-axis and the line $y = y_1$ an area B. Prove that the volumes obtained by revolving A about the x-axis and B about the y-axis are equal.

5. The area bounded by the curve $16y^2 = (x+4)^3$ and its tangent at the point $(12, 16)$ is revolved about the x-axis. Find the volume inclosed. $\qquad$ Ans. $\frac{1024}{9}\pi$.

6. The base of a solid is the area bounded by the parabola $y^2 = 2px$ and its latus rectum. Every section of the solid made by a plane at right angles to the latus rectum is a rectangle whose altitude is equal to the distance of the section from the axis of the parabola. Find the volume of the solid. $\qquad$ Ans. $\frac{1}{4}p^3$.

288   INTEGRAL CALCULUS

7. Given the ellipse $9x^2 + 25y^2 = 225$. A solid is formed about this curve in such a way that all plane sections perpendicular to the x-axis are ellipses whose foci are on the given ellipse. The major and minor axes of each section are proportional to those of the given ellipse. Find the volume of the solid.   Ans. $\frac{225}{4}\pi$.

8. Let (x, y) be a point on the curve of Art. 159, O being the origin and OA the x-axis. Show that (D) may be written

(1)   Area $= \frac{1}{2}\int (x\, dy - y\, dx)$,

by using the transformation (5), p. 4. The limits are determined by the coördinates of the extremities of the curve.

9. Derive the formula of the preceding problem directly from a figure, making use of (B) and (C), Art. 158.

The formula (1) of Problem 8 is useful for parametric equations. Find the following areas by (1).

10. The area between the involute of a circle

$$x = r\cos\theta + r\theta\sin\theta, \quad y = r\sin\theta - r\theta\cos\theta$$

and the x-axis produced to the left, in the figure of Chapter XXVI, p. 537.

11. The entire area of the hypocycloid of three cusps

$$\begin{cases} x = 2r\cos\theta + r\cos 2\theta, \\ y = 2r\sin\theta - r\sin 2\theta. \end{cases}$$

(See figure.)   Ans. $2\pi r^2$.

12. A straight uniform wire attracts a particle P according to the law of gravitation. The particle is in the line of the wire but not in the wire. Prove that the wire attracts the particle as if the mass of the wire were concentrated at a point of the wire whose distance from P is the mean proportional of the distances from P to the ends of the wire.

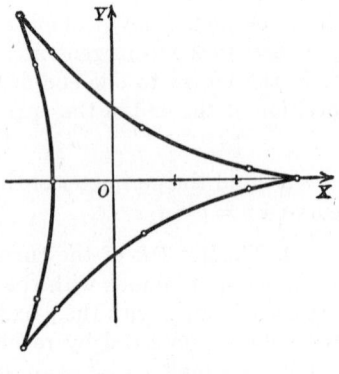

13. Find the area of the loop of the folium of Descartes, $x^3 + y^3 = 3axy$.

Hint. Let $y = tx$; then $x = \dfrac{3at}{1 + t^3}$,

$y = \dfrac{3at^2}{1 + t^3}$, and $dx = \dfrac{1 - 2t^3}{(1 + t^3)^2} 3a\, dt$.

The limits for t are 0 and ∞.

CHAPTER XVI

FORMAL INTEGRATION BY VARIOUS DEVICES

166. Introduction. Formal integration depends ultimately upon the use of a table of integrals. If, in a given case, no formula is found in the table resembling the given integral, it is often possible to transform the latter so as to make it depend upon formulas in the tables. The devices which may be used are

(a) *integration by parts* (Art. 136),
(b) *application of the theory of rational fractions*,
(c) *use of a suitable substitution*.

We proceed to discuss (b) and (c).

167. Integration of rational fractions. A rational fraction is a fraction the numerator and denominator of which are integral rational functions, that is, the variable is not affected with negative or fractional exponents. If the degree of the numerator is equal to or greater than that of the denominator, the fraction may be reduced to a mixed quantity by dividing the numerator by the denominator. For example,

$$\frac{x^4 + 3x^3}{x^2 + 2x + 1} = x^2 + x - 3 + \frac{5x + 3}{x^2 + 2x + 1}.$$

The last term is a fraction reduced to its lowest terms, having the degree of the numerator less than that of the denominator. It readily appears that the other terms are at once integrable, and hence we need consider only the fraction.

In order to integrate a differential expression involving such a fraction, it is often necessary to resolve it into simpler partial fractions, that is, to replace it by the algebraic sum of fractions of forms such that we can complete the integration. That this is always possible when the denominator can be broken up into its real prime factors is shown in algebra.*

Case I. *When the factors of the denominators are all of the first degree and none are repeated.*

---
*See Chapter XX in Hawkes's "Advanced Algebra" (Ginn and Company, Boston).

To each nonrepeated linear factor, such as $x - a$, there corresponds a partial fraction of the form

$$\frac{A}{x-a},$$

where A is a constant. The given fraction can be expressed as a sum of fractions of this form. The examples show the method.

ILLUSTRATIVE EXAMPLE. Find $\int \frac{(2x+3)dx}{x^3 + x^2 - 2x}$.

Solution. The factors of the denominator being x, $x-1$, $x+2$, we assume*

(1) $\quad \dfrac{2x+3}{x(x-1)(x+2)} = \dfrac{A}{x} + \dfrac{B}{x-1} + \dfrac{C}{x+2},$

where A, B, C are constants to be determined.
Clearing (1) of fractions, we get

(2) $\quad 2x+3 = A(x-1)(x+2) + B(x+2)x + C(x-1)x,$
$\quad 2x+3 = (A+B+C)x^2 + (A+2B-C)x - 2A.$

Since this equation is an identity, we equate the coefficients of the like powers of x in the two members according to the method of Undetermined Coefficients, and obtain three simultaneous equations

(3) $\quad \begin{cases} A + B + C = 0, \\ A + 2B - C = 2, \\ -2A = 3. \end{cases}$

Solving equations (3), we get

$$A = -\tfrac{3}{2}, \quad B = \tfrac{5}{3}, \quad C = -\tfrac{1}{6}.$$

Substituting these values in (1),

$$\frac{2x+3}{x(x-1)(x+2)} = -\frac{3}{2x} + \frac{5}{3(x-1)} - \frac{1}{6(x+2)}.$$

$$\therefore \int \frac{2x+3}{x(x-1)(x+2)} dx = -\frac{3}{2}\int \frac{dx}{x} + \frac{5}{3}\int \frac{dx}{x-1} - \frac{1}{6}\int \frac{dx}{x+2}$$

$$= -\tfrac{3}{2}\ln x + \tfrac{5}{3}\ln(x-1) - \tfrac{1}{6}\ln(x+2) + \ln c$$

$$= \ln \frac{c(x-1)^{\frac{5}{3}}}{x^{\frac{3}{2}}(x+2)^{\frac{1}{6}}}. \quad Ans.$$

A shorter method of finding the values of A, B, and C from (2) is the following:

Let factor $x = 0$; then $3 = -2A$, or $\quad A = -\tfrac{3}{2}.$
Let factor $x - 1 = 0$, or $x = 1$; then $5 = 3B$, or $\quad B = \tfrac{5}{3}.$
Let factor $x + 2 = 0$, or $x = -2$; then $-1 = 6C$, or $\quad C = -\tfrac{1}{6}.$

In every example in rational fractions the *number of constants to be determined is equal to the degree of the denominator.*

---

* In the process of decomposing the fractional part of the given differential neither the integral sign nor dx enters.

FORMAL INTEGRATION BY VARIOUS DEVICES 291

Case II. *When the factors of the denominator are all of the first degree and some are repeated.*

To every n-fold linear factor, such as $(x-a)^n$, there will correspond the sum of n partial fractions,

$$\frac{A}{(x-a)^n} + \frac{B}{(x-a)^{n-1}} + \cdots + \frac{L}{x-a},$$

in which $A, B, \cdots, L$ are constants. These partial fractions are readily integrated. For example,

$$\int \frac{A\,dx}{(x-a)^n} = A \int (x-a)^{-n}\,dx = \frac{A}{(1-n)(x-a)^{n-1}} + C.$$

ILLUSTRATIVE EXAMPLE. Find $\int \frac{x^3+1}{x(x-1)^3}\,dx$.

Solution. Since $x-1$ occurs three times as a factor, we assume

$$\frac{x^3+1}{x(x-1)^3} = \frac{A}{x} + \frac{B}{(x-1)^3} + \frac{C}{(x-1)^2} + \frac{D}{x-1}.$$

Clearing of fractions,

$x^3 + 1 = A(x-1)^3 + Bx + Cx(x-1) + Dx(x-1)^2.$
$x^3 + 1 = (A+D)x^3 + (-3A+C-2D)x^2 + (3A+B-C+D)x - A.$

Equating the coefficients of like powers of x, we get the simultaneous equations

$$A + D = 1,$$
$$-3A + C - 2D = 0,$$
$$3A + B - C + D = 0,$$
$$-A = 1.$$

Solving, $A = -1$, $B = 2$, $C = 1$, $D = 2$, and

$$\frac{x^3+1}{x(x-1)^3} = -\frac{1}{x} + \frac{2}{(x-1)^3} + \frac{1}{(x-1)^2} + \frac{2}{x-1}.$$

$$\therefore \int \frac{x^3+1}{x(x-1)^3}\,dx = -\ln x - \frac{1}{(x-1)^2} - \frac{1}{x-1} + 2\ln(x-1) + C$$

$$= -\frac{x}{(x-1)^2} + \ln \frac{(x-1)^2}{x} + C. \text{ Ans.}$$

PROBLEMS

Work out the following integrals.

1. $\int \frac{(4x-2)dx}{x^3 - x^2 - 2x} = \ln \frac{x^2 - 2x}{(x+1)^2} + C.$

2. $\int \frac{(5x^2 - 3)dx}{x^3 - x} = \ln x^3(x^2 - 1) + C.$

3. $\int \frac{(4x+3)dx}{4x^3 + 8x^2 + 3x} = -\frac{1}{2}\ln \frac{(2x+1)(2x+3)}{x^2} + C.$

292 INTEGRAL CALCULUS

4. $\int \dfrac{(4\,x^3 + 2\,x^2 + 1)dx}{4\,x^3 - x} = x + \dfrac{1}{2}\ln\dfrac{(2\,x+1)(2\,x-1)^2}{x^2} + C.$

5. $\int \dfrac{(3\,x^2 + 5\,x)dx}{(x-1)(x+1)^2} = \ln\,(x+1)(x-1)^2 - \dfrac{1}{x+1} + C.$

6. $\int \dfrac{z^2\,dz}{(z-1)^3} = \ln\,(z-1) - \dfrac{2}{z-1} - \dfrac{1}{2(z-1)^2} + C.$

7. $\int \dfrac{(y^4 - 8)dy}{y^3 + 2\,y^2} = \dfrac{y^2}{2} - 2\,y + \dfrac{4}{y} + 2\ln\,(y^2 + 2\,y) + C.$

8. $\displaystyle\int_1^2 \dfrac{(x-3)dx}{x^3 + x^2} = 4\ln\dfrac{4}{3} - \dfrac{3}{2} = -\,0.3492.$

9. $\displaystyle\int_2^4 \dfrac{(x^3 - 2)dx}{x^3 - x^2} = \dfrac{5}{2} + \ln\dfrac{4}{3} = 2.7877.$

10. $\displaystyle\int_1^3 \dfrac{(2 - x^2)dx}{x^3 + 3\,x^2 + 2\,x} = \ln\dfrac{9}{10} = -\,0.1054.$

11. $\displaystyle\int_2^3 \dfrac{(3 - x)dx}{x^3 + 4\,x^2 + 3\,x} = \ln\dfrac{81}{80} = 0.0125.$

12. $\displaystyle\int_0^1 \dfrac{(3\,x^2 + 7\,x)dx}{(x+1)(x+2)(x+3)} = \ln\dfrac{4}{3} = 0.2877.$

13. $\displaystyle\int_0^5 \dfrac{(x^2 - 3)dx}{(x+2)(x+1)^2} = \ln\dfrac{7}{2} - \dfrac{5}{3} = -\,0.4139.$

14. $\displaystyle\int_0^4 \dfrac{9\,x^2\,dx}{(2\,x+1)(x+2)^2} = 5\ln 3 - 4 = 1.4930.$

Work out each of the following integrals.

15. $\int \dfrac{8\,dx}{x^3 - 4\,x}.$

16. $\int \dfrac{(5\,x^2 - 9)dx}{x^3 - 9\,x}.$

17. $\int \dfrac{(3\,z + 7)dz}{(z+1)(z+2)(z+3)}.$

18. $\int \dfrac{(3\,x^2 + 11\,x + 2)dx}{(x+3)(x^2 - 1)}.$

19. $\int \dfrac{x^2\,dx}{(2\,x+3)(4\,x^2 - 1)}.$

20. $\int \dfrac{(t^4 + 1)dt}{t^3 - t}.$

21. $\int \dfrac{(x^2 - x - 5)dx}{x^3 + 5\,x^2}.$

22. $\int \dfrac{(5\,x^2 + 14\,x + 10)dx}{(x+2)(x+1)^2}.$

23. $\int \dfrac{(24\,y^2 + 10\,y + 5)dy}{(2\,y-1)(2\,y+1)^2}.$

24. $\int \dfrac{(x+2)dx}{x^4 + 2\,x^3 + x^2}.$

25. $\int \dfrac{(x^3 - 2\,x - 4)dx}{x^4 + 2\,x^3}.$

26. $\int \dfrac{(2\,x^2 + 1)dx}{(x-2)^3}.$

27. $\int \dfrac{(y^4 - 3\,y^3)dy}{(y^2 - 1)(y - 2)}.$

28. $\int \dfrac{(2\,t^4 + 3\,t^3 - 20\,t - 28)dt}{(t^2 - 4)(2\,t - 1)}.$

Case III. *When the denominator contains factors of the second degree but none are repeated.*

FORMAL INTEGRATION BY VARIOUS DEVICES

To every nonrepeated quadratic factor, such as $x^2 + px + q$, there corresponds a partial fraction of the form

$$\frac{Ax + B}{x^2 + px + q}.$$

The method of integration of this term is explained on page 209 (Illustrative Example 2).

If p is not zero, we complete the square in the denominator,

$$x^2 + px + \tfrac{1}{4}p^2 + q - \tfrac{1}{4}p^2 = (x + \tfrac{1}{2}p)^2 + \tfrac{1}{4}(4q - p^2). \quad (4q > p^2)$$

Let $x + \tfrac{1}{2}p = u$. Then $x = u - \tfrac{1}{2}p$, $dx = du$. Substituting these values, the new integral in terms of the variable u is readily integrated.

ILLUSTRATIVE EXAMPLE 1. Find $\int \dfrac{4 \, dx}{x^3 + 4x}$.

Solution. Assume $\dfrac{4}{x(x^2 + 4)} = \dfrac{A}{x} + \dfrac{Bx + C}{x^2 + 4}.$

Clearing of fractions, $4 = A(x^2 + 4) + x(Bx + C) = (A + B)x^2 + Cx + 4A.$
Equating the coefficients of like powers of x, we get

$$A + B = 0, \quad C = 0, \quad 4A = 4.$$

This gives $A = 1$, $B = -1$, $C = 0$, so that $\dfrac{4}{x(x^2 + 4)} = \dfrac{1}{x} - \dfrac{x}{x^2 + 4}.$

$$\therefore \int \frac{4 \, dx}{x(x^2 + 4)} = \int \frac{dx}{x} - \int \frac{x \, dx}{x^2 + 4}$$

$$= \ln x - \frac{1}{2} \ln (x^2 + 4) + \ln c = \ln \frac{cx}{\sqrt{x^2 + 4}}. \quad Ans.$$

ILLUSTRATIVE EXAMPLE 2. Prove

$$\int \frac{dx}{x^3 + 8} = \frac{1}{24} \ln \frac{(x + 2)^2}{x^2 - 2x + 4} + \frac{1}{12} \sqrt{3} \arctan \frac{x - 1}{\sqrt{3}} + C.$$

Solution. Factoring, $x^3 + 8 = (x + 2)(x^2 - 2x + 4)$.

$$\frac{1}{x^3 + 8} = \frac{Ax + B}{x^2 - 2x + 4} + \frac{C}{x + 2},$$

$$1 = (Ax + B)(x + 2) + C(x^2 - 2x + 4),$$

$$1 = (A + C)x^2 + (2A + B - 2C)x + 2B + 4C.$$

Then $\quad A = -\tfrac{1}{12}, \ B = \tfrac{1}{3}, \ C = \tfrac{1}{12}.$

Hence $\quad \int \dfrac{dx}{x^3 + 8} = \int \dfrac{-\tfrac{1}{12}x + \tfrac{1}{3}}{x^2 - 2x + 4} \, dx + \int \dfrac{\tfrac{1}{12} dx}{x + 2}.$

(4) $\quad = \dfrac{1}{12} \int \dfrac{4 - x}{x^2 - 2x + 4} \, dx + \dfrac{1}{12} \ln (x + 2) + C.$

Now $x^2 - 2x + 4 = (x - 1)^2 + 3 = u^2 + 3$, if $x - 1 = u$.
Then $x = u + 1$, $dx = du$, and

$$\int \frac{4 - x}{x^2 - 2x + 4} \, dx = \int \frac{3 - u}{u^2 + 3} \, du = \sqrt{3} \arctan \frac{u}{\sqrt{3}} - \frac{1}{2} \ln (u^2 + 3).$$

Substituting back $u = x - 1$, using (4), and reducing, we have the answer.

Case IV. *When the denominator contains factors of the second degree some of which are repeated.*

To every n-fold quadratic factor, such as $(x^2 + px + q)^n$, there will correspond the sum of n partial fractions,

$$\frac{Ax + B}{(x^2 + px + q)^n} + \frac{Cx + D}{(x^2 + px + q)^{n-1}} + \cdots + \frac{Lx + M}{x^2 + px + q}.$$

To carry out the integration, the "reduction formula"

(5) $\quad \displaystyle\int \frac{du}{(u^2 + a^2)^n} = \frac{1}{2(n-1)a^2}\left[\frac{u}{(u^2 + a^2)^{n-1}} + (2n - 3)\int \frac{du}{(u^2 + a^2)^{n-1}}\right],$

proved in the next chapter, is necessary. If $n > 2$, repeated applications of (5) are necessary. If p is not zero, we complete the square, $x^2 + px + q = (x + \frac{1}{2}p)^2 + \frac{1}{4}(4q - p^2) = u^2 + a^2$, etc., as before.

ILLUSTRATIVE EXAMPLE. Prove

$$\int \frac{2x^3 + x + 3}{(x^2 + 1)^2} dx = \ln(x^2 + 1) + \frac{1 + 3x}{2(x^2 + 1)} + \frac{3}{2} \arctan x + C.$$

Solution. Since $x^2 + 1$ occurs twice as a factor, we assume

$$\frac{2x^3 + x + 3}{(x^2 + 1)^2} = \frac{Ax + B}{(x^2 + 1)^2} + \frac{Cx + D}{x^2 + 1}.$$

Clearing of fractions,

$$2x^3 + x + 3 = Ax + B + (Cx + D)(x^2 + 1).$$

Equating the coefficients of like powers of x and solving, we get

$$A = -1, \quad B = 3, \quad C = 2, \quad D = 0.$$

Hence $\displaystyle\int \frac{2x^3 + x + 3}{(x^2 + 1)^2} dx = \int \frac{-x + 3}{(x^2 + 1)^2} dx + \int \frac{2x\, dx}{x^2 + 1}$

$$= \ln(x^2 + 1) - \int \frac{x\, dx}{(x^2 + 1)^2} + 3\int \frac{dx}{(x^2 + 1)^2}.$$

The first of these two integrals is worked out by the power formula (4), the second by (5) above, with $u = x$, $a = 1$, $n = 2$. Thus we obtain

$$\int \frac{2x^3 + x + 3}{(x^2 + 1)^2} dx = \ln(x^2 + 1) + \frac{1}{2(x^2 + 1)} + \frac{3}{2}\left[\frac{x}{x^2 + 1} + \arctan x\right] + C.$$

Reducing, we have the answer.

Conclusion. Since a rational function may always be reduced to the quotient of two integral rational functions, that is, to a rational fraction, it follows from the above discussion that any rational function whose denominator can be broken up into real quadratic and linear factors may be expressed as the algebraic sum of integral rational functions and partial fractions. The terms of this sum have forms all of which we have shown how to integrate. Hence the

FORMAL INTEGRATION BY VARIOUS DEVICES   295

Theorem. *The integral of every rational fraction whose denominator can be broken up into real quadratic and linear factors may be found, and is expressible in terms of algebraic, logarithmic, and inverse trigonometric functions, that is, in terms of the elementary functions.*

PROBLEMS

Work out the following integrals.

1. $\int \dfrac{(4\,x^2 + 6)dx}{x^3 + 3\,x} = \ln x^2(x^2 + 3) + C.$

2. $\int \dfrac{(x^2 + x)dx}{(x-1)(x^2+1)} = \ln(x-1) + \arctan x + C.$

3. $\int \dfrac{(2\,t^2 - 8\,t - 8)dt}{(t-2)(t^2+4)} = 2\ln\dfrac{t^2+4}{t-2} + C.$

4. $\int \dfrac{(x^2 + x - 10)dx}{(2\,x - 3)(x^2 + 4)} = \dfrac{1}{2}\ln\dfrac{x^2+4}{2\,x-3} + \arctan\dfrac{x}{2} + C.$

5. $\int \dfrac{(x-18)dx}{4\,x^3 + 9\,x} = \ln\dfrac{4\,x^2+9}{x^2} + \dfrac{1}{6}\arctan\dfrac{2\,x}{3} + C.$

6. $\int \dfrac{(2\,y^3 + y^2 + 2\,y + 2)dy}{y^4 + 3\,y^2 + 2} = \ln(y^2+2) + \arctan y + C.$

7. $\int \dfrac{dz}{z^4 + z^2} = -\dfrac{1}{z} - \arctan z + C.$

8. $\int \dfrac{2\,x\,dx}{(x^2+1)(x+1)^2} = \arctan x + \dfrac{1}{x+1} + C.$

9. $\int \dfrac{(x^3 + 3\,x)dx}{(x^2+1)^2} = \dfrac{1}{2}\ln(x^2+1) - \dfrac{1}{x^2+1} + C.$

10. $\int \dfrac{(x^5 + 9\,x^3 - 9\,x^2 - 9)dx}{x^3 + 9\,x} = \dfrac{x^3}{3} - \ln x(x^2+9)^4 + C.$

11. $\int \dfrac{(4\,x^2 + 2\,x + 8)dx}{x(x^2+2)^2} = \ln\dfrac{x^2}{x^2+2} + \dfrac{x}{2\,x^2+4} + \dfrac{\sqrt{2}}{4}\arctan\dfrac{x}{\sqrt{2}} + C.$

12. $\int \dfrac{t^5\,dt}{(t^2+4)^2} = \dfrac{t^2}{2} - 4\ln(t^2+4) - \dfrac{8}{t^2+4} + C.$

13. $\int \dfrac{dx}{x^3 + x^2 + x} = -\dfrac{1}{2}\ln\dfrac{x^2+x+1}{x^2} - \dfrac{\sqrt{3}}{3}\arctan\dfrac{2\,x+1}{\sqrt{3}} + C.$

14. $\int \dfrac{(x^5 + 4\,x^3)dx}{(x^2+2)^3} = \dfrac{1}{2}\ln(x^2+2) + \dfrac{1}{(x^2+2)^2} + C.$

15. $\int \dfrac{4\,dx}{x^4 - 1} = \ln\dfrac{x-1}{x+1} - 2\arctan x + C.$

16. $\int \dfrac{(2\,z^2 + 3\,z + 2)dz}{(z+2)(z^2+2\,z+2)} = 2\ln(z+2) - \arctan(z+1) + C.$

INTEGRAL CALCULUS

17. $\int \left(\dfrac{t+3}{t^2+4t+5}\right)^2 dt = \text{arc tan } (t+2) - \dfrac{1}{t^2+4t+5} + C.$

18. $\displaystyle\int_1^4 \dfrac{(5x^2+4)dx}{x^3+4x} = 3\ln 4 = 4.1589.$

19. $\displaystyle\int_0^1 \dfrac{5x\,dx}{(x+2)(x^2+1)} = \ln\dfrac{8}{9} + \dfrac{\pi}{4} = 0.667.$

20. $\displaystyle\int_0^1 \dfrac{(2x^2+x+3)dx}{(x+1)(x^2+1)} = \ln 4 + \dfrac{\pi}{4} = 2.171.$

21. $\displaystyle\int_0^1 \dfrac{(4x^2+2x)dx}{(x^2+1)(x+1)^2} = \dfrac{\pi}{4} + \dfrac{1}{2} - \ln 2 = 0.592.$

22. $\displaystyle\int_3^4 \dfrac{(5t^3-4t)dt}{t^4-16} = \ln\dfrac{12}{5} + \dfrac{3}{2}\ln\dfrac{20}{13} = 1.522.$

23. $\displaystyle\int_0^2 \dfrac{(z^3+2z^2+6z+8)dz}{(z^2+4)^2} = \dfrac{1}{2}\ln 2 + \dfrac{\pi}{4} + \dfrac{1}{8} = 1.257.$

Work out each of the following integrals.

24. $\displaystyle\int \dfrac{(6x^2+3x+4)dx}{x^3+2x}.$

25. $\displaystyle\int \dfrac{(z^4+3)dz}{(z+1)(z^2+1)}.$

26. $\displaystyle\int \dfrac{(3x^3+3x+1)dx}{x^4+3x^2}.$

27. $\displaystyle\int \dfrac{(3x^3+x^2+3)dx}{x^4+3x^2}.$

28. $\displaystyle\int \dfrac{(5x^2+12x+9)dx}{x^3+3x^2+3x}.$

29. $\displaystyle\int \dfrac{(4x^3+3x^2+18x+12)dx}{(x^2+4)^2}.$

30. $\displaystyle\int_0^{\frac{1}{2}} \dfrac{8y\,dy}{(2y+1)(4y^2+1)}.$

31. $\displaystyle\int_0^1 \dfrac{(2x^3-4)dx}{(x^2+1)(x+1)^2}.$

32. $\displaystyle\int_1^3 \dfrac{(x+10)dx}{x^3+2x^2+5x}.$

33. $\displaystyle\int_0^3 \dfrac{(2x^3+18)dx}{(x+3)(x^2+9)}.$

168. Integration by substitution of a new variable; rationalization. In the last article it was shown that all rational functions whose denominators can be broken up into real quadratic and linear factors may be integrated. Of algebraic functions which are *not rational*, that is, such as contain radicals, only a small number, relatively speaking, can be integrated in terms of elementary functions. By substituting a new variable, however, these functions can in some cases be transformed into equivalent functions that are either in the list of standard forms (Art. 128) or else rational. The method of integrating a function that is not rational by substituting for the old variable such a function of a new variable that the result is a rational function is sometimes called *integration by rationalization*. This is a very important artifice in integration, and we shall now take up some of the more important cases coming under this head.

FORMAL INTEGRATION BY VARIOUS DEVICES

Differentials containing fractional powers of x only. *Such an expression can be transformed into a rational form by means of the substitution*

$$x = z^n,$$

where n is the least common denominator of the fractional exponents of x.

For x, dx, and each radical can then be expressed rationally in terms of z.

ILLUSTRATIVE EXAMPLE 1. Prove $\displaystyle\int \frac{x^{\frac{1}{2}}\,dx}{1+x^{\frac{3}{4}}} = \frac{4}{3}x^{\frac{3}{4}} - \frac{4}{3}\ln\left(1+x^{\frac{3}{4}}\right)+C.$

Solution. Here $n = 4$. Hence let $x = z^4$.

Then $\qquad x^{\frac{1}{2}} = z^2, \quad x^{\frac{3}{4}} = z^3, \quad dx = 4\,z^3\,dz.$

Then $\displaystyle\int \frac{x^{\frac{1}{2}}\,dx}{1+x^{\frac{3}{4}}} = \int \frac{z^2}{1+z^3}\,4\,z^3\,dz = 4\int \frac{z^5}{1+z^3}\,dz$

$\qquad\qquad = 4\int\left(z^2 - \frac{z^2}{1+z^3}\right)dz = \frac{4}{3}z^3 - \frac{4}{3}\ln(1+z^3)+C.$

Substituting back $z = x^{\frac{1}{4}}$, we have the answer.

The general form of the irrational expression here treated is then

$$R\bigl(x^{\frac{1}{n}}\bigr)dx,$$

where R denotes a rational function of $x^{\frac{1}{n}}$.

Differentials containing fractional powers of $a+bx$ only. *Such an expression can be transformed into a rational form by means of the substitution*
$\qquad\qquad a + bx = z^n,$

where n is the least common denominator of the fractional exponents of the expression $a+bx$.

For x, dx, and each radical can then be expressed rationally in terms of z.

ILLUSTRATIVE EXAMPLE 2. Find $\displaystyle\int \frac{dx}{(1+x)^{\frac{3}{2}} + (1+x)^{\frac{1}{2}}}.$

Solution. Assume $\quad 1 + x = z^2.$

Then $\qquad dx = 2\,z\,dz, \quad (1+x)^{\frac{3}{2}} = z^3, \quad \text{and} \quad (1+x)^{\frac{1}{2}} = z.$

$\therefore \displaystyle\int \frac{dx}{(1+x)^{\frac{3}{2}} + (1+x)^{\frac{1}{2}}} = \int \frac{2\,z\,dz}{z^3+z} = 2\int \frac{dz}{z^2+1}$

$\qquad\qquad = 2\arctan z + C = 2\arctan(1+x)^{\frac{1}{2}} + C.$

when we substitute back the value of z in terms of x.

The general integral treated here has then the form

$$R\bigl[x,\ (a+bx)^{\frac{1}{n}}\bigr]dx,$$

where R denotes a rational function.

PROBLEMS

Work out the following integrals.

1. $\int \dfrac{(5x+9)dx}{(x-9)x^{\frac{3}{2}}} = \dfrac{2}{\sqrt{x}} + 2\ln\dfrac{\sqrt{x}-3}{\sqrt{x}+3} + C.$

2. $\int \dfrac{\sqrt{x}\,dx}{x^3+2x^2-3x} = \dfrac{1}{4}\ln\dfrac{\sqrt{x}-1}{\sqrt{x}+1} - \dfrac{\sqrt{3}}{6}\arctan\sqrt{\dfrac{x}{3}} + C.$

3. $\int \dfrac{dx}{x-x^{\frac{4}{3}}} = 3\ln\dfrac{x^{\frac{1}{3}}}{1-x^{\frac{1}{3}}} + C.$

4. $\int \dfrac{(x^{\frac{3}{2}}-x^{\frac{1}{3}})dx}{6x^{\frac{1}{4}}} = \dfrac{2}{27}x^{\frac{9}{4}} - \dfrac{2}{13}x^{\frac{13}{12}} + C.$

5. $\int \dfrac{x^2\,dx}{(4x+1)^{\frac{5}{2}}} = \dfrac{6x^2+6x+1}{12(4x+1)^{\frac{3}{2}}} + C.$

6. $\int \dfrac{dx}{x^{\frac{5}{8}}-x^{\frac{1}{8}}} = \dfrac{8x^{\frac{3}{8}}}{3} + 2\ln\dfrac{x^{\frac{1}{8}}-1}{x^{\frac{1}{8}}+1} + 4\arctan x^{\frac{1}{8}} + C.$

7. $\int \dfrac{x\,dx}{(a+bx)^{\frac{3}{2}}} = \dfrac{2(2a+bx)}{b^2\sqrt{a+bx}} + C.$

8. $\int y\sqrt[3]{a+y}\,dy = \dfrac{3}{28}(4y-3a)(a+y)^{\frac{4}{3}} + C.$

9. $\int \dfrac{(\sqrt{x+1}+1)dx}{\sqrt{x+1}-1} = x+1+4\sqrt{x+1}+4\ln(\sqrt{x+1}-1) + C.$

10. $\int \dfrac{dx}{1+\sqrt[3]{x+a}} = \dfrac{3}{2}(x+a)^{\frac{2}{3}} - 3(x+a)^{\frac{1}{3}} + 3\ln(1+\sqrt[3]{x+a}) + C.$

11. $\int \dfrac{(t+5)dt}{(t+4)\sqrt{t+2}} = 2\sqrt{t+2} + \sqrt{2}\arctan\sqrt{\dfrac{t+2}{2}} + C.$

12. $\int_0^3 \dfrac{dx}{(x+2)\sqrt{x+1}} = 2\arctan 2 - \dfrac{\pi}{2}.$

13. $\int_0^4 \dfrac{dx}{1+\sqrt{x}} = 4 - 2\ln 3.$

14. $\int_1^4 \dfrac{y\,dy}{\sqrt{2+4y}} = \dfrac{3}{2}\sqrt{2}.$

15. $\int_0^{\frac{1}{2}} \dfrac{dt}{\sqrt{2t}(9+\sqrt[3]{2t})} = 3 - 9\arctan\dfrac{1}{3}.$

16. $\int_0^1 \dfrac{x^{\frac{3}{2}}\,dx}{x+1} = \dfrac{\pi}{2} - \dfrac{4}{3}.$

17. $\int_1^{64} \dfrac{dt}{2\sqrt{t}+\sqrt[3]{t}} = 5.31.$

18. $\int_3^{29} \dfrac{(x-2)^{\frac{2}{3}}dx}{(x-2)^{\frac{2}{3}}+3} = 8 + \dfrac{3}{2}\pi\sqrt{3}.$

FORMAL INTEGRATION BY VARIOUS DEVICES 299

Work out each of the following integrals.

19. $\int \dfrac{dx}{x + 2\sqrt{x} + 5}$.

20. $\int \dfrac{dx}{x(1 - \sqrt[3]{x})}$.

21. $\int \dfrac{(x + 2) dx}{x\sqrt{x - 3}}$.

22. $\int \dfrac{y\, dy}{(2y + 3)^{\frac{4}{3}}}$.

23. $\int \dfrac{dt}{(t+1)^{\frac{1}{4}} - (t+1)^{\frac{5}{4}}}$.

24. $\int \dfrac{dx}{(x-2)^{\frac{1}{2}} - (x-2)^{\frac{3}{4}}}$.

25. $\int \dfrac{(x+3) dx}{(x+5)\sqrt{x+4}}$.

26. $\int \dfrac{(2 - \sqrt{2x+3})\, dx}{1 - 2x}$.

27. Find the area bounded by the curve $y = x + \sqrt{x+1}$, the x-axis, and the ordinates $x = 3$ and $x = 8$. Ans. $40\tfrac{1}{6}$.

28. Find the volume generated by revolving about the x-axis the area of the preceding problem.

29. Find the volume generated by revolving about the x-axis the area in the first quadrant bounded by the coördinate axes and each of the following curves.

(a) $y = 2 - \sqrt{x}$.
(b) $y = 2 - \sqrt[3]{x}$.
(c) $y = a - \sqrt{ax}$.
(d) $y = 4 - x^{\frac{2}{3}}$.

30. Find the area bounded by the curves $y = 2x + \sqrt{2x+1}$ and $y = x - \sqrt{2x+1}$ and the ordinates $x = 4$ and $x = 12$.

31. Find the area bounded by the curve
$$(x-1)y^2 = (x+1)(2y-1)$$
and the ordinates $x = 3$ and $x = 8$. Ans. $4\left[\sqrt{2} + \ln \dfrac{4 + \sqrt{2}}{4 - \sqrt{2}}\right]$.

169. Binomial differentials. A differential of the form
(1) $\qquad x^m (a + bx^n)^p\, dx$,
where a and b are any constants and the exponents m, n, p are rational numbers, is called a *binomial differential*.

Let $\qquad x = z^\alpha;\quad$ then $\quad dx = \alpha z^{\alpha - 1}\, dz$,

and $\qquad x^m (a + bx^n)^p\, dx = \alpha z^{m\alpha + \alpha - 1}(a + bz^{n\alpha})^p\, dz$.

If an integer α be chosen such that $m\alpha$ and $n\alpha$ are also integers,* we see that the given differential is equivalent to another of the same form where m and n have been replaced by integers. Also,

(2) $\qquad x^m(a + bx^n)^p\, dx = x^{m + np}(ax^{-n} + b)^p\, dx$

---

* It is always possible to choose α so that $m\alpha$ and $n\alpha$ are integers, for we can take α as the L.C.M. of the denominators of m and n.

transforms the given differential into another of the same form where the exponent n of x has been replaced by $-n$. Therefore, no matter what the algebraic sign of n may be, in one of the two differentials the exponent of x inside the parentheses will surely be positive.

When p is a positive integer the binomial may be expanded and the differential integrated termwise. In what follows p is regarded as a fraction; hence we replace it by $\dfrac{r}{s}$, where r and s are integers.*

We may then make the following statement.

Every binomial differential may be reduced to the form

$$x^m(a + bx^n)^{\frac{r}{s}}dx,$$

where m, n, r, s are integers and n is positive.

In the next section we prove that (1) can be rationalized under the following conditions.

CASE I. When $\dfrac{m+1}{n} =$ an integer or zero, by assuming

$$a + bx^n = z^s.$$

CASE II. When $\dfrac{m+1}{n} + \dfrac{r}{s} =$ an integer or zero, by assuming

$$a + bx^n = z^s x^n.$$

ILLUSTRATIVE EXAMPLE 1. $\displaystyle\int \dfrac{x^3\, dx}{(a + bx^2)^{\frac{3}{2}}} = \int x^3 (a + bx^2)^{-\frac{3}{2}} dx$

$$= \dfrac{1}{b^2}\dfrac{2a + bx^2}{\sqrt{a + bx^2}} + C.$$

Solution. $m = 3$, $n = 2$, $r = -3$, $s = 2$; and here $\dfrac{m+1}{n} = 2$, an integer. Hence this comes under Case I, and we assume

$$a + bx^2 = z^2;\ \text{whence}\ x = \left(\dfrac{z^2 - a}{b}\right)^{\frac{1}{2}},\ dx = \dfrac{z\, dz}{b^{\frac{1}{2}}(z^2 - a)^{\frac{1}{2}}},\ \text{and}\ (a + bx^2)^{\frac{3}{2}} = z^3.$$

$$\therefore \int \dfrac{x^3\, dx}{(a + bx^2)^{\frac{3}{2}}} = \int \left(\dfrac{z^2 - a}{b}\right)^{\frac{3}{2}} \cdot \dfrac{z\, dz}{b^{\frac{1}{2}}(z^2 - a)^{\frac{1}{2}}} \cdot \dfrac{1}{z^3}$$

$$= \dfrac{1}{b^2}\int (1 - az^{-2}) dz = \dfrac{1}{b^2}(z + az^{-1}) + C$$

$$= \dfrac{1}{b^2}\dfrac{2a + bx^2}{\sqrt{a + bx^2}} + C.$$

ILLUSTRATIVE EXAMPLE 2. $\displaystyle\int \dfrac{dx}{x^4\sqrt{1 + x^2}} = -\dfrac{(2x^2 - 1)(1 + x^2)^{\frac{1}{2}}}{3x^3} + C.$

Solution. $m = -4$, $n = 2$, $\dfrac{r}{s} = -\dfrac{1}{2}$; and here $\dfrac{m+1}{n} + \dfrac{r}{s} = -2$, an integer.

* The case where p is an integer is not excluded, but appears as a special case, namely, $r = p$, $s = 1$.

FORMAL INTEGRATION BY VARIOUS DEVICES

Hence this comes under Case II, and we assume

$$1 + x^2 = z^2 x^2, \quad z = \frac{(1+x^2)^{\frac{1}{2}}}{x};$$

whence $\quad x^2 = \dfrac{1}{z^2 - 1}, \quad 1 + x^2 = \dfrac{z^2}{z^2 - 1}, \quad \sqrt{1 + x^2} = \dfrac{z}{(z^2 - 1)^{\frac{1}{2}}};$

also $\quad x = \dfrac{1}{(z^2 - 1)^{\frac{1}{2}}}, \quad x^4 = \dfrac{1}{(z^2 - 1)^2}; \quad$ and $\quad dx = -\dfrac{z\,dz}{(z^2 - 1)^{\frac{3}{2}}}.$

$$\therefore \int \frac{dx}{x^4 \sqrt{1 + x^2}} = -\int \frac{\dfrac{z\,dz}{(z^2-1)^{\frac{3}{2}}}}{\dfrac{1}{(z^2-1)^2} \cdot \dfrac{z}{(z^2-1)^{\frac{1}{2}}}} = -\int (z^2 - 1)\,dz$$

$$= z - \frac{z^3}{3} + C = \frac{(2x^2 - 1)(1 + x^2)^{\frac{1}{2}}}{3\,x^3} + C.$$

PROBLEMS

Work out the following integrals.

1. $\int x^5 \sqrt{1 + x^3}\,dx = \dfrac{2(3\,x^3 - 2)(1 + x^3)^{\frac{3}{2}}}{45} + C.$

2. $\int \dfrac{x^5\,dx}{\sqrt{1 + x^3}} = \dfrac{2(x^3 - 2)\sqrt{1 + x^3}}{9} + C.$

3. $\int x^5 (8 + x^3)^{\frac{3}{2}}\,dx = \dfrac{2(5\,x^3 - 16)(8 + x^3)^{\frac{5}{2}}}{105} + C.$

4. $\int \dfrac{x^5\,dx}{(a + bx^3)^{\frac{3}{2}}} = \dfrac{2(2\,a + bx^3)}{3\,b^2 \sqrt{a + bx^3}} + C.$

5. $\int \dfrac{dx}{x^2 (1 + x^3)^{\frac{2}{3}}} = -\dfrac{(1 + x^3)^{\frac{1}{3}}}{x} + C.$

6. $\int \dfrac{dx}{x^3 (1 + x^3)^{\frac{1}{3}}} = -\dfrac{(1 + x^3)^{\frac{2}{3}}}{2\,x^2} + C.$

7. $\int \dfrac{dx}{x^2 (1 + x^4)^{\frac{3}{4}}} = -\dfrac{(1 + x^4)^{\frac{1}{4}}}{x} + C.$

8. $\int \dfrac{dx}{x^n (1 + x^n)^{\frac{1}{n}}} = -\dfrac{(1 + x^n)^{\frac{n-1}{n}}}{(n - 1)x^{n-1}} + C.$

9. $\int \dfrac{dx}{x^3 (1 + x^3)^{\frac{4}{3}}} = -\dfrac{1 + 3\,x^3}{2\,x^2 (1 + x^3)^{\frac{1}{3}}} + C.$

10. $\int \dfrac{2\sqrt{1 + x^4}\,dx}{x^3} = \ln(x^2 + \sqrt{1 + x^4}) - \dfrac{\sqrt{1 + x^4}}{x^2} + C.$

Work out each of the following integrals.

11. $\int x^5 \sqrt{1 - x^3}\, dx.$ **12.** $\int \dfrac{x^5\, dx}{\sqrt{a + bx^3}}.$ **13.** $\int x^5 (a^3 - x^3)^{\frac{3}{2}}\, dx.$

14. $\int \dfrac{(x^5 + 2\, x^2)\, dx}{(1 + x^3)^{\frac{3}{2}}}.$ **15.** $\int x(1 + x^3)^{\frac{1}{3}}\, dx.$

170. Conditions of rationalization of the binomial differential

(A) $\qquad\qquad x^m (a + bx^n)^{\frac{r}{s}}\, dx.$

CASE I. Assume $a + bx^n = z^s.$

Then $\qquad (a + bx^n)^{\frac{1}{s}} = z, \quad \text{and} \quad (a + bx^n)^{\frac{r}{s}} = z^r\,;$

also $\qquad x = \left(\dfrac{z^s - a}{b}\right)^{\frac{1}{n}}, \quad \text{and} \quad x^m = \left(\dfrac{z^s - a}{b}\right)^{\frac{m}{n}};$

hence $\qquad dx = \dfrac{s}{bn} z^{s-1} \left(\dfrac{z^s - a}{b}\right)^{\frac{1}{n} - 1} dz.$

Substituting in (A), we get

$$x^m (a + bx^n)^{\frac{r}{s}}\, dx = \dfrac{s}{bn} z^{r+s-1} \left(\dfrac{z^s - a}{b}\right)^{\frac{m+1}{n} - 1} dz.$$

The second member of this expression is rational when

$$\dfrac{m + 1}{n}$$

is an integer or zero.

CASE II. Assume $a + bx^n = z^s x^n.$

Then $\qquad x^n = \dfrac{a}{z^s - b}, \quad \text{and} \quad a + bx^n = z^s x^n = \dfrac{az^s}{z^s - b}.$

Hence $\qquad (a + bx^n)^{\frac{r}{s}} = a^{\frac{r}{s}}(z^s - b)^{-\frac{r}{s}} z^r\,;$

also $\qquad x = a^{\frac{1}{n}} (z^s - b)^{-\frac{1}{n}}, \quad x^m = a^{\frac{m}{n}} (z^s - b)^{-\frac{m}{n}};$

and $\qquad dx = -\dfrac{s}{n} a^{\frac{1}{n}} z^{s-1} (z^s - b)^{-\frac{1}{n} - 1} dz.$

Substituting in (A), we get

$$x^m (a + bx^n)^{\frac{r}{s}} dx = -\dfrac{s}{n} a^{\frac{m+1}{n} + \frac{r}{s}} (z^s - b)^{-\left(\frac{m+1}{n} + \frac{r}{s} + 1\right)} z^{r+s-1}\, dz.$$

The second member of this expression is rational when $\dfrac{m+1}{n} + \dfrac{r}{s}$ is an integer or zero.

Hence the binomial differential

$$x^m(a+bx^n)^{\frac{r}{s}}dx$$

can be rationalized in the cases given in the preceding article.

171. Transformation of trigonometric differentials.

Theorem. *A trigonometric differential involving* $\sin u$ *and* $\cos u$ *rationally only can be transformed by means of the substitution*

(1) $$\tan\dfrac{u}{2} = z,$$

or, what is the same thing, by the substitutions

(2) $$\sin u = \dfrac{2z}{1+z^2}, \quad \cos u = \dfrac{1-z^2}{1+z^2}, \quad du = \dfrac{2\,dz}{1+z^2}$$

into another differential expression which is rational in z.

Proof. From the formula for the tangent of half an angle in **(5)**, Art. 2, after squaring both members, we have

$$\tan^2\dfrac{1}{2}u = \dfrac{1-\cos u}{1+\cos u}.$$

Substituting $\tan\frac{1}{2}u = z$, and solving for $\cos u$,

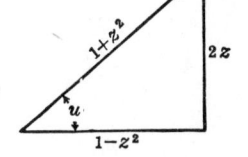

(3) $$\cos u = \dfrac{1-z^2}{1+z^2},$$

one of the formulas (2). The right triangle in the figure shows the relation (3) and gives also $\sin u$ as in (2). Finally, from (1),

$$u = 2\arctan z,$$

and hence $$du = \dfrac{2\,dz}{1+z^2}.$$

Thus the relations (2) are proved.

It is evident that if a trigonometric differential involves $\tan u$, $\operatorname{ctn} u$, $\sec u$, $\csc u$ rationally only, it will be included in the above theorem, since these four functions can be expressed rationally in terms of $\sin u$, or $\cos u$, or both. It follows, therefore, that *any rational trigonometric differential can be integrated, provided the transformed differential in terms of z can be separated into partial fractions* (see Art. 167).

INTEGRAL CALCULUS

ILLUSTRATIVE EXAMPLE. Prove $\int \dfrac{dx}{5 + 4 \sin 2x} = \dfrac{1}{3} \arctan \left(\dfrac{5 \tan x + 4}{3} \right) + C.$

Solution. Let $2x = u$. Then $x = \frac{1}{2} u$, $dx = \frac{1}{2} du$. Substituting these values, and then using (2), we have

$$\int \dfrac{dx}{5 + 4 \sin 2x} = \dfrac{1}{2} \int \dfrac{du}{5 + 4 \sin u} = \dfrac{1}{2} \int \dfrac{\dfrac{2\,dz}{1 + z^2}}{5 + \dfrac{8z}{1 + z^2}} = \int \dfrac{dz}{5 z^2 + 8 z + 5}$$

$$= \dfrac{1}{3} \arctan \left(\dfrac{5z + 4}{3} \right) + C.$$

Substituting back $z = \tan \frac{1}{2} u = \tan x$ gives the above result.

PROBLEMS

Work out the following integrals.

1. $\int \dfrac{d\theta}{1 + \sin \theta + \cos \theta} = \ln \left(1 + \tan \dfrac{\theta}{2} \right) + C.$

2. $\int \dfrac{dx}{\sin x + \tan x} = \dfrac{1}{2} \ln \tan \dfrac{x}{2} - \dfrac{1}{4} \tan^2 \dfrac{x}{2} + C.$

3. $\int \dfrac{d\phi}{5 + 4 \cos \phi} = \dfrac{2}{3} \arctan \left(\dfrac{1}{3} \tan \dfrac{\phi}{2} \right) + C.$

4. $\int \dfrac{dx}{4 + 5 \cos x} = \dfrac{1}{3} \ln \left(\dfrac{\tan \dfrac{x}{2} + 3}{\tan \dfrac{x}{2} - 3} \right) + C.$

5. $\int \dfrac{d\alpha}{3 + \cos \alpha} = \dfrac{1}{\sqrt{2}} \arctan \left(\dfrac{1}{\sqrt{2}} \tan \dfrac{\alpha}{2} \right) + C.$

6. $\int \dfrac{dx}{2 \sin x - \cos x + 3} = \arctan \left(1 + 2 \tan \dfrac{x}{2} \right) + C.$

7. $\int \dfrac{\cos \theta \, d\theta}{5 - 3 \cos \theta} = -\dfrac{\theta}{3} + \dfrac{5}{6} \arctan \left(2 \tan \dfrac{\theta}{2} \right) + C.$

8. $\int \dfrac{dx}{4 \sec x + 5} = \dfrac{2}{5} \arctan \left(\tan \dfrac{x}{2} \right) + \dfrac{4}{15} \ln \left(\dfrac{\tan \dfrac{x}{2} - 3}{\tan \dfrac{x}{2} + 3} \right) + C.$

9. $\int_0^\pi \dfrac{d\theta}{4 - 3 \cos \theta} = \dfrac{\pi}{\sqrt{7}}.$

10. $\int_0^{\frac{\pi}{2}} \dfrac{d\phi}{12 + 13 \cos \phi} = \dfrac{1}{5} \ln \dfrac{3}{2}.$

11. $\int_0^{\frac{\pi}{2}} \dfrac{dx}{2 + \sin x} = \dfrac{\pi}{3\sqrt{3}}.$

12. $\int_0^{\frac{\pi}{2}} \dfrac{d\alpha}{3 + 5 \sin \alpha} = \dfrac{1}{4} \ln 3.$

FORMAL INTEGRATION BY VARIOUS DEVICES

Work out each of the following integrals.

13. $\int \dfrac{dx}{1 + \sin x - \cos x}$.

14. $\int \dfrac{d\theta}{\operatorname{ctn} \theta + \csc \theta}$.

15. $\int \dfrac{d\phi}{13 - 5 \cos \phi}$.

16. $\int \dfrac{dt}{13 \cos t - 5}$.

17. $\int \dfrac{dx}{2 \cos x + 1}$.

18. $\int \dfrac{d\alpha}{2 + \sin \alpha}$.

19. $\int \dfrac{dx}{1 + 2 \sin x}$.

20. $\int \dfrac{\sin \theta \, d\theta}{5 + 4 \sin \theta}$.

21. $\int \dfrac{dt}{5 \sec t - 4}$.

22. $\int_0^{2\pi} \dfrac{dx}{5 + 3 \cos x}$.

23. $\int_0^{\pi} \dfrac{d\theta}{3 + 2 \cos \theta}$.

24. $\int_0^{\frac{\pi}{2}} \dfrac{d\alpha}{2 + \cos \alpha}$.

172. Miscellaneous substitutions. So far the substitutions considered have rationalized the given differential expression. In a great number of cases, however, integrations may be effected by means of substitutions which do not rationalize the given differential, but no general rule can be given, and the experience gained in working out a large number of problems must be our guide.

A very useful substitution is

$$x = \frac{1}{z}, \quad dx = -\frac{dz}{z^2},$$

called the *reciprocal substitution*. Let us use this substitution in the next example.

ILLUSTRATIVE EXAMPLE. Find $\int \dfrac{\sqrt{a^2 - x^2}}{x^4} dx$.

Solution. Making the substitution $x = \dfrac{1}{z}$, $dx = -\dfrac{dz}{z^2}$, we get

$$\int \frac{\sqrt{a^2 - x^2}}{x^4} dx = -\int (a^2 z^2 - 1)^{\frac{1}{2}} z \, dz = -\frac{(a^2 z^2 - 1)^{\frac{3}{2}}}{3 \, a^2} + C = -\frac{(a^2 - x^2)^{\frac{3}{2}}}{3 \, a^2 x^3} + C.$$

PROBLEMS

Work out the following integrals.

1. $\int \dfrac{dx}{x\sqrt{1 + x + x^2}} = \ln \left(\dfrac{cx}{2 + x + 2\sqrt{1 + x + x^2}} \right).$ Let $x = \dfrac{1}{z}$.

2. $\int \dfrac{dx}{x\sqrt{x^2 - x + 2}} = \dfrac{1}{\sqrt{2}} \ln \left(\dfrac{\sqrt{x^2 - x + 2} + x - \sqrt{2}}{\sqrt{x^2 - x + 2} + x + \sqrt{2}} \right) + C.$

   Let $\sqrt{x^2 - x + 2} = z - x.$

3. $\int \dfrac{dx}{x\sqrt{x^2 + 2x - 1}} = 2 \arctan (x + \sqrt{x^2 + 2x - 1}) + C.$

   Let $\sqrt{x^2 + 2x - 1} = z - x.$

4. $\int \dfrac{dx}{x\sqrt{2+x-x^2}} = \dfrac{1}{\sqrt{2}} \ln\left(\dfrac{\sqrt{2+2\,x}-\sqrt{2-x}}{\sqrt{2+2\,x}+\sqrt{2-x}}\right) + C.$

Let $\sqrt{2+x-x^2} = (x+1)z.$

5. $\int \dfrac{dx}{x\sqrt{5\,x-6-x^2}} = -\sqrt{\dfrac{2}{3}} \arctan\sqrt{\dfrac{2(3-x)}{3(x-2)}} + C.$

Let $\sqrt{5\,x-6-x^2} = (x-2)z.$

6. $\int \dfrac{dx}{x\sqrt{3\,x^2-2\,x-1}} = -\arcsin\left(\dfrac{1+x}{2\,x}\right) + C.$ $\qquad$ Let $x = \dfrac{1}{z}.$

7. $\int \dfrac{-dx}{x\sqrt{1+4\,x+5\,x^2}} = \ln\left(\dfrac{1+2\,x+\sqrt{1+4\,x+5\,x^2}}{x}\right) + C.$

Let $x = \dfrac{1}{z}.$

8. $\int \dfrac{dx}{x\sqrt{x^2+4\,x-4}} = -\dfrac{1}{2}\arcsin\left(\dfrac{2-x}{x\sqrt{2}}\right) + C.$ $\qquad$ Let $x = \dfrac{1}{z}.$

9. $\int \dfrac{dx}{x^2\sqrt{1+2\,x+3\,x^2}} = -\dfrac{\sqrt{1+2\,x+3\,x^2}}{x}$

$\qquad\qquad + \ln\left(\dfrac{1+x+\sqrt{1+2\,x+3\,x^2}}{x}\right) + C.$

Let $x = \dfrac{1}{z}.$

10. $\int \dfrac{dx}{x^2\sqrt{27\,x^2+6\,x-1}} = \dfrac{\sqrt{27\,x^2+6\,x-1}}{x} - 3\arcsin\left(\dfrac{1-3\,x}{6\,x}\right) + C.$

Let $x = \dfrac{1}{z}.$

11. $\int_{\frac{1}{3}}^{1} \dfrac{(x-x^3)^{\frac{1}{3}}\,dx}{x^4} = 6.$ $\qquad$ Let $x = \dfrac{1}{z}.$

12. $\int_{0}^{1} \dfrac{dx}{e^x+e^{-x}} = \arctan e - \dfrac{\pi}{4}.$ $\qquad$ Let $e^x = z.$

13. $\int_{0}^{a} \dfrac{dx}{\sqrt{ax-x^2}} = \pi.$ $\qquad$ Let $x = a\sin^2 z.$

14. $\int_{0}^{1} \sqrt{2\,t+t^2}\,dt = \sqrt{3} - \tfrac{1}{2}\ln(2+\sqrt{3}).$ $\qquad$ Let $t+1 = z.$

Work out each of the following integrals.

15. $\int \dfrac{4\,dx}{x\sqrt{x^2-2\,x+3}}.$ $\qquad$ Let $\sqrt{x^2-2\,x+3} = z - x.$

16. $\int \dfrac{4\,x\,dx}{(x^2-2\,x+3)^{\frac{3}{2}}}.$ $\qquad$ Let $\sqrt{x^2-2\,x+3} = z - x.$

17. $\int \dfrac{2\,dx}{\sqrt{5\,x-6-x^2}}.$ $\qquad$ Let $\sqrt{5\,x-6-x^2} = (x-2)z.$

18. $\int \dfrac{2\,x\,dx}{\sqrt{5\,x-6-x^2}}.$ $\qquad$ Let $\sqrt{5\,x-6-x^2} = (x-2)z.$

CHAPTER XVII

REDUCTION FORMULAS. USE OF TABLE OF INTEGRALS

173. Introduction. In this chapter formal integration is completed. The aim is eventually to lay down directions for using a table of integrals. Methods of deriving certain general formulas, called *reduction formulas*, given in all tables are developed, since these methods are typical in problems of this sort.

174. Reduction formulas for binomial differentials. When the binomial differential cannot be integrated readily by any of the methods shown so far, it is customary to employ reduction formulas deduced by the method of integration by parts. By means of these reduction formulas the given differential is expressed as the sum of two terms, one of them not affected by the sign of integration, and the other an integral of the same form as the original expression, but one which is easier to integrate. The following are the four principal reduction formulas.

(A) $\int x^m (a + bx^n)^p \, dx = \dfrac{x^{m-n+1}(a + bx^n)^{p+1}}{(np + m + 1)b}$

$\qquad - \dfrac{(m - n + 1)a}{(np + m + 1)b} \int x^{m-n}(a + bx^n)^p \, dx.$

(B) $\int x^m (a + bx^n)^p \, dx = \dfrac{x^{m+1}(a + bx^n)^p}{np + m + 1}$

$\qquad + \dfrac{anp}{np + m + 1} \int x^m(a + bx^n)^{p-1} dx.$

(C) $\int x^m (a + bx^n)^p \, dx = \dfrac{x^{m+1}(a + bx^n)^{p+1}}{(m + 1)a}$

$\qquad - \dfrac{(np + n + m + 1)b}{(m + 1)a} \int x^{m+n}(a + bx^n)^p \, dx.$

(D) $\int x^m (a + bx^n)^p \, dx = -\dfrac{x^{m+1}(a + bx^n)^{p+1}}{n(p + 1)a}$

$\qquad + \dfrac{np + n + m + 1}{n(p + 1)a} \int x^m(a + bx^n)^{p+1} dx.$

307

While it is not desirable for the student to memorize these formulas, he should know what each one will do and when each one fails. Thus:

Formula (A) diminishes m by n.   (A) fails when $np + m + 1 = 0$.
Formula (B) diminishes p by 1.   (B) fails when $np + m + 1 = 0$.
Formula (C) increases m by n.   (C) fails when $m + 1 = 0$.
Formula (D) increases p by 1.   (D) fails when $p + 1 = 0$.

I. *To derive formula* (A). The formula for integration by parts is

(1) $$\int u\, dv = uv - \int v\, du.$$ (A), Art. 136

We may apply this formula in the integration of

$$\int x^m (a + bx^n)^p\, dx$$

by placing $u = x^{m-n+1}$* and $dv = (a + bx^n)^p x^{n-1}\, dx$;
then $du = (m - n + 1) x^{m-n}\, dx$ and $v = \dfrac{(a + bx^n)^{p+1}}{nb(p+1)}$.

Substituting in (1),

(2) $$\int x^m (a + bx^n)^p\, dx = \frac{x^{m-n+1}(a + bx^n)^{p+1}}{nb(p+1)}$$
$$- \frac{m-n+1}{nb(p+1)} \int x^{m-n}(a + bx^n)^{p+1}\, dx.$$

But $\int x^{m-n}(a + bx^n)^{p+1}\, dx = \int x^{m-n}(a + bx^n)^p(a + bx^n)\, dx$

$$= a \int x^{m-n}(a + bx^n)^p\, dx$$
$$+ b \int x^m (a + bx^n)^p\, dx.$$

Substituting this in (2), we get

$$\int x^m(a + bx^n)^p\, dx = \frac{x^{m-n+1}(a+bx^n)^{p+1}}{nb(p+1)}$$
$$- \frac{(m-n+1)a}{nb(p+1)} \int x^{m-n}(a+bx^n)^p\, dx$$
$$- \frac{m-n+1}{n(p+1)} \int x^m (a+bx^n)^p\, dx.$$

Transposing the last term to the first member, combining, and solving for $\int x^m(a + bx^n)^p\, dx$, we obtain (A).

* In order to integrate dv by the power formula it is necessary that x outside the parenthesis shall have the exponent $n - 1$. Subtracting $n - 1$ from m leaves $m - n + 1$ for the exponent of x in u.

REDUCTION FORMULAS

It is seen by formula (*A*) that the integration of $x^m(a + bx^n)^p dx$ is made to depend upon the integration of another differential of the same form in which m is replaced by $m - n$. By repeated applications of formula (*A*), m may be diminished by any multiple of n.

When $np + m + 1 = 0$, formula (*A*) evidently fails (the denominator vanishing). But in that case

$$\frac{m+1}{n} + p = 0;$$

hence we can apply the method of Art. 169, and the formula is not needed.

II. *To derive formula* (*B*). Separating the factors, we may write

(3) $\quad \displaystyle\int x^m(a + bx^n)^p dx = \int x^m(a + bx^n)^{p-1}(a + bx^n) dx$

$$= a \int x^m(a + bx^n)^{p-1} dx$$

$$+ b \int x^{m+n}(a + bx^n)^{p-1} dx.$$

Now let us apply formula (*A*) to the last term of (3) by substituting in the formula $m + n$ for m, and $p - 1$ for p. This gives

$$b \int x^{m+n}(a + bx^n)^{p-1} dx = \frac{x^{m+1}(a + bx^n)^p}{np + m + 1}$$

$$- \frac{a(m+1)}{np + m + 1} \int x^m(a + bx^n)^{p-1} dx.$$

Substituting this in (3), and combining like terms, we get (*B*).

Each application of formula (*B*) diminishes p by unity. Formula (*B*) fails for the same case as (*A*).

III. *To derive formula* (*C*). Solving formula (*A*) for

$$\int x^{m-n}(a + bx^n)^p dx,$$

and substituting $m + n$ for m, we get (*C*).

Therefore each time we apply (*C*), m is replaced by $m + n$. When $m + 1 = 0$, formula (*C*) fails, but then the differential expression can be rationalized by the method of Art. 169, and the formula is not needed.

IV. *To derive formula* (*D*). Solving formula (*B*) for

$$\int x^m(a + bx^n)^{p-1} dx,$$

and substituting $p + 1$ for p, we get (*D*).

Each application of (*D*) increases p by unity. Evidently (*D*) fails when $p + 1 = 0$, but then $p = -1$ and the expression is rational.

Formula (5) of Case IV, Art. 167, is a special case of (*D*), when $m = 0$, $p = -n$, $n = 2$, $a = a^2$, $b = 1$.

ILLUSTRATIVE EXAMPLE 1. $\displaystyle\int \frac{x^3\,dx}{\sqrt{1-x^2}} = -\frac{1}{3}(x^2+2)(1-x^2)^{\frac{1}{2}} + C.$

Solution. Here $m = 3$, $n = 2$, $p = -\frac{1}{2}$, $a = 1$, $b = -1$.

We apply reduction formula (*A*) in this case because the integration of the differential would then depend on the integration of $\int x(1-x^2)^{-\frac{1}{2}}dx$, which comes under the power formula. Hence, substituting in (*A*), we obtain

$$\int x^3(1-x^2)^{-\frac{1}{2}}dx = \frac{x^{3-2+1}(1-x^2)^{-\frac{1}{2}+1}}{-1(-1+3+1)} - \frac{1(3-2+1)}{-1(-1+3+1)}\int x^{3-2}(1-x^2)^{-\frac{1}{2}}dx$$

$$= -\tfrac{1}{3}x^2(1-x^2)^{\frac{1}{2}} + \tfrac{2}{3}\int x(1-x^2)^{-\frac{1}{2}}dx$$

$$= -\tfrac{1}{3}x^2(1-x^2)^{\frac{1}{2}} - \tfrac{2}{3}(1-x^2)^{\frac{1}{2}} + C$$

$$= -\tfrac{1}{3}(x^2+2)(1-x^2)^{\frac{1}{2}} + C.$$

ILLUSTRATIVE EXAMPLE 2. $\displaystyle\int \frac{x^4\,dx}{(a^2-x^2)^{\frac{1}{2}}} = -\left(\frac{1}{4}x^3 + \frac{3}{8}a^2x\right)\sqrt{a^2-x^2}$
$$+ \frac{3}{8}a^4 \arcsin \frac{x}{a} + C.$$

HINT. Apply (*A*) twice.

ILLUSTRATIVE EXAMPLE 3. $\displaystyle\int (a^2+x^2)^{\frac{1}{2}}dx = \frac{x}{2}\sqrt{a^2+x^2}$
$$+ \frac{a^2}{2}\ln\left(x+\sqrt{a^2+x^2}\right) + C.$$

HINT. Here $m = 0$, $n = 2$, $p = \frac{1}{2}$, $a = a^2$, $b = 1$. Apply (*B*) once.

ILLUSTRATIVE EXAMPLE 4. $\displaystyle\int \frac{dx}{x^3\sqrt{x^2-1}} = \frac{(x^2-1)^{\frac{1}{2}}}{2x^2} + \frac{1}{2}\operatorname{arc\,sec} x + C.$

HINT. Apply (*C*) once.

PROBLEMS

Work out each of the following integrals.

1. $\displaystyle\int \frac{x^2\,dx}{\sqrt{a^2-x^2}} = -\frac{x}{2}\sqrt{a^2-x^2} + \frac{a^2}{2}\arcsin\frac{x}{a} + C.$

2. $\displaystyle\int \frac{x^3\,dx}{\sqrt{a^2+x^2}} = \frac{1}{3}(x^2-2a^2)\sqrt{a^2+x^2} + C.$

3. $\displaystyle\int \frac{x^5\,dx}{\sqrt{1-x^2}} = -\frac{1}{15}(3x^4+4x^2+8)\sqrt{1-x^2} + C.$

4. $\displaystyle\int x^2\sqrt{a^2-x^2}\,dx = \frac{x}{8}(2x^2-a^2)\sqrt{a^2-x^2} + \frac{a^4}{8}\arcsin\frac{x}{a} + C.$

5. $\displaystyle\int \frac{dx}{(a^2+x^2)^2} = \frac{x}{2a^2(a^2+x^2)} + \frac{1}{2a^3}\arctan\frac{x}{a} + C.$

REDUCTION FORMULAS

6. $\int \dfrac{dx}{x^3 \sqrt{a^2 - x^2}} = -\dfrac{\sqrt{a^2 - x^2}}{2 a^2 x^2} + \dfrac{1}{2 a^3} \ln \dfrac{a - \sqrt{a^2 - x^2}}{x} + C.$

7. $\int \dfrac{x^3 \, dx}{(a^2 + x^2)^{\frac{3}{2}}} = \dfrac{x^2 + 2 a^2}{\sqrt{a^2 + x^2}} + C.$

8. $\int \dfrac{dx}{(a^2 - x^2)^{\frac{5}{2}}} = \dfrac{x(3 a^2 - 2 x^2)}{3 a^4 (a^2 - x^2)^{\frac{3}{2}}} + C.$

9. $\int (x^2 + a^2)^{\frac{3}{2}} \, dx = \tfrac{1}{8} x (2 x^2 + 5 a^2) \sqrt{x^2 + a^2} + \tfrac{3}{8} a^4 \ln (x + \sqrt{x^2 + a^2}) + C.$

10. $\int x^2 \sqrt{x^2 + a^2} \, dx = \tfrac{1}{8} x (2 x^2 + a^2) \sqrt{x^2 + a^2} - \tfrac{1}{8} a^4 \ln (x + \sqrt{x^2 + a^2}) + C.$

11. $\int \dfrac{x^2 \, dx}{\sqrt{2 ax - x^2}} = -\dfrac{(x + 3 a) \sqrt{2 ax - x^2}}{2} + \dfrac{3 a^2}{2} \arccos \left(1 - \dfrac{x}{a}\right) + C.$

HINT. $\int \dfrac{x^2 \, dx}{\sqrt{2 ax - x^2}} = \int x^{\frac{3}{2}} (2 a - x)^{-\frac{1}{2}} dx.$ Apply (A) twice.

12. $\int \dfrac{y^3 \, dy}{\sqrt{4 y - y^2}} = -\tfrac{1}{3} (y^2 + 5 y + 30) \sqrt{4 y - y^2} + 20 \arccos \left(1 - \dfrac{y}{2}\right) + C.$

13. $\int \dfrac{ds}{(a^2 + s^2)^3} = \dfrac{s}{4 a^2 (a^2 + s^2)^2} + \dfrac{3 s}{8 a^4 (a^2 + s^2)} + \dfrac{3}{8 a^5} \arctan \dfrac{s}{a} + C.$

14. $\int \dfrac{y^2 \, dy}{\sqrt{9 - 4 y^2}} = -\tfrac{1}{8} y \sqrt{9 - 4 y^2} + \dfrac{9}{16} \arcsin \dfrac{2 y}{3} + C.$

15. $\int \dfrac{t^3 \, dt}{\sqrt{1 + 4 t^2}} = \dfrac{1}{24} (2 t^2 - 1) \sqrt{1 + 4 t^2} + C.$

16. $\int y^2 \sqrt{4 - 9 y^2} \, dy = \dfrac{1}{36} y (9 y^2 - 2) \sqrt{4 - 9 y^2} + \dfrac{2}{27} \arcsin \dfrac{3 y}{2} + C.$

17. $\int \dfrac{t^3 \, dt}{(1 + 9 t^2)^{\frac{3}{2}}} = \dfrac{9 t^2 + 2}{81 \sqrt{1 + 9 t^2}} + C.$

18. $\int t^2 \sqrt{1 + 4 t^2} \, dt = \tfrac{1}{32} t (1 + 8 t^2) \sqrt{1 + 4 t^2} - \tfrac{1}{64} \ln (2 t + \sqrt{1 + 4 t^2}) + C.$

Work out each of the following integrals.

19. $\int \dfrac{x^2 \, dx}{(a^2 - x^2)^2}.$

20. $\int \dfrac{dx}{x^2 (1 + x^2)^{\frac{3}{2}}}.$

21. $\int \dfrac{\sqrt{a^2 - x^2} \, dx}{x^4}.$

22. $\int \dfrac{x^8 \, dx}{\sqrt{4 - x^6}}.$

23. $\int \dfrac{\sqrt{a^2 + x^2}}{x} \, dx.$

24. $\int \dfrac{(1 - x^3)^{\frac{5}{2}} \, dx}{x}.$

25. $\int \dfrac{s^7 \, ds}{(a + bs^4)^{\frac{2}{3}}}.$

26. $\int \dfrac{z^8 \, dz}{\sqrt{5 - z^3}}.$

27. $\int \dfrac{dx}{(1 + 4 x^2)^{\frac{5}{2}}}.$

28. $\int \dfrac{dt}{t^3 \sqrt{1 - 4 t^2}}.$

29. $\int (9 y^2 + 4)^{\frac{3}{2}} \, dy.$

175. Reduction formulas for trigonometric differentials.

The method of the last article, which makes the given integral depend on another integral of the same form, is called *successive reduction*.

We shall now apply the same method to trigonometric differentials by deriving and illustrating the use of the following trigonometric reduction formulas:

(E) $$\int \sin^m x \cos^n x \, dx = \frac{\sin^{m+1} x \cos^{n-1} x}{m+n} + \frac{n-1}{m+n} \int \sin^m x \cos^{n-2} x \, dx.$$

(F) $$\int \sin^m x \cos^n x \, dx = -\frac{\sin^{m-1} x \cos^{n+1} x}{m+n} + \frac{m-1}{m+n} \int \sin^{m-2} x \cos^n x \, dx.$$

(G) $$\int \sin^m x \cos^n x \, dx = -\frac{\sin^{m+1} x \cos^{n+1} x}{n+1} + \frac{m+n+2}{n+1} \int \sin^m x \cos^{n+2} x \, dx.$$

(H) $$\int \sin^m x \cos^n x \, dx = \frac{\sin^{m+1} x \cos^{n+1} x}{m+1} + \frac{m+n+2}{m+1} \int \sin^{m+2} x \cos^n x \, dx.$$

Here the student should note that

Formula (E) *diminishes n by* 2.   (E) *fails when* $m + n = 0$.
Formula (F) *diminishes m by* 2.   (F) *fails when* $m + n = 0$.
Formula (G) *increases n by* 2.   (G) *fails when* $n + 1 = 0$.
Formula (H) *increases m by* 2.   (H) *fails when* $m + 1 = 0$.

To derive these we apply, as before, the formula for integration by parts, namely,

(1) $$\int u \, dv = uv - \int v \, du. \qquad (A), \text{ Art. } 136$$

Let $u = \cos^{n-1} x$, and $dv = \sin^m x \cos x \, dx$;

then $du = -(n-1)\cos^{n-2} x \sin x \, dx$, and $v = \dfrac{\sin^{m+1} x}{m+1}$.

Substituting in (1), we get

(2) $$\int \sin^m x \cos^n x \, dx = +\frac{\sin^{m+1} \cos^{n-1} x}{m+1} + \frac{n-1}{m+1} \int \sin^{m+2} x \cos^{n-2} x \, dx.$$

REDUCTION FORMULAS

In the same way, if we let
$$u = \sin^{m-1}x, \text{ and } dv = \cos^n x \sin x\, dx,$$
we obtain

(3) $\quad \displaystyle\int \sin^m x \cos^n x\, dx = -\frac{\sin^{m-1}x \cos^{n+1}x}{n+1}$
$\qquad\qquad\qquad\qquad + \dfrac{m-1}{n+1}\displaystyle\int \sin^{m-2}x \cos^{n+2}x\, dx.$

But $\displaystyle\int \sin^{m+2}x \cos^{n-2}x\, dx = \int \sin^m x\,(1-\cos^2 x)\cos^{n-2}x\, dx$

$\qquad\qquad\qquad = \displaystyle\int \sin^m x \cos^{n-2}x\, dx - \int \sin^m x \cos^n x\, dx.$

Substituting this in (2), combining like terms, and solving for $\displaystyle\int \sin^m x \cos^n x\, dx$, we get (E).

Making a similar substitution in (3), we get (F).

Solving formula (E) for the integral on the right-hand side, and increasing n by 2, we get (G).

In the same way we get (H) from formula (F).

Formulas (E) and (F) fail when $m+n=0$, formula (G) when $n+1=0$, and formula (H) when $m+1=0$. But in such cases we may integrate by methods which have been explained previously.

It is clear that when m and n are integers, the integral

$$\int \sin^m x \cos^n x\, dx$$

may be made to depend, by using one of the above reduction formulas, upon one of the following integrals:

$\displaystyle\int dx, \ \int \sin x\, dx, \ \int \cos x\, dx, \ \int \sin x \cos x\, dx, \ \int \dfrac{dx}{\sin x} = \int \csc x\, dx,$

$\displaystyle\int \dfrac{dx}{\cos x} = \int \sec x\, dx, \ \int \dfrac{dx}{\cos x \sin x}, \ \int \tan x\, dx, \ \int \ctn x\, dx,$

all of which we have learned how to integrate.

ILLUSTRATIVE EXAMPLE 1. Prove

$$\int \sin^2 x \cos^4 x\, dx = -\frac{\sin x \cos^5 x}{6} + \frac{\sin x \cos^3 x}{24} + \frac{1}{16}(\sin x \cos x + x) + C.$$

Solution. First applying formula (F), we get

(4) $\qquad \displaystyle\int \sin^2 x \cos^4 x\, dx = -\dfrac{\sin x \cos^5 x}{6} + \dfrac{1}{6}\int \cos^4 x\, dx.$

[Here $m=2, n=4.$]

314 INTEGRAL CALCULUS

Applying formula (*E*) to the integral in the second member of (4), we get

(5) $\quad \int \cos^4 x \, dx = \dfrac{\sin x \cos^3 x}{4} + \dfrac{3}{4} \int \cos^2 x \, dx.$

[Here $m = 0$, $n = 4$.]

Applying formula (*E*) to the second member of (5) gives

(6) $\quad \int \cos^2 x \, dx = \dfrac{\sin x \cos x}{2} + \dfrac{x}{2}.$

Now substitute the result (6) in (5), and then this result in (4). This gives the answer as above.

ILLUSTRATIVE EXAMPLE 2. Prove

$$\int \dfrac{\tan^2 2x}{\cos 2x} dx = \dfrac{1}{4} \sec 2x \tan 2x - \dfrac{1}{4} \ln (\sec 2x + \tan 2x) + C.$$

Solution. $\quad \dfrac{\tan^2 2x}{\cos 2x} = \dfrac{\sin^2 2x}{\cos^2 2x} \cdot \dfrac{1}{\cos 2x} = \dfrac{\sin^2 2x}{\cos^3 2x}.$

Let $2x = u$. Then $x = \tfrac{1}{2} u$, $dx = \tfrac{1}{2} du$, and

(7) $\quad \int \sin^2 2x \cos^{-3} 2x \, dx = \tfrac{1}{2} \int \sin^2 u \cos^{-3} u \, du.$

Apply (*G*) to the new integral in (7), with $m = 2$, $n = -3$, replacing x by u.

(8) $\quad \int \sin^2 u \cos^{-3} u \, du = -\dfrac{\sin^3 u \cos^{-2} u}{-2} + \dfrac{1}{-2} \int \sin^2 u \cos^{-1} u \, du.$

Apply (*F*) to the new integral in (8), with $m = 2$, $n = -1$.

(9) $\quad \int \sin^2 u \cos^{-1} u \, du = -\sin u + \int \cos^{-1} u \, du = -\sin u + \ln (\sec u + \tan u).$

Substituting from (9) and (8) into (7), reducing, and setting $u = 2x$, we have the answer.

PROBLEMS

Verify the following integrations.

1. $\int \sin^4 x \cos^2 x \, dx = \sin x \cos x \left[\dfrac{1}{6} \sin^4 x - \dfrac{1}{24} \sin^2 x - \dfrac{1}{16} \right] + \dfrac{x}{16} + C.$

2. $\int \tan^3 \dfrac{x}{3} \, dx = \dfrac{3}{2} \tan^2 \dfrac{x}{3} + 3 \ln \cos \dfrac{x}{3} + C.$

3. $\int \ctn^4 \theta \, d\theta = -\dfrac{\ctn^3 \theta}{3} + \ctn \theta + \theta + C.$

4. $\int \sec^3 t \, dt = \tfrac{1}{2} \sec t \tan t + \tfrac{1}{2} \ln (\sec t + \tan t) + C.$

5. $\int \csc^3 x \, dx = -\tfrac{1}{2} \csc x \ctn x + \tfrac{1}{2} \ln (\csc x - \ctn x) + C.$

6. $\int \csc^5 \theta \, d\theta = -\dfrac{\csc \theta \ctn \theta}{4} \left(\csc^2 \theta + \dfrac{3}{2} \right) + \dfrac{3}{8} \ln (\csc \theta - \ctn \theta) + C.$

7. $\int \sin^2 \phi \cos^2 \phi \, d\phi = \tfrac{1}{8} \sin \phi \cos \phi \, (2 \sin^2 \phi - 1) + \tfrac{1}{8} \phi + C.$

REDUCTION FORMULAS

8. $\int \dfrac{\operatorname{ctn}^2 2\theta\, d\theta}{\sin 2\theta} = -\dfrac{1}{4}\operatorname{ctn} 2\theta \csc 2\theta - \dfrac{1}{4}\ln(\csc 2\theta - \operatorname{ctn} 2\theta) + C.$

9. $\int \dfrac{dx}{\sin^4 x} = -\dfrac{\cos x}{3\sin^3 x} - \dfrac{2\cos x}{3\sin x} + C.$

10. $\int \cos^6\theta\, d\theta = \dfrac{\cos\theta \sin\theta}{48}[8\cos^4\theta + 10\cos^2\theta + 15] + \dfrac{5\theta}{16} + C.$

11. $\int_0^{\frac{\pi}{2}} \sin^4\theta\, d\theta = \dfrac{3\pi}{16}.$

12. $\int_0^{\pi} \cos^4 x\, dx = \dfrac{3\pi}{8}.$ **14.** $\int_0^{\pi} \sin^8\phi\, d\phi = \dfrac{35\pi}{128}.$

13. $\int_0^{\frac{\pi}{2}} \sin^6 2\theta\, d\theta = \dfrac{5\pi}{32}.$ **15.** $\int_{\frac{\pi}{4}}^{\frac{\pi}{2}} \dfrac{\cos^4 x\, dx}{\sin^2 x} = \dfrac{5}{4} - \dfrac{3\pi}{8}.$

Work out each of the following integrals:

16. $\int \sin^6 2\theta\, d\theta.$ **18.** $\int \dfrac{\sin^3 x\, dx}{\cos^5 x}.$ **20.** $\int_0^{\frac{\pi}{2}} \tan^3 \dfrac{\theta}{2}\, d\theta.$

17. $\int \csc^3 \dfrac{\theta}{2}\, d\theta.$ **19.** $\int \dfrac{d\theta}{\sin^4\theta \cos^2\theta}.$ **21.** $\int_0^{\pi} \sin^4 x\, dx.$

22. $\int_0^{\frac{\pi}{2}} \sin^3\theta \cos^3\theta\, d\theta.$ **23.** $\int_0^{\frac{\pi}{2}} (1+\sin\theta)^4\, d\theta.$

176. Use of a table of integrals. The methods of integration developed in Chapters XII, XVI, and XVII have been directed to reducing a given integral to one or more of the Standard Elementary Forms in Art. 128. Various devices have been elaborated to this end, such as

integration by parts (Art. 136);
integration by partial fractions (Art. 167);
integration by substitution of a new variable (Arts. 168–172);
use of reduction formulas (Arts. 174–175).

When, however, a more or less extensive table of integrals is available, the first step in any problem in formal integration is to search for a formula in the table by which the problem can be solved without the use of any of these devices. Such a table is given in Chapter XXVII. Some examples will now be given.

ILLUSTRATIVE EXAMPLE 1. Prove, by the Table of Integrals,

$$\int \dfrac{dx}{x^2(2+x)} = -\dfrac{1}{2x} + \dfrac{1}{4}\ln\left(\dfrac{2+x}{x}\right) + C.$$

Solution. Use 14, with $a = 2$, $b = 1$, and $u = x$.
This example, without the table, would be worked out as in Case II, Art. 167.

ILLUSTRATIVE EXAMPLE 2. Verify, by the Table of Integrals,

$$\int \frac{dx}{x(9+4x^2)} = \frac{1}{18} \ln\left(\frac{x^2}{9+4x^2}\right) + C.$$

Solution. Use 22, with $a = 3$, $b = 2$, and $u = x$.
This example, without the table, is solved as in Case III, Art. 167.

ILLUSTRATIVE EXAMPLE 3. Verify, by the Table of Integrals,

$$\int \frac{dx}{x\sqrt{4+3x}} = \frac{1}{2} \ln \frac{\sqrt{4+3x}-2}{\sqrt{4+3x}+2} + C.$$

Solution. Use 31, with $a = 4$, $b = 3$, and $u = x$.
This example, without the table, is worked out by the substitution $4 + 3x = z^2$, as shown in Art. 168.

ILLUSTRATIVE EXAMPLE 4. Verify, by the Table of Integrals,

$$\int \frac{x\,dx}{\sqrt{3x^2+4x-7}} = \frac{\sqrt{3x^2+4x-7}}{3} - \frac{2}{3\sqrt{3}} \ln\left(6x+4+2\sqrt{3}\sqrt{3x^2+4x-7}\right) + C.$$

Solution. Use 113, with $a = -7$, $b = 4$, $c = 3$, and $u = x$.
Without the table the example would be solved by completing the square as in Illustrative Example 2, p. 206.

ILLUSTRATIVE EXAMPLE 5. Verify, by the Table of Integrals,

$$\int e^{3x} \cos 2x\,dx = \frac{e^{3x}(2\sin 2x + 3\cos 2x)}{13} + C.$$

Solution. Use 154, with $a = 3$, $n = 2$, $u = x$.
Without the table the example would be solved by integration by parts. See Illustrative Example 6, Art. 136.

In many problems the given integral cannot be identified with one in the table as easily as in the preceding examples. In such cases we search for a formula in the table similar to the given integral, and such that the latter can be transformed into the former by a simple change of variable. This method has been used constantly in Chapter XII and in all integration problems hitherto.

ILLUSTRATIVE EXAMPLE 6. Verify, by the Table of Integrals,

$$\int \frac{dx}{x\sqrt{4x^2+9}} = \frac{1}{3} \ln \frac{2x}{3+\sqrt{4x^2+9}} + C.$$

Solution. Formula 47 is similar. Let $u = 2x$. Then $x = \frac{1}{2}u$, $dx = \frac{1}{2}du$, and, substituting the values in the given integral, we obtain

$$\int \frac{dx}{x\sqrt{4x^2+9}} = \int \frac{\frac{1}{2}du}{\frac{1}{2}u\sqrt{u^2+9}} = \int \frac{du}{u\sqrt{u^2+9}}.$$

Hence, applying 47, with $a = 3$, and substituting back $u = 2x$, $a = 3$, we have

$$\int \frac{dx}{x\sqrt{4x^2+9}} = -\frac{1}{3} \ln\left(\frac{3+\sqrt{4x^2+9}}{2x}\right) + C.$$

Without tables we should proceed as in Illustrative Example 2, Art. 135.

REDUCTION FORMULAS

ILLUSTRATIVE EXAMPLE 7. Verify, by the Table of Integrals,

$$\int \frac{\sqrt{9x - 4x^2}}{x^3} dx = -\frac{2}{27} \frac{(9x - 4x^2)^{\frac{3}{2}}}{x^3} + C.$$

Solution. Formula 84 is similar. Let $u = 2x$. Then $x = \frac{1}{2}u$, $dx = \frac{1}{2} du$. Substituting, we obtain

$$\int \frac{\sqrt{9x - 4x^2}}{x^3} dx = \int \frac{\sqrt{\frac{9}{2}u - u^2}}{\frac{1}{8}u^3} \frac{du}{2} = 4 \int \frac{\sqrt{\frac{9}{2}u - u^2}}{u^3} du.$$

This is now 84 with $a = \frac{9}{4}$. Hence, applying 84, and substituting back $u = 2x$, we get the required result.

If no formula from the table can be applied as in the preceding two cases, there remains the possibility that the use of one or more of the devices mentioned at the beginning of this article will lead to new integrals solvable by the table. No general directions can be given other than the rules already developed in the text for the employment of these devices.

The student should study the arrangement of the table. He will find that the Standard Forms of Art. 128 appear in their proper places. The reduction formulas of Art. 174 are given, with modifications, by 96–104. Also, the reduction formulas of Art. 175, with additional ones for various cases, are numbered 157–174. Increased power in the technique of integration will come from familiarity with the table and practice in using it.

PROBLEMS

Work out the following integrals.

1. $\int x^3 \sqrt{x^2 + 5}\, dx = \frac{1}{15}(3x^2 - 10)(x^2 + 5)^{\frac{3}{2}} + C.$

2. $\int \frac{dt}{(1 - 4t^2)^{\frac{3}{2}}} = \frac{t}{\sqrt{1 - 4t^2}} + C.$

3. $\int \frac{x^2\, dx}{\sqrt{9x^2 - 4}} = \frac{x}{18}\sqrt{9x^2 - 4} + \frac{2}{27}\ln(3x + \sqrt{9x^2 - 4}) + C.$

4. $\int \frac{d\theta}{2 - \cos 2\theta} = \frac{1}{\sqrt{3}} \arctan(\sqrt{3} \tan \theta) + C.$

5. $\int \frac{x^5\, dx}{(1 - x^4)^{\frac{3}{2}}} = \frac{x^2}{2\sqrt{1 - x^4}} - \frac{1}{2} \arcsin x^2 + C.$

6. $\int \frac{(a^2 - x^2)^{\frac{3}{2}}\, dx}{x^2} = -\frac{(x^2 + 2a^2)\sqrt{a^2 - x^2}}{2x} - \frac{3a^3}{2} \arcsin \frac{x}{a} + C.$

7. $\int e^t \sin^2 \frac{t}{2}\, dt = \frac{1}{4} e^t (2 - \sin t - \cos t) + C.$

8. $\int \dfrac{\sin 2\theta \, d\theta}{1 + \cos \theta} = 2 \ln (1 + \cos \theta) - 2 \cos \theta + C.$

9. $\int \dfrac{dx}{2 + 2x + x^2} = \arctan (x + 1) + C.$

10. $\int x^3 \sin x^2 \, dx = \tfrac{1}{2} \sin x^2 - \tfrac{1}{2} x^2 \cos x^2 + C.$

11. $\int \dfrac{dx}{\sqrt{(x-1)(2-x)}} = 2 \arcsin \sqrt{x - 1} + C.$

12. $\int \dfrac{\sqrt{9t^2 + 4}\, dt}{t} = \sqrt{9t^2 + 4} - 2 \ln \left(\dfrac{2 + \sqrt{9t^2 + 4}}{t} \right) + C.$

13. $\int \dfrac{du}{u^4 \sqrt{a^2 - u^2}} = -\dfrac{(a^2 + 2u^2)\sqrt{a^2 - u^2}}{3 a^4 u^3} + C.$

14. $\int \dfrac{dx}{x^2 \sqrt{4 - x}} = \dfrac{1}{16} \ln \left(\dfrac{\sqrt{4-x} - 2}{\sqrt{4-x} + 2} \right) - \dfrac{\sqrt{4-x}}{4x} + C.$

Work out each of the following integrals.

15. $\int \dfrac{x^5 \, dx}{5 + 4x^2}.$

16. $\int (a^2 - u^2)^{\frac{3}{2}} \, du.$

17. $\int \dfrac{dx}{x^2 + 4x + 2}.$

18. $\int \dfrac{\operatorname{ctn} t \, dt}{a + b \sin t}.$

19. $\int \sqrt{\dfrac{1 + 2y}{1 - 2y}} \, dy.$

20. $\int \dfrac{\sqrt{4x^2 - 25} \, dx}{x^2}.$

21. $\int \dfrac{x^3 \, dx}{\sqrt{a + bx^2}}.$

22. $\int \dfrac{dy}{y^2 \sqrt{y - 1}}.$

23. $\int \sqrt{\dfrac{3 + x^2}{2 + x^2}} \, x \, dx.$

24. $\int \dfrac{d\theta}{5 + 3 \sin 2\theta}.$

25. $\int \dfrac{d\theta}{3 + 5 \sin 2\theta}.$

26. $\int \dfrac{dx}{x^4 \sqrt{x^6 + a^6}}.$

27. $\int \dfrac{x \, dx}{\sqrt{x^2 + 2x + 4}}.$

28. $\int \dfrac{x \, dx}{\sqrt{4 + 2x - x^2}}.$

29. $\int \dfrac{dx}{(1 + e^x)^2}.$

30. $\int \dfrac{\sqrt{x^3 - 1} \, dx}{x}.$

31. $\int \dfrac{\sqrt{x + 4} \, dx}{x}.$

32. $\int e^t \cos^2 t \, dt.$

33. $\int \dfrac{\operatorname{ctn} \theta \, d\theta}{4 + \sin^2 \theta}.$

Evaluate each of the following definite integrals.

34. $\int_0^3 \dfrac{x \, dx}{(1 + x)^2} = 0.636.$

35. $\int_1^{\frac{5}{3}} \dfrac{dx}{x^2 \sqrt{25 - 9x^2}} = \dfrac{4}{25}.$

36. $\int_0^2 \dfrac{dx}{(4x^2 + 9)^{\frac{3}{2}}} = \dfrac{2}{45}.$

37. $\int_1^2 \dfrac{dt}{t(5 - t^2)} = 0.277.$

38. $\int_0^2 (4x^2 + 9)^{\frac{3}{2}} \, dx = 112.9.$

39. $\int_{-a}^a \sqrt{\dfrac{a - u}{a + u}} \, du = \pi a.$

REDUCTION FORMULAS

40. $\int_1^2 \dfrac{x^2\,dx}{\sqrt{9-2x^2}} = 1.338.$ 42. $\int_1^2 \dfrac{\sqrt{9-2x^2}\,dx}{x} = 1.467.$

41. $\int_1^2 \dfrac{\sqrt{9-2x^2}\,dx}{x^2} = 1.129.$ 43. $\int_0^1 t^2 e^{-t}\,dt = 0.1605.$

44. $\int_1^2 \dfrac{dy}{y^3\sqrt{4y^2+5}}.$ 46. $\int_0^2 \dfrac{x^2\,dx}{(4x^2+9)^{\frac{3}{2}}}.$ 48. $\int_1^3 \dfrac{dx}{x\sqrt{4x^2+1}}.$

45. $\int_0^{\frac{\pi}{2}} \cos^2\theta \sin^4\theta\,d\theta.$ 47. $\int_0^1 e^{-\frac{t}{5}} \cos \tfrac{1}{2}\pi t\,dt.$ 49. $\int_0^{\pi} \phi \cos \tfrac{1}{3}\phi\,d\phi.$

ADDITIONAL PROBLEMS

1. Verify the following results.

(a) $\int_1^{e^2} \dfrac{dx}{x(1+\ln x)} = \ln 3;$ (b) $\int_1^{e^2} \dfrac{dx}{x(1+\ln x)^2} = \dfrac{2}{3}.$

2. A parabola with its axis parallel to the y-axis passes through the origin and the point (1, 2). Find its equation if the area between the parabola and the x-axis is a maximum or a minimum.

Ans. $y = 6x - 4x^2$ gives a minimum.

3. Sketch the curve $y\sqrt{x} = \ln x$. Find the volume of the solid of revolution formed by revolving about the x-axis the area bounded by the curve, the x-axis, and two ordinates, one through the maximum point and the other through the point of inflection. *Ans.* $\dfrac{296}{81}\pi.$

4. A solid right circular cone of metal is formed so that the density at any point P is $20(5-r)$ lb. per cubic foot, where r is the distance in feet of the point P from the axis of the cone. Find the weight of the cone if its altitude and the radius of the base are each 3 ft. *Ans.* $630\,\pi$ lb.

NOTE. The weight of an element of uniform density is its volume times its density.

5. A hollow metal sphere has an inside radius of 6 in. and an outside radius of 10 in. The density of the metal at any point varies inversely as the distance of the point from the center of the sphere, and at the outside surface the density is 2 oz. per cubic inch. Find the weight of the sphere.

Ans. $2560\,\pi$ oz.

6. If n is an even integer, show that

$$\int_0^{\frac{\pi}{2}} \sin^n x\,dx = \int_0^{\frac{\pi}{2}} \cos^n x\,dx = \dfrac{(n-1)(n-3)\cdots(1)}{n(n-2)\cdots(2)}\dfrac{\pi}{2}.$$

7. If n is an odd integer, find the value of

$$\int_0^{\frac{\pi}{2}} \sin^n x\,dx.$$

CHAPTER XVIII

CENTROIDS, FLUID PRESSURE, AND OTHER APPLICATIONS

177. Moment of area; centroids. The *centroid* of a plane area is defined in the following manner.

A piece of stiff, flat cardboard will balance in a horizontal position if supported at a point directly under its center of gravity. This point of support is the *centroid* of the area of the flat surface of the cardboard.

For certain areas considered in elementary geometry the centroids are obvious. For a rectangle or a circle the centroid coincides with the geometrical center. In fact, if a plane figure possesses a center of symmetry, that point is the centroid. Furthermore, if a plane figure has an axis of symmetry, the centroid will lie on that axis.

The following considerations lead to the determination of the centroid by mathematical means. It is beyond the purpose of this book to justify the argument by mechanics.

Consider the area $AMPNB$ of the figure. Divide it into n rectangles, each with base Δx, as heretofore. The figure shows one of these rectangles. Let dA be its area, and $C(h, k)$ its centroid. Then

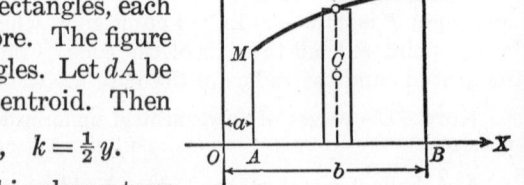

(1) $\quad dA = y\,dx, \quad h = x, \quad k = \tfrac{1}{2}y.$

The *moment of area* of this elementary rectangle about OX (or OY) is the product of its area by the perpendicular distance of its centroid from OX (or OY). If these moments are, respectively, dM_x and dM_y, then

(A) $\qquad dM_x = k\,dA, \quad dM_y = h\,dA.$

The *moment of area for the figure* $AMPNB$ is obtained by applying the Fundamental Theorem (Art. 156) to the sum of the moments of area of the n elementary rectangles. Thus we obtain

(B) $\qquad M_x = \int k\,dA, \quad M_y = \int h\,dA.$

CENTROIDS, FLUID PRESSURE

Finally, if $(\bar{x}, \bar{y})$ is the centroid of the area $AMPNB$, and A its area, then the relations between the moments of area (B) and $\bar{x}$ and $\bar{y}$ are given by

(C) $\qquad A\bar{x} = M_y, \quad A\bar{y} = M_x.$

To calculate $(\bar{x}, \bar{y})$, find the moments of area M_x and M_y. From (1) and (B), these are, for the above figure,

(2) $\qquad M_x = \tfrac{1}{2}\int_a^b y^2\,dx, \quad M_y = \int_a^b xy\,dx,$

in which the value of y in terms of x must be substituted from the equation of the curve MPN.

If the area A is known, we have, from (C),

(3) $\qquad \bar{x} = \dfrac{M_y}{A}, \quad \bar{y} = \dfrac{M_x}{A}.$

If A is not known, it may be found by integration, as in Art. 145.

ILLUSTRATIVE EXAMPLE 1. Find the centroid of the area under one arch of the sine curve

(4) $\qquad y = \sin x.$

Solution. Constructing an elementary rectangle, we have

(5) $\qquad dA = y\,dx = \sin x\,dx,$
$dM_x = k\,dA = \tfrac{1}{2} y^2\,dx = \tfrac{1}{2}\sin^2 x\,dx, \quad dM_y = h\,dA = xy\,dx = x\sin x\,dx.$

The limits are $x = 0$, $x = \pi$. Hence

(6) $A = \int_0^\pi \sin x\,dx = 2, \quad M_x = \tfrac{1}{2}\int_0^\pi \sin^2 x\,dx = \tfrac{1}{4}\pi, \quad M_y = \int_0^\pi x\sin x\,dx = \pi.$

Then, from (3), $\bar{x} = \tfrac{1}{2}\pi$, $\bar{y} = \tfrac{1}{8}\pi$. Ans.

The value of $\bar{x}$ might have been anticipated, since the line $x = \tfrac{1}{2}\pi$ is an axis of symmetry.

ILLUSTRATIVE EXAMPLE 2. In the figure the curve OPA is an arc of the parabola $y^2 = 2\,px$. Find the centroid of the area $OPAB$.

Solution. Draw an elementary rectangle, as in the figure, and mark its centroid (h, k). Then

$dA = x\,dy, \quad h = \tfrac{1}{2}x, \quad k = y.$

Using (A), $dM_x = k\,dA = xy\,dy,$
$dM_y = h\,dA = \tfrac{1}{2} x^2\,dy.$

Finding x in terms of y from $y^2 = 2\,px$, and integrating between the limits $y = 0$, $y = b$, we find

$A = \dfrac{b^3}{6\,p}, \quad M_x = \dfrac{b^4}{8\,p}, \quad M_y = \dfrac{b^5}{40\,p^2}.$

Hence $\bar{x} = \dfrac{3\,b^2}{20\,p}, \quad \bar{y} = \dfrac{3}{4}b.$ But $x = a$, $y = b$ satisfy the equation $y^2 = 2\,px$. Hence $b^2 = 2\,pa$, and $\bar{x} = \tfrac{3}{10}a$. The centroid is therefore $(\tfrac{3}{10}a, \tfrac{3}{4}b)$. Ans.

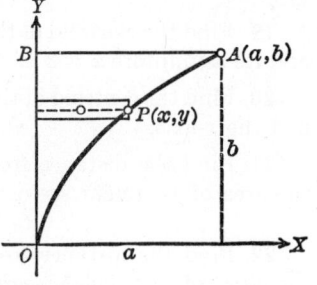

PROBLEMS

Find the centroid of each of the areas bounded by the following curves.

1. $y^2 = 2\ px$, $x = h$.   Ans. $(\frac{3}{5} h, 0)$.
2. $y = x^3$, $x = 2$, $y = 0$. $(\frac{8}{5}, \frac{16}{7})$.
3. $y = x^3$, $y = 4\ x$. (First quadrant.) $(\frac{16}{15}, \frac{64}{21})$.
4. $x = 4\ y - y^2$, $y = x$. $(\frac{12}{5}, \frac{3}{2})$.
5. $y^2 = 4\ x$, $2\ x - y = 4$. $(\frac{8}{5}, 1)$.
6. $y = x^2$, $y = 2\ x + 3$. $(1, \frac{17}{5})$.
7. $y = x^2 - 2\ x - 3$, $y = 6\ x - x^2 - 3$. $(2, 1)$.
8. $y = x^3$, $y = 8$, $x = 0$.
9. $y = 6\ x - x^2$, $y = x$.
10. $y = 4\ x - x^2$, $y = 2\ x - 3$.
11. $y = x^3 - 3\ x$, $y = x$. (First quadrant.)
12. $y^2 = a^2 - ax$, $x = 0$, $y = 0$. (First quadrant.)
13. $\dfrac{x^2}{a^2} - \dfrac{y^2}{b^2} = 1$, $y = 0$, $x = 2\ a$. (First quadrant.)

14. Find the centroid of the area bounded by the coördinate axes and the parabola $\sqrt{x} + \sqrt{y} = \sqrt{a}$.   Ans. $\bar{x} = \bar{y} = \frac{1}{5} a$.

15. Find the centroid of the area bounded by the loop of the curve $y^2 = 4\ x^2 - x^3$.   Ans. $\bar{x} = \frac{16}{7}$, $\bar{y} = 0$.

16. Find the centroid of the portion in the first quadrant of the ellipse $\dfrac{x^2}{a^2} + \dfrac{y^2}{b^2} = 1$.   Ans. $\bar{x} = \dfrac{4\ a}{3\ \pi}$, $\bar{y} = \dfrac{4\ b}{3\ \pi}$.

17. Find the centroid of the area bounded by the parabola $y^2 = 2\ px$ and the line $y = mx$.   Ans. $\bar{x} = \dfrac{4\ p}{5\ m^2}$, $\bar{y} = \dfrac{p}{m}$.

18. Find the centroid of the area included by the parabolas $y^2 = ax$ and $x^2 = by$.   Ans. $\bar{x} = \frac{9}{20} a^{\frac{1}{3}} b^{\frac{2}{3}}$, $\bar{y} = \frac{9}{20} a^{\frac{2}{3}} b^{\frac{1}{3}}$.

19. Find the centroid of the area bounded by the cissoid $y^2(2\ a - x) = x^3$ and its asymptote $x = 2\ a$.   Ans. $\bar{x} = \frac{5}{3} a$, $\bar{y} = 0$.

20. Find the centroid of the area bounded by the witch $x^2 y = 4\ a^2 (2\ a - y)$ and the x-axis.   Ans. $\bar{x} = 0$, $\bar{y} = \frac{1}{2} a$.

21. Find the distance from the center of the circle to the centroid of the area of a circular sector of angle $2\ \theta$.   Ans. $\dfrac{2\ r \sin \theta}{3\ \theta}$.

22. Find the distance from the center of the circle to the centroid of the area of a circular segment the chord of which subtends a central angle $2\ \theta$.   Ans. $\dfrac{2\ r \sin^3 \theta}{3(\theta - \sin \theta \cos \theta)}$.

CENTROIDS, FLUID PRESSURE 323

23. Find the centroid of the area bounded by the cardioid
$$\rho = a(1 + \cos \theta). \qquad Ans. \; \bar{x} = \tfrac{5}{6} a, \; \bar{y} = 0.$$

24. Find the centroid of the area bounded by one loop of the curve $\rho = a \cos 2\theta$.
$$Ans. \; \text{Distance from origin} = \frac{128 \, a\sqrt{2}}{105 \, \pi}.$$

25. Find the centroid of the area bounded by one loop of the curve $\rho = a \cos 3\theta$.
$$Ans. \; \text{Distance from origin} = \frac{81 \, a\sqrt{3}}{80 \, \pi}.$$

178. Centroid of a solid of revolution. The center of gravity of a homogeneous solid is identical with the centroid of that body considered as a geometrical solid. The centroid will lie in any plane of symmetry which the solid may possess.

To achieve a mathematical definition of the centroid of a solid of revolution, it is necessary to modify the discussion of the preceding article only in the details.

Let OX be the geometrical axis of the solid. The centroid will then lie on this axis. Let dV be an element of volume, that is, a cylinder of revolution with altitude Δx and radius y. Then $dV = \pi y^2 \, \Delta x$.

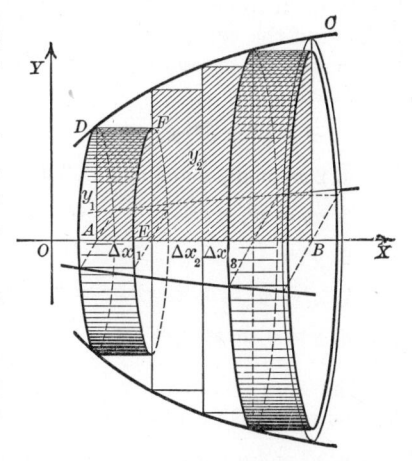

The *moment of volume* of this cylinder with respect to the plane through OY perpendicular to OX is

(1) $\qquad dM_y = x \, dV = \pi x y^2 \, \Delta x.$

The *moment of volume for the solid* is then found by the Fundamental Theorem, and $\bar{x}$ is given from

(2) $\qquad V\bar{x} = M_y = \int \pi x y^2 \, dx.$

ILLUSTRATIVE EXAMPLE. Find the centroid of a solid cone of revolution.

Solution. The equation of the element OB is
$$\frac{y}{x} = \frac{AB}{OA} = \frac{r}{h}, \quad \text{or} \quad y = \frac{rx}{h}.$$

Hence $\quad M_y = \int_0^h \pi x \frac{r^2 x^2}{h^2} \, dx = \tfrac{1}{4} \pi r^2 h^2.$

Since $V = \tfrac{1}{3} \pi r^2 h$, $\bar{x} = \tfrac{3}{4} h$. *Ans.*

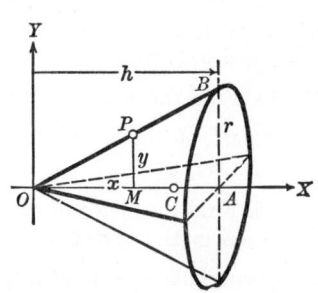

PROBLEMS

Find the centroid for each of the following solids.

1. Hemisphere. (See figure.)   Ans. $\bar{x} = \frac{3}{8} r$.

2. Paraboloid of revolution. (See figure.)   Ans. $\bar{x} = \frac{2}{3} h$.

The area bounded by OX and each curve given below is revolved about OX. Find the centroid of the solid of revolution generated.

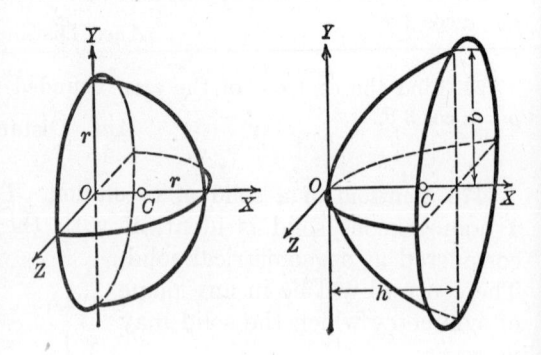

3. $x^2 - y^2 = a^2$, $x = 2a$.

4. $2xy = a^2$, $x = \frac{1}{2} a$, $x = 2a$.

5. $ay = x^2$, $x = a$.   Ans. $\bar{x} = \frac{5}{6} a$.

6. $y^2 = 4x$, $x = 1$, $x = 4$.

7. $x^2 + y^2 = 4$, $x = 0$, $x = 1$. $\bar{x} = \frac{21}{44}$.

8. $y = a \sin x$, $x = \frac{1}{2} \pi$.

The area bounded by OY and each of the curves given below is revolved about OY. Find the centroid of the solid of revolution generated.

9. $y^2 = 4ax$, $y = b$.   Ans. $\bar{y} = \frac{5}{6} b$.

10. $x^2 - y^2 = 1$, $y = 0$, $y = 1$. $\bar{y} = \frac{9}{16}$.

11. $ay^2 = x^3$, $y = a$.

12. The radii of the upper and lower bases of a frustum of a cone of revolution are, respectively, 3 in. and 6 in., and the altitude is 8 in. Locate the centroid.

13. Find the centroid of the solid formed by revolving about the y-axis the area in the first quadrant bounded by the lines $y = 0$, $x = a$, and the parabola $y^2 = 4ax$.   Ans. $\bar{y} = \frac{5}{6} a$.

14. Find the centroid of the solid formed by revolving about the x-axis that part of the area of the ellipse $\dfrac{x^2}{a^2} + \dfrac{y^2}{b^2} = 1$ which lies in the first quadrant.   Ans. $\bar{x} = \frac{3}{8} a$.

15. Find the centroid of the solid formed by revolving about the x-axis the area in the first quadrant bounded by the lines $y = 0$, $x = 2a$, and the hyperbola $\dfrac{x^2}{a^2} - \dfrac{y^2}{b^2} = 1$.

CENTROIDS, FLUID PRESSURE

16. Find the centroid of the solid formed by revolving about the x-axis the area bounded by the lines $x = 0$, $x = a$, $y = 0$, and the hyperbola $\dfrac{x^2}{a^2} - \dfrac{y^2}{b^2} + 1 = 0$.

17. Find the centroid of the solid formed by revolving about the x-axis the area bounded by the lines $y = 0$, $x = \dfrac{\pi}{4}$, and the curve $y = \sin 2x$.

18. Find the centroid of the solid formed by revolving about the x-axis the area bounded by the lines $x = 0$, $x = a$, $y = 0$, and the curve $y = e^x$.

19. The area bounded by a parabola, its axis, and its latus rectum is revolved about the latus rectum. Find the centroid of the solid generated.

Ans. Distance from focus $= \tfrac{5}{32}$ of latus rectum.

179. Fluid pressure. We will now take up the study of *fluid pressure* and learn how to calculate the pressure of a fluid on a vertical wall.

Let $ABDC$ represent part of the area of the vertical surface of one wall of a reservoir. It is desired to determine the total fluid pressure on this area. Draw the axes as in the figure, the y-axis lying in the surface of the fluid. Divide AB into n subintervals and construct horizontal rectangles within the area. Then the area of one rectangle (as EP) is $y \, \Delta x$. If this rectangle was horizontal at the depth x, the fluid pressure on it would be

$$Wxy \, \Delta x,$$

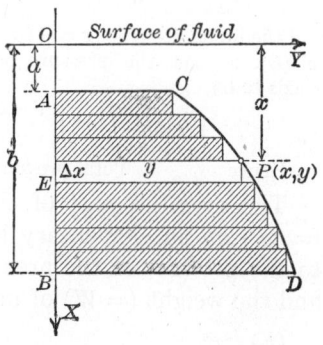

⎡ The pressure of a fluid on any given horizontal surface equals the weight ⎤
⎢ of a column of the fluid standing on that surface as a base and of height ⎥
⎣ equal to the distance of this surface below the surface of the fluid. ⎦

where $W =$ the weight of a unit volume of the fluid. Since fluid pressure is the same in all directions, it follows that $Wxy \, \Delta x$ will be approximately the pressure on the rectangle EP in its vertical position. Hence the sum

$$\sum_{i=1}^{n} Wx_i y_i \Delta x_i$$

represents approximately the pressure on all the rectangles. The pressure on the area $ABDC$ is evidently the limit of this sum. Hence, by the Fundamental Theorem,

$$\lim_{n \to \infty} \sum_{i=1}^{n} Wx_i y_i \Delta x_i = \int Wxy \, dx.$$

Hence the fluid pressure on a vertical submerged surface bounded by a curve, the x-axis, and the two horizontal lines $x = a$ and $x = b$ is given by the formula

(D) $$\text{Fluid pressure} = W \int_a^b yx\,dx,$$

where the value of y in terms of x must be substituted from the equation of the given curve.

We shall assume 62 lb. ($= W$) as the weight of a cubic foot of water.

ILLUSTRATIVE EXAMPLE 1. A circular water main 6 ft. in diameter is half full of water. Find the pressure on the gate that closes the main.

Solution. The equation of the circle is $x^2 + y^2 = 9$.

Hence $y = \sqrt{9 - x^2}$,

$W = 62$,

and the limits are from $x = 0$ to $x = 3$. Substituting in (D), we get the pressure on the right of the x-axis to be

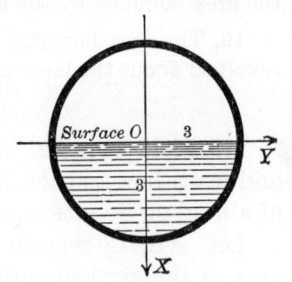

$$\text{Pressure} = 62 \int_0^3 \sqrt{9-x^2} \cdot x\,dx = [-\tfrac{62}{3}(9-x^2)^{\frac{3}{2}}]_0^3 = 558.$$

Hence Total pressure $= 2 \times 558 = 1116$ lb. *Ans.*

The essential part of the above reasoning is that the pressure ($= dP$) on an elementary horizontal strip is equal (approximately) to the product of the area of the strip ($= dA$) by its depth ($= h$) and the weight ($= W$) of unit volume of the fluid. That is,

(E) $$dP = Wh\,dA.$$

With this in mind, the axes of coördinates may be chosen in any convenient position.

ILLUSTRATIVE EXAMPLE 2. A trapezoidal gate in a dam is shown in the figure. Find the pressure on the gate when the surface of the water is 4 ft. above the top of the gate.

Solution. Choosing axes OX and OY as shown, and drawing an elementary horizontal strip, we have, using (E),

$dA = 2x\,dy$,
$h = 8 - y$,
$dP = W(8 - y)2x\,dy$.

The equation of AB is $y = 2x - 8$. Solving this equation for x, and substituting, the result is

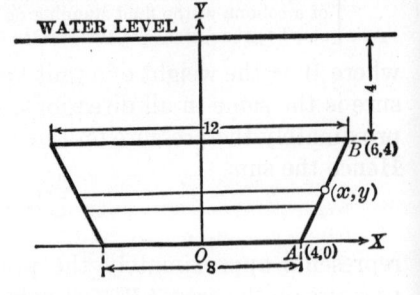

$$dP = W(8 - y)(y + 8)dy = W(64 - y^2)dy.$$

Integrating with limits $y = 0$ and $y = 4$, we obtain

$$P = W \int_0^4 (64 - y^2)dy = \tfrac{704}{3} W = 14{,}549 \text{ lb.} \; Ans.$$

CENTROIDS, FLUID PRESSURE

PROBLEMS

In the following problems the y-axis is directed vertically upward, and the x-axis is at the surface level of a liquid. Denoting the weight of a cubic unit of the liquid by W, calculate the pressure on the areas formed by joining with straight lines each set of points in the order given.

1. $(0, 0)$, $(3, 0)$, $(0, -6)$, $(0, 0)$.   *Ans.* $18\ W$.
2. $(0, 0)$, $(3, -6)$, $(0, -6)$, $(0, 0)$. $36\ W$.
3. $(0, 0)$, $(2, -2)$, $(0, -4)$, $(-2, -2)$, $(0, 0)$. $16\ W$.

4. Calculate the pressure on the lower half of an ellipse whose semi-axes are 2 and 3 units respectively, (a) when the major axis lies in the surface of the liquid; (b) when the minor axis lies in the surface.
Ans. (a) $8\ W$; (b) $12\ W$.

5. Each end of a horizontal oil tank is an ellipse of which the horizontal axis is 12 ft. long and the vertical axis 6 ft. long. Calculate the pressure on one end when the tank is half full of oil weighing 60 lb. per cubic foot.
Ans. 2160 lb.

6. The vertical end of a vat is a segment of a parabola (with vertex at the bottom) 8 ft. across the top and 16 ft. deep. Calculate the pressure on this end when the vat is full of a liquid weighing 70 lb. per cubic foot.
Ans. 38,229 lb.

7. The vertical end of a water trough is an isosceles right triangle of which each leg is 8 ft. Calculate the pressure on the end when the trough is full of water ($W = 62.5$).   *Ans.* 3771 lb.

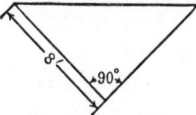

8. The vertical end of a water trough is an isosceles triangle 5 ft. across the top and 5 ft. deep. Calculate the pressure on the end when the trough is full of water.   *Ans.* 1302 lb.

9. A horizontal cylindrical tank of diameter 8 ft. is half full of oil weighing 60 lb. per cubic foot. Calculate the pressure on one end.
Ans. 2560 lb.

10. Calculate the pressure on one end if the tank of Problem 9 is full.

11. A rectangular gate in a vertical dam is 10 ft. wide and 6 ft. deep. Find (a) the pressure when the level of the water ($W = 62.5$) is 8 ft. above the top of the gate; (b) how much higher the water must rise to double the pressure found in (a).   *Ans.* (a) 41,250 lb.; (b) 11 ft.

12. Show that the pressure on any vertical surface is the product of the weight of a cubic unit of the liquid, the area of the surface, and the depth of the centroid of the area.

13. A vertical cylindrical tank, of diameter 30 ft. and height 50 ft., is full of water. Find the pressure on the curved surface.   *Ans.* 3682 tons.

180. Work. In mechanics the work done by a constant force F causing a displacement d is the product Fd. When F is variable, this definition leads to an integral. Two examples will be considered here.

Work done in pumping out a tank. Let us now consider the problem of finding the work done in emptying reservoirs of the form of solids of revolution with their axes vertical. It is convenient to assume the x-axis of the revolved curve as vertical, and the y-axis as on a level with the top of the reservoir.

Consider a reservoir such as the one shown; we wish to calculate the work done in emptying it of a fluid from the depth a to the depth b.

Divide AB into n subintervals, pass planes perpendicular to the axis of revolution through these points of division, and construct cylinders of revolution, as in Art. 160. The volume of any such cylinder will be $\pi y^2 \Delta x$ and its weight $W \pi y^2 \Delta x$, where $W =$ weight of a cubic unit of the fluid. The work done in lifting this cylinder of the fluid out of the reservoir (through the height x) will be

$$W \pi y^2 x \Delta x.$$

[Work done in lifting equals the weight multiplied by the vertical height.]

The work done in lifting all such cylinders to the top is the sum

$$\sum_{i=1}^{n} W \pi y_i^2 x_i \Delta x_i.$$

The work done in emptying that part of the reservoir will evidently be the limit of this sum. Hence, by the Fundamental Theorem,

$$\lim_{n \to \infty} \sum_{i=1}^{n} W \pi y_i^2 x_i \Delta x_i = \int W \pi y^2 x \, dx.$$

Therefore the work done in emptying a reservoir in the form of a solid of revolution from the depth a to the depth b is given by the formula

(F) $$\text{Work} = W \pi \int_a^b y^2 x \, dx,$$

where the value of y in terms of x must be substituted from the equation of the revolved curve.

CENTROIDS, FLUID PRESSURE

ILLUSTRATIVE EXAMPLE 1. Calculate the work done in pumping out the water filling a hemispherical reservoir 10 ft. deep.

Solution. The equation of the circle is

$$x^2 + y^2 = 100.$$

Hence $\quad y^2 = 100 - x^2,$
$\quad\quad W = 62,$

and the limits are from $x = 0$ to $x = 10$.
Substituting in (F), we get

$$\text{Work} = 62\,\pi \int_0^{10} (100 - x^2)x\,dx = 155{,}000\,\pi \text{ ft.-lb.}$$

The essential principle in the above reasoning is that the element of work ($= dw$) done in lifting an elementary volume ($= dV$) through a height ($= h$) is

$$dw = Wh\,dV,$$

where $W = $ weight of unit volume of the fluid. With this in mind, the axes of coördinates may be chosen in any convenient manner.

ILLUSTRATIVE EXAMPLE 2. A conical cistern is 20 ft. across the top and 15 ft. deep. If the surface of the water is 5 ft. below the top, find the work done in pumping the water to the top of the cistern.

Solution. Take axes OX and OY as in the figure. Then

$$dV = \pi x^2\,dy,$$
$$h = 15 - y,$$
$$dw = W(15 - y)\pi x^2\,dy.$$

The equation of the element OA is $x = \tfrac{2}{3}y$. Substituting,

$$dw = \pi W(15 - y)\tfrac{4}{9}y^2\,dy = \tfrac{4}{9}\pi W(15\,y^2 - y^3)dy.$$

The limits are $y = 0$ and $y = 10$, since the water is 10 ft. deep. Integrating,

$$w = \tfrac{4}{9}\pi W\int_0^{10}(15\,y^2 - y^3)dy = 216{,}421 \text{ ft.-lb. } Ans.$$

Work done by an expanding gas. If a gas in a cylinder expands against a piston head from volume v_0 cu. ft. to v_1 cu. ft., the external work done in foot-pounds is

(G) $\quad\quad\quad\quad \text{Work} = \displaystyle\int_{v_0}^{v_1} p\,dv,$

where $p = $ pressure in pounds per square foot.

Proof. Let the volume increase from v to $v + dv$.
Let $c = $ area of cross section of the cylinder.
Then $\dfrac{dv}{c} = $ distance the piston moves.

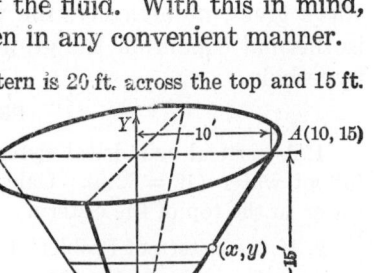

Since $pc =$ force causing the expansion dv,

$$\text{Element of work done} = pc \cdot \frac{dv}{c} = p\, dv.$$

Then (*G*) follows by the Fundamental Theorem. To use (*G*), the relation between p and v during the expansion must be known. This relation has the form

(1) $$pv^n = \text{constant},$$

the exponent n being a constant.

Isothermal expansion occurs when the temperature remains constant. Then $n = 1$, and the pressure-volume relation is

(2) $$pv = p_0 v_0 = p_1 v_1.$$

If a graph of (1) is made (*pressure-volume diagram*), plotting volumes as abscissas and pressures as ordinates, the area under this curve gives, numerically, the work done, as calculated by (*G*). In isothermal expansion the graph of (2) is a rectangular (equilateral) hyperbola.

PROBLEMS

1. A vertical cylindrical cistern of diameter 16 ft. and depth 20 ft. is full of water ($W = 62.5$). Calculate the work necessary to pump the water to the top of the cistern. *Ans.* 800,000 π ft. lb.

2. If the cistern of Problem 1 is half full, calculate the work necessary to pump the water to the top.

3. A conical cistern 20 ft. across the top and 20 ft. deep is full of water ($W = 62.5$). Calculate the work necessary to pump the water to a height of 15 ft. above the top of the cistern. *Ans.* $\dfrac{2,500,000\,\pi}{3}$ ft. lb.

4. A hemispherical tank of diameter 10 ft. is full of oil weighing 60 lb. per cubic foot. Calculate the work necessary to pump the oil to the top of the tank. *Ans.* 9375 π ft. lb.

5. A hemispherical tank of diameter 20 ft. is full of oil weighing 60 lb. per cubic foot. The oil is pumped to a height of 10 ft. above the top of the tank by an engine of $\frac{1}{2}$ H.P. (that is, the engine can do work at the rate of 16,500 ft. lb. per minute). How long will it take the engine to empty the tank?

6. Find the work done in pumping out a semi-elliptical reservoir full of water ($W = 62$). The top is a circle of diameter 6 ft., and the depth is 5 ft. *Ans.* $3487\frac{1}{2}$ π ft. lb.

7. A conical reservoir 12 ft. deep is filled with a liquid weighing 80 lb. per cubic foot. The top of the reservoir is a circle 8 ft. in diameter. Calculate the work necessary to pump the liquid to the top of the reservoir. *Ans.* 15,360 π ft. lb.

CENTROIDS, FLUID PRESSURE

8. A water tank is in the form of a hemisphere, 24 ft. in diameter, surmounted by a cylinder of the same diameter and 10 ft. high. Find the work done in pumping it out when it is filled within 2 ft. of the top.

9. A bucket of weight M is to be lifted from the bottom of a shaft h ft. deep. The weight of the rope used to hoist it is m lb. per foot. Find the work done.

10. A quantity of air with an initial volume of 200 cu. ft. and pressure of 15 lb. per square inch is compressed to 80 lb. per square inch. Determine the final volume and the work done if the isothermal law holds, that is, $pv = C$. *Ans.* 37.5 cu. ft.; 723,000 ft. lb.

11. Determine the final volume and work done in Problem 10 if the adiabatic law holds, that is $pv^n = C$, assuming $n = 1.4$.
Ans. 60 cu. ft.; 648,000 ft. lb.

12. Air at pressure of 15 lb. per square inch is compressed from 200 cu. ft. to 50 cu. ft. Determine the final pressure and the work done if the law is $pv = C$. *Ans.* 60 lb. per square inch; 599,000 ft. lb.

13. Solve Problem 12 if the law is $pv^n = C$, assuming $n = 1.4$.
Ans. 104.5 lb. per square inch; 801,000 ft. lb.

14. A quantity of gas with an initial volume of 16 cu. ft. and pressure of 60 lb. per square inch expands until the pressure is 30 lb. per square inch. Determine the final volume and the work done by the gas if the law is $pv = C$. *Ans.* 32 cu. ft.; 95,800 ft. lb.

15. Solve Problem 14 if the law is $pv^n = C$, assuming $n = 1.2$.
Ans. 28.5 cu. ft.; 75,600 ft. lb.

16. A quantity of air with an initial volume of 200 cu. ft. and pressure of 15 lb. per square inch is compressed to 30 cu. ft. Determine the final pressure and the work done if the law is $pv = C$.

17. Solve Problem 16 if the law is $pv^n = C$, assuming $n = 1.4$.

18. A gas expands from an initial pressure of 80 lb. per square inch and volume of 2.5 cu. ft. to a volume of 9 cu. ft. Find the work done if the law is $pv^n = C$, assuming $n = 1.0646$.

19. Solve Problem 18 if $n = 1.131$.

20. Determine the amount of attraction exerted by a thin, straight, homogeneous rod of uniform thickness, of length l, and of mass M upon a material point P of mass m situated at a distance of a from one end of the rod in its line of direction.

Solution. Suppose the rod to be divided into equal infinitesimal portions (elements) of length dx.
$$\frac{M}{l} = \text{mass of a unit length of rod};$$

hence $\qquad \dfrac{M}{l} dx = \text{mass of any element}.$

332   INTEGRAL CALCULUS

Newton's law for measuring the attraction between any two masses is

$$\text{Force of attraction} = \frac{\text{product of masses}}{(\text{distance between them})^2};$$

therefore the force of attraction between the particle at P and an element of the rod is

$$\frac{\frac{M}{l} m \, dx}{(x+a)^2},$$

which is then an *element of the force of attraction required*. The total attraction between the particle at P and the rod being the limit of the sum of all such elements between $x = 0$ and $x = l$, we have

$$\text{Force of attraction} = \int_0^l \frac{\frac{M}{l} m \, dx}{(x+a)^2} = \frac{Mm}{l} \int_0^l \frac{dx}{(x+a)^2} = +\frac{Mm}{a(a+l)}. \quad Ans.$$

21. Determine the amount of attraction in the last example if P lies in the perpendicular bisector of the rod at the distance a from it.

$$Ans. \quad \frac{2 \, mM}{a\sqrt{4 \, a^2 + l^2}}.$$

22. A vessel in the form of a right circular cone is filled with water. If h is its height and r the radius of the base, what time will it require to empty itself through an orifice of area a at the vertex?

Solution. Neglecting all hurtful resistances, it is known that the velocity of discharge through an orifice is that acquired by a body falling freely from a height equal to the depth of the water. If, then, x denotes the depth of the water,

$$v = \sqrt{2 \, gx}.$$

Denote by dQ the volume of water discharged in time dt, and by dx the corresponding fall of surface. The volume of water discharged through the orifice in a unit of time is

$$a\sqrt{2 \, gx},$$

being measured as a right cylinder of area of base a and altitude $v \, (= \sqrt{2 \, gx})$. Therefore in time dt

(1) $\qquad dQ = a\sqrt{2 \, gx} \, dt.$

Denoting by S the area of the surface of the water when the depth is x, we have, from geometry,

$$\frac{S}{\pi r^2} = \frac{x^2}{h^2}, \quad \text{or} \quad S = \frac{\pi r^2 x^2}{h^2}.$$

But the volume of water discharged in time dt may also be considered as the volume of a cylinder AB of area of base S and altitude dx; hence

(2) $\qquad dQ = S \, dx = \frac{\pi r^2 x^2 \, dx}{h^2}.$

Equating (1) and (2) and solving for dt,

$$dt = \frac{\pi r^2 x^2 \, dx}{ah^2 \sqrt{2 \, gx}}.$$

Therefore $\qquad t = \int_0^h \frac{\pi r^2 x^2 \, dx}{ah^2 \sqrt{2 \, gx}} = \frac{2 \, \pi r^2 \sqrt{h}}{5 \, a\sqrt{2 \, g}}. \quad Ans.$

CENTROIDS, FLUID PRESSURE

181. Mean value of a function. The *arithmetic mean* (or average value) of n numbers $y_1, y_2, \cdots, y_n$, is

(1) $\quad \bar{y} = \dfrac{1}{n}(y_1 + y_2 + \cdots + y_n).$

We proceed to establish the formula

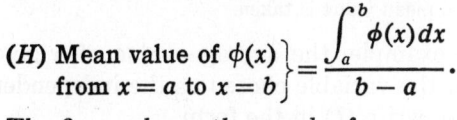

(H) Mean value of $\phi(x)$ from $x = a$ to $x = b$ $\Bigg\} = \dfrac{\int_a^b \phi(x)\,dx}{b-a}.$

The figure shows the graph of

(2) $\quad y = \phi(x).$

The mean value ($= \bar{y}$) of the ordinates of the arc PQ is to be defined. Divide AB into n equal parts each equal to Δx and let $y_1, y_2, \cdots, y_n$, be the ordinates at the n points of division. Then (1) will give an approximate value for the mean value required. Multiply numerator and denominator of the right-hand member of (1) by Δx. Then, since $n\,\Delta x = b - a$, we get

(3) $\quad \bar{y}$ (approximately) $= \dfrac{y_1\,\Delta x + y_2\,\Delta x + \cdots + y_n\,\Delta x}{b-a}.$

But the numerator in (3) is, approximately, the area $APRQB$. The average value of y (or $\phi(x)$) is defined as the limit of the right-hand member in (3) when $n \to \infty$. This gives (H).

In the figure the mean value of $\phi(x)$ equals CR if area rectangle $ABML$ = area $ABQRP$.

Taking y as the function (dependent variable), then (H) becomes

(I) $\quad \bar{y} = \dfrac{\int_a^b y\,dx}{b-a}.$

ILLUSTRATIVE EXAMPLE. Given the circle

(4) $\quad x^2 + y^2 = r^2.$

Find the average value of the ordinates in the first quadrant

(a) when y is expressed as a function of the abscissa x;

(b) when y is expressed as a function of the angle $\theta = \angle MOP$.

Solution. (a) Since $y = \sqrt{r^2 - x^2}$, the numerator in (I) is

$\int_0^r \sqrt{r^2 - x^2}\,dx = \tfrac{1}{4}\pi r^2.$ Then $\bar{y} = \tfrac{1}{4}\pi r = 0.785\,r.$ *Ans.*

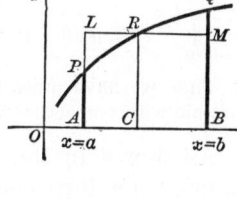

334 INTEGRAL CALCULUS

(b) Since $y = r \sin \theta$, and the limits are $\theta = 0 = a$, $\theta = \frac{1}{2}\pi = b$, the numerator in (*I*) is

$$\int_0^{\frac{1}{2}\pi} r \sin \theta \, d\theta = r.$$ Since $b - a = \frac{1}{2}\pi$, we have $\bar{y} = \frac{2\,r}{\pi} = 0.637\, r$. *Ans.*

Thus we have quite different values of y, depending upon the independent variable with respect to which the mean value is taken.

As shown in the above example, the average value of a given function y will depend upon the variable chosen as the independent variable. For this reason, we write (*I*) in the form

(5) $$\bar{y}_x = \frac{\int_a^b y \, dx}{b - a},$$

in order to indicate explicitly the variable with respect to which y is averaged.

Thus, in the Illustrative Example, we have $\bar{y}_x = 0.785\, r$, and $\bar{y}_\theta = 0.637\, r$.

PROBLEMS

1. Find the average value of $y = x^2$ from $x = 0$ to $x = 10$. *Ans.* $33\frac{1}{3}$.

2. Find the average value of the ordinates of $y^2 = 4\, x$ from $(0, 0)$ to $(4, 4)$ taken uniformly along the x-axis. *Ans.* $2\frac{2}{3}$.

3. Find the average value of the abscissas of $y^2 = 4\, x$ from $(0, 0)$ to $(4, 4)$ when uniformly distributed along the y-axis. *Ans.* $1\frac{1}{3}$.

4. Find the average value of $\sin x$ between $x = 0$ and $x = \pi$. *Ans.* $\frac{2}{\pi}$.

5. Find the average value of $\sin^2 x$ between $x = 0$ and $x = \pi$. (This average value is frequently used in the theory of alternating currents.)
Ans. $\frac{1}{2}$.

6. If a particle in a vacuum were thrown downward with an initial velocity of v_0 ft. per second, the velocity after t sec. would be given by

(1) $$v = v_0 + gt.$$

The velocity after falling s ft. would be given by

(2) $$v = \sqrt{v_0^2 + 2\, gs}.$$ (Take $g = 32$)

Find the average value of v

(a) during the first 5 sec., starting from rest; *Ans.* 80 ft. per second.

(b) during the first 5 sec., starting with an initial velocity of 36 ft. per second; *Ans.* 116 ft. per second.

(c) during the first $2\frac{1}{2}$ sec., starting from rest; *Ans.* 40 ft. per second.

(d) during the first 100 ft., starting from rest; *Ans.* $53\frac{1}{3}$ ft. per sec.

(e) during the first 100 ft., starting with an initial velocity of 60 ft. per second. *Ans.* $81\frac{2}{3}$ ft. per second.

CENTROIDS, FLUID PRESSURE

7. In simple harmonic motion $s = a \cos nt$. Find the average value of the velocity during one quarter of a period (a) as to the time; (b) as to the distance.

8. Show that in simple harmonic motion the average kinetic energy with respect to the time for any multiple of a quarter period is half the maximum kinetic energy.

9. A point is taken at random on a straight line of length a. Prove (a) that the average area of the rectangle whose sides are the two segments is $\frac{1}{6} a^2$; (b) that the average value of the sum of the squares on the two segments is $\frac{2}{3} a^2$.

10. If a point moves with constant acceleration, the average as to the time of the square of the velocity is $\frac{1}{3}(v_0^2 + v_0 v_1 + v_1^2)$, where v_0 is the initial and v_1 the final velocity.

11. Show that the average horizontal range of a particle projected with a given velocity at an arbitrary elevation is 0.6366 of the maximum horizontal range.

HINT. Take $\alpha = 0$ in the formula of Problem 35, p. 114.

The formulas

(6) $$\bar{x}_s = \frac{\int x \, ds}{\int ds}, \quad \bar{y}_s = \frac{\int y \, ds}{\int ds},$$

where (x, y) is any point on a curve for which ds is the element of arc, define the *centroid of the arc*. They give, respectively, the average values of the abscissas and ordinates of points on the curve when distributed uniformly along it. (Compare Art. 177.)

12. Show that the area of the curved surface generated by revolving an arc of a plane curve about a line in its plane not cutting the arc equals the length of the arc times the circumference of the circle described by its centroid (6). (Theorem of Pappus. Compare Art. 250.)

HINT. Use (L), Art. 164.

13. Find the centroid of the arc of the parabola $y^2 = 4x$ from $(0, 0)$ to $(4, 4)$. Ans. $\bar{x} = 1.64$, $\bar{y} = 2.29$.

14. Find the centroid of an arc of the circle $\rho = a$ between $-\theta$ and $+\theta$.

Ans. $\bar{x} = \dfrac{a \sin \theta}{\theta}$.

15. Find the centroid of the perimeter of the cardioid $\rho = a(1 + \cos \theta)$.

Ans. $\bar{x} = \frac{4}{5} a$, $\bar{y} = 0$.

16. Find by the Theorem of Pappus the centroid of the arc of the circle $x^2 + y^2 = r^2$ which lies in the first quadrant.

Ans. $\bar{x} = \bar{y} = \dfrac{2 r}{\pi}$.

17. Find by the Theorem of Pappus the surface of the torus generated by revolving the circle $(x - b)^2 + y^2 = a^2$ $(b > a)$ about the y-axis.

18. A rectangle is revolved about an axis which lies in its plane and is perpendicular to a diagonal at its extremity. Find the area of the surface generated.

19. If a particle falls from rest in a vacuum for t seconds, show that the average velocity during this time is one half the velocity at the end of the fall.

20. If a particle falls from rest in a vacuum through a distance h, show that the average velocity over this distance is two thirds the velocity at the end of the fall.

21. Rectangles with sides parallel to the coördinate axes are inscribed in the circle $x^2 + y^2 = a^2$. Find the average value of their areas if the vertical sides are equally spaced along the x-axis. *Ans.* $\frac{4}{3} a^2$.

22. Rectangles with sides parallel to the coördinate axes are inscribed in the ellipse $b^2 x^2 + a^2 y^2 = a^2 b^2$. Find the average value of their areas if the horizontal sides are equally spaced along the y-axis. *Ans.* $\frac{4}{3} ab$.

ADDITIONAL PROBLEMS

1. An area is bounded by the lines $y = x^2$, $x + y = 6$, $y = 0$, and $x = 3$. Find its centroid. *Ans.* $\bar{x} = \frac{76}{37}, \bar{y} = \frac{281}{185}$.

2. The abscissa of the centroid of the area bounded by the curve $2y = x^2$ and a certain line through the origin is 1. Find the ordinate of the centroid. *Ans.* $\frac{4}{5}$.

3. Find the centroid of the area bounded by $y = x^n (n > 0)$, the x-axis, and $x = 1$. Discuss the locus of the centroid as n varies.

Ans. $\bar{x} = \dfrac{n+1}{n+2}, \bar{y} = \dfrac{n+1}{2(2n+1)}$.

4. Find the equation of the locus of the centroid of the area bounded by the x-axis and the parabola $y = cx - x^2$ when c varies. *Ans.* $5y = 2x^2$.

5. Given the parabola $x^2 = 2py$ and any oblique line $y = mx + b$ meeting the parabola in the points A and B. Through C, the midpoint of AB, draw a line parallel to the axis of the curve meeting the parabola at D. Prove that (a) the tangent to the parabola at D is parallel to the line AB; (b) the centroid of the area $ACBD$ lies on the line CD.

6. Let P be a point on the parabola $y = x^2$, and let C be the centroid of the area bounded by the parabola, the x-axis, and the ordinate through P. Find the position of P so that the angle OPC is a maximum.

Ans. Ordinate $= \frac{5}{14}$.

CENTROIDS, FLUID PRESSURE

7. A cistern has the form of a solid generated by revolving about its vertical axis a parabolic segment cut off by a chord 8 ft. long, perpendicular to the axis and at a distance of 8 ft. from the vertex. The cistern is filled with water weighing 62.5 lb. per cubic foot. Find the amount of work required to pump over the top of the cistern one half the volume of water it contains.   *Ans.* $16,000(\sqrt{2}-1)\dfrac{\pi}{3} = 6937$ ft.-lb.

8. A hemispherical cistern of radius r is full of water. Two men, A and B, are to pump it out, each doing half the work. If A starts first, what will be the depth d of the water when he has finished his share of the work?

$$Ans.\ \frac{d}{r} = 1 - \sqrt{\frac{2-\sqrt{2}}{2}} = 0.459.$$

9. A tank in the shape of an inverted circular cone is full of water. Two men are to pump the water to the top of the tank, each doing half the work. When the first man has finished his share of the work, let z denote the ratio of the depth of water left in the tank to the original depth. Show that z is determined by the equation $6z^4 - 8z^3 + 1 = 0$. Calculate the value of z to two decimals.   *Ans.* 0.61.

10. A well is 100 ft. deep. A bucket, weighing 3 lb., has a volume of 2 cu. ft. The bucket is filled with water at the bottom of the well and is then raised at a constant rate of 5 ft. per second to the top. Neglecting the weight of the rope, find the work done in raising the bucket if it is discovered that the water is leaking out at a constant rate of 0.01 cu. ft. per second. (A cubic foot of water weighs 62.4 lb.)   *Ans.* 12,156 ft.-lb.

11. The area OAB is divided into elements such as OPQ by lines from O. Show that the area A and the moments of area M_x and M_y are given by

$$A = \tfrac{1}{2}\int (xy' - y)dx, \qquad M_x = \tfrac{1}{3}\int y(xy' - y)dx,$$

$$M_y = \tfrac{1}{3}\int x(xy' - y)dx.$$

(The centroid of a triangle is on any median at two thirds of the distance from the vertex to the opposite side.)

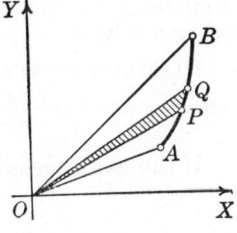

12. Find the centroid of the hyperbolic sector bounded by the equilateral hyperbola $x = a\sec\theta$, $y = a\tan\theta$ and radii from the origin to the points $(a, 0)$ and (x, y).

$$Ans.\ \bar{x} = \frac{2}{3}a\frac{\tan\theta}{\ln(\sec\theta + \tan\theta)},\ \bar{y} = \frac{2}{3}a\frac{\sec\theta - 1}{\ln(\sec\theta + \tan\theta)}.$$

CHAPTER XIX

SERIES

182. Definitions. A *sequence* is a succession of terms formed according to some fixed rule or law.

For example,   1, 4, 9, 16, 25

and $1, -x, \dfrac{x^2}{2}, -\dfrac{x^3}{3}, \dfrac{x^4}{4}, -\dfrac{x^5}{5}$

are sequences.

A *series* is the indicated sum of the terms of a sequence. Thus from the above sequences we obtain the series

$$1 + 4 + 9 + 16 + 25$$

and $1 - x + \dfrac{x^2}{2} - \dfrac{x^3}{3} + \dfrac{x^4}{4} - \dfrac{x^5}{5}.$

When the number of terms is limited, the sequence or series is said to be *finite*. When the number of terms is unlimited, the sequence or series is called an *infinite sequence* or *infinite series*.

The *general term*, or nth term, is an expression which indicates the law of formation of the terms.

ILLUSTRATIVE EXAMPLE 1. In the first example given above, the general term, or nth term, is n^2. The first term is obtained by setting $n = 1$, the tenth term by setting $n = 10$, etc.

ILLUSTRATIVE EXAMPLE 2. In the second example given above, the nth term, except for $n = 1$, is $\dfrac{(-x)^{n-1}}{n-1}$.

If the sequence is infinite, this fact is indicated by the use of dots, as

$$1, 4, 9, \cdots, n^2, \cdots.$$

Factorial numbers. An expression which occurs frequently in connection with series is a product of successive integers, beginning with 1. Thus, $1 \times 2 \times 3 \times 4 \times 5$ is called 5 *factorial* and is indicated by $\underline{|5}$ or 5!

In general, $\underline{|n} = 1 \times 2 \times 3 \times \cdots \times (n-1) \times n$

is called *n factorial*. It is understood that n is a positive integer. The expression $\underline{|n}$ has no meaning if n is not a positive integer.

183. The geometric series. For the geometric series of n terms,

(1) $$S_n = a + ar + ar^2 + \cdots + ar^{n-1},$$

it is shown in elementary algebra that

(2) $$S_n = \frac{a(1-r^n)}{1-r}, \quad \text{or} \quad S_n = \frac{a(r^n-1)}{r-1},$$

the first form being generally used if $|r| < 1$, and the second form if $|r| > 1$.

If $|r| < 1$, then r^n decreases in numerical value as n increases and

$$\lim_{n \to \infty} (r^n) = 0.$$

From formula (2) we see, therefore, that (Art. 16)

(3) $$\lim_{n \to \infty} S_n = \frac{a}{1-r}.$$

Hence if $|r| < 1$ the sum S_n of a geometric series approaches a limit as the number of terms is increased indefinitely. In this case the series is said to be **convergent**.

If $|r| > 1$, then r^n will become infinite as n increases indefinitely (Art. 18). Hence, from the second formula in (2), the sum S_n will become infinite. In this case the series is said to be **divergent**.

A peculiar situation presents itself if $r = -1$. The series then becomes

(4) $$a - a + a - a + a - a \cdots.$$

If n is even the sum is zero. If n is odd the sum is a. As n increases indefinitely the sum does not increase indefinitely and it does not approach a limit. Such a series is called an **oscillating series**.

ILLUSTRATIVE EXAMPLE. Consider the geometric series with

$$a = 1, \quad r = \tfrac{1}{2},$$

(5) $$S_n = 1 + \frac{1}{2} + \frac{1}{4} + \cdots + \frac{1}{2^{n-1}}.$$

We find, by (2), that $$S_n = \frac{1 - \frac{1}{2^n}}{1 - \frac{1}{2}} = 2 - \frac{1}{2^{n-1}}.$$

Then

(6) $\lim_{n \to \infty} S_n = 2$, which agrees with (3), when $a = 1, r = \tfrac{1}{2}$.

It is interesting to discuss (5) geometrically. To do this, lay off successive values of S_n on a straight line, as in the figure.

n	1	2	3	4,	etc.
S_n	1	$1\tfrac{1}{2}$	$1\tfrac{3}{4}$	$1\tfrac{7}{8}$	etc.

Each point thus determined bisects the segment between the preceding point and the point 2. Hence (6) is obvious.

PROBLEMS

In each of the following series (a) discover by inspection the law of formation; (b) write three more terms; (c) find the nth, or *general*, term.

1. $2 + 4 + 8 + 16 + \cdots$.   *Ans.* nth term $= 2^n$.

2. $1 - \frac{1}{2} + \frac{1}{3} - \frac{1}{4} + \cdots$. $\dfrac{(-1)^{n-1}}{n}$.

3. $-\frac{1}{2} + 0 + \frac{1}{4} + \frac{2}{5} + \frac{3}{6} + \cdots$. $\dfrac{n-2}{n+1}$.

4. $x + \dfrac{x^2}{1} + \dfrac{x^3}{1 \cdot 2} + \dfrac{x^4}{1 \cdot 2 \cdot 3} + \cdots$. $\dfrac{nx^n}{\lfloor n}$

5. $\dfrac{\sqrt{x}}{2} + \dfrac{x}{2 \cdot 4} + \dfrac{x\sqrt{x}}{2 \cdot 4 \cdot 6} + \dfrac{x^2}{2 \cdot 4 \cdot 6 \cdot 8} + \cdots$. $\dfrac{x^{\frac{n}{2}}}{2^n \lfloor n}$

6. $\dfrac{a^2}{3} - \dfrac{a^3}{5} + \dfrac{a^4}{7} - \dfrac{a^5}{9} + \cdots$. $\dfrac{(-a)^{n+1}}{2n+1}$.

Write the first four terms of the series whose nth, or *general*, term is given below.

7. $\dfrac{2^{n-1}}{\sqrt{n}}$.   *Ans.* $1 + \dfrac{2}{\sqrt{2}} + \dfrac{4}{\sqrt{3}} + \dfrac{8}{\sqrt{4}} + \cdots$.

8. $\dfrac{n+2}{2n-1}$. $3 + \dfrac{4}{3} + \dfrac{5}{5} + \dfrac{6}{7} + \cdots$.

9. $\dfrac{n}{3^{n-1}}$. $1 + \dfrac{2}{3} + \dfrac{3}{9} + \dfrac{4}{27} + \cdots$.

10. $\dfrac{x^{n-1}}{\sqrt{n}}$. $1 + \dfrac{x}{\sqrt{2}} + \dfrac{x^2}{\sqrt{3}} + \dfrac{x^3}{\sqrt{4}} + \cdots$.

11. $\dfrac{(-1)^{n-1} x^{2n-1}}{\lfloor 2n-1}$. $x - \dfrac{x^3}{\lfloor 3} + \dfrac{x^5}{\lfloor 5} - \dfrac{x^7}{\lfloor 7} + \cdots$.

12. $\dfrac{(x-a)^{n-1}}{\lfloor n}$.   13. $\dfrac{(y+n)^{2n-1}}{2^n \lfloor n}$.   14. $\dfrac{2^{\frac{n+1}{2}}}{\sqrt{n+2}}$.   15. $\dfrac{3^{n-1} \theta^n}{2^n \lfloor n-1}$.

184. Convergent and divergent series. In the series

$$S_n = u_1 + u_2 + u_3 + \cdots + u_n,$$

the variable S_n is a function of n. If we now let the number of terms ($= n$) increase without limit, one of two things may happen.

CASE I. S_n *approaches a limit, say* u, *indicated by*

(1) $\lim_{n \to \infty} S_n = u.$

The infinite series is now said to be *convergent* and to converge to the value u, or to have the value u.

CASE II. S_n approaches no limit. The infinite series is now said to be *divergent*.

Examples of divergent series are
$$1 + 2 + 3 + 4 + 5 + \cdots,$$
$$1 - 1 + 1 - 1 + \cdots.$$

As stated above, in a convergent series the *value of the series* is the number u (sometimes called the sum) defined by (1). No value is assigned to a divergent series.

In the applications of infinite series, convergent series are of major importance. Thus it is essential to have means of testing a given series for convergence or divergence.

185. General theorems. Before developing special methods for testing series, attention is called to the following theorems, of which proofs are omitted.

Theorem I. *If S_n is a variable that always increases as n increases but never exceeds some definite fixed number A, then as n increases without limit, S_n will approach a limit u which is not greater than A.*

The figure illustrates the statement. The points determined by the values S_1, S_2, S_3, etc. approach the point u, where

$$\lim_{n \to \infty} S_n = u,$$

and u is less than or equal to A.

ILLUSTRATIVE EXAMPLE. Show that the infinite series

(1) $$1 + 1 + \frac{1}{1 \cdot 2} + \frac{1}{1 \cdot 2 \cdot 3} + \cdots + \frac{1}{\underline{|n}} + \cdots$$

is convergent.

Solution. Neglect the first term, and write

(2) $$S_n = 1 + \frac{1}{1 \cdot 2} + \frac{1}{1 \cdot 2 \cdot 3} + \cdots + \frac{1}{1 \cdot 2 \cdot 3 \cdots n}.$$

Consider the variable s_n defined by

(3) $$s_n = 1 + \frac{1}{2} + \frac{1}{2 \cdot 2} + \cdots + \frac{1}{2^{n-1}},$$

in which we have replaced all integers in the denominators of (2), except 1, by 2. Obviously $S_n < s_n$. Also, in (3) we have a geometric series with $r = \frac{1}{2}$ and $s_n < 2$ no matter how large n may be (see Art. 183). Hence S_n as defined by (2) is a variable which always increases as n increases but remains less than 2. Hence S_n approaches a limit as n becomes infinite, and this limit is less than 2. Therefore the infinite series (1) is convergent, and its value is less than 3.

We shall see later that the value of (1) is the constant $e = 2.71828 \cdots$, the natural base (Art. 61).

Theorem II. *If S_n is a variable that always decreases as n increases but is never less than some definite fixed number B, then as n increases without limit, S_n will approach a limit which is not less than B.*

Consider now a convergent series
$$S_n = u_1 + u_2 + u_3 + \cdots + u_n$$
for which $\lim\limits_{n \to \infty} S_n = u$.

Let the points determined by the values S_1, S_2, S_3, etc. be plotted on a directed line. Then these points, as n increases, will approach the point determined by u (terms in S_n all of the same sign) or cluster about this point. Thus it is evident that

(A) $$\lim_{n \to \infty} u_n = 0.$$

That is, in a convergent series, the terms must approach zero as a limit.

On the other hand, if the general (or nth) term of a series does not approach zero as n becomes infinite, we know at once that the series is divergent.

(*A*) is not, however, a sufficient condition for convergence; that is, even if the nth term does approach zero, we cannot state positively that the series is convergent. For, consider the harmonic series
$$1 + \frac{1}{2} + \frac{1}{3} + \frac{1}{4} + \cdots + \frac{1}{n}.$$

Here $$\lim_{n \to \infty}(u_n) = \lim_{n \to \infty}\left(\frac{1}{n}\right) = 0;$$

that is, condition (*A*) is fulfilled. Yet we shall show in Art. 186 that the series is divergent.

We shall now proceed to deduce special tests which, as a rule, are easier to apply than the above theorems.

186. Comparison tests. In many cases it is easy to determine whether or not a given series is convergent by comparing it term by term with another series whose character is known.

Test for convergence. *Let*

(1) $$u_1 + u_2 + u_3 + \cdots$$

be a series of positive terms which it is desired to test for convergence. If a series of positive terms already known to be convergent, namely,

(2) $$a_1 + a_2 + a_3 + \cdots,$$

can be found whose terms are never less than the corresponding terms in the series (1) *to be tested, then* (1) *is a convergent series and its value does not exceed that of* (2).

Proof. Let $\quad s_n = u_1 + u_2 + u_3 + \cdots + u_n,$
and $\quad\quad\quad\quad S_n = a_1 + a_2 + a_3 + \cdots + a_n,$
and suppose that $\quad \lim_{n \to \infty} S_n = A.$

Then, since $\quad S_n < A$ and $s_n \leqq S_n,$
it follows that $s_n < A$. Hence, by Theorem I, Art. 185, s_n approaches a limit and the series (1) is convergent and its value is not greater than A.

ILLUSTRATIVE EXAMPLE 1. Test the series

(3) $\quad\quad\quad\quad 1 + \dfrac{1}{2^2} + \dfrac{1}{3^3} + \dfrac{1}{4^4} + \dfrac{1}{5^5} + \cdots.$

Solution. Compare with the geometric series

(4) $\quad\quad\quad\quad 1 + \dfrac{1}{2} + \dfrac{1}{2^2} + \dfrac{1}{2^3} + \dfrac{1}{2^4} + \cdots,$

which is known to be convergent. The terms of (4) are never less than the corresponding terms of (3). Hence (3) also is convergent.

Following a line of reasoning similar to that applied to (1) and (2), we may prove the

Test for divergence. *Let*

(5) $\quad\quad\quad\quad u_1 + u_2 + u_3 + \cdots$

be a series of positive terms to be tested, which are never less than the corresponding terms of a series of positive terms, namely,

(6) $\quad\quad\quad\quad b_1 + b_2 + b_3 + \cdots,$

known to be divergent. Then (5) *is a divergent series.*

ILLUSTRATIVE EXAMPLE 2. Show that the harmonic series

(7) $\quad\quad\quad\quad 1 + \tfrac{1}{2} + \tfrac{1}{3} + \tfrac{1}{4} + \cdots$

is divergent.

Solution. Rewrite (7) as below and compare with the series written under it. The square brackets are introduced to aid in the comparison.

(8) $\quad 1 + \tfrac{1}{2} + [\tfrac{1}{3} + \tfrac{1}{4}] + [\tfrac{1}{5} + \tfrac{1}{6} + \tfrac{1}{7} + \tfrac{1}{8}] + [\tfrac{1}{9} + \cdots + \tfrac{1}{16}] + \cdots.$
(9) $\quad \tfrac{1}{2} + \tfrac{1}{2} + [\tfrac{1}{4} + \tfrac{1}{4}] + [\tfrac{1}{8} + \tfrac{1}{8} + \tfrac{1}{8} + \tfrac{1}{8}] + [\tfrac{1}{16} + \cdots + \tfrac{1}{16}] + \cdots.$

We observe the following facts. The terms in (8) are never less than the corresponding terms in (9).

But (9) is divergent. For the sum of the terms in each square bracket is $\tfrac{1}{2}$, and S_n will increase indefinitely as n becomes infinite.

Hence (8) is divergent.

ILLUSTRATIVE EXAMPLE 3. Test the series

$$1 + \frac{1}{\sqrt{2}} + \frac{1}{\sqrt{3}} + \frac{1}{\sqrt{4}} + \cdots$$

Solution. This series is divergent, since its terms are greater than the corresponding terms of the harmonic series (7) which is divergent.

The "p series,"

(10) $$1+\frac{1}{2^p}+\frac{1}{3^p}+\frac{1}{4^p}+\cdots+\frac{1}{n^p}+\cdots,$$

is useful in applying the comparison test.

Theorem. *The p series is convergent when $p > 1$, and divergent for other values of p.*

Proof. Rewrite (10) as below and compare with the series written under it. The square brackets are used to aid in the comparison.

(11) $$1+\left[\frac{1}{2^p}+\frac{1}{3^p}\right]+\left[\frac{1}{4^p}+\frac{1}{5^p}+\frac{1}{6^p}+\frac{1}{7^p}\right]+\left[\frac{1}{8^p}+\cdots+\frac{1}{15^p}\right]+\cdots,$$

(12) $$1+\left[\frac{1}{2^p}+\frac{1}{2^p}\right]+\left[\frac{1}{4^p}+\frac{1}{4^p}+\frac{1}{4^p}+\frac{1}{4^p}\right]+\left[\frac{1}{8^p}+\cdots+\frac{1}{8^p}\right]+\cdots.$$

If $p > 1$, the terms in (12) are never less than the corresponding terms in (11). But, in (12), the sums within the brackets are

$$\frac{1}{2^p}+\frac{1}{2^p}=\frac{2}{2^p}=\frac{1}{2^{p-1}};\quad \frac{1}{4^p}+\frac{1}{4^p}+\frac{1}{4^p}+\frac{1}{4^p}=\frac{4}{4^p}=\frac{2^2}{2^{2p}}=\frac{1}{2^{2(p-1)}};$$

$$\frac{1}{8^p}+\cdots+\frac{1}{8^p}=\frac{8}{8^p}=\frac{2^3}{2^{3p}}=\frac{1}{2^{3(p-1)}},$$

and so on. Hence, to test (12) for convergence, we may consider the series

(13) $$1+\frac{1}{2^{p-1}}+\left(\frac{1}{2^{p-1}}\right)^2+\left(\frac{1}{2^{p-1}}\right)^3+\cdots.$$

When $p > 1$, series (13) is a geometric series with the common ratio less than unity, and is therefore convergent. Therefore (10) is also convergent. When $p = 1$, series (10) is the harmonic series and is divergent. When $p < 1$, the terms of series (10) will, after the first, be greater than the corresponding terms of the harmonic series; hence (10) is now divergent.   Q.E.D.

ILLUSTRATIVE EXAMPLE 4. Show that the series

(14) $$\frac{2}{2\cdot 3\cdot 4}+\frac{4}{3\cdot 4\cdot 5}+\frac{6}{4\cdot 5\cdot 6}+\cdots+\frac{2n}{(n+1)(n+2)(n+3)}+\cdots$$

is convergent.

Solution. In (14), $u_n < \frac{2n}{n^3}$, or $\frac{1}{2}u_n < \frac{1}{n^2}$;

that is, $\frac{1}{2}u_n$ is less than the general term of the p series when $p = 2$. Hence the series each of whose terms is half the corresponding term in (14) is convergent and therefore (14) also is convergent.

SERIES

PROBLEMS

Test each of the following series.

1. $\dfrac{1}{1} + \dfrac{1}{\sqrt{2^3}} + \dfrac{1}{\sqrt{3^3}} + \cdots \dfrac{1}{\sqrt{n^3}} + \cdots.$ Ans. Convergent.

2. $\dfrac{1}{1} + \dfrac{1}{\sqrt[3]{2}} + \dfrac{1}{\sqrt[3]{3}} + \cdots \dfrac{1}{\sqrt[3]{n}} + \cdots.$ Divergent.

3. $\dfrac{2}{1} + \dfrac{2}{2^2} + \dfrac{2}{3^3} + \cdots \dfrac{2}{n^n} + \cdots.$ Convergent.

4. $\dfrac{3}{1\cdot 2} + \dfrac{3}{2\cdot 3} + \dfrac{3}{3\cdot 4} + \cdots \dfrac{3}{n(n+1)} + \cdots.$ Convergent.

5. $\dfrac{4}{2\cdot 3} + \dfrac{8}{3\cdot 4} + \dfrac{12}{4\cdot 5} + \cdots \dfrac{4n}{(n+1)(n+2)} + \cdots.$ Divergent.

6. $\dfrac{3}{2\cdot 3\cdot 4} + \dfrac{5}{3\cdot 4\cdot 5} + \dfrac{7}{4\cdot 5\cdot 6} + \cdots \dfrac{2n+1}{(n+1)(n+2)(n+3)} + \cdots.$ Convergent.

7. $\dfrac{1}{5} + \dfrac{1}{10} + \dfrac{1}{15} + \cdots \dfrac{1}{5n} + \cdots.$ Divergent.

8. $\dfrac{1}{4} + \dfrac{1}{5} + \dfrac{1}{6} + \cdots \dfrac{1}{n+3} + \cdots.$ Divergent.

9. $\dfrac{1}{4} + \dfrac{1}{10} + \dfrac{1}{28} + \cdots \dfrac{1}{3^n + 1} + \cdots.$ Convergent.

10. $\dfrac{1}{3} + \dfrac{1}{\sqrt{3}} + \dfrac{1}{\sqrt[3]{3}} + \cdots \dfrac{1}{\sqrt[n]{3}} + \cdots.$ Divergent.

11. $\dfrac{1}{4} + \dfrac{1}{7} + \dfrac{1}{12} + \cdots \dfrac{1}{n^2 + 3} + \cdots.$ Convergent.

12. $\dfrac{1}{2} + \dfrac{1}{8} + \dfrac{1}{26} + \cdots \dfrac{1}{3^n - 1} + \cdots.$ Convergent.

13. $\dfrac{1}{2\cdot 2\cdot 3} + \dfrac{2}{2\cdot 3\cdot 4} + \dfrac{3}{2\cdot 4\cdot 5} + \cdots \dfrac{n}{2(n+1)(n+2)} + \cdots.$ Divergent.

14. $1 + \tfrac{1}{2} + \tfrac{1}{6} + \tfrac{1}{12} + \tfrac{1}{20} + \tfrac{1}{30} + \tfrac{1}{42} + \cdots.$ Convergent.

15. $\dfrac{2}{9} + \dfrac{2^2}{28} + \dfrac{2^3}{65} + \dfrac{2^4}{126} + \cdots.$ 17. $\dfrac{1}{\log 2} + \dfrac{1}{\log 3} + \dfrac{1}{\log 4} + \cdots.$

16. $\tfrac{2}{3} + \tfrac{2}{6} + \tfrac{2}{9} + \tfrac{2}{12} + \cdots.$ 18. $\tfrac{1}{2} + \tfrac{1}{6} + \tfrac{1}{10} + \tfrac{1}{14} + \cdots.$

187. Cauchy's test-ratio test. In the infinite geometric series

$$a + ar + ar^2 + \cdots + ar^n + ar^{n+1} + \cdots,$$

the ratio of the consecutive general terms ar^n and ar^{n+1} is the common ratio r. Moreover, we know that the series is convergent when $|r| < 1$ and divergent for other values. We now explain a ratio test which may be applied to any series.

Theorem. Let

(1) $$u_1 + u_2 + u_3 + \cdots + u_n + u_{n+1} + \cdots$$

be an infinite series of positive terms. Consider consecutive general terms u_n and u_{n+1}, and form the **test ratio**.

$$\text{Test ratio} = \frac{u_{n+1}}{u_n}.$$

Find the limit of this test ratio when n becomes infinite. Let this be

$$\rho = \lim_{n \to \infty} \frac{u_{n+1}}{u_n}.$$

I. *When $\rho < 1$, the series is convergent.*
II. *When $\rho > 1$, the series is divergent.*
III. *When $\rho = 1$, the test fails.*

Proof. I. *When $\rho < 1$.* By the definition of a limit (Art. 14) we can choose n so large, say $n = m$, that when $n \geq m$ the ratio $\dfrac{u_{n+1}}{u_n}$ will differ from ρ by as little as we please, and therefore be less than a proper fraction r. Hence

$$u_{m+1} < u_m r; \quad u_{m+2} < u_{m+1} r < u_m r^2; \quad u_{m+3} < u_m r^3;$$

and so on. Therefore, after the term u_m, each term of the series (1) is less than the corresponding term of the geometric series

(2) $$u_m r + u_m r^2 + u_m r^3 + \cdots.$$

But since $r < 1$, the series (2), and therefore also the series (1), is convergent (Art. 186).

II. *When $\rho > 1$ (or $\rho = \infty$).* Following the same line of reasoning as in I, the series (1) may be shown to be divergent.

III. *When $\rho = 1$*, the series may be either convergent or divergent; that is, the test fails. For, consider the p series, namely,

$$1 + \frac{1}{2^p} + \frac{1}{3^p} + \frac{1}{4^p} + \cdots + \frac{1}{n^p} + \frac{1}{(n+1)^p} + \cdots.$$

The test ratio is $\dfrac{u_{n+1}}{u_n} = \left(\dfrac{n}{n+1}\right)^p = \left(1 - \dfrac{1}{n+1}\right)^p;$

and $\quad \lim\limits_{n \to \infty} \left(\dfrac{u_{n+1}}{u_n}\right) = \lim\limits_{n \to \infty} \left(1 - \dfrac{1}{n+1}\right)^p = (1)^p = 1 \; (= \rho).$

Hence $\rho = 1$, no matter what value p may have. But in Art. 186 we showed that

when $p > 1$, the series converges, and
when $p \leqq 1$, the series diverges.

SERIES   347

Thus it appears that ρ can equal unity both for convergent and for divergent series. There are other tests to apply in cases like this, but the scope of our book does not admit of their consideration.

For convergence it is not enough that the test ratio is less than unity for all values of n. This test requires that the *limit* of the test ratio shall be less than unity. For instance, in the harmonic series the test ratio is always less than unity. The *limit*, however, equals unity.

The rejection of a group of terms at the beginning of a series will affect the *value* but not the *existence* of the limit.

188. Alternating series. This is the name given to a series whose terms are alternately positive and negative.

Theorem. *If* $\quad u_1 - u_2 + u_3 - u_4 + \cdots$

is an alternating series in which each term is numerically less than the one which precedes it, and if $\lim\limits_{n \to \infty} u_n = 0$,

then the series is convergent.

Proof. When n is even, S_n may be written in the two forms

(1) $\quad S_n = (u_1 - u_2) + (u_3 - u_4) + \cdots + (u_{n-1} - u_n)$,
(2) $\quad S_n = u_1 - (u_2 - u_3) - \cdots - (u_{n-2} - u_{n-1}) - u_n$.

Each expression in parentheses is positive. Hence when n increases through even values, (1) shows that S_n increases and (2) shows that S_n is always less than u_1; therefore, by Theorem I, Art. 185, S_n approaches a limit l. But S_{n+1} also approaches this limit l, since $S_{n+1} = S_n + u_{n+1}$ and $\lim u_{n+1} = 0$. Hence when n increases through all integral values, $S_n \to l$ and the series is convergent.

ILLUSTRATIVE EXAMPLE. Test the alternating series $1 - \frac{1}{2} + \frac{1}{3} - \frac{1}{4} + \cdots$.

Solution. Each term is less in numerical value than the preceding one, and $u_n = \dfrac{1}{n}$.
Hence $\lim\limits_{n \to \infty} (u_n) = 0$, and the series is convergent.

An important consequence of the above proof is the statement:

The error made by breaking a convergent alternating series off at any term does not exceed numerically the value of the first of the terms discarded.

Thus, the sum of ten terms in the above Illustrative Example is 0.646, and the value of the series differs from this by less than one eleventh.

In the above statement it is assumed, however, that the series has been carried far enough so that the terms are decreasing numerically.

189. Absolute convergence. A series is said to be *absolutely* or *unconditionally* convergent when the series formed from it by making all its terms positive is convergent. Other convergent series are said to be *conditionally convergent*.

For example, the series $1 - \dfrac{1}{2^2} + \dfrac{1}{3^3} - \dfrac{1}{4^4} + \dfrac{1}{5^5} - \cdots$

is *absolutely convergent*, since the series (3), Art. 186, is convergent. The alternating series

$$1 - \frac{1}{2} + \frac{1}{3} - \frac{1}{4} + \frac{1}{5} - \cdots$$

is *conditionally* convergent, since the harmonic series is divergent.

A series with some positive and some negative terms is convergent if the series deduced from it by making all the signs positive is convergent.

The proof of this theorem is omitted.

190. Summary. Assuming that the test-ratio test of Art. 187 holds without placing any restriction on the signs of the terms, we may summarize our results in the following

General directions for testing the series

$$u_1 + u_2 + u_3 + u_4 + \cdots + u_n + u_{n+1} + \cdots.$$

When it is an alternating series whose terms never increase in numerical value, and if $\quad \lim\limits_{n \to \infty} u_n = 0,$

then the series is convergent.

In any series in which the above conditions are not satisfied, we determine the form of u_n *and* u_{n+1}, *form the test ratio, and calculate*

$$\lim_{n \to \infty} \left(\frac{u_{n+1}}{u_n} \right) = \rho.$$

I. *When* $|\rho| < 1$, *the series is absolutely convergent.*
II. *When* $|\rho| > 1$, *the series is divergent.*
III. *When* $|\rho| = 1$, *the test fails, and we compare the series with some series which we know to be convergent, as*

$$a + ar + ar^2 + ar^3 + \cdots; \quad (r < 1) \quad \text{(geometric series)}$$

$$1 + \frac{1}{2^p} + \frac{1}{3^p} + \frac{1}{4^p} + \cdots; \quad (p > 1) \qquad (p \text{ series})$$

or compare the given series with some series which is known to be divergent, as

$$1 + \frac{1}{2} + \frac{1}{3} + \frac{1}{4} + \cdots; \qquad \text{(harmonic series)}$$

$$1 + \frac{1}{2^p} + \frac{1}{3^p} + \frac{1}{4^p} + \cdots. \quad (p < 1) \qquad (p \text{ series})$$

SERIES

ILLUSTRATIVE EXAMPLE 1. Test the series

$$1 + \frac{1}{\underline{|1}} + \frac{1}{\underline{|2}} + \frac{1}{\underline{|3}} + \frac{1}{\underline{|4}} + \cdots.$$

Solution. Here $\quad u_n = \dfrac{1}{\underline{|n-1}}, \quad u_{n+1} = \dfrac{1}{\underline{|n}}.$

$$\frac{u_{n+1}}{u_n} = \frac{\underline{|n-1}}{\underline{|n}} = \frac{1}{n}.$$

$$\rho = \lim_{n \to \infty} \frac{1}{n} = 0,$$

and the series is convergent.

ILLUSTRATIVE EXAMPLE 2. Test the series

$$\frac{\underline{|1}}{10} + \frac{\underline{|2}}{10^2} + \frac{\underline{|3}}{10^3} + \cdots.$$

Solution. Here $\quad u_n = \dfrac{\underline{|n}}{10^n}, \quad u_{n+1} = \dfrac{\underline{|n+1}}{10^{n+1}}.$

$$\frac{u_{n+1}}{u_n} = \frac{\underline{|n+1}}{10^{n+1}} \cdot \frac{10^n}{\underline{|n}} = \frac{n+1}{10}.$$

$$\rho = \lim_{n \to \infty} \frac{n+1}{10} = \infty,$$

and the series is divergent.

ILLUSTRATIVE EXAMPLE 3. Test the series

$$\frac{1}{1 \cdot 2} + \frac{1}{3 \cdot 4} + \frac{1}{5 \cdot 6} + \cdots.$$

Solution. Here $\quad u_n = \dfrac{1}{(2n-1)2n}, \quad u_{n+1} = \dfrac{1}{(2n+1)(2n+2)}.$

$$\frac{u_{n+1}}{u_n} = \frac{(2n-1)2n}{(2n+1)(2n+2)} = \frac{4n^2 - 2n}{4n^2 + 6n + 2}.$$

$$\rho = \lim_{n \to \infty} \frac{4n^2 - 2n}{4n^2 + 6n + 2} = 1,$$

by the rule in Art. 18. Hence the test-ratio test fails.
But if we compare the given series with the p series, when $p = 2$, namely,

$$1 + \frac{1}{2^2} + \frac{1}{3^2} + \frac{1}{4^2} + \cdots,$$

we see that it must be convergent, since its terms are less than the corresponding terms of this p series, which was proved convergent.

PROBLEMS

Test each of the following series.

1. $\dfrac{3}{4} + 2\left(\dfrac{3}{4}\right)^2 + 3\left(\dfrac{3}{4}\right)^3 + 4\left(\dfrac{3}{4}\right)^4 + \cdots.$ *Ans.* Convergent.

2. $\dfrac{3}{2} + \dfrac{4}{2^2} + \dfrac{5}{2^3} + \dfrac{6}{2^4} + \cdots.$ Convergent.

350 DIFFERENTIAL AND INTEGRAL CALCULUS

3. $\dfrac{3}{2} + \dfrac{3^2}{2\cdot 2^2} + \dfrac{3^3}{3\cdot 2^3} + \dfrac{3^4}{4\cdot 2^4} + \cdots$. *Ans.* Divergent.

4. $\dfrac{1}{3} + \dfrac{1\cdot 2}{3\cdot 5} + \dfrac{1\cdot 2\cdot 3}{3\cdot 5\cdot 7} + \cdots \dfrac{\underline{|n}}{3\cdot 5\cdot 7 \cdots (2n+1)} + \cdots$. Convergent.

5. $\dfrac{1}{1} + \dfrac{1\cdot 3}{1\cdot 4} + \dfrac{1\cdot 3\cdot 5}{1\cdot 4\cdot 7} + \cdots \dfrac{1\cdot 3\cdot 5 \cdots (2n-1)}{1\cdot 4\cdot 7 \cdots (3n-2)} \cdots$. Convergent.

6. $\dfrac{1}{1} + \dfrac{1\cdot 2}{1\cdot 3} + \dfrac{1\cdot 2\cdot 3}{1\cdot 3\cdot 5} + \dfrac{1\cdot 2\cdot 3\cdot 4}{1\cdot 3\cdot 5\cdot 7} + \cdots$. Convergent.

7. $\dfrac{5}{1} + \dfrac{5^2}{\underline{|2}} + \dfrac{5^3}{\underline{|3}} + \dfrac{5^4}{\underline{|4}} + \cdots$. Convergent.

8. $\dfrac{1}{9} + \dfrac{\underline{|2}}{9^2} + \dfrac{\underline{|3}}{9^3} + \dfrac{\underline{|4}}{9^4} + \cdots$. Divergent.

9. $\dfrac{1}{3} + \dfrac{3}{3^2} + \dfrac{5}{3^3} + \dfrac{7}{3^4} + \cdots$.

11. $\dfrac{1}{4\cdot 3} + \dfrac{1}{5\cdot 3^2} + \dfrac{1}{6\cdot 3^3} + \dfrac{1}{7\cdot 3^4} + \cdots$.

10. $1 + \dfrac{2\underline{|2}}{5} + \dfrac{2^2\underline{|3}}{10} + \dfrac{2^3\underline{|4}}{17} + \cdots$.

12. $\dfrac{3}{5} + \dfrac{4}{5^2} + \dfrac{5}{5^3} + \dfrac{6}{5^4} + \cdots$.

13. $\dfrac{1}{3} + \dfrac{1\cdot 3}{3\cdot 6} + \dfrac{1\cdot 3\cdot 5}{3\cdot 6\cdot 9} + \dfrac{1\cdot 3\cdot 5\cdot 7}{3\cdot 6\cdot 9\cdot 12} + \cdots$.

191. Power series. A series whose terms are monomials in ascending positive integral powers of a variable, say x, of the form

(1) $\qquad a_0 + a_1 x + a_2 x^2 + a_3 x^3 + \cdots$,

where the coefficients $a_0, a_1, a_2, \cdots$ are independent of x, is called a *power series in x*. Such series are of prime importance in the study of calculus.

A power series in x may converge for all values of x, or for no value except $x = 0$; or it may converge for some values of x different from 0 and be divergent for other values.

We shall examine (1) only for the case when the coefficients are such that
$$\lim_{n\to\infty}\left(\dfrac{a_{n+1}}{a_n}\right) = L,$$
where L is a definite number. To see the reason for this, form the test ratio (Art. 187) for (1), omitting the first term. Then we have
$$\dfrac{u_{n+1}}{u_n} = \dfrac{a_{n+1}x^{n+1}}{a_n x^n} = \dfrac{a_{n+1}}{a_n} x.$$

Hence for any fixed value of x,
$$\rho = \lim_{n\to\infty}\left(\dfrac{a_{n+1}}{a_n} x\right) = x \lim_{n\to\infty}\left(\dfrac{a_{n+1}}{a_n}\right) = xL.$$

SERIES

We have two cases:

I. If $L = 0$, the series (1) will converge for all values of x, since $\rho = 0$.

II. If L is not zero, the series will converge when $xL (= \rho)$ is numerically less than 1, that is, when x lies in the interval

$$-\frac{1}{|L|} < x < \frac{1}{|L|},$$

and will diverge for values of x outside this interval.

The end points of the interval, called the *interval of convergence*, must be examined separately. In any given series the test ratio should be formed and the interval of convergence determined by Art. 187.

ILLUSTRATIVE EXAMPLE 1. Find the interval of convergence for the series

(2) $$x - \frac{x^2}{2^2} + \frac{x^3}{3^2} - \frac{x^4}{4^2} + \cdots.$$

Solution. The test ratio here is

$$\frac{u_{n+1}}{u_n} = -\frac{n^2}{(n+1)^2} x. \text{ Also, } \lim_{n \to \infty} \frac{n^2}{(n+1)^2} = 1,$$

by Art. 18. Hence $\rho = -x$, and the series converges when x is numerically less than 1 and diverges when x is numerically greater than 1.

Now examine the end points. Substituting $x = 1$ in (2), we get

$$1 - \frac{1}{2^2} + \frac{1}{3^2} - \frac{1}{4^2} + \cdots,$$

which is an alternating series that converges.

Substituting $x = -1$ in (2), we get

$$-1 - \frac{1}{2^2} - \frac{1}{3^2} - \frac{1}{4^2} - \cdots,$$

which is convergent by comparison with the p series ($p > 1$).

The series in the above example has $[-1, 1]$ as the *interval of convergence*. This may be written $-1 \leq x \leq 1$, or indicated graphically as follows:

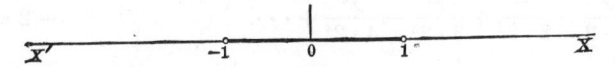

ILLUSTRATIVE EXAMPLE 2. Determine the interval of convergence for the series

$$1 + \frac{x^2}{\underline{|2}} + \frac{x^4}{\underline{|4}} + \cdots + \frac{x^{2n}}{\underline{|2n}} + \frac{x^{2n+2}}{\underline{|2n+2}} + \cdots.$$

Solution. Omitting the first term, the test ratio is

$$\frac{u_{n+1}}{u_n} = \frac{\underline{|2n}}{\underline{|2n+2}} x^2 = \frac{1}{(2n+1)(2n+2)} x^2.$$

Also $\lim_{n \to \infty} \frac{1}{(2n+1)(2n+2)} = 0$. Hence the series converges for all values of x.

PROBLEMS

For what values of the variable are the following series convergent? *Graphical representations of intervals of convergence**

1. $1 + x + x^2 + x^3 + \cdots$.   Ans. $-1 < x < 1$.

2. $x - \dfrac{x^2}{2} + \dfrac{x^3}{3} - \dfrac{x^4}{4} + \cdots$.   Ans. $-1 < x \leqq 1$.

3. $x + x^4 + x^9 + x^{16} + \cdots$.   Ans. $-1 < x < 1$.

4. $x + \dfrac{x^2}{\sqrt{2}} + \dfrac{x^3}{\sqrt{3}} + \cdots$.   Ans. $-1 \leqq x < 1$.

5. $1 + x + \dfrac{x^2}{\lfloor 2} + \dfrac{x^3}{\lfloor 3} + \cdots$.   Ans. All values of x.

6. $1 - \dfrac{\theta^2}{\lfloor 2} + \dfrac{\theta^4}{\lfloor 4} - \dfrac{\theta^6}{\lfloor 6} + \cdots$.   Ans. All values of θ.

7. $\phi - \dfrac{\phi^3}{\lfloor 3} + \dfrac{\phi^5}{\lfloor 5} - \dfrac{\phi^7}{\lfloor 7} + \cdots$.   Ans. All values of ϕ.

8. $x + \dfrac{1}{2} \cdot \dfrac{x^3}{3} + \dfrac{1 \cdot 3}{2 \cdot 4} \cdot \dfrac{x^5}{5} + \dfrac{1 \cdot 3 \cdot 5}{2 \cdot 4 \cdot 6} \cdot \dfrac{x^7}{7} + \cdots$.   Ans. $-1 \leqq x \leqq 1$.

9. $x + 2x^2 + 3x^3 + 4x^4 + \cdots$.   Ans. $-1 < x < 1$.

10. $1 - x + \dfrac{x^2}{2^2} - \dfrac{x^3}{3^2} + \cdots$. $-1 \leqq x \leqq 1$.

11. $\dfrac{x}{1 \cdot 2} - \dfrac{x^2}{2 \cdot 2^2} + \dfrac{x^3}{3 \cdot 2^3} - \dfrac{x^4}{4 \cdot 2^4} + \cdots$. $-2 < x \leqq 2$.

12. $x + \dfrac{2x^2}{\lfloor 2} + \dfrac{3x^3}{\lfloor 3} + \dfrac{4x^4}{\lfloor 4} + \cdots$.   All values.

13. $1 + \dfrac{x^2}{2 \cdot 2^2} + \dfrac{1 \cdot 3 x^4}{2 \cdot 4 \cdot 2^4} + \dfrac{1 \cdot 3 \cdot 5 x^6}{2 \cdot 4 \cdot 6 \cdot 2^6} + \cdots$. $-2 < x < 2$.

14. $\dfrac{ax}{2} + \dfrac{a^2 x^2}{5} + \dfrac{a^3 x^3}{10} + \cdots + \dfrac{a^n x^n}{n^2 + 1} + \cdots$. $(a > 0)$ $-\dfrac{1}{a} \leqq x \leqq \dfrac{1}{a}$.

* End points that are not included in the interval of convergence have circles drawn about them.

SERIES

15. $1 + \dfrac{x}{a} + \dfrac{x^2}{2\,a^2} + \dfrac{x^3}{3\,a^3} + \cdots.$ $(a > 0)$ Ans. $-a \leqq x < a.$

16. $1 + \dfrac{2^2\,x}{\underline{|2}} + \dfrac{3^2\,x^2}{\underline{|3}} + \dfrac{4^2\,x^3}{\underline{|4}} + \cdots.$ All values.

17. $\dfrac{1}{3} + \dfrac{2\,x}{2 \cdot 3^2} + \dfrac{3\,x^2}{2^2 \cdot 3^3} + \dfrac{4\,x^3}{2^3 \cdot 3^4} + \cdots.$ $-6 < x < 6.$

18. $x + 4\,x^2 + 9\,x^3 + 16\,x^4 + \cdots.$

19. $\dfrac{x}{1 \cdot 2} + \dfrac{2\,x^2}{2 \cdot 2 \cdot 3} + \dfrac{3\,x^3}{2^2 \cdot 3 \cdot 4} + \dfrac{4\,x^4}{2^3 \cdot 4 \cdot 5} + \cdots.$

20. $\dfrac{1}{2} + \dfrac{x}{3} + \dfrac{x^2}{4} + \dfrac{x^3}{5} + \cdots.$ 22. $1 - \dfrac{x}{10} + \dfrac{\underline{|2}\,x^2}{100} - \dfrac{\underline{|3}\,x^3}{1000} + \cdots.$

21. $\dfrac{10\,x}{1} + \dfrac{100\,x^2}{\underline{|2}} + \dfrac{1000\,x^3}{\underline{|3}} + \cdots.$ 23. $x + \dfrac{x^3}{3} + \dfrac{x^5}{5} + \dfrac{x^7}{7} + \cdots.$

24. $1 - \dfrac{x^2}{2^2} + \dfrac{x^4}{4^2} - \dfrac{x^6}{6^2} + \cdots.$

192. The binomial series. This important series is

(1) $\quad 1 + mx + \dfrac{m(m-1)}{1 \cdot 2}\,x^2 + \dfrac{m(m-1)(m-2)}{1 \cdot 2 \cdot 3}\,x^3 + \cdots$
$\qquad + \dfrac{m(m-1)(m-2) \cdots (m-n+1)}{\underline{|n}}\,x^n + \cdots$

where m is a constant.

If m is a positive integer, (1) is a finite series of $m+1$ terms, since all terms following that containing x^m have the factor $m-m$ in the numerator and vanish. In this case (1) is the result obtained by raising $1+x$ to the mth power. If m is not a positive integer, the series is an infinite series.

Now test (1) for convergence. We have

$$u_n = \dfrac{m(m-1)(m-2) \cdots (m-n+2)}{1 \cdot 2 \cdot 3 \cdots (n-1)}\,x^{n-1};$$

and $\quad u_{n+1} = \dfrac{m(m-1)(m-2) \cdots (m-n+2)(m-n+1)}{1 \cdot 2 \cdot 3 \cdots (n-1)n}\,x^n.$

Hence $\quad \dfrac{u_{n+1}}{u_n} = \dfrac{m-n+1}{n}\,x = \left(\dfrac{m+1}{n} - 1\right)x.$

Then, since $\lim\limits_{n \to \infty} \left(\dfrac{m+1}{n} - 1\right) = -1$, we see that $\rho = -x$,

and the series is convergent if x is numerically less than 1, and divergent when x is numerically greater than 1.

Maclaurin's Series, Art. 194, implies the following statement.

Assuming that m is not a positive integer and that $|x| < 1$, the value of the binomial series is precisely the value of $(1+x)^m$. That is,

(2) $\quad (1+x)^m = 1 + mx + \dfrac{m(m-1)}{1 \cdot 2} x^2 + \cdots.$ $\quad (|x|<1)$

If m is a positive integer, the series is finite and equals the value of the left-hand member for *all values* of x.

Equation (2) expresses the *special binomial theorem*. We may also write

(3) $\quad (a+b)^m = a^m(1+x)^m,$ if $x = \dfrac{b}{a}.$

Thus the left-hand member of (3) may also be expressed as a power series.

Examples of approximate computation by the binomial series are given below.

ILLUSTRATIVE EXAMPLE. Find $\sqrt{630}$ approximately, using the binomial series.

Solution. The perfect square nearest to 630 is 625. Hence we write

$$\sqrt{630} = \sqrt{625+5} = 25(1 + \tfrac{1}{125})^{\frac{1}{2}}.$$

Now write out (2) with $m = \tfrac{1}{2}$. The result is

$$(1+x)^{\frac{1}{2}} = 1 + \tfrac{1}{2}x - \tfrac{1}{8}x^2 + \tfrac{1}{16}x^3 - \tfrac{5}{128}x^4 + \cdots.$$

In this example, $x = \tfrac{1}{125} = 0.008$. Hence

$$(1 + \tfrac{1}{125})^{\frac{1}{2}} = 1 + 0.004 - 0.000008 + 0.000000032 + \cdots.$$

(4) $25(1 + \tfrac{1}{125})^{\frac{1}{2}} = 25 + 0.1 - 0.0002 + 0.0000008 = 25.099801$ (to the nearest figure in the sixth decimal place). *Ans.*

The series in (4) is an alternating series, and the error in the answer is less than 0.0000008.

193. Another type of power series. We shall frequently use series of the form

(1) $\quad b_0 + b_1(x-a) + b_2(x-a)^2 + \cdots + b_n(x-a)^n + \cdots$

in which a and the coefficients $b_0, b_1, \cdots, b_n, \cdots$ are constants. Such a series is called a *power series in* $(x-a)$.

Let us apply the test-ratio test to (1), as in Art. 191. Then, if

$$\lim_{n \to \infty} \dfrac{b_{n+1}}{b_n} = M,$$

we shall have, for any fixed value of x,

$$\rho = \lim_{n \to \infty} \frac{u_{n+1}}{u_n} = (x-a)M.$$

We have two cases:
I. If $M = 0$, series (1) is convergent for all values of x.
II. If M is not zero, series (1) will converge for the interval

$$a - \frac{1}{|M|} < x < a + \frac{1}{|M|}.$$

A convergent power series in x is adapted for computation when x is near zero. Series (1), if convergent, is useful when x is near the fixed value a, given in advance.

ILLUSTRATIVE EXAMPLE. Test the infinite series

$$1 - (x-1) + \frac{(x-1)^2}{2} - \frac{(x-1)^3}{3} + \cdots$$

for convergence.

Solution. Neglecting the first term, we have

$$\frac{u_{n+1}}{u_n} = -\frac{n}{n+1}(x-1).$$

Also,

$$\lim_{n \to \infty}\left(\frac{n}{n+1}\right) = 1.$$

Hence $|\rho| = |x-1|$, and the series will converge when x lies between 0 and 2. The end point $x = 2$ may be included.

PROBLEMS

1. Using the binomial series, show that

$$\frac{1}{1+x} = 1 - x + x^2 - x^3 + \cdots.$$

Verify the answer by direct division.

Using the binomial series, find approximately the values of the following numbers.

2. $\sqrt{98}$.

3. $\sqrt[3]{120}$.

4. $\sqrt[4]{630}$.

5. $\sqrt[5]{35}$.

6. $\dfrac{1}{412}$.

7. $\dfrac{1}{\sqrt{412}}$.

8. $\dfrac{1}{\sqrt[3]{990}}$.

9. $\dfrac{1}{\sqrt[4]{15}}$.

10. $\dfrac{1}{\sqrt[5]{30}}$.

11. $\sqrt{\dfrac{26}{25}}$.

12. $\sqrt[3]{\dfrac{128}{125}}$.

13. $\sqrt[4]{\dfrac{17}{16}}$.

DIFFERENTIAL AND INTEGRAL CALCULUS

For what values of the variable are the following series convergent?

14. $(x+1) - \dfrac{(x+1)^2}{2} + \dfrac{(x+1)^3}{3} - \dfrac{(x+1)^4}{4} + \cdots$.   Ans. $-2 < x \leqq 0$.

15. $(x-1) + \dfrac{(x-1)^2}{\sqrt{2}} + \dfrac{(x-1)^3}{\sqrt{3}} + \dfrac{(x-1)^4}{\sqrt{4}} + \cdots$. $0 \leqq x < 2$.

16. $2(2x+1) + \dfrac{3(2x+1)^2}{\underline{|2}} + \dfrac{4(2x+1)^3}{\underline{|3}} + \cdots$.   All values.

17. $1 + (x-2) + \dfrac{(x-2)^2}{2^2} + \dfrac{(x-2)^3}{3^2} + \dfrac{(x-2)^4}{4^2} + \cdots$.

18. $1 - 2(2x-3) + 3(2x-3)^2 - 4(2x-3)^3 + \cdots$.

19. $\dfrac{x-3}{1 \cdot 3} + \dfrac{(x-3)^2}{2 \cdot 3^2} + \dfrac{(x-3)^3}{3 \cdot 3^3} + \dfrac{(x-3)^4}{4 \cdot 3^4} + \cdots$.

20. Using the binomial series, find the first three terms of the power series in $(t-1)$ for $\sqrt{1+t}$.

Solution. $\sqrt{1+t} = \sqrt{2+(t-1)} = \sqrt{2}\left[1 + \dfrac{t-1}{2}\right]^{\frac{1}{2}}$.

Setting $m = \tfrac{1}{2}$, $x = \dfrac{t-1}{2}$ in (1), Art. 192, the result is

$$\sqrt{1+t} = \sqrt{2}\left[1 + \dfrac{(t-1)}{4} - \dfrac{(t-1)^2}{32} + \cdots\right]. \text{ Ans.}$$

21. Find the first three terms of the power series in $(t-1)$ for $\sqrt{3+t}$. Using this result, find approximately the value of $\sqrt{3.5}$.   Ans. 1.871.

22. Find the first three terms of the power series in $(t-3)$ for $\sqrt{6+t}$. Using this result, find approximately the value of $\sqrt{7}$.   Ans. 2.648.

23. Find the first three terms of the power series in $(t-1)$ for $\sqrt[3]{7+t}$. Using this result, find approximately the value of $\sqrt[3]{9}$.   Ans. 2.080.

CHAPTER XX

EXPANSION OF FUNCTIONS

194. Maclaurin's series. In this chapter the question of representing a function by a power series, or, otherwise expressed, developing (expanding) the function in a power series, will be discussed.

A convergent power series in x is obviously a function of x for all values in the interval of convergence. Thus we may write

(1) $\qquad f(x) = a_0 + a_1 x + a_2 x^2 + \cdots + a_n x^n + \cdots.$

If, then, a function is represented by a power series, what must be the form of the coefficients a_0, a_1, $\cdots$, a_n, etc.? To answer this question we proceed thus:

Set $x = 0$ in (1). Then we must have

(2) $\qquad\qquad\qquad f(0) = a_0.$

Hence the first coefficient a_0 in (1) is determined. Now *assume* that the series in (1) may be differentiated term by term, and that this differentiation may be continued. Then we shall have

(3) $\begin{cases} f'(x) = a_1 + 2\,a_2 x + 3\,a_3 x^2 + \cdots + na_n x^{n-1} + \cdots \\ f''(x) = 2\,a_2 + 6\,a_3 x + \cdots + n(n-1)a_n x^{n-2} + \cdots \\ f'''(x) = 6\,a_3 + \cdots + n(n-1)(n-2)a_n x^{n-3} + \cdots \end{cases}$

etc.

Letting $x = 0$, the results are

(4) $\quad f'(0) = a_1, \quad f''(0) = \underline{|2}\,a_2, \quad f'''(0) = \underline{|3}\,a_3, \cdots, \quad f^{(n)}(0) = \underline{|n}a_n.$

Solving (4) for $a_1, a_2, \cdots, a_n$, etc., and substituting in (1), we obtain

(A) $\quad f(x) = f(0) + f'(0)\dfrac{x}{\underline{|1}} + f''(0)\dfrac{x^2}{\underline{|2}} + \cdots + f^{(n)}(0)\dfrac{x^n}{\underline{|n}} + \cdots.$

This formula expresses $f(x)$ as a power series. We say, "the function $f(x)$ is developed (or expanded) in a power series in x." This is Maclaurin's series (or formula).*

---
* Named after Colin Maclaurin (1698–1746), and first published in his "Treatise of Fluxions" (Edinburgh, 1742). The series is really due to Stirling (1692–1770).

approximate e^x [handwritten]

It is now necessary to examine (*A*) critically. For this purpose refer to (*G*), Art. 124, and rewrite it, letting $a = 0$, $b = x$. The result is

$$(5) \quad f(x) = f(0) + f'(0)\frac{x}{\lfloor 1} + f''(0)\frac{x^2}{\lfloor 2} + \cdots + f^{(n-1)}(0)\frac{x^{n-1}}{\lfloor n-1} + R,$$

where
$$R = f^{(n)}(x_1)\frac{x^n}{\lfloor n}. \qquad (0 < x_1 < x)$$

The term R is called the *remainder after n terms*. The right-hand member of (5) agrees with Maclaurin's series (*A*) up to n terms. If we denote this sum by S_n, then (5) is

$$f(x) = S_n + R, \quad \text{or} \quad f(x) - S_n = R.$$

Now assume that, for a fixed value $x = x_0$, R approaches zero as a limit when n becomes infinite. Then S_n will approach $f(x_0)$ as a limit (Art. 14). That is, Maclaurin's series (*A*) will converge for $x = x_0$ and its value is $f(x_0)$. Thus we have the following result:

Theorem. *In order that the series (A) should converge and represent the function $f(x)$ it is necessary and sufficient that*

$$(6) \qquad \lim_{n \to \infty} R = 0.$$

It is usually easier to determine the interval of convergence (as in the preceding chapter) than that for which (6) holds. But in simple cases the two are identical.

To represent a function $f(x)$ by the power series (*A*), it is obviously *necessary* that the function and its derivatives of all orders should be finite. This is, however, not *sufficient*.

Examples of functions that cannot be represented by a Maclaurin's series are

$$\ln x \quad \text{and} \quad \text{ctn } x,$$

since both become infinite when x is zero.

The student should not fail to note the importance of such an expansion as (*A*). In all practical computations results correct to a certain number of decimal places are sought, and since the process in question replaces a function perhaps difficult to calculate by *an ordinary polynomial with constant coefficients*, it is very useful in simplifying such computations. Of course we must use terms enough to give the desired degree of accuracy.

In the case of an alternating series (Art. 188) the error made by stopping at any term is numerically less than that term.

EXPANSION OF FUNCTIONS

ILLUSTRATIVE EXAMPLE 1. Expand $\cos x$ into an infinite power series and determine for what values of x it converges.

Solution. Differentiating first and then placing $x = 0$, we get

$$f(x) = \cos x, \qquad f(0) = 1,$$
$$f'(x) = -\sin x, \qquad f'(0) = 0,$$
$$f''(x) = -\cos x, \qquad f''(0) = -1,$$
$$f'''(x) = \sin x, \qquad f'''(0) = 0,$$
$$f^{iv}(x) = \cos x, \qquad f^{iv}(0) = 1,$$
$$f^{v}(x) = -\sin x, \qquad f^{v}(0) = 0,$$
$$f^{vi}(x) = -\cos x, \qquad f^{vi}(0) = -1,$$
$$\text{etc.}, \qquad \text{etc.}$$

Substituting in (A),

(7) $$\cos x = 1 - \frac{x^2}{\lfloor 2} + \frac{x^4}{\lfloor 4} - \frac{x^6}{\lfloor 6} + \cdots.$$

Comparing with Problem 6, Art. 191, we see that the series converges for all values of x.

In the same way for $\sin x$,

(8) $$\sin x = x - \frac{x^3}{\lfloor 3} + \frac{x^5}{\lfloor 5} - \frac{x^7}{\lfloor 7} + \cdots,$$

which converges for all values of x (Problem 7, Art. 191).

In (7) and (8) it is not difficult to show that the remainder R approaches zero as a limit as n becomes infinite, when x has *any* fixed value. Consider (7). Here we may write the nth derivative in the form

$$f^{(n)}(x) = \cos\left(x + \frac{n\pi}{2}\right).$$

Hence $$R = \cos\left(x_1 + \frac{n\pi}{2}\right)\frac{x^n}{\lfloor n}.$$

Now $\cos\left(x_1 + \frac{n\pi}{2}\right)$ never exceeds 1 in numerical value. Also, the second factor of R is the nth term of the series

$$x + \frac{x^2}{\lfloor 2} + \frac{x^3}{\lfloor 3} + \cdots + \frac{x^n}{\lfloor n} + \cdots,$$

which is convergent for all values of x. Therefore it approaches zero as n becomes infinite (see (A), Art. 185). Hence (6) holds.

From the above example we see that

$$\lim_{n \to \infty} \frac{x^n}{\lfloor n} = 0.$$

Also, on the page preceding, we had

$$R = f^{(n)}(x_1)\frac{x^n}{\lfloor n}. \qquad (0 < x_1 < x)$$

Hence $\lim_{n \to \infty} R = 0$ if $f^{(n)}(x_1)$ remains finite when n becomes infinite.

360 DIFFERENTIAL AND INTEGRAL CALCULUS

ILLUSTRATIVE EXAMPLE 2. Using the series (8) found in the last example, calculate sin 1 correct to four decimal places.

Solution. Here $x = 1$ radian; that is, the angle is expressed in circular measure. Therefore, substituting $x = 1$ in (8) of the last example,

$$\sin 1 = 1 - \frac{1}{\lfloor 3} + \frac{1}{\lfloor 5} - \frac{1}{\lfloor 7} + \frac{1}{\lfloor 9} \cdots$$

Summing up the positive and negative terms separately,

$$1 = 1.00000 \cdots \qquad \frac{1}{\lfloor 3} = 0.16667 \cdots$$

$$\frac{1}{\lfloor 5} = 0.00833 \cdots \qquad \frac{1}{\lfloor 7} = 0.00020 \cdots$$

$$\overline{1.00833 \cdots} \qquad \overline{0.16687 \cdots}$$

Hence $\quad \sin 1 = 1.00833 - 0.16687 = 0.84146 \cdots,$

which is correct to five decimal places, since the error made must be less than $\frac{1}{\lfloor 9}$; that is, less than 0.000003. Obviously the value of sin 1 may be calculated to any desired degree of accuracy by simply including a sufficient number of additional terms.

PROBLEMS

Verify the following expansions of functions by Maclaurin's series and determine for what values of the variable they are convergent.

1. $e^x = 1 + x + \dfrac{x^2}{\lfloor 2} + \dfrac{x^3}{\lfloor 3} + \cdots + \dfrac{x^{n-1}}{\lfloor n-1} + \cdots.$ *Ans.* All values.

2. $\sin x = x - \dfrac{x^3}{\lfloor 3} + \dfrac{x^5}{\lfloor 5} - \cdots + \dfrac{(-1)^{n-1} x^{2n-1}}{\lfloor 2n-1} + \cdots.$ All values.

3. $\ln(1+x) = x - \dfrac{x^2}{2} + \dfrac{x^3}{3} - \dfrac{x^4}{4} + \cdots + \dfrac{(-1)^{n-1} x^n}{n} + \cdots.$ $-1 < x \leqq 1.$

4. $\ln(1-x) = -x - \dfrac{x^2}{2} - \dfrac{x^3}{3} - \dfrac{x^4}{4} - \cdots - \dfrac{x^n}{n} - \cdots.$ $-1 \leqq x < 1.$

5. $\arcsin x = x + \dfrac{1 \cdot x^3}{2 \cdot 3} + \dfrac{1 \cdot 3 \cdot x^5}{2 \cdot 4 \cdot 5} + \cdots$

$\qquad + \dfrac{1 \cdot 3 \cdots (2n-3) x^{2n-1}}{2 \cdot 4 \cdots (2n-2)(2n-1)} + \cdots.$ $-1 \leqq x \leqq 1.$

6. $\arctan x = x - \dfrac{x^3}{3} + \dfrac{x^5}{5} - \cdots + \dfrac{(-1)^{n-1} x^{2n-1}}{2n-1} + \cdots.$ $-1 \leqq x \leqq 1.$

7. $\sin\left(\dfrac{\pi}{4} + x\right) = \dfrac{1}{\sqrt{2}}\left(1 + x - \dfrac{x^2}{\lfloor 2} - \dfrac{x^3}{\lfloor 3} + \dfrac{x^4}{\lfloor 4} + \dfrac{x^5}{\lfloor 5} - \dfrac{x^6}{\lfloor 6} - \dfrac{x^7}{\lfloor 7} + \cdots\right).$

Ans. All values.

8. $\ln(a+x) = \ln a + \dfrac{x}{a} - \dfrac{x^2}{2 a^2} + \dfrac{x^3}{3 a^3} - \cdots$

$\qquad + \dfrac{(-1)^n x^{n-1}}{(n-1) a^{n-1}} + \cdots.$ $-a < x \leqq a.$

EXPANSION OF FUNCTIONS

Verify the following expansions.

9. $\tan x = x + \dfrac{x^3}{3} + \dfrac{2\,x^5}{15} + \dfrac{17\,x^7}{315} + \cdots$.

10. $\sec x = 1 + \dfrac{x^2}{2} + \dfrac{5\,x^4}{24} + \dfrac{61\,x^6}{720} + \cdots$.

11. $\sin\left(\dfrac{\pi}{3} + x\right) = \dfrac{1}{2}\left(\sqrt{3} + x - \dfrac{\sqrt{3}\,x^2}{\underline{|2}} - \dfrac{x^3}{\underline{|3}} + \dfrac{\sqrt{3}\,x^4}{\underline{|4}} + \dfrac{x^5}{\underline{|5}} \cdots\right)$.

12. $\tan\left(\dfrac{\pi}{4} + x\right) = 1 + 2\,x + 2\,x^2 + \dfrac{8\,x^3}{3} + \cdots$.

13. $\operatorname{arc\,tan}\dfrac{1}{x} = \dfrac{\pi}{2} - x + \dfrac{x^3}{3} - \dfrac{x^5}{5} + \cdots$.

14. $\tfrac{1}{2}(e^x + e^{-x}) = 1 + \dfrac{x^2}{\underline{|2}} + \dfrac{x^4}{\underline{|4}} + \dfrac{x^6}{\underline{|6}} + \cdots$.

15. $\ln(x + \sqrt{1+x^2}) = x - \dfrac{x^3}{\underline{|3}} + \dfrac{9\,x^5}{\underline{|5}} \cdots$.

16. $\ln \cos x = -\dfrac{x^2}{2} - \dfrac{x^4}{12} - \dfrac{x^6}{45} \cdots$.

Find three terms of the expansion in powers of x of each of the following functions.

17. $\cos\left(x - \dfrac{\pi}{4}\right)$.

18. $\sin(x + 1)$.

19. $e^{\sin x}$.

20. $\tfrac{1}{2}(e^x - e^{-x})$.

Compute the values of the following functions by substituting directly in the equivalent power series, taking terms enough to make the results agree with those given below.

21. $e = 2.7182 \cdots$.

Solution. Let $x = 1$ in the series of Problem 1; then

$$e = 1 + 1 + \dfrac{1}{\underline{|2}} + \dfrac{1}{\underline{|3}} + \dfrac{1}{\underline{|4}} + \dfrac{1}{\underline{|5}} + \cdots.$$

First term $= 1.00000$
Second term $= 1.00000$
Third term $= 0.50000$
Fourth term $= 0.16667 \cdots$ (Dividing third term by 3.)
Fifth term $= 0.04167 \cdots$ (Dividing fourth term by 4.)
Sixth term $= 0.00833 \cdots$ (Dividing fifth term by 5.)
Seventh term $= 0.00139 \cdots$ (Dividing sixth term by 6.)
Eighth term $= \underline{0.00020} \cdots$, etc. (Dividing seventh term by 7.)
    Adding, $e = 2.71826 \cdots$ Ans.

22. $\operatorname{arc\,tan}(\tfrac{1}{5}) = 0.1973 \cdots$; use series in Problem 6.

23. $\cos 1 = 0.5403 \cdots$; use series in (7), Illustrative Example 1.

362 DIFFERENTIAL AND INTEGRAL CALCULUS

24. $\cos 10° = 0.9848 \cdots$; use series in (7), Illustrative Example 1.

25. $\sin 0.1 = 0.0998 \cdots$; use series in Problem 2.

26. $\arcsin 1 = 1.5708 \cdots$; use series in Problem 5.

27. $\sin \dfrac{\pi}{4} = 0.7071 \cdots$; use series in Problem 2.

28. $\sin 0.5 = 0.4794 \cdots$; use series in Problem 2.

29. $e^2 = 1 + 2 + \dfrac{2^2}{\lfloor 2} + \dfrac{2^3}{\lfloor 3} + \cdots = 7.3890$.

30. $\sqrt{e} = 1 + \dfrac{1}{2} + \dfrac{1}{2^2 \lfloor 2} + \dfrac{1}{2^3 \lfloor 3} + \cdots = 1.6487$.

195. Operations with infinite series. One can carry out many of the operations of algebra and the calculus with convergent series just as one can with polynomials. The following statements are given without proof.

Let $\qquad a_0 + a_1 x + a_2 x^2 + \cdots + a_n x^n + \cdots$

and $\qquad b_0 + b_1 x + b_2 x^2 + \cdots + b_n x^n + \cdots$

be convergent power series. Then we obtain new convergent power series from them as follows:

1. *By adding (or subtracting) term by term.*

$(a_0 \pm b_0) + (a_1 \pm b_1)x + \cdots + (a_n \pm b_n)x^n + \cdots$.

2. *By multiplication and grouping terms.*

$a_0 b_0 + (a_0 b_1 + a_1 b_0)x + (a_0 b_2 + a_1 b_1 + a_2 b_0)x^2 + \cdots$.

ILLUSTRATIVE EXAMPLE 1. *Computation of logarithms.* From the series (Problems 3 and 4, Art. 194)

$\ln(1+x) = x - \tfrac{1}{2}x^2 + \tfrac{1}{3}x^3 - \tfrac{1}{4}x^4 + \cdots$,
$\ln(1-x) = -x - \tfrac{1}{2}x^2 - \tfrac{1}{3}x^3 - \tfrac{1}{4}x^4 - \cdots$,

we obtain, by subtraction of corresponding terms, and using (2), Art. 1, the new series

(1) $\qquad \ln \dfrac{1+x}{1-x} = 2(x + \tfrac{1}{3}x^3 + \tfrac{1}{5}x^5 + \tfrac{1}{7}x^7 + \cdots)$.

This series converges when $|x| < 1$.

To transform (1) into a form better adapted to computation, let N be a positive number. Then, if we set

(2) $\qquad x = \dfrac{1}{2N+1}$, whence $\dfrac{1+x}{1-x} = \dfrac{N+1}{N}$,

then $|x| < 1$ for all values of N. Substituting in (1), we get the formula

(3) $\qquad \ln(N+1) = \ln N + 2\left[\dfrac{1}{2N+1} + \dfrac{1}{3}\dfrac{1}{(2N+1)^3} + \dfrac{1}{5}\dfrac{1}{(2N+1)^5} + \cdots\right]$.

EXPANSION OF FUNCTIONS

This series converges for all positive values of N, and is well adapted to computation. For example, let $N = 1$. Then

$$\ln(N+1) = \ln 2, \quad \frac{1}{2N+1} = \frac{1}{3}.$$

Substituting in (3), the result is $\ln 2 = 0.69315$.
Placing $N = 2$ in (3), we get

$$\ln 3 = \ln 2 + 2\left[\frac{1}{5} + \frac{1}{3}\cdot\frac{1}{5^3} + \frac{1}{5}\cdot\frac{1}{5^5} + \cdots\right] = 1.09861\cdots.$$

It is only necessary to compute the logarithms of prime numbers in this way, the logarithms of composite numbers being then found by using formulas (2), Art. 1. Thus, $\ln 8 = \ln 2^3 = 3 \ln 2 = 2.07944\cdots,$
$\ln 6 = \ln 3 + \ln 2 = 1.79176\cdots.$

All the above are *Napierian*, or *natural*, *logarithms*, that is, the base is $e = 2.71828\cdots$. If we wish to find *Briggs's*, or *common, logarithms*, where the base 10 is employed, all we need to do is to change the base by means of the formula

$$\log n = \frac{\ln n}{\ln 10}.$$

Thus, $\quad \log 2 = \dfrac{\ln 2}{\ln 10} = \dfrac{0.69315}{2.30258} = 0.3010\cdots.$

In the actual computation of a table of logarithms only a few of the tabulated values are calculated from series, all the rest being found by employing theorems in the theory of logarithms and various ingenious devices designed for the purpose of saving work.

ILLUSTRATIVE EXAMPLE 2. Find the power series for $e^x \sin x$.

Solution. From the series

$$\sin x = x - \frac{x^3}{6} + \frac{x^5}{120} - \cdots \qquad \text{Problem 2, Art. 194}$$

and $\qquad e^x = 1 + x + \dfrac{x^2}{2} + \dfrac{x^3}{6} + \dfrac{x^4}{24} + \dfrac{x^5}{120} \qquad$ Problem 1, Art. 194

we obtain, by multiplication,

$$e^x \sin x = x + x^2 + \frac{x^3}{3} - \frac{x^5}{30} + \text{terms in } x^6 \text{ etc.} \quad Ans.$$

3. By division. A special case is shown in the example below.

ILLUSTRATIVE EXAMPLE 3. Find the series for $\sec x$ from the series for $\cos x$ (see (7), Art. 194).

(4) $\qquad \cos x = 1 - \dfrac{x^2}{\underline{|2}} + \dfrac{x^4}{\underline{|4}} - \dfrac{x^6}{\underline{|6}} + \cdots.$

Solution. From the formula $\sec x = \dfrac{1}{\cos x}$, we see that we have to carry through the division of 1 by the series (4). This is best done as follows.
Write (4) in the form $\cos x = 1 - z$, where

(5) $\qquad z = \dfrac{x^2}{2} - \dfrac{x^4}{24} + \dfrac{x^6}{720} - \cdots.$

Then

(6) $\qquad \sec x = \dfrac{1}{1-z} = 1 + z + z^2 + z^3 + \cdots,$

if $|z| < 1$ (Problem 1, Art. 193).

From (5), we have the series

$$z^2 = \frac{x^4}{4} - \frac{x^6}{24} + \text{terms of higher degree},$$

$$z^3 = \frac{x^6}{8} + \cdots.$$

Substituting in (6), the result is

$$\sec x = 1 + \frac{1}{2} x^2 + \frac{5}{24} x^4 + \frac{61}{720} x^6 + \cdots. \quad Ans.$$

PROBLEMS

Given $\ln 2 = 0.69315$, $\ln 3 = 1.09861$, calculate the following natural logarithms by the method of the example above.

1. $\ln 5 = 1.60944$.
2. $\ln 7 = 1.94591$.
3. $\ln 11 = 2.39790$.
4. $\ln 13 = 2.56495$.

Work out the following series.

5. $e^{-t} \cos t = 1 - t + \frac{1}{3} t^3 - \frac{1}{6} t^4 + \cdots$.

6. $\dfrac{e^x}{1-x} = 1 + 2x + \dfrac{5}{2} x^2 + \dfrac{8}{3} x^3 + \dfrac{65}{24} x^4 + \cdots$.

7. $\dfrac{\cos x}{\sqrt{1+x}} = 1 - \dfrac{1}{2} x - \dfrac{1}{8} x^2 - \dfrac{1}{16} x^3 + \dfrac{49}{384} x^4 + \cdots$.

8. $\dfrac{\sin \frac{1}{2}\theta}{1 - \frac{1}{2}\theta} = \dfrac{1}{2}\theta + \dfrac{1}{4}\theta^2 + \dfrac{5}{48}\theta^3 + \dfrac{5}{96}\theta^4 + \cdots$.

9. $\dfrac{\sqrt{1-x^2}}{\cos x} = 1 - \dfrac{1}{6} x^4 + \cdots$.

10. $e^x \tan x = x + x^2 + \dfrac{5}{6} x^3 + \dfrac{1}{2} x^4 + \cdots$.

11. $e^{-x} \sec x = 1 - x + x^2 - \dfrac{2}{3} x^3 + \dfrac{1}{2} x^4 + \cdots$.

12. $e^{-\frac{t}{2}} \sin 2t = 2t - t^2 - \dfrac{13}{12} t^3 + \dfrac{5}{8} t^4 + \cdots$.

13. $(1+x) \cos \sqrt{x} = 1 + \dfrac{1}{2} x - \dfrac{11}{24} x^2 + \dfrac{29}{720} x^3 - \dfrac{11}{8064} x^4 + \cdots$.

14. $(1 + 2x) \arcsin x = x + 2x^2 + \dfrac{1}{6} x^3 + \dfrac{1}{3} x^4 + \cdots$.

15. $\sqrt{1-x} \arctan x = x - \dfrac{1}{2} x^2 - \dfrac{11}{24} x^3 + \dfrac{5}{48} x^4 + \cdots$.

16. $\sqrt{1 - \tan x} = 1 - \dfrac{1}{2} x - \dfrac{1}{8} x^2 - \dfrac{11}{48} x^3 - \dfrac{47}{384} x^4 + \cdots$.

EXPANSION OF FUNCTIONS

17. $\sqrt{\sec x} = 1 + \dfrac{1}{4} x^2 + \dfrac{7}{96} x^4 + \cdots$.

18. $\dfrac{\ln (1 + x)}{1 + \sin x} = x - \dfrac{3}{2} x^2 + \dfrac{11}{6} x^3 - \dfrac{23}{12} x^4 + \cdots$.

19. $\dfrac{1}{\sqrt{5 - e^x}} = \dfrac{1}{2} + \dfrac{1}{16} x + \dfrac{11}{256} x^2 + \dfrac{151}{6144} x^3 + \cdots$.

20. $\sqrt{4 + \sin \phi} = 2 + \dfrac{1}{4} \phi - \dfrac{1}{64} \phi^2 - \dfrac{61}{1536} \phi^3 + \cdots$.

For the following functions find all terms of the series which involve powers of x less than x^5.

21. $e^{-\frac{x}{5}} \sin x$.

22. $e^x \cos \frac{1}{2}\sqrt{x}$.

23. $\dfrac{\sin x}{\cos 2 x}$.

24. $\sqrt{3 + e^{-x}}$.

25. $\dfrac{\ln (1 + x)}{\sqrt{1 + x}}$.

26. $\sqrt{5 - \cos x}$.

196. Differentiation and integration of power series. A convergent power series

(1) $\qquad a_0 + a_1 x + a_2 x^2 + a_3 x^3 + \cdots + a_n x^n + \cdots$

may be differentiated term by term for any value of x within the interval of convergence, and the resulting series is also convergent.

For example, from the series

$$\sin x = x - \dfrac{x^3}{\lfloor 3} + \dfrac{x^5}{\lfloor 5} - \dfrac{x^7}{\lfloor 7} + \cdots$$

we obtain, by differentiation, the new series

$$\cos x = 1 - \dfrac{x^2}{\lfloor 2} + \dfrac{x^4}{\lfloor 4} - \dfrac{x^6}{\lfloor 6} + \cdots.$$

Both series converge for all values of x (see Problems 6 and 7, Art. 191).

Again, the series (1) may be integrated term by term if the limits lie within the interval of convergence, and the resulting series will converge.

ILLUSTRATIVE EXAMPLE 1. Find the series for $\ln (1 + x)$ by integration.

Solution. Since $\dfrac{d}{dx} \ln (1 + x) = \dfrac{1}{1 + x}$, we have

$\qquad \ln (1 + x) = \displaystyle\int_0^x \dfrac{dx}{1 + x}$.

Now $\qquad \dfrac{1}{1 + x} = 1 - x + x^2 - x^3 + x^4 - \cdots$, when $|x| < 1$ (Art. 192). Substituting in (2) and integrating the right-hand member term by term, we obtain the result

$$\ln(1+x) = x - \tfrac{1}{2}x^2 + \tfrac{1}{3}x^3 - \tfrac{1}{4}x^4 + \cdots.$$

This series also converges when $|x| < 1$ (see Problem 2, Art. 191).

ILLUSTRATIVE EXAMPLE 2. Find the power series for arc sin x by integration.

Solution. Since $\dfrac{d}{dx}\arcsin x = \dfrac{1}{\sqrt{1-x^2}}$, we have

$$\text{(3)} \qquad \arcsin x = \int_0^x \frac{dx}{\sqrt{1-x^2}}. \qquad (x^2 < 1)$$

By the binomial series ((2), Art. 192), letting $m = -\tfrac{1}{2}$, and replacing x by $-x^2$, we have

$$\frac{1}{\sqrt{1-x^2}} = 1 + \frac{1}{2}x^2 + \frac{1\cdot 3}{2\cdot 4}x^4 + \frac{1\cdot 3\cdot 5}{2\cdot 4\cdot 6}x^6 + \cdots.$$

This series converges when $|x| < 1$. Substituting in (3) and integrating term by term, we get

$$\arcsin x = x + \frac{1}{2}\frac{x^3}{3} + \frac{1\cdot 3}{2\cdot 4}\frac{x^5}{5} + \frac{1\cdot 3\cdot 5}{2\cdot 4\cdot 6}\frac{x^7}{7} + \cdots. \quad Ans.$$

This series converges also when $|x| < 1$ (see Problem 8, Art. 191).

By this series, the value of π is readily computed. For, since the series converges for values of x between -1 and $+1$, we may let $x = \tfrac{1}{2}$, giving

$$\frac{\pi}{6} = \frac{1}{2} + \frac{1}{2}\cdot\frac{1}{3}\left(\frac{1}{2}\right)^3 + \frac{1\cdot 3}{2\cdot 4}\cdot\frac{1}{5}\left(\frac{1}{2}\right)^5 + \cdots,$$

or $\qquad \pi = 3.1415\cdots.$

Evidently we might have used the series of Problem 6, Art. 194, instead. Both of these series converge rather slowly, but there are other series, found by more elaborate methods, by means of which the correct value of π to a large number of decimal places may easily be calculated.

ILLUSTRATIVE EXAMPLE 3. Using series, find approximately the value of $\int_0^1 \sin x^2\, dx$.

Solution. Let $z = x^2$. Then

$$\sin z = z - \frac{z^3}{\underline{|3}} + \frac{z^5}{\underline{|5}} - \cdots. \qquad \text{Problem 2, Art. 194}$$

Hence $\qquad \sin x^2 = x^2 - \dfrac{x^6}{\underline{|3}} + \dfrac{x^{10}}{\underline{|5}}\cdots,$

and $\qquad \displaystyle\int_0^1 \sin x^2\, dx = \int_0^1 \left(x^2 - \frac{x^6}{\underline{|3}} + \frac{x^{10}}{\underline{|5}}\right)dx,$ approximately,

$$= \left[\frac{x^3}{3} - \frac{x^7}{42} + \frac{x^{11}}{1320}\right]_0^1 = 0.3333 - 0.0238 + 0.0008$$

$$= 0.3103. \quad Ans.$$

EXPANSION OF FUNCTIONS 367

PROBLEMS

1. Find the series for arc tan x by integration.
2. Find the series for ln $(1-x)$ by integration.
3. Find the series for $\sec^2 x$ by differentiating the series for tan x.
4. Find the series for ln cos x by integrating the series for tan x.

Using series, find approximately the values of the following integrals.

5. $\int_0^{\frac{1}{2}} \dfrac{\cos x \, dx}{1+x}$.   Ans. 0.3914.   9. $\int_0^{\frac{1}{4}} \dfrac{\ln(1+x) dx}{\cos x}$.   Ans. 0.0295.

6. $\int_0^{\frac{1}{2}} \dfrac{\sin x \, dx}{1-x}$.   0.185.   10. $\int_0^1 e^{-x^2} dx$.

7. $\int_0^{\frac{1}{3}} e^x \ln(1+x) dx$.   0.0628.   11. $\int_0^{\frac{1}{9}} \ln(1+\sqrt{x}) \, dx$.

8. $\int_0^{\frac{1}{2}} \dfrac{e^{-x^2} dx}{\sqrt{1-x^2}}$.   0.4815.   12. $\int_0^1 e^x \sin \sqrt{x} \, dx$.

13. $\int_0^1 \sqrt{4-x^3} \, dx$.   14. $\int_0^1 e^{-x} \cos \sqrt{x} \, dx$.   15. $\int_0^1 \sqrt{2-\sin x} \, dx$.

197. Approximate formulas derived from Maclaurin's series. By using a few terms of the power series by which a function is represented, we obtain for the function an approximate formula which possesses some degree of accuracy. Such approximate formulas are widely used in applied mathematics.

For example, taking the binomial series ((2), Art. 192), we may write down at once the following approximate formulas.

$$\begin{array}{ccc} & \textit{First approximation} & \textit{Second approximation} \\ (1+x)^m & = \ 1+mx \ = & 1+mx+\tfrac{1}{2}m(m-1)x^2; \\ \dfrac{1}{(1+x)^m} & = \ 1-mx \ = & 1-mx+\tfrac{1}{2}m(m+1)x^2. \end{array}$$

In these $|x|$ is small and m is positive.

Again, consider the sine series

(1) $$\sin x = x - \dfrac{x^3}{\underline{|3}} + \dfrac{x^5}{\underline{|5}} - \cdots.$$

Then

(2) $\quad\quad\quad \sin x = x,$

(3) $\quad\quad\quad \sin x = x - \dfrac{x^3}{6},$

etc.

are approximate formulas. Let us examine the first.

368 DIFFERENTIAL AND INTEGRAL CALCULUS

In the series in (1), assume values of x such that the terms decrease. If the first term only is retained, the value of the remaining series is numerically less than its first term $\frac{1}{6} x^3$ (Art. 188). That is,

$$\sin x = x, \text{ with } |\text{error}| < |\tfrac{1}{6} x^3|.$$

We may inquire, For what range of values of x will (2) hold to three decimal places? Then

$$|\tfrac{1}{6} x^3| < 0.0005,$$

that is, $\quad |x| < \sqrt[3]{0.003} < 0.1443 \text{ rad.}$

We then conclude that formula (2) is correct to three decimal places for values of x between -0.1443 and $+0.1443$, or, in degrees, for values between $-8°.2$ and $+8°.2$.

PROBLEMS

1. How accurate is the approximate formula $\sin x = x - \dfrac{x^3}{6}$ when (a) $x = 30°$? (b) $x = 60°$? (c) $x = 90°$?

Ans. (a) Error < 0.00033; (b) error < 0.01; (c) error < 0.08.

2. How accurate is the approximate formula $\cos x = 1 - \dfrac{x^2}{2}$ when (a) $x = 30°$? (b) $x = 60°$? (c) $x = 90°$?

Ans. (a) Error < 0.0032; (b) error < 0.05; (c) error < 0.25.

3. How accurate is the approximate formula $e^{-x} = 1 - x$ when (a) $x = 0.1$? (b) $x = 0.5$?

4. How accurate is the approximate formula $\arctan x = x - \dfrac{x^3}{3}$ when (a) $x = 0.1$? (b) $x = 0.5$? (c) $x = 1$?

Ans. (a) Error < 0.000002; (b) error < 0.006; (c) error < 0.2.

5. How many terms of the series $\sin x = x - \dfrac{x^3}{\lfloor 3} + \dfrac{x^5}{\lfloor 5} - \cdots$ must be taken to give $\sin 45°$ correct to five decimals? *Ans.* Four.

6. How many terms of the series $\cos x = 1 - \dfrac{x^2}{\lfloor 2} + \dfrac{x^4}{\lfloor 4} - \cdots$ must be taken to give $\cos 60°$ correct to five decimals?

7. How many terms of the series $\ln (1 + x) = x - \dfrac{x^2}{2} + \dfrac{x^3}{3} - \cdots$ must be taken to give $\log 1.2$ correct to five decimals? *Ans.* Six.

Verify the following approximate formulas:

8. $\dfrac{\sin x}{1 - x} = x + x^2.$

9. $\dfrac{\cos x}{1 - x^2} = 1 + \dfrac{x^2}{2}.$

10. $e^{-\theta} \cos \theta = 1 - \theta + \dfrac{\theta^3}{3}.$

11. $\int \cos \sqrt{x}\, dx = C + x - \dfrac{x^2}{4} + \dfrac{x^3}{72}.$

12. $\int e^{-x^2} dx = C + x - \dfrac{x^3}{3} + \dfrac{x^5}{10}.$

13. $\int \ln (1 - x) dx = C - \dfrac{x^2}{2} - \dfrac{x^3}{6}.$

14. $\int \arcsin x\, dx = C + \dfrac{x^2}{2} + \dfrac{x^4}{24}.$

15. $\int e^\theta \sin \theta\, d\theta = C + \dfrac{\theta^2}{2} + \dfrac{\theta^3}{3}.$

198. Taylor's series. A convergent power series in x is well adapted for calculating the value of the function which it represents for *small* values of x (near zero). We now derive an expansion in powers of $x - a$ (see Art. 193), where a is a fixed number. The series thus obtained is adapted for calculation of the function represented by it for values of x near a.

Assume that

(1) $f(x) = b_0 + b_1(x-a) + b_2(x-a)^2 + \cdots + b_n(x-a)^n + \cdots$,

and that the series represents the function. The *necessary form* of the coefficients b_0, b_1, etc. is obtained as in Art. 194. That is, we differentiate (1) with respect to x, assuming that this is possible, and continue the process. Thus we have

$f'(x) = b_1 + 2\,b_2(x-a) + \cdots + nb_n(x-a)^{n-1} + \cdots$,
$f''(x) = 2\,b_2 + \cdots + n(n-1)b_n(x-a)^{n-2} + \cdots$,

etc.

Substituting $x = a$ in these equations and in (1), and solving for b_0, b_1, b_2, $\cdots$, we obtain

$$b_0 = f(a), \quad b_1 = f'(a), \quad b_2 = \frac{f''(a)}{\underline{|2}}, \quad \cdots, \quad b_n = \frac{f^{(n)}(a)}{\underline{|n}}, \quad \cdots.$$

Replacing these values in (1), the result is

(B) $\quad f(x) = f(a) + f'(a)\,\dfrac{(x-a)}{\underline{|1}} + f''(a)\,\dfrac{(x-a)^2}{\underline{|2}} + \cdots$

$\qquad\qquad + f^{(n)}(a)\,\dfrac{(x-a)^n}{\underline{|n}} + \cdots.$

The series is called *Taylor's series* (or *formula*).*

Let us now examine (B) critically. Referring to (G), Art. 124, and letting $b = x$, the result may be written thus:

(2) $f(x) = f(a) + f'(a)\,\dfrac{(x-a)}{\underline{|1}} + \cdots + f^{(n-1)}(a)\,\dfrac{(x-a)^{n-1}}{\underline{|n-1}} + R,$

where $\qquad\qquad R = f^{(n)}(x_1)\,\dfrac{(x-a)^n}{\underline{|n}}. \qquad\qquad (a < x_1 < x)$

The term R is called the *remainder after n terms*.

Now the series in the right-hand member of (2) agrees with Taylor's series (B) up to n terms. Denote the sum of these terms by S_n. Then, from (2), we have

$$f(x) = S_n + R, \quad \text{or} \quad f(x) - S_n = R.$$

* Published by Dr. Brook Taylor (1685–1731) in his "Methodus Incrementorum" (London. 1715).

Now assume that, for a fixed value $x = x_0$, the remainder R approaches zero as a limit when n becomes infinite. Then

(3) $$\lim_{n \to \infty} S_n = f(x_0),$$

and **(B)** converges for $x = x_0$ and its value is $f(x_0)$.

Theorem. *The infinite series* **(B)** *represents the function for those values of x, and those only, for which the remainder approaches zero as the number of terms increases without limit.*

If the series converges for values of x for which the remainder does not approach zero as n increases without limit, then for such values of x the series does not represent the function $f(x)$.

It is usually easier to determine the interval of convergence of the series than that for which the remainder approaches zero; but in simple cases the two intervals are identical.

When the values of a function and its successive derivatives are known and finite for some fixed value of the variable, as $x = a$, then **(B)** is used for finding the value of the function for values of x near a, and **(B)** is also called *the expansion of $f(x)$ in the neighborhood of $x = a$.*

ILLUSTRATIVE EXAMPLE 1. Expand $\ln x$ in powers of $(x - 1)$.

Solution. $f(x) = \ln x,$ $f(1) = 0,$

$f'(x) = \dfrac{1}{x},$ $f'(1) = 1,$

$f''(x) = -\dfrac{1}{x^2},$ $f''(1) = -1,$

$f'''(x) = \dfrac{2}{x^3},$ $f'''(1) = 2,$

etc., etc.

Substituting in **(B)**, $\ln x = x - 1 - \frac{1}{2}(x-1)^2 + \frac{1}{3}(x-1)^3 - \cdots.$ *Ans.*

This converges for values of x between 0 and 2 and is the *expansion of $\ln x$ in the vicinity of $x = 1$.* See Illustrative Example, Art. 193.

ILLUSTRATIVE EXAMPLE 2. Expand $\cos x$ in powers of $\left(x - \dfrac{\pi}{4}\right)$ to four terms.

Solution. Here $f(x) = \cos x$ and $a = \dfrac{\pi}{4}$. Then we have

$f(x) = \cos x,$ $f\left(\dfrac{\pi}{4}\right) = \dfrac{1}{\sqrt{2}},$

$f'(x) = -\sin x,$ $f'\left(\dfrac{\pi}{4}\right) = -\dfrac{1}{\sqrt{2}},$

$f''(x) = -\cos x,$ $f''\left(\dfrac{\pi}{4}\right) = -\dfrac{1}{\sqrt{2}},$

$f'''(x) = \sin x,$ $f'''\left(\dfrac{\pi}{4}\right) = \dfrac{1}{\sqrt{2}},$

etc., etc.

EXPANSION OF FUNCTIONS

The series is, therefore,

$$\cos x = \frac{1}{\sqrt{2}} - \frac{1}{\sqrt{2}}\left(x - \frac{\pi}{4}\right) - \frac{1}{\sqrt{2}}\frac{\left(x - \frac{\pi}{4}\right)^2}{\underline{|2}} + \frac{1}{\sqrt{2}}\frac{\left(x - \frac{\pi}{4}\right)^3}{\underline{|3}} + \cdots.$$

The result may be written in the form

$$\cos x = 0.70711\left[1 - \left(x - \frac{\pi}{4}\right) - \frac{1}{2}\left(x - \frac{\pi}{4}\right)^2 + \frac{1}{6}\left(x - \frac{\pi}{4}\right)^3 \cdots\right].$$

To check this result let us calculate $\cos 50°$. Then $x - \frac{\pi}{4} = 5°$ expressed in radians, or $x - \frac{\pi}{4} = 0.08727$, $\left(x - \frac{\pi}{4}\right)^2 = 0.00762$, $\left(x - \frac{\pi}{4}\right)^3 = 0.00066$. With these values the series above gives $\cos 50° = 0.64278$. Five-place tables give $\cos 50° = 0.64279$.

199. Another form of Taylor's series. In (B), Art. 198, if we replace a by x_0 and let $x - a = h$, that is, $x = a + h = x_0 + h$, the result is

$$(C)\ f(x_0 + h) = f(x_0) + f'(x_0)\frac{h}{\underline{|1}} + f''(x_0)\frac{h^2}{\underline{|2}} + \cdots + f^{(n)}(x_0)\frac{h^n}{\underline{|n}} + \cdots.$$

In this second form the *new value* of $f(x)$ when x changes from x_0 to $x_0 + h$ is expanded in a power series in h, the increment of x.

ILLUSTRATIVE EXAMPLE. Expand $\sin x$ in a power series in h when x changes from x_0 to $x_0 + h$.

Solution. Here $f(x) = \sin x$, and $f(x_0 + h) = \sin(x_0 + h)$. Differentiate and arrange the work as below.

$f(x) = \sin x,$ $f(x_0) = \sin x_0,$
$f'(x) = \cos x,$ $f'(x_0) = \cos x_0,$
$f''(x) = -\sin x,$ $f''(x_0) = -\sin x_0,$
etc.,   etc.

Substituting in (C), we obtain

$$\sin(x_0 + h) = \sin x_0 + \cos x_0 \frac{h}{\underline{|1}} - \sin x_0 \frac{h^2}{\underline{|2}} - \cos x_0 \frac{h^3}{\underline{|6}} + \cdots.\ Ans.$$

PROBLEMS

Verify the following series by Taylor's formula.

1. $e^x = e^a\left[1 + (x - a) + \frac{(x-a)^2}{\underline{|2}} + \frac{(x-a)^3}{\underline{|3}} + \cdots\right].$

2. $\sin x = \sin a + (x - a)\cos a - \frac{(x-a)^2}{\underline{|2}}\sin a - \frac{(x-a)^3}{\underline{|3}}\cos a + \cdots.$

3. $\cos x = \cos a - (x - a)\sin a - \frac{(x-a)^2}{\underline{|2}}\cos a + \frac{(x-a)^3}{\underline{|3}}\sin a + \cdots.$

4. $\ln(a + x) = \ln a + \frac{x}{a} - \frac{x^2}{2\,a^2} + \frac{x^3}{3\,a^3} + \cdots.$

5. $\cos(a+x) = \cos a - x \sin a - \dfrac{x^2}{\underline{|2}} \cos a + \dfrac{x^3}{\underline{|3}} \sin a + \cdots .$

6. $\tan(x+h) = \tan x + h \sec^2 x + h^2 \sec^2 x \tan x + \cdots .$

7. $(x+h)^n = x^n + nx^{n-1}h + \dfrac{n(n-1)}{\underline{|2}} x^{n-2}h^2$
$\qquad + \dfrac{n(n-1)(n-2)}{\underline{|3}} x^{n-3}h^3 + \cdots .$

8. Expand $\sin x$ in powers of $\left(x - \dfrac{\pi}{4}\right)$ to four terms.

$$\text{Ans.} \quad \sin x = \dfrac{1}{\sqrt{2}}\left[1 + \left(x - \dfrac{\pi}{4}\right) - \dfrac{\left(x - \dfrac{\pi}{4}\right)^2}{\underline{|2}} - \dfrac{\left(x - \dfrac{\pi}{4}\right)^3}{\underline{|3}} + \cdots\right].$$

9. Expand $\tan x$ in powers of $\left(x - \dfrac{\pi}{4}\right)$ to three terms.

$$\text{Ans.} \quad \tan x = 1 + 2\left(x - \dfrac{\pi}{4}\right) + 2\left(x - \dfrac{\pi}{4}\right)^2 + \cdots .$$

10. Expand $\ln x$ in powers of $(x-2)$ to four terms.

11. Expand e^x in powers of $(x-1)$ to five terms.

12. Expand $\sin\left(\dfrac{\pi}{6} + x\right)$ in powers of x to four terms.

13. Expand $\operatorname{ctn}\left(\dfrac{\pi}{4} + x\right)$ in powers of x to three terms.

200. Approximate formulas derived from Taylor's series. Such formulas are provided by using some terms of series (*B*) or (*C*).

For example, if $f(x) = \sin x$, we have (see Problem 2, Art. 199)

(1) $\qquad \sin x = \sin a + \cos a (x - a)$

as a first approximation.

A second approximation results by taking three terms of the series. This is

(2) $\qquad \sin x = \sin a + \cos a (x-a) - \sin a \dfrac{(x-a)^2}{\underline{|2}}.$

From (1), transposing $\sin a$ and dividing by $x - a$, we get

(3) $\qquad \dfrac{\sin x - \sin a}{x - a} = \cos a.$

Since $\cos a$ is constant, this means that (approximately)

The change in the value of the sine is proportional to the change in the angle for values of the angle near a.

Formula (3) illustrates *interpolation by proportional parts.*

EXPANSION OF FUNCTIONS

ILLUSTRATIVE EXAMPLE 1. For example, let $a = 30° = 0.5236$ radian, and suppose it is required to calculate the sines of 31° and 32° by the approximate formula (1). Then, since $x - a = 1° = 0.01745$ radian,

$$\sin 31° = \sin 30° + \cos 30° \, (0.01745)$$
$$= 0.5000 + 0.8660 \times 0.01745$$
$$= 0.5000 + 0.0151 = 0.5151.$$

Similarly, $\sin 32° = \sin 30° + \cos 30° \, (0.03490) = 0.5302.$

These values by (1) are correct to three decimal places only. If greater accuracy is desired, we may use (2).

Then
$$\sin 31° = \sin 30° + \cos 30° \, (0.01745) - \frac{\sin 30°}{2}(0.01745)^2$$
$$= 0.50000 + 0.01511 - 0.00008$$
$$= 0.51503.$$

$$\sin 32° = \sin 30° + \cos 30° \, (0.03490) - \frac{\sin 30°}{2}(0.03490)^2$$
$$= 0.50000 + 0.03022 - 0.00030$$
$$= 0.52992.$$

These results are correct to four decimal places.

From (*C*) we derive approximate formulas for the increment of $f(x)$ when x changes from x_0 to $x_0 + h$. For, transposing the first term of the right-hand member, we get

(4) $$f(x_0 + h) - f(x_0) = f'(x_0)h + f''(x_0)\frac{h^2}{\lfloor 2} + \cdots.$$

The right-hand member expresses the increment of $f(x)$ as a power series in the increment of x ($= h$).

From (4) we derive, as a *first approximation*,

(5) $$f(x_0 + h) - f(x_0) = f'(x_0)h.$$

This formula was used in Art. 92. For the left-hand member is the value of the differential of $f(x)$ for $x = x_0$ and $\Delta x = h$.

As a *second approximation*, we have

(6) $$f(x_0 + h) - f(x_0) = f'(x_0)h + f''(x_0)\frac{h^2}{2}.$$

ILLUSTRATIVE EXAMPLE 2. Calculate the increment of $\tan x$, approximately, when x changes from 45° to 46°, by (5) and by (6).

Solution. From Problem 6, Art. 199, if $x = x_0$, we have

$$\tan (x_0 + h) = \tan x_0 + \sec^2 x_0 \, h + \sec^2 x_0 \tan x_0 \, h^2 + \cdots.$$

In this example $x_0 = 45°$, and $\tan x_0 = 1$, $\sec^2 x_0 = 2$.

Also, $h = 1°$ expressed in radians $= 0.01745$. Hence, by (5),

$$\tan 46° - \tan 45° = 2(0.01745) = 0.0349;$$

by (6), $\tan 46° - \tan 45° = 0.0349 + 2(0.01745)^2 = 0.0349 + 0.0006 = 0.0355.$

From the second approximation we get $\tan 46° = 1.0355$, which is correct to four places of decimals.

PROBLEMS

1. Verify the approximate formula
$$\ln(10 + x) = 2.303 + \frac{x}{10}.$$

Calculate the value of the function from this formula and compare your result with the tables, when (a) $x = -0.5$; (b) $x = -1$.

Ans. (a) Formula, 2.253; tables, 2.251.
(b) Formula, 2.203; tables, 2.197.

2. Verify the approximate formula
$$\sin\left(\frac{\pi}{6} + x\right) = 0.5 + 0.8660\, x.$$

Use the formula to calculate $\sin 27°$, $\sin 33°$, $\sin 40°$, and compare your results with the tables.

3. Verify the approximate formula
$$\tan\left(\frac{\pi}{4} + x\right) = 1 + 2x + 2x^2.$$

Use the formula to calculate $\tan 46°$, $\tan 50°$, and compare your results with the tables.

4. Verify the approximate formula
$$\cos x = \cos a - (x - a)\sin a.$$

Given
$\cos 30° = \sin 60° = 0.8660,$
$\cos 45° = \sin 45° = 0.7071,$

and
$\cos 60° = \sin 30° = 0.5,$

use the formula to calculate $\cos 32°$, $\cos 47°$, $\cos 62°$, and compare your results with the tables.

ADDITIONAL PROBLEMS

1. Given the definite integral $\int_0^{\frac{1}{2}} x^5 \ln(1 + x)\,dx$.

(a) Obtain its value by series correct to four decimal places. *Ans.* 0.0009.

(b) Obtain its value by direct computation and compare with the approximate value derived in (a).

(c) Prove that if n terms of the series are used in the computation the error is less than $\dfrac{1}{2^{n+7}(n+1)(n+7)}$.

2. Given $f(x) = e^{-\frac{x}{2}} \cos \dfrac{x}{2}$.

(a) Show that $f^{(4)}(x) = -\frac{1}{4} f(x)$.

(b) Expand $f(x)$ in a Maclaurin series to six terms.

(c) What is the coefficient of x^{12} in this series? *Ans.* $-\dfrac{1}{64\lfloor 12}$.

CHAPTER XXI

ORDINARY DIFFERENTIAL EQUATIONS *

201. Differential equations — order and degree. A differential equation is an equation involving derivatives or differentials. Differential equations have been frequently employed in this book. The illustrative examples of Art. 139 afford simple examples. Thus, from the differential equation (Illustrative Example 1)

(1) $$\frac{dy}{dx} = 2x,$$

we found, by integrating,

(2) $$y = x^2 + C.$$

Again (Illustrative Example 2), integration of the differential equation

(3) $$\frac{dy}{dx} = -\frac{x}{y}$$

led to the solution

(4) $$x^2 + y^2 = 2C.$$

Equations (1) and (3) are examples of ordinary differential equations of the *first order*, and (2) and (4) are, respectively, the **complete solutions**.

Another example is

(5) $$\frac{d^2y}{dx^2} + y = 0.$$

This is a differential equation of the *second order*, so named from the order of the derivative.

The *order of a differential equation* is the same as that of the derivative of highest order appearing in it.

The derivative of highest order appearing in a differential equation may be affected with exponents. The largest exponent gives the *degree* of the differential equation.

Thus, the differential equation

(6) $$y''^2 = (1 + y'^2)^3.$$

---

* A few types only of differential equations are treated in this chapter, namely, such types as the student is likely to encounter in elementary work in mechanics and physics.

where y' and y'' are, respectively, the first and second derivatives of y with respect to x, is of the second degree and second order.

202. Solutions of differential equations. Constants of integration. A *solution* or *integral* of a differential equation is a relation between the variables involved by which the equation is satisfied. Thus

(1) $$y = a \sin x$$

is a solution of the differential equation

(2) $$\frac{d^2y}{dx^2} + y = 0.$$

For, differentiating (1),

(3) $$\frac{d^2y}{dx^2} = -a \sin x.$$

Now, if we substitute from (1) and (3) in (2), we get

$$-a \sin x + a \sin x = 0,$$

and (2) is satisfied. Here a is an arbitrary constant. In the same manner

(4) $$y = b \cos x$$

may be shown to be a solution of (2) for any value of b. The relation

(5) $$y = c_1 \sin x + c_2 \cos x$$

is a still more general solution of (2). In fact, by giving particular values to c_1 and c_2 it is seen that the solution (5) includes the solutions (1) and (4).

The arbitrary constants c_1 and c_2 appearing in (5) are called *constants of integration*. A solution such as (5), which contains a number of arbitrary essential constants equal to the order of the equation (in this case two), is called the *complete* or *general solution*.* Solutions obtained therefrom by giving particular values to the constants are called *particular solutions*. In practice, a particular solution is obtained from the complete solution by given conditions to be satisfied by the particular solution.

ILLUSTRATIVE EXAMPLE. The complete solution of the differential equation

(1) $$y'' + y = 0$$

is $y = c_1 \cos x + c_2 \sin x$ (see (5) above).

Find a particular solution such that

(2) $$y = 2, \ y' = -1, \text{ when } x = 0.$$

* It is shown in works on differential equations that the general solution has n arbitrary constants when the differential equation is of the nth order.

ORDINARY DIFFERENTIAL EQUATIONS

Solution. From the complete solution

(3) $$y = c_1 \cos x + c_2 \sin x,$$

by differentiation, we obtain

(4) $$y' = - c_1 \sin x + c_2 \cos x.$$

Substituting in (3) and (4) from (2), we find $c_1 = 2$, $c_2 = -1$. Putting these values in (3) gives the particular solution required, $y = 2 \cos x - \sin x$. *Ans.*

The solution of a differential equation is considered as having been effected when it has been reduced to an expression involving integrals, whether the actual integration can be effected or not.

203. Verification of the solutions of differential equations. Before taking up the problem of solving differential equations, we show how to verify a given solution.

ILLUSTRATIVE EXAMPLE 1. Show that

(1) $$y = c_1 x \cos \ln x + c_2 x \sin \ln x + x \ln x$$

is a solution of the differential equation

(2) $$x^2 \frac{d^2y}{dx^2} - x \frac{dy}{dx} + 2y = x \ln x.$$

Solution. Differentiating (1), we get

(3) $$\frac{dy}{dx} = (c_2 - c_1) \sin \ln x + (c_2 + c_1) \cos \ln x + \ln x + 1,$$

(4) $$\frac{d^2y}{dx^2} = - (c_2 + c_1) \frac{\sin \ln x}{x} + (c_2 - c_1) \frac{\cos \ln x}{x} + \frac{1}{x}.$$

Substituting from (1), (3), (4), in (2), we find that the equation is identically satisfied.

ILLUSTRATIVE EXAMPLE 2. Show that

(5) $$y^2 - 4x = 0$$

is a particular solution of the differential equation

(6) $$xy'^2 - 1 = 0.$$

Solution. Differentiating (5), the result is

$$yy' - 2 = 0, \quad \text{whence} \quad y' = \frac{2}{y}.$$

Substituting this value of y' in (6) and reducing, we obtain $4x - y^2 = 0$, which is true by (5).

PROBLEMS

Verify the following solutions of the corresponding differential equations.

Differential equations   *Solutions*

1. $\dfrac{d^2y}{dx^2} - \dfrac{1}{x}\dfrac{dy}{dx} + \dfrac{2}{x} = 0.$ $y = c_1 + 2x + c_2 x^2.$

2. $\dfrac{d^2V}{dr^2} + \dfrac{2}{r}\dfrac{dV}{dr} = 0.$ $V = \dfrac{c_1}{r} + c_2.$

3. $\dfrac{d^2s}{dt^2} - \dfrac{ds}{dt} - 6s = 0.$ $s = c_1 e^{-2t} + c_2 e^{3t}.$

4. $\dfrac{d^3x}{dt^3} + 2\dfrac{d^2x}{dt^2} - \dfrac{dx}{dt} - 2x = 0.$ $x = c_1 e^t + c_2 e^{-t} + c_3 e^{-2t}.$

5. $\left(\dfrac{dy}{dx}\right)^3 - 4xy\dfrac{dy}{dx} + 8y^2 = 0.$ $y = c(x - c)^2.$

6. $x\dfrac{d^2y}{dx^2} + 2\dfrac{dy}{dx} - xy = 0.$ $xy = 2e^x - 3e^{-x}.$

7. $\dfrac{d^2s}{dt^2} + 4s = 0.$ $s = c_1 \cos(2t + c_2).$

8. $\dfrac{d^2x}{dt^2} - 6\dfrac{dx}{dt} + 13x = 39.$ $x = e^{3t}\cos 2t + 3.$

9. $y\dfrac{d^2y}{dx^2} - \left(\dfrac{dy}{dx}\right)^2 + \dfrac{dy}{dx} = 0.$ $y = ae^{\frac{x}{b}} - b.$

10. $xy\dfrac{d^2y}{dx^2} + x\left(\dfrac{dy}{dx}\right)^2 - y\dfrac{dy}{dx} = 0.$ $\dfrac{x^2}{c_1} + \dfrac{y^2}{c_2} = 1.$

11. $\dfrac{du}{dv} = \dfrac{1 + u^2}{1 + v^2}.$ $u = \dfrac{c + v}{1 - cv}.$

12. $\dfrac{d^2s}{dt^2} + 4s = 8t.$ $s = 2\sin 2t + \cos 2t + 2t.$

13. $\dfrac{d^2y}{dx^2} - 2\dfrac{dy}{dx} - 3y = e^{2x}.$ $y = c_1 e^{3x} + c_2 e^{-x} - \tfrac{1}{3}e^{2x}.$

14. $\dfrac{d^2x}{dt^2} + 9x = 5\cos 2t.$ $x = \cos 2t + 2\cos 3t + 3\sin 3t.$

15. $\dfrac{d^2x}{dt^2} + 9x = 3\cos 3t.$ $x = c_1 \cos 3t + c_2 \sin 3t + \tfrac{1}{2} t\sin 3t.$

16. $\dfrac{dy}{dx} + xy = x^3 y^3.$ $\dfrac{1}{y^2} = x^2 + 1 + ce^{x^2}.$

204. Differential equations of the first order and of the first degree. Such an equation may be brought into the form

(A) $M\,dx + N\,dy = 0,$

ORDINARY DIFFERENTIAL EQUATIONS 379

in which M and N are functions of x and y. The more common differential equations coming under this head may be divided into four types.

Type I. Variables separable. When the terms of a differential equation can be so arranged that it takes on the form

(1) $\qquad f(x)dx + F(y)dy = 0,$

where $f(x)$ is a function of x alone and $F(y)$ is a function of y alone, the process is called *separation of the variables*, and the solution is obtained by direct integration. Thus, integrating (1), we get the general solution

(2) $\qquad \int f(x)dx + \int F(y)dy = c,$

where c is an arbitrary constant.

Equations which are not given in the simple form (1) may often be brought into that form by means of the following rule for separating the variables.

FIRST STEP. *Clear of fractions; and if the equation involves derivatives, multiply through by the differential of the independent variable.*

SECOND STEP. *Collect all the terms containing the same differential into a single term. If the equation then takes on the form*

$$XY\,dx + X'Y'dy = 0,$$

where X, X' are functions of x alone, and Y, Y' are functions of y alone, it may be brought to the form (1) by dividing through by $X'Y$.

THIRD STEP. *Integrate each part separately, as in (2).*

ILLUSTRATIVE EXAMPLE 1. Solve the equation

$$\frac{dy}{dx} = \frac{1+y^2}{(1+x^2)xy}.$$

Solution. *First Step.* $(1+x^2)xy\,dy = (1+y^2)dx.$

Second Step. $(1+y^2)dx - x(1+x^2)y\,dy = 0.$

To separate the variables we now divide by $x(1+x^2)(1+y^2)$, giving

$$\frac{dx}{x(1+x^2)} - \frac{y\,dy}{1+y^2} = 0.$$

Third Step. $\int \frac{dx}{x(1+x^2)} - \int \frac{y\,dy}{1+y^2} = C,$

$\int \frac{dx}{x} - \int \frac{x\,dx}{1+x^2} - \int \frac{y\,dy}{1+y^2} = C,$ Art. 167

$\ln x - \tfrac{1}{2}\ln(1+x^2) - \tfrac{1}{2}\ln(1+y^2) = C,$

$\ln(1+x^2)(1+y^2) = 2\ln x - 2C.$

This result may be written in more compact form if we replace $-2C$ by $\ln c$, that is, give a new form to the arbitrary constant. Our solution then becomes

$$\ln(1+x^2)(1+y^2) = \ln x^2 + \ln c,$$
$$\ln(1+x^2)(1+y^2) = \ln cx^2,$$
$$(1+x^2)(1+y^2) = cx^2. \quad Ans.$$

ILLUSTRATIVE EXAMPLE 2. Solve the equation

$$a\left(x\frac{dy}{dx} + 2y\right) = xy\frac{dy}{dx}.$$

Solution. *First Step.* $\qquad ax\,dy + 2ay\,dx = xy\,dy.$
Second Step. $\qquad 2ay\,dx + x(a-y)dy = 0.$

To separate the variables we divide by xy.

$$\frac{2a\,dx}{x} + \frac{(a-y)dy}{y} = 0.$$

Third Step. $\qquad 2a\int\frac{dx}{x} + a\int\frac{dy}{y} - \int dy = C,$

$$2a\ln x + a\ln y - y = C,$$
$$a\ln x^2 y = C + y,$$
$$\ln x^2 y = \frac{C}{a} + \frac{y}{a}.$$

By passing from logarithms to exponentials this result may be written in the form

$$x^2 y = e^{\frac{C}{a} + \frac{y}{a}},$$

or $\qquad x^2 y = e^{\frac{C}{a}} \cdot e^{\frac{y}{a}}.$

Denoting the constant $e^{\frac{C}{a}}$ by c, we get our solution in the form

$$x^2 y = ce^{\frac{y}{a}}. \quad Ans.$$

Type II. Homogeneous equations. The differential equation

(A) $\qquad M\,dx + N\,dy = 0$

is said to be homogeneous when M and N are homogeneous functions of x and y of the same degree.* Such differential equations may be solved by making the substitution

(3) $\qquad y = vx.$

This will give a differential equation in v and x in which the variables are separable, and hence we may follow the rule under Type I.

---

* A function of x and y is said to be *homogeneous* in the variables if the result of replacing x and y by λx and λy (λ being arbitrary) reduces to the original function multiplied by some power of λ. This power of λ is called the *degree* of the original function.

In fact, from (**A**) we obtain

(4) $$\frac{dy}{dx} = -\frac{M}{N}.$$

Also, from (3),

(5) $$\frac{dy}{dx} = x\frac{dv}{dx} + v.$$

The right-hand member of (4) will become a function of v only when the substitution (3) is used. Hence, by using (5) and (3), we shall obtain from (4)

(6) $$x\frac{dv}{dx} + v = f(v),$$

and the variables x and v may be separated.

ILLUSTRATIVE EXAMPLE. Solve the equation

$$y^2 + x^2\frac{dy}{dx} = xy\frac{dy}{dx}.$$

Solution. $\qquad y^2\,dx + (x^2 - xy)dy = 0.$

Here $M = y^2$, $N = x^2 - xy$, and both are homogeneous and of the second degree in x and y. Also we have

$$\frac{dy}{dx} = \frac{y^2}{xy - x^2}.$$

Substitute $y = vx$. The result is

$$x\frac{dv}{dx} + v = -\frac{v^2}{1-v},$$

or $\qquad v\,dx + x(1-v)dv = 0.$

To separate the variables, divide by vx. This gives

$$\frac{dx}{x} + \frac{(1-v)dv}{v} = 0,$$

$$\int\frac{dx}{x} + \int\frac{dv}{v} - \int dv = C,$$

$$\ln x + \ln v - v = C,$$

$$\ln vx = C + v,$$

$$vx = e^{C+v} = e^C \cdot e^v,$$

$$vx = ce^v.$$

But $v = \frac{y}{x}$. Hence the complete solution is

$$y = ce^{\frac{y}{x}}. \quad Ans.$$

PROBLEMS

Find the complete solution of each of the following differential equations.

1. $(2+y)dx - (3-x)dy = 0.$ $\qquad$ Ans. $(2+y)(3-x) = c.$
2. $xy\,dx - (1+x^2)dy = 0.$ $\qquad\qquad\;\; cy^2 = 1 + x^2.$
3. $x(x+3)dy - y(2x+3)dx = 0.$ $\qquad y = cx(x+3).$

4. $\sqrt{1+x^2}\,dy - \sqrt{1-y^2}\,dx = 0.$ Ans. $\arcsin y = \ln c(x+\sqrt{1+x^2}).$
5. $d\rho + \rho \tan\theta\, d\theta = 0.$ $\rho = c\cos\theta.$
6. $(1-x)dy - y^2\,dx = 0.$ $y \ln c(1-x) = 1.$
7. $(x+2y)dx + (2x-3y)dy = 0.$ $x^2 + 4xy - 3y^2 = c.$
8. $(3x+5y)dx + (4x+6y)dy = 0.$ $(x+y)^2(x+2y) = c.$
9. $2(x+y)dx + y\,dy = 0.$

Ans. $\tfrac{1}{2}\ln(2x^2 + 2xy + y^2) - \arctan\left(\dfrac{x+y}{x}\right) = c.$

10. $(8y+10x)dx + (5y+7x)dy = 0.$ $(x+y)^2(2x+y)^3 = c.$
11. $(2x+y)dx + (x+3y)dy = 0.$ $2x^2 + 2xy + 3y^2 = c.$
12. $\sqrt{1-4t^2}\,ds + 2\sqrt{1-s^2}\,dt = 0.$ $s\sqrt{1-4t^2} + 2t\sqrt{1-s^2} = c.$
13. $2z(3z+1)dw + (1-2w)dz = 0.$ $(2w-1)(1+3z) = 3cz.$
14. $2x\,dz - 2z\,dx = \sqrt{x^2+4z^2}\,dx.$ $1 + 4cz - c^2 x^2 = 0.$
15. $(x+4y)dx + 2x\,dy = 0.$ $x^3 + 6x^2 y = c.$
16. $(2x^2+y^2)dx + (2xy+3y^2)dy = 0.$ $2x^3 + 3xy^2 + 3y^3 = c.$
17. $\dfrac{du}{dv} = \dfrac{1+u^2}{1+v^2}.$ $u = \dfrac{v+c}{1-cv}.$
18. $(3+2y)x\,dx + (x^2-2)dy = 0.$
19. $2(1+y)dx - (1-x)dy = 0.$
20. $(1+y)x\,dx - (1+x)y\,dy = 0.$
21. $(ax+b)dy - y^2\,dx = 0.$
22. $(3x+y)dx + (x+y)dy = 0.$
23. $xy(y+2)dx - (y+1)dy = 0.$
24. $(1+x^2)dy - (1-y^2)dx = 0.$
25. $(x-2y)dx - (2x+y)dy = 0.$
26. $(3x+2y)dx + x\,dy = 0.$
27. $3(5x+3y)dx + (11x+5y)dy = 0.$
28. $(x^2+y^2)dx + (2xy+y^2)dy = 0.$
29. $2y\,dx - (2x-y)dy = 0.$

In each of the following problems find the particular solution which is determined by the given values of x and y.

30. $\dfrac{dx}{y} + \dfrac{4\,dy}{x} = 0;\ x=4,\ y=2.$ Ans. $x^2 + 4y^2 = 32.$
31. $(x^2+y^2)dx = 2xy\,dy;\ x=1,\ y=0.$ $y^2 = x^2 - x.$
32. $x\,dy - y\,dx = \sqrt{x^2+y^2}\,dx;\ x=\tfrac{1}{2},\ y=0.$ $1 + 4y - 4x^2 = 0.$
33. $(1+y^2)dy = y\,dx;\ x=2,\ y=2.$

34. Find the equation of the curve whose slope at any point is equal to $-\left(1+\dfrac{y}{x}\right)$ and which passes through the point $(2, 1)$.   Ans. $x^2 + 2xy = 8.$

35. Find the equation of the curve whose slope at any point is equal to $\dfrac{y-1}{x^2+x}$ and which passes through the point $(1, 0)$.   Ans. $y(1+x) = 1-x.$

Type III. Linear equations. The linear differential equation of the first order in y is of the form

(B) $$\frac{dy}{dx} + Py = Q,$$

where P and Q are functions of x alone, or constants.

Similarly, the equation

(C) $$\frac{dx}{dy} + Hx = J,$$

where H and J are functions of y or constants, is a linear differential equation.

To integrate (B), let

(7) $$y = uz,$$

where z and u are functions of x to be determined. Differentiating (7),

(8) $$\frac{dy}{dx} = u\frac{dz}{dx} + z\frac{du}{dx}.$$

Substituting from (8) and (7) in (B), the result is

$$u\frac{dz}{dx} + z\frac{du}{dx} + Puz = Q, \text{ or}$$

(9) $$u\frac{dz}{dx} + \left(\frac{du}{dx} + Pu\right)z = Q.$$

We now determine u by integrating

(10) $$\frac{du}{dx} + Pu = 0,$$

in which the variables x and u are separable. Using the value of u thus obtained, we find z by solving

(11) $$u\frac{dz}{dx} = Q,$$

in which x and z can be separated. Obviously, the values of u and z thus found will satisfy (9), and the solution of (B) is then given by (7). The following examples show the details.

ILLUSTRATIVE EXAMPLE 1. Solve the equation

(12) $$\frac{dy}{dx} - \frac{2y}{x+1} = (x+1)^{\frac{5}{2}}.$$

Solution. This is evidently in the linear form (B), where

$$P = -\frac{2}{x+1} \text{ and } Q = (x+1)^{\frac{5}{2}}.$$

Let $y = uz$; then

$$\frac{dy}{dx} = u\frac{dz}{dx} + z\frac{du}{dx}.$$

Substituting in the given equation (12), we get

$$u\frac{dz}{dx} + z\frac{du}{dx} - \frac{2\,uz}{1+x} = (x+1)^{\frac{5}{2}}, \text{ or}$$

(13) $$u\frac{dz}{dx} + \left(\frac{du}{dx} - \frac{2\,u}{1+x}\right)z = (x+1)^{\frac{5}{2}}.$$

To determine u we place the coefficient of z equal to zero. This gives

$$\frac{du}{dx} - \frac{2\,u}{1+x} = 0,$$

$$\frac{du}{u} = \frac{2\,dx}{1+x}.$$

Integrating, we get $\quad \ln u = 2\ln(1+x) = \ln(1+x)^2.$

(14) $\quad\therefore u = (1+x)^2.$*

Equation (13) now becomes, since the term in z drops out,

$$u\frac{dz}{dx} = (x+1)^{\frac{5}{2}}.$$

Replacing u by its value from (14),

$$\frac{dz}{dx} = (x+1)^{\frac{1}{2}}.$$

Integrating, we get $\quad dz = (x+1)^{\frac{1}{2}}dx.$

(15) $$z = \frac{2(x+1)^{\frac{3}{2}}}{3} + C.$$

Substituting from (15) and (14) in $y = uz$, we get the complete solution

$$y = \frac{2(x+1)^{\frac{7}{2}}}{3} + C(x+1)^2. \text{ Ans.}$$

ILLUSTRATIVE EXAMPLE 2. Derive a formula for the complete solution of (B).

Solution. Solving (10), we have

$$\ln u + \int P\,dx = \ln k,$$

where $\ln k$ is the constant of integration;

whence $$u = ke^{-\int P\,dx}.$$

Substituting this value of u in (11), and separating the variables z and x, the result is

$$dz = \frac{Q}{k}e^{\int P\,dx}dx.$$

Integrating, and substituting in (7), we obtain

$$y = e^{-\int P\,dx}\left(\int Qe^{\int P\,dx}dx + C\right). \text{ Ans.}$$

It should be observed that the constant k cancels out of the final result. For this reason it is customary to omit the constant of integration in solving (10).

* For the sake of simplicity we have assumed the particular value zero for the constant of integration. Otherwise we should have $u = c(1+x)^2$. But in the work that follows c finally cancels out. (See Illustrative Example 2.)

Type IV. Equations reducible to the linear form. Some equations that are not linear can be reduced to the linear form by means of a suitable transformation. One type of such equations is

(D) $$\frac{dy}{dx} + Py = Qy^n,$$

where P and Q are functions of x alone or constants. Equation (D) may be reduced to the linear form (B), Type III, by means of the substitution $z = y^{-n+1}$. Such a reduction, however, is not necessary if we employ the same method for finding the solution as that given under Type III. Let us illustrate this by means of an example.

ILLUSTRATIVE EXAMPLE. Solve the equation

(16) $$\frac{dy}{dx} + \frac{y}{x} = a \ln x \cdot y^2.$$

Solution. This is evidently in the form (D), where

$$P = \frac{1}{x}, \quad Q = a \ln x, \quad n = 2.$$

Let $y = uz$; then $\quad \dfrac{dy}{dx} = u\dfrac{dz}{dx} + z\dfrac{du}{dx}.$

Substituting in (16), we get

$$u\frac{dz}{dx} + z\frac{du}{dx} + \frac{uz}{x} = a \ln x \cdot u^2 z^2,$$

(17) $$u\frac{dz}{dx} + \left(\frac{du}{dx} + \frac{u}{x}\right)z = a \ln x \cdot u^2 z^2.$$

To determine u we place the coefficient of z equal to zero. This gives

$$\frac{du}{dx} + \frac{u}{x} = 0,$$

$$\frac{du}{u} = -\frac{dx}{x}.$$

Integrating, we get $\quad \ln u = -\ln x = \ln \dfrac{1}{x},$

(18) $$u = \frac{1}{x}.$$

Since the term in z drops out, equation (17) now becomes

$$u\frac{dz}{dx} = a \ln x \cdot u^2 z^2,$$

$$\frac{dz}{dx} = a \ln x \cdot u z^2.$$

Replacing u by its value from (18),

$$\frac{dz}{dx} = a \ln x \cdot \frac{z^2}{x},$$

$$\frac{dz}{z^2} = a \ln x \cdot \frac{dx}{x}.$$

Integrating, we get $\quad -\dfrac{1}{z} = \dfrac{a(\ln x)^2}{2} + C,$

(19) $\qquad z = -\dfrac{2}{a(\ln x)^2 + 2C}.$

Substituting from (19) and (18) in $y = uz$, we get the complete solution

$$y = -\dfrac{1}{x} \cdot \dfrac{2}{a(\ln x)^2 + 2C},$$

or $\qquad xy[a(\ln x)^2 + 2C] + 2 = 0.$ Ans.

PROBLEMS

Find the complete solution of each of the following differential equations.

1. $x\dfrac{dy}{dx} - 2y = 2x.$ \qquad Ans. $y = cx^2 - 2x.$

2. $x\dfrac{dy}{dx} - 2y = -x.$ \qquad $y = x + cx^2.$

3. $\dfrac{dy}{dx} - 2y = 1 - 2x.$ \qquad $y = x + ce^{2x}.$

4. $x\dfrac{dy}{dx} - 3y = -2nx.$ \qquad $y = nx + cx^3.$

5. $\dfrac{dy}{dx} - y = -2e^{-x}.$ \qquad $y = e^{-x} + ce^x.$

6. $\dfrac{ds}{dt} - s\,\mathrm{ctn}\,t = 1 - (t+2)\,\mathrm{ctn}\,t.$ \qquad $s = t + 2 + c\sin t.$

7. $\dfrac{dy}{dx} + \dfrac{2y}{x} = 2y^2.$ \qquad $cx^2 y + 2xy - 1 = 0.$

8. $\dfrac{ds}{dt} + s\tan t = 2t + t^2 \tan t.$ \qquad $s = t^2 + c\cos t.$

9. $x\dfrac{dy}{dx} - y = (x-1)e^x.$ \qquad $y = e^x + cx.$

10. $\dfrac{dy}{dx} + \dfrac{y}{x} = y^3.$ \qquad $cx^2 y^2 + 2xy^2 - 1 = 0.$

11. $\dfrac{ds}{dt} + \dfrac{s}{t} = \cos t + \dfrac{\sin t}{t}.$ \qquad $s = \sin t + \dfrac{c}{t}.$

12. $nx\dfrac{dy}{dx} + 2y = xy^{n+1}.$ \qquad $cx^2 y^n + xy^n - 1 = 0.$

13. $\dfrac{ds}{dt} + s = \cos t - \sin t.$ \qquad $s = \cos t + ce^{-t}.$

14. $\dfrac{ds}{dt} - s\,\mathrm{ctn}\,t = e^t(1 - \mathrm{ctn}\,t).$ \qquad $s = e^t + c\sin t.$

ORDINARY DIFFERENTIAL EQUATIONS

15. $x \dfrac{dy}{dx} - 2y + 3x = 0$.

16. $\dfrac{dy}{dx} + y = 2 + 2x$.

17. $x \dfrac{dy}{dx} + y = (1+x)e^x$.

18. $\dfrac{dy}{dx} - y = 1 - 2x$.

19. $x \dfrac{dy}{dx} + y + x^2 y^2 = 0$.

20. $\dfrac{ds}{dt} - s \operatorname{ctn} t + \csc t = 0$.

21. $2 \dfrac{dy}{dx} + y = (x-1) y^3$.

22. $x \dfrac{dy}{dx} - y = x \cos x - \sin x$.

23. $n \dfrac{dy}{dx} - y + (x^2 + 2x) y^{n+1} = 0$.

24. $\dfrac{ds}{dt} + s \tan t = e^{-t} (\tan t - 1)$.

In each of the following problems find the particular solution which is determined by the given values of x and y.

25. $\dfrac{dy}{dx} - \dfrac{2y}{x} = x^2 e^x$; $x = 1, y = 0$.   *Ans.* $y = x^2(e^x - e)$.

26. $\dfrac{dy}{dx} + \dfrac{2y}{x} = \dfrac{1}{x^2}$; $x = 1, y = 2$. $y = \dfrac{x+1}{x^2}$.

27. $\dfrac{dy}{dx} + y \tan x = \sec x$; $x = 0, y = -1$. $y = \sin x - \cos x$.

28. $\dfrac{dy}{dx} - \dfrac{2y}{x+1} = (x+1)^3$; $x = 0, y = 1$. $2y = (x+1)^4 + (x+1)^2$.

29. Find the equation of the curve whose slope at any point is equal to $\dfrac{2y + x + 1}{x}$ and which passes through the point $(1, 0)$.
   Ans. $2y = 3x^2 - 2x - 1$.

30. Find the equation of the curve whose slope at any point is equal to $\dfrac{y^2 \ln x - y}{x}$ and which passes through the point $(1, 1)$. *Ans.* $y(1 + \ln x) = 1$.

205. Two special types of differential equations of higher order. The differential equations discussed in this article occur frequently.

(E) $$\dfrac{d^n y}{dx^n} = X,$$

where X is a function of x alone or a constant.

To integrate, first multiply both members by dx. Then, integrating, we have

$$\dfrac{d^{n-1} y}{dx^{n-1}} = \int \dfrac{d^n y}{dx^n} \, dx = \int X \, dx + c_1.$$

Repeat the process $(n-1)$ times. Then the complete solution containing n arbitrary constants will be obtained.

388   DIFFERENTIAL AND INTEGRAL CALCULUS

ILLUSTRATIVE EXAMPLE. Solve $\dfrac{d^3y}{dx^3} = xe^x$.

Solution. Multiplying both members by dx and integrating,

$$\dfrac{d^2y}{dx^2} = \int xe^x\, dx + C_1,$$

or

$$\dfrac{d^2y}{dx^2} = xe^x - e^x + C_1.$$

Repeating the process,

$$\dfrac{dy}{dx} = \int xe^x\, dx - \int e^x\, dx + \int C_1\, dx + C_2,$$

or

$$\dfrac{dy}{dx} = xe^x - 2\,e^x + C_1 x + C_2.$$

$$y = \int xe^x\, dx - \int 2\,e^x\, dx + \int C_1 x\, dx + \int C_2\, dx + C_3$$

$$= xe^x - 3\,e^x + \dfrac{C_1 x^2}{2} + C_2 x + C_3.$$

Hence $\quad y = xe^x - 3\,e^x + c_1 x^2 + c_2 x + c_3.$ *Ans.*

A second type of much importance is

(F) $$\dfrac{d^2y}{dx^2} = Y,$$

where Y is a function of y alone.

To integrate, proceed thus. Write the equation

$$dy' = Y\, dx,$$

and multiply both members by y'. The result is

$$y'\, dy' = Y y'\, dx.$$

But $y'\, dx = dy$, and the preceding equation becomes

$$y' dy' = Y\, dy.$$

The variables y and y' are now separated. Integrating, the result is

$$\tfrac{1}{2}\, y'^2 = \int Y\, dy + C_1.$$

The right-hand member is a function of y. Extract the square root, separate the variables, x and y, and integrate again. The following example illustrates the method.

ILLUSTRATIVE EXAMPLE. Solve $\dfrac{d^2y}{dx^2} + a^2 y = 0$.

Solution. Here $\dfrac{dy'}{dx} = \dfrac{d^2y}{dx^2} = -a^2 y$, and hence the equation belongs to type (F).

Multiplying both members by $y'dx$ and proceeding as above, we get

$$y'\, dy' = -a^2 y\, dy.$$

Integrating, $\quad \tfrac{1}{2}\, y'^2 = -\tfrac{1}{2}\, a^2 y^2 + C.$

$$y' = \sqrt{2C - a^2 y^2}.$$

$$\dfrac{dy}{dx} = \sqrt{C_1 - a^2 y^2}.$$

ORDINARY DIFFERENTIAL EQUATIONS

setting $2C = C_1$ and taking the positive sign of the radical. Separating the variables, we get

$$\frac{dy}{\sqrt{C_1 - a^2 y^2}} = dx.$$

Integrating, $\quad \dfrac{1}{a} \arcsin \dfrac{ay}{\sqrt{C_1}} = x + C_2,$

or $\qquad\qquad \arcsin \dfrac{ay}{\sqrt{C_1}} = ax + aC_2.$

This is the same as $\dfrac{ay}{\sqrt{C_1}} = \sin(ax + aC_2)$

$\qquad\qquad = \sin ax \cos aC_2 + \cos ax \sin aC_2, \quad$ (4), Art. 2

or $\qquad\qquad y = \dfrac{\sqrt{C_1}}{a} \cos aC_2 \cdot \sin ax + \dfrac{\sqrt{C_1}}{a} \sin aC_2 \cdot \cos ax.$

Hence $\qquad y = c_1 \sin ax + c_2 \cos ax.$ Ans.

PROBLEMS

Find the complete solution of each of the following differential equations.

1. $\dfrac{d^2 x}{dt^2} = t^2.$ $\qquad$ Ans. $x = \dfrac{t^4}{12} + c_1 t + c_2.$

2. $\dfrac{d^2 x}{dt^2} = x.$ $\qquad\qquad x = c_1 e^t + c_2 e^{-t}.$

3. $\dfrac{d^2 x}{dt^2} = 4 \sin 2t.$ $\qquad x = -\sin 2t + c_1 t + c_2.$

4. $\dfrac{d^2 x}{dt^2} = e^{2t}.$ $\qquad\qquad x = \dfrac{e^{2t}}{4} + c_1 t + c_2.$

5. $\dfrac{d^2 s}{dt^2} = \dfrac{1}{(s+1)^3}.$ $\qquad c_1(s+1)^2 = (c_1 t + c_2)^2 + 1.$

6. $\dfrac{d^2 s}{dt^2} = \dfrac{1}{\sqrt{as}}.$ $\qquad 3t = 2 a^{\frac{1}{4}} (s^{\frac{1}{2}} - 2 c_1)(s^{\frac{1}{2}} + c_1)^{\frac{1}{2}} + c_2.$

7. $\dfrac{d^2 y}{dt^2} = \dfrac{a}{y^3}.$ $\qquad c_1 y^2 = a + (c_1 t + c_2)^2.$

8. $\dfrac{d^2 y}{dx^2} + \dfrac{a^2}{y^2} = 0.$
 Ans. $\sqrt{c_1 y^2 + y} - \dfrac{1}{\sqrt{c_1}} \ln(\sqrt{c_1 y} + \sqrt{1 + c_1 y}) = ac_1 \sqrt{2}\, x + c_2.$

9. $\dfrac{d^2 s}{dt^2} + \dfrac{k}{s^2} = 0.$ Find t, having given that $s = a$, $\dfrac{ds}{dt} = 0$, when $t = 0$.

 Ans. $t = \sqrt{\dfrac{a}{2k}} \left\{ \sqrt{as - s^2} + a \arcsin \sqrt{\dfrac{a-s}{a}} \right\}.$

10. $\dfrac{d^3 y}{dx^3} = x + \sin x.$ $\qquad$ 11. $\dfrac{d^2 s}{dt^2} = a \cos nt.$ $\qquad$ 12. $\dfrac{d^2 y}{dx^2} = 4y.$

206. Linear differential equations of the second order with constant coefficients. Equations of the form

(G) $$\frac{d^2y}{dx^2} + p\frac{dy}{dx} + qy = 0,$$

where p and q are constants, are important in applied mathematics.

To obtain a particular solution of (G), let us try to determine the value of the constant r so that (G) will be satisfied by

(1) $$y = e^{rx}.$$

Differentiating (1), we obtain

(2) $$\frac{dy}{dx} = re^{rx}, \quad \frac{d^2y}{dx^2} = r^2 e^{rx}.$$

Substituting from (1) and (2) in (G) and dividing out the factor e^{rx}, the result is

(3) $$r^2 + pr + q = 0,$$

a quadratic equation whose roots are the values of r required. Equation (3) is called the *auxiliary equation* for (G). If (3) has distinct roots r_1 and r_2, then

(4) $$y = e^{r_1 x} \quad \text{and} \quad y = e^{r_2 x}$$

are distinct particular solutions of (G), and the complete solution is

(5) $$y = c_1 e^{r_1 x} + c_2 e^{r_2 x}.$$

In fact, (5) contains two essential arbitrary constants, and (G) is satisfied by this relation.

ILLUSTRATIVE EXAMPLE 1. Solve

(6) $$\frac{d^2y}{dx^2} - 2\frac{dy}{dx} - 3y = 0.$$

Solution. The auxiliary equation is

(7) $$r^2 - 2r - 3 = 0.$$

Solving (7), the roots are 3 and -1, and by (5) the complete solution is

$$y = c_1 e^{3x} + c_2 e^{-x}. \quad Ans.$$

Check. Substituting this value of y in (6), the equation is satisfied.

Roots of (3) *imaginary*. If the roots of the auxiliary equation (3) are imaginary, the exponents in (5) will also be imaginary. A real complete solution may be found, however, by choosing imaginary values of c_1 and c_2 in (5). In fact, let

(8) $$r_1 = a + b\sqrt{-1}, \quad r_2 = a - b\sqrt{-1}$$

be the pair of conjugate imaginary roots of (3). Then

(9) $$e^{r_1 x} = e^{(a+b\sqrt{-1})x} = e^{ax} e^{bx\sqrt{-1}}, \quad e^{r_2 x} = e^{(a-b\sqrt{-1})x} = e^{ax} e^{-bx\sqrt{-1}}.$$

Substituting these values in (5), we obtain

(10) $$y = e^{ax}(c_1 e^{bx\sqrt{-1}} + c_2 e^{-bx\sqrt{-1}}).$$

In the algebra of imaginary numbers it is shown that*

$$e^{bx\sqrt{-1}} = \cos bx + \sqrt{-1}\sin bx, \qquad e^{-bx\sqrt{-1}} = \cos bx - \sqrt{-1}\sin bx.$$

When these values are substituted in (10), the complete solution may be written

(11) $$y = e^{ax}(A \cos bx + B \sin bx),$$

if the new arbitrary constants A and B are determined from c_1 and c_2 by $A = c_1 + c_2$, $B = (c_1 - c_2)\sqrt{-1}$. That is, we now take for c_1 and c_2 in (5) the imaginary values $c_1 = \frac{1}{2}(A - B\sqrt{-1})$, $c_2 = \frac{1}{2}(A + B\sqrt{-1})$.

By giving to A and B in (11) the values 1 and 0, and 0 and 1, in turn, we see that

(12) $$y = e^{ax}\cos bx \quad \text{and} \quad y = e^{ax}\sin bx$$

are real particular solutions of (G).

ILLUSTRATIVE EXAMPLE 2. Solve

(13) $$\frac{d^2y}{dx^2} + k^2 y = 0.$$

Solution. The auxiliary equation (3) is now

$$r^2 + k^2 = 0. \quad \therefore r = \pm k\sqrt{-1}.$$

Comparing with (8), we see that $a = 0$, $b = k$. Hence, by (11), the complete solution is
$$y = A \cos kx + B \sin kx.$$

Check. When this value of y is substituted in (13), the equation is satisfied. Compare this method with that used for the same example in Art. 205 ($k = a$).

REMARK. A different form for the solution is obtained by setting $A = C \cos \alpha$, $B = C \sin \alpha$ in the above value of y. Then $y = C \cos(kx - \alpha)$. (By (4), p. 3.)

---

* Let $i = \sqrt{-1}$, and assume that the series for e^x in Problem 1, Art. 194, represents the function when x is replaced by ibx. Then, since $i^2 = -1$, $i^3 = -i$, $i^4 = 1$, etc., we have

(14) $$e^{ibx} = 1 + ibx - \frac{b^2 x^2}{\underline{|2}} - i\frac{b^3 x^3}{\underline{|3}} + \frac{b^4 x^4}{\underline{|4}} + i\frac{b^5 x^5}{\underline{|5}} - \cdots.$$

Also, replacing x by bx in (7) and (8), Art. 194,

$$\cos bx = 1 - \frac{b^2 x^2}{\underline{|2}} + \frac{b^4 x^4}{\underline{|4}} - \cdots, \qquad \sin bx = bx - \frac{b^3 x^3}{\underline{|3}} + \frac{b^5 x^5}{\underline{|5}} - \cdots.$$

Then, by Art. 195,

(15) $$\cos bx + i \sin bx = 1 + ibx - \frac{b^2 x^2}{\underline{|2}} - i\frac{b^3 x^3}{\underline{|3}} + \frac{b^4 x^4}{\underline{|4}} + i\frac{b^5 x^5}{\underline{|5}} - \cdots,$$

assuming that the series represents the function. The right-hand members of (14) and (15) are identical. Therefore $e^{ibx} = \cos bx + i \sin bx$.

Roots of (3) *real and equal.* The roots of the auxiliary equation (3) will be equal if $p^2 = 4\,q$. Then (3) may be written, by substituting $q = \tfrac{1}{4}\,p^2$,

(14) $\qquad\qquad r^2 + pr + \tfrac{1}{4}\,p^2 = (r + \tfrac{1}{2}\,p)^2 = 0,$

and $r_1 = r_2 = -\tfrac{1}{2}\,p$. In this case

(15) $\qquad\qquad y = e^{r_1 x} \quad \text{and} \quad y = xe^{r_1 x}$

are distinct particular solutions, and

(16) $\qquad\qquad y = e^{r_1 x}(c_1 + c_2 x)$

is the complete solution.

To corroborate this statement, it is only necessary to prove that the second equation in (15) gives a solution. But we have, by differentiating,

(17) $\quad y = xe^{r_1 x}, \quad \dfrac{dy}{dx} = e^{r_1 x}(1 + r_1 x), \quad \dfrac{d^2 y}{dx^2} = e^{r_1 x}(2\,r_1 + r_1^2 x).$

Substituting from (17) into the left-hand member of (*G*), the result is, after canceling $e^{r_1 x}$,

(18) $\qquad\qquad (r_1^2 + pr_1 + q)x + 2\,r_1 + p.$

This vanishes, since r_1 satisfies (3) and equals $-\tfrac{1}{2}\,p$.

ILLUSTRATIVE EXAMPLE 3. Solve

(19) $\qquad\qquad \dfrac{d^2 s}{dt^2} + 2\,\dfrac{ds}{dt} + s = 0.$

Find the particular solution such that

$$s = 4 \text{ and } \dfrac{ds}{dt} = -2 \text{ when } t = 0.$$

Solution. The auxiliary equation is

$$r^2 + 2\,r + 1 = 0, \quad \text{or} \quad (r+1)^2 = 0.$$

Hence the roots are equal, $r_1 = -1$, and, by (16),

(20) $\qquad\qquad s = e^{-t}(c_1 + c_2 t).$

This is the complete solution.

To find the required particular solution, we must find values for the constants c_1 and c_2 such that the given conditions,

$$s = 4 \text{ and } \dfrac{ds}{dt} = -2 \text{ when } t = 0$$

are satisfied.

Substituting in the complete solution (20) the given values $s = 4$, $t = 0$, we have $4 = c_1$, and hence

(21) $\qquad\qquad s = e^{-t}(4 + c_2 t).$

ORDINARY DIFFERENTIAL EQUATIONS 393

Now differentiate (21) with respect to t. We get

$$\frac{ds}{dt} = e^{-t}(c_2 - 4 - c_2 t).$$

By the given conditions, $\frac{ds}{dt} = -2$ when $t = 0$.

Substituting, the result is $-2 = c_2 - 4$, and hence $c_2 = 2$. Then the particular solution required is $s = e^{-t}(4 + 2\,t)$. *Ans.*

PROBLEMS

Find the complete solution of each of the following differential equations.

1. $\dfrac{d^2x}{dt^2} - \dfrac{dx}{dt} - 2\,x = 0.$ *Ans.* $x = c_1 e^{-t} + c_2 e^{2t}.$

2. $\dfrac{d^2y}{dx^2} - 4\dfrac{dy}{dx} + 3\,y = 0.$ $y = c_1 e^x + c_2 e^{3x}.$

3. $\dfrac{d^2s}{dt^2} - 2\dfrac{ds}{dt} + s = 0.$ $s = c_1 e^t + c_2 t e^t.$

4. $\dfrac{d^2x}{dt^2} + 16\,x = 0.$ $x = c_1 \cos 4\,t + c_2 \sin 4\,t.$

5. $\dfrac{d^2s}{dt^2} - 4\,s = 0.$ $s = c_1 e^{2t} + c_2 e^{-2t}.$

6. $\dfrac{d^2y}{dx^2} + 4\dfrac{dy}{dx} = 0.$ $y = c_1 + c_2 e^{-4x}.$

7. $\dfrac{d^2s}{dt^2} + 2\dfrac{ds}{dt} + 2\,s = 0.$ $s = e^{-t}(c_1 \cos t + c_2 \sin t).$

8. $\dfrac{d^2s}{dt^2} - 2\dfrac{ds}{dt} + 5\,s = 0.$ $s = e^t(c_1 \cos 2\,t + c_2 \sin 2\,t).$

9. $\dfrac{d^2\theta}{dt^2} - 5\dfrac{d\theta}{dt} + 4\,\theta = 0.$ 13. $\dfrac{d^2s}{dt^2} - 3\,s = 0.$

10. $\dfrac{d^2y}{dx^2} + 6\dfrac{dy}{dx} + 9\,y = 0.$ 14. $\dfrac{d^2y}{dx^2} - n\dfrac{dy}{dx} = 0.$

11. $\dfrac{d^2y}{dx^2} + 5\dfrac{dy}{dx} + 6\,y = 0.$ 15. $\dfrac{d^2s}{dt^2} - 6\dfrac{ds}{dt} + 25\,s = 0.$

12. $\dfrac{d^2s}{dt^2} + 3\,s = 0.$ 16. $\dfrac{d^2x}{dt^2} + 2\dfrac{dx}{dt} + 10\,x = 0.$

In the following problems find the particular solution which satisfies the given conditions.

17. $\dfrac{d^2s}{dt^2} + 3\dfrac{ds}{dt} + 2\,s = 0$; $s = 0, \dfrac{ds}{dt} = 1,$ when $t = 0.$ *Ans.* $s = e^{-t} - e^{-2t}.$

18. $\dfrac{d^2x}{dt^2} + n^2 x = 0$; $x = a, \dfrac{dx}{dt} = 0,$ when $t = 0.$ $x = a \cos nt.$

19. $\dfrac{d^2x}{dt^2} - n^2x = 0$; $x = 2, \dfrac{dx}{dt} = 0$, when $t = 0$.   Ans. $x = e^{nt} + e^{-nt}$.

20. $\dfrac{d^2y}{dt^2} + 2\dfrac{dy}{dt} - 8y = 0$; $y = 0, \dfrac{dy}{dt} = 24$, when $t = 0$.
                                        Ans. $y = 4(e^{2t} - e^{-4t})$.

21. $\dfrac{d^2s}{dt^2} - 8\dfrac{ds}{dt} + 16s = 0$; $s = 0, \dfrac{ds}{dt} = 1$, when $t = 0$. $s = te^{4t}$.

22. $\dfrac{d^2x}{dt^2} - a\dfrac{dx}{dt} = 0$; $x = 0, \dfrac{dx}{dt} = a$, when $t = 0$. $x = e^{at} - 1$.

23. $\dfrac{d^2s}{dt^2} + 8\dfrac{ds}{dt} + 25s = 0$; $s = 4, \dfrac{ds}{dt} = -16$, when $t = 0$.
                                        Ans. $s = 4e^{-4t}\cos 3t$.

24. $\dfrac{d^2x}{dt^2} - 6\dfrac{dx}{dt} + 10x = 0$; $x = 1, \dfrac{dx}{dt} = 4$, when $t = 0$.
                                        Ans. $x = e^{3t}(\cos t + \sin t)$.

25. $\dfrac{d^2s}{dt^2} + 4s = 0$; $s = 0, \dfrac{ds}{dt} = 4$, when $t = 0$.

26. $\dfrac{d^2x}{dt^2} - 4x = 0$; $x = 10, \dfrac{dx}{dt} = 0$, when $t = 0$.

27. $\dfrac{d^2y}{dt^2} + 4\dfrac{dy}{dt} = 0$; $y = 1, \dfrac{dy}{dt} = 2$, when $t = 0$.

28. $\dfrac{d^2x}{dt^2} - 4\dfrac{dx}{dt} + 4x = 0$; $x = 2, \dfrac{dx}{dt} = 5$, when $t = 0$.

29. $\dfrac{d^2x}{dt^2} - 4\dfrac{dx}{dt} + 13x = 0$; $x = 2, \dfrac{dx}{dt} = 4$, when $t = 0$.

30. $\dfrac{d^2s}{dt^2} + 4\dfrac{ds}{dt} + 8s = 0$; $s = 0, \dfrac{ds}{dt} = 8$, when $t = 0$.

To solve the differential equation

(H) $$\dfrac{d^2y}{dx^2} + p\dfrac{dy}{dx} + qy = X,$$

where p and q are constants and X is a function of the independent variable x or a constant, three steps are necessary.

FIRST STEP. *Solve the equation* (G). *Let the complete solution be*

(22) $y = u$.

Then u is called the **complementary function** for (H).

SECOND STEP. *Obtain a particular solution*

(23) $y = v$

of (H) *by trial*.

THIRD STEP. *The complete solution of* (H) *is now*

(24) $y = u + v$.

ORDINARY DIFFERENTIAL EQUATIONS 395

In fact, when the value of y from (24) is substituted in (H), it is seen that the equation is satisfied, and (24) contains two essential arbitrary constants.

To determine the particular solution (23), the following directions are useful (see also Art. 208). In the formulas all letters except x, the independent variable, are constants.

General case. If $y = X$ is not a particular solution of (G), if

Form of X		Form of v
$X = a + bx,$	assume	$y = v = A + Bx;$
$X = ae^{bx},$	assume	$y = v = Ae^{bx};$
$X = a_1 \cos bx + a_2 \sin bx,$	assume	$y = v = A_1 \cos bx + A_2 \sin bx.$

Special case. If $y = X$ is a particular solution of (G), assume for v the above form multiplied by x (the independent variable).

The method consists in substituting $y = v$, as given above, in (H), and determining the constants A, B, A_1, A_2, so that (H) is satisfied.

ILLUSTRATIVE EXAMPLE 4. Solve

(25) $$\frac{d^2y}{dx^2} - 2\frac{dy}{dx} - 3y = 2x.$$

Solution. *First Step.* The complementary function u is found from the complete solution of

(26) $$\frac{d^2y}{dx^2} - 2\frac{dy}{dx} - 3y = 0.$$

By Illustrative Example 1, above, therefore

(27) $$y = u = c_1 e^{3x} + c_2 e^{-x}.$$

Second Step. Since $y = X = 2x$ is not a particular solution of (26), assume for a particular solution of (25)

(28) $$y = v = A + Bx.$$

Substituting this value in (25) and collecting terms, the result is

(29) $$-2B - 3A - 3Bx = 2x.$$

Equating coefficients of like powers of x, we get

$$-2B - 3A = 0, \quad -3B = 2.$$

Solving, $A = \frac{4}{9}$, $B = -\frac{2}{3}$, and substituting in (28), we obtain the particular solution

(30) $$y = v = \frac{4}{9} - \frac{2}{3}x.$$

Third Step. Then, from (27) and (30), the complete solution is

$$y = u + v = c_1 e^{3x} + c_2 e^{-x} + \frac{4}{9} - \frac{2}{3}x. \quad Ans.$$

ILLUSTRATIVE EXAMPLE 5. Solve

(31) $$\frac{d^2y}{dx^2} - 2\frac{dy}{dx} - 3y = 2e^{-x}.$$

Solution. *First Step.* The complementary function is (27), or

(32) $$y = u = c_1 e^{3x} + c_2 e^{-x}.$$

Second Step. Here $y = X = 2e^{-x}$ is a particular solution of (26), for it is obtained from the complete solution (32) by letting $c_1 = 0$, $c_2 = 2$. Hence assume for a particular solution v of (31),

(33) $$y = v = Axe^{-x}.$$

Differentiating (33), we obtain

(34) $$\frac{dy}{dx} = Ae^{-x}(1-x), \quad \frac{d^2y}{dx^2} = Ae^{-x}(x-2).$$

Substituting from (33) and (34) in (31), we obtain

(35) $$Ae^{-x}(x-2) - 2Ae^{-x}(1-x) - 3Axe^{-x} = 2e^{-x}.$$

Simplifying, we get $-4Ae^{-x} = 2e^{-x}$, and hence $A = -\tfrac{1}{2}$. Substituting in (33) we obtain

(36) $$y = v = -\tfrac{1}{2} xe^{-x}.$$

Third Step. The complete solution of (31) is, therefore,

$$y = u + v = c_1 e^{3x} + c_2 e^{-x} - \tfrac{1}{2} xe^{-x}. \quad Ans.$$

ILLUSTRATIVE EXAMPLE 6. Determine the particular solution of

(37) $$\frac{d^2s}{dt^2} + 4s = 2\cos 2t,$$

such that $s = 0$ and $\dfrac{ds}{dt} = 2$ when $t = 0$.

Solution. Find the complete solution first.

First Step. Solving

(38) $$\frac{d^2s}{dt^2} + 4s = 0,$$

we find the complementary function

(39) $$s = u = c_1 \cos 2t + c_2 \sin 2t.$$

Second Step. Considering the right-hand member in (37), we observe that $s = 2 \cos 2t$ is a particular solution of (38) resulting from (39) when $c_1 = 2$, $c_2 = 0$. Hence for a particular solution $s = v$ of (37) assume

(40) $$s = v = t(A_1 \cos 2t + A_2 \sin 2t).$$

Differentiating (40), we obtain

(41) $$\begin{cases} \dfrac{ds}{dt} = A_1 \cos 2t + A_2 \sin 2t - 2t(A_1 \sin 2t - A_2 \cos 2t). \\ \dfrac{d^2s}{dt^2} = -4A_1 \sin 2t + 4A_2 \cos 2t - 4t(A_1 \cos 2t + A_2 \sin 2t). \end{cases}$$

ORDINARY DIFFERENTIAL EQUATIONS 397

Substituting from (40) and (41) in (37), and simplifying, the result is

(42) $\qquad -4 A_1 \sin 2t + 4 A_2 \cos 2t = 2 \cos 2t.$

This equation becomes an identity when $A_1 = 0$, $A_2 = \frac{1}{2}$. Substituting in (40), we get

(43) $\qquad s = v = \frac{1}{2} t \sin 2t.$

Third Step. By (39) and (43) the complete solution of (37) is

(44) $\qquad s = c_1 \cos 2t + c_2 \sin 2t + \frac{1}{2} t \sin 2t.$

We must now determine c_1 and c_2 so that

(45) $\qquad s = 0$ and $\dfrac{ds}{dt} = 2$ when $t = 0$.

Differentiating (44),

(46) $\qquad \dfrac{ds}{dt} = -2 c_1 \sin 2t + 2 c_2 \cos 2t + \frac{1}{2} \sin 2t + t \cos 2t.$

Substituting the given conditions (45) in (44) and (46), the results are

$$0 = c_1, \quad 2 = 2 c_2. \quad \therefore c_1 = 0, \; c_2 = 1.$$

Putting these values back in (44), the particular solution required is

(47) $\qquad s = \sin 2t + \frac{1}{2} t \sin 2t.$ *Ans.*

PROBLEMS

Find the complete solution of each of the following differential equations.

1. $\dfrac{d^2x}{dt^2} + x = at + b.$ *Ans.* $x = c_1 \cos t + c_2 \sin t + at + b.$

2. $\dfrac{d^2x}{dt^2} + x = ae^{bt}.$ $x = c_1 \cos t + c_2 \sin t + \dfrac{ae^{bt}}{b^2 + 1}.$

3. $\dfrac{d^2x}{dt^2} + x = 4 \cos t.$ $x = c_1 \cos t + c_2 \sin t + 2 t \sin t.$

4. $\dfrac{d^2x}{dt^2} + x = 4 \sin 2t.$ $x = c_1 \cos t + c_2 \sin t - \frac{4}{3} \sin 2t.$

5. $\dfrac{d^2s}{dt^2} - 4s = at + b.$ $s = c_1 e^{2t} + c_2 e^{-2t} - \frac{1}{4}(at + b).$

6. $\dfrac{d^2s}{dt^2} - 4s = 2 e^t.$ $s = c_1 e^{2t} + c_2 e^{-2t} - \frac{2}{3} e^t.$

7. $\dfrac{d^2s}{dt^2} - 4s = e^{2t}.$ $s = c_1 e^{2t} + c_2 e^{-2t} + \frac{1}{4} t e^{2t}.$

8. $\dfrac{d^2s}{dt^2} - 4s = 2 \cos 2t.$ $s = c_1 e^{2t} + c_2 e^{-2t} - \frac{1}{4} \cos 2t.$

9. $\dfrac{d^2y}{dx^2} + 9y = 5 x^2.$ $y = c_1 \cos 3x + c_2 \sin 3x + \frac{5}{9} x^2 - \frac{10}{81}.$

398 DIFFERENTIAL AND INTEGRAL CALCULUS

10. $\dfrac{d^2x}{dt^2} - \dfrac{dx}{dt} - 2x = 4t.$ Ans. $x = c_1 e^{-t} + c_2 e^{2t} + 1 - 2t.$

11. $\dfrac{d^2x}{dt^2} - 2\dfrac{dx}{dt} + x = 8.$ $x = c_1 e^t + c_2 t e^t + 8.$

12. $\dfrac{d^2s}{dt^2} - 4\dfrac{ds}{dt} + 3s = 6 e^{2t}.$ $s = c_1 e^t + c_2 e^{3t} - 6 e^{2t}.$

13. $\dfrac{d^2s}{dt^2} + 2\dfrac{ds}{dt} + 2s = 8 e^{2t}.$ $s = e^{-t}(c_1 \cos t + c_2 \sin t) + \tfrac{4}{5} e^{2t}.$

14. $\dfrac{d^2x}{dt^2} - 4\dfrac{dx}{dt} + 3x = 4 e^t.$ $x = c_1 e^t + c_2 e^{3t} - 2 t e^t.$

15. $\dfrac{d^2y}{dt^2} - 2\dfrac{dy}{dt} + 5y = 3 \cos t.$ $y = e^t(c_1 \cos 2t + c_2 \sin 2t) + \tfrac{3}{5} \cos t - \tfrac{3}{10} \sin t.$

16. $\dfrac{d^2y}{dt^2} - 2\dfrac{dy}{dt} + 5y = 3 \sin 2t.$ $y = e^t(c_1 \cos 2t + c_2 \sin 2t) + \tfrac{12}{17} \cos 2t + \tfrac{3}{17} \sin 2t.$

17. $\dfrac{d^2s}{dt^2} + 9s = 3 \cos 2t.$

18. $\dfrac{d^2x}{dt^2} + 4x = 2 \sin 2t.$

19. $\dfrac{d^2y}{dt^2} - y = 2 + e^t.$

20. $\dfrac{d^2z}{dt^2} - 4z = t - e^t.$

21. $\dfrac{d^2x}{dt^2} + 2x = t^2 - 2.$

22. $4\dfrac{d^2s}{dt^2} + s = 5 \cos \dfrac{t}{2}.$

23. $\dfrac{d^2s}{dt^2} + 3\dfrac{ds}{dt} + 2s = 2 \sin t.$

24. $\dfrac{d^2x}{dt^2} - 8\dfrac{dx}{dt} + 16x = 4 - 8t.$

25. $\dfrac{d^2y}{dt^2} - 8\dfrac{dy}{dt} + 25y = 5 \cos 2t.$

26. $\dfrac{d^2s}{dt^2} + 6\dfrac{ds}{dt} + 10s = 5 \sin 2t.$

In the following problems find the particular solution which satisfies the given conditions.

27. $\dfrac{d^2s}{dt^2} + 9s = t + \dfrac{1}{2};$ $s = \dfrac{1}{18},$ $\dfrac{ds}{dt} = \dfrac{1}{9},$ when $t = 0.$ Ans. $s = \dfrac{1}{9} t + \dfrac{1}{18}.$

28. $\dfrac{d^2s}{dt^2} + 9s = 9 e^{3t};$ $s = 1,$ $\dfrac{ds}{dt} = \dfrac{3}{2},$ when $t = 0.$

Ans. $s = \tfrac{1}{2}(\cos 3t + e^{3t}).$

29. $\dfrac{d^2s}{dt^2} + 9s = 5 \cos 2t;$ $s = 1,$ $\dfrac{ds}{dt} = 3,$ when $t = 0.$

Ans. $s = \sin 3t + \cos 2t.$

30. $\dfrac{d^2s}{dt^2} + 9s = 3 \cos 3t;$ $s = 0,$ $\dfrac{ds}{dt} = 6,$ when $t = 0.$

Ans. $s = 2 \sin 3t + \tfrac{1}{2} t \sin 3t.$

31. $\dfrac{d^2x}{dt^2} - 2\dfrac{dx}{dt} - 3x = 2t + 1;$ $x = \dfrac{1}{3},$ $\dfrac{dx}{dt} = -\dfrac{4}{9},$ when $t = 0.$

Ans. $x = \tfrac{1}{9}(e^{3t} + e^{-t} - 6t + 1).$

32. $\dfrac{d^2x}{dt^2} - 6\dfrac{dx}{dt} + 13x = 39$; $x = 4$, $\dfrac{dx}{dt} = 3$, when $t = 0$.

Ans. $x = e^{3t} \cos 2t + 3$.

33. $\dfrac{d^2s}{dt^2} + 9s = 4 - 3t$; $s = 0$, $\dfrac{ds}{dt} = 0$, when $t = 0$.

34. $\dfrac{d^2s}{dt^2} - 9s = 6t$; $s = 0$, $\dfrac{ds}{dt} = 0$, when $t = 0$.

35. $\dfrac{d^2y}{dx^2} - 2\dfrac{dy}{dx} = 2x$; $y = 2$, $\dfrac{dy}{dx} = 0$, when $x = 0$.

36. $\dfrac{d^2x}{dt^2} + x = 2\cos 2t$; $x = 0$, $\dfrac{dx}{dt} = 2$, when $t = 0$.

37. $\dfrac{d^2s}{dt^2} + 4s = 2\cos 2t$; $s = 0$, $\dfrac{ds}{dt} = 2$, when $t = 0$.

38. $\dfrac{d^2x}{dt^2} - 2\dfrac{dx}{dt} + 2x = 2\sin t$; $x = 0$, $\dfrac{dx}{dt} = 0$, when $t = 0$.

39. $\dfrac{d^2y}{dx^2} + 5\dfrac{dy}{dx} + 4y = 2e^x$; $y = 1$, $\dfrac{dy}{dx} = 0$, when $x = 0$.

40. $\dfrac{d^2y}{dx^2} + 4\dfrac{dy}{dx} + 4y = 4e^{2x}$; $y = 0$, $\dfrac{dy}{dx} = 2$, when $x = 0$.

207. Applications. Compound-interest law. A simple application of differential equations is afforded by problems in which the rate of change of the function with respect to the variable (Art. 50) for any value of the variable is proportional to the corresponding value of the function. That is, if $y = f(x)$,

(1) $$\dfrac{dy}{dx} = ky,$$

where k is a constant. In this differential equation the variables are separable as in Type I, p. 379. Solving, we get

(2) $$y = ce^{kx},$$

where c is an arbitrary constant. Thus the function y is an exponential function (Art. 62). Conversely, given (2), it is easily shown by differentiation that y satisfies (1). The connection of (1) with the name "compound-interest law" is shown as follows:

Let $y = $ a sum of money in dollars accumulating at compound interest;
$i = $ interest in dollars on one dollar for a year;
$\Delta t = $ an interval of time measured in years;
$\Delta y = $ the interest of y dollars for the interval of time Δt.

Then $\Delta y = iy\,\Delta t$. Therefore

(3) $$\dfrac{\Delta y}{\Delta t} = iy.$$

Equation (3) states that the average rate of change of y (Art. 50) for the period of time Δt is proportional to y itself. In business, interest is added to the principal at stated times only, — yearly, quarterly, etc. In other words, y changes discontinuously with t. But in nature, changes proceed on the whole in a continuous manner. So that to adapt equation (3) to natural phenomena we must imagine the sum y to accumulate continuously; that is, assume the interval of time Δt to be infinitesimal. Then equation (3) becomes

$$\frac{dy}{dt} = iy,$$

and the rate of change of y is proportional to y, agreeing with (1) if $k = i$.

In (1) the function y is said to change according to the compound-interest law.

A second example is afforded by the complete solution of the equation

(4) $$\frac{dy}{dx} = ky + c,$$

where k and c are constants not equal to zero. For let $c = ak$. Then (4) may be written

(5) $$\frac{d}{dx}(y + a) = k(y + a).$$

This equation states that the function $y + a$ changes according to the compound-interest law. The differential equation (4), or (5), is solved as in Type I, p. 379. The solution is

(6) $$y = ce^{kx} - a.$$

ILLUSTRATIVE EXAMPLE 1. The function y of x changes according to the compound-interest law. When $x = 1$, $y = 4$; when $x = 2$, $y = 12$. Find the law.

Solution. By (1) we have

(7) $$\frac{dy}{dx} = ky.$$

Separating the variables and integrating, we get

$$\ln y = kx + C.$$

We have to find the values of k and C. Substitute the given values of x and y. Then
$$\ln 4 = k + C, \quad \ln 12 = 2k + C.$$
Solving, $k = \ln 12 - \ln 4 = \ln 3 = 1.0986$, $C = \ln 4 - \ln 3 = \ln \frac{4}{3}$.
Therefore $\ln y = 1.0986\, x + \ln \frac{4}{3}$, and $y = \frac{4}{3} e^{1.0986 x}$. *Ans.*

ILLUSTRATIVE EXAMPLE 2. *Washing down a solution.* Water is run into a tank containing a saline (or acid) solution with the purpose of reducing its strength. The volume v of the mixture in the tank is kept constant. If $s =$ quantity of salt (or acid) in the tank at any time, and $x =$ amount of water which has run through, show that the rate of decrease of s with respect to x varies as s, and, in fact, that $\frac{ds}{dx} = -\frac{s}{v}$.

Solution. Since $s =$ quantity of salt in the mixture of total volume v, the quantity of salt in any other volume u of the mixture is $\frac{s}{v} u$.

ORDINARY DIFFERENTIAL EQUATIONS

Suppose a volume Δx of the mixture is dipped out of the tank. The amount of salt thus dipped out will be $\frac{s}{v} \Delta x$, and hence the change in the amount of salt in the tank is given by

(8) $$\Delta s = -\frac{s}{v} \Delta x.$$

Suppose now that a volume of water Δx is added to fill the tank to its original volume v. Then from (8) the ratio of the amount of salt removed to the volume of water added is given by

$$\frac{\Delta s}{\Delta x} = -\frac{s}{v}.$$

When $\Delta x \to 0$ we obtain the instantaneous rate of change of s with respect to x, namely,

$$\frac{ds}{dx} = -\frac{s}{v}. \quad Ans.$$

Hence s changes according to the compound-interest law.

PROBLEMS

1. The rate of change of a function y with respect to x equals $\frac{1}{3} y$, and $y = 4$ when $x = -1$. Find the law connecting x and y. *Ans.* $y = 5.58\, e^{\frac{1}{3}x}$.

2. The rate of change of a function y with respect to x equals $2 - y$, and $y = 8$ when $x = 0$. Find the law. *Ans.* $y = 6\, e^{-x} + 2$.

3. In Illustrative Example 2, if $v = 10{,}000$ gal., how much water must be run through to wash down 50 per cent of the salt? *Ans.* 6931 gal.

4. *Newton's law of cooling.* If the excess temperature of a body above the temperature of the surrounding air is x degrees, the time-rate of decrease of x is proportional to x. If this excess temperature was at first 80 degrees, and after 1 min. is 70 degrees, what will it be after 2 min.? In how many minutes will it decrease 20 degrees?

5. Atmospheric pressure p at points above the earth's surface as a function of the altitude h above sea level changes according to the compound-interest law. Assuming $p = 15$ lb. per sq. in. when $h = 0$, and 10 lb. when $h = 10{,}000$ ft., find p (a) when $h = 5000$ ft.; (b) when $h = 15{,}000$ ft. *Ans.* (a) 12.2 lb.; (b) 8.15 lb.

6. The velocity of a chemical reaction in which x is the amount transformed in time t is the time-rate of change of x.

Reaction of the first order. Let $a =$ concentration at the beginning of the experiment. Then $\frac{dx}{dt} = k(a - x)$, since the rate of change is proportional to the concentration at that instant. (Note that $a - x$, the concentration, changes according to the compound-interest law.)

Prove that k, the velocity constant, is equal to $\frac{1}{t} \ln \frac{a}{a - x}$.

7. In the inversion of raw sugar, the time-rate of change varies as the amount of raw sugar remaining. If, after 10 hr., 1000 lb. of raw sugar have been reduced to 800 lb., how much raw sugar will remain at the expiration of 24 hr.? *Ans.* 586 lb.

8. In an electric circuit with given voltage E and current i (amperes), the voltage E is consumed in overcoming (1) the resistance R (ohms) of the circuit, and (2) the inductance L, the equation being

$$E = Ri + L\frac{di}{dt}, \quad \text{or} \quad \frac{di}{dt} = \frac{1}{L}(E - Ri).$$

This process therefore comes under equation (4) above, E, R, L being constants. Given $L = 640$, $R = 250$, $E = 500$, and $i = 0$ when $t = 0$, show that the current will approach 2 amperes as t increases. Also find in how many seconds i will reach 90 per cent of its maximum value. *Ans.* 5.9 sec.

9. In a condenser discharging electricity, the time-rate of change of the voltage e is proportional to e, and e decreases with the time. Given $k = \frac{1}{40}$, find t if e decreases to 10 per cent of its original value. *Ans.* 92 sec.

10. *Building up a saline (or acid) solution* by adding salt (or acid), maintaining constant volume, leads to the equation $\dfrac{dy}{dx} = \dfrac{1}{v}(v - y)$, where $v =$ the constant volume, $y =$ salt (or acid) in the tank at any moment, $x =$ salt (or acid) added from the beginning. Derive this result and compare with Illustrative Example 2 above.

208. Applications to problems in mechanics. Many important problems in mechanics and physics are solved by the methods explained in this chapter. For example, problems in rectilinear motion often lead to differential equations of the first or second order, and the solution of the problems depends upon solving these equations.

Before giving illustrative examples it is to be recalled (Arts. 51 and 59) that

(1) $$v = \frac{ds}{dt}, \quad a = \frac{d^2s}{dt^2} = \frac{dv}{dt} = v\frac{dv}{ds},$$

where v and a are, respectively, the velocity and acceleration at any instant of time $(= t)$, and s equals the distance of the moving point at this time from a fixed origin on the linear path.

ILLUSTRATIVE EXAMPLE 1. In a rectilinear motion the acceleration is inversely proportional to the square of the distance s, and equals -1 when $s = 2$. That is,

(2) $$\text{Acceleration} = a = -\frac{4}{s^2}.$$

Also, $v = 5$, $s = 8$, when $t = 0$.

(a) Find v when $s = 24$.

Solution. From (2), using the last form for a, we obtain

(3) $$v\frac{dv}{ds} = -\frac{4}{s^2}.$$

Multiplying both members by ds and integrating, the result is

(4) $$\frac{v^2}{2} = \frac{4}{s} + C, \quad \text{or} \quad v^2 = \frac{8}{s} + C'.$$

ORDINARY DIFFERENTIAL EQUATIONS

Substituting in (4) the above conditions $v = 5$, $s = 8$, we find $C' = 24$. Hence (4) becomes

(5) $$v^2 = \frac{8}{s} + 24.$$

From this equation, if $s = 24$, $v = \tfrac{1}{3}\sqrt{219} = 4.93$. *Ans.*

(b) Find the time which elapses when the point moves from $s = 8$ to $s = 24$.

Solution. Solving (5) for v, we get

(6) $$\frac{ds}{dt} = v = \sqrt{8}\,\frac{\sqrt{s + 3\,s^2}}{s}.$$

Separating the variables s and t and solving for t with limits as given, $s = 8$, $s = 24$, we find, for the elapsed time,

(7) $$t = \frac{1}{2\sqrt{2}} \int_8^{24} \frac{s\,ds}{\sqrt{s + 3\,s^2}} = 3.23.\ \textit{Ans.}$$

NOTE. Using the first form in (1) for a, (2) is

$$\frac{d^2s}{dt^2} = -\frac{4}{s^2},$$

which is of the form *(F)*, Art. 205. The method of integration here is the same as in Art. 205.

An important type of rectilinear motion is that in which the acceleration and the distance are in a constant ratio and differ in sign. Then we may write

(8) $$a = -k^2 s,$$

where $k^2 =$ magnitude of a at unit distance.

Remembering that a force and the acceleration caused by it differ in magnitude only, we see in the above case that the effective force is directed always toward the point $s = 0$ and is, in magnitude, directly proportional to the distance s. The motion is called *simple harmonic vibration*.

From (8), we have, using (1),

(9) $$\frac{d^2s}{dt^2} + k^2 s = 0,$$

a linear equation in s and t of the second order with constant coefficients. Integrating (see Illustrative Example 2, Art. 206), we obtain the complete solution,

(10) $$s = c_1 \cos kt + c_2 \sin kt.$$

From (10), by differentiation,

(11) $$v = k(-c_1 \sin kt + c_2 \cos kt).$$

It is easy to see that the motion defined by (10) is a periodic oscillation between the extreme positions $s = b$, and $s = -b$, determined by

(12) $$b = \sqrt{c_1{}^2 + c_2{}^2}, \quad \text{period} = \frac{2\,\pi}{k}.$$

In fact, we may replace the constants c_1 and c_2 in (10) by other constants b and A, such that

(13) $$c_1 = b \sin A, \quad c_2 = b \cos A.$$

Substituting these values in (10), it reduces to

(14) $$s = b \sin(kt + A), \quad \text{by (4), Art. 2}$$

and now the truth of the above statement is manifest.

In the following examples we give cases when the simple harmonic motion is disturbed by other forces. In all cases the problem depends upon the solution of an equation of the form (*G*) and (*H*), discussed above.

ILLUSTRATIVE EXAMPLE 2. In a rectilinear motion

(15) $$a = -\tfrac{5}{4} s - v.$$

Also, $v = 2$, $s = 0$, when $t = 0$.
(a) Find the equation of motion (s in terms of t).

Solution. Using (1), we have, from (15),

(16) $$\frac{d^2 s}{dt^2} + \frac{ds}{dt} + \frac{5}{4} s = 0,$$

an equation of the form (*G*). The roots of the auxiliary equation $r^2 + r + \tfrac{5}{4} = 0$ are

$$r_1 = -\tfrac{1}{2} + \sqrt{-1}, \quad r_2 = -\tfrac{1}{2} - \sqrt{-1}.$$

Hence the complete solution of (16) is

(17) $$s = e^{-\tfrac{1}{2}t}(c_1 \cos t + c_2 \sin t).$$

By the given conditions, $s = 0$ when $t = 0$. Substituting these values in (17), we find $c_1 = 0$, and hence

(18) $$s = c_2 e^{-\tfrac{1}{2}t} \sin t.$$

Differentiating to find v, we get

(19) $$v = c_2 e^{-\tfrac{1}{2}t}(-\tfrac{1}{2} \sin t + \cos t).$$

Substituting the given values $v = 2$ when $t = 0$, we have $2 = c_2$. With this value of c_2, (18) becomes

(20) $$s = 2 e^{-\tfrac{1}{2}t} \sin t. \quad \text{Ans.}$$

(b) For what values of t will $v = 0$?

Solution. When $v = 0$, the expression in the parenthesis of the right-hand member of (19) must vanish. Setting this equal to zero, we readily obtain

(21) $$\tan t = 2.$$

For any value of t satisfying (21), v will vanish. These values are

(22) $$t = 1.10 + n\pi. \quad (n = \text{an integer}). \quad \textbf{Ans.}$$

ORDINARY DIFFERENTIAL EQUATIONS 405

Successive values of t from (22) differ by the constant interval of time π.

Discussion. This example illustrates *damped harmonic vibration.* In fact, in (15) the acceleration is the sum of two components

(23) $\quad\quad\quad\quad\quad\quad a_1 = -\frac{5}{4} s, \quad a_2 = -v.$

The simple harmonic vibration corresponding to the component a_1 is now disturbed by a *damping force* with the acceleration a_2, that is, by a force proportional to the velocity and opposite to the direction of motion. The effects of this damping force are twofold.

First, the interval of time between successive positions of the point where $v = 0$ is *lengthened* by the damping force. In fact, for the simple harmonic vibration

(24) $\quad\quad\quad\quad\quad\quad a_1 = -\frac{5}{4} s,$

we have, by comparison with (8), $k = \frac{1}{2}\sqrt{5} = 1.12$, and the half-period, by (12), is $0.89\ \pi$. As we have seen above, for the damped harmonic vibration the corresponding interval is π.

Second, the values of s for the successive extreme positions where $v = 0$, instead of being equal, now form a decreasing geometric progression. Proof is omitted.

ILLUSTRATIVE EXAMPLE 3. In a rectilinear motion

(25) $\quad\quad\quad\quad\quad a = -4s + 2\cos 2t.$

Also, $s = 0, v = 2,$ when $t = 0$.

(a) Find the equation of motion.

Solution. By (1), we have, from (25),

(26) $\quad\quad\quad\quad\quad \dfrac{d^2s}{dt^2} + 4s = 2\cos 2t.$

The particular solution required was found in Illustrative Example 6, Art. 206, and is given by equation (47), p. 397. Hence

(27) $\quad s = \sin 2t + \frac{1}{2} t \sin 2t.$ *Ans.*

(b) For what values of t will $v = 0$?

Solution. Differentiating (27) to find v, and setting the result equal to zero, we get

(28) $\quad (2 + t)\cos 2t + \frac{1}{2} \sin 2t = 0;$

or, dividing (28) through by $\cos 2t$,

(29) $\quad \frac{1}{2} \tan 2t + 2 + t = 0.$

The roots of this equation may be found as explained in Arts. 87–89. The figure shows the curves (see Art. 88)

(30) $\quad\quad\quad y = \frac{1}{2}\tan 2t, \quad y = -2 - t,$

and the abscissas of the points of intersection are, approximately,

$\quad\quad\quad\quad\quad t = 0.88, 2.36,$ etc. *Ans.*

Discussion. This example illustrates *forced harmonic vibration.* In fact, in (25) the acceleration is the sum of two components

(31) $\quad\quad\quad\quad a_1 = -4s, \quad a_2 = 2\cos 2t.$

The simple harmonic vibration corresponding to the component a_1 with the period π is now disturbed by a force with the acceleration a_2, that is, by a periodic force whose period ($=\pi$) is *the same* as the period of the undisturbed simple harmonic vibration. The effects of this disturbing force are twofold.

First, the interval of time between successive positions of the point where $v=0$ is no longer constant, but decreases and approaches $\frac{1}{2}\pi$. This is clear from the above figure.

Second, the values of s for the successive extreme positions where $v=0$ now increase and eventually become, in numerical value, indefinitely great.

PROBLEMS

In each of the following problems the acceleration and initial conditions are given. Find the equation of motion.

1. $a = -k^2 s$; $s=0, v=v_0$, when $t=0$. *Ans.* $s = \dfrac{v_0}{k}\sin kt$.

2. $a = -k^2 s$; $s=s_0, v=0$, when $t=0$. $s = s_0 \cos kt$.

3. $a = -k^2 s$; $s=s_0, v=v_0$, when $t=0$.

4. $a = 6-s$; $s=0, v=0$, when $t=0$. $s = 6(1-\cos t)$.

5. $a = \sin 2t - s$; $s=0, v=0$, when $t=0$. $s = \tfrac{2}{3}\sin t - \tfrac{1}{3}\sin 2t$.

6. $a = 2\cos t - s$; $s=2, v=0$, when $t=0$. $s = 2\cos t + t\sin t$.

7. $a = -2v - 2s$; $s=3, v=-3$, when $t=0$. $s = 3 e^{-t} \cos t$.

8. $a = -k^2 s + b$; $s=0, v=0$, when $t=0$.

9. $a = -nv$; $s=0, v=n$, when $t=0$.

10. $a = 8t - 4s$; $s=0, v=4$, when $t=0$.

11. $a = 4\sin t - 4s$; $s=0, v=0$, when $t=0$.

12. $a = 2\sin 2t - 4s$; $s=0, v=0$, when $t=0$.

13. $a = -2v - 5s$; $s=1, v=1$, when $t=0$.

14. Given $a = 8 - 4s$, and $v=0$, $s=0$, when $t=0$. Show that the motion is a simple harmonic vibration with the center at $s=2$, with an amplitude 2 and a period π.

15. The acceleration of a particle is given by

$$a = 5\cos 2t - 9s.$$

(a) If the particle starts from rest at the origin, find its equation of motion. *Ans.* $s = \cos 2t - \cos 3t$.
What is the greatest distance from the origin reached by the particle?

(b) If the particle starts from the origin with velocity $v=6$, find its equation of motion. *Ans.* $s = \cos 2t + 2\sin 3t - \cos 3t$.
What is the greatest distance from the origin reached by the particle?

16. Answer the questions of the preceding problem if the acceleration is given by
$$a = 3\cos 3t - 9s.$$

Ans. (a) $s = \tfrac{1}{2} t \sin 3t$: (b) $s = \tfrac{1}{2} t \sin 3t + 2\sin 3t$.

17. A body falls from rest under the action of its weight and a small resistance which varies as the velocity. Prove the following relations:

$$a = g - kv.$$

$$v = \frac{g}{k}(1 - e^{-kt}).$$

$$s = \frac{g}{k^2}(kt + e^{-kt} - 1).$$

$$ks + v + \frac{g}{k}\ln\left(1 - \frac{kv}{g}\right) = 0.$$

18. A body falls from rest a distance of 80 ft. Assuming $a = 32 - v$, find the time. *Ans.* 3.47 sec.

19. A boat moving in still water is subject to a retardation proportional to its velocity at any instant. Show that the velocity t sec. after the power is shut off is given by the formula $v = ce^{-kt}$, where c is the velocity at the instant the power is shut off.

20. At a certain instant a boat drifting in still water has a velocity of 4 mi. per hour. One minute later the velocity is 2 mi. per hour. Find the distance moved.

21. Under certain conditions the equation defining the swing of a galvanometer is

$$\frac{d^2\theta}{dt^2} + 2\mu\frac{d\theta}{dt} + k^2\theta = 0.$$

Show that it will not swing through the zero point if $\mu > k$. Find the complete solution if $\mu < k$.

209. Linear differential equations of the nth order with constant coefficients. The solution of the linear differential equation

(*I*) $$\frac{d^n y}{dx^n} + p_1\frac{d^{n-1}y}{dx^{n-1}} + p_2\frac{d^{n-2}y}{dx^{n-2}} + \cdots + p_n y = 0,$$

in which the coefficients $p_1, p_2, \cdots, p_n$ are constants, will now be discussed.

The substitution of e^{rx} for y in the first member gives

$$(r^n + p_1 r^{n-1} + p_2 r^{n-2} + \cdots + p_n)e^{rx}.$$

This expression vanishes for all values of r which satisfy the equation

(1) $$r^n + p_1 r^{n-1} + p_2 r^{n-2} + \cdots + p_n = 0;$$

and therefore for each of these values of r, e^{rx} is a solution of (*I*). Equation (1) is called the *auxiliary equation* of (*I*). We observe that the coefficients are the same in both, the exponents in (1) corresponding to the order of the derivatives in (*I*), and y being replaced by 1.

From the roots of the equation we may write down particular solutions of the differential equation (*I*). The results are those of Art. 206 extended to cases when the order exceeds two, and are proved in more advanced textbooks.

Rule to solve the equation (*I*)

FIRST STEP. *Write down the corresponding auxiliary equation*

(1) $\quad r^n + p_1 r^{n-1} + p_2 r^{n-2} + \cdots + p_n = 0.$

SECOND STEP. *Solve completely the auxiliary equation.*

THIRD STEP. *From the roots of the auxiliary equation write down the corresponding particular solutions of the differential equation as follows:*

AUXILIARY EQUATION	DIFFERENTIAL EQUATION
(a) Each distinct real root r_1	gives a particular solution $e^{r_1 x}$.
(b) Each distinct pair of imaginary roots $a \pm bi$	gives two particular solutions $e^{ax} \cos bx$, $e^{ax} \sin bx$.
(c) A multiple root occurring s times	gives s (or $2s$) particular solutions obtained by multiplying the particular solutions (a) (or (b)) by $1, x, x^2, \cdots, x^{s-1}$.

FOURTH STEP. *Multiply each of the n* independent solutions thus found by an arbitrary constant and add the results. This result set equal to y gives the complete solution.*

ILLUSTRATIVE EXAMPLE 1. Solve $\dfrac{d^3 y}{dx^3} - 3 \dfrac{d^2 y}{dx^2} + 4 y = 0.$

Solution. Follow the above rule.

First Step. $r^3 - 3 r^2 + 4 = 0$, auxiliary equation.
Second Step. Solving, the roots are $-1, 2, 2$.
Third Step. (a) The root -1 gives the solution e^{-x}.
(c) The double root 2 gives the two solutions e^{2x}, xe^{2x}.
Fourth Step. The complete solution is

$$y = c_1 e^{-x} + c_2 e^{2x} + c_3 x e^{2x}. \quad Ans.$$

ILLUSTRATIVE EXAMPLE 2. Solve $\dfrac{d^4 y}{dx^4} - 4 \dfrac{d^3 y}{dx^3} + 10 \dfrac{d^2 y}{dx^2} - 12 \dfrac{dy}{dx} + 5 y = 0.$

Solution. Follow the above rule.

First Step. $r^4 - 4 r^3 + 10 r^2 - 12 r + 5 = 0$, auxiliary equation.
Second Step. Solving, the roots are $1, 1, 1 \pm 2 i$.

---

* A check on the accuracy of the work is found in the fact that the first three steps must give n independent solutions.

Third Step. (b) The pair of imaginary roots $1 \pm 2i$ gives the two solutions

$$e^x \cos 2x, \; e^x \sin 2x. \qquad (a=1, b=2)$$

(c) The double root 1 gives the two solutions e^x, xe^x.

Fourth Step. The complete solution is

$$y = c_1 e^x + c_2 x e^x + c_3 e^x \cos 2x + c_4 e^x \sin 2x,$$

or $\qquad y = (c_1 + c_2 x + c_3 \cos 2x + c_4 \sin 2x)e^x.$ *Ans.*

The linear differential equation

$$(J) \qquad \frac{d^n y}{dx^n} + p_1 \frac{d^{n-1} y}{dx^{n-1}} + p_2 \frac{d^{n-2} y}{dx^{n-2}} + \cdots + p_n y = X,$$

in which $p_1, p_2, \cdots, p_n$ are constants, and X is a function of x or a constant, is solved by methods like those used in Art. 206 for equation (H). The three steps described on page 394 are to be followed here also. That is, we solve first the equation (I), obtaining

$$(2) \qquad y = u,$$

the complete solution of (I). Then u is the complementary function for (J).

Next we find in any manner a particular solution of (J),

$$(3) \qquad y = v.$$

Then the complete solution of (J) is

$$(4) \qquad y = u + v.$$

In finding (3), methods of trial may be used analogous to those given on page 395 for $n = 2$. The rules given there for the general case apply also for any value of n. In any case we may follow the

Rule to find a particular solution of (J)

FIRST STEP. *Differentiate successively the given equation (J) and obtain, either directly or by elimination, a differential equation of a higher order of Type (I).*

SECOND STEP. *Solving this new equation by the rule on page 408, we get its complete solution*

$$y = u + v,$$

where the part u is the complementary function of (J) already found in the first step,* and v is the sum of the additional terms found.

THIRD STEP. *To find the values of the constants of integration in the particular solution v, substitute*

$$y = v$$

* From the method of derivation it is obvious that every solution of the original equation must also be a solution of the derived equation.

410 DIFFERENTIAL AND INTEGRAL CALCULUS

and its derivatives in the given equation (*J*). *In the resulting identity equate the coefficients of like terms, solve for the constants of integration, and substitute their values back in*

$$y = u + v,$$

giving the complete solution of (*J*).

This method will now be illustrated by means of examples.

NOTE. The solution of the auxiliary equation of the new derived differential equation is facilitated by observing that the left-hand member of that equation is exactly divisible by the left-hand member of the auxiliary equation used in finding the complementary function.

ILLUSTRATIVE EXAMPLE. Solve

(5) $\qquad y'' - 3\,y' + 2\,y = xe^x.$

Solution. We first find the complementary function u. Solving

(6) $\qquad y'' - 3\,y' + 2\,y = 0,$

the result is

(7) $\qquad y = u = c_1 e^{2x} + c_2 e^x.$

First Step. We now differentiate (5), obtaining

(8) $\qquad y''' - 3\,y'' + 2\,y' = xe^x + e^x.$

Subtracting (5) from (8), the result is

(9) $\qquad y''' - 4\,y'' + 5\,y' - 2\,y = e^x.$

Differentiating (9), we obtain

(10) $\qquad y^{\text{iv}} - 4\,y''' + 5\,y'' - 2\,y' = e^x.$

Subtracting (9) from (10), we get

(11) $\qquad y^{\text{iv}} - 5\,y''' + 9\,y'' - 7\,y' + 2\,y = 0,$

an equation of Type (*I*).

Second Step. Solve (11). The auxiliary equation is

(12) $\qquad r^4 - 5\,r^3 + 9\,r^2 - 7\,r + 2 = 0.$

The left-hand member must be divisible by $r^2 - 3\,r + 2$, since the auxiliary equation for (6) is $r^2 - 3\,r + 2 = 0$. In fact, we find that (12) may be written

(13) $\qquad (r^2 - 3\,r + 2)(r - 1)^2 = 0.$

The roots are $r = 1, 1, 1, 2$. Hence the complete solution of (11) is

(14) $\qquad y = c_1 e^{2x} + e^x(c_2 + c_3 x + c_4 x^2).$

Third Step. Comparing (7) and (14), we see that

(15) $\qquad y = v = e^x(c_3 x + c_4 x^2)$

will be a particular solution of (5) for suitable values of the constants c_3 and c_4.

Differentiating (15), we obtain

(16) $\qquad y' = e^x(c_3 + (c_3 + 2\,c_4)x + c_4 x^2),$
$\qquad\qquad y'' = e^x(2(c_3 + c_4) + (c_3 + 4\,c_4)x + c_4 x^2).$

ORDINARY DIFFERENTIAL EQUATIONS

Substituting in (5) from (15) and (16), dividing both members by e^x and reducing, the result is

(17) $\qquad 2\,c_4 - c_3 - 2\,c_4 x = x.$

Equating coefficients of like powers of x, we obtain $-2\,c_4 = 1$, $2\,c_4 - c_3 = 0$, whence we find $c_3 = -1$, $c_4 = -\frac{1}{2}$. Substituting these values in (15), the particular solution is

(18) $\qquad y = v = e^x(-x - \frac{1}{2}x^2),$

and the complete solution is

$$y = u + v = c_1 e^{2x} + c_2 e^x - e^x(x + \tfrac{1}{2}x^2). \quad Ans.$$

PROBLEMS

Find the complete solution of each of the following differential equations.

1. $\dfrac{d^3y}{dx^3} + 4\dfrac{dy}{dx} = 0.$ $\qquad$ Ans. $y = c_1 + c_2 \cos 2x + c_3 \sin 2x.$

2. $\dfrac{d^4y}{dx^4} + 4\dfrac{d^2y}{dx^2} = 0.$ $\qquad y = c_1 + c_2 x + c_3 \cos 2x + c_4 \sin 2x.$

3. $\dfrac{d^5y}{dx^5} - \dfrac{dy}{dx} = 0.$ $\qquad y = c_1 + c_2 e^x + c_3 e^{-x} + c_4 \cos x + c_5 \sin x.$

4. $\dfrac{d^4s}{dt^4} + 3\dfrac{d^2s}{dt^2} - 4s = 0.$ $\qquad s = c_1 e^t + c_2 e^{-t} + c_3 \cos 2t + c_4 \sin 2t.$

5. $\dfrac{d^4y}{dx^4} - \dfrac{d^3y}{dx^3} + 9\dfrac{d^2y}{dx^2} - 9\dfrac{dy}{dx} = 0.$
$\qquad$ Ans. $y = c_1 + c_2 e^x + c_3 \cos 3x + c_4 \sin 3x.$

6. $\dfrac{d^4s}{dt^4} + 8\dfrac{d^2s}{dt^2} + 16s = 0.$ Ans. $s = (c_1 + c_2 t)\cos 2t + (c_3 + c_4 t)\sin 2t.$

7. $\dfrac{d^3x}{dt^3} + 6\dfrac{d^2x}{dt^2} + 12\dfrac{dx}{dt} + 8x = 0.$ $\qquad$ Ans. $x = e^{-2t}(c_1 + c_2 t + c_3 t^2).$

8. $\dfrac{d^4s}{dt^4} - s = t^3 + 3t.$ Ans. $s = c_1 e^t + c_2 e^{-t} + c_3 \cos t + c_4 \sin t - t^3 - 3t.$

9. $\dfrac{d^3y}{dx^3} - 4\dfrac{dy}{dx} = 2x^2.$ $\qquad y = c_1 + c_2 e^{2x} + c_3 e^{-2x} - \tfrac{1}{6}x^3 - \tfrac{1}{4}x.$

10. $\dfrac{d^4y}{dx^4} - \dfrac{d^2y}{dx^2} = 4x.$ $\qquad y = c_1 + c_2 x + c_3 e^x + c_4 e^{-x} - \tfrac{2}{3}x^3.$

11. $\dfrac{d^2y}{dx^2} - 3\dfrac{dy}{dx} + 2y = xe^{3x}.$ $y = c_1 e^x + c_2 e^{2x} - \tfrac{3}{4}e^{3x} + \tfrac{1}{2}xe^{3x}.$

12. $\dfrac{d^2s}{dt^2} - 9\dfrac{ds}{dt} + 20s = t^2 e^{3t}.$ Ans. $s = c_1 e^{4t} + c_2 e^{5t} + \dfrac{e^{3t}(7 + 6t + 2t^2)}{4}.$

13. $\dfrac{d^2s}{dt^2} + 4s = t\sin^2 t.$
$\qquad$ Ans. $s = c_1 \cos 2t + c_2 \sin 2t + \dfrac{t}{8} - \dfrac{t\cos 2t}{32} - \dfrac{t^2 \sin 2t}{16}.$

14. $\dfrac{d^4y}{dx^4} - 9\dfrac{d^2y}{dx^2} = 0.$

15. $\dfrac{d^3x}{dt^3} + 8\,x = 0.$

16. $\dfrac{d^4y}{dx^4} - 13\dfrac{d^2y}{dx^2} + 36\,y = 0.$

17. $\dfrac{d^4y}{dx^4} + 5\dfrac{d^2y}{dx^2} - 36\,y = 0.$

18. $\dfrac{d^4s}{dt^4} + 13\dfrac{d^2s}{dt^2} + 36\,s = 18\,t - 36.$

19. $\dfrac{d^4x}{dt^4} + 2\dfrac{d^2x}{dt^2} + x = t + 2.$

20. $\dfrac{d^3s}{dt^3} - 2\dfrac{d^2s}{dt^2} + \dfrac{ds}{dt} = e^t.$

MISCELLANEOUS PROBLEMS

Find the complete solution of each of the following differential equations.

1. $8\left(\dfrac{dy}{dt}\right)^3 = 27\,y.$ Ans. $y = (t+c)^{\frac{3}{2}}.$

2. $\left(\dfrac{dy}{dx}\right)^3 - 27\,y^2 = 0.$ $y = (x+c)^3.$

3. $4\left(\dfrac{dy}{dx}\right)^2 = 9\,x.$ $y = x^{\frac{3}{2}} + c.$

4. $(1 + x^2)dy = \sqrt{1 - y^2}\,dx.$ $\dfrac{y}{\sqrt{1-y^2}} = \dfrac{x+c}{1-cx}.$

5. $(x+y)dx = (x-y)dy.$ $\ln(x^2 + y^2) - 2\,\text{arc tan}\,\dfrac{y}{x} = c.$

6. $\dfrac{dy}{dx} + y = e^{-x}.$ $y = (x+c)e^{-x}.$

7. $\dfrac{d^2s}{dt^2} - 4\dfrac{ds}{dt} + 3\,s = 0.$ $s = c_1 e^t + c_2 e^{3t}.$

8. $\dfrac{d^2s}{dt^2} - 4\dfrac{ds}{dt} + 4\,s = 0.$ $s = c_1 e^{2t} + c_2 t e^{2t}.$

9. $\dfrac{d^2s}{dt^2} - 4\dfrac{ds}{dt} + 8\,s = 0.$ Ans. $s = e^{2t}(c_1 \cos 2\,t + c_2 \sin 2\,t).$

10. $\dfrac{d^2y}{dt^2} - 2\dfrac{dy}{dt} - 3\,y = e^{2t}.$ $y = c_1 e^{3t} + c_2 e^{-t} - \tfrac{1}{3} e^{2t}.$

11. $\dfrac{d^2x}{dt^2} + k^2 x = at + b.$ $x = c_1 \cos kt + c_2 \sin kt + \dfrac{at+b}{k^2}.$

12. $\dfrac{d^2x}{dt^2} + k^2 x = ae^{bt}.$ $x = c_1 \cos kt + c_2 \sin kt + \dfrac{ae^{bt}}{b^2 + k^2}.$

13. $\dfrac{d^2x}{dt^2} - k^2 x = a \cos kt.$ $x = c_1 e^{kt} + c_2 e^{-kt} - \dfrac{a}{2\,k^2}\cos kt.$

14. $\dfrac{d^2x}{dt^2} + k^2 x = a \sin kt.$ $x = c_1 \cos kt + c_2 \sin kt - \dfrac{a}{2\,k}\,t\cos kt.$

15. $(x^2 - 2\,y^2)dx + 2\,xy\,dy = 0.$ $y^2 + x^2 \ln cx = 0.$

16. $\dfrac{dy}{dx} + \dfrac{4\,xy}{x^2+1} = \dfrac{1}{(x^2+1)^3}.$ $y(x^2+1)^2 = \text{arc tan}\,x + c.$

17. $\dfrac{d^4s}{dt^4} - 5\dfrac{d^2s}{dt^2} + 4s = 0.$ Ans. $s = c_1 e^t + c_2 e^{-t} + c_3 e^{2t} + c_4 e^{-2t}.$

18. $\dfrac{d^4s}{dt^4} + 5\dfrac{d^2s}{dt^2} + 4s = 0.$
  Ans. $s = c_1 \cos t + c_2 \sin t + c_3 \cos 2t + c_4 \sin 2t.$

19. $xy^2\, dy = (x^3 + y^3) dx.$

20. $dy + xy(1 - x^2 y^2) dx = 0.$

21. $\dfrac{d^2x}{dt^2} - 8\dfrac{dx}{dt} + 25 x = 0.$

22. $\dfrac{d^2x}{dt^2} + 4x = 8t + 2.$

23. $\dfrac{d^2x}{dt^2} + 4x = e^{-t}.$

24. $\dfrac{d^2x}{dt^2} + 4x = 6 \cos 3t.$

25. $\dfrac{d^2x}{dt^2} + 4x = 2 \cos 2t.$

Solve each of the following differential equations by making the transformation suggested.

26. $t^2 \dfrac{ds}{dt} - 2st - s^3 = 0.$ Let $s = \dfrac{t^2}{v}.$ Ans. $\dfrac{t^4}{2s^2} + \dfrac{t^3}{3} = c.$

27. $(t^2 + t)ds = (t^2 + 2st + s)dt.$ Let $s = vt.$ Ans. $s = ct(1 + t) - t.$

28. $(3 + 2st)s\, dt = (3 - 2st)t\, ds.$ Let $st = v.$

29. $(x + y)^2 \dfrac{dy}{dx} = 2x + 2y + 5.$ Let $x + y = v.$

ADDITIONAL PROBLEMS

1. For a certain curve the area bounded by the curve, the x-axis, and any two ordinates is k times the length of arc intercepted between the ordinates, and the curve passes through the point $(0, k)$. Show that the curve must be a catenary.

2. The acceleration of a man dropping in a parachute from a stationary balloon is $32 - \tfrac{1}{8} v^2$ ft. per second per second, where v is the velocity in feet per second. If he reaches the ground in one minute, prove that the height of the balloon is a little more than 950 ft.

3. A point moving on the x-axis is subject to an acceleration directed toward the origin and proportional to its distance from the origin and to a retardation proportional to its velocity. Given that the differential equation for x is of the form

$$\dfrac{d^2x}{dt^2} + m\dfrac{dx}{dt} + nx = 0,$$

where m and n are positive, and given the initial conditions $x = 10$, $\dfrac{dx}{dt} = 0$, when $t = 0$, find in each of the following cases x and $\dfrac{dx}{dt}$ and discuss the motion.

  (a) $m = 4, n = 5$; (b) $m = 4, n = 4$; (c) $m = 4, n = 3$.

CHAPTER XXII

HYPERBOLIC FUNCTIONS

210. Hyperbolic sine and cosine. Certain simple expressions involving exponential functions (Art. 62) occur frequently in applied mathematics. They are called *hyperbolic functions*. The justification for this name is brought out below in Art. 215. Two of these functions, the *hyperbolic sine* and *hyperbolic cosine* of a variable v, written, respectively, sinh v and cosh v, are defined by the equations

$$(A) \qquad \sinh v = \frac{e^v - e^{-v}}{2}, \quad \cosh v = \frac{e^v + e^{-v}}{2},$$

where e is, as usual, the Napierian base (Art. 61). These functions are not, however, independent, for we have from (A)

$$(B) \qquad \cosh^2 v - \sinh^2 v = 1.$$

$$\left[\text{From } (A), \text{ squaring, } \cosh^2 v = \frac{e^{2v} + 2 + e^{-2v}}{4}, \sinh^2 v = \frac{e^{2v} - 2 + e^{-2v}}{4}.\right.$$
$$\left.\text{Hence, by subtraction, } \cosh^2 v - \sinh^2 v = 1.\right]$$

From (A), by solving for the exponential functions, we get

$$(1) \qquad e^v = \cosh v + \sinh v, \quad e^{-v} = \cosh v - \sinh v.$$

ILLUSTRATIVE EXAMPLE. Show that the complete solution of the differential equation

$$(2) \qquad \frac{d^2y}{dx^2} - a^2 y = 0$$

may be written $\qquad y = A \sinh ax + B \cosh ax,$

where A and B are constants.

Solution. By Art. 206 the auxiliary equation for (2) is $r^2 - a^2 = 0$, whose roots are a and $-a$. Therefore the complete solution of (2) is

$$y = c_1 e^{ax} + c_2 e^{-ax}.$$

The values of e^{ax} and e^{-ax} are found from (1) by taking $v = ax$. Hence

$$e^{ax} = \cosh ax + \sinh ax, \ e^{-ax} = \cosh ax - \sinh ax,$$
$$y = c_1(\cosh ax + \sinh ax) + c_2(\cosh ax - \sinh ax)$$
$$= (c_1 + c_2)\cosh ax + (c_1 - c_2)\sinh ax.$$

Letting $c_1 - c_2 = A$, $c_1 + c_2 = B$, we obtain the desired form.
The result should be compared with Illustrative Example 2, p. 391.

HYPERBOLIC FUNCTIONS 415

211. Other hyperbolic functions. The *hyperbolic tangent*, tanh v, is defined by

(C) $$\tanh v = \frac{\sinh v}{\cosh v} = \frac{e^v - e^{-v}}{e^v + e^{-v}}.$$

The equations

(1) $$\operatorname{ctnh} v = \frac{1}{\tanh v}, \quad \operatorname{sech} v = \frac{1}{\cosh v}, \quad \operatorname{csch} v = \frac{1}{\sinh v}$$

define, respectively, the *hyperbolic cotangent, hyperbolic secant,* and *hyperbolic cosecant.* The ratios used in (C) and (1) are the same as those in (2), p. 2, for the corresponding trigonometric functions.

The following relations hold:

(2) $$1 - \tanh^2 v = \operatorname{sech}^2 v, \quad \operatorname{ctnh}^2 v - 1 = \operatorname{csch}^2 v,$$

analogous to formulas in (2), p. 2. The proof of the first formula is given below.

The following statements hold for the values of the hyperbolic functions. They should be verified.

$\sinh v$ can have any value; $\cosh v$, any positive value not less than 1; $\operatorname{sech} v$, any positive value not exceeding 1; $\tanh v$, any value numerically less than 1; $\operatorname{ctnh} v$, any value numerically greater than 1; $\operatorname{csch} v$, any value except zero. Also, from the definitions, we have

(3) $$\begin{aligned}\sinh(-v) &= -\sinh v, & \operatorname{csch}(-v) &= -\operatorname{csch} v, \\ \cosh(-v) &= \cosh v, & \operatorname{sech}(-v) &= \operatorname{sech} v, \\ \tanh(-v) &= -\tanh v, & \operatorname{ctnh}(-v) &= -\operatorname{ctnh} v.\end{aligned}$$

ILLUSTRATIVE EXAMPLE. Given $\tanh x = \frac{4}{5}$. Find the values of the other hyperbolic functions.

Solution. In (B) divide each term by $\cosh^2 x$. Then we get

$$1 - \frac{\sinh^2 x}{\cosh^2 x} = \frac{1}{\cosh^2 x}.$$

Therefore $1 - \tanh^2 x = \operatorname{sech}^2 x.$ By (C) and (1)

Since $\tanh x = \frac{4}{5}$, this equation gives $\operatorname{sech} x = \frac{3}{5}$, the negative value being inadmissible. Then

$$\cosh x = \frac{1}{\operatorname{sech} x} = \frac{5}{3}, \qquad \text{by (1)}$$

$$\sinh x = \cosh x \tanh x = \frac{4}{3}, \qquad \text{by (C)}$$

$$\operatorname{ctnh} x = \tfrac{5}{4}, \text{ and } \operatorname{csch} x = \tfrac{3}{4}. \qquad \text{By (1)}$$

212. Table of values of the hyperbolic sine, cosine, and tangent. Graphs. A table giving the values, to four significant figures, of $\sinh v$, $\cosh v$, $\tanh v$ for values of v from 0 to 5.9 is shown on p. 416. For negative values of v, use the relations (3), Art. 211.

HYPERBOLIC FUNCTIONS

v	Sinh v	Cosh v	Tanh v	v	Sinh v	Cosh v	Tanh v	v	Sinh v	Cosh v	Tanh v
.00	.0000	1.000	.0000	.50	.5211	1.128	.4621	1.0	1.175	1.543	.7616
.01	.0100	1.000	.0100	.51	.5324	1.133	.4700	1.1	1.336	1.669	.8005
.02	.0200	1.000	.0200	.52	.5438	1.138	.4777	1.2	1.509	1.811	.8337
.03	.0300	1.000	.0300	.53	.5552	1.144	.4854	1.3	1.698	1.971	.8617
.04	.0400	1.001	.0400	.54	.5666	1.149	.4930	1.4	1.904	2.151	.8854
.05	.0500	1.001	.0500	.55	.5782	1.155	.5005	1.5	2.129	2.352	.9052
.06	.0600	1.002	.0599	.56	.5897	1.161	.5080	1.6	2.376	2.577	.9217
.07	.0701	1.002	.0699	.57	.6014	1.167	.5154	1.7	2.646	2.828	.9354
.08	.0801	1.003	.0798	.58	.6131	1.173	.5227	1.8	2.942	3.107	.9468
.09	.0901	1.004	.0898	.59	.6248	1.179	.5299	1.9	3.268	3.418	.9562
.10	.1002	1.005	.0997	.60	.6367	1.185	.5370	2.0	3.627	3.762	.9640
.11	.1102	1.006	.1096	.61	.6485	1.192	.5441	2.1	4.022	4.144	.9705
.12	.1203	1.007	.1194	.62	.6605	1.198	.5511	2.2	4.457	4.568	.9757
.13	.1304	1.008	.1293	.63	.6725	1.205	.5581	2.3	4.937	5.037	.9801
.14	.1405	1.010	.1391	.64	.6846	1.212	.5649	2.4	5.466	5.557	.9837
.15	.1506	1.011	.1489	.65	.6967	1.219	.5717	2.5	6.050	6.132	.9866
.16	.1607	1.013	.1587	.66	.7090	1.226	.5784	2.6	6.695	6.769	.9890
.17	.1708	1.014	.1684	.67	.7213	1.233	.5850	2.7	7.406	7.473	.9910
.18	.1810	1.016	.1781	.68	.7336	1.240	.5915	2.8	8.192	8.253	.9926
.19	.1911	1.018	.1878	.69	.7461	1.248	.5980	2.9	9.060	9.115	.9940
.20	.2013	1.020	.1974	.70	.7586	1.255	.6044	3.0	10.02	10.07	.9951
.21	.2115	1.022	.2070	.71	.7712	1.263	.6107	3.1	11.08	11.12	.9960
.22	.2218	1.024	.2165	.72	.7838	1.271	.6169	3.2	12.25	12.29	.9967
.23	.2320	1.027	.2260	.73	.7966	1.278	.6231	3.3	13.54	13.57	.9973
.24	.2423	1.029	.2355	.74	.8094	1.287	.6291	3.4	14.97	15.00	.9978
.25	.2526	1.031	.2449	.75	.8223	1.295	.6352	3.5	16.54	16.57	.9982
.26	.2629	1.034	.2543	.76	.8353	1.303	.6411	3.6	18.29	18.31	.9985
.27	.2733	1.037	.2636	.77	.8484	1.311	.6469	3.7	20.21	20.24	.9988
.28	.2837	1.039	.2729	.78	.8615	1.320	.6527	3.8	22.34	22.36	.9990
.29	.2941	1.042	.2821	.79	.8748	1.329	.6584	3.9	24.69	24.71	.9992
.30	.3045	1.045	.2913	.80	.8881	1.337	.6640	4.0	27.29	27.31	.9993
.31	.3150	1.048	.3004	.81	.9015	1.346	.6696	4.1	30.16	30.18	.9995
.32	.3255	1.052	.3095	.82	.9150	1.355	.6751	4.2	33.34	33.35	.9996
.33	.3360	1.055	.3185	.83	.9286	1.365	.6805	4.3	36.84	36.86	.9996
.34	.3466	1.058	.3275	.84	.9423	1.374	.6858	4.4	40.72	40.73	.9997
.35	.3572	1.062	.3364	.85	.9561	1.384	.6911	4.5	45.00	45.01	.9998
.36	.3678	1.066	.3452	.86	.9700	1.393	.6963	4.6	49.74	49.75	.9998
.37	.3785	1.069	.3540	.87	.9840	1.403	.7014	4.7	54.97	54.98	.9998
.38	.3892	1.073	.3627	.88	.9981	1.413	.7064	4.8	60.75	60.76	.9999
.39	.4000	1.077	.3714	.89	1.012	1.423	.7114	4.9	67.14	67.15	.9999
.40	.4108	1.081	.3800	.90	1.027	1.433	.7163	5.0	74.20	74.21	.9999
.41	.4216	1.085	.3885	.91	1.041	1.443	.7211	5.1	82.01	82.01	.9999
.42	.4325	1.090	.3969	.92	1.055	1.454	.7259	5.2	90.63	90.64	.9999
.43	.4434	1.094	.4053	.93	1.070	1.465	.7306	5.3	100.17	100.17	1.0000
.44	.4543	1.098	.4136	.94	1.085	1.475	.7352	5.4	110.70	110.71	1.0000
.45	.4653	1.103	.4219	.95	1.099	1.486	.7398	5.5	122.34	122.35	1.0000
.46	.4764	1.108	.4301	.96	1.114	1.497	.7443	5.6	135.21	135.22	1.0000
.47	.4875	1.112	.4382	.97	1.129	1.509	.7487	5.7	149.43	149.44	1.0000
.48	.4986	1.117	.4462	.98	1.145	1.520	.7531	5.8	165.15	165.15	1.0000
.49	.5098	1.122	.4542	.99	1.160	1.531	.7574	5.9	182.52	182.52	1.0000

HYPERBOLIC FUNCTIONS 417

When $v \to +\infty$, sinh v and cosh v become infinite, while tanh v approaches unity as a limit.

Graphs of sinh x, cosh x, and tanh x (Figs. 1, 2, 3) are easily drawn by making use of the table.

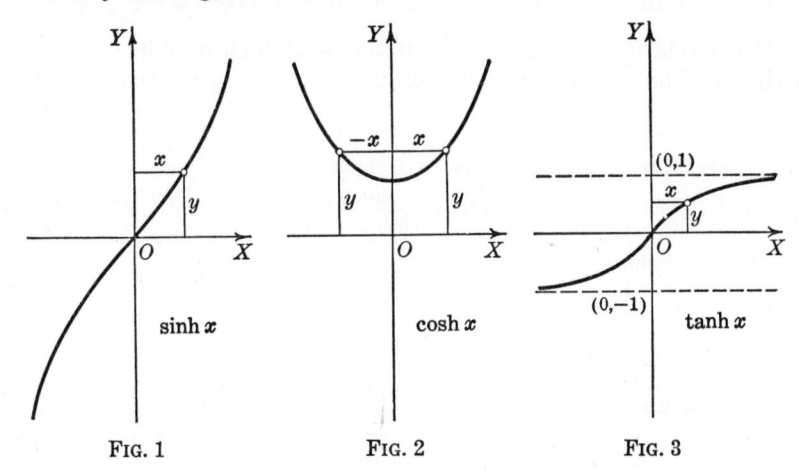

sinh x cosh x tanh x

FIG. 1   FIG. 2   FIG. 3

213. Hyperbolic functions of $v + w$. Formulas for hyperbolic functions, corresponding to two of (4), p. 3, are

(D)   sinh $(v + w) =$ sinh v cosh $w +$ cosh v sinh w,
(E)   cosh $(v + w) =$ cosh v cosh $w +$ sinh v sinh w.

Proof of (D). From the definition (A), replacing v by $v + w$, we have

(1) $$\sinh(v+w) = \frac{e^{v+w} - e^{-v-w}}{2},$$

(2) $$\cosh(v+w) = \frac{e^{v+w} + e^{-v-w}}{2}.$$

The right-hand member of (1) is transformed as follows, making use of (1), Art. 210.

$$\frac{e^{v+w} - e^{-v-w}}{2} = \frac{e^v e^w - e^{-v} e^{-w}}{2}$$

$$= \frac{(\cosh v + \sinh v)(\cosh w + \sinh w) - (\cosh v - \sinh v)(\cosh w - \sinh w)}{2}.$$

Multiplying out and reducing, we get (D). Formula (E) is proved in the same way.

If we set $w = v$ in (D) and (E), we have

(3)   sinh $2v = 2$ sinh v cosh v,
(4)   cosh $2v = \cosh^2 v + \sinh^2 v$.

418   DIFFERENTIAL AND INTEGRAL CALCULUS

These are analogous to the formulas for $\sin 2x$ and $\cos 2x$, respectively, of (5), p. 3. From (B) and (4), we get results which correspond to the formulas for $\sin^2 x$ and $\cos^2 x$ in (5), p. 3. These are

(5) $\quad \sinh^2 v = \tfrac{1}{2}\cosh 2v - \tfrac{1}{2}, \quad \cosh^2 v = \tfrac{1}{2}\cosh 2v + \tfrac{1}{2}.$

Other relations for hyperbolic functions, which may be compared with those on page 3 for trigonometric functions, are given in the problems.

ILLUSTRATIVE EXAMPLE. Derive the formula

(6) $\quad\quad\quad\quad\quad\quad \tanh v = \dfrac{\sinh 2v}{\cosh 2v + 1}.$

Solution. From (5), by division, we get

(7) $\quad\quad\quad\quad\quad\quad \tanh^2 v = \dfrac{\cosh 2v - 1}{\cosh 2v + 1}.$

Now $\quad \dfrac{\cosh 2v - 1}{\cosh 2v + 1} \cdot \dfrac{\cosh 2v + 1}{\cosh 2v + 1} = \dfrac{\cosh^2 2v - 1}{(\cosh 2v + 1)^2}.$

By (B), $\cosh^2 2v - 1 = \sinh^2 2v$. Hence (7) becomes

(8) $\quad\quad\quad\quad\quad\quad \tanh^2 v = \dfrac{\sinh^2 2v}{(\cosh 2v + 1)^2},$

and therefore $\quad\quad \tanh v = \pm \dfrac{\sinh 2v}{\cosh 2v + 1}.$

The sign before the right-hand member must now be examined. From (3) we have

$$\sinh 2v = \dfrac{2\sinh v}{\cosh v}\cosh^2 v = 2\tanh v \cosh^2 v.$$

Therefore $\sinh 2v$ and $\tanh v$ will always agree in sign. Also $\cosh 2v + 1$ is always positive. Hence the positive sign must be used, and we get (6).

If v is replaced by $\tfrac{1}{2}v$, (6) becomes

(9) $\quad\quad\quad\quad\quad\quad \tanh \dfrac{v}{2} = \dfrac{\sinh v}{\cosh v + 1}.$

PROBLEMS

1. The value of one hyperbolic function is given. Find the values of the others and check as far as possible by the table on page 416.

   (a) $\cosh x = 1.25.$ \quad\quad (c) $\sinh x = 10.$
   (b) $\operatorname{csch} x = -0.75.$ \quad (d) $\operatorname{ctnh} x = -2.5.$

Prove each of the formulas in Problems 2–7, and compare with the corresponding formula (if any) in (2), (4)–(6), pp. 2, 3.

2. $1 - \operatorname{ctnh}^2 v = -\operatorname{csch}^2 v.$

3. $\sinh(v - w) = \sinh v \cosh w - \cosh v \sinh w,$
   $\cosh(v - w) = \cosh v \cosh w - \sinh v \sinh w.$

HYPERBOLIC FUNCTIONS

4. $\tanh(v \pm w) = \dfrac{\tanh v \pm \tanh w}{1 \pm \tanh v \tanh w}$.

5. $\sinh \dfrac{v}{2} = \pm \sqrt{\dfrac{\cosh v - 1}{2}}$, $\cosh \dfrac{v}{2} = + \sqrt{\dfrac{\cosh v + 1}{2}}$.

6. $\sinh v + \sinh w = 2 \sinh \tfrac{1}{2}(v+w) \cosh \tfrac{1}{2}(v-w)$,
$\cosh v + \cosh w = 2 \cosh \tfrac{1}{2}(v+w) \cosh \tfrac{1}{2}(v-w)$.

7. $\tanh \tfrac{1}{2}(v-w) = \dfrac{\sinh v - \sinh w}{\cosh v + \cosh w}$.

8. Show that the equation of the catenary (figure, p. 532) may be written $y = a \cosh \dfrac{x}{a}$.

9. Solve the differential equation $\dfrac{d^2y}{dx^2} - y = 0$ in terms of hyperbolic functions, given that $y = 3$ when $x = 0$, and $y = 0$ when $\tanh x = \tfrac{4}{5}$.

$Ans.$ $y = 3 \cosh x - 3.75 \sinh x$.

10. Show that $\mathrm{sech}\,(-x) = \mathrm{sech}\, x$. Draw the graph and prove $\lim\limits_{x \to \infty} \mathrm{sech}\, x = 0$.

11. Show that $\mathrm{ctnh}\,(-x) = -\mathrm{ctnh}\, x$. Draw the graph and prove $\lim\limits_{x \to +\infty} \mathrm{ctnh}\, x = 1$.

12. Show that $\mathrm{csch}\,(-x) = -\mathrm{csch}\, x$. Draw the graph and prove $\lim\limits_{x \to \infty} \mathrm{csch}\, x = 0$.

13. Prove (a) $\sinh 3u = 3 \sinh u + 4 \sinh^3 u$;
(b) $\cosh 3u = 4 \cosh^3 u - 3 \cosh u$.

14. Show that $(\sinh x + \cosh x)^n = \sinh nx + \cosh nx$. ($n$ any positive integer.)

15. Prove that $\sinh^2 x - \sinh^2 y = \sinh(x+y) \sinh(x-y)$.

16. Simplify $\dfrac{\cosh 2u + \cosh 4v}{\sinh 2u + \sinh 4v}$. $Ans.$ $\mathrm{ctnh}\,(u + 2v)$.

17. Parametric equations for the *tractrix* may be written

$$x = t - a \tanh \dfrac{t}{a}, \quad y = a\, \mathrm{sech}\, \dfrac{t}{a}.$$

The parameter is t, and a is a constant. Plot the curve when $a = 4$. (The tractrix is the curve for which the length of the tangent (Art. 43) is constant and equal to a. Figure in Chapter XXVI.)

18. Solve $\dfrac{d^2y}{dx^2} = n^2(y - mx^2)$.

$Ans.$ $y = A \cosh nx + B \sinh nx + mx^2 + \dfrac{2m}{n^2}$.

214. Derivatives. The formulas, in which v is a function of x, are as follows.

$$\text{XXVII} \qquad \frac{d}{dx} \sinh v = \cosh v \, \frac{dv}{dx}.$$

$$\text{XXVIII} \qquad \frac{d}{dx} \cosh v = \sinh v \, \frac{dv}{dx}.$$

$$\text{XXIX} \qquad \frac{d}{dx} \tanh v = \operatorname{sech}^2 v \, \frac{dv}{dx}.$$

$$\text{XXX} \qquad \frac{d}{dx} \operatorname{ctnh} v = - \operatorname{csch}^2 v \, \frac{dv}{dx}.$$

$$\text{XXXI} \qquad \frac{d}{dx} \operatorname{sech} v = - \operatorname{sech} v \tanh v \, \frac{dv}{dx}.$$

$$\text{XXXII} \qquad \frac{d}{dx} \operatorname{csch} v = - \operatorname{csch} v \operatorname{ctnh} v \, \frac{dv}{dx}.$$

Proof of XXVII. By (A), $\sinh v = \dfrac{e^v - e^{-v}}{2}.$

Then
$$\frac{d}{dx} \sinh v = \frac{e^v \dfrac{dv}{dx} + e^{-v} \dfrac{dv}{dx}}{2}$$

$$= \frac{e^v + e^{-v}}{2} \frac{dv}{dx}$$

$$= \cosh v \, \frac{dv}{dx},$$

using (A).

Formula **XXVIII** is proved in a similar manner. The proof of **XXIX** is analogous to that given in Art. 72 for the derivative of $\tan v$. To prove **XXX–XXXII**, differentiate the forms as given in (1), Art. 211. The details are left as exercises.

215. Relations to the equilateral hyperbola. The curve for which

(1) $$x = a \cosh v, \quad y = a \sinh v$$

are parametric equations is the equilateral hyperbola $x^2 - y^2 = a^2$. For, eliminating the parameter v by squaring and subtracting, we have
$$x^2 - y^2 = a^2(\cosh^2 v - \sinh^2 v) = a^2. \qquad \text{By } (B)$$

Fig. 2, p. 421, shows a hyperbolic sector OAP_1 bounded by the arc AP_1 of (1), the semitransverse axis OA, and the radius vector OP_1. At P_1, $v = v_1$

HYPERBOLIC FUNCTIONS

Theorem. *The area of the hyperbolic sector OAP_1 equals $\frac{1}{2} a^2 v_1$.*

Proof. Let (ρ, θ) be the polar coördinates of any point on the arc AP_1. Then the element of area is (Art. 159) $dA = \frac{1}{2} \rho^2 d\theta$.

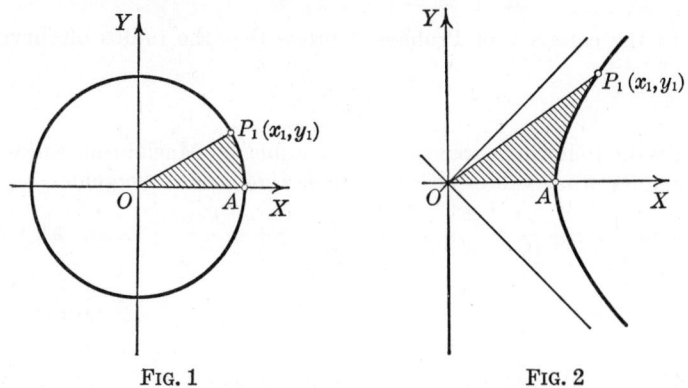

Fig. 1   Fig. 2

But $\qquad \rho^2 = x^2 + y^2 = a^2(\cosh^2 v + \sinh^2 v)$. $\qquad$ Using (1)

Also by **(5)**, p. 4,

$$\theta = \arctan \frac{y}{x} = \arctan (\tanh v). \qquad \text{By (1)}$$

Therefore $\qquad \dfrac{d\theta}{dv} = \dfrac{\operatorname{sech}^2 v}{1 + \tanh^2 v}.$

By **XXII**, p. 87, and **XXIX**

Using **(C)** and (1), Art. 211, we get

$$d\theta = \frac{dv}{\cosh^2 v + \sinh^2 v},$$

and therefore $\qquad dA = \frac{1}{2} a^2 dv$.

The theorem follows by integration, since $v = 0$ at A. $\qquad$ **Q.E.D.**

The parametric equations of the circle in Fig. 1 are

$$x = r \cos t, \quad y = r \sin t. \qquad \text{Art. 81}$$

The parameter t equals t_1 at P_1, and t_1 is the measure of the central angle AOP_1 in radians. Hence the area of the circular sector AOP_1 is $\frac{1}{2} r^2 t_1$.

Let $r = a = 1$. Then in Fig. 1, for $P(x, y)$,

$$x = \cos t, \quad y = \sin t, \quad \tfrac{1}{2} t = \text{area } AOP.$$

In Fig. 2, for $P(x, y)$,

$$x = \cosh v, \quad y = \sinh v, \quad \tfrac{1}{2} v = \text{area } AOP.$$

Hyperbolic functions, therefore, have the same relations to the equilateral hyperbola as the trigonometric functions do to the circle.

PROBLEMS

1. Show that the element of length of arc for the catenary $y = a \cosh \dfrac{x}{a}$ is given by $ds = \cosh \dfrac{x}{a} dx$.

2. In the catenary of Problem 1 prove that the radius of curvature equals $\dfrac{y^2}{a}$.

Verify the following expansions of functions by Maclaurin's series, and determine for what values of the variable they are convergent.

3. $\sinh x = x + \dfrac{x^3}{\underline{|3}} + \dfrac{x^5}{\underline{|5}} + \cdots + \dfrac{x^{2n-1}}{\underline{|2n-1}} + \cdots$.   *Ans.* All values.

4. $\cosh x = 1 + \dfrac{x^2}{\underline{|2}} + \dfrac{x^4}{4} + \cdots + \dfrac{x^{2n}}{\underline{|2n}} + \cdots$.   *Ans.* All values.

Verify the following expansions, using the series in Problems 3 and 4 and the methods explained in Art. 195.

5. $\operatorname{sech} x = 1 - \tfrac{1}{2} x^2 + \tfrac{5}{24} x^4 - \tfrac{61}{720} x^6 + \cdots$.

6. $\tanh x = x - \tfrac{1}{3} x^3 + \tfrac{2}{15} x^5 - \tfrac{17}{315} x^7 + \cdots$.

7. Test the function $5 \cosh x + 4 \sinh x$ for maximum or minimum values.   *Ans.* Minimum value, 3.

8. Test the function $A \sinh x + B \cosh x$ for maximum or minimum values.   *Ans.* If $B^2 > A^2$, a maximum value $-\sqrt{B^2 - A^2}$ if $B < 0$; a minimum value $+\sqrt{B^2 - A^2}$ if $B > 0$.

9. Derive the series in Problems 3 and 4 from the series for e^x and e^{-x} by subtraction and addition. (Use (A) and Art. 195.)

10. Let $ds =$ length of the element of arc; let $\rho = \sqrt{x^2 + y^2} =$ radius vector of $P(x, y)$ for the circle or equilateral hyperbola of Art. 215, and take limits of integration for the arc AP_1 in Figs. 1 and 2, p. 421. Prove

(a) $\displaystyle\int \dfrac{ds}{\rho} = t_1$ for the circle; (b) $\displaystyle\int \dfrac{ds}{\rho} = v_1$ for the hyperbola.

11. Prove $\lim\limits_{x \to +\infty} (\cosh x - \sinh x) = 0$.

12. Evaluate each of the following indeterminate forms.

(a) $\lim\limits_{x \to 0} \dfrac{\sinh x}{x}$.   (b) $\lim\limits_{x \to 0} \dfrac{\tanh x}{x}$.

(c) $\lim\limits_{x \to 0} \dfrac{1 - \cosh x}{x^2}$.   *Ans.* $-\tfrac{1}{2}$.

13. Given $\tan \phi = \sinh x$. Prove $\dfrac{d\phi}{dx} = \operatorname{sech} x$.

HYPERBOLIC FUNCTIONS

14. Derive the expansion
$$\arctan(\sinh x) = x - \tfrac{1}{6}x^3 + \tfrac{1}{24}x^5 - \tfrac{61}{5040}x^7 + \cdots$$
by integration as in Art. 196, using the result in Problem 5.

15. Prove the following theorems for the tractrix (see figure)
$$x = t - a\tanh\frac{t}{a}, \quad y = a\operatorname{sech}\frac{t}{a}.$$

(a) The parameter t equals the intercept of the tangent on the x-axis.
(b) The constant a equals the length of the tangent (Art. 43).
(c) The evolute is the catenary $\beta = a\cosh\dfrac{\alpha}{a}$.
(d) The radius of curvature PC is $a\sinh\dfrac{t}{a}$.

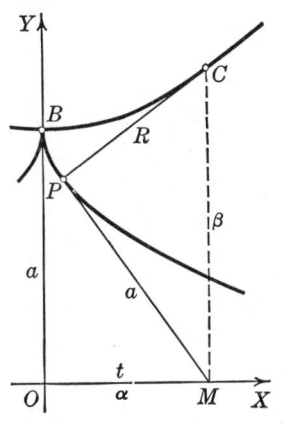

216. Inverse hyperbolic functions. The relation
$$(1) \qquad y = \sinh v$$
is also written
$$(2) \qquad v = \sinh^{-1} y,$$
and read "v equals the inverse hyperbolic sine of y." Therefore $\sinh v$ and $\sinh^{-1} y$ are inverse functions (Art. 39). The same notation and nomenclature are used for the other inverse hyperbolic functions, $\cosh^{-1} v$ ("inverse hyperbolic cosine of v"), etc.

The curves
$$(3) \qquad y = \sinh x, \quad y = \cosh x, \quad y = \tanh x$$
are shown again on page 424. Assume now that y is given.

In Fig. 1, y may have any value, positive or negative, and then the value of x is uniquely determined.

In Fig. 2, y may have any positive value not less than 1. When $y > 1$, x has two values equal numerically and differing in sign.

In Fig. 3, y may have any value numerically less than 1, and then the value of x is uniquely determined.

Summarized, the results are

The function $\sinh^{-1} v$ is uniquely determined for any value of v. Also $\sinh^{-1}(-v) = -\sinh^{-1} v$.

The function $\cosh^{-1} v$, when $v > 1$, has two values differing only in sign. Also $\cosh^{-1} 1 = 0$.

The function $\tanh^{-1} v$ is uniquely determined when $v^2 < 1$. Also $\tanh^{-1}(-v) = -\tanh^{-1} v$.

424 DIFFERENTIAL AND INTEGRAL CALCULUS

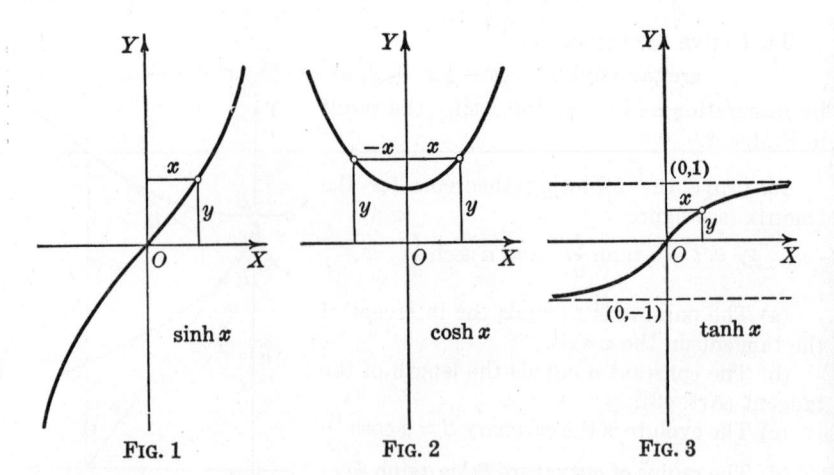

FIG. 1 — sinh x
FIG. 2 — cosh x
FIG. 3 — tanh x

Hyperbolic functions were defined in Art. 210 in terms of exponential functions. The inverse hyperbolic functions are expressible in terms of logarithmic functions. The relations are

(F) $\quad\quad \sinh^{-1} x = \ln(x + \sqrt{x^2 + 1}).$ $\quad\quad$ (Any x)

(G) $\quad\quad \cosh^{-1} x = \ln(x \pm \sqrt{x^2 - 1}).$ $\quad\quad$ ($x \geq 1$)

(H) $\quad\quad \tanh^{-1} x = \frac{1}{2} \ln \frac{1+x}{1-x}.$ $\quad\quad$ ($x^2 < 1$)

Proof of (F). Let $v = \sinh^{-1} x$. Then

(4) $\quad\quad x = \sinh v = \dfrac{e^v - e^{-v}}{2}.$ $\quad\quad$ By (A)

To solve (4) for v, write it as follows:

$$e^v - \frac{1}{e^v} - 2x = 0, \quad \text{or} \quad e^{2v} - 2xe^v - 1 = 0.$$

This is a quadratic equation in e^v. Solving, $e^v = x \pm \sqrt{x^2 + 1}$.

Since e^v is always positive, the negative sign before the radical must be discarded. Hence, using Napierian logarithms, we have (F).

Proof of (G). Let $v = \cosh^{-1} x$. Then

(5) $\quad\quad x = \cosh v = \dfrac{e^v + e^{-v}}{2}.$ $\quad\quad$ By (A)

Clearing and reducing, we have $e^{2v} - 2xe^v + 1 = 0$.
Solving, $\quad\quad e^v = x \pm \sqrt{x^2 - 1}.$

Both values must be retained. Taking logarithms gives (G).

HYPERBOLIC FUNCTIONS

Proof of (H). Let $v = \tanh^{-1} x$. Then

(6) $$x = \tanh v = \frac{e^v - e^{-v}}{e^v + e^{-v}}.$$ **By (C)**

Clearing of fractions and simplifying, the result is

$$(x-1)e^v + (x+1)e^{-v} = 0. \text{ Hence } e^{2v} = \frac{1+x}{1-x}.$$

Taking logarithms, we have (H).

ILLUSTRATIVE EXAMPLE. Transform

(7) $$5 \cosh x + 4 \sinh x$$

into the form $C \cosh (x + a)$, where C and a are constants, and find C and a.

Solution. By (E), Art. 213, we have

(8) $\qquad C \cosh (x + a) = C \cosh x \cosh a + C \sinh x \sinh a.$

Hence (7) will have the desired form if C and a satisfy the equations

(9) $\qquad C \cosh a = 5, \quad C \sinh a = 4.$

Squaring, subtracting, and using (B), Art. 210, we get $C^2 = 9$. Then $C = +3$, since $\cosh a$ must be positive. Also, by division, $\tanh a = \frac{4}{5}$. Hence

$$a = \tanh^{-1} 0.8 = \tfrac{1}{2} \ln 9. \qquad \text{By } (H)$$

Therefore $a = 1.099$ and

(10) $\qquad 5 \cosh x + 4 \sinh x = 3 \cosh (x + 1.099).$

The graph of the function $5 \cosh x + 4 \sinh x$ may be obtained from the graph of $3 \cosh x$ by translating the y-axis to the new origin $(1.099, 0)$. (Compare with Illustrative Example 2, p. 391.)

When x is given, the values of $\sinh^{-1} x$, $\cosh^{-1} x$, or $\tanh^{-1} x$ can be determined by the table on page 416 to not more than three significant figures. For example, $\sinh^{-1} 0.25 = 0.247$; $\cosh^{-1} 3 = \pm 1.76$. For greater accuracy (F), (G), or (H) may be used if tables of Napierian logarithms are at hand.*

217. Derivatives (continued). The formulas, in which v is a function of x, are as follows.

XXXIII $\qquad \dfrac{d}{dx} \sinh^{-1} v = \dfrac{\dfrac{dv}{dx}}{\sqrt{v^2 + 1}}.$ (Any v)

XXXIV $\qquad \dfrac{d}{dx} \cosh^{-1} v = \dfrac{\dfrac{dv}{dx}}{\pm \sqrt{v^2 - 1}}.$ ($v > 1$)

XXXV $\qquad \dfrac{d}{dx} \tanh^{-1} v = \dfrac{\dfrac{dv}{dx}}{1 - v^2}.$ ($v^2 < 1$)

---

* The Smithsonian Mathematical Tables, "Hyperbolic Functions" (1909), give the values of $\sinh u$, $\cosh u$, $\tanh u$, $\operatorname{ctnh} u$ to five significant figures. Values of the corresponding inverse functions to five significant figures may be found from these tables.

Proof of XXXIII. (Compare Art. 75.) Let

$$y = \sinh^{-1} v;$$

then
$$v = \sinh y.$$

Differentiating with respect to y, by **XXVII**,

$$\frac{dv}{dy} = \cosh y;$$

therefore
$$\frac{dy}{dv} = \frac{1}{\cosh y}. \qquad \text{By } (C), \text{ Art. 39}$$

Since v is a function of x, this may be substituted in (A), Art. 38, giving

$$\frac{dy}{dx} = \frac{1}{\cosh y}\frac{dv}{dx} = \frac{1}{\sqrt{v^2+1}}\frac{dv}{dx}.$$

$[\cosh y = \sqrt{\sinh^2 y + 1} = \sqrt{v^2+1}, \text{ by } (B).]$

The proofs of **XXXIV** and **XXXV** are similar. Other formulas are the following.

(*I*) $\qquad \text{ctnh}^{-1} x = \frac{1}{2} \ln \frac{x+1}{x-1}.$ $\qquad (x^2 > 1)$

(*J*) $\qquad \text{sech}^{-1} x = \ln \left(\frac{1}{x} \pm \sqrt{\frac{1}{x^2} - 1} \right).$ $\qquad (0 < x \leqq 1)$

(*K*) $\qquad \text{csch}^{-1} x = \ln \left(\frac{1}{x} + \sqrt{\frac{1}{x^2} + 1} \right).$ $\qquad (x^2 > 0)$

XXXVI $\qquad \dfrac{d}{dx} \text{ctnh}^{-1} v = \dfrac{-\dfrac{dv}{dx}}{v^2 - 1}.$ $\qquad (v^2 > 1)$

XXXVII $\qquad \dfrac{d}{dx} \text{sech}^{-1} v = \dfrac{-\dfrac{dv}{dx}}{\pm v\sqrt{1 - v^2}}.$ $\qquad (0 < v < 1)$

XXXVIII $\qquad \dfrac{d}{dx} \text{csch}^{-1} v = \dfrac{-\dfrac{dv}{dx}}{v^2 \sqrt{1 + \dfrac{1}{v^2}}}.$ $\qquad (v^2 > 0)$

Details of the proofs are called for in Problems 5–8 on the next page.

HYPERBOLIC FUNCTIONS

PROBLEMS

1. Show that the two values of $\cosh^{-1} x$ in (G) differ only in sign.

2. Draw the graph of $y = \frac{1}{2} \sinh^{-1} x$. Check in the figure the values of y and y' when $x = 2$. *Ans.* $y = 0.72$, $y' = 0.2236$.

3. Prove **XXXIII** directly by differentiating (F).

4. Draw the graph of each of the following and check in the figure the values of y and y' for the given value of x.
   (a) $y = \cosh^{-1} x$; $x = 2$.
   (b) $y = \tanh^{-1} x$; $x = -0.75$.

5. Prove **XXXIV** and **XXXV**.

6. Derive (I) and **XXXVI**.

7. Derive (J) and **XXXVII**.

8. Derive (K) and **XXXVIII**.

9. Derive the expansion

$$\tanh^{-1} x = x + \frac{x^3}{3} + \frac{x^5}{5} + \cdots$$

by Art. 195.

10. Given $\sinh x = \tan \phi$. Prove
    (a) $x = \ln (\sec \phi + \tan \phi)$; (b) $\dfrac{dx}{d\phi} = \sec \phi$.

11. Show that $\operatorname{csch}^{-1} v = \sinh^{-1} \dfrac{1}{v}$. Derive **XXXVIII** from **XXXIII**, using this relation.

12. Evaluate $\lim\limits_{x=\infty} x \operatorname{ctnh}^{-1} x$. *Ans.* 1.

13. Evaluate $\lim\limits_{x=\infty} x \operatorname{csch}^{-1} x$.

14. Derive the expansion

$$\sinh^{-1} x = x - \frac{1}{6} x^3 + \frac{1 \cdot 3}{2 \cdot 4} \frac{x^5}{5} - \cdots.$$

15. Evaluate $\lim\limits_{x \to +\infty} (\sinh^{-1} x - \ln x)$. *Ans.* $\ln 2$.

16. Show that $\operatorname{ctnh}^{-1} v = \tanh^{-1} \dfrac{1}{v}$, $\operatorname{sech}^{-1} v = \cosh^{-1} \dfrac{1}{v}$, and verify **XXXVI** and **XXXVII** from these relations.

17. Prove $\dfrac{d}{dx} \tanh^{-1} \dfrac{\tanh a \tan x}{\operatorname{sech} a + \sec x} = \dfrac{\sinh a}{1 + \cosh a \cos x}$.

18. Draw the graphs of (a) $y = \operatorname{ctnh}^{-1} x$; (b) $y = \operatorname{sech}^{-1} x$; (c) $y = \operatorname{csch}^{-1} x$, using the theorem of Problem 28, p. 41.

218. Telegraph line. Assume in a telegraph line that a "steady state" of flow of electricity from A, *the home end*, to B, *the receiving end*, has been established, with perfect insulation and uniform linear leakage. P is any intermediate point. We need to consider:

the *electromotive force* (volts), e.m.f., E_A at A, E_B at B, E at P;
the *current strength* (amperes), I_A at A, I_B at B, I at P;
the *characteristic constants* α and r_0, whose values depend upon the linear resistance and leakage. They are positive numbers.

Let $x = AP$. Then it is shown in books on electrical engineering that E and I are functions of x such that

(1) $$\frac{d^2E}{dx^2} - \alpha^2 E = 0,$$

(2) $$r_0 \alpha I = -\frac{dE}{dx}.$$

We wish to find the e.m.f. and current strength at P. They are

(3) $$E = E_A \cosh \alpha x - r_0 I_A \sinh \alpha x,$$

(4) $$I = I_A \cosh \alpha x - \frac{E_A}{r_0} \sinh \alpha x.$$

Proof. The complete solution of (1) is (Illustrative Example, Art. 210)

(5) $$E = A \cosh \alpha x + B \sinh \alpha x.$$

Substituting in (2), the result is

(6) $$r_0 I = - A \sinh \alpha x - B \cosh \alpha x.$$

But $E = E_A$, $I = I_A$ when $x = 0$. Therefore $A = E_A$, $B = -r_0 I_A$, and (5) and (6) become (3) and (4) respectively.

For the solution in terms of the e.m.f. and current strength at the receiving end, see Problem 2 below.

PROBLEMS

All refer to a telegraph line in a "steady state," and $L = AB$.

1. Given $E_A = 200$ volts, $L = 500$ kilometers, $r_0 = 4000$ ohms, $\alpha = 0.0025$, $I_B = 0$. Find I_A and E_B.

Ans. $I_A = 0.05 \tanh 1.25 = 0.04238$ ampere;
$E_B = 200 \operatorname{sech} 1.25 = 105.8$ volts $= 0.53\, E_A$.

HYPERBOLIC FUNCTIONS

2. If $y = PB =$ distance of P from the receiving end, show that

$$E = E_B \cosh \alpha y + r_0 I_B \sinh \alpha y, \quad I = I_B \cosh \alpha y + \frac{E_B}{r_0} \sinh \alpha y.$$

3. Given $E_A = 200$ volts, $I_A = 0.04$ ampere, $r_0 = 4000$ ohms, $\alpha = 0.0025$. Show that

$$E = 120 \cosh (1.099 - 0.0025 \, x), \quad I = 0.03 \sinh (1.099 - 0.0025 \, x).$$

(See the Illustrative Example, Art. 216. Thus E tends towards a minimum value of 120 volts and I approaches 0 as x approaches 439.6.)

4. Given $E_A = 160$ volts, $I_A = 0.05$ ampere, $r_0 = 4000$ ohms, $\alpha = 0.0025$. Show that

$$E = 120 \sinh (1.099 - 0.0025 \, x), \quad I = 0.03 \cosh (1.099 - 0.0025 \, x).$$

(See the Illustrative Example, Art. 216. Thus E approaches zero and I decreases to a minimum value of 0.03 ampere when x approaches 439.6.)

5. Prove that $\dfrac{d^2 I}{dx^2} - \alpha^2 I = 0$. (Thus E and I are solutions of the same linear differential equation, which has the form $y'' - \alpha^2 y = 0$.)

6. Given $E_A = r_0 I_A$. Prove

(a) $E = E_A e^{-\alpha x}, \quad I = I_A e^{-\alpha x}$;
(b) $E = r_0 I$;
(c) $E \to 0$ when x becomes infinite.

(For example, if $r_0 = 4000$, and the impressed e.m.f. at the home end of the line is 4000 times the current strength, then at *every point* of the line the e.m.f. is 4000 times the current strength and diminishes towards zero as the length of the line is indefinitely increased.)

7. In Problem 6 show that the electromotive force at unit distance along the line from P equals $Ee^{-\alpha}$, where e is the Napierian base.

8. Prove the following.

(a) If $I_B = 0$, then $E_A = r_0 I_A \ctnh \alpha L$.
(b) If $E_B = 0$, then $E_A = r_0 I_A \tanh \alpha L$.

ADDITIONAL PROBLEMS

Derive the following relations.

1. If $E_A > r_0 I_A$ and $\tau = \tanh^{-1} \dfrac{r_0 I_A}{E_A}$, then

$$E = E_A \sech \tau \cosh (\tau - \alpha x), \quad I = I_A \csch \tau \sinh (\tau - \alpha x).$$

2. If $E_A < r_0 I_A$ and $\tau = \tanh^{-1} \dfrac{E_A}{r_0 I_A}$, then

$$E = E_A \csch \tau \sinh (\tau - \alpha x), \quad I = I_A \sech \tau \cosh (\tau - \alpha x).$$

219. Integrals. A list of elementary integrals involving hyperbolic functions and supplementing Art. 128 is given here.

(24) $$\int \sinh v \, dv = \cosh v + C.$$

(25) $$\int \cosh v \, dv = \sinh v + C.$$

(26) $$\int \tanh v \, dv = \ln \cosh v + C.$$

(27) $$\int \operatorname{ctnh} v \, dv = \ln \sinh v + C.$$

(28) $$\int \operatorname{sech}^2 v \, dv = \tanh v + C.$$

(29) $$\int \operatorname{csch}^2 v \, dv = -\operatorname{ctnh} v + C.$$

(30) $$\int \operatorname{sech} v \tanh v \, dv = -\operatorname{sech} v + C.$$

(31) $$\int \operatorname{csch} v \operatorname{ctnh} v \, dv = -\operatorname{csch} v + C.$$

The proofs follow immediately from **XXVII–XXXII**, except for (26) and (27). To prove (26), we have

$$\int \tanh v \, dv = \int \frac{\sinh v}{\cosh v} \, dv \qquad \text{by (C)}$$

$$= \int \frac{d(\cosh v)}{\cosh v} = \ln \cosh v + C.$$

The proof of (27) is similar.

ILLUSTRATIVE EXAMPLE. Derive the formulas

(1) $$\int \operatorname{sech} v \, dv = \arctan (\sinh v) + C;$$

(2) $$\int \operatorname{csch} v \, dv = \ln \tanh \frac{v}{2} + C.$$

Solution. Since $\operatorname{sech} v = \dfrac{1}{\cosh v} = \dfrac{\cosh v}{\cosh^2 v} = \dfrac{\cosh v}{1 + \sinh^2 v}$, by (B)

we have $\int \operatorname{sech} v \, dv = \int \dfrac{\cosh v \, dv}{1 + \sinh^2 v} = \int \dfrac{d(\sinh v)}{1 + \sinh^2 v}$

$$= \int d[\arctan (\sinh v)] = \arctan (\sinh v) + C.$$

HYPERBOLIC FUNCTIONS

To derive (2) we have (compare Art. 131)

$$\operatorname{csch} v = \operatorname{csch} v \, \frac{\operatorname{csch} v + \operatorname{ctnh} v}{\operatorname{csch} v + \operatorname{ctnh} v}$$

$$= \frac{\operatorname{csch}^2 v + \operatorname{csch} v \operatorname{ctnh} v}{\operatorname{ctnh} v + \operatorname{csch} v}.$$

$$\int \operatorname{csch} v \, dv = -\int \frac{-\operatorname{csch}^2 v - \operatorname{csch} v \operatorname{ctnh} v}{\operatorname{ctnh} v + \operatorname{csch} v} \, dv$$

$$= -\int \frac{d(\operatorname{ctnh} v + \operatorname{csch} v)}{\operatorname{ctnh} v + \operatorname{csch} v}$$

$$= -\ln (\operatorname{ctnh} v + \operatorname{csch} v) + C$$

$$= -\ln \left(\frac{\cosh v}{\sinh v} + \frac{1}{\sinh v} \right) + C \qquad \text{by (1), Art. 211}$$

$$= -\ln (\cosh v + 1) + \ln \sinh v + C = \ln \frac{\sinh v}{\cosh v + 1} + C$$

$$= \ln \tanh \frac{v}{2} + C. \qquad \text{By (9), Art. 213}$$

PROBLEMS

Work out the following integrals.

1. $\int \sinh^2 v \, dv = \frac{1}{4} \sinh 2v - \frac{1}{2} v + C.$

2. $\int \cosh^2 v \, dv = \frac{1}{4} \sinh 2v + \frac{1}{2} v + C.$

3. $\int \tanh^2 v \, dv = v - \tanh v + C.$

4. $\int \operatorname{ctnh}^2 v \, dv = v - \operatorname{ctnh} v + C.$

5. $\int \sinh^3 v \, dv = \frac{1}{3} \cosh^3 v - \cosh v + C.$

6. $\int \cosh^3 v \, dv = \frac{1}{3} \sinh^3 v + \sinh v + C.$

7. $\int \tanh^3 v \, dv = \ln \cosh v - \frac{1}{2} \tanh^2 v + C.$

8. $\int \tanh^4 v \, dv = v - \tanh v - \frac{1}{3} \tanh^3 v + C.$

9. $\int \operatorname{csch}^3 v \, dv = -\frac{1}{2} \operatorname{csch} v \operatorname{ctnh} v - \frac{1}{2} \ln \tanh \frac{v}{2} + C.$

10. $\int x \sinh x \, dx = x \cosh x - \sinh x + C.$

11. $\int \cos x \sinh x \, dx = \frac{1}{2} (\cos x \cosh x + \sin x \sinh x) + C.$

12. $\int \sinh (mx) \sinh (nx) = \frac{1}{m^2 - n^2} [m \sinh (nx) \cosh (mx) - n \cosh (nx) \sinh (mx)] + C.$

DIFFERENTIAL AND INTEGRAL CALCULUS

Work out each of the following integrals.

13. $\int \sinh^4 x \, dx.$

14. $\int \text{sech}^4 2x \, dx.$

15. $\int \sin x \cosh x \, dx.$

16. $\int x \cosh x \, dx.$

17. $\int x^2 \cosh x \, dx.$

18. $\int e^x \sinh x \, dx.$

19. $\int e^{ax} \cosh x \, dx.$

Work out each of the following, using the hyperbolic substitution indicated. (Compare Art. 135.)

20. $\int \sqrt{x^2 - 4} \, dx$; $x = 2 \cosh v.$ Ans. $\frac{1}{2} x \sqrt{x^2 - 4} - 2 \cosh^{-1} \frac{1}{2} x + C.$

21. $\int \dfrac{du}{(a^2 - u^2)^{\frac{3}{2}}}$; $u = a \tanh z.$

22. $\int \dfrac{(x+2)dx}{x^2 + 2x + 5}$; $x = 2 \sinh z - 1.$

23. The arc of the catenary $y = a \cosh \dfrac{x}{a}$ from $(0, a)$ to (x, y) is revolved about the y-axis. Find the area of the curved surface generated, using hyperbolic functions.

24. Find the centroid of the hyperbolic sector OAP_1 in Fig. 2, Art. 215. (Compare Problem 12, p. 337.)   Ans. $\bar{x} = \dfrac{2}{3} a \dfrac{\sinh v_1}{v_1}$, $\bar{y} = \dfrac{2}{3} a \dfrac{\cosh v_1 - 1}{v_1}.$

220. Integrals (continued). From XXXIII–XXXVIII we may derive integrals. Some of them we have already met in Art. 128. Their values are now expressible in terms of inverse hyperbolic functions.

(32) $\quad \int \dfrac{dv}{\sqrt{v^2 + a^2}} = \sinh^{-1} \dfrac{v}{a} + C.$ $\hfill (v \text{ any value})$

(33) $\quad \int \dfrac{dv}{\sqrt{v^2 - a^2}} = \cosh^{-1} \dfrac{v}{a} + C.$ $\hfill (v \geq a)$

(34) $\quad \int \dfrac{dv}{a^2 - v^2} = \dfrac{1}{a} \tanh^{-1} \dfrac{v}{a} + C.$ $\hfill (v^2 < a^2)$

(35) $\quad \int \dfrac{dv}{v^2 - a^2} = -\dfrac{1}{a} \text{ctnh}^{-1} \dfrac{v}{a} + C.$ $\hfill (v^2 > a^2)$

(36) $\quad \int \dfrac{dv}{v\sqrt{a^2 - v^2}} = -\dfrac{1}{a} \text{sech}^{-1} \dfrac{v}{a} + C.$ $\hfill (0 < v < a)$

(37) $\quad \int \dfrac{dv}{v\sqrt{v^2 + a^2}} = -\dfrac{1}{a} \text{csch}^{-1} \dfrac{v}{a} + C.$ $\hfill (v \text{ any value})$

HYPERBOLIC FUNCTIONS

(38) $\quad \int \sqrt{v^2 + a^2}\, dv = \dfrac{v}{2}\sqrt{v^2 + a^2} + \dfrac{a^2}{2}\sinh^{-1}\dfrac{v}{a} + C.$

(39) $\quad \int \sqrt{v^2 - a^2}\, dv = \dfrac{v}{2}\sqrt{v^2 - a^2} - \dfrac{a^2}{2}\cosh^{-1}\dfrac{v}{a} + C.$

In (33) and (39) the positive value of the inverse hyperbolic cosine must be used, and in (36) the positive value of the inverse hyperbolic secant.

Proofs of (32) and (33). Let $x = \dfrac{v}{a}$ in (*F*). Then

$$\sinh^{-1}\dfrac{v}{a} = \ln\left(\dfrac{v}{a} + \sqrt{\dfrac{v^2}{a^2} + 1}\right) = \ln(v + \sqrt{v^2 + a^2}) - \ln a.$$

Hence

(1) $\qquad \ln(v + \sqrt{v^2 + a^2}) = \sinh^{-1}\dfrac{v}{a} + \ln a.$

In the same way, from (*G*) we get

(2) $\qquad \ln(v + \sqrt{v^2 - a^2}) = \cosh^{-1}\dfrac{v}{a} + \ln a.$

Using these results in the right-hand member of (21), p. 193, we get (32) and (33).

Proofs of (34) and (35). Let $x = \dfrac{v}{a}$ in (*H*). Then

(3) $\qquad \dfrac{1}{2}\ln\dfrac{a+v}{a-v} = \tanh^{-1}\dfrac{v}{a}.$

Then (34) follows from (3) and (19 *a*), p. 192.
In the same way, from (*I*) and (19), p. 192, we get (35).

$$\left[\text{In (19)},\ \ln\dfrac{v-a}{v+a} = -\ln\dfrac{v+a}{v-a}.\right]$$

Proof of (36). Since

$$d\left(-\dfrac{1}{a}\operatorname{sech}^{-1}\dfrac{v}{a}\right) = \dfrac{\dfrac{1}{a}d\left(\dfrac{v}{a}\right)}{\pm\dfrac{v}{a}\sqrt{1 - \dfrac{v^2}{a^2}}} = \dfrac{dv}{\pm v\sqrt{a^2 - v^2}},\ \text{by XXXVII}$$

we have $\displaystyle\int \dfrac{dv}{v\sqrt{a^2 - v^2}} = -\dfrac{1}{a}\operatorname{sech}^{-1}\dfrac{v}{a}$ if the positive sign before the radical is chosen. The proof of (37) is similar.

Formulas (38) and (39) follow from (23), p. 193, using (1) and (2).

REMARK. Since

$$\operatorname{ctnh}^{-1}\dfrac{v}{a} = \tanh^{-1}\dfrac{a}{v},\quad \operatorname{sech}^{-1}\dfrac{v}{a} = \cosh^{-1}\dfrac{a}{v},\quad \operatorname{csch}^{-1}\dfrac{v}{a} = \sinh^{-1}\dfrac{a}{v},$$

the integrals (35)–(37) may also be expressed in terms of the functions most convenient for use of the table of Art. 212.

ILLUSTRATIVE EXAMPLE. Derive (37) by means of the substitution $v = a \operatorname{csch} z$. (Compare Art. 135.)

Solution. We have

$$\sqrt{v^2 + a^2} = \sqrt{a^2 \operatorname{csch}^2 z + a^2} = a \operatorname{ctnh} z. \quad \text{By (2), Art. 211}$$

Also $\quad dv = -a \operatorname{csch} z \operatorname{ctnh} z \, dz.$ By XXXII

Therefore $\quad \displaystyle\int \frac{dv}{v\sqrt{v^2 + a^2}} = \int \frac{-a \operatorname{csch} z \operatorname{ctnh} z \, dz}{a \operatorname{csch} z \cdot a \operatorname{ctnh} z} = -\frac{1}{a} z + C.$

Since $z = \operatorname{csch}^{-1} \dfrac{v}{a}$, we have (37).

PROBLEMS

1. In the figure the curve is the equilateral hyperbola $x^2 - y^2 = a^2$. Using Art. 142, prove that
(a) area $AMP = $ triangle $OMP - \dfrac{1}{2} a^2 \cosh^{-1} \dfrac{x}{a}$;

(b) sector $OAP = \dfrac{1}{2} a^2 \cosh^{-1} \dfrac{x}{a} = \dfrac{1}{2} a^2 v$,

if $x = a \cosh v$. (Thus we have an alternative proof of the theorem of Art. 215.)

2. Derive each of the following power series by integration (Art. 196).

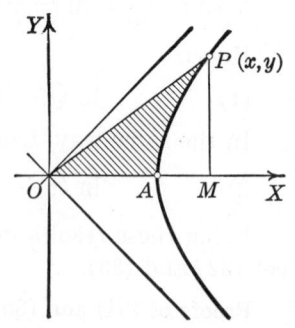

(a) $\tanh^{-1} x = x + \dfrac{1}{3} x^3 + \dfrac{1}{5} x^5 + \cdots + \dfrac{x^{2n-1}}{2n-1} + \cdots \qquad (x^2 < 1)$

(b) $\sinh^{-1} x = x - \dfrac{1}{2} \dfrac{x^3}{3} + \dfrac{1 \cdot 3}{2 \cdot 4} \dfrac{x^5}{5} - \dfrac{1 \cdot 3 \cdot 5}{2 \cdot 4 \cdot 6} \dfrac{x^7}{7} + \cdots \qquad (x^2 < 1)$

Work out the following integrals.

3. $\displaystyle\int \sinh^{-1} x \, dx = x \sinh^{-1} x - \sqrt{1 + x^2} + C.$

4. $\displaystyle\int \tanh^{-1} x \, dx.$ 	**5.** $\displaystyle\int x \cosh^{-1} x \, dx.$

Work out, using hyperbolic functions.

6. $\displaystyle\int_0^1 \frac{dx}{\sqrt{16 x^2 + 9}}.$ 	**7.** $\displaystyle\int_0^2 \frac{dx}{25 - 4 x^2}.$ 	**8.** $\displaystyle\int_3^5 \sqrt{x^2 - 9} \, dx.$

9. Find the length of arc for the parabola $x^2 = 4y$ from $(0, 0)$ to $(4, 4)$, using hyperbolic functions. *Ans.* $\sqrt{20} + \sinh^{-1} 2 = 5.92.$

10. Find the area bounded by the catenary $y = a \cosh \dfrac{x}{a}$ and the line $y = 2a$.

11. The downward acceleration a of a falling body is given by $a = 32 - \tfrac{1}{2} v^2$, and $v = 0$, $s = 0$, when $t = 0$. Find v and s.
Ans. $v = 8 \tanh 4t$, $s = 2 \ln \cosh 4t$.

HYPERBOLIC FUNCTIONS

221. The gudermannian. The function arc tan (sinh v), which occurs frequently in mathematics (for example, in (1), Illustrative Example, Art. 219), is called the *gudermannian** of v. The symbol used is gd v (read " gudermannian of v "). Thus

(1) $\qquad$ gd v = arc tan (sinh v).

The derivative is

XXXIX $\qquad \dfrac{d}{dx}$ gd v = sech $v \dfrac{dv}{dx}$,

assuming v to be a function of x.

Proof. Differentiating (1), we get

$$\frac{d}{dv} \text{gd } v = \frac{\cosh v}{1 + \sinh^2 v}.$$ By **XXII** and **XXVII**

But $\qquad 1 + \sinh^2 v = \cosh^2 v$, $\qquad$ by (*B*)

and $\qquad \dfrac{1}{\cosh v} =$ sech v. $\qquad$ By (1), Art. 211

Then $\qquad \dfrac{d}{dx}$ gd v = sech $v \dfrac{dv}{dx}$. $\qquad$ By (*A*), Art. 38

From the definition (1) and Art. 77, we have

(2) gd (0) = 0 ; gd ($-v$) = $-$gd v ; gd ($+\infty$) = $\tfrac{1}{2}\pi$; gd ($-\infty$) = $-\tfrac{1}{2}\pi$.

When v increases, gd v increases (since sech $v > 0$). Its value lies between $-\tfrac{1}{2}\pi$ and $+\tfrac{1}{2}\pi$. Some values are given in the accompanying table. By (1), Art. 219,

(40) $\qquad \displaystyle\int$ sech $v \, dv =$ gd $v + C$.

To find the inverse function (Art. 39) let

(3) $\quad \phi =$ arc tan (sinh v), $\quad (-\tfrac{1}{2}\pi < \phi < \tfrac{1}{2}\pi)$

and solve for v. The result is

(4) $\qquad v = \sinh^{-1}$ (tan ϕ).

v	gd v
0.5	0.480
1.0	0.864
1.5	1.132
2.0	1.302
2.5	1.407
3.0	1.471
3.5	1.510
4.0	1.534
4.5	1.549
5.0	1.557

From (3) we have tan $\phi =$ sinh v. Since $\cosh^2 v = 1 + \sinh^2 v$, by (*B*), the trigonometric functions of ϕ, when $v > 0$, can be read off from the accompanying right triangle. Thus sin $\phi =$ tanh v, cos $\phi =$ sech v, etc.

The inverse function † (4) may be written

(5) $\qquad v = \ln (\sec \phi + \tan \phi)$.

---
* Named after the mathematician Gudermann. His papers were published in 1830.
† The symbol gd$^{-1}\phi$ is used by some writers ($v =$ gd$^{-1}\phi$).

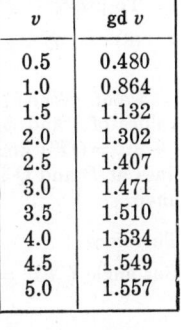

Proof. Replace x in (F), Art. 216, by $\tan \phi$, and note that $1 + \tan^2 \phi$ is equal to $\sec^2 \phi$, by (2), p. 2.

Conversely, given (5), then $\phi = \text{gd } v$.

Proof. Changing to exponentials, (5) becomes

$$\sec \phi + \tan \phi = e^v, \quad \text{or} \quad \tan \phi - e^v = -\sec \phi.$$

Squaring both members, substituting $\sec^2 \phi = 1 + \tan^2 \phi$, and reducing, the result is
$$-2 \tan \phi \, e^v + e^{2v} = 1.$$

Solving for $\tan \phi$, we get

$$\tan \phi = \frac{e^v - e^{-v}}{2} = \sinh v. \qquad \text{By } (A)$$

Hence $\qquad \phi = \arctan (\sinh v) = \text{gd } v.$ $\qquad$ Q.E.D.

ILLUSTRATIVE EXAMPLE. In the tractrix let
$a = $ length of the tangent PT (constant by definition);
$t = $ intercept of the tangent on the x-axis;
$\phi = $ angle between the tangent line directed upward and the y-axis;
$\phi = 0$ when $t = 0$.
Then $B(0, a)$ is on the curve.
To prove
(6) $\qquad \phi = \text{gd}\left(\dfrac{t}{a}\right).$

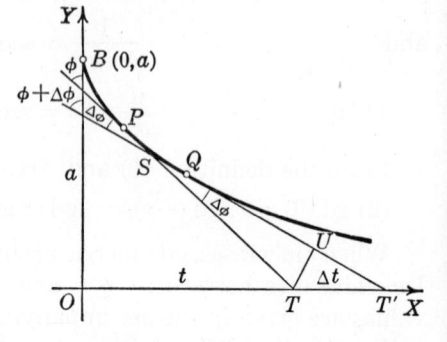

Proof. When t is given, ϕ is determined. Hence ϕ is a function of t. Let the values of t and ϕ for the tangent line at Q, a point near P, be, respectively, $t + \Delta t (= OT')$ and $\phi + \Delta \phi$. Draw TU perpendicular to QT'. Let the tangent lines at P and Q intersect at S. Then in the right triangles UTT' and STU we have $\qquad TU = TT' \cos UTT'; \quad TU = TS \sin TSU.$

Therefore $\qquad TT' \cos UTT' = TS \sin TSU.$

But angle $UTT' = \phi + \Delta \phi$, angle $TSU = \Delta \phi$, $TT' = \Delta t$. Hence

$$\Delta t \cos (\phi + \Delta \phi) = TS \sin \Delta \phi.$$

Let Q move along the curve towards P, and let $\Delta \phi \to 0$. Then Δt and $\Delta \phi$ are infinitesimals. Also S approaches P and $TS \to a$. Hence, by the Replacement Theorem, Art. 98, and (B), Art. 68, we have

$$dt \cos \phi = a \, d\phi, \quad \text{or} \quad d\frac{t}{a} = \sec \phi \, d\phi.$$

Integrating, and remembering that $\phi = 0$ when $t = 0$, we get

$$\frac{t}{a} = \ln (\sec \phi + \tan \phi).$$

Therefore, by (5), $\qquad \phi = \text{gd}\left(\dfrac{t}{a}\right).$ $\qquad$ Q.E.D.

HYPERBOLIC FUNCTIONS

PROBLEMS

1. The figure shows the circle $x^2 + y^2 = 1$ and equilateral hyperbola $x^2 - y^2 = 1$ in the first quadrant. From M, the foot of the ordinate MP of any point P on the hyperbola, draw MT tangent to the circle. Let $\tfrac{1}{2}v=$ area of the hyperbolic sector OAP (Art. 215), and $\phi =$ angle AOT. Prove $\phi = \text{gd } v$.

2. Prove

(a) $\text{gd } v = 2 \arctan e^v - \tfrac{1}{2}\pi$;

(b) $\int \sinh v \tanh v \, dv = \sinh v - \text{gd } v + C$.

3. Draw the graph of $y = \text{gd } x$. Calculate y and y' when $x = 1$. See figure.

Ans. $y = 0.86$, $y' = 0.65$.

4. In the Illustrative Example, p. 436, if P is (x, y), prove that

$$x = t - a \sin \phi, \quad y = a \cos \phi.$$

From these and (6) derive the parametric equations

$$x = t - a \tanh \frac{t}{a}, \quad y = a \text{ sech } \frac{t}{a}$$

for the tractrix. Find the rectangular equation also.

5. Derive $\int \text{sech}^3 v \, dv = \tfrac{1}{2} \text{ sech } v \tanh v + \tfrac{1}{2} \text{ gd } v + C$.

6. If the length of the tangent of a curve (Art. 43) is constant $(= a)$,

(a) prove $\dfrac{dy}{dx} = -\dfrac{y}{\sqrt{a^2 - y^2}}$.

(b) Integrate by the hyperbolic substitution $y = a \text{ sech } \dfrac{t}{a}$ and the condition $x = 0$ when $t = 0$, and in this way derive the equations of the tractrix in Problem 4.

7. Evaluate each of the following by differentiation.

(a) $\lim\limits_{x \to 0} \dfrac{\text{gd } x - x}{x^3}$;

(b) $\lim\limits_{x \to 0} \dfrac{\text{gd } x - \sin x}{x^5}$.

Ans. (a) $-\tfrac{1}{6}$; (b) $\tfrac{1}{30}$.

8. Using the expansion of Problem 14, Art. 215, we have

$$\text{gd } x = x - \tfrac{1}{6} x^3 + \tfrac{1}{24} x^5 - \tfrac{61}{5040} x^7 + \cdots.$$

Calculate the value of gd 0.5 to four places of decimals. Ans. 0.4804.

9. The equation (5), p. 435, may be written

$$v = \ln \tan (\tfrac{1}{4}\pi + \tfrac{1}{2}\phi).$$

Prove this statement, making use of (2), (4), and (5), pp. 2, 3.

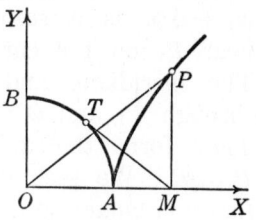

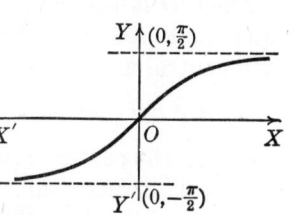

222. Mercator's Chart. The figure shows a portion (one eighth) of a sphere representing the earth. North Pole N, equator EF, longitude θ_1 and latitude ϕ_1 of the point P_1 are indicated. Q, with longitude $\theta_1 + \Delta\theta$, latitude $\phi_1 + \Delta\phi$, is a second point, near P_1 on the curve P_1QV. The meridians and parallels through P_1 and Q are shown. They form the quadrilateral P_1SQR. We seek expressions for the circular arcs RQ and P_1R.

Since O is the center of the equal arcs RQ and P_1S, each with central angle $\Delta\phi$, we have

(1) $\quad$ arc $RQ = $ arc $P_1S = a\,\Delta\phi$.

Also, C is the center of arc P_1R, with central angle $\Delta\theta$. Hence arc $P_1R = CP_1 \cdot \Delta\theta$. But, in the right triangle OP_1C (right angle at C), $CP_1 = a\cos\phi_1$. Hence

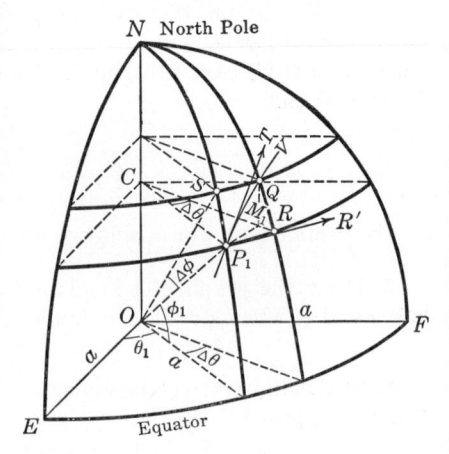

$$\text{arc } P_1R = a\cos\phi_1\,\Delta\theta.$$

The line P_1R' is tangent at P_1 to the parallel P_1R. The line P_1T is tangent* at P_1 to the curve P_1QV. The angle at P_1 between the curve and the parallel is the angle $R'P_1T$. Then

(2) $\quad \tan R'P_1T = \sec\phi_1 \left(\dfrac{d\phi}{d\theta}\right)_1,$

the value of the derivative being found from the equation

(3) $\quad \theta = f(\phi)$

satisfied by the latitude and longitude of each point of the curve P_1QV.

Proof of (2). By the Replacement Theorem, Art. 98, it can be shown † that

(4) $\quad \tan R'P_1T = \lim\limits_{\Delta\theta \to 0} \dfrac{\text{arc } RQ}{\text{arc } P_1R}.$

Substituting the values from (1), we get (2).

---

* Defined as in Art. 28 as the limiting position of the secant through P_1 and Q when Q approaches P_1 along the curve P_1Q.

† The details are indicated in the Additional Problems (p. 443). Note that arc RQ and arc P_1R are, respectively, opposite and adjacent to the angle at P_1 in the curvilinear triangle P_1RQ.

HYPERBOLIC FUNCTIONS

On Mercator's* Chart of the earth's surface the point with latitude ϕ, longitude θ, is represented by the point (x, y) such that

(5) $\quad\quad\quad x = \theta, \quad y = \ln(\sec\phi + \tan\phi)$,

or, inversely,

(6) $\quad\quad\quad \theta = x, \quad \phi = \operatorname{gd} y.$ $\quad\quad$ By Art. 221

In (5) and (6), θ and ϕ are expressed in radians. Meridians (θ = constant) are represented on the chart by lines parallel to the y-axis, parallels (ϕ = constant) by lines parallel to the x-axis. The curve (3) is given by the parametric equations

(7) $\quad\quad\quad x = f(\phi), \quad y = \ln(\sec\phi + \tan\phi).$

Theorem. *The angle between a curve on the sphere and an intersecting parallel is unchanged by mapping.*

Proof. Let (x_1, y_1) be the point on (7) where $\phi = \phi_1$. The parallel becomes the line $y = y_1$ on the chart. Hence we have to prove that the curve (7) is such that

(8) $\quad\quad\quad \left(\dfrac{dy}{dx}\right)_1 = \sec\phi_1\left(\dfrac{d\phi}{d\theta}\right)_1.$ $\quad\quad$ Using (2)

From (7) and (3), we get

$$\frac{dy}{d\phi} = \sec\phi, \quad \frac{dx}{d\phi} = f'(\phi) = \frac{d\theta}{d\phi}.$$

Then (8) follows from (A), Art. 81, and (C), Art. 39. $\quad\quad$ Q.E.D.

Two important corollaries follow.

Cor. I. The angle at P_1 on the sphere formed by two curves, P_1QR and $P_1Q'R'$, will equal the angle on the chart at (x_1, y_1) formed by the corresponding curves. Hence angles remain unchanged by mapping.

Cor. II. A straight line on the chart with slope $\tan\alpha$ corresponds to a curve on the sphere cutting all parallels under the same angle α. This curve is called a *rhumb line* (or *loxodrome*).

Along a rhumb line

(9) $\quad\quad\quad \phi = \operatorname{gd}(\theta\tan\alpha + b).$

This follows from (6) and $y = x\tan\alpha + b$. The course of a ship proceeding always in the same direction lies along a rhumb line. In the representation (5), θ, and therefore x, has values from $-\pi$ to $+\pi$, inclusive. On the other hand, y may have any value (Art. 221). Hence the entire surface of the earth is mapped on the strip of the xy-plane determined by the lines $x = -\pi$ and $x = +\pi$.

---

* Gerardus Mercator (1512–1594), a noted cartographer, published his Chart of the World in 1569. His name is the Latinized form of Gerhard Kremer.

440   DIFFERENTIAL AND INTEGRAL CALCULUS

By the table of Art. 221 we may find the latitude in degrees of the parallels which are given on the chart by the lines $y =$ constant.

y	0	0.5	1.0	1.5	2	3	4	5
lat.	0°	27° 31'	49° 36'	64° 51'	74° 35'	84° 18'	87° 54'	89° 14'

A rhumb line is represented on the map by a series of parallel line segments such as AA_1, BB_1, CC_1, etc. in the figure, where BA_1, CB_1, etc. are parallel to the x-axis. The representation is "conformal"; that is, the *shape* of small areas is preserved. This follows from Cor. I. For example, a triangular figure * on the earth's surface bounded by rhumb lines will be a triangle on the map, and corresponding angles in the two figures will be equal. But the distortion of an area on the earth's surface by mapping depends upon its distance from the equator. Problem 4, p. 442, brings out this point.

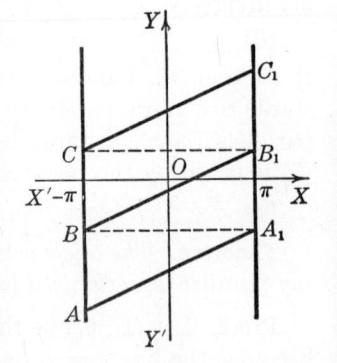

223. Relations between trigonometric and hyperbolic functions. Let the exponent v of the exponential function e^v be a *complex number* $x + iy$ (x and y real numbers; $i = \sqrt{-1}$). Then we assume as a definition

(1) $\qquad e^{x+iy} = e^x e^{iy} = e^x(\cos y + i \sin y).$

If $x = 0$, we have (see p. 391)

(2) $\qquad e^{iy} = \cos y + i \sin y.$

Change y to $-y$. Then (2) becomes

(3) $\qquad e^{-iy} = \cos y - i \sin y.$

Solving (2) and (3) for $\sin y$ and $\cos y$, the results are

(4) $\qquad \sin y = \dfrac{e^{iy} - e^{-iy}}{2i}, \quad \cos y = \dfrac{e^{iy} + e^{-iy}}{2}.$

Thus the sine and cosine of a real variable are expressed in terms of exponential functions with imaginary exponents.

Formulas (4) and (A) suggest definitions of the functions concerned when the variable is any complex number z. These definitions are

(5) $\qquad \sin z = \dfrac{e^{iz} - e^{-iz}}{2i}, \quad \cos z = \dfrac{e^{iz} + e^{-iz}}{2},$

$\qquad \sinh z = \dfrac{e^z - e^{-z}}{2}, \quad \cosh z = \dfrac{e^z + e^{-z}}{2}.$

---

* The lines $x = -\pi$ and $x = +\pi$ represent the same meridian (180° W. or 180° E.). It is assumed that this meridian does not cross the curvilinear triangle. In the figure, A_1 and B represent the same point on the earth, as do also B_1 and C.

HYPERBOLIC FUNCTIONS

The other trigonometric and hyperbolic functions of z are defined by the same ratios as are used when the variable is a real number. From (5) we may prove the following:

(L) $\qquad \sinh iz = i \sin z, \quad \cosh iz = \cos z.$

$\left[\sinh iz = \dfrac{e^{iz} - e^{-iz}}{2} = i \sin z, \text{ using (5); etc.} \right]$

From (L), by division, we get

(6) $\qquad \tanh iz = i \tan z.$

The similarity of many formulas in this chapter to others for trigonometric functions is explained by the relations (L) and (6) (see Illustrative Example 2). The right-hand members of (5) are expressible as complex numbers whose real parts involve only trigonometric and hyperbolic functions of real variables. This appears below in Illustrative Example 1.

ILLUSTRATIVE EXAMPLE 1. Derive the formula

(7) $\qquad \sinh(x + iy) = \sinh x \cos y + i \cosh x \sin y.$

Solution. By (5), if $z = x + iy$, we have

(8) $\qquad \sinh(x + iy) = \dfrac{e^{x+iy} - e^{-x-iy}}{2}$

(9) $\qquad = \dfrac{e^x(\cos y + i \sin y) - e^{-x}(\cos y - i \sin y)}{2}.$ By (1)

By (1), Art. 210, if $v = x$, we have

$e^x = \cosh x + \sinh x, \quad e^{-x} = \cosh x - \sinh x.$

Substitute these values in (9), and reduce. The result is (7). Changing i to $-i$, (7) becomes

$\sinh(x - iy) = \sinh x \cos y - i \cosh x \sin y.$

The form of the right-hand member here and in (7) should be noticed.

ILLUSTRATIVE EXAMPLE 2. Prove directly by (5) the relations

$\sin^2 z + \cos^2 z = 1, \quad \cosh^2 z - \sinh^2 z = 1.$

Solution. The details are the same as in the proof of (B), Art. 210. The first relation may be derived from the second as follows:
Let $z = iv$. Then $\cosh^2 iv - \sinh^2 iv = 1.$ But, by (L), $\cosh iv = \cos v$, $\sinh iv = i \sin v$. Hence $\cos^2 v + \sin^2 v = 1.$

PROBLEMS

1. Using differentials, show that the distance apart on Mercator's Chart of the lines parallel to the x-axis which represent the parallels at latitudes ϕ_1 and $\phi_1 + \Delta\phi$, respectively, varies as sec ϕ_1.

2. Along a rhumb line $\phi = \text{gd}\,(\theta \tan \alpha + b)$. Prove by differentiation that $\tan \alpha = \sec \phi \, \dfrac{d\phi}{d\theta}.$

442   DIFFERENTIAL AND INTEGRAL CALCULUS

3. The altitude h of the zone on the sphere bounded by the parallels $\phi = \phi_2$, $\phi = \phi_1$ ($\phi_2 > \phi_1$) is $a(\sin \phi_2 - \sin \phi_1)$ (see figure, p. 438). If $y = y_2$, $y = y_1$ are the corresponding parallels on the map, prove the following.

(a) $h = a(\tanh y_2 - \tanh y_1)$;

(b) $dy = \dfrac{1}{a} \sec^2 \phi_1 \, dh$, if $\phi_2 = \phi_1 + d\phi$.

4. Using (b), Problem 3, show that equal zones of small altitude whose lower bases are parallels at latitudes 0, 30°, 45°, 60°, respectively, map into rectangles whose areas are as 3 : 4 : 6 : 12. (The area of a zone equals its altitude times the circumference of a great circle.)

5. Describe the direction of a curve on the sphere (a) if $\dfrac{d\phi}{d\theta} = 0$; (b) if $\dfrac{d\phi}{d\theta}$ becomes infinite.

6. Derive each of the following formulas by the method used in Illustrative Example 1.

(a) $\cosh(x + iy) = \cosh x \cos y + i \sinh x \sin y$;
(b) $\sin(x + iy) = \sin x \cosh y + i \cos x \sinh y$;
(c) $\cos(x + iy) = \cos x \cosh y - i \sin x \sinh y$.

From these write the values of $\cosh(x - iy)$, $\sin(x - iy)$, $\cos(x - iy)$.

7. Prove   (a) $\sinh\left(i\dfrac{\pi}{2} \pm x\right) = i \cosh x$;

(b) $\cosh\left(i\dfrac{\pi}{2} \pm x\right) = \pm i \sinh x$.

8. Evaluate each of the following to two places of decimals.

(a) $\sinh(1.5 + i)$;   (b) $\cosh(1 - i)$;
(c) $\cos(0.8 + 0.5\,i)$;   (d) $\sin(0.5 + 0.8\,i)$.

Ans. (a) $1.15 + 1.98\,i$; (c) $0.78 - 0.37\,i$.

9. Evaluate each of the following indeterminate forms.

(a) $\lim\limits_{x \to 0} \dfrac{\sinh nx}{\sinh x}$.   *Ans.* n.

(b) $\lim\limits_{x \to 0} \dfrac{\sinh^2 x}{x^2}$. 1.

(c) $\lim\limits_{x \to 0} \dfrac{\operatorname{sech} x - 1}{x^2}$. $-\tfrac{1}{2}$.

(d) $\lim\limits_{x \to 0} \dfrac{\sinh x - x}{x^3}$. $\tfrac{1}{6}$.

(e) $\lim\limits_{x \to 0} \dfrac{x - \tanh x}{x^3}$. $\tfrac{1}{3}$.

HYPERBOLIC FUNCTIONS

ADDITIONAL PROBLEMS

1. In the figure of Art. 222, P_1M_1 is drawn perpendicular to CR, and therefore perpendicular to the plane of the meridian NQR. Then triangle P_1QM_1 is a right triangle (the chord P_1Q is not shown), and $\tan M_1P_1Q = \dfrac{M_1Q}{P_1M_1}$. When $\Delta\theta \to 0$, the line P_1M_1 (produced) approaches the tangent P_1R', and angle M_1P_1Q approaches angle $R'P_1T$. Therefore

(10) $$\tan R'P_1T = \lim_{\Delta\theta \to 0} \frac{M_1Q}{P_1M_1}.$$

Compare with (4), Art. 222, and show that

(a) $\lim\limits_{\Delta\theta \to 0} \dfrac{P_1M_1}{\text{arc } P_1R} = 1$ (see Fig. 1);

(b) $\lim\limits_{\Delta\theta \to 0} \dfrac{M_1Q}{\text{arc } RQ} = 1$ (See Fig. 2, which shows the plane of the meridian NQR). In triangle M_1QR show that M_1R is an infinitesimal of higher order than QR when $\Delta\theta$ and $\Delta\phi$ are of the same order (Art. 99). Then see Problem, p. 148.

FIG. 1

FIG. 2

Using (a) and (b) and the Replacement Theorem, Art. 98, (10) becomes (4), Art. 222.

2. If ds_1 is the element of the length of arc for a curve on the sphere of Art. 222, prove that $ds_1{}^2 = a^2(d\phi^2 + \cos^2\phi\, d\theta^2)$. (In the figure of Art. 222, $(\text{chord } P_1Q)^2 = \overline{P_1M_1}^2 + \overline{M_1Q}^2$ and $\lim \dfrac{\text{chord } P_1Q}{\text{arc } P_1Q} = 1$.)

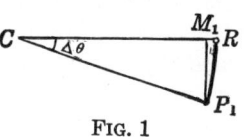

3. If ds is the differential of the arc of a curve on Mercator's Chart, show that $ds^2 = \sec^2\phi(d\phi^2 + \cos^2\phi\, d\theta^2)$. (Comparing with Problem 2, we have $ds_1{}^2 = a^2 \cos^2\phi\, ds^2$.)

4. Find the length of a rhumb line between points whose difference of latitude is $\Delta\phi$.   Ans. $a \csc \alpha\, \Delta\phi$. ($a =$ radius of the earth.)

5. Prove that the first four formulas in (4), p. 3, and (D), (E), Art. 213, hold when x, y, v, w are replaced by complex numbers. (Use the definitions (5).)

6. Prove the formulas of Problem 6, p. 442, by using the results in Additional Problem 5 and (L).

7. Prove that $\tanh(x + iy) = \dfrac{\sinh 2x + i \sin 2y}{\cosh 2x + \cos 2y}$.

8. Derive the formula for $\tan(x + iy)$ from the result in the preceding problem.

CHAPTER XXIII

PARTIAL DIFFERENTIATION

224. Functions of several variables. Continuity. The preceding chapters have been devoted to applications of the calculus to functions of one variable. We now turn to functions of more than one independent variable. Simple examples of such functions are afforded by formulas from elementary mathematics. Thus, in the relation for the volume v of a right circular cylinder,

(1) $$v = \pi x^2 y,$$

v is a function of the two independent variables x ($=$ radius) and y ($=$ altitude). Again, in the formula for the area u of an oblique plane triangle,

(2) $$u = \tfrac{1}{2} xy \sin \alpha,$$

u is a function of the three independent variables x, y, and α, representing, respectively, two sides and the included angle.

Obviously, in (1), as well as in (2), the values which can be assigned to the variables in the right-hand member are entirely independent of one another.

The relation

(3) $$z = f(x, y)$$

can be represented graphically by a surface, the locus of the equation (3) obtained by interpreting x, y, z as rectangular coördinates, as in solid analytic geometry. This surface is the graph of the function of two variables, $f(x, y)$.

A function $f(x, y)$ of two independent variables x and y is defined as *continuous for* $x = a$, $y = b$, when

(A) $$\lim_{\substack{x \to a \\ y \to b}} f(x, y) = f(a, b),$$

no matter in what way x and y approach their respective limits a and b.

This definition is sometimes roughly summed up in the statement that *a very small change in one or both of the variables produces a very small change in the value of the function.**

---

* This will be better understood if the student again reads over Art. 17 on continuous functions of a single variable.

We may illustrate this geometrically by considering the surface represented by the equation

(3) $$z = f(x, y).$$

Consider a fixed point P on the surface where $x = a$ and $y = b$. Denote by Δx and Δy the increments of the variables x and y, and by Δz the corresponding increment of the function z, the coördinates of P' being

$$(x + \Delta x, y + \Delta y, z + \Delta z).$$

At P the value of the function is

$$z = f(a, b) = MP.$$

If the function is continuous at P, then however Δx and Δy may approach zero as a limit, Δz will also approach zero as a limit. That is, $M'P'$ will approach coincidence with MP, the point P' approaching the point P on the surface from any direction whatever.

A similar definition holds for a continuous function of more than two variables.

In what follows, only values of the variables are considered for which a function is continuous.

225. Partial derivatives. In the relation

(1) $$z = f(x, y),$$

we may hold y fast and let x alone vary. Then z becomes a function of one variable x, and we may form its derivative in the usual manner. The notation is

$$\frac{\partial z}{\partial x} = \text{partial derivative of } z \text{ with respect to } x \text{ } (y \text{ remains constant}).*$$

Similarly,

$$\frac{\partial z}{\partial y} = \text{partial derivative of } z \text{ with respect to } y \text{ } (x \text{ remains constant}).*$$

Corresponding symbols are used for partial derivatives of functions of three or more variables.

In order to avoid confusion the round ∂ † has been generally adopted to indicate partial differentiation.

---
* The constant values are substituted in the function before differentiating.
† Introduced by Jacobi (1804–1851).

ILLUSTRATIVE EXAMPLE 1. Find the partial derivatives of $z = ax^2 + 2bxy + cy^2$.

Solution. $\dfrac{\partial z}{\partial x} = 2ax + 2by$, treating y as a constant,

$\dfrac{\partial z}{\partial y} = 2bx + 2cy$, treating x as a constant.

ILLUSTRATIVE EXAMPLE 2. Find the partial derivatives of $u = \sin(ax + by + cz)$.

Solution. $\dfrac{\partial u}{\partial x} = a\cos(ax + by + cz)$, treating y and z as constants,

$\dfrac{\partial u}{\partial y} = b\cos(ax + by + cz)$, treating x and z as constants,

$\dfrac{\partial u}{\partial z} = c\cos(ax + by + cz)$, treating y and x as constants.

Referring to (1), we have, in the notations commonly used,

$$\frac{\partial z}{\partial x} = \frac{\partial}{\partial x}f(x, y) = \frac{\partial f}{\partial x} = f_x(x, y) = f_x = z_x;$$

$$\frac{\partial z}{\partial y} = \frac{\partial}{\partial y}f(x, y) = \frac{\partial f}{\partial y} = f_y(x, y) = f_y = z_y.$$

Similar notations are used for functions of any number of variables. Referring to Art. 24, we shall have

(2) $\qquad f_x(x, y_0) = \lim\limits_{\Delta x \to 0} \dfrac{f(x + \Delta x, y_0) - f(x, y_0)}{\Delta x}$,

(3) $\qquad f_y(x_0, y) = \lim\limits_{\Delta y \to 0} \dfrac{f(x_0, y + \Delta y) - f(x_0, y)}{\Delta y}$.

226. Partial derivatives interpreted geometrically. Let the equation of the surface shown in the figure be

$z = f(x, y)$.

Pass a plane $EFGH$ through the point P (where $x = a$ and $y = b$) on the surface parallel to the XOZ-plane. Since the equation of this plane is

$y = b$,

the equation of the curve of intersection JPK with the surface is

$z = f(x, b)$,

if we consider EF as the axis of Z and EH as the axis of X. In this plane $\dfrac{\partial z}{\partial x}$ means the same as $\dfrac{dz}{dx}$, and we have

(1) $\qquad \dfrac{\partial z}{\partial x} = \tan MTP =$ slope of curve of intersection JK at P.

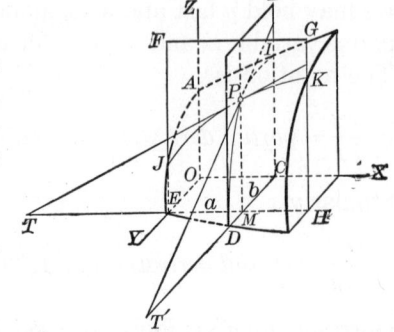

PARTIAL DIFFERENTIATION

Similarly, if we pass the plane BCD through P parallel to the YOZ-plane, its equation is
$$x = a,$$
and for the curve of intersection DPI, $\dfrac{\partial z}{\partial y}$ means the same as $\dfrac{dz}{dy}$. Hence

(2) $\dfrac{\partial z}{\partial y} = -\tan MT'P =$ slope of curve of intersection DI at P.

ILLUSTRATIVE EXAMPLE. Given the ellipsoid $\dfrac{x^2}{24} + \dfrac{y^2}{12} + \dfrac{z^2}{6} = 1$; find the slope of the curve of intersection of the ellipsoid made (a) by the plane $y = 1$ at the point where $x = 4$ and z is positive; (b) by the plane $x = 2$ at the point where $y = 3$ and z is positive.

Solution. Considering y as constant,
$$\frac{2x}{24} + \frac{2z}{6}\frac{\partial z}{\partial x} = 0, \quad \text{or} \quad \frac{\partial z}{\partial x} = -\frac{x}{4z}.$$
When x is constant, $\dfrac{2y}{12} + \dfrac{2z}{6}\dfrac{\partial z}{\partial y} = 0,$ or $\dfrac{\partial z}{\partial y} = -\dfrac{y}{2z}.$

(a) When $y = 1$ and $x = 4$, $z = \sqrt{\dfrac{3}{2}}$. $\therefore \dfrac{\partial z}{\partial x} = -\sqrt{\dfrac{2}{3}}$. Ans.

(b) When $x = 2$ and $y = 3$, $z = \dfrac{1}{\sqrt{2}}$. $\therefore \dfrac{\partial z}{\partial y} = -\dfrac{3}{2}\sqrt{2}$. Ans.

PROBLEMS

Work out the following partial derivatives.

1. $z = Ax^2 + Bxy + Cy^2 + Dx + Ey + F.$
   Ans. $\dfrac{\partial z}{\partial x} = 2Ax + By + D$; $\dfrac{\partial z}{\partial y} = Bx + 2Cy + E.$

2. $f(x, y) = Ax^3 + 3Bx^2y + 3Cxy^2 + Dy^3.$
   Ans. $f_x(x, y) = 3(Ax^2 + 2Bxy + Cy^2)$;
   $f_y(x, y) = 3(Bx^2 + 2Cxy + Dy^2).$

3. $f(x, y) = \dfrac{Ax + By}{Cx + Dy}.$ Ans. $\dfrac{\partial f}{\partial x} = \dfrac{(AD - BC)y}{(Cx + Dy)^2}$; $\dfrac{\partial f}{\partial y} = \dfrac{(BC - AD)x}{(Cx + Dy)^2}.$

4. $u = xy + yz + zx.$ Ans. $u_x = y + z$; $u_y = x + z$; $u_z = x + y.$

5. $f(x, y) = (x + y)\sin(x - y).$
   Ans. $f_x(x, y) = \sin(x - y) + (x + y)\cos(x - y)$;
   $f_y(x, y) = \sin(x - y) - (x + y)\cos(x - y).$

6. $\rho = \sin 2\theta \cos 3\phi.$ Ans. $\dfrac{\partial \rho}{\partial \theta} = 2\cos 2\theta \cos 3\phi$;
   $\dfrac{\partial \rho}{\partial \phi} = -3\sin 2\theta \sin 3\phi.$

7. $\rho = e^{\theta + \phi}\cos(\theta - \phi).$ Ans. $\dfrac{\partial \rho}{\partial \theta} = e^{\theta + \phi}\{\cos(\theta - \phi) - \sin(\theta - \phi)\}$;
   $\dfrac{\partial \rho}{\partial \phi} = e^{\theta + \phi}\{\cos(\theta - \phi) + \sin(\theta - \phi)\}.$

DIFFERENTIAL AND INTEGRAL CALCULUS

Find the partial derivatives of the following functions.

8. $f(x, y) = 3x^4 - 4x^3y + 6x^2y^2$.
11. $f(x, y) = (x + 2y) \tan(2x + y)$.

9. $u = \dfrac{x + 2y}{y + 2z}$.
12. $\rho = \tan 2\theta \operatorname{ctn} 4\phi$.

10. $z = e^{\frac{x}{y}} \ln \dfrac{y}{x}$.
13. $\rho = e^{-\theta} \cos \dfrac{\phi}{\theta}$.

14. If $f(x, y) = 2x^2 - 3xy + 4y^2$, show that $f_x(2, 3) = -1$, $f_y(2, 3) = 18$.

15. If $f(x, y) = \dfrac{2x}{x - y}$, show that $f_x(3, 1) = -\tfrac{1}{2}$, $f_y(3, 1) = \tfrac{3}{2}$.

16. If $f(x, y) = e^{-x} \sin(x + 2y)$, show that $f_x\left(0, \dfrac{\pi}{4}\right) = -1$, $f_y\left(0, \dfrac{\pi}{4}\right) = 0$.

17. If $u = Ax^4 + 2Bx^2y^2 + Cy^4$, show that $x \dfrac{\partial u}{\partial x} + y \dfrac{\partial u}{\partial y} = 4u$.

18. If $u = \dfrac{x^2 y^2}{x + y}$, show that $x \dfrac{\partial u}{\partial x} + y \dfrac{\partial u}{\partial y} = 3u$.

19. If $u = x^2 y + y^2 z + z^2 x$, show that $\dfrac{\partial u}{\partial x} + \dfrac{\partial u}{\partial y} + \dfrac{\partial u}{\partial z} = (x + y + z)^2$.

20. If $u = \dfrac{Ax^n + By^n}{Cx^2 + Dy^2}$, show that $x \dfrac{\partial u}{\partial x} + y \dfrac{\partial u}{\partial y} = (n - 2)u$.

21. The area of a triangle is given by the formula $K = \tfrac{1}{2} bc \sin A$. Given $b = 10$ in., $c = 20$ in., $A = 60°$.
(a) Find the area.
(b) Find the rate of change of the area with respect to the side b if c and A remain constant.
(c) Find the rate of change of the area with respect to the angle A if b and c remain constant.
(d) Using the rate found in (c), calculate approximately the change in area if the angle is increased by one degree.
(e) Find the rate of change of c with respect to b if the area and the angle remain constant.

22. The law of cosines for any triangle is $a^2 = b^2 + c^2 - 2bc \cos A$. Given $b = 10$ in., $c = 15$ in., $A = 60°$.
(a) Find a.
(b) Find the rate of change of a with respect to b if c and A remain constant.
(c) Using the rate found in (b), calculate approximately the change in a if b is decreased by 1 in.
(d) Find the rate of change of a with respect to A if b and c remain constant.
(e) Find the rate of change of c with respect to A if a and b remain constant.

PARTIAL DIFFERENTIATION

227. The total differential. We have already considered the differential of a function of one variable in Art. 91. Thus, if $y = f(x)$, we defined and proved

(1) $$dy = f'(x)\,\Delta x = \frac{dy}{dx}\,\Delta x = \frac{dy}{dx}\,dx.$$

We shall next consider a function of two variables

(2) $$u = f(x, y)$$

and begin by proving the following theorem.

Theorem. *Let u and its first partial derivatives be continuous in some neighborhood of the point (x, y). Then there are quantities ϵ and ϵ' such that*
$$\Delta u = f_x(x, y)\,\Delta x + f_y(x, y)\,\Delta y + \epsilon\Delta x + \epsilon'\Delta y$$
and
$$\lim_{\Delta x \to 0,\, \Delta y \to 0} \epsilon = 0, \qquad \lim_{\Delta x \to 0,\, \Delta y \to 0} \epsilon' = 0.$$

Proof. Let Δx and Δy be the increments of x and y respectively, and let Δu be the corresponding increment of u. Then

(3) $$\Delta u = f(x + \Delta x, y + \Delta y) - f(x, y)$$

is called the *total increment* of u.

Adding and subtracting $f(x, y + \Delta y)$ in the second member,

(4) $$\Delta u = [f(x + \Delta x, y + \Delta y) - f(x, y + \Delta y)] \\ + [f(x, y + \Delta y) - f(x, y)].$$

Applying the Theorem of Mean Value (*D*), Art. 116, to each of the two differences in (4), we get, for the first difference,

(5) $$f(x + \Delta x, y + \Delta y) - f(x, y + \Delta y) = f_x(x + \theta_1\,\Delta x, y + \Delta y)\Delta x.$$

$\left[\begin{array}{l}a = x,\ \Delta a = \Delta x,\ \text{and since } x \text{ varies while } y + \Delta y \text{ remains} \\ \text{constant, we get the partial derivative with respect to } x.\end{array}\right]$

For the second difference,

(6) $$f(x, y + \Delta y) - f(x, y) = f_y(x, y + \theta_2\,\Delta y)\Delta y.$$

$\left[\begin{array}{l}a = y,\ \Delta a = \Delta y,\ \text{and since } y \text{ varies while } x \text{ remains constant,} \\ \text{we get the partial derivative with respect to } y.\end{array}\right]$

Substituting from (5) and (6) in (4) gives

(7) $$\Delta u = f_x(x + \theta_1\,\Delta x, y + \Delta y)\Delta x + f_y(x, y + \theta_2\,\Delta y)\Delta y.$$

Since $f_x(x, y)$ and $f_y(x, y)$ are continuous near (x, y), we may write

(8) $$\epsilon = f_x(x + \theta_1\Delta x, y + \Delta y) - f_x(x, y),$$
(9) $$\epsilon' = f_y(x, y + \theta_2\Delta y) - f_y(x, y);$$

and, from the definition of a continuous function, we know that

$$\lim_{\Delta x \to 0,\, \Delta y \to 0} \epsilon = 0, \qquad \lim_{\Delta x \to 0,\, \Delta y \to 0} \epsilon' = 0.$$

450   DIFFERENTIAL AND INTEGRAL CALCULUS

Solving (8) and (9) for $f_x(x + \theta_1 \Delta x, y + \Delta y)$ and $f_y(x, y + \theta_2 \Delta y)$, and substituting in (7), we have

(10) $\Delta u = f_x(x, y) \Delta x + f_y(x, y) \Delta y + \epsilon \Delta x + \epsilon' \Delta y.$ Q.E.D.

We then define as the *total differential* ($= du$) of u

(11) $du = f_x(x, y) \Delta x + f_y(x, y) \Delta y.$

The right-hand member in (11) is the "principal part" of the right-hand member of (10), that is, du is a close approximate value of Δu for small values of Δx and Δy (compare Art. 92). Obviously, if $u = x$, (11) becomes $dx = \Delta x$. If $u = y$, (11) becomes $dy = \Delta y$. Substituting, then, in (11) for Δx and Δy their corresponding differentials, we obtain the important formula

(B) $du = f_x(x, y)\, dx + f_y(x, y)\, dy = \dfrac{\partial u}{\partial x} dx + \dfrac{\partial u}{\partial y} dy = \dfrac{\partial f}{\partial x} dx + \dfrac{\partial f}{\partial y} dy,$

which should be compared with (1) at the beginning of this article. If u is a function of three variables, its total differential is

(C) $du = \dfrac{\partial u}{\partial x} dx + \dfrac{\partial u}{\partial y} dy + \dfrac{\partial u}{\partial z} dz;$

and so on for any number of variables.

A geometric interpretation of (B) is given in Art. 238.

ILLUSTRATIVE EXAMPLE 1. Compute Δu and du for the function

(12) $u = 2\,x^2 + 3\,y^2,$

when $x = 10$, $y = 8$, $\Delta x = 0.2$, $\Delta y = 0.3$, and compare the results.

Solution. From (12) we get (compare Art. 27)

$u + \Delta u = 2(x + \Delta x)^2 + 3(y + \Delta y)^2$
$= 2\,x^2 + 3\,y^2 + 4\,x\,\Delta x + 6\,y\,\Delta y + 2(\Delta x)^2 + 3(\Delta y)^2.$

(13) $\Delta u = 4\,x\,\Delta x + 6\,y\,\Delta y + 2(\Delta x)^2 + 3(\Delta y)^2.$

Differentiating (12), we find $\dfrac{\partial u}{\partial x} = 4\,x$, $\dfrac{\partial u}{\partial y} = 6\,y$. Then from (B)

(14) $du = 4\,x\,dx + 6\,y\,dy.$

Remembering that $\Delta x = dx$, $\Delta y = dy$, we see that the right-hand member in (14) is the "principal part" of the right-hand member in (13), for the additional terms are of the second degree in Δx or Δy. This statement illustrates (10) and (11) above (namely, $\epsilon = 2\,\Delta x$, $\epsilon' = 3\,\Delta y$).

Substituting the given values in (13) and (14), we get

(15) $\Delta u = 8 + 14.4 + 0.08 + 0.27 = 22.75;$

(16) $du = 8 + 14.4 = 22.4.$

Then $\Delta u - du = 0.35 = 1.6\%$ of Δu. Ans.

PARTIAL DIFFERENTIATION

ILLUSTRATIVE EXAMPLE 2. Given $u = \arctan \dfrac{y}{x}$, find du.

Solution. $\quad \dfrac{\partial u}{\partial x} = -\dfrac{y}{x^2 + y^2}, \quad \dfrac{\partial u}{\partial y} = \dfrac{x}{x^2 + y^2}.$

Substituting in (B), $\quad du = \dfrac{x\,dy - y\,dx}{x^2 + y^2}.$ Ans.

PROBLEMS

Find the total differential of each of the following functions.

1. $z = 2x^3 - 4xy^2 + 3y^3$. Ans. $dz = (6x^2 - 4y^2)dx + (9y^2 - 8xy)dy$.

2. $u = \dfrac{Ax + By}{Cx + Dy}$. $\qquad du = \dfrac{(AD - BC)(y\,dx - x\,dy)}{(Cx + Dy)^2}$.

3. $u = xy^2z^3$. $\qquad du = y^2z^3\,dx + 2xyz^3\,dy + 3xy^2z^2\,dz$.

4. $u = x^2 \cos 2y$. $\qquad$ 5. $\theta = \arctan \dfrac{y}{x}$. $\qquad$ 6. $u = (x - y)\ln(x + y)$.

7. If $x^2 + y^2 + z^2 = a^2$, show that $dz = -\dfrac{x\,dx + y\,dy}{z}$.

8. Find dz if $4x^2 - 9y^2 - 16z^2 = 100$.

9. Compute Δu and du for the function $u = x^2 - 3xy + 2y^2$ when $x = 2, y = -3, \Delta x = -0.3, \Delta y = 0.2.$ $\quad$ Ans. $\Delta u = -7.15, du = -7.5$.

10. Compute du for the function $u = (x + y)\sqrt{x - y}$ when $x = 6, y = 2$, $dx = \frac{1}{4}, dy = -\frac{1}{2}.$ $\qquad\qquad$ Ans. 1.

11. Compute Δu and du for the function $u = xy + 2x - 4y$ when $x = 2, y = 3, \Delta x = 0.4, \Delta y = -0.2.$

12. Compute $d\rho$ for the function $\rho = e^{\frac{\theta}{2}} \sin(\theta - \phi)$ when $\theta = 0$, $\phi = \frac{1}{2}\pi, d\theta = 0.2, d\phi = -0.2.$

228. Approximation of the total increment. Small errors. Formulas (B) and (C) are used to calculate Δu approximately. Also, when the values of x and y are determined by measurement or experiment, and hence subject to small errors Δx and Δy, a close approximation to the error in u can be found by (B). (Compare Arts. 92, 93.)

ILLUSTRATIVE EXAMPLE 1. Find, approximately, the volume of tin in a thin cylindrical can without a top if the inside diameter and height are, respectively, 6 in. and 8 in., and the thickness is $\frac{1}{8}$ in.

Solution. The volume v of a solid right circular cylinder with diameter x and height y is

(1) $\qquad\qquad v = \frac{1}{4}\pi x^2 y.$

Obviously, the exact volume of the can is the difference Δv between the volumes of two solid cylinders for which $x = 6\frac{1}{4}, y = 8\frac{1}{8}$, and $x = 6, y = 8$, respectively. Since only an approximate value is required, we calculate dv instead of Δv.

Differentiating (1), and using (B), we get

(2) $\quad dv = \frac{1}{2}\pi xy\, dx + \frac{1}{4}\pi x^2\, dy.$

Substituting in (2) $x = 6$, $y = 8$, $dx = \frac{1}{4}$, $dy = \frac{1}{8}$, the result is

$$dv = 7\tfrac{1}{8}\,\pi = 22.4 \text{ cu. in. } Ans.$$

The exact value is $\Delta v = 23.1$ cu. in.

ILLUSTRATIVE EXAMPLE 2. Two sides of an oblique plane triangle measured, respectively, 63 ft. and 78 ft., and the included angle measured 60°. These measurements were subject to errors whose maximum values are 0.1 ft. in each length and 1° in the angle. Find the approximate maximum error and the percentage error made in calculating the third side from these measurements.

Solution. Using the law of cosines ((7), Art. 2),

(3) $\quad u^2 = x^2 + y^2 - 2\,xy\cos\alpha,$

where x, y are the given sides, α the included angle, and u the third side. The given data are

(4) $x = 63,\quad y = 78,\quad \alpha = 60° = \dfrac{\pi}{3},\quad dx = dy = 0.1,\quad d\alpha = 0.01745$ (radian).

Differentiating (3), we get

$$\frac{\partial u}{\partial x} = \frac{x - y\cos\alpha}{u},\quad \frac{\partial u}{\partial y} = \frac{y - x\cos\alpha}{u},\quad \frac{\partial u}{\partial \alpha} = \frac{xy\sin\alpha}{u}.$$

Hence, using (C),

$$du = \frac{(x - y\cos\alpha)dx + (y - x\cos\alpha)dy + xy\sin\alpha\, d\alpha}{u}.$$

Substituting the values from (4), we find

$$du = \frac{2.4 + 4.65 + 74.25}{71.7} = 1.13 \text{ ft. } Ans.$$

The percentage error $= 100\,\dfrac{du}{u} = 1.6\%$. *Ans.*

PROBLEMS

1. The legs of a right triangle measured 6 ft. and 8 ft. respectively, with maximum errors in each of 0.1 ft. Find approximately the maximum error and percentage error in calculating (a) the area, (b) the hypotenuse, from these measurements. *Ans.* (a) 0.7 sq. ft., 2.9%; (b) 0.14 ft., 1.4%.

2. In the preceding problem find, from the given dimensions, the angle opposite the longer side, and calculate the approximate maximum error in that angle in radians and degrees.

3. The radii of the bases of a frustum of a right circular cone measure 5 in. and 11 in. respectively, and the slant height measures 12 in. The maximum error in each measurement is 0.1 in. Find the approximate error and percentage error in calculating from these measurements (a) the altitude; (b) the volume (see (12), Art. 1).

Ans. (a) 0.23 in., 2.2%; (b) 24.4 π cu. in., $3\tfrac{1}{2}\%$.

PARTIAL DIFFERENTIATION

4. One side of a triangle measures 2000 ft., and the adjacent angles measure 30° and 60° respectively, with a maximum error in each angle of 30′. The maximum error in the measurement of the side is ±1 ft. Find the approximate maximum error and percentage error in calculating from these measurements (a) the altitude on the given side; (b) the area of the triangle. *Ans.* (a) 17.88 ft., 2.1%.

5. The diameter and altitude of a right circular cylinder are found by measurement to be 12 in. and 8 in. respectively. If there is a probable error of 0.2 in. in each measurement, what is approximately the greatest possible error in the computed volume? *Ans.* 16.8 π cu. in.

6. The dimensions of a box are found by measurement to be 6 ft., 8 ft., 12 ft. If there is a probable error of 0.05 ft., (a) what is approximately the greatest possible error in the computed volume? (b) What is the percentage error? *Ans.* (a) 10.8 cu. ft.; (b) $\frac{15}{8}$%.

7. Given the surface $z = \dfrac{x-y}{x+y}$. If, at the point where $x=4$, $y=2$, x and y are each increased by $\frac{1}{10}$, what is the approximate change in z? *Ans.* $-\frac{1}{90}$.

8. The specific gravity of a solid is given by the formula $s = \dfrac{P}{w}$, where P is the weight in a vacuum and w is the weight of an equal volume of water. How is the computed specific gravity affected by an error of $\pm \frac{1}{10}$ in weighing P and $\pm \frac{1}{20}$ in weighing w, assuming $P=8$ and $w=1$ in the experiment, (a) if both errors are positive? (b) if one error is negative? (c) What is approximately the largest percentage error? *Ans.* (a) 0.3; (b) 0.5; (c) 6¼%.

9. The diameter and slant height of a right circular cone are found by measurement to be 10 in. and 20 in. respectively. If there is a probable error of 0.2 in. in each measurement, what approximately is the greatest possible error in the computed value of (a) the volume? (b) the curved surface? *Ans.* (a) $\dfrac{37\,\pi\sqrt{15}}{18} = 25$ cu. in.; (b) $3\pi = 9.42$ sq. in.

10. Two sides of a triangle are found by measurement to be 63 ft. and 78 ft. and the included angle to be 60°. If there is a probable error of 0.5 ft. in measuring the sides and of 2° in measuring the angle, what is approximately the greatest possible error in the computed value of the area? (See (7), Art. 2.) *Ans.* 73.6 sq. ft.

11. If specific gravity is determined by the formula $s = \dfrac{A}{A-W}$, where A is the weight in air and W the weight in water, what is (a) approximately the largest error in s if A can be read within 0.01 lb. and W within 0.02 lb., the actual readings being $A=9$ lb., $W=5$ lb.? (b) the largest relative error? *Ans.* (a) 0.0144; (b) $\frac{23}{3600}$.

454   DIFFERENTIAL AND INTEGRAL CALCULUS

12. The resistance of a circuit was found by using the formula $C = \dfrac{E}{R}$, where $C =$ current and $E =$ electromotive force. If there is an error of $\frac{1}{10}$ ampere in reading C and $\frac{1}{20}$ volt in reading E, (a) what is the approximate error in R if the readings are $C = 15$ amperes and $E = 110$ volts? (b) What is the percentage error?   *Ans.* (a) 0.0522 ohms; (b) $\frac{47}{66}$%.

13. If the formula $\sin(x + y) = \sin x \cos y + \cos x \sin y$ were used to calculate $\sin(x + y)$, what approximate error would result if an error of 0.1° were made in measuring both x and y, the measurements of the two acute angles giving $\sin x = \frac{3}{5}$ and $\sin y = \frac{5}{13}$?   *Ans.* 0.0018.

14. The acceleration of a particle down an inclined plane is given by $a = g \sin i$. If g varies by 0.1 ft. per second per second, and i, which is measured as 30°, may be in error 1°, what is the approximate error in the computed value of a? Take the normal value of g to be 32 ft. per second per second.   *Ans.* 0.534 ft. per second per second.

15. The period of a pendulum is $P = 2\pi\sqrt{\dfrac{l}{g}}$. (a) What is the greatest approximate error in the period if there is an error of $\pm\,0.1$ ft. in measuring a 10-foot suspension and g, taken as 32 ft. per second per second, may be in error by 0.05 ft. per second per second? (b) What is the percentage error?   *Ans.* (a) 0.0204 sec.; (b) $\frac{37}{64}$%.

16. The dimensions of a cone are radius of base $= 4$ in., altitude $= 6$ in. What is the approximate error in volume and in total surface if there is a shortage of 0.01 in. per inch in the measure used?
Ans. $dV = 3.0159$ cu. in.; $dS = 2.818$ sq. in.

17. The length l and the period P of a simple pendulum are connected by the equation $4\pi^2 l = P^2 g$. If l is calculated assuming $P = 1$ sec. and $g = 32$ ft. per second per second, what is approximately the error in l if the true values are $P = 1.02$ sec. and $g = 32.01$ ft. per second per second? What is the percentage error?

18. A solid is in the form of a cylinder capped at each end with a hemisphere of the same radius as the cylinder. Its measured dimensions are diameter $= 8$ in. and total length $= 20$ in. What is approximately the error in volume and surface if the tape used in measuring has stretched uniformly $\frac{1}{2}$% beyond its proper length?

19. Assuming the characteristic equation of a perfect gas to be $vp = Rt$, where $v =$ volume, $p =$ pressure, $t =$ absolute temperature, and R a constant, what is the relation between the differentials dv, dp, dt?
Ans. $v\,dp + p\,dv = R\,dt$.

20. Using the result in the last example as applied to air, suppose that in a given case we have found by actual experiment that $t = 300°$ C., $p = 2000$ lb. per square foot, $v = 14.4$ cu. ft. Find the change in p, assuming it to be uniform, when t changes to 301° C., and v to 14.5 cu. ft. $R = 96$.   *Ans.* -7.22 lb. per square foot.

PARTIAL DIFFERENTIATION

229. Total derivatives. Rates. Turn now to the case where x and y in

(1) $$u = f(x, y)$$

are not independent variables. Assume, for example, that both are functions of a third variable t, namely,

(2) $$x = \phi(t), \quad y = \psi(t).$$

When these values are substituted in (1), u becomes a function of one variable t, and its derivative may be found in the usual manner. We now have

(3) $$du = \frac{du}{dt} dt, \quad dx = \frac{dx}{dt} dt, \quad dy = \frac{dy}{dt} dt.$$

Formula **(B)** was established with the assumption that x and y were independent variables. We may easily show that it holds also in the present case. To this end, return to (10), Art. 227, and divide both members by Δt. Changing the notation, this may be written

(4) $$\frac{\Delta u}{\Delta t} = \frac{\partial u}{\partial x} \frac{\Delta x}{\Delta t} + \frac{\partial u}{\partial y} \frac{\Delta y}{\Delta t} + \left(\epsilon \frac{\Delta x}{\Delta t} + \epsilon' \frac{\Delta y}{\Delta t} \right).$$

Now when $\Delta t \to 0$, $\Delta x \to 0$ and $\Delta y \to 0$. Hence (see Art. 227)

$$\lim_{\Delta t \to 0} \epsilon = 0, \quad \lim_{\Delta t \to 0} \epsilon' = 0.$$

Therefore, when $\Delta t \to 0$, (4) becomes

(D) $$\frac{du}{dt} = \frac{\partial u}{\partial x} \frac{dx}{dt} + \frac{\partial u}{\partial y} \frac{dy}{dt}.$$

Multiplying both members by dt and using (3), we obtain **(B)**. That is, **(B)** *holds also when x and y are functions of a third variable t.*

In the same way, if $$u = f(x, y, z),$$

and x, y, z are all functions of t, we get

(E) $$\frac{du}{dt} = \frac{\partial u}{\partial x} \frac{dx}{dt} + \frac{\partial u}{\partial y} \frac{dy}{dt} + \frac{\partial u}{\partial z} \frac{dz}{dt},$$

and so on for any number of variables.

In **(D)** we may suppose $t = x$; then y is a function of x, and u is really a function of the one variable x, giving

(F) $$\frac{du}{dx} = \frac{\partial u}{\partial x} + \frac{\partial u}{\partial y} \frac{dy}{dx}.$$

In the same way, from **(E)** we have, when y and z are functions of x,

(G) $$\frac{du}{dx} = \frac{\partial u}{\partial x} + \frac{\partial u}{\partial y} \frac{dy}{dx} + \frac{\partial u}{\partial z} \frac{dz}{dx}.$$

456   DIFFERENTIAL AND INTEGRAL CALCULUS

The student should observe that $\dfrac{\partial u}{\partial x}$ and $\dfrac{du}{dx}$ have quite different meanings. The partial derivative $\dfrac{\partial u}{\partial x}$ is formed on the supposition that the *particular variable x alone varies*, all other variables being held fast. But

$$\frac{du}{dx} = \lim_{\Delta x \to 0} \left(\frac{\Delta u}{\Delta x}\right),$$

where Δu is the *total increment of u* due to changes in *all the variables* caused by the change Δx in the independent variable. In contradistinction to partial derivatives, $\dfrac{du}{dt}$, $\dfrac{du}{dx}$ are called *total derivatives* with respect to t and x respectively.

It should be observed that $\dfrac{\partial u}{\partial x}$ has a perfectly definite value for any point (x, y), while $\dfrac{du}{dx}$ depends not only on the point (x, y) but also on the particular direction chosen to reach that point.

ILLUSTRATIVE EXAMPLE 1. Given $u = \sin \dfrac{x}{y}$, $x = e^t$, $y = t^2$; find $\dfrac{du}{dt}$.

Solution. $\dfrac{\partial u}{\partial x} = \dfrac{1}{y}\cos\dfrac{x}{y}$, $\dfrac{\partial u}{\partial y} = -\dfrac{x}{y^2}\cos\dfrac{x}{y}$; $\dfrac{dx}{dt} = e^t$, $\dfrac{dy}{dt} = 2t$.

Substituting in (D), $\dfrac{du}{dt} = (t-2)\dfrac{e^t}{t^3}\cos\dfrac{e^t}{t^2}.$ *Ans.*

ILLUSTRATIVE EXAMPLE 2. Given $u = e^{ax}(y - z)$, $y = a\sin x$, $z = \cos x$; find $\dfrac{du}{dx}$.

Solution. $\dfrac{\partial u}{\partial x} = ae^{ax}(y - z)$, $\dfrac{\partial u}{\partial y} = e^{ax}$, $\dfrac{\partial u}{\partial z} = -e^{ax}$; $\dfrac{dy}{dx} = a\cos x$, $\dfrac{dz}{dx} = -\sin x$.

Substituting in (G),

$$\frac{du}{dx} = ae^{ax}(y - z) + ae^{ax}\cos x + e^{ax}\sin x = e^{ax}(a^2 + 1)\sin x. \; Ans.$$

NOTE. In examples like the above, u could, by substitution, be found explicitly in terms of the independent variable and then differentiated directly; but generally this process would be longer and in many cases could not be used at all.

Formulas (D) and (E) are useful in all applications involving time-rates of change of functions of two or more variables. The process is practically the same as that outlined in the rule given in Art. 52, except that, instead of differentiating with respect to t (Third Step), we find the partial derivatives and substitute in (D) or (E). Let us illustrate by an example.

PARTIAL DIFFERENTIATION 457

ILLUSTRATIVE EXAMPLE 3. The altitude of a circular cone is 100 in., and decreases at the rate of 10 in. per second; and the radius of the base is 50 in., and increases at the rate of 5 in. per second. At what rate is the volume changing?

Solution. Let $x =$ radius of base, $y =$ altitude; then

$$u = \frac{1}{3} \pi x^2 y = \text{volume}, \quad \frac{\partial u}{\partial x} = \frac{2}{3} \pi xy, \quad \frac{\partial u}{\partial y} = \frac{1}{3} \pi x^2.$$

Substituting in (D), $\dfrac{du}{dt} = \dfrac{2}{3} \pi xy \dfrac{dx}{dt} + \dfrac{1}{3} \pi x^2 \dfrac{dy}{dt}.$

But $x = 50, \quad y = 100, \quad \dfrac{dx}{dt} = 5, \quad \dfrac{dy}{dt} = -10.$

$\therefore \dfrac{du}{dt} = \dfrac{2}{3} \pi \cdot 5000 \cdot 5 - \dfrac{1}{3} \pi \cdot 2500 \cdot 10 = 15.15$ cu. ft. per second, increasing. *Ans.*

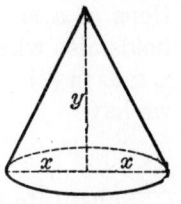

230. Change of variables. If the variables in

(1) $\qquad u = f(x, y)$

are changed by the transformation

(2) $\qquad x = \phi(r, s), \quad y = \psi(r, s),$

the partial derivatives of u with respect to the new variables r and s can be obtained by (D). For, if we hold s fast, then x and y in (2) are functions of r only. Hence we have

(3) $\qquad \dfrac{\partial u}{\partial r} = \dfrac{\partial u}{\partial x} \dfrac{\partial x}{\partial r} + \dfrac{\partial u}{\partial y} \dfrac{\partial y}{\partial r},$

all derivatives with respect to r now being partial.

In the same way,

(4) $\qquad \dfrac{\partial u}{\partial s} = \dfrac{\partial u}{\partial x} \dfrac{\partial x}{\partial s} + \dfrac{\partial u}{\partial y} \dfrac{\partial y}{\partial s}.$

In particular, let the transformation be

(5) $\qquad x = x' + h, \quad y = y' + k,$

the new variables being x' and y', and h, k being constants. Then

$$\dfrac{\partial x}{\partial x'} = 1, \quad \dfrac{\partial x}{\partial y'} = 0, \quad \dfrac{\partial y}{\partial x'} = 0, \quad \dfrac{\partial y}{\partial y'} = 1.$$

Then we obtain, from (3) and (4),

(6) $\qquad \dfrac{\partial u}{\partial x} = \dfrac{\partial u}{\partial x'}, \quad \dfrac{\partial u}{\partial y} = \dfrac{\partial u}{\partial y'}.$

Hence the transformation (5) *leaves the value of the partial derivatives unchanged.*

If the values of x and y in (5) are substituted in (1), the result is

(7) $\qquad u = f(x, y) = F(x', y').$

458   DIFFERENTIAL AND INTEGRAL CALCULUS

The results in (6) may now be written

(8) $\quad f_x(x, y) = F_{x'}(x', y'), \quad f_y(x, y) = F_{y'}(x', y')$.

In Art. 229 it was shown that (B) is true when x and y are functions of a single independent variable t. We prove now that (B) holds also when x and y are functions of two independent variables r, s, as in (2). For by (B), when r and s are the independent variables, we have

$$dx = \frac{\partial x}{\partial r} dr + \frac{\partial x}{\partial s} ds, \quad dy = \frac{\partial y}{\partial r} dr + \frac{\partial y}{\partial s} ds.$$

Substitute these values in the expression

(9) $\quad \dfrac{\partial u}{\partial x} dx + \dfrac{\partial u}{\partial y} dy$

and reduce by (3) and (4). The result is

(10) $\quad \dfrac{\partial u}{\partial r} dr + \dfrac{\partial u}{\partial s} ds.$

But, by (1) and (2), u becomes a function of the independent variables r and s. Then, by (B), (10) equals du. Hence (9) also equals du.

Therefore (B) holds when x and y are functions of one or of two independent variables. In the same way, it can be shown that (C) holds when x, y, z are functions of one, two, or three independent variables.

231. Differentiation of implicit functions. The equation

(1) $\quad\quad\quad\quad f(x, y) = 0$

defines either x or y as an implicit function of the other. It represents an equation containing x and y when all its terms have been transposed to the first member. Let

(2) $\quad\quad\quad\quad u = f(x, y)$;

then $\quad\quad\quad\quad \dfrac{du}{dx} = \dfrac{\partial f}{\partial x} + \dfrac{\partial f}{\partial y} \dfrac{dy}{dx},$ $\quad\quad$ by (F)

and y is an arbitrary function of x. Now let y be the function of x satisfying (1). Then $u = 0$ and $du = 0$, and hence

(3) $\quad\quad\quad\quad \dfrac{\partial f}{\partial x} + \dfrac{\partial f}{\partial y} \dfrac{dy}{dx} = 0.$

Solving, we get

(H) $\quad\quad\quad\quad \dfrac{dy}{dx} = -\dfrac{\dfrac{\partial f}{\partial x}}{\dfrac{\partial f}{\partial y}}.$ $\quad\quad \left(\dfrac{\partial f}{\partial y} \neq 0\right)$

PARTIAL DIFFERENTIATION 459

Thus we have a formula for differentiating implicit functions. This formula in the form (3) is equivalent to the process employed in Art. 41 for differentiating implicit functions, and all the examples on pages 40 and 41 may be solved by it.

When the equation of a curve is in the form (1), formula (*H*) affords an easy way of getting the slope.

ILLUSTRATIVE EXAMPLE 1. Given $x^2y^4 + \sin y = 0$, find $\dfrac{dy}{dx}$.

Solution. Let $f(x, y) = x^2y^4 + \sin y$.

$$\frac{\partial f}{\partial x} = 2\,xy^4, \quad \frac{\partial f}{\partial y} = 4\,x^2y^3 + \cos y.$$

Therefore, from (*H*), $\quad \dfrac{dy}{dx} = -\dfrac{2\,xy^4}{4\,x^2y^3 + \cos y}.\quad$ Ans.

ILLUSTRATIVE EXAMPLE 2. If x increases at the rate of 2 in. per second as it passes through the value $x = 3$ in., at what rate must y change when $y = 1$ in., in order that the function $2\,xy^2 - 3\,x^2y$ shall remain constant?

Solution. Let $u = 2\,xy^2 - 3\,x^2y$; then, since u remains constant, $\dfrac{du}{dt} = 0$. Substituting this value in the left-hand member of (*D*), transposing, and solving for $\dfrac{dy}{dt}$, we get

(4) $\qquad\qquad \dfrac{dy}{dt} = -\dfrac{\frac{\partial u}{\partial x}}{\frac{\partial u}{\partial y}}\dfrac{dx}{dt}.$

Also, $\qquad\qquad \dfrac{\partial u}{\partial x} = 2\,y^2 - 6\,xy, \quad \dfrac{\partial u}{\partial y} = 4\,xy - 3\,x^2.$

Now, substituting in (4), $\quad \dfrac{dy}{dt} = -\dfrac{2\,y^2 - 6\,xy}{4\,xy - 3\,x^2}\dfrac{dx}{dt}.$

But $\qquad\qquad x = 3, \quad y = 1, \quad \dfrac{dx}{dt} = 2.$

Therefore, $\qquad\qquad \dfrac{dy}{dt} = -2\tfrac{2}{15}$ in. per second. Ans.

In like manner, the equation

(5) $\qquad\qquad F(x, y, z) = 0$

defines z as an implicit function of the two independent variables x and y. To find the partial derivatives of z with respect to x and to y, proceed as follows.

Let $\qquad\qquad u = f(x, y, z).$

Then $\qquad\qquad du = \dfrac{\partial F}{\partial x}\,dx + \dfrac{\partial F}{\partial y}\,dy + \dfrac{\partial F}{\partial z}\,dz,$

by (*B*), and this holds no matter what the independent variables are (Art. 230). Now let z be chosen as that function of the independent variables x and y which satisfies (5). Then $u = 0$, $du = 0$, and we have

(6) $$\frac{\partial F}{\partial x} dx + \frac{\partial F}{\partial y} dy + \frac{\partial F}{\partial z} dz = 0.$$

But now $$dz = \frac{\partial z}{\partial x} dx + \frac{\partial z}{\partial y} dy. \quad \text{By } (B)$$

Substituting this value in (6) and simplifying gives

$$\left(\frac{\partial F}{\partial x} + \frac{\partial F}{\partial z}\frac{\partial z}{\partial x}\right)dx + \left(\frac{\partial F}{\partial y} + \frac{\partial F}{\partial z}\frac{\partial z}{\partial y}\right)dy = 0.$$

Here $dx \, (= \Delta x)$ and $dy \, (= \Delta y)$ are independent increments. We may therefore set $dy = 0$, $dx \neq 0$, divide through by dx, and solve for $\frac{\partial z}{\partial x}$. The result is

(I) $$\frac{\partial z}{\partial x} = -\frac{\frac{\partial F}{\partial x}}{\frac{\partial F}{\partial z}}.$$

Proceeding in a similar manner, we may prove

(J) $$\frac{\partial z}{\partial y} = -\frac{\frac{\partial F}{\partial y}}{\frac{\partial F}{\partial z}}.$$

Formulas (I) and (J) are to be interpreted as follows: In the left-hand members z is the function of x and y satisfying (5). In the right-hand members F is the function of three variables, x, y, z, given in the left-hand member of (5).

The generalization of (H), (I), (J) to an implicit function u of any number of variables is now obvious.

ILLUSTRATIVE EXAMPLE. By the equation

$$\frac{x^2}{24} + \frac{y^2}{12} + \frac{z^2}{6} - 1 = 0,$$

z is defined as an implicit function of x and y. Find the partial derivatives of this function.

Solution. $\quad F = \frac{x^2}{24} + \frac{y^2}{12} + \frac{z^2}{6} - 1.$

Hence $\quad \frac{\partial F}{\partial x} = \frac{x}{12}, \quad \frac{\partial F}{\partial y} = \frac{y}{6}, \quad \frac{\partial F}{\partial z} = \frac{z}{3}.$

Substituting in (I) and (J), we get

$$\frac{\partial z}{\partial x} = -\frac{x}{4\,z}, \quad \frac{\partial z}{\partial y} = -\frac{y}{2\,z}. \quad Ans.$$

(Compare with the Illustrative Example in Art. 226.)

PARTIAL DIFFERENTIATION

PROBLEMS

In Problems 1–5 find $\dfrac{du}{dt}$.

1. $u = x^2 - 3xy + 2y^2$; $x = \cos t$, $y = \sin t$.

   Ans. $\dfrac{du}{dt} = \sin 2t - 3 \cos 2t$.

2. $u = x + 4\sqrt{xy} - 3y$; $x = t^3$, $y = \dfrac{1}{t}$. $\quad \dfrac{du}{dt} = 3t^2 + 4 + \dfrac{3}{t^2}$.

3. $u = e^x \sin y + e^y \sin x$; $x = \tfrac{1}{2}t$, $y = 2t$.

   Ans. $\dfrac{du}{dt} = e^{\tfrac{1}{2}t}(\tfrac{1}{2} \sin 2t + 2 \cos 2t) + e^{2t}(2 \sin \tfrac{1}{2}t + \tfrac{1}{2} \cos \tfrac{1}{2}t)$.

4. $u = 2x^2 - xy + y^2$; $x = \cos 2t$, $y = \sin t$.

5. $u = xy + yz + zx$; $x = \dfrac{1}{t}$, $y = e^t$, $z = e^{-t}$.

In Problems 6–10 find $\dfrac{dy}{dx}$ by formula (H).

6. $Ax^2 + 2Bxy + Cy^2 + 2Dx + 2Ey + F = 0$.

   Ans. $\dfrac{dy}{dx} = -\dfrac{Ax + By + D}{Bx + Cy + E}$.

7. $x^3 + y^3 - 3axy = 0$. $\quad \dfrac{dy}{dx} = \dfrac{ay - x^2}{y^2 - ax}$.

8. $e^x \sin y - e^y \cos x = 1$. $\quad \dfrac{dy}{dx} = \dfrac{e^y \sin x + e^x \sin y}{e^y \cos x - e^x \cos y}$.

9. $x^4 - x^2 y^2 - x^2 + 2y^2 = 8$.

10. $Ax^4 + 2Bx^2 y^2 + Cy^4 = (x^2 + y^2)^2$.

In Problems 11–15 verify that the given values of x and y satisfy the equation, and find the corresponding value of $\dfrac{dy}{dx}$.

11. $x^2 + 2xy + 2y = 22$; $x = 2$, $y = 3$. $\quad$ Ans. $\dfrac{dy}{dx} = -\dfrac{5}{3}$.

12. $x^3 - y^3 + 4xy = 0$; $x = 2$, $y = -2$. $\quad \dfrac{dy}{dx} = 1$.

13. $Ax + By + Ce^{xy} = C$; $x = 0$, $y = 0$. $\quad \dfrac{dy}{dx} = -\dfrac{A}{B}$.

14. $2x - \sqrt{2xy} + y = 4$; $x = 2$, $y = 4$.

15. $e^x \cos y + e^y \sin x = 1$; $x = 0$, $y = 0$.

In Problems 16–20 find $\dfrac{\partial z}{\partial x}$ and $\dfrac{\partial z}{\partial y}$.

16. $Ax^2 + By^2 + Cz^2 = D$. $\quad$ Ans. $\dfrac{\partial z}{\partial x} = -\dfrac{Ax}{Cz}$; $\dfrac{\partial z}{\partial y} = -\dfrac{By}{Cz}$.

17. $Axy + Byz + Czx = D$. $\quad \dfrac{\partial z}{\partial x} = -\dfrac{Ay + Cz}{Cx + By}$; $\dfrac{\partial z}{\partial y} = -\dfrac{Ax + Bz}{Cx + By}$.

18. $x + 2y + z - 2\sqrt{xyz} = 10$. Ans. $\dfrac{\partial z}{\partial x} = \dfrac{yz - \sqrt{xyz}}{\sqrt{xyz} - xy}$; $\dfrac{\partial z}{\partial y} = \dfrac{xz - 2\sqrt{xyz}}{\sqrt{xyz} - xy}$.

19. $x^3 + y^3 + z^3 - 3\,axyz = 0$.

20. $Ax^2 + By^2 + Cz^2 + 2\,Dxy + 2\,Eyz + 2\,Fzx = G$.

21. A point is moving on the curve of intersection of the sphere $x^2 + y^2 + z^2 = 49$ and the plane $y = 2$. When x is 6 and is increasing 4 units per second, find (a) the rate at which z is changing and (b) the speed with which the point is moving.

Ans. (a) 8 units per second; (b) $4\sqrt{5}$ units per second.

22. A point is moving on the curve of intersection of the surface $x^2 + xy + y^2 - z^2 = 0$ and the plane $x - y + 2 = 0$. When x is 3 and is increasing 2 units per second, find (a) the rate at which y is changing, (b) the rate at which z is changing, (c) the speed with which the point is moving. Ans. (a) 2 units per second; (b) $\tfrac{24}{7}$ units per second; (c) 4.44 units per second.

23. The characteristic equation of a perfect gas is $R\theta = pv$, where θ is the temperature, p the pressure, v the volume, and R a constant. At a certain instant a given amount of gas has a volume of 15 cu. ft. and is under a pressure of 25 lb. per square inch. Assuming $R = 96$, find the temperature and the rate at which the temperature is changing if the volume is increasing at the rate of $\tfrac{1}{2}$ cu. ft. per second and the pressure is decreasing at the rate of $\tfrac{1}{10}$ lb. per square inch per second.

Ans. Temperature is increasing at the rate of $\tfrac{11}{96}$ degrees per second.

24. A triangle ABC is being transformed so that the angle A changes at a uniform rate from $0°$ to $90°$ in 10 sec., while side AC decreases 1 in. per second and side AB increases 1 in. per second. If at the time of observation $A = 60°$, $AC = 16$ in., and $AB = 10$ in., (a) how fast is BC changing? (b) how fast is the area of ABC changing?

Ans. (a) 0.911 in. per second; (b) 8.88 sq. in. per second.

232. Derivatives of higher order. If

(1) $$u = f(x, y),$$

then

(2) $$\dfrac{\partial u}{\partial x} = f_x(x, y), \quad \dfrac{\partial u}{\partial y} = f_y(x, y)$$

are themselves functions of x and y, and can, in turn, be differentiated. Thus, taking the first function and differentiating, we have,

(3) $$\dfrac{\partial^2 u}{\partial x^2} = f_{xx}(x, y), \quad \dfrac{\partial^2 u}{\partial y\,\partial x} = f_{yx}(x, y).$$

In the same manner, from the second function in (2), we obtain

(4) $$\dfrac{\partial^2 u}{\partial x\,\partial y} = f_{xy}(x, y), \quad \dfrac{\partial^2 u}{\partial y^2} = f_{yy}(x, y).$$

PARTIAL DIFFERENTIATION

In (3) and (4) there are apparently four derivatives of the second order. It is shown below that

(K) $$\frac{\partial^2 u}{\partial y\, \partial x} = \frac{\partial^2 u}{\partial x\, \partial y},$$

provided, merely, that the derivatives concerned are continuous. That is, *the order of differentiating successively with respect to x and y is immaterial.* Thus $f(x, y)$ has only three partial derivatives of the second order, namely,

(5) $\quad f_{xx}(x, y), \quad f_{xy}(x, y) = f_{yx}(x, y), \quad f_{yy}(x, y).$

This may be easily extended to higher derivatives. For instance, since (K) is true,

$$\frac{\partial^3 u}{\partial x^2\, \partial y} = \frac{\partial}{\partial x}\left(\frac{\partial^2 u}{\partial x\, \partial y}\right) = \frac{\partial^3 u}{\partial x\, \partial y\, \partial x} = \frac{\partial^2}{\partial x\, \partial y}\left(\frac{\partial u}{\partial x}\right) = \frac{\partial^2}{\partial y\, \partial x}\left(\frac{\partial u}{\partial x}\right) = \frac{\partial^3 u}{\partial y\, \partial x^2}.$$

Similar results hold for functions of three or more variables.

ILLUSTRATIVE EXAMPLE. Given $u = x^3 y - 3\, x^2 y^3$; verify $\dfrac{\partial^2 u}{\partial y\, \partial x} = \dfrac{\partial^2 u}{\partial x\, \partial y}$.

Solution. $\quad \dfrac{\partial u}{\partial x} = 3\, x^2 y - 6\, xy^3, \quad \dfrac{\partial^2 u}{\partial y\, \partial x} = 3\, x^2 - 18\, xy^2,$

$\dfrac{\partial u}{\partial y} = x^3 - 9\, x^2 y^2, \quad \dfrac{\partial^2 u}{\partial x\, \partial y} = 3\, x^2 - 18\, xy^2.$

Hence the formula is verified.

Proof of (K). Consider the expression

(6) $\quad F = f(x + \Delta x, y + \Delta y) - f(x + \Delta x, y) - f(x, y + \Delta y) + f(x, y).$

Introduce the function

(7) $\quad \phi(u) = f(u, y + \Delta y) - f(u, y),$

where u is an auxiliary variable. Then

$\phi(x + \Delta x) = f(x + \Delta x, y + \Delta y) - f(x + \Delta x, y),$

(8) $\quad \phi(x) = f(x, y + \Delta y) - f(x, y).$

Hence (6) may be written

(9) $\quad F = \phi(x + \Delta x) - \phi(x).$

Applying the Theorem of Mean Value (D), Art. 116,

(10) $\quad F = \Delta x\, \phi'(x + \theta_1 \Delta x). \qquad (0 < \theta_1 < 1)$

$[f(x) = \phi(u), a = x, \Delta a = \Delta x.]$

464   DIFFERENTIAL AND INTEGRAL CALCULUS

The value of $\phi'(x + \theta_1 \Delta x)$ is obtained from (8) by taking the partial derivative with respect to x and replacing x by $x + \theta_1 \Delta x$. Thus (10) becomes

(11) $F = \Delta x (f_x(x + \theta_1 \Delta x, y + \Delta y) - f_x(x + \theta_1 \Delta x, y))$.

Now apply (D), Art. 116, to $f_x(x + \theta_1 \Delta x, v)$, regarding v as the independent variable. Then

(12) $F = \Delta x \, \Delta y \, f_{yx}(x + \theta_1 \Delta x, y + \theta_2 \Delta y)$. $(0 < \theta_2 < 1)$

If the second and third terms of the right-hand member of (6) are interchanged, a similar procedure will give

(13) $F = \Delta y \, \Delta x f_{xy}(x + \theta_3 \Delta x, y + \theta_4 \Delta y)$. $(0 < \theta_3 < 1, 0 < \theta_4 < 1)$

Hence from (12) and (13),

(14) $f_{yx}(x + \theta_1 \Delta x, y + \theta_2 \Delta y) = f_{xy}(x + \theta_3 \Delta x, y + \theta_4 \Delta y)$.

Taking the limit of both sides as Δx and Δy approach zero as limits, we have

(15) $f_{yx}(x, y) = f_{xy}(x, y)$,

since these functions are assumed to be continuous.

PROBLEMS

Find the second partial derivatives of each of the following functions.

1. $f(x, y) = Ax^2 + 2\,Bxy + Cy^2$.
   Ans. $f_{xx}(x, y) = 2\,A$; $f_{xy}(x, y) = 2\,B$; $f_{yy}(x, y) = 2\,C$.

2. $f(x, y) = Ax^3 + Bx^2y + Cxy^2 + Dy^3$.
Ans. $f_{xx}(x, y) = 6\,Ax + 2\,By$; $f_{xy}(x, y) = 2\,Bx + 2\,Cy$; $f_{yy}(x, y) = 2\,Cx + 6\,Dy$.

3. $f(x, y) = Ax + By + Ce^{xy}$.
   Ans. $f_{xx}(x, y) = Cy^2 e^{xy}$; $f_{xy} = C(1 + xy)e^{xy}$; $f_{yy}(x, y) = Cx^2 e^{xy}$.

4. $f(x, y) = \dfrac{Ax + By}{Cx + Dy}$.

5. $f(x, y) = x^2 \cos y + y^2 \sin x$.

6. If $f(x, y) = x^3 + 3\,x^2y + 6\,xy^2 - y^3$, show that
   $f_{xx}(2, 3) = 30, f_{xy}(2, 3) = 48, f_{yy}(2, 3) = 6$.

7. If $f(x, y) = x^4 - 4\,x^3y + 8\,xy^3 - y^4$, show that
   $f_{xx}(2, -1) = 96, f_{xy}(2, -1) = -24, f_{yy}(2, -1) = -108$.

8. If $f(x, y) = 2\,x^4 - 3\,x^2y^2 + y^4$, find the values of
   $f_{xx}(2, -2), f_{xy}(2, -2), f_{yy}(2, -2)$.

PARTIAL DIFFERENTIATION

9. If $u = Ax^4 + Bx^3y + Cx^2y^2 + Dxy^3 + Ey^4$, verify the following results.

$$\frac{\partial^3 u}{\partial x^3} = 24\ Ax + 6\ By, \qquad \frac{\partial^3 u}{\partial x^2 \partial y} = 6\ Bx + 4\ Cy,$$

$$\frac{\partial^3 u}{\partial x\ \partial y^2} = 4\ Cx + 6\ Dy, \qquad \frac{\partial^3 u}{\partial y^3} = 6\ Dx + 24\ Ey.$$

10. If $u = (ax^2 + by^2 + cz^2)^3$, show that $\dfrac{\partial^3 u}{\partial x^2\ \partial y} = \dfrac{\partial^3 u}{\partial x\ \partial y\ \partial x} = \dfrac{\partial^3 u}{\partial y\ \partial x^2}$.

11. If $u = \dfrac{xy}{x+y}$, show that $x^2 \dfrac{\partial^2 u}{\partial x^2} + 2\ xy \dfrac{\partial^2 u}{\partial x\ \partial y} + y^2 \dfrac{\partial^2 u}{\partial y^2} = 0$.

12. If $u = \ln \sqrt{x^2 + y^2}$, show that $\dfrac{\partial^2 u}{\partial x^2} + \dfrac{\partial^2 u}{\partial y^2} = 0$.

13. If $v = \dfrac{1}{\sqrt{x^2 + y^2 + z^2}}$, show that $\dfrac{\partial^2 u}{\partial x^2} + \dfrac{\partial^2 u}{\partial y^2} + \dfrac{\partial^2 u}{\partial z^2} = 0$.

ADDITIONAL PROBLEMS

1. A circular hill has a central vertical section in the form of the curve whose equation is $x^2 + 160\ y - 1600 = 0$, where the unit is 1 yd. The top is being cut down in horizontal layers at the constant rate of 100 cu. yd. per day. How fast is the area of the horizontal cross section increasing when the top has been cut down a vertical distance of 4 yd.?

Ans. 25 sq. yd. per day.

2. If $u = \dfrac{e^{xy}}{e^x + e^y}$, show that $\dfrac{\partial u}{\partial x} + \dfrac{\partial u}{\partial y} = (x + y - 1)u$.

3. If $u = \dfrac{1}{r}$, where $r = \sqrt{x^2 + y^2 + z^2}$, show that

$$\left(\frac{\partial u}{\partial x}\right)^2 + \left(\frac{\partial u}{\partial y}\right)^2 + \left(\frac{\partial u}{\partial z}\right)^2 = \frac{1}{r^4}.$$

4. If $z = x^2 \arctan \dfrac{y}{x} - y^2 \arctan \dfrac{x}{y}$, show that $\dfrac{\partial^2 z}{\partial x\ \partial y} = \dfrac{x^2 - y^2}{x^2 + y^2}$.

5. If $u = z \arctan \dfrac{x}{y}$, show that $\dfrac{\partial^2 u}{\partial x^2} + \dfrac{\partial^2 u}{\partial y^2} + \dfrac{\partial^2 u}{\partial z^2} = 0$.

6. If $u = \ln(e^x + e^y + e^z)$, show that $\dfrac{\partial^3 u}{\partial x\ \partial y\ \partial z} = 2\ e^{x+y+z-3u}$.

7. If $u = f(x, y)$ and $x = r \cos \theta$, $y = r \sin \theta$, show that

$$\frac{\partial u}{\partial x} = \cos \theta\ \frac{\partial u}{\partial r} - \frac{\sin \theta}{r}\ \frac{\partial u}{\partial \theta},$$

$$\frac{\partial u}{\partial y} = \sin \theta\ \frac{\partial u}{\partial r} + \frac{\cos \theta}{r}\ \frac{\partial u}{\partial \theta}.$$

8. Let $u = (x_1^2 + x_2^2 + \cdots + x_n^2)^k$. What values of k will satisfy the equation $\dfrac{\partial^2 u}{\partial x_1^2} + \dfrac{\partial^2 u}{\partial x_2^2} + \cdots + \dfrac{\partial^2 u}{\partial x_n^2} = 0$? *Ans.* $k = 1 - \dfrac{n}{2}\ (n > 2)$.

CHAPTER XXIV

APPLICATIONS OF PARTIAL DERIVATIVES

233. Envelope of a family of curves. The equation of a curve generally involves, besides the variables x and y, certain constants upon which the size, shape, and position of that particular curve depend. For example, the locus of the equation

$$(x - \alpha)^2 + y^2 = r^2$$

is a circle whose center lies on the x-axis at a distance of α from the origin, its size depending on the radius r. Suppose α to take on a series of values while r is held fast; then we shall have a corresponding series of circles of equal radius differing in their distances from the origin, as shown in the figure.

Any system of curves formed in this way is called a *family of curves*, and the quantity α, which is constant for any one curve, but changes in passing from one curve to another, is called a *variable parameter*. To indicate that α enters as a variable parameter it is usual to insert it in the functional symbol, thus:

$$f(x, y, \alpha) = 0.$$

The curves of a family may be tangent to the same curve or group of curves, as in the above figure. In that case the name *envelope* of the family is applied to the curve or group of curves. We shall now explain a method for finding the equation of the envelope of a family of curves. Suppose that the curve whose parametric equations are

(1) $\qquad x = \phi(\alpha), \quad y = \psi(\alpha)$

is tangent to each curve of the family

(2) $\qquad f(x, y, \alpha) = 0,$

the parameter α being the same in both cases. For any common value of α equations (1) will satisfy (2). Hence, by (E), Art. 229, since $u = f(x, y, \alpha)$, $du = df = 0$, and z is replaced by α, we have

(3) $\qquad f_x(x, y, \alpha)\phi'(\alpha) + f_y(x, y, \alpha)\psi'(\alpha) + f_\alpha(x, y, \alpha) = 0.$

466

APPLICATIONS OF PARTIAL DERIVATIVES

The slope of (1) at any point is

(4) $$\frac{dy}{dx} = \frac{\psi'(\alpha)}{\phi'(\alpha)},$$ (*A*), Art. 81

and the slope of (2) at any point is

(5) $$\frac{dy}{dx} = -\frac{f_x(x, y, \alpha)}{f_y(x, y, \alpha)}.$$ (*H*), Art. 231

Hence if the curves (1) and (2) are tangent, the slopes at a point of tangency will be equal, giving

$$\frac{\psi'(\alpha)}{\phi'(\alpha)} = -\frac{f_x(x, y, \alpha)}{f_y(x, y, \alpha)}, \text{ or}$$

(6) $$f_x(x, y, \alpha)\phi'(\alpha) + f_y(x, y, \alpha)\psi'(\alpha) = 0.$$

Comparing (6) and (3) gives

(7) $$f_\alpha(x, y, \alpha) = 0.$$

Therefore the coördinates of the point of tangency satisfy

(8) $$f(x, y, \alpha) = 0 \text{ and } f_\alpha(x, y, \alpha) = 0;$$

that is, the parametric equations of the envelope, in case an envelope exists, may be found by solving these equations for x and y in terms of the parameter α.

General rule for finding the parametric equations of the envelope

FIRST STEP. *Write the equation of the family of curves in the form $f(x, y, \alpha) = 0$ and derive the equation $f_\alpha(x, y, \alpha) = 0$.*

SECOND STEP. *Solve these two equations for x and y in terms of the parameter α.*

The rectangular equation may be found either by eliminating α between the equations (8) or from the parametric equations (Art. 81).

ILLUSTRATIVE EXAMPLE 1. For the family of circles at the beginning of this article,
$$f(x, y, \alpha) = (x - \alpha)^2 + y^2 - r^2 = 0.$$

Hence $$f_\alpha(x, y, \alpha) = (x - \alpha) = 0.$$

Eliminating α, the result is $y^2 - r^2 = 0$, or $y = r$, $y = -r$, and these are the equations of the lines AB and CD in the figure.

ILLUSTRATIVE EXAMPLE 2. Find the envelope of the family of straight lines $x \cos \alpha + y \sin \alpha = p$, α being the variable parameter.

Solution. (9) $$f(x, y, \alpha) = x \cos \alpha + y \sin \alpha - p = 0.$$

First Step. Differentiating with respect to α,

(10) $\qquad f_\alpha(x, y, \alpha) = -x \sin \alpha + y \cos \alpha = 0.$

Second Step. Multiplying (9) by $\cos \alpha$ and (10) by $\sin \alpha$ and subtracting, we get
$$x = p \cos \alpha.$$
Similarly, eliminating x between (9) and (10),
$$y = p \sin \alpha.$$
The *parametric equations of the envelope* are therefore

(11) $\qquad \begin{cases} x = p \cos \alpha, \\ y = p \sin \alpha, \end{cases}$

α being the parameter. Squaring equations (11) and adding, we get $\qquad x^2 + y^2 = p^2,$

the rectangular equation of the envelope, a circle.

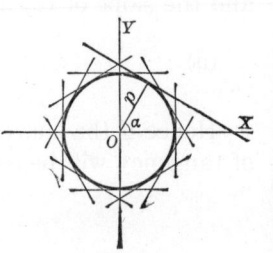

ILLUSTRATIVE EXAMPLE 3. Find the envelope of a line of constant length a whose extremities move along two fixed rectangular axes.

Solution. Let $AB = a$ in length, and let

(12) $\qquad x \cos \alpha + y \sin \alpha - p = 0$

be its equation. Now as AB moves, both α and p will vary. But p may be found in terms of α. For $AO = AB \cos \alpha = a \cos \alpha$, and also $p = AO \sin \alpha = a \sin \alpha \cos \alpha$. Substituting in (12), we get

(13) $\qquad x \cos \alpha + y \sin \alpha - a \sin \alpha \cos \alpha = 0,$

where α is the variable parameter. This equation is in the form $f(x, y, \alpha) = 0$. Differentiating with respect to α, the equation $f_\alpha(x, y, \alpha) = 0$ is

(14) $\qquad -x \sin \alpha + y \cos \alpha + a \sin^2 \alpha - a \cos^2 \alpha = 0.$

Solving (13) and (14) for x and y in terms of α, the result is

(15) $\qquad \begin{cases} x = a \sin^3 \alpha, \\ y = a \cos^3 \alpha, \end{cases}$

the parametric equations of the envelope, a hypocycloid. The corresponding rectangular equation is found from equations (15) by eliminating α as follows:

$$x^{\frac{2}{3}} = a^{\frac{2}{3}} \sin^2 \alpha,$$

$$y^{\frac{2}{3}} = a^{\frac{2}{3}} \cos^2 \alpha.$$

Adding, $\qquad x^{\frac{2}{3}} + y^{\frac{2}{3}} = a^{\frac{2}{3}},$ the rectangular equation of the hypocycloid.

Many problems occur in which it is convenient to use two parameters connected by an equation of condition. By using the latter, one parameter may be eliminated from the equation of the family of curves. It is, however, often better to proceed as in the following example.

APPLICATIONS OF PARTIAL DERIVATIVES 469

ILLUSTRATIVE EXAMPLE 4. Find the envelope of the family of ellipses whose axes coincide and whose area is constant.

Solution. (16) $\dfrac{x^2}{a^2} + \dfrac{y^2}{b^2} = 1$

is the equation of the ellipse, where a and b are the variable parameters connected by the equation

(17) $\quad \pi ab = k,$

πab being the area of an ellipse whose semiaxes are a and b. Differentiating, regarding a and b as variables and x and y as constants, we have, using differentials,

$$\frac{x^2 da}{a^3} + \frac{y^2 db}{b^3} = 0, \text{ from (16)},$$

and $\quad b\,da + a\,db = 0,$ from (17).

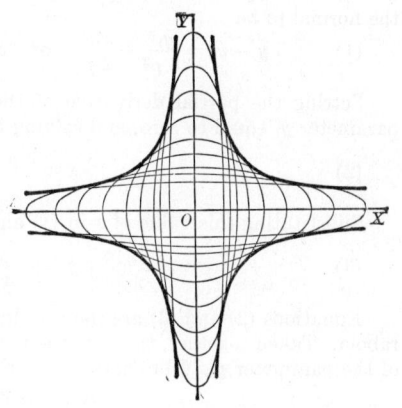

Transposing one term in each to the second member and dividing, the result is

$$\frac{x^2}{a^2} = \frac{y^2}{b^2}.$$

Therefore, using (16), $\quad \dfrac{x^2}{a^2} = \dfrac{1}{2}$ and $\dfrac{y^2}{b^2} = \dfrac{1}{2},$

whence $\quad a = \pm x\sqrt{2} \quad$ and $\quad b = \pm y\sqrt{2}.$

Substituting these values in (17), we get the envelope $xy = \pm \dfrac{k}{2\,\pi}$, a pair of conjugate rectangular hyperbolas (see figure).

234. The evolute of a given curve considered as the envelope of its normals. Since the normals to a curve are all tangent to the evolute (Art. 110), it is evident that *the evolute of a curve may be defined as the envelope of its normals*. It is also interesting to notice that if we find the parametric equations of the envelope by the method of the previous article, we get the coördinates x and y of the center of curvature; so that we have here *a second method for finding the coördinates of the center of curvature*. If we eliminate the variable parameter, we obtain the rectangular equation of the evolute.

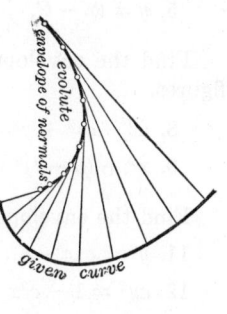

ILLUSTRATIVE EXAMPLE. Find the evolute of the parabola $y^2 = 4\,px$ considered as the envelope of its normals.

Solution. The equation of the normal at any point (x_1, y_1) is

$$y - y_1 = -\frac{y_1}{2\,p}(x - x_1)$$

470 DIFFERENTIAL AND INTEGRAL CALCULUS

from (2), Art. 43. As we are considering the normals all along the curve, both x_1 and y_1 will vary. Eliminating x_1 by means of $y_1^2 = 4\,px_1$, we find the equation of the normal to be

(1) $\qquad y - y_1 = \dfrac{y_1^3}{8\,p^2} - \dfrac{xy_1}{2\,p},\quad$ or $\quad xy_1 + 2\,py - 2\,py_1 - \dfrac{y_1^3}{4\,p} = 0.$

Setting the partial derivative of the left-hand member with respect to the parameter y_1 equal to zero, and solving for x, we find

(2) $\qquad\qquad\qquad x = \dfrac{3\,y_1^2 + 8\,p^2}{4\,p}.$

Substituting this value of x in (1) and solving for y,

(3) $\qquad\qquad\qquad y = -\dfrac{y_1^3}{4\,p^2}.$

Equations (2) and (3) are the coördinates of the center of curvature of the parabola. Taken together, they are the parametric equations of the evolute in terms of the parameter y_1. Eliminating y_1, we obtain

$$27\,py^2 = 4(x - 2\,p)^3,$$

the rectangular equation of the evolute of the parabola. This is the same result we obtained in Illustrative Example 1, Art. 109, by the first method.

PROBLEMS

Find the envelopes of the following systems of straight lines and draw the figures.

1. $y = mx + m^2.$ $\qquad\qquad\qquad$ Ans. $x^2 + 4\,y = 0.$

2. $y = \dfrac{x}{m} + m^2.$ $\qquad\qquad\qquad$ $27\,x^2 = 4\,y^3.$

3. $y = m^2x - 2\,m^3.$ $\qquad\qquad\qquad$ $27\,y = x^3.$

4. $y = 2\,mx + m^4.$ $\qquad\qquad\qquad$ $16\,y^3 + 27\,x^4 = 0.$

5. $y = tx - t^2.$ $\qquad$ 6. $y = t^2x + t.$ $\qquad$ 7. $y = mx - 2\,m^2.$

Find the envelopes of the following systems of circles and draw the figures.

8. $(x - c)^2 + y^2 = 4\,c.$ $\qquad\qquad$ Ans. $y^2 = 4\,x + 4.$

9. $x^2 + (y - t)^2 = 2\,t.$ $\qquad\qquad$ 10. $(x - t)^2 + (y + t)^2 = t^2.$

Find the envelopes of the following systems of parabolas.

11. $y^2 = c(x - c).$ $\qquad\qquad\qquad$ Ans. $2\,y = \pm\,x.$

12. $cy^2 = 1 - c^2x.$

13. Find the evolute of the ellipse $b^2x^2 + a^2y^2 = a^2b^2$, taking the equation of the normal in the form $by = ax\tan\phi - (a^2 - b^2)\sin\phi$, the eccentric angle ϕ being the parameter.

Ans. $x = \dfrac{a^2 - b^2}{a}\cos^3\phi,\; y = \dfrac{b^2 - a^2}{b}\sin^3\phi\,;\;$ or $(ax)^{\frac{2}{3}} + (by)^{\frac{2}{3}} = (a^2 - b^2)^{\frac{2}{3}}.$

APPLICATIONS OF PARTIAL DERIVATIVES

14. Find the evolute of the hypocycloid $x^{\frac{2}{3}} + y^{\frac{2}{3}} = a^{\frac{2}{3}}$, the equation of whose normal is
$$y \cos \tau - x \sin \tau = a \cos 2\tau,$$
τ being the parameter. *Ans.* $(x+y)^{\frac{2}{3}} + (x-y)^{\frac{2}{3}} = 2 a^{\frac{2}{3}}$.

15. Find the envelope of the circles which pass through the origin and have their centers on the hyperbola $x^2 - y^2 = c^2$.
Ans. The lemniscate $(x^2 + y^2)^2 = 4 c^2 (x^2 - y^2)$.

16. Find the envelope of a line such that the sum of its intercepts on the axes equals c. *Ans.* The parabola $x^{\frac{1}{2}} + y^{\frac{1}{2}} = c^{\frac{1}{2}}$.

17. Find the envelope of the family of ellipses $b^2 x^2 + a^2 y^2 = a^2 b^2$ when the sum of its semiaxes equals c. *Ans.* The hypocycloid $x^{\frac{2}{3}} + y^{\frac{2}{3}} = c^{\frac{2}{3}}$.

18. Projectiles are fired from a gun with an initial velocity v_0. Supposing the gun can be given any elevation and is kept always in the same vertical plane, what is the envelope of all possible trajectories, the resistance of the air being neglected?

HINT. The equation of any trajectory is
$$y = x \tan \alpha - \frac{gx^2}{2 v_0^2 \cos^2 \alpha},$$
α being the variable parameter.
Ans. The parabola $y = \frac{v_0^2}{2g} - \frac{gx^2}{2 v_0^2}$.

19. If the family
$$t^2 f(x, y) + t g(x, y) + h(x, y) = 0$$
has an envelope, show that it is
$$g^2(x, y) - 4 f(x, y) h(x, y) = 0.$$

235. Tangent line and normal plane to a skew curve. The student is already familiar with the parametric representation of a plane curve (Art. 81). In order to extend this notion to curves in space, let the coördinates of any point $P(x, y, z)$ on a skew curve be given as functions of some fourth variable which we shall denote by t; thus,

(1) $x = \phi(t), \quad y = \psi(t), \quad z = \chi(t).$

The elimination of the parameter t between these equations two by two will give us the equations of the projecting cylinders of the curve on the coördinate planes.

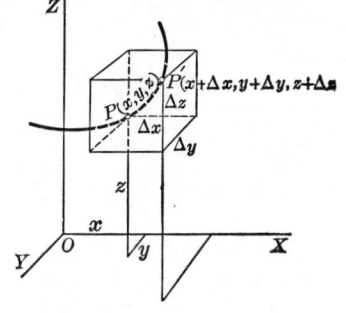

472   DIFFERENTIAL AND INTEGRAL CALCULUS

Let the point $P(x, y, z)$ correspond to the value t of the parameter, and the point $P'(x + \Delta x, y + \Delta y, z + \Delta z)$ correspond to the value $t + \Delta t$ where $\Delta x, \Delta y, \Delta z$ are the increments of x, y, z due to the increment Δt as found from equations (1). From analytic geometry of three dimensions, we know that the direction cosines of the secant (diagonal) PP' are proportional to

$$\Delta x, \quad \Delta y, \quad \Delta z;$$

or, dividing through by Δt and denoting the direction angles of the secant by α', β', γ',

(2) $$\frac{\cos \alpha'}{\frac{\Delta x}{\Delta t}} = \frac{\cos \beta'}{\frac{\Delta y}{\Delta t}} = \frac{\cos \gamma'}{\frac{\Delta z}{\Delta t}}.$$

Now let P' approach P along the curve. Then Δt, and therefore also $\Delta x, \Delta y, \Delta z$, will approach zero as a limit, and the secant PP' will approach the tangent line to the curve at P as a limiting position.

Now $$\lim_{\Delta t \to 0} \frac{\Delta x}{\Delta t} = \frac{dx}{dt} = \phi'(t), \text{ etc.}$$

Hence, for the tangent line,

(A) $$\frac{\cos \alpha}{\frac{dx}{dt}} = \frac{\cos \beta}{\frac{dy}{dt}} = \frac{\cos \gamma}{\frac{dz}{dt}}.$$

When the point of contact is $P_1(x_1, y_1, z_1)$, we use the notation

(3) $$\left|\frac{dx}{dt}\right|_1 = \text{value of } \frac{dx}{dt} \text{ when } x = x_1, y = y_1, z = z_1,$$

and similar notation for the other derivatives.

Hence, by (2) and (4), p. 5, we have the following result.

The equations of the tangent line to the curve whose equations are

(1) $$x = \phi(t), \quad y = \psi(t), \quad z = \chi(t)$$

at the point $P_1(x_1, y_1, z_1)$ *are*

(B) $$\frac{x - x_1}{\left|\frac{dx}{dt}\right|_1} = \frac{y - y_1}{\left|\frac{dy}{dt}\right|_1} = \frac{z - z_1}{\left|\frac{dz}{dt}\right|_1}.$$

The *normal plane* of a skew curve at a point $P_1(x_1, y_1, z_1)$ is the plane which passes through P_1 and is perpendicular to the tangent line at P_1. The denominators in (B) are the direction numbers of the tangent line at P_1. Hence we have the following result.

APPLICATIONS OF PARTIAL DERIVATIVES

The equation of the normal plane to the curve (1) *at* P_1 (x_1, y_1, z_1) *is*

(C) $\quad \left|\dfrac{dx}{dt}\right|_1 (x - x_1) + \left|\dfrac{dy}{dt}\right|_1 (y - y_1) + \left|\dfrac{dz}{dt}\right|_1 (z - z_1) = 0.$

ILLUSTRATIVE EXAMPLE. Find the equations of the tangent line and the equation of the normal plane to the circular helix (θ being the parameter)

(4) $\quad \begin{cases} x = a \cos \theta, \\ y = a \sin \theta, \\ z = b\theta, \end{cases}$

(a) at any point (x_1, y_1, z_1); (b) when $\theta = 2\pi$.

Solution. $\dfrac{dx}{d\theta} = -a \sin \theta = -y, \quad \dfrac{dy}{d\theta} = a \cos \theta = x, \quad \dfrac{dz}{d\theta} = b.$

Substituting in (B) and (C), we get, at (x_1, y_1, z_1),

(5) $\quad \dfrac{x - x_1}{-y_1} = \dfrac{y - y_1}{x_1} = \dfrac{z - z_1}{b},$ tangent line,

and
$$-y_1(x - x_1) + x_1(y - y_1) + b(z - z_1) = 0,$$
normal plane.

When $\theta = 2\pi$, the point on the curve is T $(a, 0, 2b\pi)$, giving

$$\dfrac{x - a}{0} = \dfrac{y - 0}{a} = \dfrac{z - 2b\pi}{b},$$

or $\quad x = a, \quad by = az - 2ab\pi,$
the equations of the tangent line, and

$$ay + bz - 2b^2\pi = 0,$$

the equation of the normal plane.

REMARK. For the tangent line (5) we have by (2) and (4), p. 5,

$$\cos \gamma = \dfrac{b}{\sqrt{x_1^2 + y_1^2 + b^2}} = \dfrac{b}{\sqrt{a^2 + b^2}} = \text{a constant.}$$

That is, the helix cuts all elements of the cylinder $x^2 + y^2 = a^2$ under the same angle.

236. Length of arc of a skew curve. From the figure of the preceding article we have

(1) $\quad \dfrac{(\text{Chord } PP')^2}{\Delta t^2} = \left(\dfrac{\Delta x}{\Delta t}\right)^2 + \left(\dfrac{\Delta y}{\Delta t}\right)^2 + \left(\dfrac{\Delta z}{\Delta t}\right)^2.$

Let arc $PP' = \Delta s$. Proceeding as in Art. 95, we easily prove

(2) $\quad \left(\dfrac{ds}{dt}\right)^2 = \left(\dfrac{dx}{dt}\right)^2 + \left(\dfrac{dy}{dt}\right)^2 + \left(\dfrac{dz}{dt}\right)^2.$

From this we obtain

(D) $\quad s = \displaystyle\int_{t_0}^{t_1} (dx^2 + dy^2 + dz^2)^{\frac{1}{2}},$

where $x = \phi(t), y = \psi(t), z = \chi(t)$, as in (1), Art. 235.

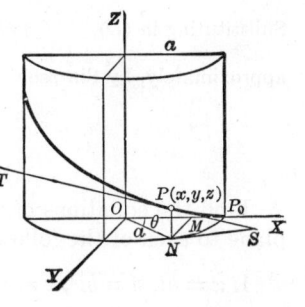

474 DIFFERENTIAL AND INTEGRAL CALCULUS

The direction cosines of the tangent line can now be given a simple form. For, from (A) of the preceding article, and by the above equation (2), using formulas in (2), p. 5, we have

(3) $$\cos \alpha = \frac{dx}{ds}, \quad \cos \beta = \frac{dy}{ds}, \quad \cos \gamma = \frac{dz}{ds}.$$

ILLUSTRATIVE EXAMPLE. Find the length of arc of the skew cubic

(4) $$x = t, \quad y = \tfrac{1}{2} t^2, \quad z = \tfrac{1}{3} t^3$$

between the points where $t = 0$ and $t = 4$.

Solution. Differentiating (4), we obtain

$$dx = dt, \quad dy = t\, dt, \quad dz = t^2\, dt.$$

Substituting in (D), $\quad s = \int_0^4 \sqrt{1 + t^2 + t^4}\, dt = 23.92,$

approximately, by Simpson's Rule, letting $n = 8$.

PROBLEMS

Find the equations of the tangent line and the equation of the normal plane to each of the following skew curves at the point indicated.

1. $x = at,\ y = bt^2,\ z = ct^3;\ t = 1.$

   Ans. $\dfrac{x-a}{a} = \dfrac{y-b}{2\,b} = \dfrac{z-c}{3\,c};\ ax + 2\,by + 3\,cz = a^2 + 2\,b^2 + 3\,c^2.$

2. $x = 2\,t,\ y = t^2,\ z = 4\,t^4;\ t = 1.$

   Ans. $\dfrac{x-2}{2} = \dfrac{y-1}{2} = \dfrac{z-4}{16};\ x + y + 8\,z = 35.$

3. $x = t^2 - 1,\ y = t + 1,\ z = t^3;\ t = 2.$

   Ans. $\dfrac{x-3}{4} = \dfrac{y-3}{1} = \dfrac{z-8}{12};\ 4\,x + y + 12\,z = 111.$

4. $x = t^3 - 1,\ y = t^2 + t,\ z = 4\,t^3 - 3\,t + 1;\ t = 1.$

   Ans. $\dfrac{x}{3} = \dfrac{y-2}{3} = \dfrac{z-2}{9};\ x + y + 3\,z = 8.$

5. $x = 2\,t - 3,\ y = 5 - t^2,\ z = \dfrac{2}{t};\ t = 2.$

6. $x = a \cos t,\ y = b \sin t,\ z = t;\ t = \tfrac{1}{6}\pi.$

7. $x = t,\ y = e^t,\ z = e^{-t};\ t = 0.$

8. $x = \cos t,\ y = \sin t,\ z = \tan t;\ t = 0.$

9. Find the length of arc of the circular helix

   $$x = a \cos \theta, \quad y = a \sin \theta, \quad z = b\theta$$

   between the points where $\theta = 0$ and $\theta = 2\,\pi$. Ans. $2\,\pi\sqrt{a^2 + b^2}.$

APPLICATIONS OF PARTIAL DERIVATIVES 475

10. Find the length of arc of the curve
$$x = 3\,\theta\cos\theta, \quad y = 3\,\theta\sin\theta, \quad z = 4\,\theta$$
between the points where $\theta = 0$ and $\theta = 4$.   *Ans.* $26 + \frac{25}{6}\ln 5 = 32.70$.

11. Find the length of arc of the curve
$$x = 2\,t, \quad y = t^2 - 2, \quad z = 1 - t^2$$
between the points where $t = 0$ and $t = 2$.

12. Given the two curves
(5) $\qquad x = t, \quad y = 2\,t^2, \quad z = -\dfrac{1}{t};$
(6) $\qquad x = 1 - \theta, \quad y = 2\cos\theta, \quad z = \sin\theta - 1.$
(a) Show that the two curves intersect at the point $A(1, 2, -1)$.
(b) Find the direction cosines of the tangent line to (5) at A.
$$\textit{Ans.} \quad \frac{1}{\sqrt{18}},\ \frac{4}{\sqrt{18}},\ \frac{1}{\sqrt{18}}.$$
(c) Find the direction cosines of the tangent line to (6) at A.
(d) Find the angle of intersection of the curves at A.   *Ans.* $90°$.

13. Given the two curves
$$x = 2 - t, \quad y = t^2 - 4, \quad z = t^3 - 8;$$
$$x = \sin\theta, \quad y = \theta, \quad z = 1 - \cos\theta.$$
(a) Show that the two curves intersect at the origin O.
(b) Find the direction cosines of the tangent line to each curve at O.
(c) Find the angle of intersection of the curves at O.

14. (a) If OF, OE, ON in the first figure of Art. 222 are chosen as axes of coördinates OX, OY, OZ, respectively, and if $P(x, y, z)$ is a point on the sphere, prove that $x = a\cos\phi\sin\theta$, $y = a\cos\phi\cos\theta$, $z = a\sin\phi$, if ϕ and θ are, respectively, the latitude and longitude of P.

(b) Using (3), and (3) on page 5, find the angle α at P between a curve on the sphere for which $\theta = f(\phi)$ and the parallel through P.
$$\textit{Ans.} \quad \tan\alpha = \sec\phi\,\frac{d\phi}{d\theta}, \text{ as in Art. 222.}$$

237. Normal line and tangent plane to a surface. A straight line is said to be *tangent to a surface* at a point P if it is the limiting position of a secant line through P and a neighboring point P' on the surface when P' is made to approach P along a curve on the surface. We now proceed to establish a theorem of fundamental importance.

Theorem. *All tangent lines to a surface at a given point lie in a plane.*

Proof. Let
(1) $\qquad F(x, y, z) = 0$
be the equation of the given surface, and let $P(x, y, z)$ be the given point on the surface. If now P' be made to approach P along a curve

C lying on the surface and passing through P and P', then evidently the secant line PP' approaches the position of a tangent line to the curve C at P. Now let the equations of the curve C be

(2) $\qquad x = \phi(t), \quad y = \psi(t), \quad z = \chi(t).$

Then the equation (1) must be satisfied identically by these values. Hence, if $u = F(x, y, z)$, then $u = 0$, $du = 0$, and by (E), Art. 229,

(3) $\qquad \dfrac{\partial F}{\partial x}\dfrac{dx}{dt} + \dfrac{\partial F}{\partial y}\dfrac{dy}{dt} + \dfrac{\partial F}{\partial z}\dfrac{dz}{dt} = 0.$

This equation (see (3), Art. 4) shows that the tangent line to (2), whose direction cosines are proportional to

$$\dfrac{dx}{dt}, \ \dfrac{dy}{dt}, \ \dfrac{dz}{dt},$$

is perpendicular to a line whose direction cosines are proportional to

(4) $\qquad \dfrac{\partial F}{\partial x}, \ \dfrac{\partial F}{\partial y}, \ \dfrac{\partial F}{\partial z}.$ $\qquad$ By (3), p. 5

Let $P_1(x_1, y_1, z_1)$ be a point on the surface and

(5) $\qquad \left|\dfrac{\partial F}{\partial x}\right|_1, \ \left|\dfrac{\partial F}{\partial y}\right|_1, \ \left|\dfrac{\partial F}{\partial z}\right|_1$

the values of the partial derivatives in (4) when $x = x_1$, $y = y_1$, $z = z_1$. The line passing through P_1 whose direction numbers are given by (5) is called the *normal* line to the surface at P_1. Hence we have the following result:

The equations of the normal line to the surface

(1) $\qquad F(x, y, z) = 0$

at $P_1(x_1, y_1, z_1)$ *are*

(E) $\qquad \dfrac{x - x_1}{\left|\dfrac{\partial F}{\partial x}\right|_1} = \dfrac{y - y_1}{\left|\dfrac{\partial F}{\partial y}\right|_1} = \dfrac{z - z_1}{\left|\dfrac{\partial F}{\partial z}\right|_1}.$

The preceding argument shows that *all* tangent lines to the surface (1) at P_1 are perpendicular to the normal line at P_1. Hence they lie in a plane. Thus the theorem is proved.

This plane is called the *tangent plane* at (P_1).

We may now state the following result.

The equation of the tangent plane to the surface (1) *at the point of contact* $P_1(x_1, y_1, z_1)$ *is*

(F) $\qquad \left|\dfrac{\partial F}{\partial x}\right|_1 (x - x_1) + \left|\dfrac{\partial F}{\partial y}\right|_1 (y - y_1) + \left|\dfrac{\partial F}{\partial z}\right|_1 (z - z_1) = 0.$

APPLICATIONS OF PARTIAL DERIVATIVES 477

REMARK. If all the denominators in (E) vanish, the normal line and tangent plane are indeterminate. Such points are called *singular points* and are excluded here.

In case the equation of the surface is given in the form $z = f(x, y)$, let

(6) $\qquad F(x, y, z) = f(x, y) - z = 0.$

Then $\qquad \dfrac{\partial F}{\partial x} = \dfrac{\partial f}{\partial x} = \dfrac{\partial z}{\partial x}, \quad \dfrac{\partial F}{\partial y} = \dfrac{\partial f}{\partial y} = \dfrac{\partial z}{\partial y}, \quad \dfrac{\partial F}{\partial z} = -1.$

Hence, by (E), we have the following result.

The equations of the normal line to the surface $z = f(x, y)$ at (x_1, y_1, z_1) are

(G) $\qquad \dfrac{x - x_1}{\left|\dfrac{\partial z}{\partial x}\right|_1} = \dfrac{y - y_1}{\left|\dfrac{\partial z}{\partial y}\right|_1} = \dfrac{z - z_1}{-1}.$

Also, from (F), we obtain

(H) $\qquad \left|\dfrac{\partial z}{\partial x}\right|_1 (x - x_1) + \left|\dfrac{\partial z}{\partial y}\right|_1 (y - y_1) - (z - z_1) = 0,$

which is then *the formula for the equation of a plane tangent at (x_1, y_1, z_1) to a surface whose equation is given in the form $z = f(x, y)$*.

238. Geometric interpretation of the total differential. We are now in a position to discuss formula (B), Art. 227, by geometry, in a manner entirely analogous to that in Art. 91.

Consider the surface

(1) $\qquad\qquad z = f(x, y),$

and the point (x_1, y_1, z_1) on it. Then the total differential of (1) is, when $\qquad\qquad x = x_1, \quad y = y_1,$

(2) $\qquad\qquad dz = \left|\dfrac{\partial z}{\partial x}\right|_1 \Delta x + \left|\dfrac{\partial z}{\partial y}\right|_1 \Delta y,$

using (B), Art. 227, and replacing dx and dy by their equivalents, Δx and Δy, respectively. Let us find the z-coördinate of the point in the tangent plane at P_1 where

$\qquad\qquad x = x_1 + \Delta x, \quad y = y_1 + \Delta y.$

Substituting these values in (H) of Art. 237, we find

(3) $\qquad\qquad z - z_1 = \left|\dfrac{\partial z}{\partial x}\right|_1 \Delta x + \left|\dfrac{\partial z}{\partial y}\right|_1 \Delta y.$

Comparing (2) and (3), we get $dz = z - z_1$. Hence the

Theorem. *The total differential of a function $f(x, y)$ corresponding to the increments Δx and Δy equals the corresponding increment of the z-coördinate of the tangent plane to the surface $z = f(x, y)$.*

Thus, in the figure, PP' is the plane tangent to surface PQ at $P(x, y, z)$.

Let $\quad AB = \Delta x$

and $\quad CD = \Delta y$;

then $\quad dz = z - z_1 = DP' - DE = EP'$.

Notice also that $\quad \Delta z = DQ - DE = EQ$.

ILLUSTRATIVE EXAMPLE. Find the equation of the tangent plane and the equations of the normal line to the sphere $x^2 + y^2 + z^2 = 14$ at the point (1, 2, 3).

Solution. Let $F(x, y, z) = x^2 + y^2 + z^2 - 14$;

then $\quad \dfrac{\partial F}{\partial x} = 2x, \quad \dfrac{\partial F}{\partial y} = 2y, \quad \dfrac{\partial F}{\partial z} = 2z; \quad x_1 = 1, \quad y_1 = 2, \quad z_1 = 3.$

Therefore $\quad \left|\dfrac{\partial F}{\partial x}\right|_1 = 2, \quad \left|\dfrac{\partial F}{\partial y}\right|_1 = 4, \quad \left|\dfrac{\partial F}{\partial z}\right|_1 = 6.$

Substituting in (F), $\quad 2(x-1) + 4(y-2) + 6(z-3) = 0$,
or $\quad x + 2y + 3z = 14$, the tangent plane.

Substituting in (E), $\quad \dfrac{x-1}{2} = \dfrac{y-2}{4} = \dfrac{z-3}{6}$,

giving $\quad z = 3x$ and $2z = 3y$, equations of the normal line. *Ans.*

PROBLEMS

Find the equation of the tangent plane and the equations of the normal line to each of the following surfaces at the point indicated.

1. $x^2 + y^2 + z^2 = 49$; (6, 2, 3).
   Ans. $6x + 2y + 3z = 49$; $\dfrac{x-6}{6} = \dfrac{y-2}{2} = \dfrac{z-3}{3}$.

2. $z = x^2 + y^2 - 1$; (2, 1, 4).
   Ans. $4x + 2y - z = 6$; $\dfrac{x-2}{4} = \dfrac{y-1}{2} = \dfrac{z-4}{-1}$.

APPLICATIONS OF PARTIAL DERIVATIVES

3. $x^2 + xy^2 + y^3 + z + 1 = 0$; $(2, -3, 4)$.
   Ans. $13x + 15y + z + 15 = 0$; $\dfrac{x-2}{13} = \dfrac{y+3}{15} = \dfrac{z-4}{1}$.

4. $x^2 + 2xy + y^2 + z - 7 = 0$; $(1, -2, 6)$.
   Ans. $2x + 2y - z + 8 = 0$; $\dfrac{x-1}{2} = \dfrac{y+2}{2} = \dfrac{z-6}{-1}$.

5. $x^2y^2 + xz - 2y^3 - 10 = 0$; $(2, 1, 4)$.
   Ans. $4x + y + z - 13 = 0$; $\dfrac{x-2}{4} = \dfrac{y-1}{1} = \dfrac{z-4}{1}$.

6. $x^2 - y^2 - z^2 = 1$; $(3, 2, 2)$.

7. $x^2 + y^2 - z^2 = 25$; $(5, 5, 5)$.

8. $2x^2 + 3y^2 + 4z^2 = 6$; $(1, 1, \tfrac{1}{2})$.

9. $x + y - z^2 = 3$; $(3, 4, 2)$.

10. Find the equation of the tangent plane to the hyperboloid of two sheets $\dfrac{x^2}{a^2} - \dfrac{y^2}{b^2} - \dfrac{z^2}{c^2} = 1$ at (x_1, y_1, z_1).   Ans. $\dfrac{x_1 x}{a^2} - \dfrac{y_1 y}{b^2} - \dfrac{z_1 z}{c^2} = 1$.

11. Find the equation of the tangent plane at the point (x_1, y_1, z_1) on the surface $ax^2 + by^2 + cz^2 + d = 0$.   Ans. $ax_1 x + by_1 y + cz_1 z + d = 0$.

12. Show that the equation of the plane tangent to the sphere
$$x^2 + y^2 + z^2 + 2Lx + 2My + 2Nz + D = 0$$
at the point (x_1, y_1, z_1) is
$$x_1 x + y_1 y + z_1 z + L(x + x_1) + M(y + y_1) + N(z + z_1) + D = 0.$$

13. Find the equation of the tangent plane at any point of the surface
$$x^{\tfrac{2}{3}} + y^{\tfrac{2}{3}} + z^{\tfrac{2}{3}} = a^{\tfrac{2}{3}},$$
and show that the sum of the squares of the intercepts on the axes made by the tangent plane is constant.

14. Prove that the tetrahedron formed by the coördinate planes and any tangent plane to the surface $xyz = a^3$ is of constant volume.

15. The surface $x^2 - 4y^2 - 4z = 0$ is cut by the curve $x = \dfrac{t^2}{2},\ y = \dfrac{4}{t}$, $z = \dfrac{t - 2t^2}{2}$ at the point $(2, 2, -3)$. What is the angle of intersection?
   Ans. $90° - \arccos \dfrac{19}{3\sqrt{138}} = 32°\,37'$.

16. The surface $x^2 + y^2 + 3z^2 = 25$ and the curve $x = 2t$, $y = \dfrac{3}{t}$, $z = -2t^2$ intersect at the point on the curve given by $t = 1$. What is the angle of intersection?   Ans. $90° - \arccos \dfrac{19}{7\sqrt{29}} = 30°\,16'$.

17. The ellipsoid $x^2 + 2y^2 + 3z^2 = 20$ and the skew curve $x = \tfrac{3}{2}(t^2+1)$, $y = t^4 + 1$, $z = t^3$ meet at the point $(3, 2, 1)$. Show that the curve cuts the surface orthogonally.

480   DIFFERENTIAL AND INTEGRAL CALCULUS

239. Another form of the equations of the tangent line and normal plane to a skew curve. If the curve in question be the curve of intersection AB of the two surfaces $F(x, y, z) = 0$ and $G(x, y, z) = 0$, the tangent line PT at $P(x_1, y_1, z_1)$ is the intersection of the tangent planes CD and CE at that point, for it is also tangent to both surfaces and hence must lie in both tangent planes. The equations of the two tangent planes at P are, from (F),

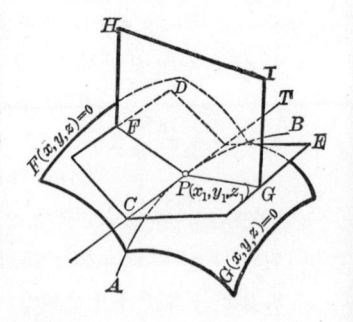

(1)
$$\left|\frac{\partial F}{\partial x}\right|_1 (x - x_1) + \left|\frac{\partial F}{\partial y}\right|_1 (y - y_1) + \left|\frac{\partial F}{\partial z}\right|_1 (z - z_1) = 0,$$

$$\left|\frac{\partial G}{\partial x}\right|_1 (x - x_1) + \left|\frac{\partial G}{\partial y}\right|_1 (y - y_1) + \left|\frac{\partial G}{\partial z}\right|_1 (z - z_1) = 0.$$

Taken simultaneously, these equations are the equations of the tangent line PT to the skew curve AB.

If A, B, C are direction numbers for the line of intersection of the planes (1), then, by (6), Art. 4,

(2)
$$A = \left|\frac{\partial F}{\partial y}\right|_1 \left|\frac{\partial G}{\partial z}\right|_1 - \left|\frac{\partial F}{\partial z}\right|_1 \left|\frac{\partial G}{\partial y}\right|_1, \quad B = \left|\frac{\partial F}{\partial z}\right|_1 \left|\frac{\partial G}{\partial x}\right|_1 - \left|\frac{\partial F}{\partial x}\right|_1 \left|\frac{\partial G}{\partial z}\right|_1,$$

$$C = \left|\frac{\partial F}{\partial x}\right|_1 \left|\frac{\partial G}{\partial y}\right|_1 - \left|\frac{\partial F}{\partial y}\right|_1 \left|\frac{\partial G}{\partial x}\right|_1.$$

Then the equations of the tangent line CPT are

(3)
$$\frac{x - x_1}{A} = \frac{y - y_1}{B} = \frac{z - z_1}{C}.$$

The equation of the normal plane PHI is

(4) $$A(x - x_1) + B(y - y_1) + C(z - z_1) = 0.$$

ILLUSTRATIVE EXAMPLE 1. Find the equations of the tangent line and the equation of the normal plane at $(r, r, r\sqrt{2})$ to the curve of intersection of the sphere and cylinder whose equations are, respectively, $x^2 + y^2 + z^2 = 4r^2$, $x^2 + y^2 = 2rx$.

Solution. Let $F = x^2 + y^2 + z^2 - 4r^2$ and $G = x^2 + y^2 - 2rx$.

$$\left|\frac{\partial F}{\partial x}\right|_1 = 2r, \quad \left|\frac{\partial F}{\partial y}\right|_1 = 2r, \quad \left|\frac{\partial F}{\partial z}\right|_1 = 2r\sqrt{2};$$

$$\left|\frac{\partial G}{\partial x}\right|_1 = 0, \quad \left|\frac{\partial G}{\partial y}\right|_1 = 2r, \quad \left|\frac{\partial G}{\partial z}\right|_1 = 0.$$

APPLICATIONS OF PARTIAL DERIVATIVES

Substituting in (2), we find

$$A = -4r^2\sqrt{2}, \quad B = 0, \quad C = 4r^2.$$

Hence, by (3), we have

$$\frac{x-r}{-\sqrt{2}} = \frac{y-r}{0} = \frac{z-r\sqrt{2}}{1};$$

or $\quad y = r, \quad x + \sqrt{2}\,z = 3\,r,$

the equations of the tangent PT at P to the curve of intersection.
Substituting in (4), we get the equation of the normal plane,

$$-\sqrt{2}(x-r) + 0(y-r) + (z-r\sqrt{2}) = 0,$$

or $\quad \sqrt{2}\,x - z = 0.$

ILLUSTRATIVE EXAMPLE 2. Find the angle of intersection of the surfaces in the preceding example at the point given.

Solution. The angle of intersection equals the angle between the tangent planes or normal lines. We have found direction numbers for these lines above in Illustrative Example 1 (see (*E*), Art. 237).

These are $\quad a = 2\,r, \; b = 2\,r, \; c = 2\,r\sqrt{2}.$

$\quad\quad\quad\quad\quad\; a' = 0, \; b' = 2\,r, \; c' = 0.$

Hence, by (6), Art. 4

$$\cos\theta = \frac{4\,r^2}{8\,r^2} = \frac{1}{2}. \quad \theta = 60°. \text{ Ans.}$$

PROBLEMS

Find the equations of the tangent line and the equation of the normal plane to each of the following curves at the point indicated.

1. $x^2 + y^2 + z^2 = 49,\; x^2 + y^2 = 13;\; (3, 2, -6).$

   Ans. $\dfrac{x-3}{2} = \dfrac{y-2}{-3},\; z + 6 = 0;\; 2\,x - 3\,y = 0.$

2. $z = x^2 + y^2 - 1,\; 3\,x^2 + 2\,y^2 + z^2 = 30;\; (2, 1, 4).$

   Ans. $\dfrac{x-2}{5} = \dfrac{y-1}{-11} = \dfrac{z-4}{-2};\; 5\,x - 11\,y - 2\,z + 9 = 0.$

3. $x^2 + y^2 - z^2 = 16,\; x^2 + 4\,y^2 + 4\,z^2 = 84;\; (2, 4, 2).$

   Ans. $\dfrac{x-2}{16} = \dfrac{y-4}{-5} = \dfrac{z-2}{6};\; 16\,x - 5\,y + 6\,z = 24.$

4. $x^2 + y^2 + 3\,z^2 = 32,\; 2\,x^2 + y^2 - z^2 = 0;\; (2, 1, 3).$

   Ans. $\dfrac{x-2}{6} = \dfrac{y-1}{-21} = \dfrac{z-3}{1};\; 6\,x - 21\,y + z + 6 = 0.$

5. $x^2 - y^2 - z^2 = 1,\; x^2 - y^2 + z^2 = 9;\; (3, 2, 2).$

6. $x^2 + 4\,y^2 - 4\,z^2 = 0,\; 2\,x + y + z - 24 = 0;\; (8, 3, 5).$

7. The equations of a helix (spiral) are
$$x^2 + y^2 = r^2,$$
$$y = x \tan \frac{z}{c}.$$

Show that at the point (x_1, y_1, z_1) the equations of the tangent line are
$$c(x - x_1) + y_1(z - z_1) = 0,$$
$$c(y - y_1) - x_1(z - z_1) = 0;$$
and the equation of the normal plane is
$$y_1 x - x_1 y - c(z - z_1) = 0.$$

8. The surfaces $x^2 y^2 + 2x + z^3 = 16$ and $3x^2 + y^2 - 2z = 9$ intersect in a curve which passes through the point $(2, 1, 2)$. What are the equations of the respective tangent planes to the two surfaces at this point?
Ans. $3x + 4y + 6z = 22$; $6x + y - z = 11$.

9. Show that the ellipsoid $x^2 + 3y^2 + 2z^2 = 9$ and the sphere $x^2 + y^2 + z^2 - 8x - 8y - 6z + 24 = 0$ are tangent to each other at the point $(2, 1, 1)$.

10. Show that the paraboloid $3x^2 + 2y^2 - 2z = 1$ and the sphere $x^2 + y^2 + z^2 - 4y - 2z + 2 = 0$ cut orthogonally at the point $(1, 1, 2)$.

240. Law of the Mean. The applications of partial derivatives to be given now depend upon the Law of the Mean for functions of several variables. The result to be derived is based upon the discussion in Art. 116. We proceed to establish the formula

(1) $f(x_0 + h, y_0 + k) = f(x_0, y_0) + h f_x(x_0 + \theta h, y_0 + \theta k)$
$\qquad\qquad\qquad\qquad\qquad + k f_y(x_0 + \theta h, y_0 + \theta k).$ $(0 < \theta < 1)$

To this end let

(2) $\qquad\qquad F(t) = f(x_0 + ht, y_0 + kt).$

Apply **(D)**, Art. 116, to $F(t)$, with $a = 0$, and $\Delta a = 1$. Then we have

(3) $\qquad\qquad F(1) = F(0) + F'(\theta).$ $\qquad\qquad (0 < \theta < 1)$

But from (2), by **(D)**, Art. 229, since $x = x_0 + ht$, $y = y_0 + kt$,

(4) $\qquad F'(t) = h f_x(x_0 + ht, y_0 + kt) + k f_y(x_0 + ht, y_0 + kt).$

Then, from (2), we get

(5) $\qquad F(1) = f(x_0 + h, y_0 + k)$, $\quad F(0) = f(x_0, y_0),$

and, from (4),

(6) $\qquad F'(\theta) = h f_x(x_0 + \theta h, y_0 + \theta k) + k f_y(x_0 + \theta h, y_0 + \theta k).$

When these results are substituted in (3), we obtain (1).

APPLICATIONS OF PARTIAL DERIVATIVES 483

If we desire a formula analogous to *(F)*, Art. 124, we must form $F''(t)$. Applying again *(D)*, Art. 229, we get

$$\frac{d}{dt} f_x(x_0 + ht, y_0 + kt) = hf_{xx}(x_0 + ht, y_0 + kt) + kf_{yx}(x_0 + ht, y_0 + kt);$$

$$\frac{d}{dt} f_y(x_0 + ht, y_0 + kt) = hf_{xy}(x_0 + ht, y_0 + kt) + kf_{yy}(x_0 + ht, y_0 + kt).$$

Hence from (4), we have by differentiating with respect to t,

(7) $F''(t) = h^2 f_{xx}(x_0 + ht, y_0 + kt) + 2 hk f_{xy}(x_0 + ht, y_0 + kt)$
$\qquad + k^2 f_{yy}(x_0 + ht, y_0 + kt).$

From *(F)*, Art. 124, letting $b = 1$, $a = 0$, $x_2 = \theta$, we get

(8) $\qquad F(1) = F(0) + F'(0) + \frac{1}{\lfloor 2} F''(\theta).$

We may easily prove now the extended Law of the Mean for a function of two variables by substituting in (8) from (5), (4), and (7). Thus we get

(9) $f(x_0 + h, y_0 + k) = f(x_0, y_0) + hf_x(x_0, y_0) + kf_y(x_0, y_0)$
$\qquad + \frac{1}{\lfloor 2} [h^2 f_{xx}(x_0 + \theta h, y_0 + \theta k) + 2 hk f_{xy}(x_0 + \theta h, y_0 + \theta k)$
$\qquad + k^2 f_{yy}(x_0 + \theta h, y_0 + \theta k)].$ $\qquad (0 < \theta < 1).$

There is no difficulty in establishing the corresponding formulas for functions of more than two variables, nor in extending the laws in a manner analogous to that at the end of Art. 124.

241. Maxima and minima of functions of several variables. In Art. 46, and again in Art. 125, were derived necessary and sufficient conditions for maximum and minimum values of a function of one variable. We now take up this problem when several independent variables are present.

The function $f(x, y)$ is said to be a *maximum* at $x = a$, $y = b$ when $f(a, b)$ is greater than $f(x, y)$ for all values of x and y in the neighborhood of a and b. Similarly, $f(x, y)$ is said to be a *minimum* at $x = a$, $y = b$ when $f(a, b)$ is less than $f(x, y)$ for all values of x and y in the neighborhood of a and b.

These definitions may be stated in analytical form as follows:

If, for all values of h and k numerically less than some small positive quantity,

(1) $\qquad f(a + h, b + k) - f(a, b) = $ a negative number,

then $f(a, b)$ is a *maximum* value of $f(x, y)$. If

(2) $\qquad f(a + h, b + k) - f(a, b) = $ a positive number,

then $f(a, b)$ is a *minimum* value of $f(x, y)$.

These statements may be interpreted geometrically as follows. A point P on the surface
$$z = f(x, y)$$
is a maximum point when it is "higher" than all other points on the surface in its neighborhood, the coördinate plane XOY being assumed horizontal. Similarly, P' is a minimum point on the surface when it is "lower" than all other points on the surface in its neighborhood.

Hence if
$$z_1 = f(a, b)$$
is a maximum or minimum, the tangent plane at (a, b, z_1) must be horizontal, that is, parallel to XOY. But the tangent plane (H), Art. 237, is parallel to XOY when the coefficients of x and y are zero. Hence we have the following result.

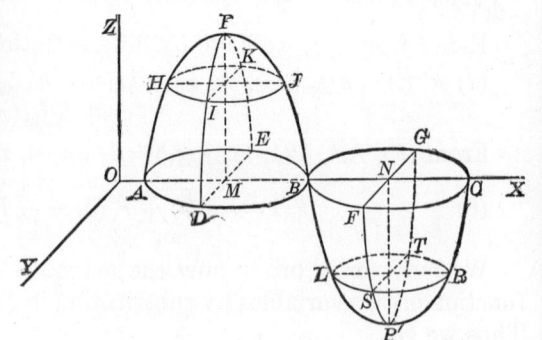

A necessary condition that $f(a, b)$ shall be a maximum or minimum value of $f(x, y)$ is that the equations

(3) $$\frac{\partial f}{\partial x} = 0, \quad \frac{\partial f}{\partial y} = 0$$

shall be satisfied by $x = a$, $y = b$.

The conditions (3) may be obtained without use of the tangent plane. For, when $y = b$, the function $f(x, b)$ can neither increase nor decrease when x passes through a (see Art. 45). Hence follows the first of equations (3). The same statement applies to the function $f(a, y)$. Thus we have the second equation in (3).

The method just expounded applies to a function of three variables $f(x, y, z)$. That is, a *necessary* condition that $f(a, b, c)$ shall be a maximum or a minimum value is that the equations

(4) $$\frac{\partial f}{\partial x} = 0, \quad \frac{\partial f}{\partial y} = 0, \quad \frac{\partial f}{\partial z} = 0$$

shall have the common solution $x = a$, $y = b$, $z = c$.

For necessary and sufficient conditions the problem is much more difficult (see below). But in many applied problems the existence of a maximum or minimum value is known in advance, and no test is necessary.

APPLICATIONS OF PARTIAL DERIVATIVES

ILLUSTRATIVE EXAMPLE 1. A long piece of tin 24 in. wide is to be made into a trough by bending up two sides. Find the width and inclination of each side if the carrying capacity is a maximum.

Solution. The area of the cross section shown in the figure must be a maximum. The cross section is a trapezoid of upper base $24 - 2x + 2x \cos \alpha$, lower base $24 - 2x$, and altitude $x \sin \alpha$. The area A is given by

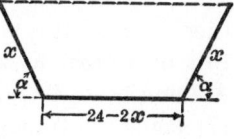

(5) $\qquad A = 24 x \sin \alpha - 2 x^2 \sin \alpha + x^2 \sin \alpha \cos \alpha.$

By differentiation we have

$$\frac{\partial A}{\partial x} = 24 \sin \alpha - 4 x \sin \alpha + 2 x \sin \alpha \cos \alpha.$$

$$\frac{\partial A}{\partial \alpha} = 24 x \cos \alpha - 2 x^2 \cos \alpha + x^2(\cos^2 \alpha - \sin^2 \alpha).$$

Setting the partial derivatives equal to zero, we have the two equations

$$2 \sin \alpha(12 - 2x + x \cos \alpha) = 0.$$

$$x[24 \cos \alpha - 2 x \cos \alpha + x(\cos^2 \alpha - \sin^2 \alpha)] = 0.$$

One solution of this system is $\alpha = 0$, $x = 0$, which has no meaning in the physical problem. Assuming $\alpha \neq 0$, $x \neq 0$, and solving the equations, we get $\cos \alpha = \frac{1}{2}$, $x = 8$.

A consideration of the physical problem shows that there must exist a maximum value of the area. Hence this maximum value occurs when $\alpha = 60°$ and $x = 8''$.

We now establish a sufficient condition. Assuming that equations (3) hold, we obtain from (9), Art. 240, substituting $x_0 = a$, $y_0 = b$, and transposing,

(6) $\qquad f(a+h, b+k) - f(a, b) = \frac{1}{2}[h^2 f_{xx}(x, y) + 2 hk f_{xy}(x, y) + k^2 f_{yy}(x, y)],$

where we have set $x = a + \theta h$, $y = b + \theta k$. By (1) and (2), $f(a, b)$ will be a maximum (or a minimum) if the right-hand member is negative (or positive) for all values of h and k sufficiently small in numerical value (zero excluded). Set

(7) $\qquad A = f_{xx}(x, y), \quad B = f_{xy}(x, y), \quad C = f_{yy}(x, y),$

and consider the identity

(8) $\quad Ah^2 + 2 Bhk + Ck^2 = \frac{1}{A}[(Ah + Bk)^2 + (AC - B^2)k^2].$

The expression within the square brackets in the right-hand member in (8) is always positive if

(9) $\qquad\qquad\qquad AC - B^2 > 0.$

and the left-hand member therefore has the same sign as A (or C, since, by (9), A and C must agree in sign). The question now is, therefore, to interpret the criterion (9) for the right-hand member in (6), in which, as already stated, h and k are numerically small. Assume that (9) holds when $x = a$, $y = b$. Then, the derivatives in (7) being continuous, it will hold also for values of x, y near a, b. Also, the sign of A (or C) will be the same as the sign of $f_{xx}(a, b)$ (or $f_{yy}(a, b)$). Thus we have established the following **rule for finding maximum and minimum values of a function $f(x, y)$**.

FIRST STEP. *Solve the simultaneous equations*

$$\frac{\partial f}{\partial x} = 0, \quad \frac{\partial f}{\partial y} = 0.$$

SECOND STEP. *Calculate for these values of x and y the value of*

$$\Delta = \frac{\partial^2 f}{\partial x^2} \frac{\partial^2 f}{\partial y^2} - \left(\frac{\partial^2 f}{\partial x \, \partial y}\right)^2.$$

THIRD STEP. *The function will have*

a maximum value if $\Delta > 0$ *and* $\dfrac{\partial^2 f}{\partial x^2} \left(\text{or } \dfrac{\partial^2 f}{\partial y^2}\right) < 0$;

a minimum value if $\Delta > 0$ *and* $\dfrac{\partial^2 f}{\partial x^2} \left(\text{or } \dfrac{\partial^2 f}{\partial y^2}\right) > 0$.

If Δ is *negative*, it is not difficult to see that $f(x, y)$ will have *neither a maximum nor a minimum value.*

The student should notice that this rule does not necessarily give *all* maximum and minimum values. For a pair of values of x and y determined by the First Step may cause Δ to vanish, and may lead to a maximum or a minimum or neither. Further investigation is therefore necessary for such values. The rule is, however, sufficient for solving many important examples.

The question of maxima and minima of functions of three or more independent variables must be left to more advanced treatises.

ILLUSTRATIVE EXAMPLE 2. Examine the function $3\,axy - x^3 - y^3$ for maximum and minimum values.

Solution. $\quad f(x, y) = 3\,axy - x^3 - y^3$.

First Step. $\quad \dfrac{\partial f}{\partial x} = 3\,ay - 3\,x^2 = 0, \quad \dfrac{\partial f}{\partial y} = 3\,ax - 3\,y^2 = 0.$

Solving these two simultaneous equations, we get

$$x = 0, \quad x = a,$$
$$y = 0, \quad y = a,$$

APPLICATIONS OF PARTIAL DERIVATIVES 487

Second Step. $\dfrac{\partial^2 f}{\partial x^2} = -6x, \quad \dfrac{\partial^2 f}{\partial x \partial y} = 3a, \quad \dfrac{\partial^2 f}{\partial y^2} = -6y;$

$$\Delta = \dfrac{\partial^2 f}{\partial x^2} \dfrac{\partial^2 f}{\partial y^2} - \left(\dfrac{\partial^2 f}{\partial x \partial y}\right)^2 = 36\,xy - 9\,a^2.$$

Third Step. When $x = 0$ and $y = 0$, $\Delta = -9\,a^2$, and there can be neither a maximum nor a minimum at $(0, 0)$.

When $x = a$ and $y = a$, $\Delta = +27\,a^2$; and since $\dfrac{\partial^2 f}{\partial x^2} = -6a$, we have the conditions for a maximum value of the function fulfilled at (a, a). Substituting $x = a$, $y = a$ in the given function, we get its maximum value equal to a^3.

ILLUSTRATIVE EXAMPLE 3. Divide a into three parts such that their product shall be a maximum.

Solution. Let $x =$ first part, $y =$ second part; then $a - (x + y) = a - x - y =$ third part, and the function to be examined is

$$f(x, y) = xy(a - x - y).$$

First Step. $\dfrac{\partial f}{\partial x} = ay - 2\,xy - y^2 = 0, \quad \dfrac{\partial f}{\partial y} = ax - 2\,xy - x^2 = 0.$

Solving simultaneously, we get as one pair of values $x = \dfrac{a}{3}, y = \dfrac{a}{3}.$

Second Step. $\dfrac{\partial^2 f}{\partial x^2} = -2y, \quad \dfrac{\partial^2 f}{\partial x \partial y} = a - 2x - 2y, \quad \dfrac{\partial^2 f}{\partial y^2} = -2x;$

$$\Delta = 4\,xy - (a - 2x - 2y)^2.$$

Third Step. When $x = \dfrac{a}{3}$ and $y = \dfrac{a}{3}$, $\Delta = \dfrac{a^2}{3}$; and since $\dfrac{\partial^2 f}{\partial x^2} = -\dfrac{2a}{3}$, it is seen that our product is a maximum when $x = \dfrac{a}{3}, y = \dfrac{a}{3}$. Therefore the third part is also $\dfrac{a}{3}$, and the maximum value of the product is $\dfrac{a^3}{27}$.

PROBLEMS

Discuss for maxima and minima the following functions.

1. $x^2 + xy + y^2 - 6x + 2.$ Ans. $x = 4, y = -2$ gives min.
2. $4x + 2y - x^2 + xy - y^2.$ $x = \tfrac{10}{3}, y = \tfrac{8}{3}$ gives max.
3. $2x^2 - 2xy + y^2 + 5x - 3y.$ $x = -1, y = \tfrac{1}{2}$ gives min.
4. $x^3 - 3\,axy + y^3.$ $x = y = a$ gives min.
5. $\sin x + \sin y + \sin(x + y).$ $x = y = \dfrac{\pi}{3}$ gives max.

                                                        $x = y = \dfrac{5\pi}{3}$ gives min.

6. $x^2 - xy + y^2 + ax + by + c.$

7. $xy + \dfrac{a^3}{x} + \dfrac{b^3}{y}.$

8. Show that the maximum value of $\dfrac{(ax + by + c)^2}{x^2 + y^2 + 1}$ is $a^2 + b^2 + c^2.$

488   DIFFERENTIAL AND INTEGRAL CALCULUS

9. Find the rectangular parallelepiped of maximum volume which has three faces in the coördinate planes and one vertex in the plane $\frac{x}{a} + \frac{y}{b} + \frac{z}{c} = 1$.   *Ans.* Volume $= \frac{abc}{27}$.

10. Find the volume of the largest rectangular parallelepiped that can be inscribed in the ellipsoid $\frac{x^2}{a^2} + \frac{y^2}{b^2} + \frac{z^2}{c^2} = 1$.   *Ans.* $\frac{8\,abc}{3\sqrt{3}}$.

11. A pentagon is composed of a rectangle surmounted by an isosceles triangle. If the perimeter of the pentagon has a given value P, find the dimensions for maximum area.

Ans. $\alpha = 30°$, $2x = \dfrac{P}{2 + 2\sec\alpha - \tan\alpha}$,

$y = \dfrac{P}{2} - x(1 + \sec\alpha)$.

12. Find the shortest distance between the lines $x = \frac{y}{2} = \frac{z}{3}$ and $x = y - 3 = z$.

13. A manufacturer produces two lines of candy at constant average costs of 50 cents and 60 cents per pound respectively. If the selling price of the first line is x cents per pound and of the second line is y cents per pound, the number of pounds which can be sold each week is given by the formulas $N_1 = 250(y - x)$, $N_2 = 32{,}000 + 250(x - 2y)$.

Show that for maximum profit the selling prices should be fixed at 89 cents and 94 cents per pound respectively.

14. A solid consists of a cylinder surmounted by a cone. Let $r =$ radius, $a =$ altitude of the cylinder, $h =$ altitude of the cone, $S =$ surface, and $V =$ volume of the solid. Show that the minimum value of S for a given value of V is given by

$$r^3 = \frac{3\,V}{\pi(3 + \sqrt{5})}, \quad h = \frac{2\,r}{\sqrt{5}}, \quad S = \frac{3\,V}{r}.$$

242. Taylor's theorem for functions of two or more variables. The expansion of $f(x, y)$ is found by using the methods and results of Arts. 194 and 240. We consider

(1) $F(t) = f(x + ht, y + kt)$,

and expand $F(t)$ as in (5), Art. 194. The result is

(2) $F(t) = F(0) + F'(0)\dfrac{t}{\underline{|1}} + F''(0)\dfrac{t^2}{\underline{|2}} + \cdots + F^{(n-1)}(0)\dfrac{t^{n-1}}{\underline{|n-1}} + R.$

We obtain the values of $F(0)$, $F'(0)$, $F''(0)$, by substituting $t = 0$ in (2), (4), (7), Art. 240. By differentiating (7), and putting $t = 0$, the expressions for $F'''(0)$ etc. will result. These are omitted here. Note, however, that $F'''(0)$ is homogeneous and of the third degree

APPLICATIONS OF PARTIAL DERIVATIVES 489

in h and k. A similar property holds for higher derivatives. If these values are substituted in (2), and we set $t = 1$, the result is

(3) $f(x+h, y+k) = f(x,y) + hf_x(x,y) + kf_y(x,y)$
$$+ \frac{1}{\lfloor 2} [h^2 f_{xx}(x,y) + 2 hk f_{xy}(x,y) + k^2 f_{yy}(x,y)] + \cdots + R.$$

The expression for R is complicated and will be omitted from this point on.

In (3) write $x = a$, $y = b$, and then replace h by $(x - a)$ and k by $(y - b)$. The result is *Taylor's theorem for a function of two variables*,

(*I*) $f(x,y) = f(a,b) + f_x(a,b)(x-a) + f_y(a,b)(y-b)$
$$+ \frac{1}{\lfloor 2} [f_{xx}(a,b)(x-a)^2 + 2f_{xy}(a,b)(x-a)(y-b)$$
$$+ f_{yy}(a,b)(y-b)^2] + \cdots.$$

Finally, setting $a = b = 0$, we obtain an expansion corresponding to Maclaurin's series (*A*), Art. 194,

(*J*) $f(x,y) = f(0,0) + f_x(0,0)x + f_y(0,0)y$
$$+ \frac{1}{\lfloor 2} [f_{xx}(0,0)x^2 + 2f_{xy}(0,0)xy + f_{yy}(0,0)y^2] + \cdots.$$

The right-hand member in (*J*) may be written as the infinite series

(4) $$u_0 + \frac{u_1}{\lfloor 1} + \frac{u_2}{\lfloor 2} + \cdots,$$

where $u_0 = f(0,0)$,
$u_1 = f_x(0,0)x + f_y(0,0)y$,
$u_2 = f_{xx}(0,0)x^2 + 2f_{xy}(0,0)xy + f_{yy}(0,0)y^2$,
etc.

These terms in (4) are homogeneous polynomials in (x, y). The degree of each is equal to the subscript. That is, by (*J*) the function is expanded into a sum of polynomials homogeneous in (x, y) and of ascending degree. Similarly, in (*I*) the terms in the expansion are polynomials homogeneous in $(x - a, y - b)$.

Formula (*I*) is called *the expansion of $f(x, y)$ at the point (a, b)*.

Reference must be made to more advanced treatises for proof of the problem of determining those values of (x, y) for which the expansions (*I*) and (*J*) hold.

By breaking off series (4) at any term, an approximate formula for $f(x, y)$ is obtained for values near (a, b) or $(0, 0)$. Compare Art. 200.

490 DIFFERENTIAL AND INTEGRAL CALCULUS

ILLUSTRATIVE EXAMPLE. Expand
$$xy^2 + \sin xy$$
at the point $(1, \tfrac{1}{2}\pi)$ up to terms of the third degree.

Solution. Here $a = 1, b = \tfrac{1}{2}\pi$,
and
$f(x, y) = xy^2 + \sin xy,$
$f_x(x, y) = y^2 + y \cos xy,$
$f_y(x, y) = 2xy + x \cos xy,$
$f_{xx}(x, y) = -y^2 \sin xy,$
$f_{xy}(x, y) = 2y + \cos xy - xy \sin xy,$
$f_{yy}(x, y) = 2x - x^2 \sin xy.$

Substituting $x = 1$, $y = \tfrac{1}{2}\pi$, the results are
$f(1, \tfrac{1}{2}\pi) = \tfrac{1}{4}\pi^2 + 1,$
$f_x(1, \tfrac{1}{2}\pi) = \tfrac{1}{4}\pi^2,$
$f_y(1, \tfrac{1}{2}\pi) = \pi,$
$f_{xx}(1, \tfrac{1}{2}\pi) = -\tfrac{1}{4}\pi^2,$
$f_{xy}(1, \tfrac{1}{2}\pi) = \tfrac{1}{2}\pi,$
$f_{yy}(1, \tfrac{1}{2}\pi) = 1.$

Substituting in (I), we get
$$xy^2 + \sin xy = 1 + \tfrac{1}{4}\pi^2 + \tfrac{1}{4}\pi^2(x-1) + \pi(y - \tfrac{1}{2}\pi)$$
$$+ \frac{1}{\underline{|2}}\left[-\tfrac{1}{4}\pi^2(x-1)^2 + \pi(x-1)\left(y - \tfrac{1}{2}\pi\right) + \left(y - \tfrac{1}{2}\pi\right)^2\right] + \cdots \quad Ans.$$

Formulas for expanding a function of three variables $f(x, y, z)$ are readily derived, and are left as problems.

PROBLEMS

1. From (1) above, show that
$$F'''(0) = h^3 \left|\frac{\partial^3 f}{\partial x^3}\right|_0 + 3h^2k \left|\frac{\partial^3 f}{\partial x^2 \partial y}\right|_0 + 3hk^2 \left|\frac{\partial^3 f}{\partial x \partial y^2}\right|_0 + k^3 \left|\frac{\partial^3 f}{\partial y^3}\right|_0.$$

2. Verify the following expansion.
$$\cos x \cos y = 1 - \frac{x^2 + y^2}{\underline{|2}} + \frac{x^4 + 6x^2y^2 + y^4}{\underline{|4}}$$
$$- \frac{x^6 + 15 x^4 y^2 + 15 x^2 y^4 + y^6}{\underline{|6}} + \cdots.$$

3. Expand $\sin x \sin y$ in powers of x and y.

4. Verify the following expansion.
$$a^x \ln (1+y) = y + \tfrac{1}{2}(2 xy \ln a - y^2 + x^2 y \ln^2 a - xy^2 \ln a) + \tfrac{1}{3} y^3 + \cdots.$$

5. Expand $x^3 + xy^2$ at the point $(1, 2)$.

6. Verify the following expansion.
$$\sin (x + y) = x + y - \frac{x^3 + 3x^2y + 3xy^2 + y^3}{\underline{|3}} + \cdots.$$

Verify the following approximate formulas for small values of x and y.

7. $e^x \sin y = y + xy.$

8. $e^x \ln (1 + y) = y + xy.$

9. $\sqrt{\dfrac{1+x}{1+y}} = 1 + \tfrac{1}{2}(x - y).$

CHAPTER XXV

MULTIPLE INTEGRALS

243. Partial and successive integration. Corresponding to *partial differentiation* in the differential calculus we have the inverse process of *partial integration* in the integral calculus. As may be inferred from the connection, partial integration means that, having given a differential expression involving two or more independent variables, we integrate it, considering first *a single one only* as varying and all the rest constant. Then we integrate the result, considering another one as varying and the others constant, and so on. Such integrals are called *double, triple*, etc., according to the number of variables, and are known as *multiple integrals*.

In the solution of this problem the only new feature is that the constant of integration has a new form. We shall illustrate this by means of examples. Thus, suppose we wish to find u, having given

$$\frac{\partial u}{\partial x} = 2x + y + 3.$$

Integrating this with respect to x, considering y as constant, we have

$$u = x^2 + xy + 3x + \phi,$$

where ϕ denotes the constant of integration. But since y was regarded as constant during this integration, ϕ may involve y. We shall then indicate this dependence of ϕ on y by replacing ϕ by the symbol $\phi(y)$. Hence the most general form of u is

$$u = x^2 + xy + 3x + \phi(y),$$

where $\phi(y)$ denotes an arbitrary function of y.

As another problem let us find

$$u = \iint (x^2 + y^2) dy\, dx.$$

This means that we wish to find u, having given

$$\frac{\partial^2 u}{\partial x\, \partial y} = x^2 + y^2.$$

492 DIFFERENTIAL AND INTEGRAL CALCULUS

Integrating first with respect to y, regarding x as constant, we get

$$\frac{\partial u}{\partial x} = x^2 y + \frac{y^3}{3} + \psi(x),$$

where $\psi(x)$ is an arbitrary function of x.

Now integrating this result with respect to x, regarding y as constant, we have

$$u = \frac{x^3 y}{3} + \frac{xy^3}{3} + \Psi(x) + \Phi(y),$$

where $\Phi(y)$ is an arbitrary function of y, and

$$\Psi(x) = \int \psi(x) dx.$$

244. Definite double integral. Geometric interpretation. Let $f(x, y)$ be a continuous and single-valued function of x and y. Geometrically,

(1) $$z = f(x, y)$$

is the equation of a surface, as KL. Take some area S in the XOY-plane and construct upon S as a base the right cylinder whose elements are parallel to OZ. Let this cylinder inclose the area S' on KL. Let us now find the volume V of the solid bounded by S, S', and the cylindrical surface. We proceed as follows:

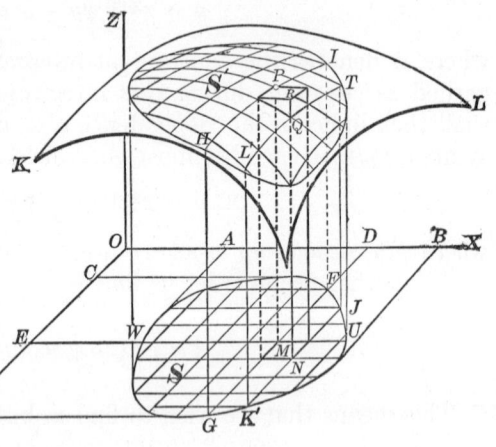

At equal distances apart ($= \Delta x$) in the area S draw a set of lines parallel to OY, and then a second set parallel to OX at equal distances apart ($= \Delta y$). Through these lines pass planes parallel to YOZ and XOZ respectively. Then within the areas S and S' we have a network of lines, as in the figure, that in S being composed of rectangles, each of area $\Delta x \, \Delta y$. This construction divides the cylinder into a number of vertical columns, such as $MNPQ$, whose upper and lower bases are corresponding portions of the networks in S' and S respectively. As the upper bases of these columns are curvilinear, we of course cannot calculate the volume of the columns directly. Let us replace these columns by prisms whose upper bases are found thus: each column is cut through by a plane parallel to XOY passed through that vertex of the upper base for which x and y have the least numerical values. Thus the column $MNPQ$ is replaced by the right prism $MNPR$, the upper base being in a plane through P parallel to the XOY-plane.

If the coördinates of P are (x, y, z), then $MP = z = f(x, y)$, and therefore

(2) \qquad Volume of $MNPR = f(x, y)\Delta y\, \Delta x$.

Calculating the volume of each of the other prisms formed in the same way by replacing x and y in (2) by corresponding values, and adding the results, we obtain a volume V' approximately equal to V; that is,

(3) $$V' = \sum \sum f(x, y)\Delta y\, \Delta x;$$

where the double summation sign $\sum \sum$ indicates that values of *two variables* x, y must be taken account of in the quantity to be summed up.

If now in the figure we increase the number of divisions of the network in S indefinitely by letting Δx and Δy diminish indefinitely, and calculate in each case the double sum (3), then obviously V' will approach V as a limit, and hence we have the fundamental result

(4) $$V = \lim_{\substack{\Delta x \to 0 \\ \Delta y \to 0}} \sum \sum f(x, y)\Delta y\, \Delta x.$$

We show now that this limit can be found by successive integration.

The required volume may be found as follows: Consider any one of the slices into which the solid is divided by two successive planes parallel to YOZ; for example, the slice whose faces are $FIHG$ and $JTL'K'$. The thickness of this slice is Δx. Now the values of z along the curve HI are found by writing $x = OD$ in the equation $z = f(x, y)$; that is, along HI

$$z = f(OD, y).$$

Hence \qquad Area $FIHG = \displaystyle\int_{DF}^{DG} f(OD, y)dy$.

The volume of the slice under discussion is approximately equal to that of a prism with base $FIHG$ and altitude Δx; that is, equal to

$$\Delta x \cdot \text{area } FIHG = \Delta x \int_{DF}^{DG} f(OD, y)dy.$$

The required volume of the whole solid is evidently the limit of the sum of all prisms constructed in like manner, as $x\ (= OD)$ varies from OA to OB; that is,

$$(5) \qquad V = \int_{OA}^{OB} dx \int_{DF}^{DG} f(x, y) dy.$$

Similarly, it may be shown that

$$(6) \qquad V = \int_{OC}^{OV} dy \int_{EW}^{EU} f(x, y) dx.$$

The integrals (5) and (6) are also written in the more compact form

$$\int_{OA}^{OB} \int_{DF}^{DG} f(x, y) dy\, dx \quad \text{and} \quad \int_{OC}^{OV} \int_{EW}^{EU} f(x, y) dx\, dy.$$

In (5) the limits DF and DG are functions of x, since they are found by solving the equation of the boundary curve of the base of the solid for y.

Similarly, in (6) the limits EW and EU are functions of y. Now comparison of (4), (5), and (6) gives the result

$$(A) \quad V = \lim_{\substack{\Delta x \to 0 \\ \Delta y \to 0}} \sum \sum f(x, y)\, \Delta y \cdot \Delta x = \int_{a_2}^{a_1} \int_{u_2}^{u_1} f(x, y) dy\, dx$$

$$= \int_{b_2}^{b_1} \int_{v_2}^{v_1} f(x, y) dx\, dy,$$

where v_1 and v_2 are, in general, functions of y, and u_1 and u_2 functions of x. The second integral sign in each case applies to the first differential.

Equation (A) is an extension of the Fundamental Theorem of Art. 156 to double sums.

Our result may be stated in the following form.

The definite double integral

$$\int_{a_2}^{a_1} \int_{u_2}^{u_1} f(x, y) dy\, dx$$

may be interpreted as that portion of the volume of a right cylinder which is included between the plane XOY and the surface

$$z = f(x, y),$$

the base of the cylinder being the area in the XOY-plane bounded by the curves
$$y = u_1, \quad y = u_2, \quad x = a_1, \quad x = a_2.$$

A similar statement holds for the second integral.

It is instructive to look upon the above process of finding the volume of the solid as follows.

MULTIPLE INTEGRALS 495

Consider a column with rectangular base $dy\,dx$ and of altitude z as an element of the volume. Summing up all such elements from $y = DF$ to $y = DG$, x in the meanwhile being constant (say $= OD$), gives the volume of a thin slice having $FGHI$ as one face. The volume of the whole solid is then found by summing up all such slices from $x = OA$ to $x = OB$.

In successive integration involving two variables the order of integration denotes that the limits on the second integral sign correspond to the variable whose differential is written first, the differentials of the variables and their corresponding limits being written in the reverse order. Before attempting to apply successive integration to practical problems it is best that the student should acquire by practice some facility in evaluating definite multiple integrals.

ILLUSTRATIVE EXAMPLE 1. Find the value of the definite double integral

$$\int_0^a \int_0^{\sqrt{a^2-x^2}} (x+y)\,dy\,dx.$$

Solution. $\int_0^a \int_0^{\sqrt{a^2-x^2}} (x+y)\,dy\,dx$

$= \int_0^a \left[\int_0^{\sqrt{a^2-x^2}} (x+y)\,dy \right] dx$

$= \int_0^a \left[xy + \frac{y^2}{2} \right]_0^{\sqrt{a^2-x^2}} dx$

$= \int_0^a \left(x\sqrt{a^2-x^2} + \frac{a^2-x^2}{2} \right) dx$

$= \frac{2\,a^3}{3}.$ Ans.

Interpreting this result geometrically, we have found the volume of the solid of cylindrical shape standing on OAB as base and bounded at the top by the surface (plane) $z = x + y$.

The solid here stands on a base in the XOY-plane bounded by

$\left.\begin{array}{l} y = 0 \text{ (line } OB) \\ y = \sqrt{a^2-x^2} \text{ (quadrant of circle } AB) \end{array}\right\}$ from y limits;

$\left.\begin{array}{l} x = 0 \text{ (line } OA) \\ x = a \text{ (line } BE) \end{array}\right\}$ from x limits.

ILLUSTRATIVE EXAMPLE 2. Verify $\int_b^{2b} \int_0^a (a-y)x^2\,dy\,dx = \frac{7\,a^2b^3}{6}.$

Solution. $\int_b^{2b} \int_0^a (a-y)x^2\,dy\,dx = \int_b^{2b} \left[ay - \frac{y^2}{2} \right]_0^a x^2\,dx = \int_b^{2b} \frac{a^2}{2} x^2\,dx = \frac{7\,a^2b^3}{6}.$

ILLUSTRATIVE EXAMPLE 3. Verify $\int_0^a \int_{-\sqrt{a^2-x^2}}^{\sqrt{a^2-x^2}} x\,dy\,dx = \frac{2\,a^3}{3}.$

Solution. $\int_0^a \int_{-\sqrt{a^2-x^2}}^{\sqrt{a^2-x^2}} x\,dy\,dx = \int_0^a \left[xy \right]_{-\sqrt{a^2-x^2}}^{\sqrt{a^2-x^2}} dx$

$= \int_0^a 2\,x\sqrt{a^2-x^2}\,dx = \left[-\frac{2}{3}\left(a^2-x^2\right)^{\frac{3}{2}} \right]_0^a = \frac{2}{3}\,a^3.$

496 DIFFERENTIAL AND INTEGRAL CALCULUS

In successive integration involving three variables the order of integration is denoted in the same way as for two variables; that is, the order of the limits on the integral signs, reading from the inside to the left, is the same as the order of the corresponding variables whose differentials are read from the inside to the right.

ILLUSTRATIVE EXAMPLE 4. Verify $\int_2^3 \int_1^2 \int_2^5 xy^2 \, dz \, dy \, dx = \frac{35}{2}$.

Solution. $\int_2^3 \int_1^2 \int_2^5 xy^2 \, dz \, dy \, dx = \int_2^3 \int_1^2 \left[\int_2^5 xy^2 \, dz \right] dy \, dx = \int_2^3 \int_1^2 \left[xy^2 z \right]_2^5 dy \, dx$

$= 3 \int_2^3 \int_1^2 xy^2 \, dy \, dx = 3 \int_2^3 \left[\int_1^2 xy^2 \, dy \right] dx$

$= 3 \int_2^3 \left[\frac{xy^3}{3} \right]_1^2 dx = 7 \int_2^3 x \, dx = \frac{35}{2}.$

In Problems 1–10 in the following list the solid whose volume equals the value of the integral should be described.

PROBLEMS

Work out the following definite integrals.

1. $\int_0^1 \int_0^2 (x + 2) dy \, dx = 5.$

2. $\int_0^4 \int_0^x y \, dy \, dx = \frac{32}{3}.$

3. $\int_0^a \int_0^{\sqrt{x}} dy \, dx = \frac{2}{3} a^{\frac{3}{2}}.$

4. $\int_1^2 \int_0^y y \, dx \, dy = \frac{7}{6}.$

5. $\int_0^2 \int_0^{x^2} y \, dy \, dx = \frac{16}{5}.$

6. $\int_1^2 \int_y^{y^2} (x + 2y) dx \, dy = \frac{143}{30}.$

7. $\int_0^{-1} \int_{y+1}^{2y} xy \, dx \, dy = \frac{11}{24}.$

8. $\int_{-1}^1 \int_0^{x^2} (x + y) dy \, dx = \frac{1}{5}.$

9. $\int_0^2 \int_0^x (x^2 + y^2) dy \, dx = \frac{16}{3}.$

10. $\int_0^1 \int_0^{x^2} e^{\frac{y}{x}} \, dy \, dx = \frac{1}{2}.$

11. $\int_b^a \int_\beta^\alpha \rho^2 \sin \theta \, d\theta \, d\rho = \frac{1}{3}(a^3 - b^3)(\cos \beta - \cos \alpha).$

12. $\int_0^\pi \int_0^{a \cos \theta} \rho \sin \theta \, d\rho \, d\theta = \frac{1}{3} a^2.$

13. $\int_0^\pi \int_0^{a(1 + \cos \theta)} \rho^2 \sin \theta \, d\rho \, d\theta = \frac{4}{3} a^3.$

14. $\int_0^{\frac{\pi}{2}} \int_{a \cos \theta}^a \rho^4 \, d\rho \, d\theta = (\pi - \frac{16}{15}) \frac{a^5}{10}.$

15. $\int_b^a \int_0^b \int_a^{2a} x^2 y^2 z \, dz \, dy \, dx = \frac{1}{6} a^2 b^3 (a^3 - b^3).$

16. $\int_0^a \int_0^x \int_a^y x^3 y^2 z \, dz \, dy \, dx = \frac{1}{96} a^9.$

MULTIPLE INTEGRALS

17. $\int_0^1 \int_{y^2}^1 \int_0^{1-x} x\, dz\, dx\, dy = \frac{4}{35}$.

18. $\int_0^1 \int_0^{1-x} \int_0^{1-y^2} z\, dz\, dy\, dx = \frac{11}{60}$.

19. $\int_1^2 \int_0^z \int_0^{z\sqrt{3}} \left(\frac{x}{x^2+y^2}\right) dy\, dx\, dz = \frac{1}{2}\pi$.

20. $\int_0^1 \int_0^x \int_0^{x+y} e^{x+y+z} dz\, dy\, dx = \frac{1}{8}e^4 - \frac{3}{4}e^2 + e - \frac{3}{8}$.

245. Value of a definite double integral taken over a region S. In the last article the definite double integral appeared as a volume. This does not necessarily mean that every definite double integral is a volume, for the physical interpretation of the result depends on the nature of the quantities represented by x, y, z. If x, y, z are the coordinates of a point in space, then the result is indeed a volume. In order to give the definite double integral in question an interpretation not necessarily involving the geometric concept of volume, we observe that the variable z does not occur explicitly in the integral, and therefore we may confine ourselves to the XOY-plane. In fact, let us consider simply a region S in the XOY-plane, and a given function $f(x, y)$. Within this region construct rectangular elements of area by drawing a network of lines, as in Art. 244. Choose a point (x, y) of the rectangular element of area $\Delta x\, \Delta y$, *either within the rectangle or on its perimeter*. Form the product

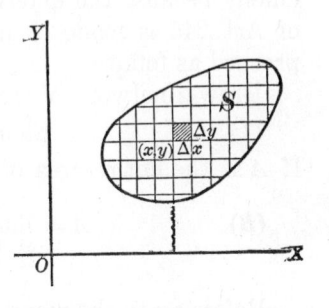

$$f(x, y)\Delta x\, \Delta y,$$

and similar products for all other rectangular elements. Sum up these products. The result is

$$\sum\sum f(x, y)\Delta x\, \Delta y.$$

Finally let $\Delta x \to 0$, and $\Delta y \to 0$.

We write the result

(1) $\qquad \lim\limits_{\substack{\Delta x \to 0 \\ \Delta y \to 0}} \sum\sum f(x, y)\Delta x\, \Delta y = \iint_S f(x, y) dx\, dy,$

and call it *the double integral of the function* $f(x, y)$ *taken over the region S*.

By (A) the value of the left-hand member in (1) was found by successive integration when $f(x, y)$ had no negative values for the region S. The reasoning of Art. 244 will hold, however, if the portion

S' of the surface $z = f(x, y)$ lies below the plane XOY. The limit of the double sum will then be the volume with a negative sign. The integrals in (A) will give the same negative number. Finally, if $f(x, y)$ is sometimes positive, sometimes negative for points of S, we may divide S into subregions in which $f(x, y)$ will be either always positive or always negative. The reasoning will hold for each subregion and therefore for the combined region S. Hence the conclusion: *the double integral in* (1) *may be evaluated in all cases by successive integration.*

It remains to explain the method of determining the *limits* of integration. This is done in the next article.

246. Plane area as a definite double integral. Rectangular coördinates. The problem of plane areas has been solved by single integration in Art. 145. The discussion using double integration is useful chiefly because the determination of limits for the general problem of Art. 245 is made clear. To set up the desired double integral, proceed as follows.

Draw a network of rectangles as before. Then, in the figure,

(1) Element of area $= \Delta x \, \Delta y$.

If A is the entire area of the region S, obviously, by (1), Art. 245,

(B) $$A = \lim_{\substack{\Delta x \to 0 \\ \Delta y \to 0}} \sum\sum \Delta x \, \Delta y = \iint_S dx \, dy.$$

Referring to the result stated in Art. 245, we may say:

The area of any region is the value of the double integral of the function $f(x, y) = 1$ taken over that region.

Or, also: *The area equals numerically the volume of a right cylinder of unit height erected on the base S.* (Art. 244.)

The examples show how the limits of integration are found.

ILLUSTRATIVE EXAMPLE 1. Calculate that portion of the area above OX which is bounded by the semicubical parabola $y^2 = x^3$ and the straight line $y = x$.

Solution. The order of integration is indicated in the figure. Integrate first with respect to x. That is, sum up first the elements $dx \, dy$ in a horizontal strip. Then we have

$$\int_{AB}^{AC} dx \, dy = dy \int_{AB}^{AC} dx = \text{area of a horizontal strip of altitude } dy.$$

Next, integrate this result with respect to y. This corresponds to summing up all horizontal strips. In this way we obtain

$$A = \int_0^{OD} \int_{AB}^{AC} dx \, dy.$$

MULTIPLE INTEGRALS

The limits AB and AC are found by solving each of the equations of the bounding curves for x. Thus from the equation of the line, $x = AB = y$, and from the equation of the curve, $x = AC = y^{\frac{2}{3}}$. To determine OD, solve the two equations *simultaneously* to find the point of intersection E. This gives the point $(1, 1)$; hence $OD = 1$. Therefore

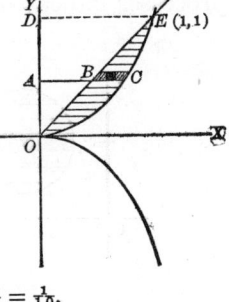

$$A = \int_0^1 \int_y^{y^{\frac{2}{3}}} dx\, dy = \int_0^1 (y^{\frac{2}{3}} - y)dy = \left[\tfrac{3}{5} y^{\frac{5}{3}} - \tfrac{1}{2} y^2\right]_0^1$$
$$= \tfrac{3}{5} - \tfrac{1}{2} = \tfrac{1}{10}. \text{ Ans.}$$

Or we may begin by summing up the elements $dx\, dy$ in a vertical strip, and then sum up these strips. We shall then have

$$A = \int_0^1 \int_{x^{\frac{3}{2}}}^x dy\, dx = \int_0^1 \left(x - x^{\frac{3}{2}}\right)dx = \tfrac{1}{2} - \tfrac{2}{5} = \tfrac{1}{10}.$$

In this example either order of integration may be chosen. This is not always the case, as the following example shows.

ILLUSTRATIVE EXAMPLE 2. Find the area in the first quadrant bounded by the x-axis and the curves $\quad x^2 + y^2 = 10, \quad y^2 = 9x$.

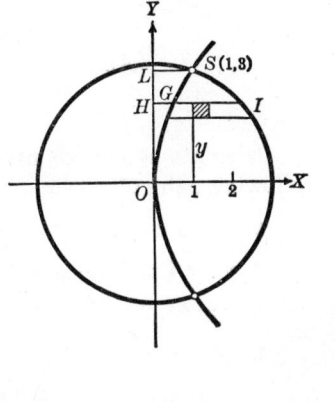

Solution. Here we first integrate with respect to x to cover a horizontal strip, that is, from the parabola to the circle. We then have, for the entire area,

$$A = \int_0^3 \int_{HG}^{HI} dx\, dy,$$

since the point of intersection S is $(1, 3)$. To find HG, solve $y^2 = 9x$ for x. Then

$$x = HG = \tfrac{1}{9} y^2.$$

To find HI, solve $x^2 + y^2 = 10$ for x. We get

$$x = HI = +\sqrt{10 - y^2}.$$

Hence

$$A = \int_0^3 \int_{\frac{1}{9}y^2}^{\sqrt{10-y^2}} dx\, dy = \left[\tfrac{y}{2} \sqrt{10 - y^2} + 5 \arcsin \tfrac{y}{\sqrt{10}} - \tfrac{1}{27} y^3\right]_0^3 = 6.75. \text{ Ans.}$$

If we integrate first with respect to y, using vertical strips, two integrals are necessary. Then

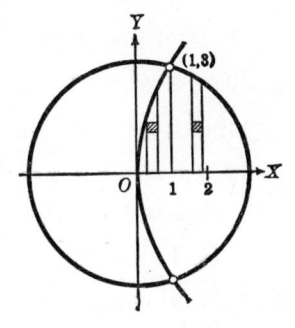

$$A = \int_0^1 \int_0^{3\sqrt{x}} dy\, dx + \int_1^{\sqrt{10}} \int_0^{\sqrt{10-x^2}} dy\, dx = 6.75.$$

The order of integration should be such that the area is given by one integral, if this is possible.

The examples above show that we set

$$A = \iint dx\, dy \quad \text{or} \quad A = \iint dy\, dx$$

according to the nature of the curves bounding the area. The figures below illustrate, in a general way, the difference in the summation processes indicated by the two integrals.

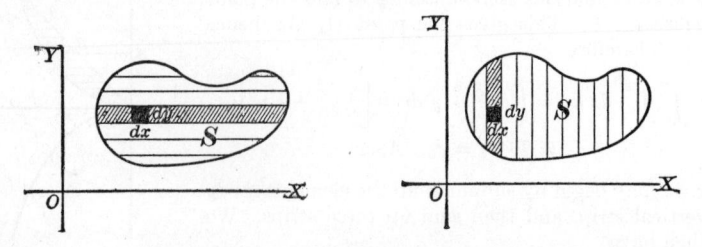

PROBLEMS

1. Find by double integration the area between the two parabolas $3y^2 = 25x$ and $5x^2 = 9y$, (a) by integrating first with respect to y; (b) by integrating first with respect to x.

$$\text{Ans. (a)} \int_0^3 \int_{\frac{5x^2}{9}}^{\sqrt{\frac{25x}{3}}} dy\, dx = 5; \quad \text{(b)} \int_0^5 \int_{\frac{3y^2}{25}}^{\sqrt{\frac{9y}{5}}} dx\, dy = 5.$$

Calculate by double integration the finite area bounded by each of the following pairs of curves.

2. $y = 4x - x^2,\ y = x.$ Ans. $4\frac{1}{2}$.

3. $y^2 = 4x,\ 2x - y = 4.$ 9.

4. $y = x^2,\ 2x - y + 3 = 0.$ $\frac{32}{3}$.

5. $y^2 = 2x,\ x^2 = 6y.$ 4.

6. $y^2 = 4x,\ x = 12 + 2y - y^2.$ $\frac{4096}{75}$.

7. $y^2 = 2x,\ x^2 + y^2 = 4x.$ $\pi - \frac{8}{3}$.

8. $y^2 = 9 + x,\ y^2 = 9 - 3x.$ 48.

9. $(x^2 + 4a^2)y = 8a^3,\ 2y = x,\ x = 0.$ $a^2(\pi - 1)$.

10. $x^{\frac{1}{2}} + y^{\frac{1}{2}} = a^{\frac{1}{2}},\ x + y = a.$ $\frac{1}{3}a^2$.

11. $x^{\frac{2}{3}} + y^{\frac{2}{3}} = a^{\frac{2}{3}},\ x + y = a.$ $\frac{1}{32}(16 - 3\pi)a^2$.

12. $y = x^3 - 2x,\ y = 6x - x^3.$ 16.

13. $x = 6y - y^2,\ y = x.$ 16. $x^2 + y^2 = 25,\ 27y^2 = 16x^3.$

14. $4y^2 = x^3,\ y = x.$ 17. $(2a - x)y^2 = x^3,\ y^2 = ax.$

15. $y^2 = x + 4,\ y^2 = 4 - 2x.$ 18. $x^2 - y^2 = 14,\ x^2 + y^2 = 36.$

247. Volume under a surface. In Art. 244 we discussed the volume of a solid bounded by a surface

(1) $$z = f(x, y),$$

the XOY-plane, and a cylinder. The elements of the cylinder were parallel to OZ, and its base was a region S in the XOY-plane. The volume of this solid is, by (A),

(2) $$V = \iint_S z \, dx \, dy = \iint_S f(x, y) dx \, dy.$$

The order of integration and the limits are the same as for the area of the region S. The volume of a solid of this type is the "volume under the surface (1)." The analogous problem for the plane, "area under a curve," has been treated in Chapter XIV. As a special case the volume may be bounded by the surface and the XOY-plane itself.

Note that the element of volume in (2) is a right prism with base $dx \, dy$ and altitude z.

ILLUSTRATIVE EXAMPLE 1. Find the volume bounded by the elliptic paraboloid

(3) $$4z = 16 - 4x^2 - y^2$$

and the XOY-plane.

Solution. Solving (3) for z, we get

(4) $$z = 4 - x^2 - \tfrac{1}{4} y^2.$$

Letting $z = 0$, we obtain

(5) $$4x^2 + y^2 = 16,$$

which is the equation of the perimeter of the base of the solid in the XOY-plane. Hence by (2), using the value of z in (4),

(6) $$V = 4 \int_0^2 \int_0^{2\sqrt{4-x^2}} (4 - x^2 - \tfrac{1}{4} y^2) dy \, dx = 16 \pi. \text{ Ans.}$$

The limits are taken for the area OAB of the ellipse (5) lying in the first quadrant.

ILLUSTRATIVE EXAMPLE 2. Find the volume of the solid bounded by the paraboloid of revolution

(7) $$x^2 + y^2 = az,$$

the XOY-plane, and the cylinder

(8) $$x^2 + y^2 = 2ax.$$

Solution. Solving (7) for z, and finding the limits for the area of the base of the cylinder (8) in the XOY-plane, we get, using (2),

$$V = 2 \int_0^{2a} \int_0^{\sqrt{2ax - x^2}} \frac{x^2 + y^2}{a} dy \, dx = \tfrac{3}{2} \pi a^3. \text{ Ans.}$$

For the area ONA (see figure), $MN = \sqrt{2ax - x^2}$, (solving (8) for y), and $OA = 2a$. These are the limits.

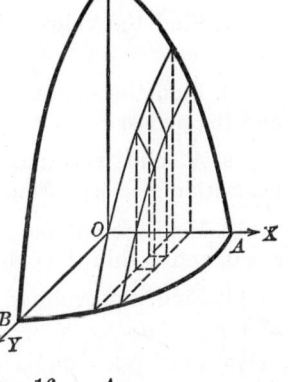

PROBLEMS

1. Find the volume under $z = 4 - x^2$, above $z = 0$, and within $y^2 = 4x$.

   Ans. $V = 2\int_0^2 \int_0^{2\sqrt{x}} (4 - x^2)dy\, dx = 17.24$.

2. Find the volume under the plane $x + z = 2$, above $z = 0$, and within $x^2 + y^2 = 4$.

   Ans. $V = 2\int_{-2}^2 \int_0^{\sqrt{4-x^2}} (2 - x)dy\, dx = 8\pi$.

3. Find the volume bounded by the plane $\frac{x}{a} + \frac{y}{b} + \frac{z}{c} = 1$ and the coordinate planes. Ans. $\frac{1}{6} abc$.

4. Find the volume bounded above by $x + z = 4$, below by $z = 0$, and laterally by $y^2 = 4x$. Ans. $\frac{512}{15}$.

5. Find the volume of the solid bounded above by $y^2 = a^2 - az$, below by $z = 0$, and within $x^2 + y^2 = a^2$. Ans. $\frac{3}{4} \pi a^3$.

6. Find the volume under the elliptic paraboloid $z = 1 - \frac{1}{4}x^2 - \frac{1}{9}y^2$ and above $z = 0$. Ans. 3π.

7. Find the volume under the plane $x + y + z = 8$, and bounded by the planes $x + 2y = 8$, $x - 2y = 8$, $x = 0$, $z = 0$. Ans. $170\frac{2}{3}$.

8. Find the volume bounded by the cylindrical surface $x^2 + az = a^2$ and the planes $x + y = a$, $y = 0$, $z = 0$. Ans. $\frac{3}{4} a^3$.

9. A solid is bounded by the surfaces $y^2 + z^2 = 4ax$, $x = 3a$, and lies within $y^2 = ax$. Find the volume. Ans. $(6\pi + 9\sqrt{3})a^3$.

10. Find the volume below the cylindrical surface $y^2 = a^2 - az$, above $z = 0$, and within the cylindrical surface $x^2 + y^2 = ax$. Ans. $\frac{15}{64} \pi a^3$.

11. Find the volume below $z = 2x + a$, above $z = 0$, and within $x^2 + y^2 = 2ax$. Ans. $3\pi a^3$.

12. Find the volume under $y^2 + z = 4$, above $z = 0$, and within the cylindrical surfaces $y^2 - 2x = 0$, $y^2 = 8 - 2x$. Ans. $\frac{512}{15}$.

13. A solid is bounded by the paraboloid $x^2 + y^2 = az$, the cylindrical surface $y^2 = a^2 - ax$, and the planes $x = 0$, $z = 0$. Find the volume. Ans. $\frac{4}{7} a^3$.

14. Find the volume under $4z = 16 - 4x^2 - y^2$, above $z = 0$, and within $x^2 + y^2 = 2x$. Ans. $\frac{43}{16} \pi$.

15. The axes of two circular cylindrical surfaces intersect at right angles and their radii are equal ($= r$). Find the common volume. Ans. $\frac{16}{3} r^3$.

16. Find the volume of the closed surface $x^{\frac{2}{3}} + y^{\frac{2}{3}} + z^{\frac{2}{3}} = a^{\frac{2}{3}}$. (The trace on each coördinate plane is the astroid, Chapter XXVI.) Ans. $\frac{4}{35} \pi a^3$.

17. Find the volume common to $y^2 + z^2 = 4ax$ and $x^2 + y^2 = 2ax$.

   Ans. $(2\pi + \frac{16}{3})a^3$.

MULTIPLE INTEGRALS

248. Directions for setting up a double integral. We shall now state a rule for forming the double integral which will give a required property. Applications are made in the following articles. For single integration the corresponding rule is given in Art. 156.

FIRST STEP. *Draw the curves which bound the region, or area, concerned.*

SECOND STEP. *At any point $P(x, y)$ within the area construct the rectangular element of area $\Delta x\, \Delta y$.*

THIRD STEP. *Work out the function $f(x, y)$, which, when multiplied by $\Delta x\, \Delta y$, gives the required property for the rectangular element of area.*

FOURTH STEP. *The required integral is*

$$\iint f(x, y) dx\, dy$$

taken over the given region, or area. The order of integration and limits are determined in the same manner as in finding the area itself.

249. Moment of area and centroids. This problem is treated in Art. 177 by single integration. Double integration is often more convenient.

We follow the rule of the preceding article. The moments of area for the rectangular element of area are, respectively,

$x\, \Delta x\, \Delta y$, with respect to OY,

$y\, \Delta x\, \Delta y$, with respect to OX.

Hence for the entire area, using the notation of Art. 177, we have

(C) $\quad M_x = \iint y\, dx\, dy, \quad M_y = \iint x\, dx\, dy.$

The centroid of the area is given by

(D) $\quad \bar{x} = \dfrac{M_y}{\text{area}}, \quad \bar{y} = \dfrac{M_x}{\text{area}}.$

In (C) the integrals give the values of the double integrals of the functions

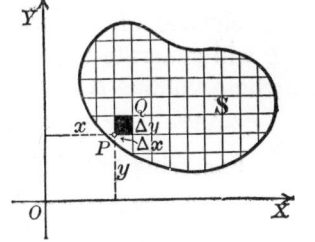

$f(x, y) = y \quad \text{and} \quad f(x, y) = x,$

respectively, taken over the area. (Art. 245.)

For an area bounded by a curve, the x-axis, and two ordinates (the "area under a curve"), we derive from (C)

(1) $\quad M_x = \displaystyle\int_a^b \int_0^y y\, dy\, dx = \tfrac{1}{2}\int_a^b y^2\, dx,$

$\quad M_y = \displaystyle\int_a^b \int_0^y x\, dy\, dx = \int_a^b xy\, dx.$

These agree with (2), Art. 177. Note that y in (1) is the ordinate of a point on the curve, and its value in terms of x must be found from the equation of the curve and substituted in the integrand before integration.

ILLUSTRATIVE EXAMPLE. Find the centroid of the area in the first quadrant bounded by the semicubical parabola $y^2 = x^3$ and the straight line $y = x$.

Solution. The order and the limits of integration were found in Illustrative Example 1, Art. 246. Hence, using (C),

$$M_x = \int_0^1 \int_y^{y^{\frac{2}{3}}} y\, dx\, dy = \int_0^1 (y^{\frac{5}{3}} - y^2) dy = \tfrac{1}{24}.$$

$$M_y = \int_0^1 \int_y^{y^{\frac{2}{3}}} x\, dx\, dy = \tfrac{1}{2} \int_0^1 (y^{\frac{4}{3}} - y^2) dy = \tfrac{1}{21}.$$

Since $A = \text{area} = \tfrac{1}{10}$,

we have, from (D), $\bar{x} = \tfrac{10}{21} = 0.48$, $\bar{y} = \tfrac{5}{12} = 0.42$. Ans.

250. Theorem of Pappus. A useful relation between centroids and volumes of solids of revolution is expressed in the following theorem.

If a plane area is revolved about an axis lying in its plane and not crossing it, the volume of the solid of revolution thus generated is equal to the product of the plane area by the circumference described by its centroid.

Proof. Let the area in the figure be revolved about the x-axis. The rectangular element of area within the region S at P will generate a hollow circular cylinder whose volume ΔV is given by

$$\Delta V = \pi(y + \Delta y)^2\, \Delta x - \pi y^2\, \Delta x.$$

Factoring and simplifying, we get

$$\Delta V = 2\, \pi(y + \tfrac{1}{2} \Delta y)\, \Delta x\, \Delta y.$$

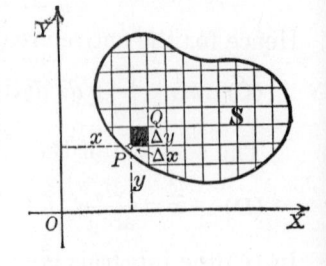

Now, in (1), Art. 245, (x, y) in $f(x, y)$ is a point " either within the rectangle PQ or on its perimeter." But $(x, y + \tfrac{1}{2} \Delta y)$ is a point on the perimeter of PQ. Therefore let $f(x, y) = 2\, \pi y$. Then ΔV has the form $f(x, y)\, \Delta x\, \Delta y$, and, by (1), Art. 245, and (C),

(1) $\qquad V_x = 2\, \pi \iint_S y\, \Delta x\, \Delta y = 2\, \pi M_x.$

Finally, using (D), we get

(2) $\qquad V_x = 2\, \pi \bar{y} \cdot A.$

MULTIPLE INTEGRALS

where A is the area of the region S. The right-hand member is the product of the area by the circumference described by its centroid. Hence the theorem is proved. We write the result

(3) $$V = 2\pi\bar{y} \cdot A.$$

If two of the quantities V, $\bar{y}$, A are known, the other can be found by (3).

ILLUSTRATIVE EXAMPLE. Find the centroid of the trapezoid $OMPB$ of the figure by the Theorem of Pappus.

Solution. Area $OMPB = \frac{1}{2}(3+5)8 = 32$. Revolving the figure about OX, the solid formed is a frustum of a cone of revolution. Hence, by (12), Art. 1, since $a = 8$, $R = 5$, $r = 3$,

$$V_x = \frac{8\pi}{3}(25 + 9 + 15) = \frac{392}{3}\pi.$$

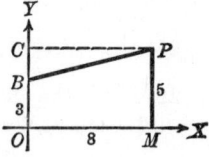

Hence, by (3), $\bar{y} = \dfrac{V_x}{2\pi A} = \dfrac{392}{192} = 2.04.$

Revolving the figure about OY, the volume generated is the difference of the volumes of the cylinder generated by $OCPM$ and the cone generated by the triangle BCP. Hence

$$V_y = 320\pi - \frac{128\pi}{3} = \frac{832}{3}\pi.$$

Hence, by the theorem, $\bar{x} = \dfrac{V_y}{2\pi A} = \dfrac{832}{192} = 4\frac{1}{3}.$

The centroid is $(4\frac{1}{3}, 2.04)$. *Ans.*

PROBLEMS

Find the centroid of the area bounded by each of the following curves:

1. $y = x^3$, $y = 4x$. (Area in first quadrant.)   *Ans.* $(\frac{16}{15}, \frac{64}{21})$.
2. $y = 6x - x^2$, $y = x$. $(\frac{5}{2}, 5)$.
3. $y = 4x - x^2$, $y = 2x - 3$. $(1, \frac{3}{5})$.
4. $x^2 = 4y$, $x - 2y + 4 = 0$. $(1, \frac{8}{5})$.
5. $y = x^2$, $2x - y + 3 = 0$. $(1, \frac{17}{5})$.
6. $y = x^2 - 2x - 3$, $y = 2x - 3$. $(2, -\frac{3}{5})$.
7. $y^2 = x$, $x + y = 2$, $y = 0$. (First quadrant.) $(\frac{32}{35}, \frac{5}{14})$.
8. $y^2 = x$, $x + y = 2$, $x = 0$. $(\frac{8}{25}, \frac{11}{10})$.
9. $y^3 = x^2$, $2y = x$. $(\frac{10}{3}, \frac{40}{21})$.
10. $4y = 3x^2$, $2y^2 = 9x$. $(\frac{9}{10}, \frac{27}{20})$.
11. $y^2 = 2x$, $y = x - x^2$. $(\frac{14}{15}, -\frac{11}{15})$.
12. $y^2 = 8x$, $x + y = 6$. $(\frac{34}{5}, -4)$.
13. $y^2 = 4x$, $y^2 = 5 - x$. $(\frac{11}{5}, 0)$.
14. $y = 6x - x^2$, $x + y = 6$. $(\frac{7}{2}, 5)$.

15. $x = 4y - y^2$, $y = x$.   Ans. $(\frac{12}{5}, \frac{3}{2})$.

16. $y = 4x - x^2$, $y = 5 - 2x$. $(3, \frac{3}{5})$.

17. $y^2 = 4x$, $2x - y = 4$. $(\frac{8}{5}, 1)$.

18. $y = x^2 - 2x - 3$, $y = 6x - x^2 - 3$. $(2, 1)$.

19. $x^2 + y^2 = 1$, $x + y = 1$. $(0.585, 0.585)$.

20. $x^2 + y^2 = 32$, $y^2 = 4x$.

21. $y^2 = 4x$, $2x + y = 4$.

22. $x^2 + y^2 - 10x = 0$, $x^2 = y$.

23. $x^2 = y$, $2y = 6x - x^2$.

24. $x^{\frac{2}{3}} + y^{\frac{2}{3}} = a^{\frac{2}{3}}$. (Area in first quadrant.) $\left(\dfrac{256\,a}{315\,\pi}, \dfrac{256\,a}{315\,\pi}\right)$.

25. $x^{\frac{1}{2}} + y^{\frac{1}{2}} = a^{\frac{1}{2}}$, $x = 0$, $y = 0$. $\left(\dfrac{a}{5}, \dfrac{a}{5}\right)$.

26. Find the centroid of the area under one arch of the cycloid $x = a(\theta - \sin\theta)$, $y = a(1 - \cos\theta)$.   Ans. $\left(\pi a, \dfrac{5\,a}{6}\right)$.

27. Using the Theorem of Pappus, find the centroid of a semicircle.

     Ans. Distance from diameter $= \dfrac{4\,r}{3\,\pi}$.

28. Using the Theorem of Pappus, find the centroid of the area of the ellipse $\dfrac{x^2}{a^2} + \dfrac{y^2}{b^2} = 1$ which lies in the first quadrant.   Ans. $\left(\dfrac{4\,a}{3\,\pi}, \dfrac{4\,b}{3\,\pi}\right)$.

29. Using the Theorem of Pappus find the volume of the torus generated by revolving the circle $(x - b)^2 + y^2 = a^2$ ($b > a$) about the y-axis.   Ans. $2\,\pi^2 a^2 b$.

30. A rectangle is revolved about an axis which lies in its plane and is perpendicular to a diagonal at its extremity. Find the volume of the solid generated.

251. Center of fluid pressure. The problem of calculating the pressure of a fluid on a vertical wall was discussed in Art. 179.

The pressures on the rectangular elements of the figure constitute a system of parallel forces, since they are perpendicular to the plane of the area XOY. The resultant of this system of forces is the total fluid pressure P, given by (D), Art. 179.

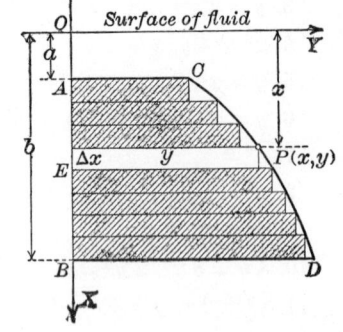

(1) $P = \overline{W} \displaystyle\int_a^b yx\,dx$.

MULTIPLE INTEGRALS 507

The point of application of P is called the *center of fluid pressure*. We wish to find the x-coördinate ($= x_0$) of this point.

To this end we use the *principle of force moments*. This may be stated thus:

The sum of the turning moments of a system of parallel forces about an axis is equal to the turning moment of their resultant about this axis.

Now the fluid pressure dP on the rectangular element EP is, by Art. 179,

(2) $$dP = \overline{W}xy\,\Delta x.$$

The turning moment of this force about the axis OY is the product of dP by its lever arm OE ($= x$), or, using (2),

(3) Turning moment of $dP = x\,dP = \overline{W}x^2y\,\Delta x.$

Hence we have, for the entire turning moment for the distributed fluid pressure,

(4) Total turning moment $= \int_a^b \overline{W}x^2y\,dx.$

But the turning moment of the resultant fluid pressure P is $x_0 P$. Hence

(5) $$x_0 P = \overline{W}\int_a^b x^2 y\,dx.$$

Solving for x_0 and using (1), we get the formula for the *depth of the center of pressure*

(6) $$x_0 = \frac{\int_a^b x^2\,dA}{\int_a^b x\,dA},$$

where dA = element of area = $y\,dx$.

The denominator in (6) is the moment of area of $ABCD$ with respect to OY (see Art. 177). The numerator is an integral not met with hitherto. It is called the *moment of inertia of the area ABCD about OY*.

The letter I is commonly used for moment of inertia about an axis, and a subscript is attached to designate the axis. Thus (6) becomes

(7) $$x_0 = \frac{I_y}{M_y}.$$

The usual notation for moment of inertia about an axis l is

(8) $$I_l = \int r^2\, dA,$$

in which

(9) $r =$ distance of the element dA from the axis l.

The problem of this article is one of many which lead to moments of inertia. In the following section the calculation of moments of inertia by double and single integration is explained. Applications are also given.

252. Moment of inertia of an area. In mechanics the moment of inertia of an area about an axis is an important concept. The calculation of moments of inertia will now be explained. We follow the rule of Art. 248.

For the elementary rectangle PQ at $P(x, y)$ the moment of inertia about OX is defined as

(1) $$y^2 \Delta x\, \Delta y,$$

and about the y-axis it is

(2) $$x^2 \Delta x\, \Delta y.$$

Then, if I_x and I_y are the corresponding moments of inertia for the entire area, we have (compare (8), Art. 251)

(E) $$I_x = \iint y^2\, dx\, dy, \quad I_y = \iint x^2\, dx\, dy.$$

The radii of gyration r_x and r_y are given by

(F) $$r_x^2 = \frac{I_x}{\text{area}}, \quad r_y^2 = \frac{I_y}{\text{area}}.$$

In (E) the functions whose integrals are taken over the area are, respectively, $f(x, y) = y^2$, and $f(x, y) = x^2$.

Formulas (E) become simple for an area "under a curve," that is, an area bounded by a curve, the x-axis and two ordinates. Thus we obtain

(3) $$I_x = \int_a^b \int_0^y y^2\, dy\, dx = \tfrac{1}{3} \int_a^b y^3\, dx,$$
$$I_y = \int_a^b \int_0^y x^2\, dy\, dx = \int_a^b x^2 y\, dx.$$

In these equations y is the ordinate of a point on the curve, and its value in terms of x must be found from the equation of this curve and substituted in the integrand.

Formulas for moments of inertia I are written in the form

(G) $$I = Ar^2,$$

where A = area and r = radius of gyration. Solving (F) for I_x and I_y will give this form.

Dimensions. If the linear unit is 1 in., the moment of inertia has the dimensions in.^{4}. By (F), r_x and r_y are lengths, in inches.

ILLUSTRATIVE EXAMPLE 1. Find $I_x, I_y,$ and the corresponding radii of gyration for the area of Illustrative Example 1, Art. 246.

Solution. Using the same order of integration and the same limits as before, we have, by (E),

$$I_x = \int_0^1 \int_y^{y^{\frac{2}{3}}} y^2 \, dx \, dy = \int_0^1 (y^{\frac{8}{3}} - y^3) \, dy = \tfrac{1}{44}.$$

$$I_y = \int_0^1 \int_y^{y^{\frac{2}{3}}} x^2 \, dx \, dy = \tfrac{1}{3} \int_0^1 (y^2 - y^3) \, dy = \tfrac{1}{36}.$$

Since A = area = $\tfrac{1}{10}$, we find, by (F),

$$r_x = 0.48, \ r_y = 0.53. \ Ans.$$

ILLUSTRATIVE EXAMPLE 2. Find I_x and I_y for the parabolic segment BOC in the figure.

Solution. With the axes of coördinates as drawn, the equation of the bounding parabola is

(4) $$y^2 = 2 \, px.$$

Since $B(a, b)$ is a point on the curve, we get, by substituting $x = a, y = b$ in (4), $b^2 = 2 \, pa$. Solving this equation for $2 \, p$ and substituting its value in (4), we obtain

(5) $$y^2 = \frac{b^2 x}{a}, \quad \text{or} \quad y = \frac{bx^{\frac{1}{2}}}{a^{\frac{1}{2}}}.$$

The moments of inertia for the area under the parabola OPB in the first quadrant will be half the required moments. Hence, using (3), and substituting the value of y from (5), we get

$$\tfrac{1}{2} I_x = \tfrac{1}{3} \int_0^a \frac{b^3}{a^{\frac{3}{2}}} x^{\frac{3}{2}} \, dx = \frac{2}{15} ab^3. \quad \therefore I_x = \frac{4}{15} ab^3.$$

$$\tfrac{1}{2} I_y = \int_0^a x^2 \frac{b}{a^{\frac{1}{2}}} x^{\frac{1}{2}} \, dx = \frac{2}{7} a^3 b. \quad \therefore I_y = \frac{4}{7} a^3 b.$$

For the area of the segment, we find

$$\tfrac{1}{2} A = \int_0^a y \, dx = \int_0^a \frac{b}{a^{\frac{1}{2}}} x^{\frac{1}{2}} \, dx = \frac{2}{3} ab. \quad \therefore A = \frac{4}{3} ab.$$

Hence, by (F), $\quad r_x^2 = \dfrac{I_x}{A} = \dfrac{1}{5} b^2,$ and $\quad I_x = \dfrac{1}{5} Ab^2,$

$$r_y^2 = \frac{I_y}{A} = \frac{3}{7} a^2, \quad \text{and} \quad I_y = \frac{3}{7} Aa^2.$$

The results are in the form (G). *Ans.*

In the figure on page 325 the axis OY lies in the surface of the fluid. If we denote this axis in any figure by s, then the depth of the center of pressure, is by (7), Art. 251

(6) $$x_0 = \frac{I_s}{M_s} = \frac{r_s^2}{h_s},$$

if $\quad r_s = $ radius of gyration about the axis s,
and $\quad h_s = $ depth of centroid below the axis s.

ILLUSTRATIVE EXAMPLE 3. Find the depth of the center of pressure on the trapezoidal water gate of the figure. Compare Illustrative Example 2, Art. 179.

Solution. Choose axes OX and OY as shown, and draw an elementary horizontal strip. Let the distance of this strip from the axis s at the water level be r. Then
$$r = 8 - y, \quad dA = 2x\,dy.$$

Hence, by (8), Art. 251, and by the definition of moment of area (Art. 177), we have

(7) $\quad I_s = \int r^2 dA = \int (8-y)^2 2x\,dy,$

(8) $\quad M_s = \int r\,dA = \int (8-y) 2x\,dy.$

The equation of AB is $y = 2x - 8$.
Solving this for x, substituting in (7) and (8), and integrating with limits $y = 0$, $y = 4$, we obtain

$$I_s = \int_0^4 (8-y)^2(8+y)dy = 1429\tfrac{1}{3},$$

$$M_s = \int_0^4 (64 - y^2)dy = 234\tfrac{2}{3}.$$

Hence, by (7), Art. 251, $x_0 = 6.09$. Ans.

253. Polar moment of inertia. The moment of inertia of the elementary rectangle PQ about the origin O is the product of the area and $\overline{OP}^2$, that is,

(1) $\quad (x^2 + y^2)\Delta x\,\Delta y.$

Hence, by Art. 248, for the entire area

(2) $\quad I_0 = \iint (x^2 + y^2)dx\,dy.$

We may, however, write the right-hand member as the sum of two integrals, for (2) is clearly the same as

(3) $\quad I_0 = \iint x^2\,dx\,dy + \iint y^2\,dx\,dy = I_x + I_y.$

Hence we have the following theorem.

The moment of inertia of an area about the origin equals the sum of its moments of inertia about the x-axis and the y-axis.

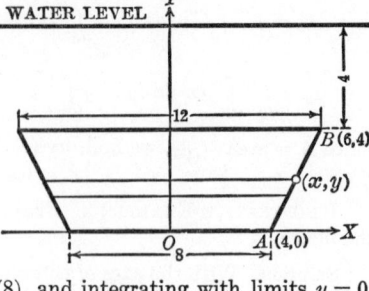

PROBLEMS

Find I_x, I_y, and I_0 for each of the areas described below.

1. The semicircle which is to the right of the y-axis and which is bounded by $x^2 + y^2 = r^2$. *Ans.* $I_x = I_y = \dfrac{Ar^2}{4}$.

2. The isosceles triangle of height h and base a whose vertices are $(0, 0), \left(h, \dfrac{a}{2}\right), \left(h, -\dfrac{a}{2}\right)$. *Ans.* $I_x = \dfrac{Aa^2}{24}, I_y = \dfrac{Ah^2}{2}$.

3. The right triangle whose vertices are $(0, 0)$, (b, a), $(b, 0)$. *Ans.* $I_x = \dfrac{Aa^2}{6}, I_y = \dfrac{Ab^2}{2}$.

4. The ellipse $\dfrac{x^2}{a^2} + \dfrac{y^2}{b^2} = 1$. *Ans.* $I_x = \dfrac{Ab^2}{4}, I_y = \dfrac{Aa^2}{4}$.

5. The area in the first quadrant bounded by $y^2 = 4x$, $x = 4$, $y = 0$. *Ans.* $I_x = \dfrac{16\,A}{5}, I_y = \dfrac{48\,A}{7}$.

6. The area included between the ellipse $\dfrac{x^2}{9} + \dfrac{y^2}{4} = 1$ and the circle $x^2 + y^2 = 2y$. *Ans.* $I_x = \dfrac{19\,A}{20}, I_y = \dfrac{53\,A}{20}$.

7. The area included between the ellipses $\dfrac{x^2}{16} + \dfrac{y^2}{9} = 1$ and $x^2 + \dfrac{y^2}{4} = 1$. *Ans.* $I_x = \dfrac{5\,A}{2}, I_y = \dfrac{19\,A}{4}$.

8. The area included between the circle $x^2 + y^2 = 16$ and the circle $x^2 + (y + 2)^2 = 1$. *Ans.* $I_x = \dfrac{239\,A}{60}, I_y = \dfrac{17\,A}{4}$.

9. The area included between the circle $x^2 + y^2 = 36$ and the circle $x^2 + (y + 3)^2 = 4$. *Ans.* $I_x = \dfrac{71\,A}{8}, I_y = 10\,A$.

10. The area between the circle $x^2 + y^2 = 4$ and the ellipse $\dfrac{x^2}{36} + \dfrac{y^2}{16} = 1$. *Ans.* $I_x = \dfrac{23\,A}{5}, I_y = \dfrac{53\,A}{5}$.

11. The entire area bounded by $x^{\frac{2}{3}} + y^{\frac{2}{3}} = a^{\frac{2}{3}}$. *Ans.* $I_x = I_y = \dfrac{7\,Aa^2}{64}$.

12. Find the depth of the center of pressure on a triangular water gate having its vertex below the base, which is horizontal and on a level with the surface of the water.

13. Find the depth of the center of pressure on a rectangular water gate 8 ft. wide and 4 ft. deep when the level of the water is 5 ft. above the top of the gate. *Ans.* **7.19 ft. below the surface of the water.**

14. Find the depth of the center of pressure on the end of a horizontal cylindrical oil tank of diameter 5 ft. when the depth of oil is (a) 2.5 ft.; (b) 4 ft.; (c) 6 ft.   *Ans.* (a) $\dfrac{15\,\pi}{32} = 1.47$ ft.; (b) approximately 2.4 ft.; (c) $\dfrac{221}{56} = 3.95$ ft.

254. Polar coördinates. Plane area. When the equations of the curves bounding an area are given in polar coördinates, certain modifications are necessary.

The area is now divided into elementary portions, as follows:

Draw arcs of circles with the common center O with successive radii differing by $\Delta\rho$. Thus, in Fig. 1, $OP = \rho$, $OS = \rho + \Delta\rho$. Then

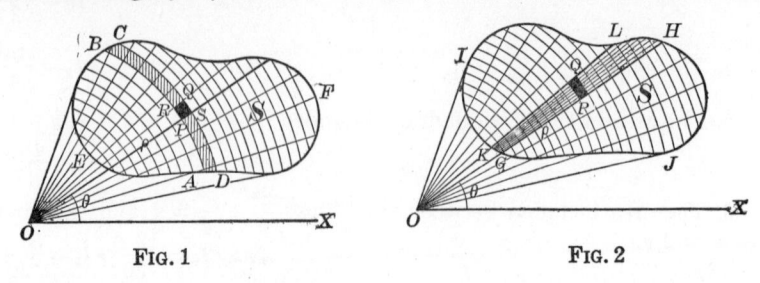

FIG. 1   FIG. 2

draw radial lines from O such that the angle between any two consecutive lines is the same and equal to $\Delta\theta$. Thus, in Fig. 1, angle $POR = \Delta\theta$.

The area will now contain a large number of rectangular portions, such as $PSQR$ in Fig. 1.

Let $PSQR = \Delta A$. Now ΔA is the difference of the areas of the circular sectors POR and SOQ. Hence

(1) $\quad \Delta A = \tfrac{1}{2}(\rho + \Delta\rho)^2\,\Delta\theta - \tfrac{1}{2}\rho^2\,\Delta\theta = \rho\,\Delta\rho\,\Delta\theta + \tfrac{1}{2}\Delta\rho^2\,\Delta\theta$.

The function $f(x, y)$ of Art. 245 is to be replaced by a function using polar coördinates. Let this be $F(\rho, \theta)$. Then, proceeding as in Art. 245, we choose a point (ρ, θ) of ΔA, form the product

$$F(\rho, \theta)\Delta A$$

for each ΔA within the region S, add these products, and finally let $\Delta\rho \to 0$ and $\Delta\theta \to 0$. It is shown in Art. 258 that the limiting value of this double sum may be found by successive integration. We now write (compare (1), Art. 245)

(2) $\quad \displaystyle\lim_{\substack{\Delta\rho \to 0 \\ \Delta\theta \to 0}} \sum\sum F(\rho, \theta)\,\Delta A = \iint_S F(\rho, \theta)\rho\,d\rho\,d\theta,$

MULTIPLE INTEGRALS 513

and call it *the double integral of the function $F(\rho, \theta)$ taken over the region S.*

Note in (2) that the value of ΔA in (1) has been replaced in the integral by $\rho\, d\rho\, d\theta$.

The simplest case of (2) is that of finding the area of the region S. We then have

(H) $\qquad A = \iint \rho\, d\rho\, d\theta = \iint \rho\, d\theta\, d\rho.$

These are easily remembered if we think of the elements (checks) as being rectangles with dimensions $\rho\, d\theta$ and $d\rho$, and hence of area $\rho\, d\theta\, d\rho$.

The figures below illustrate, in a general way, the difference in the processes indicated by the two integrals.

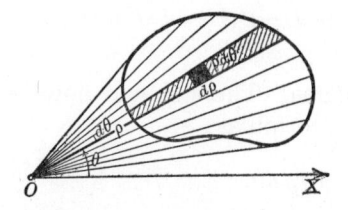

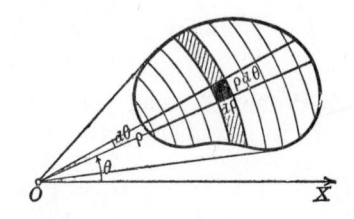

In the first, we integrate first with respect to ρ, since $d\rho$ precedes $d\theta$, keeping θ constant. This process will cover the radial strip $KGHL$ in Fig. 2, p. 512. The limits for ρ are $\rho = OG$ and $\rho = OH$, found by solving the equation (or equations) of the bounding curve (or curves) for ρ in terms of θ. Then integrate varying θ, the limits being $\theta = \angle JOX$ and $\theta = \angle IOX$.

The second integral in (2) is worked out by integrating with respect to θ, ρ remaining constant. This step covers the circular strip $ABCD$ in Fig. 1, p. 512, between two consecutive circular arcs. Then integrate varying ρ.

When the area is bounded by a curve and two of its radii vectores (area swept over by the radius vector), we obtain from the first form in (H)

$$A = \int_\alpha^\beta \int_0^\rho \rho\, d\rho\, d\theta = \tfrac{1}{2} \int_\alpha^\beta \rho^2\, d\theta,$$

which agrees with (D), Art. 159.

Double integrals in polar coördinates have one of the forms

(3) $\qquad \iint F(\rho, \theta)\rho\, d\rho\, d\theta \quad \text{or} \quad \iint F(\rho, \theta)\rho\, d\theta\, d\rho.$

514   DIFFERENTIAL AND INTEGRAL CALCULUS

ILLUSTRATIVE EXAMPLE 1. Find the limits for the double integral giving some required property related to the area inside the circle $\rho = 2r\cos\theta$ and outside the circle $\rho = r$.

Solution. The points of intersection are $A\left(r, \frac{\pi}{3}\right)$, and $B\left(r, -\frac{\pi}{3}\right)$. Use the first form in (3).

The limits for ρ are
$$\rho = OG = r,$$
$$\rho = OH = 2r\cos\theta;$$
for θ they are $\frac{\pi}{3}$ and $-\frac{\pi}{3}$. Ans.

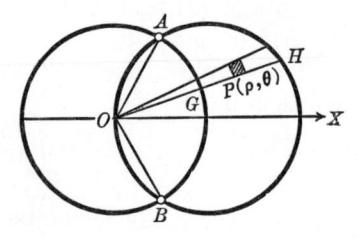

ILLUSTRATIVE EXAMPLE 2. Find the area inside the circle $\rho = 2r\cos\theta$ and outside the circle $\rho = r$.

Solution. From Illustrative Example 1 above, we have

$$A = \int_{-\frac{\pi}{3}}^{\frac{\pi}{3}} \int_{r}^{2r\cos\theta} \rho\, d\rho\, d\theta = \int_{-\frac{\pi}{3}}^{\frac{\pi}{3}} \tfrac{1}{2}(4r^2\cos^2\theta - r^2)d\theta = r^2(\tfrac{1}{3}\pi + \tfrac{1}{2}\sqrt{3}) = 1.91\, r^2.\ \text{Ans.}$$

255. Problems using polar coördinates. There should now be no difficulty in establishing the following formulas:

(1) $$M_x = \iint \rho^2 \sin\theta\, d\rho\, d\theta.$$

(2) $$M_y = \iint \rho^2 \cos\theta\, d\rho\, d\theta.$$

(3) $$I_x = \iint \rho^3 \sin^2\theta\, d\rho\, d\theta.$$

(4) $$I_y = \iint \rho^3 \cos^2\theta\, d\rho\, d\theta.$$

(5) $$I_0 = \iint \rho^3\, d\rho\, d\theta.$$

The order of the differentials will have to be changed if integration with respect to θ is performed first.

ILLUSTRATIVE EXAMPLE 1. On account of important applications the moments of inertia of a circle are now worked out.

Let $a =$ radius. Then, by (5), the polar moment of inertia with respect to the center is

(6) $$I_0 = \int_0^a \left[\int_0^{2\pi} d\theta\right] \rho^3\, d\rho = \frac{\pi a^4}{2} = \frac{A}{2}a^2,$$

where $A =$ area of the circle.

Also, since $I_x = I_y$, by symmetry, we have, by (3), Art. 253,

(7) $$I_x = \tfrac{1}{2} I_0 = \frac{A}{4} a^2.$$

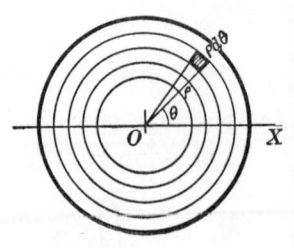

MULTIPLE INTEGRALS

In words: *The polar moment of inertia of a circle with respect to its center equals the product of half the area and the square of the radius; the polar moment of inertia with respect to any diameter equals the product of one fourth the area and the square of the radius.*

ILLUSTRATIVE EXAMPLE 2. Find the centroid of a loop of the lemniscate

$$\rho^2 = a^2 \cos 2\theta.$$

Solution. Since OX is an axis of symmetry, we have $\bar{y} = 0$.

$$\tfrac{1}{2} A = \int_0^{\frac{1}{4}\pi} \int_0^{a\sqrt{\cos 2\theta}} \rho \, d\rho \, d\theta$$

$$= \frac{a^2}{2} \int_0^{\frac{1}{4}\pi} \cos 2\theta \, d\theta = \frac{a^2}{4}.$$

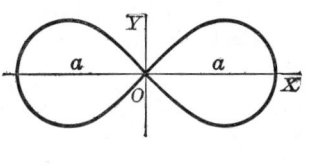

$$\tfrac{1}{2} M_y = \int_0^{\frac{1}{4}\pi} \int_0^{a\sqrt{\cos 2\theta}} \rho^2 \cos\theta \, d\rho \, d\theta = \tfrac{1}{3} a^3 \int_0^{\frac{1}{4}\pi} (\cos 2\theta)^{\frac{3}{2}} \cos\theta \, d\theta.$$

$$= \tfrac{1}{3} a^3 \int_0^{\frac{1}{4}\pi} (1 - 2\sin^2\theta)^{\frac{3}{2}} \cos\theta \, d\theta \qquad \text{by (5), Art. 2}$$

$$= \frac{\sqrt{2}}{6} a^3 \int_0^1 (1 - z^2)^{\frac{3}{2}} dz \left(\text{if } \sin\theta = \tfrac{1}{2} z\sqrt{2} \right) = \frac{\pi}{32} a^3 \sqrt{2}.$$

Hence $\quad \bar{x} = \dfrac{M_y}{A} = \dfrac{\pi}{8} a\sqrt{2} = 0.55\, a.$ Ans.

ILLUSTRATIVE EXAMPLE 3. Find I_0 over the region bounded by the circle $\rho = 2 r \cos\theta$.

Solution. Summing up for the elements in the triangular-shaped strip OP, the ρ-limits are zero and $2 r \cos\theta$ (found from the equation of the circle).

Summing up for all such strips, the θ-limits are $-\dfrac{\pi}{2}$ and $\dfrac{\pi}{2}$. Hence, by (5),

$$I_0 = \int_{-\frac{\pi}{2}}^{\frac{\pi}{2}} \int_0^{2r\cos\theta} \rho^3 \, d\rho \, d\theta = \frac{3\,\pi r^4}{2}. \quad \text{Ans.}$$

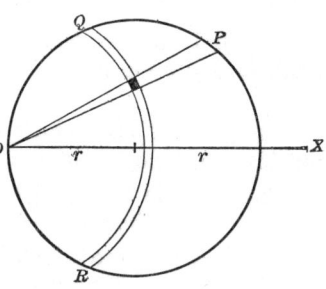

Or, summing up first for the elements in a circular strip (as QR), we have

$$I_0 = \int_0^{2r} \int_{-\arccos\frac{\rho}{2r}}^{\arccos\frac{\rho}{2r}} \rho^3 \, d\theta \, d\rho = \frac{3\,\pi r^4}{2}. \quad \text{Ans.}$$

PROBLEMS

1. Find the area inside the circle $\rho = \tfrac{3}{2}$ and to the right of the line $4 \rho \cos\theta = 3$.
Ans. $\dfrac{3(4\,\pi - 3\sqrt{3})}{16}.$

2. Find the area which is inside the circle $\rho = 3 \cos\theta$ and outside the circle $\rho = \tfrac{3}{2}$.
Ans. $\dfrac{3(2\,\pi + 3\sqrt{3})}{8}.$

3. Find the area which is inside the circle $\rho = 3 \cos \theta$ and outside the circle $\rho = \cos \theta$. *Ans.* 2π.

4. Find the area inside the cardioid $\rho = 1 + \cos \theta$ and to the right of the line $4 \rho \cos \theta = 3$.
Ans. $\dfrac{\pi}{2} + \dfrac{9\sqrt{3}}{16}$.

5. Find the area which is inside the cardioid $\rho = 1 + \cos \theta$ and outside the circle $\rho = 1$.
Ans. $\dfrac{\pi}{4} + 2$.

6. Find the area which is inside the circle $\rho = 1$ and outside the cardioid $\rho = 1 + \cos \theta$.
Ans. $2 - \dfrac{\pi}{4}$.

7. Find the area which is inside the circle $\rho = 3 \cos \theta$ and outside the cardioid $\rho = 1 + \cos \theta$.
Ans. π.

8. Find the area which is inside the circle $\rho = 1$ and outside the parabola $\rho(1 + \cos \theta) = 1$.
Ans. $\dfrac{\pi}{2} - \dfrac{2}{3}$.

9. Find the area which is inside the cardioid $\rho = 1 + \cos \theta$ and outside the parabola $\rho(1 + \cos \theta) = 1$.
Ans. $\dfrac{3\pi}{4} + \dfrac{4}{3}$.

10. Find the area which is inside the circle $\rho = \cos \theta + \sin \theta$ and outside the circle $\rho = 1$.
Ans. $\dfrac{1}{2}$.

11. Find the area which is inside the circle $\rho = \sin \theta$ and outside the cardioid $\rho = 1 - \cos \theta$.
Ans. $1 - \dfrac{\pi}{4}$.

12. Find the area which is inside the lemniscate $\rho^2 = 2 a^2 \cos 2\theta$ and outside the circle $\rho = a$.
Ans. $0.684 a^2$.

13. Find the area which is inside the cardioid $\rho = 4(1 + \cos \theta)$ and outside the parabola $\rho(1 - \cos \theta) = 3$.
Ans. 5.504.

14. Find the area which is inside the circle $\rho = 2a \cos \theta$ and outside the circle $\rho = a$. Find the centroid of the area and I_x and I_y.

Ans. $A = \left(\dfrac{\pi}{3} + \dfrac{\sqrt{3}}{2}\right) a^2$, $\bar{x} = \dfrac{(8\pi + 3\sqrt{3})a}{2(2\pi + 3\sqrt{3})}$,

$I_x = \left(\dfrac{\pi}{12} + \dfrac{3\sqrt{3}}{16}\right) a^4$, $I_y = \left(\dfrac{3\pi}{4} + \dfrac{11\sqrt{3}}{16}\right) a^4$.

15. Find the centroid of the area bounded by the cardioid
$\rho = a(1 + \cos \theta)$.
Ans. $\bar{x} = \dfrac{5a}{6}$.

16. Find the centroid of the area bounded by a loop of the curve $\rho = a \cos 2\theta$.
Ans. $\bar{x} = \dfrac{128\sqrt{2}\,a}{105 \pi}$.

17. Find the centroid of the area bounded by a loop of the curve $\rho = a \cos 3\theta$.
Ans. $\bar{x} = \dfrac{81\sqrt{3}\,a}{80 \pi}$.

18. Find I_y for the lemniscate $\rho^2 = a^2 \cos 2\theta$. Ans. $\dfrac{A}{48}(3\pi+8)a^2$.

19. Find I_x for the cardioid $\rho = a(1+\cos\theta)$.

20. Find I_x and I_y for one loop of the curve $\rho = a\cos 2\theta$.

21. Prove from (1), Art. 254, that

$$\lim_{\substack{\Delta\rho\to 0 \\ \Delta\theta\to 0}} \frac{\Delta A}{\rho\,\Delta\rho\,\Delta\theta} = 1,$$

and therefore ΔA "differs from $\rho\,\Delta\rho\,\Delta\theta$ by an infinitesimal of higher order" (Art. 99). Then ΔA in the left-hand member of (2), Art. 254, may be replaced by $\rho\,\Delta\rho\,\Delta\theta$. (Proof omitted.)

256. General method for finding the areas of curved surfaces. The method given in Art. 164 applied only to the area of a surface of revolution. We shall now give a more general method. Let

(1) $$z = f(x, y)$$

be the equation of the surface KL in the figure, and suppose it is required to calculate the area of the region S' lying on the surface.

Denote by S the region on the XOY-plane which is the orthogonal projection of S' on that plane. Now pass planes parallel to YOZ and XOZ at common distances Δx and Δy respectively. As in Art. 244, these planes form truncated prisms (as PB) bounded at the top by a portion (as PQ) of the given surface whose projection on the XOY-plane is a rectangle of area $\Delta x\,\Delta y$ (as AB). This rectangle also forms the lower base of the prism. The coördinates of P are (x, y, z).

Now consider the plane tangent to the surface KL at P. Evidently the same rectangle AB is the projection on the XOY-plane of that portion of the tangent plane (PR) which is intercepted by the prism PB. Assuming γ as the angle the tangent plane makes with the XOY-plane, we have

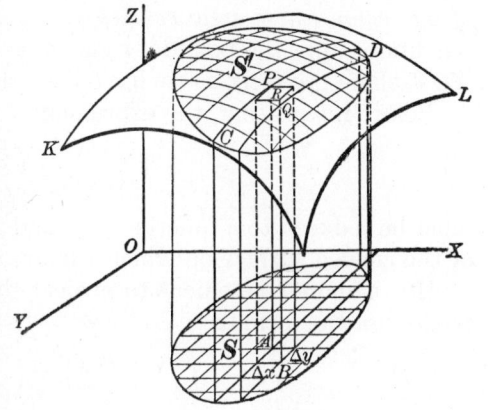

$$\text{Area } AB = \text{area } PR \cdot \cos\gamma,$$

[The projection of a plane area upon a second plane is equal to the area of the portion projected multiplied by the cosine of the angle between the planes.]

or $$\Delta y\,\Delta x = \text{area } PR \cdot \cos\gamma.$$

Now γ is equal to the angle between OZ and a line from O perpendicular to the tangent plane. Hence from (*H*), Art. 237, and (2) and (3), Art. 4, we have

$$\cos \gamma = \frac{1}{\left[1+\left(\frac{\partial z}{\partial x}\right)^2+\left(\frac{\partial z}{\partial y}\right)^2\right]^{\frac{1}{2}}}.$$

Then $\quad$ Area $PR = \left[1+\left(\frac{\partial z}{\partial x}\right)^2+\left(\frac{\partial z}{\partial y}\right)^2\right]^{\frac{1}{2}} \Delta y\, \Delta x.$

This we take as the element of area of the region S'. We then define the area of the region S' as

$$\lim_{\substack{\Delta x \to 0 \\ \Delta y \to 0}} \sum \sum \left[1+\left(\frac{\partial z}{\partial x}\right)^2+\left(\frac{\partial z}{\partial y}\right)^2\right]^{\frac{1}{2}} \Delta y\, \Delta x,$$

the summation extending over the region S, as in Art. 245. Denoting by A the area of the region S', we have

(*I*) $\quad\quad A = \iint_S \left[1+\left(\frac{\partial z}{\partial x}\right)^2+\left(\frac{\partial z}{\partial y}\right)^2\right]^{\frac{1}{2}} dy\, dx,$

the limits of integration depending *on the projection on the XOY-plane of the region whose area we wish to calculate.* Thus, for (*I*) we choose our limits from the boundary curve or curves of the region S in the XOY-plane precisely as we have been doing in the previous sections.

Before integrating, the expression

$$1+\left(\frac{\partial z}{\partial x}\right)^2+\left(\frac{\partial z}{\partial y}\right)^2$$

must be reduced to a function of x and y only, by using the equation of the curved surface on which the area lies.

If it is more convenient to project the required area on the XOZ-plane, use the formula

(*J*) $\quad\quad A = \iint_S \left[1+\left(\frac{\partial y}{\partial x}\right)^2+\left(\frac{\partial y}{\partial z}\right)^2\right]^{\frac{1}{2}} dz\, dx,$

where the limits are found from the boundary of the region S, which is now the projection of the required area on the XOZ-plane.

Similarly, we may use

(*K*) $\quad\quad A = \iint_S \left[1+\left(\frac{\partial x}{\partial y}\right)^2+\left(\frac{\partial x}{\partial z}\right)^2\right]^{\frac{1}{2}} dz\, dy,$

the limits being found from the projection of the required area on the YOZ-plane.

In some problems it is required to find the area of a portion of one surface intercepted by a second surface. In such cases the partial derivatives required for substitution in the formula should be found from the equation of the surface whose partial area is wanted.

Since the limits are found by projecting the required area on one of the coördinate planes, it should be remembered that

To find the projection of the area required on the XOY-plane, eliminate z between the equations of the surfaces whose intersections form the boundary of the area.

Similarly, we eliminate y to find the projection on the XOZ-plane, and x to find it on the YOZ-plane.

This area of a curved surface gives a further illustration of *integration of a function over a given area*. Thus in (*I*) we integrate the function
$$\left[1+\left(\frac{\partial z}{\partial x}\right)^2+\left(\frac{\partial z}{\partial y}\right)^2\right]^{\frac{1}{2}}$$
over the projection of the required curved surface on the XOY-plane.

As remarked above, (*J*) and (*K*) must be reduced to
$$\iint f(x, z)\,dz\,dx \quad \text{and} \quad \iint f(y, z)\,dy\,dz,$$
respectively, by means of the equation of the surface on which the required curved surface lies.

ILLUSTRATIVE EXAMPLE 1. Find the area of the surface of the sphere
$$x^2 + y^2 + z^2 = r^2$$
by double integration.

Solution. Let *ABC* in the figure be one eighth of the surface of the sphere. Here
$$\frac{\partial z}{\partial x} = -\frac{x}{z}, \quad \frac{\partial z}{\partial y} = -\frac{y}{z},$$
and $\quad 1+\left(\frac{\partial z}{\partial x}\right)^2+\left(\frac{\partial z}{\partial y}\right)^2 = 1 + \frac{x^2}{z^2} + \frac{y^2}{z^2} = \frac{x^2+y^2+z^2}{z^2} = \frac{r^2}{r^2-x^2-y^2}.$

The projection of the area required on the XOY-plane is *AOB*, a region bounded by $x=0 (=OB)$; $y=0 (=OA)$; $x^2+y^2=r^2 (=BA)$.

Integrating first with respect to y, we sum up all the elements along a strip (as *DEGF*) which is also projected on the XOY-plane in a strip (as *MNGF*); that is, our y-limits are zero and MF $(=\sqrt{r^2-x^2})$. Then integrating with respect to x sums up all such strips composing the surface *ABC*; that is, our x-limits are zero and $OA (=r)$. Substituting in (*I*), we get

$$\frac{A}{8} = \int_0^r \int_0^{\sqrt{r^2-x^2}} \frac{r\,dy\,dx}{\sqrt{r^2-x^2-y^2}} = \frac{\pi r^2}{2},$$

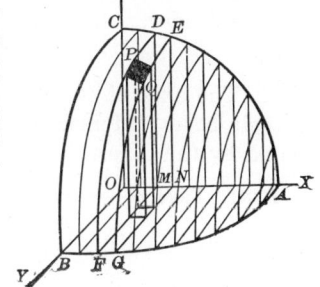

or $\quad A = 4\pi r^2.$ Ans.

520   DIFFERENTIAL AND INTEGRAL CALCULUS

ILLUSTRATIVE EXAMPLE 2. The center of a sphere of radius r is on the surface of a right cylinder, the radius of whose base is $\frac{r}{2}$. Find the area of the surface of the cylinder intercepted by the sphere.

Solution. Taking the origin at the center of the sphere, an element of the cylinder for the z-axis, and a diameter of a right section of the cylinder for the x-axis, the equation of the sphere is

$$x^2 + y^2 + z^2 = r^2,$$

and of the cylinder $x^2 + y^2 = rx$.

$ODAPB$ is evidently one fourth of the cylindrical surface required. Since this area projects into the semicircular arc ODA on the XOY-plane, there is no region S from which to determine our limits in this plane; hence we shall project our area on, say, the XOZ-plane. Then the region S over which we integrate is $OACB$, which is bounded by $z = 0$ $(= OA)$, $x = 0$ $(= OB)$, and $z^2 + rx = r^2$ $(= ACB)$, the last equation being found by eliminating y between the equations of the two surfaces. Integrating first with respect to z means that we sum up all the elements in a vertical strip (as PD), the z-limits being zero and $\sqrt{r^2 - rx}$. Then, on integrating with respect to x, we sum up all such strips, the x-limits being zero and r.

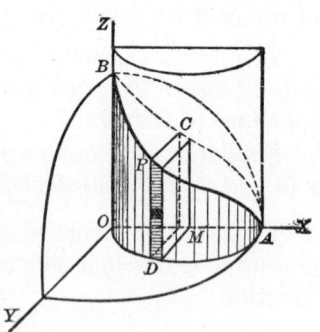

Since the required surface lies on the cylinder, the partial derivatives required for formula (*J*) must be found from the equation of the cylinder.

Hence
$$\frac{\partial y}{\partial x} = \frac{r - 2x}{2y}, \quad \frac{\partial y}{\partial z} = 0.$$

Substituting in (*J*),

$$\frac{A}{4} = \int_0^r \int_0^{\sqrt{r^2 - rx}} \left[1 + \left(\frac{r - 2x}{2y}\right)^2\right]^{\frac{1}{2}} dz\, dx.$$

Substituting the value of y in terms of x from the equation of the cylinder,

$$A = 2r \int_0^r \int_0^{\sqrt{r^2 - rx}} \frac{dz\, dx}{\sqrt{rx - x^2}} = 2r \int_0^r \frac{\sqrt{r^2 - rx}}{\sqrt{rx - x^2}} dx = 2r \int_0^r \sqrt{\frac{r}{x}}\, dx = 4\, r^2.$$

PROBLEMS

1. In the preceding example find the surface of the sphere intercepted by the cylinder.   *Ans.* $4r \int_0^r \int_0^{\sqrt{rx - x^2}} \frac{dy\, dx}{\sqrt{r^2 - x^2 - y^2}} = 2(\pi - 2)r^2.$

2. The axes of two equal right circular cylinders, r being the radius of their bases, intersect at right angles. Find the surface of one intercepted by the other.

HINT. Take $x^2 + z^2 = r^2$ and $x^2 + y^2 = r^2$ as the equations of the cylinders.

Ans. $8r \int_0^r \int_0^{\sqrt{r^2 - x^2}} \frac{dy\, dx}{\sqrt{r^2 - x^2}} = 8\, r^2.$

MULTIPLE INTEGRALS 521

3. Find the area of that portion of the sphere $x^2 + y^2 + z^2 = 2\,ay$ cut out by one nappe of the cone $x^2 + z^2 = y^2$. *Ans.* $2\,\pi a^2$.

4. Find the surface of the cylinder $x^2 + y^2 = r^2$ included between the plane $z = mx$ and the XOY-plane. *Ans.* $4\,r^2 m$.

5. Find the area of that part of the plane $\dfrac{x}{a} + \dfrac{y}{b} + \dfrac{z}{c} = 1$ which is intercepted by the coördinate planes.
Ans. $\tfrac{1}{2}\sqrt{b^2 c^2 + c^2 a^2 + a^2 b^2}$.

6. Find the area of the portion of the sphere $x^2 + y^2 + z^2 = 2\,ay$ which lies within the paraboloid $by = x^2 + z^2$. *Ans.* $2\,\pi ab$.

7. In the preceding example find the area of the portion of the paraboloid which lies within the sphere.

8. Find the area of the surface of the paraboloid $y^2 + z^2 = 4\,ax$ intercepted by the parabolic cylinder $y^2 = ax$ and the plane $x = 3\,a$.
Ans. $\tfrac{56}{9}\pi a^2$.

9. In the preceding problem find the area of the surface of the cylinder intercepted by the paraboloid and plane. *Ans.* $(13\sqrt{13} - 1)\dfrac{a^2}{\sqrt{3}}$.

10. Find the surface of the cylinder $z^2 + (x\cos\alpha + y\sin\alpha)^2 = r^2$ which is situated in the positive compartment of coördinates.

HINT. The axis of this cylinder is the line $z = 0$, $x\cos\alpha + y\sin\alpha = 0$; and the radius of the base is r. *Ans.* $\dfrac{r^2}{\sin\alpha\cos\alpha}$.

11. Find the area of that portion of the surface of the cylinder $y^{\frac{2}{3}} + z^{\frac{2}{3}} = a^{\frac{2}{3}}$ bounded by a curve whose projection on the XY-plane is $x^{\frac{2}{3}} + y^{\frac{2}{3}} = a^{\frac{2}{3}}$. *Ans.* $\tfrac{12}{5} a^2$.

12. Find by integration the area of that portion of the surface of the sphere $x^2 + y^2 + z^2 = 100$ which lies between the parallel planes $x = -8$ and $x = 6$.

257. Volumes found by triple integration. In many cases the volume of a solid bounded by surfaces whose equations are given may be calculated by means of three successive integrations, the process being merely an extension of the methods employed in the preceding articles of this chapter (see also Art. 247).

Suppose the solid in question is divided by planes parallel to the coördinate planes into rectangular parallelepipeds having the dimensions Δz, Δy, Δx. The volume of one of these parallelepipeds is

$$\Delta z \cdot \Delta y \cdot \Delta x,$$

and we choose it as the element of volume.

Now sum up all such elements within the region R bounded by the given surfaces by first summing up all the elements in a column parallel to one of the coördinate axes, then summing up all such columns in a slice parallel to one of the coördinate planes containing that axis, and finally summing up all such slices within the region in question. The volume V of the solid will then be the limit of this triple sum as Δz, Δy, Δx each approach zero as a limit. That is,

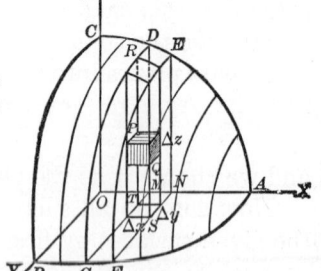

(1) $$V = \lim_{\substack{\Delta x \to 0 \\ \Delta y \to 0 \\ \Delta z \to 0}} \sum \sum \sum_R \Delta z \, \Delta y \, \Delta x,$$

the summations being extended over the entire region R bounded by the given surfaces. This limit is denoted by

(L) $$V = \iiint_R dz \, dy \, dx.$$

By extension of the principle of Art. 245, we speak of (L) as the *triple integral of the function* $f(x, y, z) = 1$ *throughout the region* R. Many problems require the integration of a *variable* function of x, y, and z throughout a given region. The notation is

(2) $$\iiint_R f(x, y, z) dz \, dy \, dx,$$

which is, of course, the limit of a triple sum analogous to the double sums we have already discussed. In more advanced treatises it is shown that the triple integral (2) is evaluated by successive integration. The limits are found in the same manner as for (L).

Simple examples of (2) are afforded by the formulas for the centroid $(\bar{x}, \bar{y}, \bar{z})$ (center of gravity) of a homogeneous solid, namely,

$$V\bar{x} = \iiint x \, dx \, dy \, dz, \quad V\bar{y} = \iiint y \, dx \, dy \, dz, \quad V\bar{z} = \iiint z \, dx \, dy \, dz.$$

They are obtained by reasoning as in Art. 249, using moments of volume. In the integrands, (x, y, z) is an interior point. The centroid will lie in any plane of symmetry.

ILLUSTRATIVE EXAMPLE 1. Find the volume of that portion of the ellipsoid

$$\frac{x^2}{a^2} + \frac{y^2}{b^2} + \frac{z^2}{c^2} = 1$$

which lies in the first octant.

Solution. Let O-ABC be that portion of the ellipsoid whose volume is required, the equations of the bounding surfaces being

(3) $\quad \dfrac{x^2}{a^2} + \dfrac{y^2}{b^2} + \dfrac{z^2}{c^2} = 1 \; (= ABC),$
(4) $\quad z = 0 \; (= OAB),$
(5) $\quad y = 0 \; (= OAC),$
(6) $\quad x = 0 \; (= OBC).$

PQ is an element, being one of the rectangular parallelepipeds with dimensions Δz, Δy, Δx into which the planes parallel to the coördinate planes have divided the region.

Integrating first with respect to z, we sum up all such elements in a column (as RS), the z-limits being zero (from (4)) and $TR = c\sqrt{1 - \frac{x^2}{a^2} - \frac{y^2}{b^2}}$ (from (3) by solving for z).

Integrating next with respect to y, we sum up all such columns in a slice (as $DEMNGF$), the y-limits being zero (from (5)) and $MG = b\sqrt{1 - \frac{x^2}{a^2}}$ (from the equation of the curve AGB, namely $\frac{x^2}{a^2} + \frac{y^2}{b^2} = 1$, by solving for y).

Lastly, integrating with respect to x, we sum up all such slices within the entire region O-ABC, the x-limits being zero (from (6)) and $OA = a$.

Hence
$$V = \int_0^a \int_0^{b\sqrt{1-\frac{x^2}{a^2}}} \int_0^{c\sqrt{1-\frac{x^2}{a^2}-\frac{y^2}{b^2}}} dz\, dy\, dx$$
$$= \frac{\pi cb}{4\, a^2} \int_0^a (a^2 - x^2) dx = \frac{\pi abc}{6}.$$

Therefore the volume of the entire ellipsoid is $\frac{4\, \pi abc}{3}$.

ILLUSTRATIVE EXAMPLE 2. Find the volume of the solid bounded by the surfaces

(7) $\quad z = 4 - x^2 - \tfrac{1}{4} y^2$,
(8) $\quad z = 3\, x^2 + \tfrac{1}{4} y^2$.

Solution. The surfaces are the elliptic paraboloids of the figure. Eliminating z between (7) and (8), we find

(9) $\quad 4\, x^2 + \tfrac{1}{2} y^2 = 4$,

which is the equation of the cylinder $ABCD$ (see figure) that passes through the curve of intersection of (7) and (8) and has its elements parallel to OZ.

We have

(10) $\quad V = 4 \int_0^1 \int_0^{2\sqrt{2(1-x^2)}} \int_{3x^2+\frac{1}{4}y^2}^{4-x^2-\frac{1}{4}y^2} dz\, dy\, dx.$

The limits are found as follows:

Integrating with respect to z, we sum up the elements of volume $dz\, dy\, dx$ in a column of base $dy\, dx$ from the surface (8) to the surface (7) (MP to MQ in figure). The limits for z are, then, given by the right-hand members in these equations. Thus we find

(11) $\quad V = 4 \int_0^1 \int_0^{2\sqrt{2(1-x^2)}} (4 - 4\, x^2 - \tfrac{1}{2}\, y^2) dy\, dx.$

The limits on this double integral are those for the region OAB, the portion of the area of the base of the cylinder (9) which lies in the first quadrant. Working out (11), we find $V = 4\, \pi\sqrt{2} = 17.77$ cubic units. *Ans.*

The problem given may be such that the first integration should be performed with respect to x or y, and not with respect to z, as above. The limits must then be determined in accordance with the preceding discussion.

258. Volumes, using cylindrical coördinates. In many problems involving integration the work is much simplified by employing cylindrical coördinates (ρ, θ, z) as defined in (7), p. 6. The cylindrical equation of any one of the bounding surfaces may often be written down directly from its definition. In any case it may be found from its rectangular equation by the substitution

(1) $$x = \rho \cos \theta, \quad y = \rho \sin \theta.$$

Cylindrical coördinates are especially useful when a bounding surface is a surface of revolution. For the equation of such a surface, when the axis is OZ, will have the form $z = f(\rho)$; that is, the coördinate θ will be absent.

Volume under a surface. Let

(2) $$z = F(\rho, \theta)$$

be the cylindrical equation of a surface, as KL in the figure. We wish to find the volume of the solid bounded above by this surface, below by the plane XOY, and laterally by the cylindrical surface whose right section by the plane XOY is the region S. This cylindrical surface intercepts on the surface (2) the region S'.

Divide the solid into elements of volume as follows: Divide S into elements of area ΔA by drawing radial lines from O and arcs of circles about O, as in Art. 254. Pass planes through the radial lines and OZ. Pass cylindrical surfaces of revolution about OZ standing on the circular arcs within S. Then the solid is divided into columns such as $MNPQ$, where area $MN = \Delta A$, and $MP = z$. The element of volume is then a right prism with base ΔA and altitude z. Hence

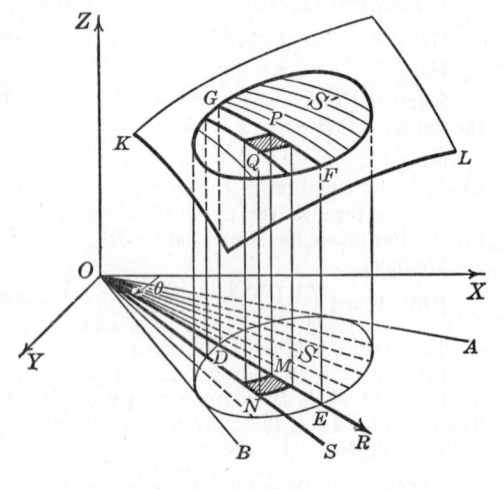

(3) $$\Delta V = z \, \Delta A.$$

The volume V is found by summing up the prisms (3) whose bases lie within S and finding the limit of this sum when the radial lines

MULTIPLE INTEGRALS 525

and circular arcs within S increase in number so that $\Delta\rho \to 0$ and $\Delta\theta \to 0$. That is,

(4) $$V = \lim_{\substack{\Delta\rho \to 0 \\ \Delta\theta \to 0}} \sum\sum z \, \Delta A.$$

We now show that the double limit in (4) may be found by successive integration. (Compare Art. 244.) This is done by finding the volume, approximately, of a slice of the solid included between two radial planes such as ROZ and SOZ, and then taking the limit of the sum of these slices.

Let $DEFG$ be the section of the solid in the plane ROZ. The values of z along the curve GPF are given by (2) when θ ($=$ angle XOR) is held fast. In the plane ROZ take OR and OZ as rectangular axes and (ρ, z) as coördinates. Let $(\bar{\rho}, \bar{z})$ be the centroid of area $DEFG$. Then by (2) and (3), Art. 177,

$$\bar{\rho} \cdot \text{area } DEFG = \int_{OD}^{OE} \rho z \, d\rho = \int_{OD}^{OE} \rho F(\rho, \theta) d\rho.$$

The integral will be a function of θ.

Now revolve area $DEFG$ about OZ. By Art. 250, the volume of the solid of revolution thus generated is $2\pi\bar{\rho} \cdot$ area $DEFG$. The planes ROZ and SOZ cut out a wedge from this solid of revolution whose volume is $\Delta\theta\bar{\rho} \cdot$ area $DEFG$, since angle $ROS = \Delta\theta$ (radians). Therefore

(5) $$\Delta\theta \int_{OD}^{OE} \rho F(\rho, \theta) d\rho$$

is equal, approximately, to the volume of the slice of the solid included between the planes ROZ and SOZ. The limit of the sum of the wedges (5) when $\Delta\theta \to 0$ is the exact volume.

Hence

(6) $$V = \int_\alpha^\beta \int_{\rho_1}^{\rho_2} F(\rho, \theta) \rho \, d\rho \, d\theta,$$

where $\alpha = \angle XOA$, $\beta = \angle XOB$, $\rho_1 = OD = f_1(\theta)$, $\rho_2 = OE = f_2(\theta)$, values to be found from the polar equations of the curves bounding S.

The element of the integral in (6), namely,

$$F(\rho, \theta) \rho \, d\rho \, d\theta = z\rho \, d\rho \, d\theta,$$

may be thought of as the volume of a right prism of altitude z and base of area $\rho \, d\rho \, d\theta$. Thus ΔA in (3) is replaced by $\rho \, \Delta\rho \, \Delta\theta$, as in Art. 254.

We have now the formula*

(M) $$V = \iint_S z\rho \, d\rho \, d\theta = \iint_S F(\rho, \theta)\rho \, d\rho \, d\theta$$

―――――
*The *order* of integration is immaterial. Proof is omitted.

for the volume under the surface (2), and the limits are found as in Art. 254 for the area of the region S.

From (M) and (4) we may derive (2), Art. 254.

ILLUSTRATIVE EXAMPLE 1. Show that the volume of the solid bounded by the ellipsoid of revolution $b^2(x^2 + y^2) + a^2z^2 = a^2b^2$ and the cylindrical surface $x^2 + y^2 - ax = 0$ is given by

(7) $$V = 4\frac{b}{a}\int_0^{\frac{1}{2}\pi}\int_0^{a\cos\theta} \sqrt{a^2 - \rho^2}\, \rho\, d\rho\, d\theta.$$

Evaluate this integral.

Solution. By (1) the cylindrical equation of the ellipsoid is $b^2\rho^2 + a^2z^2 = a^2b^2$. Hence

(8) $$z = \frac{b}{a}\sqrt{a^2 - \rho^2}.$$

The polar equation of the circle $x^2 + y^2 - ax = 0$ in the XY-plane bounding S is, by (1),

(9) $\rho = a\cos\theta.$

For the semicircle the limits for ρ are zero and $a\cos\theta$, when θ is held fast, and for θ, zero and $\frac{1}{2}\pi$. Substituting in (M) the value of z from (8) and the above limits, we get (7). Integrating,

$$V = \tfrac{2}{9} a^2 b(3\pi - 4) = 1.206\, a^2b.$$

Volume by triple integration. The element of volume ΔV will now be an *element* of the right prism used above in (3), that is, a right prism with base ΔA and altitude Δz. The solid is divided into such elements by passing through it the planes and cylindrical surfaces used in the figure at the beginning of this article and also planes parallel to the plane XOY at distances apart equal to Δz. We now have

(10) $\quad\quad\quad \Delta V = \Delta z\, \Delta A.$

By summation and taking the limit when $\Delta z \to 0$, $\Delta\rho \to 0$, $\Delta\theta \to 0$, we have

(N) $$V = \iiint \rho\, dz\, d\rho\, d\theta,$$

for ΔA may be replaced by $\rho\, \Delta\rho\, \Delta\theta$ as before.

Formulas (3) in Art. 257 for the centroid become

$$V\bar{x} = \iiint \rho^2 \cos\theta\, dz\, d\rho\, d\theta, \quad V\bar{y} = \iiint \rho^2 \sin\theta\, dz\, d\rho\, d\theta,$$

$$V\bar{z} = \iiint \rho z\, dz\, d\rho\, d\theta,$$

when cylindrical coördinates are used.

MULTIPLE INTEGRALS

ILLUSTRATIVE EXAMPLE 2. Find the volume of the solid whose upper surface is on the sphere

(11) $\quad x^2 + y^2 + z^2 = 8$

and whose lower surface is on the paraboloid of revolution

(12) $\quad x^2 + y^2 = 2z$.

Solution. The figure shows the sphere and the paraboloid in the first octant. The curve of intersection AB lies in the plane $z = 2$. Its projection DE on the XY-plane is the circle

(13) $\quad x^2 + y^2 = 4$.

The cylindrical equations are, by (1):

(14) $\rho^2 + z^2 = 8$ (the sphere (11));

(15) $\rho^2 = 2z$ (the paraboloid (12));

(16) $\rho = 2$ (the circle (13)).

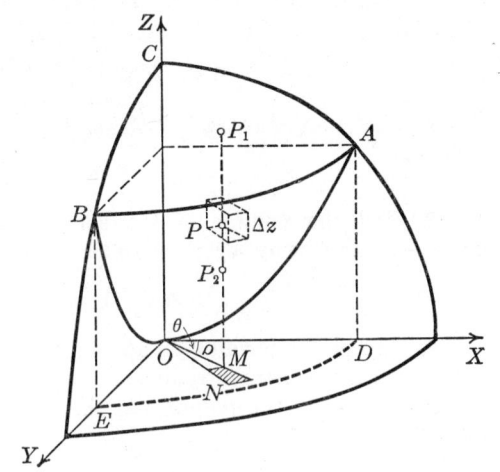

An element of area ΔA in the circle (16) is drawn at $M(\rho, \theta)$ in the figure. An element of volume ΔV is shown at $P(\rho, \theta, z)$.

We have, by (N),

(17) $$V = \int_0^{2\pi} \int_0^2 \int_{\frac{1}{2}\rho^2}^{\sqrt{8-\rho^2}} \rho \, dz \, d\rho \, d\theta.$$

The limits are found as follows: Integrating with respect to z (holding ρ and θ fast), we sum up the elements of volume (10) in a column from the surface (15) to the surface (14) (MP_2 to MP_1 in the figure). From (15), $z = MP_2 = \frac{1}{2}\rho^2$; from (14), $z = MP_1 = \sqrt{8 - \rho^2}$, the z-limits. The limits for ρ and θ are those for the area of the circle (16). Integrating with respect to ρ gives the sum of the columns in the slice included between the plane passing through OZ and OM and the plane passing through OZ and ON. The final integration sums up these slices.

Integrating in (17),

$$V = \tfrac{4}{3}\pi(8\sqrt{2} - 7) = 18.1. \text{ Ans.}$$

In the following problems, formulas (M) and (N) are to be used when the equations of bounding surfaces are in cylindrical coördinates (cylindrical equations). If the corresponding rectangular equations are needed for drawing a figure, they may be found by the transformation

(18) $\quad \rho^2 = x^2 + y^2, \quad \theta = \arctan \dfrac{y}{x}$,

to which may be added

(19) $\quad \sin \theta = \dfrac{y}{\sqrt{x^2 + y^2}}, \quad \cos \theta = \dfrac{x}{\sqrt{x^2 + y^2}}.$

DIFFERENTIAL AND INTEGRAL CALCULUS

PROBLEMS

1. Find the volume of the solid below the cylindrical surface $x^2 + z = 4$, above the plane $x + z = 2$, and included between the planes $y = 0$, $y = 3$.

$$\text{Ans.} \quad V = \int_0^3 \int_{-1}^2 \int_{2-x}^{4-x^2} dz\, dx\, dy = 13\tfrac{1}{2} \text{ cubic units.}$$

2. Work out Illustrative Example 2, Art. 247, using cylindrical coordinates.

$$\text{Ans.} \quad V = 2 \int_0^{\frac{1}{2}\pi} \int_0^{2a\cos\theta} \frac{\rho^3}{a} d\rho\, d\theta = \frac{3}{2}\pi a^3.$$

3. Find the volume of the solid bounded above by the cylinder $z = 4 - x^2$ and below by the elliptic paraboloid $z = 3x^2 + y^2$.

$$\text{Ans.} \quad V = 4 \int_0^1 \int_0^{2\sqrt{1-x^2}} \int_{3x^2+y^2}^{4-x^2} dz\, dy\, dx = 4\pi.$$

4. Two planes forming an angle α radians with each other meet along a diameter of a sphere of radius a. Find the volume of the *spherical wedge* included between the planes and the spherical surface, using cylindrical coördinates.
 $\text{Ans.} \; \tfrac{2}{3}\alpha a^3.$

5. Find the volume below the plane $z = x$ and above the elliptic paraboloid $z = x^2 + y^2$.
 $\text{Ans.} \; \tfrac{1}{32}\pi.$

6. Work Problem 5 using cylindrical coördinates.

$$\text{Ans.} \quad V = 2 \int_0^{\frac{1}{2}\pi} \int_0^{\cos\theta} \int_{\rho^2}^{\rho\cos\theta} \rho\, dz\, d\rho\, d\theta = \tfrac{1}{32}\pi.$$

7. Find the volume bounded by the sphere $\rho^2 + z^2 = a^2$ within the cylinder $\rho = a\cos\theta$.
 $\text{Ans.} \; \tfrac{2}{3}a^3(\pi - \tfrac{4}{3}).$

8. Find the volume above $z = 0$, below the cone $z^2 = x^2 + y^2$, and within the cylinder $x^2 + y^2 = 2ax$, using cylindrical coördinates.
 $\text{Ans.} \; \tfrac{32}{9}a^3.$

9. Find the volume of the solid bounded by $z = x + 1$ and $2z = x^2 + y^2$.
 $\text{Ans.} \; \tfrac{9}{4}\pi.$

10. In Problem 3 show that integration with respect to z gives (without further integration) $V = 4A - 4I_y - I_x$, where A is the area of the ellipse $4x^2 + y^2 = 4$, and I_x and I_y are moments of inertia for this ellipse as given by (E), Art. 252.

11. Find the volume below the plane $2z = 4 + \rho\cos\theta$, above $z = 0$, and within the cylinder $\rho = 2\cos\theta$.
 $\text{Ans.} \; \tfrac{5}{2}\pi.$

12. A solid is bounded by the paraboloid of revolution $az = \rho^2$ and the plane $z = c$. Find the centroid.
 $\text{Ans.} \; (0, 0, \tfrac{2}{3}c).$

13. A solid is bounded by the hyperboloid $z^2 = a^2 + \rho^2$ and the upper nappe of the cone $z^2 = 2\rho^2$. Find the volume.
 $\text{Ans.} \; \tfrac{2}{3}\pi a^3(\sqrt{2} - 1).$

14. Find the centroid of the solid in Problem 13.
 $\text{Ans.} \; (0, 0, \tfrac{3}{8}a(\sqrt{2}+1)).$

MULTIPLE INTEGRALS

15. Find the centroid of the solid in Problem 1. *Ans.* $(\frac{1}{2}, \frac{3}{2}, \frac{12}{5})$.

16. Find the centroid of the solid in Problem 2. *Ans.* $(\frac{4}{3} a, 0, \frac{10}{9} a)$.

17. Find the centroid of the solid in Problem 8.

18. Find the volume of the solid bounded below by $z = 0$, above by the cone $z = a - \rho$, and laterally by $\rho = a \cos \theta$. *Ans.* $\frac{1}{36} a^3 (9 \pi - 16)$.

19. Find the centroid of the solid in the preceding problem.

20. Find the volume of the solid below the spherical surface $\rho^2 + z^2 = 25$ and above the upper nappe of the conical surface $z = \rho + 1$.

21. A homogeneous tetrahedron is bounded by the coördinate planes and the plane $\dfrac{x}{a} + \dfrac{y}{b} + \dfrac{z}{c} = 1$. Show that the centroid is $(\frac{1}{4} a, \frac{1}{4} b, \frac{1}{4} c)$.

22. Assume that the density of the tetrahedron in Problem 21 is $D = kx$.
(a) Show that the mass is $M = \frac{1}{24} k a^2 b c$.
(b) Show that $\bar{x} = \frac{2}{5} a$.

23. In a hemisphere of radius a the density varies as the distance from the center. $D = kr$.
(a) Show that the mass is $M = \frac{1}{2} \pi k a^4$.
(b) Show that the distance of the centroid from the base is $\frac{2}{5} a$.

24. A sphere of radius a is cut by a cone whose axis is a diameter OZ of the sphere and whose vertex O lies on the surface. The vertex angle of the cone is 60°.
(a) Show that the volume of that part of the sphere which lies within the cone is $\frac{7}{12} \pi a^3$.
(b) Show that the distance of the centroid from the vertex of the cone is $\frac{37}{28} a$.

25. Compare Illustrative Example 3, Art. 165, and Illustrative Example 1, Art. 257, and derive (N), Art. 165, from (L), Art. 257.

26. Derive formula (2), Art. 178, from the first formula in (3), Art. 257.

ADDITIONAL PROBLEMS

1. Find the volume bounded above by the sphere $\rho^2 + z^2 = r^2$, below by the cone $z = \rho \operatorname{ctn} \phi$, and included between the planes $\theta = \beta$, $\theta = \beta + \Delta\beta$, ϕ and β being acute angles. (The solid is part of a spherical wedge, like $O\text{-}SQN$ in the figure of Art. 222 when OQ is drawn.)
Ans. $\frac{1}{3} r^3 \Delta\beta (1 - \cos \phi)$.

2. Find (without integration) the volume bounded by the sphere $\rho^2 + z^2 = r^2$, the cones $z = \rho \operatorname{ctn} \phi$, $z = \rho \operatorname{ctn} (\phi + \Delta\phi)$, and the planes $\theta - \beta$, $\theta = \beta + \Delta\beta$, using the result in the preceding problem. (The solid is like $O\text{-}P_1RQS$ in the figure of Art. 222 when OR and OQ are drawn.)
Ans. $\frac{2}{3} r^3 \Delta\beta \sin (\phi + \frac{1}{2} \Delta\phi) \sin \frac{1}{2}\Delta\phi$.

530 DIFFERENTIAL AND INTEGRAL CALCULUS

3. Find (without integration) the volume bounded by $z = \rho \operatorname{ctn} \phi$, $z = \rho \operatorname{ctn}(\phi + \Delta\phi)$, $\theta = \beta$, $\theta = \beta + \Delta\beta$, and included between the spheres $\rho^2 + z^2 = r^2$, $\rho^2 + z^2 = (r + \Delta r)^2$, using the answer in Problem 2.

Ans. $2 \Delta\beta \Delta r \sin(\phi + \tfrac{1}{2}\Delta\phi) \sin \tfrac{1}{2} \Delta\phi (r^2 + r \Delta r + \tfrac{1}{3} \Delta r^2)$.

(The solid is obtained from the figure in Art. 222 by producing each of the radii OP_1, OR, OQ, OS a distance Δr to P_1', R', Q', S' on the sphere $\rho^2 + z^2 = (r + \Delta r)^2$. The cones intersect this sphere in the circular arcs $P_1'R'$ and $Q'S'$, and the planes in the arcs of great circles $P_1'S'$, $R'Q'$. The solid has the vertices P_1RQS-$P_1'R'Q'S'$.)

4. The solid of Problem 3 is the element of volume ΔV when spherical coördinates (8), p. 6, are used. Replace β by θ. Then one vertex P of ΔV has the spherical coördinates (r, ϕ, θ). Prove, from Problem 3,

$$\lim_{\substack{\Delta r \to 0 \\ \Delta \theta \to 0 \\ \Delta \phi \to 0}} \frac{\Delta V}{r^2 \sin \phi \, \Delta r \, \Delta \phi \, \Delta \theta} = 1.$$

Therefore ΔV differs from $r^2 \sin \phi \, \Delta r \, \Delta \phi \, \Delta \theta$ by an infinitesimal of higher order (Art. 99).

5. In the solid of the preceding problem prove that the edges of ΔV meeting at any vertex are mutually perpendicular, and that the lengths of those intersecting at (r, ϕ, θ) are, respectively, Δr, $r \Delta\phi$, $r \sin \phi \, \Delta\theta$.

6. Describe the three systems of surfaces (spheres, cones, planes) which must be drawn to divide a solid R into elements of volume ΔV (Problem 4) when spherical coördinates are used. Let (r, ϕ, θ) be any point of ΔV. Then we write

$$\lim_{\substack{\Delta r \to 0 \\ \Delta \theta \to 0 \\ \Delta \phi \to 0}} \sum\sum\sum F(r, \phi, \theta) \Delta V = \iiint_R F(r, \phi, \theta) r^2 \sin \phi \, dr \, d\phi \, d\theta.$$

In the left-hand member ΔV may be replaced by $r^2 \sin \phi \, \Delta r \, \Delta \phi \, \Delta \theta$ (see Problem 4), that is, by the product of the three edges in Problem 5. The right-hand member is calculated by successive integration. (Proof omitted.)

7. Work out the integral in the preceding problem if $F(r, \phi, \theta) = r$, and R is the sphere $r = 2a \cos \phi$, that is, $x^2 + y^2 + z^2 = 2az$.

Ans. $\int_0^{2\pi} \int_0^{\frac{1}{2}\pi} \int_0^{2a \cos \phi} r^3 \sin \phi \, dr \, d\phi \, d\theta = \tfrac{8}{5} \pi a^4$.

8. Work out the integral in Problem 6 if $F(r, \phi, \theta) = r^2 \cos \phi$ and R is the region $r = 2a \cos \phi$.   *Ans.* $\tfrac{64}{35} \pi a^5$.

CHAPTER XXVI

CURVES FOR REFERENCE

For the convenience of the student a number of the more common curves employed in the text are collected here.

CUBICAL PARABOLA

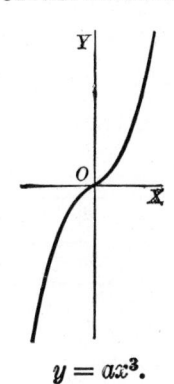

$y = ax^3.$

SEMICUBICAL PARABOLA

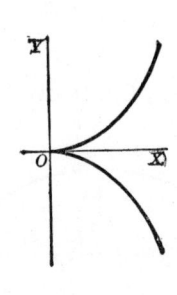

$y^2 = ax^3.$

THE WITCH OF AGNESI

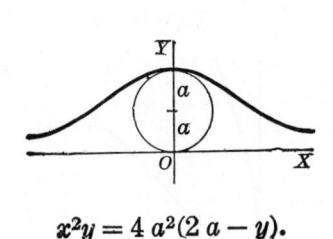

$x^2 y = 4 a^2 (2a - y).$

THE CISSOID OF DIOCLES

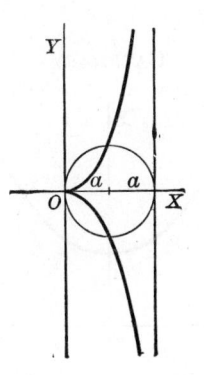

$y^2 (2a - x) = x^3.$

The Lemniscate of Bernoulli

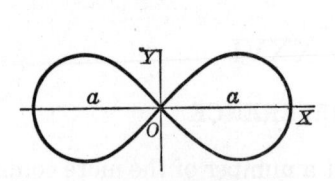

$(x^2+y^2)^2 = a^2(x^2-y^2).$
$\rho^2 = a^2 \cos 2\theta.$

The Conchoid of Nicomedes

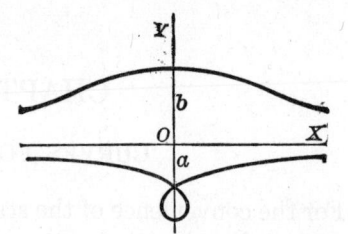

$x^2 y^2 = (y+a)^2(b^2-y^2).$
(In the figure, $b > a$.)

Cycloid, Ordinary Case

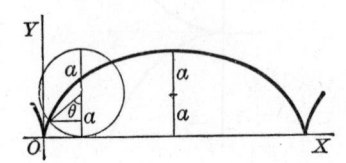

$x = a \text{ arc vers } \dfrac{y}{a} - \sqrt{2ay-y^2}.$
$\begin{cases} x = a(\theta - \sin\theta), \\ y = a(1 - \cos\theta). \end{cases}$

Cycloid, Vertex at Origin

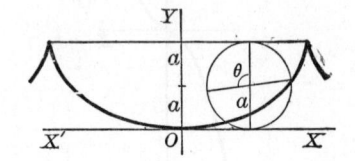

$x = a \text{ arc vers } \dfrac{y}{a} + \sqrt{2ay-y^2}$
$\begin{cases} x = a(\theta + \sin\theta), \\ y = a(1 - \cos\theta). \end{cases}$

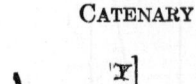

Catenary

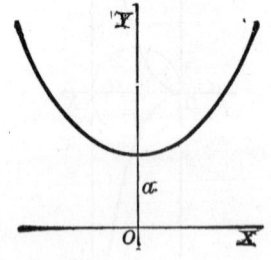

$y = \dfrac{a}{2}\left(e^{\frac{x}{a}} + e^{-\frac{x}{a}}\right) = a \cosh \dfrac{x}{a}.$

Parabola

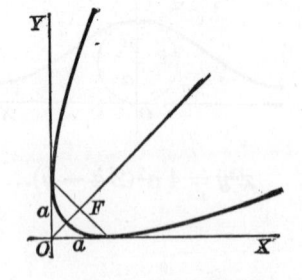

$x^{\frac{1}{2}} + y^{\frac{1}{2}} = a^{\frac{1}{2}}.$

CURVES FOR REFERENCE

Hypocycloid of Four Cusps (Astroid)

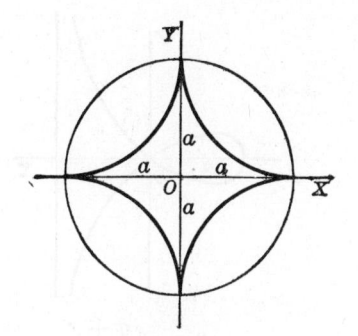

$$x^{\frac{2}{3}} + y^{\frac{2}{3}} = a^{\frac{2}{3}}.$$
$$\begin{cases} x = a \cos^3 \theta, \\ y = a \sin^3 \theta. \end{cases}$$

Evolute of Ellipse

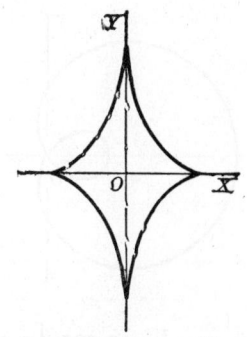

$$(ax)^{\frac{2}{3}} + (by)^{\frac{2}{3}} = (a^2 - b^2)^{\frac{2}{3}}$$
$$\begin{cases} x = A \cos^3 \theta, \, Aa = a^2 - b^2, \\ y = B \sin^3 \theta, \, Bb = b^2 - a^2. \end{cases}$$

Cardioid

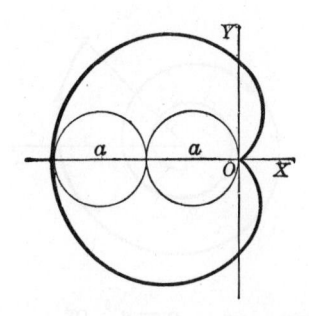

$$x^2 + y^2 + ax = a\sqrt{x^2 + y^2}.$$
$$\rho = a(1 - \cos \theta).$$

Folium of Descartes

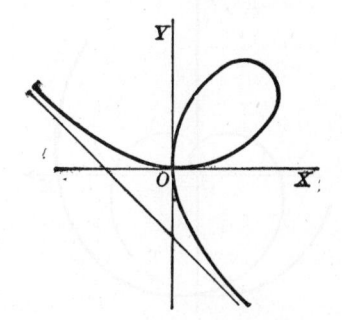

$$x^3 + y^3 - 3\,axy = 0.$$

Sine Curve

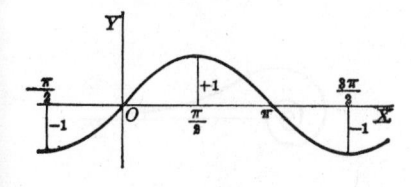

$$y = \sin x.$$

Cosine Curve

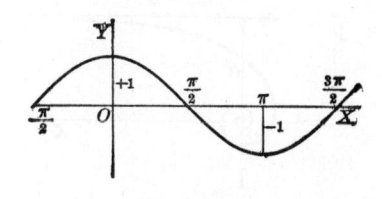

$$y = \cos x.$$

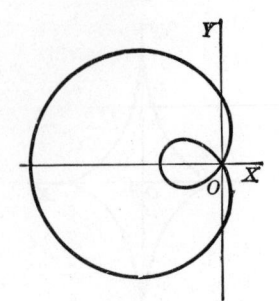

LIMAÇON

$\rho = b - a\cos\theta.$
(In the figure, $b < a$.)

STROPHOID

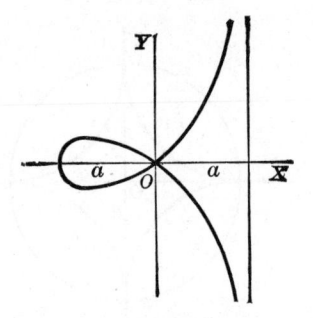

$y^2 = x^2 \dfrac{a+x}{a-x}.$

SPIRAL OF ARCHIMEDES

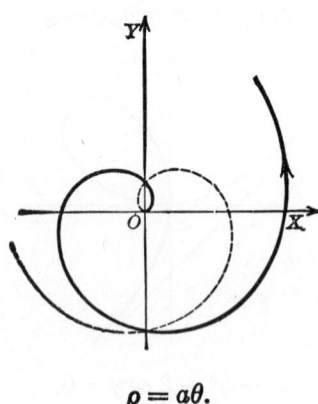

$\rho = a\theta.$

LOGARITHMIC OR EQUIANGULAR SPIRAL

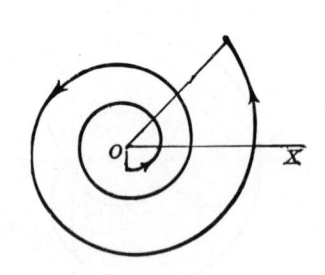

$\rho = e^{a\theta}$, or
$\log \rho = a\theta.$

HYPERBOLIC OR RECIPROCAL SPIRAL

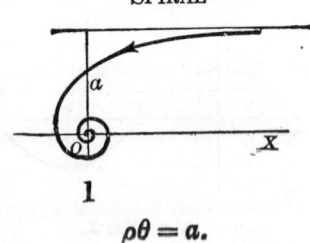

$\rho\theta = a.$

LITUUS

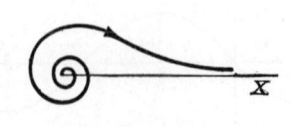

$\rho^2\theta = a^2.$

CURVES FOR REFERENCE

PARABOLIC SPIRAL

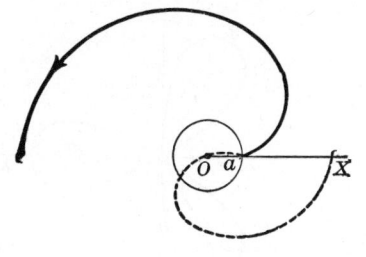

$(\rho - a)^2 = 4\,ac\theta.$

LOGARITHMIC CURVE

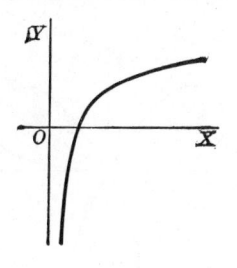

$y = \ln x.$

EXPONENTIAL CURVE

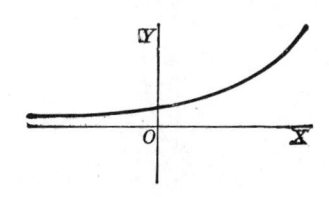

$y = e^x.$

PROBABILITY CURVE

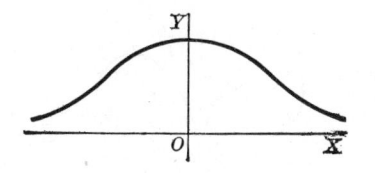

$y = e^{-x^2}.$

SECANT CURVE

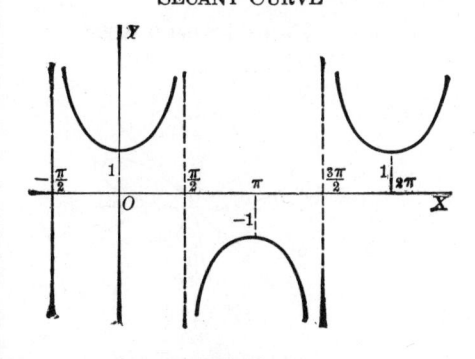

$y = \sec x.$

TANGENT CURVE

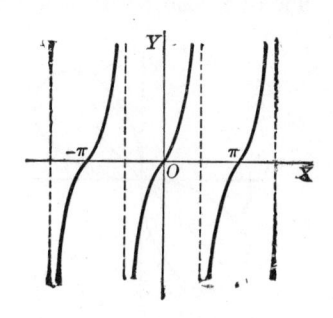

$y = \tan x.$

THREE-LEAVED ROSE

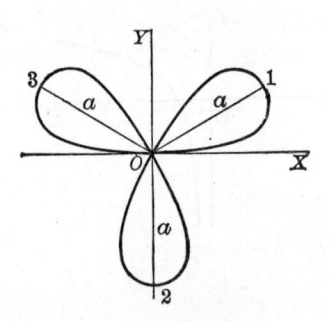

$\rho = a \sin 3\theta.$

THREE-LEAVED ROSE

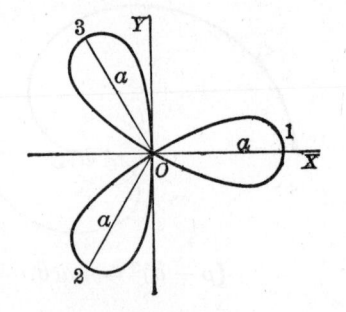

$\rho = a \cos 3\theta.$

FOUR-LEAVED ROSE

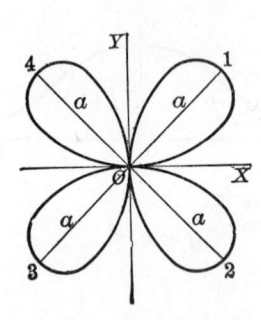

$\rho = a \sin 2\theta.$

FOUR-LEAVED ROSE

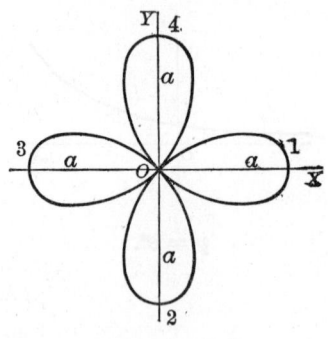

$\rho = a \cos 2\theta.$

TWO-LEAVED ROSE LEMNISCATE

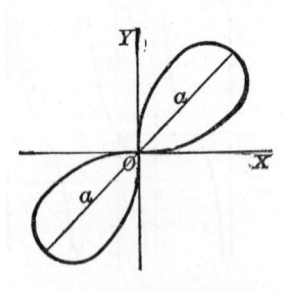

$\rho^2 = a^2 \sin 2\theta.$

EIGHT-LEAVED ROSE

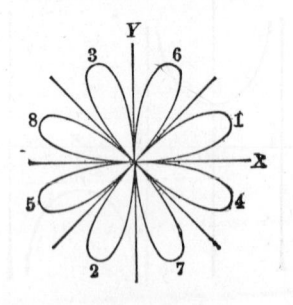

$\rho = a \sin 4\theta.$

CURVES FOR REFERENCE

Parabola

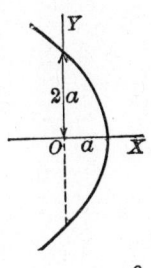

$$\rho = a \sec^2 \frac{\theta}{2}.$$

Equilateral Hyperbola

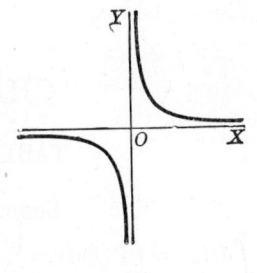

$$xy = a.$$

Involute of a Circle

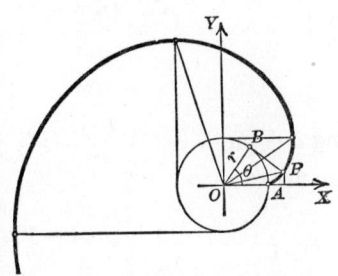

$$\begin{cases} x = r \cos \theta + r\theta \sin \theta, \\ y = r \sin \theta - r\theta \cos \theta. \end{cases}$$

Tractrix

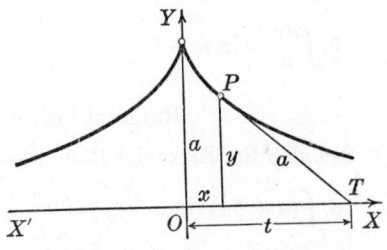

$$x = a \operatorname{sech}^{-1} \frac{y}{a} - \sqrt{a^2 - y^2}.$$

$$\begin{cases} x = t - a \tanh \frac{t}{a}, \\ y = a \operatorname{sech} \frac{t}{a}. \end{cases}$$

CHAPTER XXVII

TABLE OF INTEGRALS

Some Elementary Forms

1. $\int df(x) = \int f'(x)dx = f(x) + C.$

2. $\int a\,du = a\int du.$

3. $\int (du \pm dv \pm dw \pm \cdots) = \int du \pm \int dv \pm \int dw \pm \cdots.$

4. $\int u^n\,du = \dfrac{u^{n+1}}{n+1} + C.$ $\hfill (n \neq -1)$

5. $\int \dfrac{du}{u} = \ln u + C.$

Rational Forms containing $a + bu$

See also the Binomial Reduction Formulas 96–104.

6. $\int (a + bu)^n\,du = \dfrac{(a+bu)^{n+1}}{b(n+1)} + C.$ $\hfill (n \neq -1)$

7. $\int \dfrac{du}{a+bu} = \dfrac{1}{b}\ln(a+bu) + C.$

8. $\int \dfrac{u\,du}{a+bu} = \dfrac{1}{b^2}[a+bu - a\ln(a+bu)] + C.$

9. $\int \dfrac{u^2\,du}{a+bu} = \dfrac{1}{b^3}[\tfrac{1}{2}(a+bu)^2 - 2a(a+bu) + a^2\ln(a+bu)] + C.$

10. $\int \dfrac{u\,du}{(a+bu)^2} = \dfrac{1}{b^2}\left[\dfrac{a}{a+bu} + \ln(a+bu)\right] + C.$

11. $\int \dfrac{u^2\,du}{(a+bu)^2} = \dfrac{1}{b^3}\left[a+bu - \dfrac{a^2}{a+bu} - 2a\ln(a+bu)\right] + C.$

12. $\int \dfrac{u\,du}{(a+bu)^3} = \dfrac{1}{b^2}\left[-\dfrac{1}{a+bu} + \dfrac{a}{2(a+bu)^2}\right] + C.$

13. $\int \dfrac{du}{u(a+bu)} = -\dfrac{1}{a}\ln\left(\dfrac{a+bu}{u}\right) + C.$

14. $\int \dfrac{du}{u^2(a+bu)} = -\dfrac{1}{au} + \dfrac{b}{a^2}\ln\left(\dfrac{a+bu}{u}\right) + C.$

15. $\int \dfrac{du}{u(a+bu)^2} = \dfrac{1}{a(a+bu)} - \dfrac{1}{a^2}\ln\left(\dfrac{a+bu}{u}\right) + C.$

Rational Forms containing $a^2 \pm b^2u^2$

16. $\int \dfrac{du}{a^2 + b^2u^2} = \dfrac{1}{ab} \arctan \dfrac{bu}{a} + C.$

17. $\int \dfrac{du}{a^2 - b^2u^2} = \dfrac{1}{2\,ab} \ln \left(\dfrac{a + bu}{a - bu}\right) + C.$ \quad $(a^2 > b^2u^2)$

$\int \dfrac{du}{b^2u^2 - a^2} = \dfrac{1}{2\,ab} \ln \left(\dfrac{bu - a}{bu + a}\right) + C.$ \quad $(a^2 < b^2u^2)$

18. $\int u(a^2 \pm b^2u^2)^n du = \dfrac{(a^2 \pm b^2u^2)^{n+1}}{\pm 2\,b^2(n + 1)} + C.$ \quad $(n \neq -1)$

19. $\int \dfrac{u\,du}{a^2 \pm b^2u^2} = \dfrac{1}{\pm 2\,b^2} \ln (a^2 \pm b^2u^2) + C.$

20. $\int \dfrac{u^m\,du}{(a^2 \pm b^2u^2)^p} = \dfrac{u^{m-1}}{\pm b^2(m - 2p + 1)(a^2 \pm b^2u^2)^{p-1}}$
$ - \dfrac{a^2(m - 1)}{\pm b^2(m - 2p + 1)} \int \dfrac{u^{m-2}\,du}{(a^2 \pm b^2u^2)^p}.$

21. $\int \dfrac{u^m\,du}{(a^2 \pm b^2u^2)^p} = \dfrac{u^{m+1}}{2\,a^2(p - 1)(a^2 \pm b^2u^2)^{p-1}}$
$ - \dfrac{m - 2p + 3}{2\,a^2(p - 1)} \int \dfrac{u^m\,du}{(a^2 \pm b^2u^2)^{p-1}}.$

22. $\int \dfrac{du}{u(a^2 \pm b^2u^2)} = \dfrac{1}{2\,a^2} \ln \left(\dfrac{u^2}{a^2 \pm b^2u^2}\right) + C.$

23. $\int \dfrac{du}{u^m(a^2 \pm b^2u^2)^p} = -\dfrac{1}{a^2(m - 1)u^{m-1}(a^2 \pm b^2u^2)^{p-1}}$
$ - \dfrac{\pm b^2(m + 2p - 3)}{a^2(m - 1)} \int \dfrac{du}{u^{m-2}(a^2 \pm b^2u^2)^p}.$

24. $\int \dfrac{du}{u^m(a^2 \pm b^2u^2)^p} = \dfrac{1}{2\,a^2(p - 1)u^{m-1}(a^2 \pm b^2u^2)^{p-1}}$
$ + \dfrac{m + 2p - 3}{2\,a^2(p - 1)} \int \dfrac{du}{u^m(a^2 \pm b^2u^2)^{p-1}}.$

Forms containing $\sqrt{a + bu}$

The integrand may be rationalized by setting $a + bu = v^2$. See also the Binomial Reduction Formulas 96–104.

25. $\int u\sqrt{a + bu}\,du = -\dfrac{2(2a - 3bu)(a + bu)^{\frac{3}{2}}}{15\,b^2} + C.$

26. $\int u^2\sqrt{a + bu}\,du = \dfrac{2(8a^2 - 12abu + 15b^2u^2)(a + bu)^{\frac{3}{2}}}{105\,b^3} + C.$

27. $\int u^m\sqrt{a + bu}\,du = \dfrac{2\,u^m(a + bu)^{\frac{3}{2}}}{b(2m + 3)} - \dfrac{2\,am}{b(2m + 3)} \int u^{m-1}\sqrt{a + bu}\,du.$

28. $\int \dfrac{u\,du}{\sqrt{a + bu}} = -\dfrac{2(2a - bu)\sqrt{a + bu}}{3\,b^2} + C.$

540   DIFFERENTIAL AND INTEGRAL CALCULUS

29. $\int \dfrac{u^2 \, du}{\sqrt{a+bu}} = \dfrac{2(8a^2 - 4abu + 3b^2u^2)\sqrt{a+bu}}{15 b^3} + C.$

30. $\int \dfrac{u^m \, du}{\sqrt{a+bu}} = \dfrac{2 u^m \sqrt{a+bu}}{b(2m+1)} - \dfrac{2am}{b(2m+1)} \int \dfrac{u^{m-1} \, du}{\sqrt{a+bu}}.$

31. $\int \dfrac{du}{u\sqrt{a+bu}} = \dfrac{1}{\sqrt{a}} \ln \left(\dfrac{\sqrt{a+bu} - \sqrt{a}}{\sqrt{a+bu} + \sqrt{a}} \right) + C, \text{ for } a > 0.$

32. $\int \dfrac{du}{u\sqrt{a+bu}} = \dfrac{2}{\sqrt{-a}} \arctan \sqrt{\dfrac{a+bu}{-a}} + C, \text{ for } a < 0.$

33. $\int \dfrac{du}{u^m \sqrt{a+bu}} = -\dfrac{\sqrt{a+bu}}{a(m-1)u^{m-1}} - \dfrac{b(2m-3)}{2a(m-1)} \int \dfrac{du}{u^{m-1}\sqrt{a+bu}}.$

34. $\int \dfrac{\sqrt{a+bu} \, du}{u} = 2\sqrt{a+bu} + a \int \dfrac{du}{u\sqrt{a+bu}}.$

35. $\int \dfrac{\sqrt{a+bu} \, du}{u^m} = -\dfrac{(a+bu)^{\frac{3}{2}}}{a(m-1)u^{m-1}} - \dfrac{b(2m-5)}{2a(m-1)} \int \dfrac{\sqrt{a+bu} \, du}{u^{m-1}}.$

Forms containing $\sqrt{u^2 \pm a^2}$

In this group of formulas we may replace

$$\ln(u + \sqrt{u^2 + a^2}) \quad \text{by} \quad \sinh^{-1} \dfrac{u}{a},$$

$$\ln(u + \sqrt{u^2 - a^2}) \quad \text{by} \quad \cosh^{-1} \dfrac{u}{a},$$

$$\ln \left(\dfrac{a + \sqrt{u^2 + a^2}}{u} \right) \quad \text{by} \quad \sinh^{-1} \dfrac{a}{u}.$$

36. $\int (u^2 \pm a^2)^{\frac{1}{2}} du = \dfrac{u}{2} \sqrt{u^2 \pm a^2} \pm \dfrac{a^2}{2} \ln(u + \sqrt{u^2 \pm a^2}) + C.$

37. $\int (u^2 \pm a^2)^{\frac{n}{2}} du = \dfrac{u(u^2 \pm a^2)^{\frac{n}{2}}}{n+1} \pm \dfrac{na^2}{n+1} \int (u^2 \pm a^2)^{\frac{n}{2} - 1} du. \; (n \neq -1)$

38. $\int u(u^2 \pm a^2)^{\frac{n}{2}} du = \dfrac{(u^2 \pm a^2)^{\frac{n}{2}+1}}{n+2} + C. \hspace{2cm} (n \neq -2)$

39. $\int u^m (u^2 \pm a^2)^{\frac{n}{2}} du = \dfrac{u^{m-1}(u^2 \pm a^2)^{\frac{n}{2}+1}}{n+m+1}$
$\hspace{3cm} - \dfrac{\pm a^2(m-1)}{n+m+1} \int u^{m-2}(u^2 \pm a^2)^{\frac{n}{2}} du.$

40. $\int \dfrac{du}{(u^2 \pm a^2)^{\frac{1}{2}}} = \ln(u + \sqrt{u^2 \pm a^2}) + C.$

41. $\int \dfrac{du}{(u^2 \pm a^2)^{\frac{3}{2}}} = \dfrac{u}{\pm a^2 \sqrt{u^2 \pm a^2}} + C.$

TABLE OF INTEGRALS

42. $\displaystyle\int \frac{u\,du}{(u^2 \pm a^2)^{\frac{n}{2}}} = \frac{(u^2 \pm a^2)^{1-\frac{n}{2}}}{2-n} + C.$

43. $\displaystyle\int \frac{u^2\,du}{(u^2 \pm a^2)^{\frac{1}{2}}} = \frac{u}{2}\sqrt{u^2 \pm a^2} - \frac{\pm a^2}{2} \ln(u + \sqrt{u^2 \pm a^2}) + C.$

44. $\displaystyle\int \frac{u^2\,du}{(u^2 \pm a^2)^{\frac{3}{2}}} = -\frac{u}{\sqrt{u^2 \pm a^2}} + \ln(u + \sqrt{u^2 \pm a^2}) + C.$

45. $\displaystyle\int \frac{u^m\,du}{(u^2 \pm a^2)^{\frac{n}{2}}} = \frac{u^{m-1}}{(m-n+1)(u^2 \pm a^2)^{\frac{n}{2}-1}} - \frac{\pm a^2(m-1)}{m-n+1}\int\frac{u^{m-2}\,du}{(u^2 \pm a^2)^{\frac{n}{2}}}.$

46. $\displaystyle\int \frac{u^m\,du}{(u^2 \pm a^2)^{\frac{n}{2}}} = \frac{u^{m+1}}{\pm a^2(n-2)(u^2 \pm a^2)^{\frac{n}{2}-1}} - \frac{m-n+3}{\pm a^2(n-2)}\int\frac{u^m\,du}{(u^2 \pm a^2)^{\frac{n}{2}-1}}.$

47. $\displaystyle\int \frac{du}{u(u^2 + a^2)^{\frac{1}{2}}} = -\frac{1}{a}\ln\left(\frac{a + \sqrt{u^2 + a^2}}{u}\right) + C.$

48. $\displaystyle\int \frac{du}{u(u^2 - a^2)^{\frac{1}{2}}} = \frac{1}{a}\operatorname{arc\,sec}\frac{u}{a} + C.$

49. $\displaystyle\int \frac{du}{u^2(u^2 \pm a^2)^{\frac{1}{2}}} = -\frac{\sqrt{u^2 \pm a^2}}{\pm a^2 u} + C.$

50. $\displaystyle\int \frac{du}{u^3(u^2 + a^2)^{\frac{1}{2}}} = -\frac{\sqrt{u^2 + a^2}}{2a^2 u^2} + \frac{1}{2a^3}\ln\left(\frac{a + \sqrt{u^2 + a^2}}{u}\right) + C.$

51. $\displaystyle\int \frac{du}{u^3(u^2 - a^2)^{\frac{1}{2}}} = \frac{\sqrt{u^2 - a^2}}{2a^2 u^2} + \frac{1}{2a^3}\operatorname{arc\,sec}\frac{u}{a} + C.$

52. $\displaystyle\int \frac{du}{u^m(u^2 \pm a^2)^{\frac{n}{2}}} = -\frac{1}{\pm a^2(m-1)u^{m-1}(u^2 \pm a^2)^{\frac{n}{2}-1}} - \frac{m+n-3}{\pm a^2(m-1)}\int\frac{du}{u^{m-2}(u^2 \pm a^2)^{\frac{n}{2}}}.$

53. $\displaystyle\int \frac{du}{u^m(u^2 \pm a^2)^{\frac{n}{2}}} = \frac{1}{\pm a^2(n-2)u^{m-1}(u^2 \pm a^2)^{\frac{n}{2}-1}} + \frac{m+n-3}{\pm a^2(n-2)}\int\frac{du}{u^m(u^2 \pm a^2)^{\frac{n}{2}-1}}.$

54. $\displaystyle\int \frac{(u^2 + a^2)^{\frac{1}{2}}\,du}{u} = \sqrt{u^2 + a^2} - a\ln\left(\frac{a + \sqrt{u^2 + a^2}}{u}\right) + C.$

55. $\displaystyle\int \frac{(u^2 - a^2)^{\frac{1}{2}}\,du}{u} = \sqrt{u^2 - a^2} - a\operatorname{arc\,sec}\frac{u}{a} + C.$

542   DIFFERENTIAL AND INTEGRAL CALCULUS

56. $\int \dfrac{(u^2 \pm a^2)^{\frac{1}{2}} du}{u^2} = -\dfrac{\sqrt{u^2 \pm a^2}}{u} + \ln(u + \sqrt{u^2 \pm a^2}) + C.$

57. $\int \dfrac{(u^2 \pm a^2)^{\frac{n}{2}} du}{u^m} = -\dfrac{(u^2 \pm a^2)^{\frac{n}{2}+1}}{\pm a^2(m-1)u^{m-1}}$
$\qquad - \dfrac{m-n-3}{\pm a^2(m-1)} \int \dfrac{(u^2 \pm a^2)^{\frac{n}{2}} du}{u^{m-2}}.$

58. $\int \dfrac{(u^2 \pm a^2)^{\frac{n}{2}} du}{u^m} = \dfrac{(u^2 \pm a^2)^{\frac{n}{2}}}{(n-m+1)u^{m-1}} + \dfrac{\pm a^2 n}{n-m+1} \int \dfrac{(u^2 \pm a^2)^{\frac{n}{2}-1} du}{u^m}.$

Forms containing $\sqrt{a^2 - u^2}$

59. $\int (a^2 - u^2)^{\frac{1}{2}} du = \dfrac{u}{2}\sqrt{a^2-u^2} + \dfrac{a^2}{2} \arcsin \dfrac{u}{a} + C.$

60. $\int (a^2 - u^2)^{\frac{n}{2}} du = \dfrac{u(a^2-u^2)^{\frac{n}{2}}}{n+1} + \dfrac{a^2 n}{n+1} \int (a^2-u^2)^{\frac{n}{2}-1} du.$ $(n \neq -1)$

61. $\int u(a^2 - u^2)^{\frac{n}{2}} du = -\dfrac{(a^2-u^2)^{\frac{n}{2}+1}}{n+2} + C.$ $(n \neq -2)$

62. $\int u^m (a^2 - u^2)^{\frac{n}{2}} du = -\dfrac{u^{m-1}(a^2-u^2)^{\frac{n}{2}+1}}{n+m+1}$
$\qquad + \dfrac{a^2(m-1)}{n+m+1} \int u^{m-2}(a^2-u^2)^{\frac{n}{2}} du.$

63. $\int \dfrac{du}{(a^2-u^2)^{\frac{1}{2}}} = \arcsin \dfrac{u}{a} + C.$

64. $\int \dfrac{du}{(a^2-u^2)^{\frac{3}{2}}} = \dfrac{u}{a^2 \sqrt{a^2-u^2}} + C.$

65. $\int \dfrac{u\, du}{(a^2-u^2)^{\frac{n}{2}}} = \dfrac{(a^2-u^2)^{1-\frac{n}{2}}}{n-2} + C.$

66. $\int \dfrac{u^2\, du}{(a^2-u^2)^{\frac{1}{2}}} = -\dfrac{u}{2}\sqrt{a^2-u^2} + \dfrac{a^2}{2} \arcsin \dfrac{u}{a} + C.$

67. $\int \dfrac{u^2\, du}{(a^2-u^2)^{\frac{3}{2}}} = \dfrac{u}{\sqrt{a^2-u^2}} - \arcsin \dfrac{u}{a} + C.$

68. $\int \dfrac{u^m\, du}{(a^2-u^2)^{\frac{n}{2}}} = -\dfrac{u^{m-1}}{(m-n+1)(a^2-u^2)^{\frac{n}{2}-1}} + \dfrac{a^2(m-1)}{m-n+1} \int \dfrac{u^{m-2} du}{(a^2-u^2)^{\frac{n}{2}}}.$

69. $\int \dfrac{u^m\, du}{(a^2-u^2)^{\frac{n}{2}}} = \dfrac{u^{m+1}}{a^2(n-2)(a^2-u^2)^{\frac{n}{2}-1}} - \dfrac{m-n+3}{a^2(n-2)} \int \dfrac{u^m\, du}{(a^2-u^2)^{\frac{n}{2}-1}}.$

TABLE OF INTEGRALS 543

70. $\int \dfrac{du}{u(a^2-u^2)^{\frac{1}{2}}} = -\dfrac{1}{a} \ln\left(\dfrac{a+\sqrt{a^2-u^2}}{u}\right) + C = -\dfrac{1}{a} \cosh^{-1}\dfrac{a}{u} + C.$

71. $\int \dfrac{du}{u^2(a^2-u^2)^{\frac{1}{2}}} = -\dfrac{\sqrt{a^2-u^2}}{a^2 u} + C.$

72. $\int \dfrac{du}{u^3(a^2-u^2)^{\frac{1}{2}}} = -\dfrac{\sqrt{a^2-u^2}}{2\,a^2 u^2} - \dfrac{1}{2\,a^3} \ln\left(\dfrac{a+\sqrt{a^2-u^2}}{u}\right) + C$

$= -\dfrac{\sqrt{a^2-u^2}}{2\,a^2 u^2} - \dfrac{1}{2\,a^3} \cosh^{-1}\dfrac{a}{u} + C.$

73. $\int \dfrac{du}{u^m(a^2-u^2)^{\frac{n}{2}}} = -\dfrac{1}{a^2(m-1)u^{m-1}(a^2-u^2)^{\frac{n}{2}-1}}$
$+ \dfrac{m+n-3}{a^2(m-1)} \int \dfrac{du}{u^{m-2}(a^2-u^2)^{\frac{n}{2}}}.$

74. $\int \dfrac{du}{u^m(a^2-u^2)^{\frac{n}{2}}} = \dfrac{1}{a^2(n-2)u^{m-1}(a^2-u^2)^{\frac{n}{2}-1}}$
$+ \dfrac{m+n-3}{a^2(n-2)} \int \dfrac{du}{u^m(a^2-u^2)^{\frac{n}{2}-1}}.$

75. $\int \dfrac{(a^2-u^2)^{\frac{1}{2}}\,du}{u} = \sqrt{a^2-u^2} - a \ln\left(\dfrac{a+\sqrt{a^2-u^2}}{u}\right) + C$

$= \sqrt{a^2-u^2} - a \cosh^{-1}\dfrac{a}{u} + C.$

76. $\int \dfrac{(a^2-u^2)^{\frac{1}{2}}\,du}{u^2} = -\dfrac{\sqrt{a^2-u^2}}{u} - \arcsin\dfrac{u}{a} + C.$

77. $\int \dfrac{(a^2-u^2)^{\frac{n}{2}}\,du}{u^m} = -\dfrac{(a^2-u^2)^{\frac{n}{2}+1}}{a^2(m-1)u^{m-1}} + \dfrac{m-n-3}{a^2(m-1)} \int \dfrac{(a^2-u^2)^{\frac{n}{2}}\,du}{u^{m-2}}.$

78. $\int \dfrac{(a^2-u^2)^{\frac{n}{2}}\,du}{u^m} = \dfrac{(a^2-u^2)^{\frac{n}{2}}}{(n-m+1)u^{m-1}} + \dfrac{a^2 n}{n-m+1} \int \dfrac{(a^2-u^2)^{\frac{n}{2}-1}\,du}{u^m}.$

Forms containing $\sqrt{2\,au \pm u^2}$

The Binomial Reduction Formulas 96–104 may be applied by writing $\sqrt{2\,au \pm u^2} = u^{\frac{1}{2}}(2\,a \pm u)^{\frac{1}{2}}.$

79. $\int \sqrt{2\,au-u^2}\,du = \dfrac{u-a}{2}\sqrt{2\,au-u^2} + \dfrac{a^2}{2}\arccos\left(1-\dfrac{u}{a}\right) + C.$

80. $\int u\sqrt{2\,au-u^2}\,du = -\dfrac{3\,a^2+au-2\,u^2}{6}\sqrt{2\,au-u^2}$
$+ \dfrac{a^3}{2}\arccos\left(1-\dfrac{u}{a}\right) + C.$

81. $\int u^m \sqrt{2au - u^2}\, du = -\dfrac{u^{m-1}(2au - u^2)^{\frac{3}{2}}}{m + 2}$
$+ \dfrac{a(2m + 1)}{m + 2}\int u^{m-1}\sqrt{2au - u^2}\, du.$

82. $\int \dfrac{\sqrt{2au - u^2}\, du}{u} = \sqrt{2au - u^2} + a \arccos\left(1 - \dfrac{u}{a}\right) + C.$

83. $\int \dfrac{\sqrt{2au - u^2}\, du}{u^2} = -\dfrac{2\sqrt{2au - u^2}}{u} - \arccos\left(1 - \dfrac{u}{a}\right) + C.$

84. $\int \dfrac{\sqrt{2au - u^2}\, du}{u^3} = -\dfrac{(2au - u^2)^{\frac{3}{2}}}{3au^3} + C.$

85. $\int \dfrac{\sqrt{2au - u^2}\, du}{u^m} = -\dfrac{(2au - u^2)^{\frac{3}{2}}}{a(2m - 3)u^m} + \dfrac{m - 3}{a(2m - 3)}\int \dfrac{\sqrt{2au - u^2}\, du}{u^{m-1}}.$

86. $\int \dfrac{du}{\sqrt{2au - u^2}} = \arccos\left(1 - \dfrac{u}{a}\right) + C.$

87. $\int \dfrac{du}{\sqrt{2au + u^2}} = \ln\left(u + a + \sqrt{2au + u^2}\right) + C.$

88. $\int F(u, \sqrt{2au + u^2})\, du = \int F(z - a, \sqrt{z^2 - a^2})\, dz$, where $z = u + a$.

89. $\int \dfrac{u\, du}{\sqrt{2au - u^2}} = -\sqrt{2au - u^2} + a \arccos\left(1 - \dfrac{u}{a}\right) + C.$

90. $\int \dfrac{u^2\, du}{\sqrt{2au - u^2}} = -\dfrac{(u + 3a)\sqrt{2au - u^2}}{2} + \dfrac{3a^2}{2}\arccos\left(1 - \dfrac{u}{a}\right) + C.$

91. $\int \dfrac{u^m\, du}{\sqrt{2au - u^2}} = -\dfrac{u^{m-1}\sqrt{2au - u^2}}{m} + \dfrac{a(2m - 1)}{m}\int \dfrac{u^{m-1}\, du}{\sqrt{2au - u^2}}.$

92. $\int \dfrac{du}{u\sqrt{2au - u^2}} = -\dfrac{\sqrt{2au - u^2}}{au} + C.$

93. $\int \dfrac{du}{u^m\sqrt{2au - u^2}} = -\dfrac{\sqrt{2au - u^2}}{a(2m - 1)u^m} + \dfrac{m - 1}{a(2m - 1)}\int \dfrac{du}{u^{m-1}\sqrt{2au - u^2}}.$

94. $\int \dfrac{du}{(2au - u^2)^{\frac{3}{2}}} = \dfrac{u - a}{a^2\sqrt{2au - u^2}} + C.$

95. $\int \dfrac{u\, du}{(2au - u^2)^{\frac{3}{2}}} = \dfrac{u}{a\sqrt{2au - u^2}} + C.$

Binomial Reduction Formulas

96. $\int u^m(a + bu^q)^p\, du = \dfrac{u^{m-q+1}(a + bu^q)^{p+1}}{b(pq + m + 1)}$
$- \dfrac{a(m - q + 1)}{b(pq + m + 1)}\int u^{m-q}(a + bu^q)^p\, du.$

TABLE OF INTEGRALS 545

97. $\int u^m(a+bu^q)^p\, du = \dfrac{u^{m+1}(a+bu^q)^p}{pq+m+1}$
$+ \dfrac{apq}{pq+m+1}\int u^m(a+bu^q)^{p-1}\, du.$

98. $\int \dfrac{du}{u^m(a+bu^q)^p} = -\dfrac{1}{a(m-1)u^{m-1}(a+bu^q)^{p-1}}$
$- \dfrac{b(m-q+pq-1)}{a(m-1)}\int \dfrac{du}{u^{m-q}(a+bu^q)^p}.$

99. $\int \dfrac{du}{u^m(a+bu^q)^p} = \dfrac{1}{aq(p-1)u^{m-1}(a+bu^q)^{p-1}}$
$+ \dfrac{m-q+pq-1}{aq(p-1)}\int \dfrac{du}{u^m(a+bu^q)^{p-1}}.$

100. $\int \dfrac{du}{u(a+bu^q)} = \dfrac{1}{aq}\ln\left(\dfrac{u^q}{a+bu^q}\right)+C.$

101. $\int \dfrac{(a+bu^q)^p\, du}{u^m} = -\dfrac{(a+bu^q)^{p+1}}{a(m-1)u^{m-1}}$
$- \dfrac{b(m-q-pq-1)}{a(m-1)}\int \dfrac{(a+bu^q)^p\, du}{u^{m-q}}.$

102. $\int \dfrac{(a+bu^q)^p\, du}{u^m} = \dfrac{(a+bu^q)^p}{(pq-m+1)u^{m-1}}$
$+ \dfrac{apq}{pq-m+1}\int \dfrac{(a+bu^q)^{p-1}\, du}{u^m}.$

103. $\int \dfrac{u^m\, du}{(a+bu^q)^p} = \dfrac{u^{m-q+1}}{b(m-pq+1)(a+bu^q)^{p-1}}$
$- \dfrac{a(m-q+1)}{b(m-pq+1)}\int \dfrac{u^{m-q}\, du}{(a+bu^q)^p}.$

104. $\int \dfrac{u^m\, du}{(a+bu^q)^p} = \dfrac{u^{m+1}}{aq(p-1)(a+bu^q)^{p-1}}$
$- \dfrac{m+q-pq+1}{aq(p-1)}\int \dfrac{u^m\, du}{(a+bu^q)^{p-1}}.$

Forms containing $a+bu\pm cu^2$ $(c>0)$

The expression $a+bu+cu^2$ may be reduced to a binomial by writing
$u = z - \dfrac{b}{2c},\ k = \dfrac{b^2-4ac}{4c^2}.$
Then $\qquad a+bu+cu^2 = c(z^2-k).$
The expression $a+bu-cu^2$ may be reduced to a binomial by writing
$u = z + \dfrac{b}{2c},\ k = \dfrac{b^2+4ac}{4c^2}.$
Then $\qquad a+bu-cu^2 = c(k-z^2).$

105. $\int \dfrac{du}{a+bu+cu^2} = \dfrac{2}{\sqrt{4ac-b^2}}\arctan\left(\dfrac{2cu+b}{\sqrt{4ac-b^2}}\right)+C,$
when $b^2 < 4ac.$

546 DIFFERENTIAL AND INTEGRAL CALCULUS

106. $\int \dfrac{du}{a+bu+cu^2} = \dfrac{1}{\sqrt{b^2-4ac}} \ln\left(\dfrac{2cu+b-\sqrt{b^2-4ac}}{2cu+b+\sqrt{b^2-4ac}}\right) + C,$
   when $b^2 > 4ac$.

107. $\int \dfrac{du}{a+bu-cu^2} = \dfrac{1}{\sqrt{b^2+4ac}} \ln\left(\dfrac{\sqrt{b^2+4ac}+2cu-b}{\sqrt{b^2+4ac}-2cu+b}\right) + C.$

108. $\int \dfrac{(Mu+N)du}{a+bu\pm cu^2} = \dfrac{\pm M}{2c} \ln(a+bu\pm cu^2))$
   $+ \left(N \mp \dfrac{bM}{2c}\right) \int \dfrac{du}{a+bu\pm cu^2}.$

109. $\int \sqrt{a+bu+cu^2}\, du = \dfrac{2cu+b}{4c}\sqrt{a+bu+cu^2}$
   $- \dfrac{b^2-4ac}{8c^{\frac{3}{2}}} \ln(2cu+b+2\sqrt{c}\sqrt{a+bu+cu^2}) \mp C.$

110. $\int \sqrt{a+bu-cu^2}\, du = \dfrac{2cu-b}{4c}\sqrt{a+bu-cu^2}$
   $+ \dfrac{b^2+4ac}{8c^{\frac{3}{2}}} \arcsin\left(\dfrac{2cu-b}{\sqrt{b^2+4ac}}\right) + C.$

111. $\int \dfrac{du}{\sqrt{a+bu+cu^2}} = \dfrac{1}{\sqrt{c}} \ln(2cu+b+2\sqrt{c}\sqrt{a+bu+cu^2}) + C.$

112. $\int \dfrac{du}{\sqrt{a+bu-cu^2}} = \dfrac{1}{\sqrt{c}} \arcsin\left(\dfrac{2cu-b}{\sqrt{b^2+4ac}}\right) + C.$

113. $\int \dfrac{u\,du}{\sqrt{a+bu+cu^2}} = \dfrac{\sqrt{a+bu+cu^2}}{c}$
   $- \dfrac{b}{2c^{\frac{3}{2}}} \ln(2cu+b+2\sqrt{c}\sqrt{a+bu+cu^2}) \mp C.$

114. $\int \dfrac{u\,du}{\sqrt{a+bu-cu^2}} = -\dfrac{\sqrt{a+bu-cu^2}}{c}$
   $+ \dfrac{b}{2c^{\frac{3}{2}}} \arcsin\left(\dfrac{2cu-b}{\sqrt{b^2+4ac}}\right) + C.$

Other Algebraic Forms

115. $\int \sqrt{\dfrac{a+u}{b+u}}\, du = \sqrt{(a+u)(b+u)}$
   $+ (a-b) \log_e(\sqrt{a+u}+\sqrt{b+u}) \mp C.$

116. $\int \sqrt{\dfrac{a-u}{b+u}}\, du = \sqrt{(a-u)(b+u)}$
   $+ (a+b) \arcsin \sqrt{\dfrac{u+b}{a+b}} + C.$

117. $\int \sqrt{\dfrac{a+u}{b-u}}\, du = -\sqrt{(a+u)(b-u)} - (a+b) \arcsin \sqrt{\dfrac{b-u}{a+b}} + C.$

118. $\int \sqrt{\dfrac{1+u}{1-u}}\, du = -\sqrt{1-u^2} + \arcsin u + C.$

119. $\int \dfrac{du}{\sqrt{(u-a)(b-u)}} = 2 \arcsin \sqrt{\dfrac{u-a}{b-a}} + C.$

TABLE OF INTEGRALS 547

Exponential and Logarithmic Forms

120. $\int e^{au}\, du = \dfrac{e^{au}}{a} + C.$

121. $\int b^{au}\, du = \dfrac{b^{au}}{a \ln b} + C.$

122. $\int u e^{au}\, du = \dfrac{e^{au}}{a^2}(au - 1) + C.$

123. $\int u^n e^{au}\, du = \dfrac{u^n e^{au}}{a} - \dfrac{n}{a}\int u^{n-1} e^{au}\, du.$

124. $\int u^n b^{au}\, du = \dfrac{u^n b^{au}}{a \ln b} - \dfrac{n}{a \ln b}\int u^{n-1} b^{au}\, du + C.$

125. $\int \dfrac{b^{au}\, du}{u^n} = -\dfrac{b^{au}}{(n-1)u^{n-1}} + \dfrac{a \ln b}{n-1}\int \dfrac{b^{au}\, du}{u^{n-1}}.$

126. $\int \ln u\, du = u \ln u - u + C.$

127. $\int u^n \ln u\, du = u^{n+1}\left[\dfrac{\ln u}{n+1} - \dfrac{1}{(n+1)^2}\right] + C.$

128. $\int u^m \ln^n u\, du = \dfrac{u^{m+1}}{m+1} \ln^n u - \dfrac{n}{m+1}\int u^m \ln^{n-1} u\, du.$

129. $\int e^{au} \ln u\, du = \dfrac{e^{au} \ln u}{a} - \dfrac{1}{a}\int \dfrac{e^{au}}{u}\, du.$

130. $\int \dfrac{du}{u \ln u} = \ln(\ln u) + C.$

Trigonometric Forms

In forms involving $\tan u$, $\operatorname{ctn} u$, $\sec u$, $\csc u$, which do not appear below, first use the relations

$$\tan u = \dfrac{\sin u}{\cos u}, \quad \operatorname{ctn} u = \dfrac{\cos u}{\sin u}, \quad \sec u = \dfrac{1}{\cos u}, \quad \csc u = \dfrac{1}{\sin u}.$$

131. $\int \sin u\, du = -\cos u + C.$

132. $\int \cos u\, du = \sin u + C.$

133. $\int \tan u\, du = -\ln \cos u + C = \ln \sec u + C.$

134. $\int \operatorname{ctn} u\, du = \ln \sin u + C.$

135. $\int \sec u\, du = \int \dfrac{du}{\cos u} = \ln(\sec u + \tan u) + C$
$\qquad = \ln \tan\left(\dfrac{u}{2} + \dfrac{\pi}{4}\right) + C.$

136. $\int \csc u\, du = \int \dfrac{du}{\sin u} = \ln(\csc u - \operatorname{ctn} u) + C$
$\qquad = \ln \tan \dfrac{u}{2} + C.$

137. $\int \sec^2 u \, du = \tan u + C.$

138. $\int \csc^2 u \, du = -\ctn u + C.$

139. $\int \sec u \tan u \, du = \sec u + C.$

140. $\int \csc u \ctn u \, du = -\csc u + C.$

141. $\int \sin^2 u \, du = \tfrac{1}{2} u - \tfrac{1}{4} \sin 2u + C.$

142. $\int \cos^2 u \, du = \tfrac{1}{2} u + \tfrac{1}{4} \sin 2u + C.$

143. $\int \cos^n u \sin u \, du = -\dfrac{\cos^{n+1} u}{n+1} + C.$

144. $\int \sin^n u \cos u \, du = \dfrac{\sin^{n+1} u}{n+1} + C.$

145. $\int \sin mu \sin nu \, du = -\dfrac{\sin (m+n)u}{2(m+n)} + \dfrac{\sin (m-n)u}{2(m-n)} + C.$

146. $\int \cos mu \cos nu \, du = \dfrac{\sin (m+n)u}{2(m+n)} + \dfrac{\sin (m-n)u}{2(m-n)} + C.$

147. $\int \sin mu \cos nu \, du = -\dfrac{\cos (m+n)u}{2(m+n)} - \dfrac{\cos (m-n)u}{2(m-n)} + C.$

148. $\int \dfrac{du}{1 + \cos a \cos u} = 2 \csc a \arctan (\tan \tfrac{1}{2} a \tan \tfrac{1}{2} u) + C.$

149. $\int \dfrac{du}{\cos a + \cos u} = \csc a \ln \left(\dfrac{1 + \tan \tfrac{1}{2} a \tan \tfrac{1}{2} u}{1 - \tan \tfrac{1}{2} a \tan \tfrac{1}{2} u} \right) + C \quad (\tan^2 \tfrac{1}{2} u < \ctn^2 \tfrac{1}{2} a)$

$\quad = 2 \csc a \tanh^{-1}(\tan \tfrac{1}{2} a \tan \tfrac{1}{2} u) + C \quad (\tan^2 \tfrac{1}{2} u < \ctn^2 \tfrac{1}{2} a)$

150. $\int \dfrac{du}{1 + \cos a \sin u} = 2 \csc a \arctan (\csc a \tan \tfrac{1}{2} u + \ctn a) + C.$

151. $\int \dfrac{du}{\cos a + \sin u} = \csc a \ln \left(\dfrac{\tan a - \tan \tfrac{1}{2} u - \sec a}{\tan a + \tan \tfrac{1}{2} u + \sec a} \right) + C$

$\quad\quad [(\ctn a \tan \tfrac{1}{2} u + \csc a)^2 < 1]$

$\quad = -2 \csc a \tanh^{-1}(\ctn a \tan \tfrac{1}{2} u + \csc a) + C$

$\quad\quad [(\ctn a \tan \tfrac{1}{2} u + \csc a)^2 < 1]$

152. $\int \dfrac{du}{a^2 \cos^2 u + b^2 \sin^2 u} = \dfrac{1}{ab} \arctan \left(\dfrac{b \tan u}{a} \right) + C.$

153. $\int e^{au} \sin nu \, du = \dfrac{e^{au}(a \sin nu - n \cos nu)}{a^2 + n^2} + C.$

154. $\int e^{au} \cos nu \, du = \dfrac{e^{au}(n \sin nu + a \cos nu)}{a^2 + n^2} + C.$

155. $\int u \sin u \, du = \sin u - u \cos u + C.$

156. $\int u \cos u \, du = \cos u + u \sin u + C.$

TABLE OF INTEGRALS 549

Trigonometric Reduction Formulas

157. $\displaystyle\int \sin^n u \, du = -\frac{\sin^{n-1} u \cos u}{n} + \frac{n-1}{n} \int \sin^{n-2} u \, du.$

158. $\displaystyle\int \cos^n u \, du = \frac{\cos^{n-1} u \sin u}{n} + \frac{n-1}{n} \int \cos^{n-2} u \, du.$

159. $\displaystyle\int \frac{du}{\sin^n u} = -\frac{\cos u}{(n-1) \sin^{n-1} u} + \frac{n-2}{n-1} \int \frac{du}{\sin^{n-2} u}.$

160. $\displaystyle\int \frac{du}{\cos^n u} = \frac{\sin u}{(n-1) \cos^{n-1} u} + \frac{n-2}{n-1} \int \frac{du}{\cos^{n-2} u}.$

161. $\displaystyle\int \cos^m u \sin^n u \, du = \frac{\cos^{m-1} u \sin^{n+1} u}{m+n} + \frac{m-1}{m+n} \int \cos^{m-2} u \sin^n u \, du.$

162. $\displaystyle\int \cos^m u \sin^n u \, du = -\frac{\sin^{n-1} u \cos^{m+1} u}{m+n} + \frac{n-1}{m+n} \int \cos^m u \sin^{n-2} u \, du.$

163. $\displaystyle\int \frac{du}{\cos^m u \sin^n u} = \frac{1}{(m-1) \sin^{n-1} u \cos^{m-1} u}$
$\qquad + \frac{m+n-2}{m-1} \displaystyle\int \frac{du}{\cos^{m-2} u \sin^n u}.$

164. $\displaystyle\int \frac{du}{\cos^m u \sin^n u} = -\frac{1}{(n-1) \sin^{n-1} u \cos^{m-1} u}$
$\qquad + \frac{m+n-2}{n-1} \displaystyle\int \frac{du}{\cos^m u \sin^{n-2} u}.$

165. $\displaystyle\int \frac{\cos^m u \, du}{\sin^n u} = -\frac{\cos^{m+1} u}{(n-1) \sin^{n-1} u} - \frac{m-n+2}{n-1} \int \frac{\cos^m u \, du}{\sin^{n-2} u}.$

166. $\displaystyle\int \frac{\cos^m u \, du}{\sin^n u} = \frac{\cos^{m-1} u}{(m-n) \sin^{n-1} u} + \frac{m-1}{m-n} \int \frac{\cos^{m-2} u \, du}{\sin^n u}.$

167. $\displaystyle\int \frac{\sin^n u \, du}{\cos^m u} = \frac{\sin^{n+1} u}{(m-1) \cos^{m-1} u} - \frac{n-m+2}{m-1} \int \frac{\sin^n u \, du}{\cos^{m-2} u}.$

168. $\displaystyle\int \frac{\sin^n u \, du}{\cos^m u} = -\frac{\sin^{n-1} u}{(n-m) \cos^{m-1} u} + \frac{n-1}{n-m} \int \frac{\sin^{n-2} u \, du}{\cos^m u}.$

169. $\displaystyle\int \tan^n u \, du = \frac{\tan^{n-1} u}{n-1} - \int \tan^{n-2} u \, du.$

170. $\displaystyle\int \ctn^n u \, du = -\frac{\ctn^{n-1} u}{n-1} - \int \ctn^{n-2} u \, du.$

171. $\displaystyle\int e^{au} \cos^n u \, du = \frac{e^{au} \cos^{n-1} u (a \cos u + n \sin u)}{a^2 + n^2}$
$\qquad + \frac{n(n-1)}{a^2 + n^2} \displaystyle\int e^{au} \cos^{n-2} u \, du.$

172. $\displaystyle\int e^{au} \sin^n u \, du = \frac{e^{au} \sin^{n-1} u (a \sin u - n \cos u)}{a^2 + n^2}$
$\qquad + \frac{n(n-1)}{a^2 + n^2} \displaystyle\int e^{au} \sin^{n-2} u \, du.$

173. $\int u^m \cos au \, du = \dfrac{u^{m-1}}{a^2} (au \sin au + m \cos au)$
$\qquad - \dfrac{m(m-1)}{a^2} \int u^{m-2} \cos au \, du.$

174. $\int u^m \sin au \, du = \dfrac{u^{m-1}}{a^2} (m \sin au - au \cos au)$
$\qquad - \dfrac{m(m-1)}{a^2} \int u^{m-2} \sin au \, du.$

[Inverse Trigonometric Functions

175. $\int \operatorname{arc\,sin} u \, du = u \operatorname{arc\,sin} u + \sqrt{1-u^2} + C.$

176. $\int \operatorname{arc\,cos} u \, du = u \operatorname{arc\,cos} u - \sqrt{1-u^2} + C.$

177. $\int \operatorname{arc\,tan} u \, du = u \operatorname{arc\,tan} u - \ln \sqrt{1+u^2} + C.$

178. $\int \operatorname{arc\,ctn} u \, du = u \operatorname{arc\,ctn} u + \ln \sqrt{1+u^2} + C.$

179. $\int \operatorname{arc\,sec} u \, du = u \operatorname{arc\,sec} u - \ln (u + \sqrt{u^2-1}) + C.$
$\qquad = u \operatorname{arc\,sec} u - \cosh^{-1} u + C.$

180. $\int \operatorname{arc\,csc} u \, du = u \operatorname{arc\,csc} u + \ln (u + \sqrt{u^2-1}) + C$
$\qquad = u \operatorname{arc\,csc} u + \cosh^{-1} u + C.$

Hyperbolic Functions

181. $\int \sinh u \, du = \cosh u + C.$

182. $\int \cosh u \, du = \sinh u + C.$

183. $\int \tanh u \, du = \ln \cosh u + C.$

184. $\int \operatorname{ctnh} u \, du = \ln \sinh u + C.$

185. $\int \operatorname{sech} u \, du = \operatorname{arc\,tan} (\sinh u) + C = \operatorname{gd} u + C.$

186. $\int \operatorname{csch} u \, du = \ln \tanh \tfrac{1}{2} u + C.$

187. $\int \operatorname{sech}^2 u \, du = \tanh u + C.$

188. $\int \operatorname{csch}^2 u \, du = - \operatorname{ctnh} u + C.$

189. $\int \operatorname{sech} u \tanh u \, du = - \operatorname{sech} u + C.$

TABLE OF INTEGRALS

190. $\int \operatorname{csch} u \operatorname{ctnh} u \, du = - \operatorname{csch} u + C.$

191. $\int \sinh^2 u \, du = \tfrac{1}{4} \sinh 2u - \tfrac{1}{2} u + C.$

192. $\int \cosh^2 u \, du = \tfrac{1}{4} \sinh 2u + \tfrac{1}{2} u + C.$

193. $\int \tanh^2 u \, du = u - \tanh u + C.$

194. $\int \operatorname{ctnh}^2 u \, du = u - \operatorname{ctnh} u + C.$

195. $\int u \sinh u \, du = u \cosh u - \sinh u + C.$

196. $\int u \cosh u \, du = u \sinh u - \cosh u + C.$

197. $\int \sinh^{-1} u \, du = u \sinh^{-1} u - \sqrt{1 + u^2} + C.$

198. $\int \cosh^{-1} u \, du = u \cosh^{-1} u - \sqrt{u^2 - 1} + C.$

199. $\int \tanh^{-1} u \, du = u \tanh^{-1} u + \tfrac{1}{2} \ln(1 - u^2) + C.$

200. $\int \operatorname{ctnh}^{-1} u \, du = u \operatorname{ctnh}^{-1} u + \tfrac{1}{2} \ln(1 - u^2) + C.$

201. $\int \operatorname{sech}^{-1} u \, du = u \operatorname{sech}^{-1} u + \operatorname{gd}(\tanh^{-1} u) + C$
$\qquad = u \operatorname{sech}^{-1} u + \arcsin u + C.$

202. $\int \operatorname{csch}^{-1} u \, du = u \operatorname{csch}^{-1} u + \sinh^{-1} u + C.$

203. $\int \sinh mu \sinh nu \, du = \dfrac{\sinh(m+n)u}{2(m+n)} - \dfrac{\sinh(m-n)u}{2(m-n)} + C. \; \left(m \gtrless n\right)$

204. $\int \cosh mu \cosh nu \, du = \dfrac{\sinh(m+n)u}{2(m+n)} + \dfrac{\sinh(m-n)u}{2(m-n)} + C. \; \left(m \gtrless n\right)$

205. $\int \sinh mu \cosh nu \, du = \dfrac{\cosh(m+n)u}{2(m+n)} + \dfrac{\cosh(m-n)u}{2(m-n)} + C. \; \left(m \gtrless n\right)$

206. $\int \dfrac{du}{\cosh a + \cosh u} = 2 \operatorname{csch} a \tanh^{-1}(\tanh \tfrac{1}{2} u \tanh \tfrac{1}{2} a) + C.$

207. $\int \dfrac{du}{\cos a + \cosh u} = 2 \csc a \arctan(\tanh \tfrac{1}{2} u \tan \tfrac{1}{2} a) + C.$

208. $\int \dfrac{du}{1 + \cos a \cosh u} = 2 \csc a \tanh^{-1}(\tanh \tfrac{1}{2} u \tan \tfrac{1}{2} a) + C$
$\qquad\qquad\qquad (\tanh^2 \tfrac{1}{2} u < \operatorname{ctn}^2 \tfrac{1}{2} a)$

209. $\int e^{au} \sinh nu \, du = \dfrac{e^{au}(a \sinh nu - n \cosh nu)}{a^2 - n^2} + C.$

210. $\int e^{au} \cosh nu \, du = \dfrac{e^{au}(a \cosh nu - n \sinh nu)}{a^2 - n^2} + C.$

Differential Equations

In these formulas A, B, α denote arbitrary constants.

1. The differential equation of Harmonic Vibration.

$$\frac{d^2s}{dt^2} + k^2 s = 0.$$

The general solution may be written in the following forms.
(a) $s = Ae^{kti} + Be^{-kti}$,
(b) $s = A \cos kt + B \sin kt$,
(c) $s = A \cos (kt + \alpha)$,
(d) $s = A \sin (kt + \alpha)$.

2. The differential equation of Damped Vibration.

$$\frac{d^2s}{dt^2} + 2\mu \frac{ds}{dt} + k^2 s = 0, \ \mu < k.$$

The general solution is

$$s = e^{-\mu t}(A \cos \sqrt{k^2 - \mu^2}\, t + B \sin \sqrt{k^2 - \mu^2}\, t), \text{ or}$$
$$s = e^{-\mu t} \cos (\sqrt{k^2 - \mu^2}\, t + \alpha).$$

3. The differential equation of Forced Vibration.

(a) $\dfrac{d^2s}{dt^2} + k^2 s = L \cos nt + M \sin nt$, where $n \neq k$.

The general solution is

$$s = A \cos kt + B \sin kt + \frac{L}{k^2 - n^2} \cos nt + \frac{M}{k^2 - n^2} \sin nt.$$

(b) $\dfrac{d^2s}{dt^2} + k^2 s = L \cos kt + M \sin kt.$

The general solution is

$$s = A \cos kt + B \sin kt + \frac{L}{2k} t \sin kt - \frac{M}{2k} t \cos kt.$$

INDEX

(The numbers refer to pages)

Absolute convergence, 348
Acceleration, curvilinear motion, 121; rectilinear motion, 83
Adiabatic law, 70
Agnesi, witch of, 531
Anchor ring, 267
Angle of intersection, of plane curves, 43; polar form of, 126; of skew curves, 475; of surfaces, 481
Approximate formulas, 367, 372, 490
Arc, centroid of, 335; differential of, 142, 144, 473; length of, plane curve, 271; skew curve, 473
Archimedes, 127, 128, 155, 277, 534
Area, of a curved surface, 517; moment of, 320, 503, 514; moment of inertia of, 508, 514; plane, 241, 258, 498; in polar coördinates, 262, 512; of a surface of revolution, 277
Astroid, 119, 533
Auxiliary equation, 390

Bending, direction of, 75
Bernoulli, lemniscate of, 532
Binomial differentials, 299, 307
Binomial theorem, 1, 353
Boyle's law, 70

Calculation, of e, 361; of logarithms, 362; of π, 366
Cardioid, 117, 119, 125, 135, 145, 155, 244, 271, 275, 281, 323, 335, 516, 517, 533
Catenary, 152, 270, 276, 282, 423, 432, 434, 532
Cauchy, 345

Center of fluid pressure, 506
Centroid, of a homogeneous solid, 522, 526; of a plane area, 320, 337, 503; of a solid of revolution, 323
Change of variable, 166, 240, 457
Cissoid, 44, 46, 270, 277, 322, 531
Complementary function, 394
Complex number, 440
Compound-interest law, 399
Conchoid, 532
Conoid, 284
Constant, 7; absolute, 7; arbitrary, 7; of integration, 189, 229, 233, 376; numerical, 7
Continuity of functions, 12, 444
Convergence, 340
Coördinates, cylindrical, 6, 524–527; polar, 123; spherical, 6, 529, 530
Cosine curve, 533
Critical values, 52
Cubic, skew, 474
Curvature, 149; center of, 157, 171; circle of, 153, 170; radius of, 152
Curve-tracing, 81
Curvilinear motion, 120, 146
Cycloid, 116, 119, 144, 151, 161, 244, 270, 274, 276, 281, 532
Cylindrical coördinates, 6, 524–527

Derivative, definition, 21; interpretation of, by geometry, 25, 446; partial, 445; as a rate, 64; symbols for, 22, 445; total, 455; transformation of, 166
Descartes, folium of, 46, 119, 288, 533

553

Differential, 136; of arc, 142, 144, 473; of area, 237; formulas for, 140; geometric interpretation, 137, 477; as an infinitesimal, 146; total, 449

Differential coefficient, 21, 136

Differential equations, applications to mechanics, 402; definitions, 375; first order, 378; higher order, 387; homogeneous, 380; linear, 383, 390, 407

Differentiation, 22; formulas for, 28, 29, 86, 87, 115, 119, 420, 425, 426, 435; general rule for, 23; of implicit functions, 40, 73, 154, 458; logarithmic, 93; partial, 445, 462; successive, 73, 462

Diocles, cissoid of, 44, 46, 270, 277, 322, 531

Direction of a curve, 42

Ellipsoid, 280, 285, 522

Envelopes, 466

Equations, graphical solution, 128; interpolation, 129; of motion, 120; Newton's method, 131

Errors, 138, 451; percentage, 138; relative, 138

Evolute, 158, 469; of the cycloid, 161; of the ellipse, 160, 533; of the parabola, 159, 470; properties of, 162

Exponential curve, 89, 535

Exponential function, 89

Factorial number, 338

Family of curves, 230, 466

Fluid pressure, 325; center of, 506

Fluxions, 19, 357

Folium of Descartes, 46, 119, 288, 533

Formulas, approximate, 367, 372, 490; for reference, 1–6

Fourier, 238

Function, complementary, 394; continuity of, 12, 444; decreasing, 50; definition of, 8, 444; derived, 21; differentiable, 21, 23; discontinuous, example of, 108; exponential, 89; of a function, 37; graph of, 10, 444; hyperbolic, 414, 415; implicit, 39, 73, 458; increasing, 50; inverse, 38; inverse hyperbolic, 423; inverse trigonometric, 105; logarithmic, 89; mean value of, 333; periodic, 97; of several variables, 444; sine, 97; table of hyperbolics, 416; transcendental, 86; trigonometric, 99

Fundamental theorem of the integral calculus, 254–257

Graph of a function, 10, 444

Gravity, center of, 320, 323

Greek alphabet, 6

Gudermann, 435

Gudermannian, 435; inverse, 435

Gyration, radius of, 508

Harmonic vibration, 403, 405

Helix, circular, 473, 474, 482

Horner's method, 130

Hyperboloid, 528

Hypocycloid, 46, 119, 156, 244, 268, 270, 276, 280, 288, 468, 533

Increments, 19; approximation of, 137, 451

Indeterminate forms, 174

Inertia, moment of, 508, 514

Infinitesimals, 17; replacement theorem, 147

Infinity, 13

Inflectional points, 79

Initial conditions, 229

Integrals, 188; change in limits, 240; decomposition of interval, 250; definite, 237; discontinuous, 251; geometric representation, 244, 492; improper, 250–253; indefinite, 189; interchange of limits, 249; multiple, 491; table of, 538; use of table, 315

INDEX

Integrand, 195
Integration, 187; approximate, 245; of binomial differentials, 299; formulas for, 191–193, 430, 432, 433, 435; fundamental theorem of, 254; by miscellaneous substitutions, 221, 305, 432; by parts, 223; by rational fractions, 289; by rationalization, 221, 296; by reciprocal substitution, 305; by reduction formulas, 307, 312; successive, 491; of trigonometric forms, 213, 303, 312
Interpolation, 129, 372
Interval of a variable, 7
Involute, 163; of a circle, 156, 276, 288, 537
Isothermal expansion, 330

Jacobi, 445

Laplace, 19
Laws of the mean, 172, 182, 482
Leibnitz, 26
Lemniscate, 127, 155, 263, 515, 516, 517, 532
Length of arc, plane curves, 271; in polar coördinates, 274; of skew curves, 473
Limaçon, 534
Limit of a variable, 10
Limits, change in, 240; of an integral, 238; theorems on, 11, 17
Lituus, 534
Logarithmic curve, 89, 535
Logarithmic differentiation, 93
Logarithmic function, 89
Logarithms, common, 88; natural, 87
Loxodrome, 439

Maclaurin's series, 357, 367
Maxima and minima, 47; analytic treatment, 182; definitions, 52; first method, 53; functions of two variables, 483; second method, 76

Mean value, extended theorem of, 182; of a function, 333; theorems of, 172, 482
Mechanics, 402
Mercator, 439; chart, 439
Moment, of area, 320, 503; of inertia, 507, 508, 514; polar, 510, 514
Motion, curvilinear, 120; rectilinear, 65, 83

Napierian logarithms, 88
Newton, 19, 27, 131, 132, 133, 332, 401
Nicomedes, conchoid of, 532
Normal, to a plane curve, 43; plane to a skew curve, 471, 480; to a surface, 475

Osculating circle, 170

Pappus, theorems of, 335, 504
Parabola, 532, 537; cubical, 156, 531; semicubical, 266, 498, 531
Parabolic rule, 247
Paraboloid of revolution, 268, 521, 528
Parameter, 115, 466
Parametric equations, 115; first derivative, 115; second derivative, 119
Point of inflection, 79
Polar coördinates, 123; moment of inertia, 514; subnormal, 126; subtangent, 126
Power rule, 32
Pressure, fluid, 325; center of, 506
Probability curve, 535
Projectile, 121, 234, 471

Quadratic equation, 1

Radius, of curvature, 152; of gyration, 508
Railroad curves, 152
Rates, 64, 455

INDEX

Rectification, of plane curves, 271; in polar coördinates, 274; of skew curves, 473
Reduction formulas, 307, 312
Replacement theorem, 147
Rhumb line, 439
Rolle's theorem, 169
Roots of equations, 128
Rose-leaf curves, 536

Secant curve, 535
Sequence, 338
Series, 338; absolute convergence, 348; alternating, 347; approximate formulas from, 367, 372; binomial, 353; Cauchy's test, 345; comparison tests, 342; convergent, 340; differentiation and integration of, 365; divergent, 340; geometric, 339; harmonic, 342; Maclaurin's, 357; operations with, 362; oscillating, 339; p, 344; power, 350, 354; Taylor's, 369, 488
Simpson's rule, 247
Sine curve, 533
Skew curves, 471, 480; length of, 473
Slope of a curve, 42; parametric form, 115; polar form, 125
Solids of revolution, centroid of, 323; surface of, 277; volume of, 265, 267, 268
Speed, 121, 146
Spherical coördinates, 6, 529, 530
Spheroid, oblate, 266; prolate, 266
Spiral, of Archimedes, 127, 128, 155, 277, 534; hyperbolic or reciprocal, 119, 128, 264, 277, 534; logarithmic or equiangular, 127, 128, 534; parabolic, 535

Stirling, 357
Strophoid, 534
Subnormal, 43; polar, 126
Subtangent, 43; polar, 126
Successive differentiation, 73, 462
Successive integration, 491

Table, of hyperbolic functions, 416; of integrals, 538
Tangent, horizontal, 42, 117; to a plane curve, 43; to a skew curve, 471, 480; plane to a surface, 475; vertical, 42, 117
Tangent curve, 535
Taylor's theorem, 369, 488
Telegraph line, 428–429
Torus, 267
Tractrix, 85, 270, 282, 419, 423, 436, 537
Transformation of derivatives, 166
Transition curves, 152
Trapezoidal rule, 245
Triple integration, 521
Trisectrix, 155, 264

Variable, change of, 166, 457; definition, 7; dependent, 8; independent, 8
Velocity, curvilinear motion, 120, 146; rectilinear motion, 65
Vibration, damped harmonic, 405; forced harmonic, 405; simple harmonic, 403
Volume, of a hollow solid of revolution, 267; of a solid with known cross section, 283; of a solid of revolution, 265, 267; under a surface, 501, 524; by triple integration, 521, 526

Witch, 62, 251, 269, 322, 531
Work, 328